Water Wave Kinematics

NATO ASI Series

Advanced Science Institutes Series

A Series presenting the results of activities sponsored by the NATO Science Committee, which aims at the dissemination of advanced scientific and technological knowledge, with a view to strengthening links between scientific communities.

The Series is published by an international board of publishers in conjunction with the NATO Scientific Affairs Division

A Life Sciences B Physics	Plenum Publishing Corporation London and New York
C Mathematical and Physical Sciences D Behavioural and Social Sciences E Applied Sciences	Kluwer Academic Publishers Dordrecht, Boston and London
F Computer and Systems Sciences G Ecological Sciences H Cell Biology	Springer-Verlag Berlin, Heidelberg, New York, London, Paris and Tokyo

Series E: Applied Sciences - Vol. 178

Water Wave Kinematics

edited by

A. Tørum

Norwegian Institute of Technology /
Norwegian Hydrotechnical Laboratory,
Trondheim, Norway

and

O. T. Gudmestad

Statoil,
Stavanger, Norway

Kluwer Academic Publishers

Dordrecht / Boston / London

Published in cooperation with NATO Scientific Affairs Division

Proceedings of the NATO Advanced Research Workshop on
Water Wave Kinematics
Molde, Norway
22–25 May 1989

Library of Congress Cataloging in Publication Data

Water wave kinematics / edited by A. Tørum, O.T. Gudmestad.
        p.   cm. -- (NATO ASI series. Series E, Applied sciences ;
no.)
    "Published in cooperation with NATO Scientific Affairs Division."
    ISBN-13:978-94-010-6725-6   e-ISBN-13:978-94-009-0531-3
    DOI: 10.1007/978-94-009-0531-3

    1. Water waves--Congresses.   2. Kinematics--Congresses.
I. Tørum, A. (Alf), 1933-   . II. Gudmestad, O. T. (Ove T.), 1947-
. III. Series.
TC172.W36   1990
627--dc20
                                                    89-77044

ISBN-13:978-94-010-6725-6

Published by Kluwer Academic Publishers,
P.O. Box 17, 3300 AA Dordrecht, The Netherlands.

Kluwer Academic Publishers incorporates the publishing programmes of
D. Reidel, Martinus Nijhoff, Dr W. Junk and MTP Press.

Sold and distributed in the U.S.A. and Canada
by Kluwer Academic Publishers,
101 Philip Drive, Norwell, MA 02061, U.S.A.

In all other countries, sold and distributed
by Kluwer Academic Publishers Group,
P.O. Box 322, 3300 AH Dordrecht, The Netherlands.

Printed on acid-free paper

CONTENTS

Short contributions

Short contributions

Short contributions

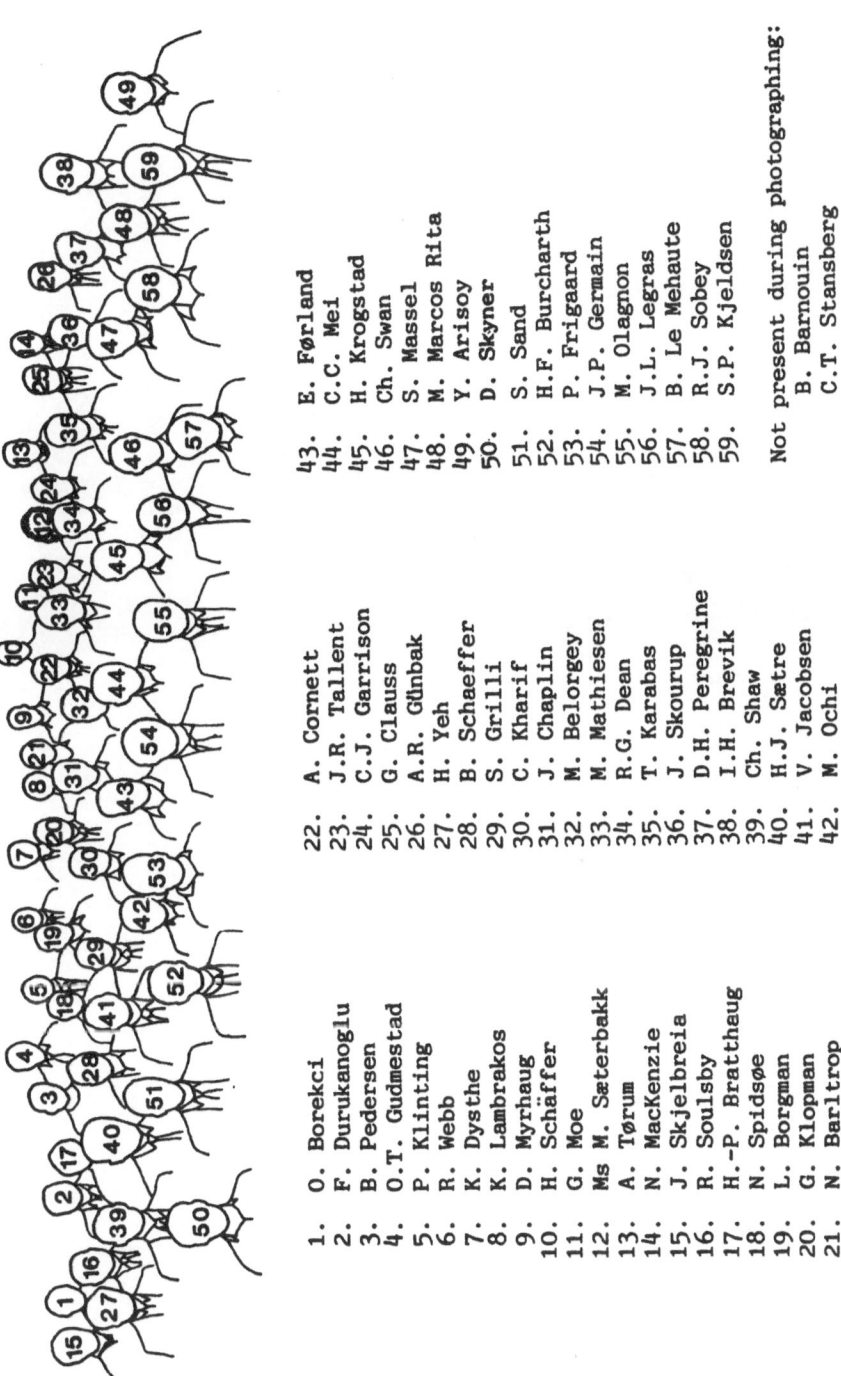

1. O. Borekci
2. F. Durukanoglu
3. B. Pedersen
4. O.T. Gudmestad
5. P. Klinting
6. R. Webb
7. K. Dysthe
8. K. Lambrakos
9. D. Myrhaug
10. H. Schäffer
11. G. Moe
12. Ms M. Sæterbakk
13. A. Tørum
14. N. MacKenzie
15. J. Skjelbreia
16. R. Soulsby
17. H.-P. Bratthaug
18. N. Spidsøe
19. L. Borgman
20. G. Klopman
21. N. Barltrop

22. A. Cornett
23. J.R. Tallent
24. C.J. Garrison
25. G. Clauss
26. A.R. Günbak
27. H. Yeh
28. B. Schaeffer
29. S. Grilli
30. C. Kharif
31. J. Chaplin
32. M. Belorgey
33. M. Mathiesen
34. R.G. Dean
35. T. Karabas
36. J. Skourup
37. D.H. Peregrine
38. I.H. Brevik
39. Ch. Shaw
40. H.J. Sætre
41. V. Jacobsen
42. M. Ochi

43. E. Førland
44. C.C. Mei
45. H. Krogstad
46. Ch. Swan
47. S. Massel
48. M. Marcos Rita
49. Y. Arisoy
50. D. Skyner
51. S. Sand
52. H.F. Burcharth
53. P. Frigaard
54. J.P. Germain
55. M. Olagnon
56. J.L. Legras
57. B. Le Mehaute
58. R.J. Sobey
59. S.P. Kjeldsen

Not present during photographing:

B. Barnouin
C.T. Stansberg

PREFACE

Water wave kinematics is a central field of study in ocean and coastal engineering. The wave forces on structures as well as sand erosion both on coastlines and in the ocean are to a large extent governed by the local distribution of velocities and accelerations of the water particles.

Our knowledge of waves has generally been derived from measurements of the water surface elevations. The reason for this is that the surface elevations have been of primary interest and fairly cheap and reliable instruments have been developed for such measurements. The water wave kinematics has then been derived from the surface elevation information by various theories. However, the different theories for the calculation of water particle velocities and acceleration have turned out to give significant differences in the calculated responses of structures. In recent years new measurement techniques have made it possible to make accurate velocity measurements. Hence, the editors deemed it to be useful to bring together a group of experts working actively as researchers in the field of water wave kinematics. These experts included theoreticians as well as experimentalists on wave kinematics. It was also deemed useful to include experts on the response of structures to have their views from a structural engineering point of view on what information is really needed on water wave kinematics.

The objectives of the workshop were:

- to summarize research related to wave kinematics
  (state-of-the-art)
- to define approaches for design of safe and cost efficient
  offshore and coastal structures
- to exchange ideas and methodologies and to serve as a
  forum for idea generation
- to define further research needs and most efficient methods to
  solve those needs.

A grant from NATO International Scientific Exchange Programme made it possible to hold a NATO Advanced Research Workshop in Molde, Norway, 22 - 25 May 1989. Additional funding was obtained from Statoil, Norway, the Royal Norwegian Council for Scientific and Indusstrial Research, Norway (NTNF) and from the Foundation for Scientific and Industrial Research at the Norwegian Institute of Technology, (SINTEF) Norway. We are thankful to these institutions for having made it possible to arrange the workshop.

The workshop was organized as a series of lectures, including invited lectures, with discussions. There was also set up five working groups on five different subjects: Deep water waves, shallow water waves, breaking and freak waves, measurements and forces.

The discussions and conclusions of the working groups were summarized in reports which are included in the proceedings.

The scientific committee for the workshop had the following members:

Professor A. Tørum   Norwegian Hydrotechnical Laboratory/
                                 Norwegian Institute of Technology,
                                 Norway.
                                 Director of the Workshop.

Dr.Scient. O.T.Gudmestad   Statoil, Norway
                                 Director of the Workshop

Professor R.G. Dean   University of Florida, USA

Professor D.H. Peregrine   University of Bristol, United Kingdom

Dr. S. Sand   Danish Maritime Institute, Denmark

The editors would like to thank the other members of the scientific committee for their help and effort to organize the workshop.

We also like to thank all the other participants in the workshop. Each one contributed in a significant way to make the workshop a rewarding event.

Finally we like to thank the invited speakers and the contributors for their effort in preparing their manuscripts and thus making these proceedings possible. We hope the proceedings will be a source for information and inspiration.

September 1989

Alf Tørum   Ove T. Gudmestad
Editor   Editor

Reports from Working Groups

Report from Working Group on Deep Water Wave Kinematics

K.F.Lambrakos, Chairman
O.T.Gudmestad, Presenter

1 INTRODUCTION AND SUMMARY

There are presently many wave kinematics procedures that
are used by the offshore engineer to calculate design
kinematics. These procedures include the highly
nonlinear, two-dimensional breaking wave methods, the
linear, random three-dimensional methods and others. In
many cases the application determines the procedure to
be used, e.g. for steel piled jacket (SPJ) static design
the Stokes type procedures have traditionally been used.
In other instances, the choice of the procedure depends
on regulatory agency requirements, familiarity,
available experimental verification, and also
simplicity. The argument can be made that all these
procedures have served the offshore industry well since
there have been no major offshore platform failures that
are directly attributed to wave kinematics inaccuracies.
This success, however, may be due to general
conservative approaches used in platform design, and to
compensating factors. All procedures represent
approximations to the real offshore kinematics problem,
and as design procedures become more sophisticated,
aiming to improve safety while reducing platform costs,
there is a two-fold need (1) for additional verification
and improvement of the various available procedures and
(2) for development of procedures that solve
satisfactorily the general problem of three-dimensional
irregular waves with currents. The additional
verification and establishment of ranges of
applicability of present procedures is of utmost
importance since some of the procedures, e.g. stretching
methods for irregular waves, give significantly
different kinematics in the wave crest region.

A. Tørum and O. T. Gudmestad (eds.), Water Wave Kinematics, 3–7.
© 1990 *Kluwer Academic Publishers.*

1.1._____Statement_of_the_problem

The general problem of wave kinematics and currents in deep water is quite complex, and needless to say it has not been solved. The physical model assumed for this discussion is deterministic (stochastic aspects of wave kinematics are not considered) and includes:

- regular or irregular finite height, non breaking waves
- wave conditions with or without energy spreading
- wave conditions with or without current.

In the case of waves only or waves with uniform current, the fluid is assumed irrotational and incompressible. For the case of nonuniform current with waves, the appropriate vorticity is assumed to be included. It is believed that any accurate solution to the analytical relationships describing this stated problem will give sufficiently accurate kinematics. Of course this will need to be substantiated with data.

1.2_____Status

A satisfactory method for calculating kinematics in the general irregular wave with energy spreading cases that satisfies to high degree of accuracy Laplace's equation and boundary conditions is not presently available. For practical application, however, there are many analytical and empirical procedures that can be used depending on application. These procedures are backed up by varying degrees of experimental and/or analytical verification.

The extension of the two-dimensional wave procedures to three-dimensional waves is not straightforward neither from the analytical or technical nor from the wave input description side. Available solutions (empirical or low order analytical) are approximate, and, no extensive comparisons with data have been carried out.

The presence of storm tide or other non-uniform currents adds to the difficulty of the general wave only problem. Analytical solutions for regular waves (Stokes) with arbitrary current profiles are available. Also, empirical solutions are available for the general case, but those solutions can benefit from additional data.

1.3_____Recommendations

Further measurements in the laboratory followed by carefully defined field tests are required. Test programs should be based on the need of the offshore industry. The industry should define the acceptable levels of accuracy in the kinematics, which would normally depend on application.

It is also required to develop analytical or numerical methods which solve the more general problem of kinematics for currents plus irregular, three-dimensional waves. In order to achieve this goal there is need for international cooperation and data sharing. Definition of standard bench-mark wave and current cases would also expedite verification of the procedures.

2   KINEMATICS MODELS

There are four major classes of wave kinematics problems that are discussed below which represent different levels of difficulty in the search for solutions. These are: two dimensional waves with and without current, three-dimensional waves with and without current.

2.1.   Two dimensional waves

There is no kinematics procedure or theory presently available for the general case which satisfies the governing equations and the boundary conditions. The relatively new boundary integral methods hold promise but at present have computational and free boundary limitations.

Simple procedures (e.g. Wheeler stretching, delta stretching, Gudmestad method etc) which represent extrapolations of linear theory to finite height irregular waves have been successfully compared to available data according to published results. Since these procedures are basically empirical, their use far outside the range of wave parameters for which they have been verified should be done with caution. Additional kinematics data (laboratory and field) are needed in the wave crest to further verify and define the range of applicability of these methods. Theoretical efforts to assess the methods are also worthwhile.

There exist analytically robust procedures (Stokes methods, boundary integral methods, etc) to treat regular or limited groups of waves with high accuracy although further testing against available data is necessary for these procedures, also. Verified methods for using these procedures with irregular waves are yet to be established.

Numerical procedures (e.g. stream function, EXVP method, etc) that treat irregular single or multiple waves can also be used. These procedures fit with high accuracy one or both boundary conditions. The applicability of these procedures is limited by the assumptions they are based on (rigid shape or changing shape waves) and the required length of the wave trace that can be treated.

Linear theory could be used for low wave steepness or for elevations beyond one wave height below the deepest wave trough. Higher order theories (2nd order) are available and are applicable for low steepness. These theories, however, require additional comparisons with data, also.

2.2   Three dimensional waves

There is no general theory presently available apart from general linear wave theory and there is no procedure under development which is computationally efficient for the nonlinear case. Some of the available two-dimensional empirical procedures (stretching methods) may in principle be extended to the three-dimensional wave case. Test data will be needed to verify the accuracy of such extensions.

The linear wave theory could be used for the low wave steepnesses. Higher order numerical procedures (e.g. Forristall's KBCF method) could be used if made computer efficient. Here also there is a need for data verification.

2.3.   Current with waves

There is no analytical procedure available for the general case. There are, however, empirical procedures which are used by the offshore industry; these procedures have not been validated with data. For two-dimensional, steady regular waves and collinear current of arbitrary profile new theories exist and insight can be gained with these theories as to how waves interact with currents. Measurements are needed in the wave zone in particular to test the goodness of these various procedures.

3   RECOMMENDATIONS

The recommendations below address both short term and long term needs. The wave kinematics problem remains still a challenging problem for the theorists and experimentalist, and will require good imagination to bring about new and improved solutions that can be put to use in offshore design.

3.1._____Data

It is recommended that the industry clearly define the required accuracy of new data needed for the different applications. Laboratory measurements on kinematics for the following wave cases are needed:

- 2D regular waves and groups of waves for a larger range of parameters to verify analytical and numerical procedures;
- 2D irregular waves to verify and extend the test data base;
- 2D current with regular and irregular waves;
- 3D irregular waves in order ot extend methods and procedures presently used for 2D waves;
- 3D irregular waves with current.

The laboratory tests should also aim towards definition of <u>field tests</u> which subsequently should be carried out.

3.2_____Analysis_methods

Research should concentrate on the following main tasks:
- Develop fully consistent 2D theory for irregular waves;
- Develop empirical procedures for current with waves;
- Develop fully consistent 2D theory for current with regular and with irregular waves;
- Extend empirical procedures from 2D to 3D;
- Develop higher order analytical and numerical procedures for 3D irregular waves with and without current;
- Clarify further the mean flow in the kinematics in waves.

4   COOPERATION

In order to successfully develop the analytical tools needed and to fully benefit from future testing in the laboratory and in the field, international cooperation is required. It is recommended that standard test cases be defined through the cooperation of the offshore industry and research institutions in order to verify procedures. For advancing the kinematics procedures it is essential that there be data sharing. Also, the offshore industry must continue to provide data for long term research projects at the institutions and the research institutions must continue to carry out basic research to solve engineering needs.

Report from Working Group on
Shallow Water Wave Kinematics

R.L.Soulsby, Chairman
G.Klopman, Presenter

INTRODUCTION

The Working Group on Shallow Water Kinematics (WGSWK)
defined "shallow water" as extending from the
conventional deepwater limit, h/Lo = 0,5, to the
shoreline. A decision was made to focus on the needs for
the next two or so decades with criteria of methodology
and data required by the designer and a generally
improved understanding of the phenomena related to
shallow water kinematics. Although of importance, wave
hindcasting was not considered as within the Group's
scope. Finally, in order to provide focus to our
recommendations, only six problem areas have been
selected for emphasis.

STATUS

The shallow water region is characterized as one of
substantial nonlinearity, and directionality with the
possibility of strong currents and large tidal
fluctuations. At present, substantial variability in
methodology ranging from ad hoc to sophistication is
employed in the prediction of shallow water wave
kinematics. The range of methodology encompasses regular
vs irregular waves, deterministic vs stochastic
representation and various methods of combining waves
and currents. Although each of them can play useful
roles, a better understanding is needed of the strength
and relatively applicability and differences to be
expected. Regulatory agencies prefer straight forward
procedures suitable for codification. A need exists to
develop and introduce procedures that have greater
realism and physics and can be readily applied by
industry.

The workshop addressed a number of problems relative to
shallow water kinematics. It was concluded that for the
case of regular waves without currents, existing wave
theories are adequate. However, more emphasis needs to
be directed to the problems of regular waves with a
current which varies over depth. Of particular concern
is the appropriateness of methods for predicting near
surface kinematics. Due, in part, to the high
nonlinearity of the surfzone and the significance of

9

A. Tørum and O. T. Gudmestad (eds.), Water Wave Kinematics, 9–15.
© 1990 Kluwer Academic Publishers.

sediment transport in this area, more effort here is needed. Other areas of reseach need are described in the following sections of this Working Group Report.

LONG-TERM CLIMATE DESCRIPTION

A Long-term climate description of a region is an important engineering activity which requires research in order to improve its reliability. In the absence of long-term widespread measurements, emphasis is thrown onto the use of recently developped shallow water codes such as -3 GWAM. The hindcasting is performed for adjacent deep water and wave energy is then radiated in towards the shallow water region. Inherent difficulties with this systematic approach centre around the influence of current-interaction, shoaling, and large variations of still water level with time. Areas of immediate research concern include the following:

- Algorithms to optimize long-term information whilst reducing the bulk of wind and pressure fields processed. (Such as sequential boot-strapping techniques and other sub-sampling procedures).

- Improving storage techniques for time-step and spatial interpolation.

- Statistical methods for addressing questions of simultaneity of water depth, current wave fields.

B Although many platforms are equipped to record seastate data, this is still done on a systematically intermittant basis (e.g. 20 minutes for each 3 hours). This appears to be satisfactory for most spectral and statistical requirements, but cannot be expected to record the more extreme events which may occur whilst the instruments are on stand-by. Without resorting to continuous recording, and thence involving storage and servicing problems, it would be highly desirable to develop methods of recording the extrements events in as much detailed as possible.

WAVE TRANSFORMATION & DETERMINISTIC MODELS

A In the analysis of nearshore wave phenomena, the starting point will most often be information (p.g., the directional wave spectrum) of the wave field in deep water. Therefore, one needs wave transformation models that can transfer this information into the shallow water area of interest and to provide input for surf zone models.

The different model employed are chosen according to the scale and distance from the shore of the area under consideration:

- Outer area:
  Depth-Current refraction models (linear)
  Ray theory, finite difference
  Mesh size 100 m - 1 km
  Size of area modelled 25 km^2 - 1000 km^2

- Intermediate area:
  Models based on mild slope equation (linear, extended to include dissipation and boundary absorbtion) the parabolic approxination (linear or weakly nonlinear and dissipative).
  Mesh size: 5m - 20 m
  Size of area modelled: 1 km^2 - 25km^2.

- Harbour areas:
  Models based on "Boussinesq equation". (weakly nonlinear).
  Hehnholz equation or 30 linear potensial solutions.
  Mesh size: 5 km - 20 m.
  Size of area modelled: < 5 km^2.

Research needs:
- Spectral model including refraction, diffraction, frequency shifts due to wave breaking
- generation of long waves by short waves, and their transformation.

B As the water depth becomes small and/or the waves become high, the analytical wave models seize to be valid: The prediction of maximum waves of constant form (with or without current) require the use of stream-function-like theories. In some cases these data can also be fitted to extreme wave data to account for the asymmetry found in nature. Asymmetry statistics or realistic methods of generalizing wave assymetry should be developed. The wave models for prediction of kinematics under waves should be evaluation with high quality data.

2D models based on potential flow can now solve fully nonlinear problems from deep to shallow water. Up to now they are however limited:

- to the first breaking wave
- lack of a radiation boundary for arbitrary waves
- cannot model friction and flow separation

Research should be directed toward:

- introducing variety
- investigating post-breaking extensions of such models
- developing a free boundary (i.e. weakly reflective boundary conditions).
- 3 D potential flow models

WAVE-CURRENT INTERACTION

Waves interact with current in a number of important, and often nonlinar ways within the upper frictionless part of the water column and through their combined friction at the bed. These strongly influence the wave propagation characteristics, the current distribution, the resulting sediment transport, and forces on structures. The State of the Art and the Research Needs on various modes of interaction are as follows:

A Refraction

At the moment, refraction of linear waves by horizontally sheared, vertical uniform currents can be calculated. It is however, necessary to extend these methods to nonlinear waves.

B Kinematics

Methods for the computation of regular nonlinear waves on condirectional vertically sheared currents have recently appeared. They should be extended to non-codirectional cases, and to random waves. Also, they should be combined with above mentioned item to predict refraction and kinematics due to horizontally and vertically sheared currents.

C Mass transport

Theories exist to calculate the mass transport of water near the bed for a laminar wave boundary layer in the absence of currents. These should be extended, first to the turbulent wave boundary layer, then to the case where an external (e.g. tidal) current is imposed.

D Bottom friction

Several theories exist to calculate mean and peak bed shear stresses due to sinuisoidal waves superimposed on a current. These prediction methods need to be simplified (without offering much of their accuracy) to be applicable in wave phase averaged numerical wave and current models. Also, extensions should be made toward irregular (frequency and direction) waves, breaking waves and to distinguish the wave propagation direction.

All the above topics require advances through theoretical and numerical methods, and laboratory and field measurements.

The topics require to be combined together to provide fully interactive 2D numerical models of wave and current fields, leading possibly (on the long term) to 3D models.

NONLINEAR DIRECTIONAL WAVE THEORY

A constant reoccuring problem in ocean engineering consists of finding the best choice for a given application between random, linear, directional wave thoery and nonlinear undirectional deterministic wave theory. Each approach seems to be optimal for some class of problems, but there are other types of problems for which neither is quite right. A nonlinear, directional, wave theory is sorely needed to bridge betwen the two extremes. In recent years, several new approaches to the problem have emerged. This, coupled with the increasing power of available computers, makes now an appropriate item to undertake the solution of this reoccurent problem, both for deep and shallow water.

One potential direction of attack is to combine waves from several directions while forcing satisfaction of the wave equation and free surface conditons. A second direction consists of a iterative procedure which starts with a linear, directional sea surface and, in a stepwise fashion, proceeds to modify the surface toward a better satisfaction of free surface conditions while maintaining the proper directional spectrum and any conditioning constraints.

There probably are other alternatives also available. The development of such a general random directional wave theory would be of great value in future ocean engineering investigations.

SPLASH ZONE KINEMATICS

The wave kinematics at or near the free water surface are very critical in determining overturning moment and other load parameters in an offshore structure. It is difficult to make field measurements of wave kinematics in this zone since the measuring devices are alternately submerged and in the air. The abrupt shock to the instrument at the moment of passing through the air-water interface typically induces transients in the measurements which are difficult to interpret. Measurement of forces in the splash zone experience

similar shock problems. Non intrusive measurement scheme
for both kinematics and forces need to be developed and
studied.

From an analytical perspective relative to random,
linear directional wave theory, the prediction of
kinematics and forces for locations above mean water
level are perplexing and create paradoxes within linear
wave theory. Linear theory assumes infinitesmal
amplitudes and so kinematics above mean water level are
contradictory. If the theory is used without
modification above mean water level, the velocities are
overestimated in a part of the force regime critical to
most structural computations. Various "stretching"
schemes have been proposed and use, but these are quite
"ad hoc" in nature and theoretically unsatisfactory. It
would be much more desireable to have some scheme which
closely satisfies the free surface boundary conditions.

Some experiments have been carried out to investigate
splash zone kinematics and forces under laboratory
conditions. Use of these results demands a knowledge of
scale effects which for these processes are poorly
understood. In field conditions also there is likely to
be a thin wind-induced surface current which may locally
enhance particle velocities and forces.

In summary there is a research need for (a) improved
instrumentation for measuring forces and kinematics in
the wave splash zone, and (b) more theoretically sound
ways to modify random linear directional wave theory for
loading points in the splash loading zone. (c) improved
understanding of scale effects and (d) of the importance
of wind-induced surface currents.

SURF ZONE HYDRODYNAMICS SEDIMENT TRANSPORT

To further our understanding of the processes that shape
our coastlines, it is essential that we understand the
hydrodynamics of the surf zone. Of particular interest
are the distribution over depth of wave induced
longshore and onshore-offshore currents and the momentum
fluxes which drive them. Advances in this area will
enable more refined studies of sediment transport modes
and rates.

As has been stressed by Dr. N. Barltrop during this
workshop it is important to determine the appropriate
wave theory to use for a particular study. Waves in the
surf zone are non-linear. The hydrodynamics of the surf
zone is further complicated by the geometry of the
bottom. Therefore it is suggested that shallow water

wave theories (for constant and variable depth) be
evaluated for applicability over various geometries
(starting with planar and simple curved geometries and
eventually extending to geometries which include bars).
In parallel with the theoretical studies numerical
approaches (such as the Stream Function Wave Theory or
the Numerical Wave Tank) should be considered.
Infragravity waves can also cause substantial velocity
fluctuations and water surface displacements in the surf
zone.

The turbulence associated with wave breaking causes the
wave height to decay which is directly related to the
momentum fluxes. Thus studies to determine the variation
of eddy viscosity over depth and across the surf zone
are essential. Another process to be further studied is
the reforming of waves as they propagate over the
troughs between bars.

Further research is also required in including the
effects of the bottom boundary layer on the wave induced
currents. In the area of sediment transport studies are
required to model sediment transport across the surf
zone and over depth. Shallow water symmetrical and
asymmetrical waves and their interactions with currents
should be considered as driving mechanisms.

Sediment transport driven by random waves is also an
area requiring research efforts.

Better models to quantify longshore and crosshore
sediment transport rates for suspended and bedload load
transport are required. The threshold of motion and
sediment pick-up rates are topics still requiring better
understanding. One area of particular interest is the
determiniation of the extent over depth of the breaking
induced mixing and its effects on sediment transport. In
this respect breaker types should be considered in
analysis. Equilibrium beach profiles also need further
research. To obtain a broader understanding of the
mechanisms shaping coastal regions the sediment
transport outside the surf zone requires attention. The
study of the transport in this zone including tidal and
storm surge currents is of importance. Surf zone and
non-surf zone models of transport need to be connected
in attempts to obtain a global model.

The verification of the results of the investigations
suggested through laboratory and especially field
testing is of utmost importance.

Report from the Working Group on
Breaking and Freak Waves

D.H.Peregrine, Chairman
S.Sand,   Presenter

Deep-water breaking

Presentations at the workshop have included discussion of waves breaking in deep-water but details are only given in the context of wave tank simulations. Conventional measurements at sea do not distinguish broken from unbroken waves.

We wish to emphasize that the physics of breaking waves can differ between wave tanks and waves at sea. In particular, the most frequent and important occurrences of deep-water breaking are in wind-driven waves. The character of wind-driven waves varies with their age, but at any time breaking is intermittent and short-crested. This makes it difficult to quantify and measure breaking events.

Statistics of breaking would be invaluable. Some statistics of white caps have been collected and are clearly related to breaking events. Efforts to quantify the relation would be welcome and should be contrasted with statistics of breaking based on spectral distributions which only represent the exceeding of a mathematical criteria. Such criteria need to be established from the physics of wind-wave interaction. An initial approach based on measurements following the mariner's type of observation is recommended, e.g. the percentage of breaking waves as a function of wind speed, the size and duration of white caps, depth of air entrainment etc. We consider relations between sea state and Beaufort number could be confirmed and compared with modern wind casting methods.

Short-crestedness implies a need to account for three-dimensional breaking in multi-directional seas, but first the behaviour of ideal regular waves needs clarification both experimentally and theoretically.

The important effects of wind, with its pressure distribution driving the waves and the vorticity that wind drift induces in the surface layer have yet to be quantified, although, their qualitative importance has been demonstrated. This research is urgently needed since wave breaking is intimately linked with wave generation, and this linkage produces the most severe states that ships and offshore structures encounter.

We propose that measurements of breaking wind-waves be first conducted in wind-wave flumes, where it should be possible to measure all significant parameters,

A. Tørum and O. T. Gudmestad (eds.), Water Wave Kinematics, 17–20.
© 1990 Kluwer Academic Publishers.

though some development of instrumentation may be needed, especially to recognise wave breaking. In particular local properties worth measuring include elevation, velocity, vorticity, turbulent intensity, in both air and water as well as air or water entrainment. General properties to be examined include the type of breaking, three-dimensional structure and the role of breaking in frequency downshifting.

Theoretical studies need to be encouraged into quantifying the effects of surface drift and of three-dimensionality. Current numerical models can describe two-dimensional irrotational waves as they break. Addition of a layer of vorticity at the surface is feasible. Three-dimensional irrotational programs are likely to be developed but combination with vorticity seems unlikely in that case.

Shallow-water breaking

Breaking induced by bottom topography is a well-studied area because of its importance in all aspects of coastal engineering. For the purposes of this working group we ignore its implications for and interactions with surf-zone processes, other than the basic hydrodynamics. The presentations in the workshop reflect current research frontiers. That is investigation of the complex but deterministic flow field that occurs after breaking is represented by Tallent's detailed description of the eddies and splashes for a wave on a beach, the numerical modelling of wave behaviour before any splash occurs is described by Grilli and by Peregrine. These are all two-dimensional studies. For shallow-water breaking this limitation is not serious since such breaking is usually longcrested.

A number of substantial field experiments in the last decade are giving a good appreciation of surf-zone kinematics. However, for many practical purposes results from regular wave trains are used, often with extremely simplified models of the surf zone.

At the research level, many detailed aspects need study as mentioned below but the only major gaps in our knowledge concern the flow once breaking starts, until, on gentle slopes, the wave has become a turbulent bore; and the flow when breaking stops as when a broken wave enters deeper water. Further study is required because of the importance of the breaking phase in processes such as sediment transport, the generation of nearshore circulation which influences dispersion of pollution etc, the determination of set-up and run-up with relevance to flooding, and the severe conditions that breaking provides for structures such as breakwaters and sea walls.

Research is needed in the experimental area to clarify the strong effects of wind which at the present time are mainly recorded in books on surfing. Onshore winds lead to a dominance of spilling breakers, whereas offshore winds encourage plunging breakers. Several aspects of breaking need to be quantified, although we can recognise spilling and plunging breakers, intermediate cases are not readily described. That is, we still lack good definition for the relative size and strength of breakers.

The present initial studies on breaker structure need to be developed and extended since for many applications this is the most significant phase of wave development. There are still distinct experimental difficulties, and instruments or measuring techniques need to be developed, e.g. to measure air-water ratios.

Experiments and characterization of irregular waves as they break are required, especially on steeper slopes where backwash has a significant influence on breaker character.

Field observations have provided substantial data banks which can be usefully studied with further analysis and interpretation. We note however that conventional measurements can be of much greater value if simultaneous film or video records are made since it is not always easy to recognise wave breaking from other records.

Theoretical study is hampered by the difficulties of describing the breaking process, though there is a good chance of progress if the simple turbulence modelling of Svendsen and Madsen (1984) is developed. Verification of the potential flow models up to breaking by comparison with a wider range of experiments is desirable, and these models can be improved in a number of ways, e.g. inclusion of vorticity layers may mimic wind drift, or a sheared backwash flow.

Mathematical models of a simpler kind, such as those using the finite–amplitude shallow–water equations could provide insight into the interaction between the dominant waves and longer period waves (infra–gravity waves) whose origin may be in the surf–zone region and which have a dominant influence on run–up on gentle beaches.

Freak waves

The workshop presentations of Kjeldsen and Sand give a good indication of the "state of the art" in this area. The following definition of freak waves has been proposed by Kjeldsen

$$H_{max}/H_s > 2.0$$

since this will in most cases (number of waves n = 1000–2000) represent a clear exceedance of the most probable maximum wave according to the Rayleigh distribution (c.f. Sand et al). On the basis of observations in the Gulf of Mexico, the North Sea, the Irish Sea, off the coast of Norway, etc. it is believed that the phenomenon appears in both deep and shallow water, and is dependent on locality.

It is proposed that the following causes be further investigated:

Wave phenomenon causing freak waves –

– wave focussing
– waves arising from different directions
– super–position of waves of different scales

Amongst the physical causes are

– topographical refraction, diffraction and reflection on different scales
– wave–current refraction
– colliding low pressure systems (e.g. collisions between polar lows and ordinary low pressures)

- combination of swell and wind seas
- response of the wave field to a turning wind field such as a hurricane

In many cases freak waves appear as a result of combinations of the phenomena mentioned above (see Kjeldsen).

Recommendations for future research are listed under the following headings:

- Full scale freak wave measurement program (selection of instrumentation, site platforms, etc.)
- Extended analysis of existing data.
- Correlation of meteorological information and freak wave occurrence.
- Further description of freak wave categories as non–breaking, spilling or plunging breakers.
- Kinematic theory of 3–D freak waves (especially above MWL).
- Ultimate and accidental limit state design guidelines and possible forecasting for ships and the oil industry.

(See also recommendations given by ISSC on this subject in August, 1988).

Report from the Working Group on
Measurement of Wave Kinematics

P.Klinting,   Chairman
J.E.Skjelbreia   Secretary

1 INTRODUCTION

It was the position of this group that the other working
groups in this workshop would provide the needs and
justification for additional wave kinematics
measurements. Consequently, our recommendations will
focus more on the means of implementing measurements and
making suggestions for improving the state-of-the-art.
Given the very different natures of field and laboratory
measurements requirements, this group felt it best to
consider the requirements independently. It is important
to note that our recommendation philosophy will not be
limited to currently available methods and techniques
but to also include technology and techniques that show
promise for future wave kinematic measurements.

2 STATUS

2.1 Field

The quantity of field measurements available is
inadequate and the quality, with respect to wave
kinematics, is, in general, low. Furthermore, available
measurements are often severly limited in usefulness due
to the measurement locations, structural influence and
lack of one or more velocity components. Instrument
reliability is very uncertain as instruments are often
improperly calibrated and maintenance and recalibration
are infrequent or non-existent. Not infrequently,
valuable information in severe storms is lost due to
equipment failure.

2.2. Laboratory

In contrast to field measurements, the quantity and
variety of laboratory measurements available is
significant, but are still lacking in the important
near-surface and bottom zones. The possibility of
excellent quality in measurements is high given: the
opportunity for assured reliable instrument operation
through frequent testing and calibration; the
disturbance to the flow, by the measuring instrument or
its support structure, can be made negligible or

A. Tørum and O. T. Gudmestad (eds.), Water Wave Kinematics, 21–23.
© 1990 Kluwer Academic Publishers.

completely eliminated altogether; and that three-component measurements are attainable. Despite this, laboratory measurements must be viewed cautiously due to the possibility of improper modelling and uncontrollable tank effects.

3 USE OF MEASUREMENTS

3.1 Field

In general measurements are made for a specific purpose. Field measurements are though focusing on topics not covering the wave kinematics. Instead oceanographic knowledge is sought in order to define statistical parameters, for example :
- wave height
- tidal and ocean currents
- wind speed

However, time series of more severe events are in general available.

3.2 Laboratory

Laboratory experiments are, in general, conducted to verify the validity of theories and extensions to theories. Therefore laboratory measurements are, for the present, limited to unidirectional problems and most often to shallow or very shallow water cases since theoretical efforts are extensive here.

Laboratory measurements, therefore, do not cover the complex situation as found in nature that we attempt to learn more about.

4 MEASURING TECHNIQUES

4.1. Field

The most frequently used velocimeters are of the electromagnetic flow meter type. A few acoustic meters are also now available, their applicability to wave kinematics is uncertain. For current measurements, mechanical propeller meters are commonly applied for short term measuring programs. They are all kept at fixed positions by mounting directly to fixed structures or suspended from buoyant bodies. In general, none of the above mentioned meter types are suitable for measurements in the splash zone.

4.2 Laboratory

Commonly used types of velocimeters are elctromagnetic, acoustic, and micropropellers meters. Optical techniques such as laser Doppler velocimetry have become increasingly used providing measurements without disturbing the flow.

5 RECOMMENDATIONS

5.1. Field

a. Equipment operation should be kept at a level securing high fidelity, i.e. sufficient calibration and verification are needed.

b. Equipment should be made durable.

c. Measurements are to be taken at sufficient distance from structural influence. All three velocity components are needed.

d. Equipment for near surface measurements needs development.

e. Three-component LDV should be developed.

5.2 Laboratory

Techniques in general should be improved to allow for a more free choice in setting up an experiment. Whole-field measurements may prove excellent for verification after improvement of the recording and processing techniques. Maintenance and reliability should be given the utmost attention.

Report from the Working Group on

Forces

A. Cornett, Chairman
N.Spidsoe, Secretary

The discussions of this working group were an example of the dialogue and international co-operation that is personally rewarding for all participants and leads to valuable consolidation of the state-of-the-art and also points to future research needs.

This has been a workshop on water wave kinematics. Forcing on ocean and coastal structures is fundamentally related to kinematics - the two subjects should not be considered independently. However, the discussions of this working group were focussed on the issue of how knowledge of kinematics can lead to predictions of forcing. The discussions began with consideration of Morison's equation, and expanded to include other topics such as freak waves, laboratory standards, and the need for an accessible complete information data base of field and laboratory data.

We have seen that the determination of force coefficients for an irregular wave case cannot be expected to follow from any regular flow situation. The case of coexisting waves and current should also be considered separately. It has been suggested that the use of random process theory to determine the variance and statistics of extreme Morison equation forcing in irregular waves leads to very encouraging results.

It is generally recognized that kinematics in the near field are more significant than those in the far field. Further study of wave effects on force coefficients is recommended.

More data is needed on force coefficients in situations typical of the real ocean environment. Continued research into force coefficients appropriate for: multidirectional seas, coexisting waves and currents, clusters of cylinders, and relative velocity situations, is recommended. The emphasis of this research should remain on understanding the physics of the forcing process, particularly the importance of kinematics in the near field.

A. Tørum and O. T. Gudmestad (eds.), Water Wave Kinematics, 25–27.

Having called for more research and more data, the working group considers it important to recommend that the data be available to researchers around the world, and that it be reported in a sufficiently complete manner so that results from various sources may be compared. Probabilistic format and methods should be more systematically used in the processing and presentation of experimental results in order to:
(a) make them objectively comparable, and
(b) provide the information needed for reliability design methods.
A common data base for force coefficient values, together with the ability to inter-compare the results from various sources, would encourage more rigorous application of the Morison equation.

The accident at the Ekofisk field in 1984 in which a control room wall was damaged by a wave crest at least 22 m above mean water level has been mentioned by both Kjeldsen and Sand et al. This event provides graphic evidence that "freak" waves are a definite hazard to ocean structures. There is a need to learn more about the processes through which these waves are formed, their kinematics, and their probability of occurence. Efforts to understand the influence of crossing seas, topography, and co-existing current, on the frequency and formation of these extreme waves should continue. More good field data is needed to extend our understanding of "freak" waves. It has been suggested that the wave rider buoy is not capable of making these measurements.

A related topic is the forcing caused by breaking waves which may not necessarily be "freaks". Breaking and very steep asymmetric waves can exert significantly larger forces on ocean structures than steady symmetric waves of similar height and period. Such unsteady waves exist in real seas, and they should be considered by the designers of ocean structures. It is recommended that research into the probability of occurrence of wave breaking in a variety of realistic multidirectional seas sould continue. It is also recommended that study of the forcing due to unsteady and breaking waves should continue.

A gap still exists betwen regulations - and therefore, engineering practice - and state-of-the-art research. There is a need for continual upgrading of design specifications to reflect advances made by the research community.

There was some discussion of the need for quality
assessment of wave basin tests. Information of
reflection characteristics, wave generator capabilities,
and the vertical distribution of kinematics measured in
the basin at various distances from the wave generator,
should be documented and made available by basin
operators.

More research on long waves and their influence on the
slow drift motions of floating structures is
recommended.

In summary, we would like to encourage the participants
of the other working groups to continue to advance on
many of the kinematics-related issues discussed at this
workshop. This will help those of us in the forcing
working group to solve some of our own problems.
Expressed in a different way, learning more about
hydrodynamic forcing on coastal and ocean structures is
intimately linked to greater understanding of wave
kinematics.

Introductory Lectures

STOCHASTIC DESCRIPTION OF OFFSHORE ENVIRONMENT

M. K. OCHI
University of Florida
336 Weil Hall, COE Department
Gainesville, Florida, 32611
U.S.A.

ABSTRACT. This paper presents a state-of-the-art review of the stochastic description of the ocean environment. Non-linearity consideration in random seas, wave breaking, wave groups, and stochastic properties of offshore currents are discussed presenting phenomena observed in the real world.

1. INTRODUCTION

Properties of wind-generated waves observed in the ocean are not readily definable on an individual basis; hence, they may best be represented from the probabilistic point of view. In other words, waves are considered to be a random process following a certain probabilistic law, from which various statistical characteristics can be evaluated. Here, the probabilistic law (probability distribution) is developed taking into consideration the physics (kinematics) of the waves. In turn, the probabilistic approach dealing with random waves may serve to aid the further advancement of knowledge of wave kinematics.

This paper presents a state-of-the-art review of the stochastic description of the ocean environment. The subjects reviewed in this paper are primarily those relevant to wave kinematics on which significant progress has been made in recent years. In reviewing the current status, specific effort is made to clarify the basic principles or basic approach for evaluating the stochastic properties of a random phenomenon, rather than to present the mathematical derivation of prediction formulations.

Non-linearity consideration in random waves is not commonly used, although the stochastic prediction method was developed in 1963 based on wave kinematics. This prediction method as well as several other prediction methods pursuing different approaches are summarized.

Wave breaking and wave groups are discussed presenting phenomena observed in the real world. Discussed also are the bases and approaches for stochastic prediction of the frequency of occurrence of

31

A. Tørum and O. T. Gudmestad (eds.), Water Wave Kinematics, 31–56.
© 1990 *Kluwer Academic Publishers*.

the phenomena in random seas.

The stochastic properties of offshore currents have not been clarified to date. Some properties which may be of interest for consideration in wave kinematics are outlined from the results of a study currently being undertaken.

2. NON-LINEAR RANDOM WAVES

For the probabilistic presentation of surface wave profiles observed in random seas, the wave profile is commonly considered to be an accumulation (linear superposition) of sinusoidal components with random phases which may be written as

$$\eta = \sum_i \alpha_i \, \xi_i \, , \tag{1}$$

where α_i = constant, ξ_i = sinusoidal wave.

Hence, waves are assumed to be a Gaussian random process based on the central limit theorem in probability theory. For a more precise presentation of the sea surface, however, the non-linear concept must be introduced in which quadratic and higher-order terms associated with the sinusoidal components should be considered. In other words, we may write the non-linear wave profile as

$$\eta = \sum_i \alpha_i \, \xi_i + \sum_{i,j} \alpha_{ij} \, \xi_i \, \xi_j \, . \tag{2}$$

The concept of the effect of non-linearity on ocean wave spectra was introduced in the early 1960s by Tick (1959, 1961), Phillips (1960, 1961), and Hasselmann (1961, 1962). The effect of non-linearity on the statistical distribution of wave profiles was first clarified by Longuet-Higgins (1963), followed by many related studies. These include Tayfun (1980, 1986), Anastasiou et al. (1982), Tung and Huang (1983), Huang et al. (1983), Srokosz and Longuet-Higgins (1986), Langley (1986), etc., among others.

The statistical properties of non-linear waves are outlined in the following discussions.

Assuming that the wave slope is small, the velocity potential, ϕ, and surface elevation, η, of non-linear waves are expanded in a perturbation series (Hasselman 1962). That is,

$$\phi = \phi_1 + \phi_2 + \text{----}$$

$$\eta = \eta_1 + \eta_2 + \text{----} \, . \tag{3}$$

Then, from consideration of the continuity equation, the kinematic and dynamic boundary conditions at the free surface, etc., the first-order solution and the second-order additional term for non-linear waves can be derived (Longuet-Higgins 1963). For unidirectional seas, we may write,

$$\eta_1 = \sum_{n=1}^{N} a_n \cos \chi_n$$

$$\eta_2 = \frac{1}{2g} \sum_{m=1}^{N} \sum_{n=1}^{N} a_n a_m \omega_n^2 \cos(\chi_n + \chi_m) - \frac{1}{2g} \sum_{m=1}^{N} \sum_{n \geq m}^{N} a_n a_m (\omega_m^2 - \omega_n^2) \cos(\chi_n - \chi_m),$$

$$(4)$$

where a_m, a_n = amplitude, $\chi_n = k_n x - \omega_n t + \xi_n$,

   k_n = wave number, ω_n = frequency, ξ_n = phase.

It is observed in Eq.(4) that the non-linear wave profile is expressed in the same form as shown in Eq.(2).

From the probabilistic view point, the wave profile η_1 (linear wave) obeys the normal distribution. Hence, it follows that the distribution applicable for the sum of η_1 and η_2 may be expressed as an extension of the normal distribution; most likely in the form of series.

In probability theory, an arbitrarily given standardized probability density function (distribution with zero mean and unit variance) can be expressed, in general, in the following form associated with the standardized normal distribution $\alpha(z)$:

$$f(z) = \alpha(z) + a_1 \alpha^{(1)}(z) + a_2 \alpha^{(2)}(z) + \text{----}$$

$$(5)$$

where $\alpha^{(n)}(z)$ = n-th derivative of $\alpha(z)$.

By applying the property that Hermite polynomials are orthogonal with respect to $\alpha(z)$, the probability density function $f(z)$ can be written in terms of moments and Hermite polynomials. That is,

$$f(z) = \frac{1}{\sqrt{2\pi}} e^{-\frac{z^2}{2}} \left[1 + \frac{m_3}{3!} H_3(z) + \frac{m_4 - 3}{4!} H_4(z) \right.$$

$$\left. + \frac{m_5 - 10m_3}{5!} H_5(z) + \frac{m_6 - 15m_4 + 30}{6!} H_6(z) + \text{----} \right],$$

$$(6)$$

where m_j = j-th moment of the standardized random variable.

$H_j(z)$ = Hermite polynomials of degree j.

The probability density function given above is called the Gram-Charlier series of Type A. This probability density function may be applied to the sum of linear and non-linear wave profiles, $(\eta_1 + \eta_2)$, by letting $z = (\eta_1 + \eta_2)/\sigma$, where σ^2 is the variance of waves which can be evaluated from analysis of either the observed time history or the spectrum.

Longuet-Higgins presented an arbitrarily given probability density function as a function of the normal distribution through the cumulant generating function, which is the logarithm of the characteristic function of the distribution given as

$$\Psi(t) = \ln \left(\int_{-\infty}^{\infty} f(x)\, e^{itx}\, dx \right) = \sum_{j=1}^{\infty} \frac{(it)^j}{j!}\, k_j , \qquad (7)$$

where k_j = cumulant, $f(x)$ = any probability density function.

Inversely, the probability density function $f(x)$ can be evaluated from

$$f(x) = \frac{1}{2\pi} \int_{-\infty}^{\infty} \exp\left\{ \Psi(t) - itx \right\} dt$$

$$= \frac{1}{2\pi} \int_{-\infty}^{\infty} \exp\left\{ (k_1 - x)it + \frac{k_2}{2!}(it)^2 + \frac{k_3}{3!}(it)^3 + ---- \right\} dt. \qquad (8)$$

By applying the property of the Hermite polynomials and by standardizing the random variable X by $Z = (X - k_1)/\sqrt{k_2}$, the following standardized probability density function can be obtained:

$$f(z) = \frac{1}{2\pi}\, e^{-\frac{z^2}{2}} \left[1 + \frac{\lambda_3}{3!} H_3(z) + \frac{\lambda_4}{4!} H_4(z) + \frac{\lambda_5}{5!} H_5(z) \right.$$

$$\left. + \frac{\lambda_6}{6!} + \frac{\lambda_3^2}{2!(3!)^2} H_6(z) + ---- \right], \qquad (9)$$

where $\lambda_j = k_j/(k_2)^{j/2}$. In particular, λ_3 = skewness,

λ_4 = kurtosis minus 3.

In reality, the probability density functions given in Eqs.(6) and
(9) can be proven to be the same. However, the significant advantage
of Longuet-Higgins' approach is that the method lends itself to the
derivation of the joint non-Gaussian probability density function for
two random variables which can be applied to displacement and velocity
of non-linear waves.

An example of the application of the probability density function to
non-linear waves is shown in Figure 1. Wave data are obtained in an
area of finite water depth and hence the wave time history shows a
definite excess of high crests and shallow troughs. The histogram of
the wave profile is not symmetric with respect to the mean value. As
can be seen in the figure, the probability density function consisting
of terms with the parameters λ_3, λ^2_3, and λ_4 agrees well with the
histogram. This is a general trend observed from many examples.
However, the probability density function becomes negative for large
negative displacement. The negative part is not included in Figure 1.
Although the magnitude of the negative portion is relatively small,
this is a drawback of the probability density function given by Eqs.
(6) and (9), in practice.

In order to circumvent this weakness of the non-Gaussian probability
density function, Tayfun (1980, 1986) expressed the non-linear wave
profile in the form of (amplitude modulated) Stokes wave under the
assumption of a narrow-banded spectrum. That is,

$$\eta = a \cos \chi + \frac{1}{2} a^2 \bar{k} \cos 2\chi .\qquad(10)$$

where   a = random amplitude, $\chi = \bar{\omega}t + \xi$, $\bar{\omega}$ = mean

frequency = m_1/m_o, $\bar{k}$ = mean wave number = $\bar{\omega}^2/g$.

The probability density function of η cannot be obtained in closed
form; instead, it is derived numerically by differentiating the
cumulative distribution function of η .

An example of a comparison between the Tayfun, Gram-Charlier, and the
normal distributions as applied to simulated data is shown in Figure 2
(Tayfun 1980). As can be seen in the figure, the Tayfun distribution
agrees well with the Gram-Charlier distribution except for the tail
portion representing the large negative values of the distribution.

Huang et al. (1983) also developed a probability density function for
non-linear waves in random seas based on Stokes expansion carried out
to the third order. The probability distribution developed by Tayfun
and that developed by Huang et al., however, are applicable for waves
with a narrow-banded spectrum in contrast to the distribution developed
by Longuet-Higgins.

Another method to analytically develop a probability distribution for
non-linear waves without limitation of the spectral band-width was
recently presented by Langley (1987). The principle of this method is
that the statistical distribution of the response of a non-linear
system which is given by the two term Volterra series expansion can be

expressed in the form given in Eq.(2) with i=j. Here, ξ_i in Eq.(2) are independent orthogonal standardized normal variates and the coefficients α_{ij} are the eigenvalues of an integral equation and α_i can be found from the eigenfunction and input spectrum (Kac and Siegert 1947). This approach has been applied for derivation of the statistical distribution of non-linear responses of marine structures by Neal (1974), Vinje (1983), Naess (1985), Langley (1987), etc. In applying this concept to non-linear waves, Langley developed the input based on second order random wave theory. Upon finding the constants α_i and α_{ij}, the probability density function of η of Eq.(2) with $\xi_i = \xi_j$ can be derived through the characteristic function of η, but the density function cannot be obtained in closed form.

3. BREAKING WAVES IN RANDOM SEAS

Wave breaking in deep water is a subject of extreme interest on which numerous papers exist in the literature. Among others, one of the most important topics for breaking is that concerned with the breaking mechanism. The criterion for breaking has been discussed from different view points by many researchers. These include Stokes (1880), Mitchel (1893), Dean (1968), Banner and Phillips (1974), Van Dorn and Pazen (1975), Nath and Ramsey (1976), Longuet-Higgins (1969, 1974, 1976, 1980), Longuet-Higgins and Cokelet (1976), Kjeldsen and Myrhaug (1978), Ochi and Tsai (1983), Snyder and Kennedy (1983), Weissman et al. (1984), Srokosz (1985), Holthuijsen and Herber (1986), Xu et al. (1986), Ramberg and Griffin (1987), etc.

The most commonly known wave breaking criterion is the Stokes limiting wave steepness. That is, breaking takes place when the wave height exceeds 14.2 percent of the Stokes limiting wave length which is 20 percent greater than that of ordinary sinusoidal waves of the same frequency. This criterion can be written in terms of wave height, H, and period, T, as

$$H \geq 0,027 \ gT^2. \tag{11}$$

Results of laboratory experiments on breaking of regular waves generated in a uniform section channel indeed demonstrate that breaking takes place when the condition given in Eq.(11) is satisfied. However, the breaking criterion applicable for irregular waves is substantially different from that for regular waves. Figure 3 shows breaking criteria obtained in three different series of laboratory experiments. The lines given in the figure are the average of the scattered data for each experiment. Indicated also in the figure are the lower and upper limits of wave heights tested. The breaking criterion obtained by Ramberg and Griffin (1987) was established from experiments in waves generated in a convergent channel. Ochi and Tsai (1983) established the criterion from breaking of irregular waves generated by a wavemaker

installed at the end of a tank. On the other hand, Xu et al. (1983)
obtained the criterion from breaking of irregular waves generated by a
blower installed in an air-sea tank. Although the experimental
conditions are quiet different, the breaking criteria obtained from
these experiments are essentially the same. As an average, the wave
height-period relationship for breaking may be written as

$$H \geq 0.020 \ gT^2. \tag{12}$$

Needless to say, breaking is associated with instability
characteristics of waves. Studies on the stability of non-linear
gravity waves in deep water have been carried out by many researchers.
These include analytical studies made by Longuet-Higgins (1978a, b),
Longuet-Higgins and Cockelet (1978), and experimental studies carried
out by Melville (1982) and Sue et al, (1982), among others. One of the
conclusions derived by Longuet-Higgins is that the instability of
regular waves occurs when the wave slope ak (where a = amplitude. k =
wave number) approaches 0.436. This can be written in terms of wave
height and period as $H = 0.022 \ gT^2$ which is very close to the
relationship shown in Figure 3.

It is noted that the wave breaking discussed above is that which
takes place along the excursion (slope) crossing the zeroline, which
may be called Type I breaking as illustrated in Figure 4. There is
another type breaking which takes place along the excursion above the
zeroline (Type II) shown in Figure 4. This is the breaking of waves
superimposed on long waves; hence, the local height where breaking
occurs is small. Xu et al. (1986) observed this type of breaking in
their experiments on wind generated waves. In their analysis for
establishing the breaking criterion, the period of the long carrier
waves was used, resulting in the breaking wave height is very low H =
$(0.005—0.010) \ gT^2$. Breaking wave heights of the same order were
reported in the results of field observations made in Lake Washington
(Weissman et al. 1984). It is believed, however, that Eq.(12) would
also hold for the case shown in Figure 4 if the excursion between local
crests are considered as breaking criteria.

For a more precise description of steep wave profiles for which
breaking is imminent, Kjeldsen and Myrhaug (1978) propose three
parameters; crest front steepness, ξ, vertical asymmetry factor λ, and
horizontal asymmetry factor μ. Definitions of these parameters are
given in Figure 5. An analysis of steep waves obtained from field data
and laboratory experiments shows the maximum values of these parameters
to be $\xi = 0.83$, $\lambda = 7.7$, and $\mu = 0.92$; however, it has not been
clarified whether or not this maximum values directly reflect breaking
conditions.

A wave breaking criterion expressed in terms of acceleration was
first proposed by Phillips (1958) as -g in connection with development
of the equilibrium range considered for wave spectra. Longuet-Higgins
(1963) gives a particle acceleration of -0.5g in a Stokes crest angle
of 120° flow. Snyder and Kennedy (1983) show a threshold vertical

acceleration of -0.5g for white capping waves. The breaking criterion
given in Eq.(12) can be converted to a limiting acceleration of -0.4g.
Longuet-Higgins (1985), however, pointed out some ambiguity in the
definition of wave acceleration used as a criterion of breaking waves.
He showed that the wave acceleration obtained from measurements by a
fixed vertical probe is an apparent acceleration (called the Eulerian
acceleration), while wave particle acceleration ideally measured by a
small floating buoy is the true acceleration (called the Lagrangian
acceleration). These accelerations are different for waves of finite
amplitude, and he has shown analytically a significant difference
between them.

The probability of occurrence of wave breaking in random seas can be
evaluated by applying the breaking criterion given in Eq.(12) to the
joint probability density function of wave height and period.

The joint probability density function of wave height and period has
been analytically developed for waves with a narrow-banded spectrum by
Longuet-Higgins (1983) and for waves with a non-narrow-banded spectrum
by Cavanié et al. (1976) and Lindgren and Rychlik (1982). The first
two distributions are given in closed form. These joint probability
density functions can be used for evaluating the probability of
occurrence of Type I excursion, while the joint distribution of
excursion and its associated time interval for Type II excursion was
developed by Ochi an Tsai (1983).

In order to compute the probability of occurrence of wave breaking
associated with Type I and Type II excursions, it is convenient to
express the joint probability density function as well as the breaking
criterion in dimensionless form. The breaking criterion given in
Eq.(12) can be written in dimensionless form as

$$\nu \geq \alpha\left(\overline{T}_m^2/\sqrt{m_o}\right) \lambda^2 \tag{13}$$

where $\alpha = 0.196$ cm/sec^2,

ν = excursion (twice the amplitude for Type I)$/\sqrt{m_o}$

λ = time interval (period for Type I)$/\overline{T}_m$

$\overline{T}_m$ = average time interval between successive positive maxima

   $= 4\pi\left\{\sqrt{1-\varepsilon^2} \ / \ (1+\sqrt{1-\varepsilon^2})\right\} \sqrt{m_o/m_2}$

ε = band-width parameter of spectrum

m_j = j-th moment of the wave spectrum.

The dimensionless joint probability density functions applicable for
Type I and Type II excursions are given in Appendix I. Here, Cavanié's
joint distribution is considered for Type I excursion.

In the computation of occurrence of wave breaking, care must be taken to evaluate the frequency of occurrence on Type II excursion relative to Type I excursion prior to evaluating the probability of breaking. From knowledge of frequency of occurrence of Type I and Type II excursions for a given wave spectrum, the conditional probability of occurrence of breaking, given either Type I or Type II excursion can be computed.

As an example of the probability of occurrence of breaking waves in random seas, computations are made for the two wave spectra shown in Figure 6. These spectra are obtained from measurements in the North Atlantic. Spectrum A represents a sea of significant wave height of 4.6 m (15.1 ft) while Spectrum B is that for a significant wave height of 10.8 m (35.4 ft).

The joint probability density functions of the Type I excursion and its associated time interval as well as the lines indicating breaking conditions are shown in Figure 7. As can be seen in the figure, the shapes of the dimensionless joint probability density functions are the same. However, the location of the lines indicating the breaking condition differs substantially and this in turn yields a difference in the probabilities of breaking; the probability being zero for Spectrum A and 0.13 for Spectrum B.

The probability of wave breaking can also be evaluated by applying the criterion given in terms of acceleration to the probability density function of the vertical acceleration of random waves. Following this approach, Srokosz (1986) shows that the probability of breaking evaluated through the acceleration criterion is almost equal to as that evaluated through the joint probability density function if the limiting acceleration for breaking is chosen to be 0.4g.

Evaluation of the probability of breaking in random seas by using the acceleration criterion is much simpler than through use of the joint probability density function. However, the latter approach provides information on dimensions of waves which break making it possible to modify the shape of the original wave spectrum due to breaking.

4. WAVE GROUPS IN RANDOM SEAS

Another interesting phenomenon observed at sea is a sequence of high waves having nearly equal periods which are commonly known as wave groups. The physical explanation of this phenomenon has yet to be clarified; however, Mollo-Christensen and Ramamonjiarisoa (1978) explain that the wave field does not consist of independently propagating Fourier components but instead consists wholly or in part of wave groups of a permanent type. As evidence, they present results of field and laboratory observations indicating that harmonic components of wind-generated waves propagate at higher phase velocities than those predicted by linear theory, and that velocities are nearly constant for frequencies greater than the modal frequency of the spectrum (see Figure 8).

Wave groups often cause serious problems for the safety of marine

systems when the period of the individual waves in the group is close
to the marine system's natural motion period. This is because the
motion is augmented due to resonance with the waves which, in turn, can
induce capsizing of the marine system. Many studies have been carried
out on this subject primarily focusing on the frequency of occurrence
of the phenomenon in random seas. These include Goda (1972), Nolte and
Hsu (1972), Ewing (1973), Rye (1974), 1979), Rye and Lervik (1981),
Kimura (1980), Battjes and Vledder (1984), Ochi and Sahinoglou (1989),
and Longuet-Higgins (1984), etc.

In order to obtain the statistical characteristics of group waves, we
may consider the envelope of the wave profile by connecting crests or
troughs as shown in Figure 9. The wave group is then defined as a
level-crossing phenomenon associated with the envelope. For evaluating
the time duration and number of waves in a group, we may assume the
waves to be a narrow-banded Gaussian random process and express the
wave profile $x(t)$ as

$$x(t) = x_c(t) \cos \bar{\omega}t - x_s(t) \sin \bar{\omega}t , \tag{14}$$

where $\quad x_c(t) = \sum a_n \cos\{(\omega_n - \bar{\omega})t + \varepsilon\}$

$\qquad\qquad x_s(t) = \sum a_n \sin\{(\omega_n - \bar{\omega})t + \varepsilon\}$.

Then, we consider the up-crossing of the envelope at time t_1 and downcrossing at time t_2 for both displacement and velocity. Hence, it is
necessary to first formulate the joint Gaussian probability density
function for a set of eight random variables. That is,

$$f(\mathbf{X}) = \frac{1}{(2\pi)^4} \cdot \frac{1}{|\Sigma|} \, e^{-\frac{1}{2}\mathbf{X}'\Sigma^{-1}\mathbf{X}} \tag{15}$$

where $\quad \mathbf{X}' = (x_{c1}, \overset{\text{'}}{x}_{c1}, x_{c2}, \overset{\text{'}}{x}_{c2}, x_{s1}, \overset{\text{'}}{x}_{s1}, x_{s2}, \overset{\text{'}}{x}_{s2})$

Σ = covariance matrix whose elements can be evaluated from
wave spectrum.

The joint probability density function $f(x)$ can be evaluated from a
given wave spectrum; however, the inverse operation of the covariance
matrix involved in Eq.(15) is extremely difficult to perform. In
addition, the follow up transformation to the joint probability density
function of the envelope is not feasible. For this reason, in almost
all studies, the number of waves involved in a wave group, the time
duration, etc., are all average values.

However, this difficulty can be circumvented by decomposing a given wave spectrum into two parts; each part being symmetric about its mean frequency (Ochi and Sahinoglou 1989a). One contains primarily the lower frequency energy components of the wave spectrum, while the other contains the higher frequencies of the spectrum as illustrated in Figure 10. The sum of the two areas is equal to the area of the original spectrum. The parameters of the two symmetric shaped spectra are determined through a least square fitting technique such that the difference between the shape of the original spectrum and the sum of the two symmetric spectra is minimal.

By applying this technique, the covariance matrix involved in Eq.(15). is drastically simplified such that all covariances between x_c and x_s in Eq.(14) are zero. Hence, the inverse operation of the matrix can be simplified and the follow up transformation to the joint probability density function at times t_1 and t_2 is feasible, and therefor the probability density function of time duration can be derived.

However, care must be taken in applying the envelope crossing concept to the wave group problem. That is, it has been customary to consider the exceedance of the envelope above a certain level to identify a wave group. Unfortunately, this is not correct, in practice; if the time duration $\tau_{\alpha+}$ shown in Figure 9 is relatively short; there may be only one wave crest (or no wave crest) in time $\tau_{\alpha+}$ ———— which obviously does not constitute a wave group even though the wave envelope exceeds the specified level. Therefore, the probability distribution of time duration associated with envelope crossings should be modified. If this modification is incorporated in the probability distribution, it is found that the majority of the level crossings of the wave envelope cannot be considered as the occurrence of a wave group.

Table 1 shows an example of the probability of occurrence of a wave group when the envelope exceeds a specified level. Computations are carried out for a two-parameter wave spectra with significant wave height of 10 m and modal frequency of 0.42 rps. As can be seen in the table, for the crossing at a level of 5 m, for example, only 15 percent of the crossings can be considered as a wave group in which two or more wave crests exist for each crossing. This implies that the probability of occurrence of wave groups computed by applying the currently available method is substantially overestimated.

The number of waves in a wave group can also be evaluated by applying the Markov chain concept (Kimura 1980). The Markov chain concept considers the transition probability which represents the probability that a given state is followed by a specified state after a one step transition (one cycle for the present case). Therefore, one of the transition probabilities deals with a large wave height followed by another large height which constitutes a group above a specified level. Kimura evaluated the transition probabilities from the joint Rayleigh probability distribution for wave height and the joint Weibull probability distribution for wave period. The latter distribution is necessary to evaluate the time interval between two successive occurrences of wave groups. Although the latter distribution does not have a theoretical background, Kimura claims that the distribution agrees well with observed data.

In order to evaluate the parameter involved in the joint Rayleigh probability distribution for a given wave spectrum, Kimura derived the relationship between the parameter and spectral band-width from simulation data. However, Longuet-Higgins (1984) considered the joint envelope distribution function whose parameters are expressed in terms of wave spectral density function, and evaluated the transition probabilities for the Markov chain. He also proved that the Markov chain approach and the envelope level-crossing approach yield the same results for waves having a sufficiently narrow-band spectrum.

5. STOCHASTIC PROPERTIES OF OFFSHORE CURRENTS

An interesting phenomenon observed in random seas is that of wave-current interactions. The effects of currents on random waves, associated water particles kinematics, and the wave energy spectrum have been discussed by many researchers. Although it is reasonable to assume that currents are uniform for the study of wave-current interactions, a problem remains as to the magnitude of the currents to be used for evaluating the effect of currents on waves, in practice.

The magnitude of current velocity measured in offshore areas fluctuates in a random fashion with frequencies covering a wide range from 0.0008 to 0.08 cycles per hour (period range from 12 to 1200 hours). Furthermore, the direction of current velocity changes with time with irregular angular frequency of rotation, although it rotates clockwise in the northern hemisphere as shown in Figure 11. Thus, current velocity is considered to be a random process with varying frequency and direction.

In order to clarify the statistical properties of offshore currents and to estimate the extreme magnitude of current velocity from measured data, Ochi and McMillen (1988) carried out analyses on 4 months of continuous current data measured at 26, 47, and 75 meters in a water depth of 84 meters. In their study, the time history of the current velocity is first decomposed into East-West and North-South directions, and then further decomposed into high and low-frequency components. The high-frequency components are considered to be the currents associated with tides, while the low-frequency components are those attributed to all other environmental conditions which may be called the residuals. A cutoff frequency of 0.035 cycles/hour (28.4 hour period) is used to separate the time histories into their respective tidal and residual components. Some significant findings derived from the results of analysis are summarized below.

Deviations from the mean value of the current velocity for the East-West components (including both tidal and residual velocity components) follow the non-Gaussian probability distribution with parameter λ_4 which is associated with the kurtosis of the distribution (see Figure 12). The same trend can also be observed for deviations from the mean in the North-South components. It is also found that the correlation coefficient evaluated for the East-West and the North-South current

velocity time histories is extremely small; hence, they are considered
to be statistically independent.

When the East-West and North-South velocity components are decomposed
into high-frequency (tidal) and low-frequency (residual) components,
the tidal and residual components of the velocity are found to be
statistically independent. Both the tidal and residual currents may be
assumed to be Gaussian random processes as shown in Figures 13 and 14,
respectively.

The area under the spectral density function agrees well with the
variance obtained from the time histories for both the East-West and
North-South components. This result substantiates the fact that the
spectral density function of the current velocity in any direction is
interrelated with the variance of current fluctuations. Hence, the
probabilistic prediction of the magnitude of current velocity may be
made through spectral analysis.

The statistical properties of current velocity obtained from analysis
of current data at deeper depths are nearly the same as those obtained
from near-surface current data. The square-root of the current energy,
which is proportional to the current velocity, decreases exponentially
with increase in water depth. As shown in Figure 15, the tidal and
residual energies both decrease substantially with increase in water
depth. However, the rate of decrease for the residual currents is much
larger than that for the tidal currents.

The magnitude of extreme current velocity (including both tidal and
residual components) may be estimated by applying order statistics to
the non-Gaussian distribution. Figure 16 shows the computed extreme
current velocities for various directions plotted in polar coordinates.
As can be seen in the figure, there is no substantial difference in the
magnitude of the extreme current velocities in various directions for
data used in the analysis. The magnitudes of the extreme current
velocities in the shaded domain in the figure are nearly equal (70.3 -
69.6 cm/sec). The magnitude of the extreme current velocity observed
in the same time period is 74.0 cm/sec in the South-West direction.

In order to more clearly define the stochastic properties of
currents, the vector sums of the spectral densities in the East-West
and the North-South directions is evaluated individually for high and
low frequency components. The results are shown in Figure 17. The
areas under these spectra represent the variances of residual and tidal
current, respectively.

By combining the high frequency (or low frequency) velocities in the
East-West and North-South directions in vector form, we can obtain the
information on tidal (or residual) current velocities. Figures 18 and
19 show examples of polar diagrams indicating the magnitude and
direction of the tidal and residual current velocities, respectively,
as a function of time. The tidal and residual currents are
statistically uncorrelated; the correlation coefficient between them is
extremely small, on the order of 0.05. Hence, the possibility of
simultaneous occurrence of large tidal current and large residual
current is almost nill.

APPENDIX I

Joint probability density function of Type I excursion and its associated period.

$$f_I(\nu, \lambda) = \frac{1}{32\sqrt{2\pi}} \frac{\left(1 + \sqrt{1-\epsilon^2}\right)^3}{\epsilon(1-\epsilon^2)} \frac{\nu^2}{\lambda^5}$$

$$\times e^{-\frac{\nu^2}{8\epsilon^2\lambda^4}\left\{\frac{1}{16}\frac{\left(1+\sqrt{1-\epsilon^2}\right)^4}{1-\epsilon^2} - \frac{1}{2}\left(1+\sqrt{1-\epsilon^2}\right)^2\lambda^2 + \lambda^4\right\}}$$

Joint probability density function of Type II excursion and its associated period.

$$f_{II}(\nu, \lambda) = \frac{k(\epsilon)}{16\sqrt{\pi}} \frac{\left(1+\sqrt{1-\epsilon^2}\right)^5}{\epsilon(1-\epsilon^2)\left(\epsilon^2 + 2\sqrt{1-\epsilon^2}\right)} \frac{\nu^2}{\lambda^5}$$

$$\times \exp\left\{-\frac{\nu^2}{2\epsilon^2}\left(\left\{1 - \beta(\epsilon,\lambda)\right\}^2 + \left\{\frac{\epsilon}{8}\frac{\left(1+\sqrt{1-\epsilon^2}\right)^2}{\sqrt{1-\epsilon^2}}\frac{1}{\lambda^2}\right\}^2\right)\right\}$$

$$\times \left[\begin{array}{l} \exp\left\{\frac{\lambda^2}{4\epsilon^2}\left(1 - \beta(\epsilon,\lambda)\right)^2\right\} \cdot \left\{1 - \Phi\left(\frac{\nu}{\sqrt{2}\epsilon}\left\{1 + \beta(\epsilon,\lambda)\right\}\right)\right\} \\[2em] + \frac{\sqrt{2(1-\epsilon^2)}}{1+\epsilon^2} \exp\left\{\frac{\nu^2}{2\epsilon^2}\left((1-\epsilon^2) - 2\beta(\epsilon,\lambda)\right)\right\} \\[2em] + \frac{2\sqrt{\pi}\sqrt{1-\epsilon^2}}{\epsilon(1+\epsilon^2)^{3/2}} \nu\left\{1 - \beta(\epsilon,\lambda)\right\} \cdot \exp\left\{\frac{\nu^2}{2(1+\epsilon^2)\epsilon^2}\left(1 - \beta(\epsilon,\lambda)\right)\right\} \\[2em] \times \left\{1 - \Phi\left(\frac{\nu}{\epsilon\sqrt{1+\epsilon^2}}\left\{\epsilon^2 + \beta(\epsilon,\lambda)\right\}\right)\right\} \end{array} \right]$$

$$0 \le \nu < \infty, \quad 0 \le \lambda < \infty$$

REFERENCES

Anastasiou,K., Tickell,R.G., and Chaplin,J.R.(1982) The Non-Linear Properties of Random Wave Kinematics. Proc. Third Inter. Conf. Behavior Offshore Structures, Vol.1, pp. 493-515.

Banner,M.L. and Phillips,O.M.(1974) On the Incipient Breaking of Small Scale Waves. Journal Fluid Mech., Vol.65, Part 4, pp. 674-656.

Battjes,J.A. and Van Vledder,G.P.(1984) Verification of Kimura's Theory for Wave Group Statistics. Proc. Int. Conf. Coastal Eng., Vol.1, pp. 642-648.

Cavanié,A., Arhan,M., and Ezraty.R.(1976) A Statistical Relationship between Individual Heights and Periods of Storm Waves. Proc. Conf. Behavior Offshore Structures, Vol.2, pp. 354-360.

Dean,R.G.(1968) Breaking Wave Criteria: A Study Employing a Numerical Wave Theory. Proc. 11th Int. Conf. on Coastal Eng., Vol.1, pp. 108-123.

Ewing,J.A.(1973) Mean Length of Runs of High Waves. Journal Geophy. Res., Vol.78, pp. 1933-1936.

Goda,Y.(1972) On Wave Groups. Proc. Conf. Behavior Offshore Structures, Vol.1, pp. 1-14.

Hasselmann,K.(1961) On the Non-Linear Energy Transfer in a Wave Spectrum. Proc. Conf. Ocean Wave Spectra, pp. 191-197.

Hasselmann,K.(1962) On the Non-Linear Energy Transfer in a Gravity-Wave Spectrum. Part 1, General Theory. Journal Fluid Mech., Vol.12, pp. 481-500.

Holthuijsen,L.H. and Herbers,T.H.C.(1986) Statistics of Breaking Waves Observed as Whitecaps in the Open Sea. Journal Phys. Oceanogr., Vol.16, pp. 290-297.

Huang,N.E., Long,S.R., Tung,C.C., Yuen,Y., and Bliven.L.F.(1983) A Non-Gaussian Statistical Model for Surface Elevation of Non-Linear Random Wave Field. Journal Geophy. Res, Vol.88, pp. 7597-7606.

Kac,M. and Siegert,A.J.F.(1947) On the Theory of Noise in Radio Receivers with Square-Law Detectors. Journal App. Phys., Vol.18, pp. 383-397.

Kimura,A.(1980) Statistical Properties of Random Wave Groups. Proc. 17th Int. Conf. on Coastal Eng., Vol.3, pp. 2955-2973.

Kjeldsen,S.P. and Myrhaug,D.(1978) Kinematics and Dynamics of Breaking Waves. River and Harbour Laboratory Report, Norwegian Int. of Technology.

Langley,R.S.(1987) A Statistical Analysis of Low Frequency Second-Order Forces and Motions. Journal App. Ocean Res., Vol.9, pp. 163-170.

Langley,R.S.(1987) A Statistical Analysis of Non-Linear Random Waves. Journal Ocean Eng., Vol.14, No.5, pp. 389-407.

Lindgren,G. and Rychlik,I.(1982) Wave Characteristic Distributions for Gaussian Waves —— Wave-Length, Amplitude and Steepness. Journal Ocean Eng., Vol.9, No.5, pp. 411-432.

Longuet-Higgins,M.S.(1963) The Effect of Non-Linearities on Statistical Disstributions in the Theory of Sea Waves. Journal Fluid Mech., Vol.17, Part 3, pp. 459-480.

Longuet-Higgins,M.S.(1963) The Generation of Capillary Waves by Steep Gravity Waves. Journal Fluid Mech., Vol.16, pp. 138-159.

Longuet-Higgins,M.S.(1969) On Wave Breaking and Equilibrium Spectrum of Wind-Generated Waves. Proc. Roy. Soc. London, Ser.A, Vol.310, pp. 151-159.

Longuet-Higgins,M.S.(1974) Breaking Waves —— In Deep or Shallow Water. Proc. 10th Symp. on Naval Hydro., pp. 597-605.

Longuet-Higgins,M.S.(1976) Recent Development in the Study of Breaking Waves. Proc. 15th Int. Conf. on Coastal Eng., Vol.1, pp. 441-460.

Longuet-Higgins,M.S.(1978a) The Instabilities of Gravity Waves of Finite Amplitude in Deep Water. I: Superharmonics. Proc. Roy. Soc. London, Ser.A, Vol.360, pp. 471-488.

Longuet-Higgins,M.S.(1978b) The Instabilities of Gravity Waves of Finite Amplitude in Deep Water. II: Subharmonics. Proc. Roy. Soc. London, Ser.A, Vol.360, pp. 489-505.

Longuet-Higgins,M.S.(1980) The Unsolved Problem of Breaking Waves. Proc. 18th Int. Conf. on Coastal Eng., Vol.1, pp. 1-48.

Longuet-Higgins,M.S.(1983) On the Joint Distribution of Wave Periods and Amplitudes in a Random Wave Field. Proc. Roy. Soc. London, Ser.A, Vol.389, pp. 241-258.

Longuet-Higgins,M.S.(1984) Statistical Properties of Wave Groups in a Random Sea State. Phil. Trans. Roy. Soc. London, Ser.A, Vol.312, pp. 219-250.

Longuet-Higgins,M.S.(1985) Accelerations in Steep Gravity Waves. Journal Phys. Oceanogr., Vol.15, pp. 1570-1579.

Longuet-Higgins,M.S. and Cokelet,E.D.(1976) The Deformation of Deep Surface Waves on Water. I: A Numerical Method of Computation. Proc. Roy. Soc. London, Ser.A, Vol.350, pp. 1-36.

Longuet-Higgins,M.S. and Cokelet,E.D.(1978) The Deformation of Steep Surface Waves on Water. II: Growth of Normal-Mode Instabilities. Proc. Roy. Soc. London, Ser.A, Vol.364, pp. 1-28.

Melville,W.K.(1982) The Instability and Breaking of Deep-Water Waves. Journal Fluid Mech., Vol.115, pp. 165-185.

Mitchell,J.H.(1893) The Highest Waves in Water. Journal Phil. Mag., Vol.36, pp. 430-437.

Mollo-Christensen,E. and Ramamonjiarisoa,A.(1978) Modelling the Presence of Wave Groups in a Random Wave Field. Journal Geophy. Res., Vol.83, No.C8, pp. 4117-4122.

Naess,A.(1985) Statistical Analysis of Second Order Response of Marine Structures. Journal Ship Res., Vol.29, No.4, pp. 270-284.

Nath,J.H. and Ramsey,F.L.(1976) Probability Distributions of Breaking Wave Heights Emphasizing the Utilization of the JONSWAP Spectrum. Journal Phys. Oceans., Vol.6, pp. 316-323.

Neal,E.(1974) Second-Order Hydrodynamic Forces due to Stochastic Excitation. Proc. 10th Sympo. on Naval Hydro., pp. 517-537.

Nolte,K.G. and Hsu,F.H.(1972) Statistics of Ocean Wave Groups. Proc. 4th Offshore Tech. Conf., Vol.2, pp. 637-644.

Ochi,M.K. and Tsai,C.H.(1983) Prediction of Breaking Waves in Deep Water. Journal Phy. Oceanography, Vol.13, No.11, pp. 2008-2019.

Ochi,M.K.(1986) Non-Gaussian Random Processes in Ocean Engineering. Journal Probabilistic Eng. Mechanics, Vol.1, No.1, pp. 28-39.

Ochi,M.K. and McMillen,R.I.(1988) Stochastic Analysis of Offshore Current. Proc. 21st Int. Conf. on Coastal Eng., Vol.3, pp. 2536-2549.

Ochi,M.K. and Sahinoglou,I.I.(1989a) Stochastic Characteristics of Wave Group in Random Seas; Part 1, Time Duration of and Number of Waves in a Wave Group. Journal App. Ocean Res., Vol.11, No.1, pp. 39-50.

Ochi,M.K. and Sahinoglou,I.I.(1989b) Stochastic Characteristics of Wave Group in Random Seas; Part 2, Frequency of Occurrence of Wave Group. Journal App. Ocean Res., Vol.11, No.2, pp. 89-99.

Phillips,O.M.(1958) The Equilibrium Range in Spectrum of Wind Generated Waves. Journal Fluid Mech., Vol.4, pp. 785-790.

Phillips,O.M.(1960) On the Dynamics of Unsteady Gravity Waves of Finite Amplitude. Part 1, The Elementary Interactions. Journal Fluid Mech., Vol.9, pp. 193-217.

Phillips,O.M.(1961) On the Dynamics of Unsteady Gravity Waves of Finite Amplitude. Part 2. Local Properties of a Random Wave Field. Journal Fluid Mech., Vol.11, pp. 143-155.

Ramberg,S.E. and Griffin,M.(1987) Laboratory Study of Steep and Breaking Deep Water Waves. Journal, Waterway, Port, Coastal, and Ocean Eng., ASCE, Vol.113, No.5, pp. 493-506.

Rye,H.(1974) Wave Group Formation Among Storm Waves. Proc. 14th Int. Conf. on Coastal Eng., Vol.1, pp. 164-183.

Rye,H.(1979) Wave Parameter Studies and Wave Groups. Proc. Int. Conf. on Sea Climatology, pp. 89-123.

Rye,H. and Lervik,E.(1981) Wave Grouping Studies by Means of Correlation Techniques. Norwegian Mar. Res., Vol.4, pp. 12-21.

Snyder,R.L. and Kennedy,R.M.(1983) On the Formation of Whitecaps by a Threshold Mechanism. Part 1: Basic Formlulation. Journal Phys. Oceanogr., Vol.13, pp. 1482-1492.

Srokosz.M.A.(1985) On the Probability of Wave Breaking in Deep Water. Journal Phys. Oceanogr., Vol.16, pp. 382-385.

Srokosz,M.A. and Longuet-Higgins,M.S.(1986) On the Skewness of Sea-Surface Elevation. Journal Fluid Mech., Vol.164, pp. 487-497.

Stokes,G.G.(1880) On the Theory of Oscillatory Waves. Math. Phys. Pap., Vol.1, pp. 225-228, Cambridge Univ. Press.

Sue,M.Y., Bergin,M., Marler,P. and Myrick,R.(1982) Experiments on Non-Linear Instabilities and Evolution of Steep Gravity-Wave Trains. Journal Fluid Mech., Vol.124, pp. 45-72.

Tayfun,M.A.(1980) Narrow-Band Nonlinear Sea Waves. Journal Geophy. Res., Vol.85, No.C3, pp. 1548-1552.

Tayfun,M.A.(1986) On Narrow-Band Representation of Ocean Waves. Journal Geophy. Res., Vol.91, No.C6, pp. 7743-7752.

Tick.L.J.(1959) A Non-Linear Random Model of Gravity Waves. Journal Math. and Mechanics, Vol.8, No.5, pp. 643-651.

Tick.L.J.(1961) Non-Linear Probability Models of Ocean Waves. Proc. Conf. Ocean Wave Spectra, pp. 163-169.

Tung,C.C. and Huang,N.E.(1984) Statistical Properties of the Kinematics and Dynamics of Nonlinear Waves. Journal Phys. Oceano., Vol.14, No.3, pp. 594-600.

Van Dorn,W.G. and Pazan,S.E.(1975) Laboratory Investigation of Wave Breaking. Part 2: Deep Water Waves. Advanced Ocean Eng. Lab. Rep. 71, Scripps Inst. Oceano.

48

Vinje,T.(1983) On the Statistical Distribution of Second-Order Forces
and Motion. Inter. Shipbuild. Prog., Vol.30, pp. 58-68.
Weissman,M.A., Atakturk,S.S. and Katsaros,K.B.(1984) Detection of
Breaking Events in a Wind-Generated Wave Field. Journal Phy.
Oceanography, Vol.14, pp. 1608-1619.
Xu,D., Hwang,P.A., and Wu,J.(1986) Breaking of Wind-Generated Waves.
Journal Phys. Oceanogr., Vol.16, pp. 2172-2178.

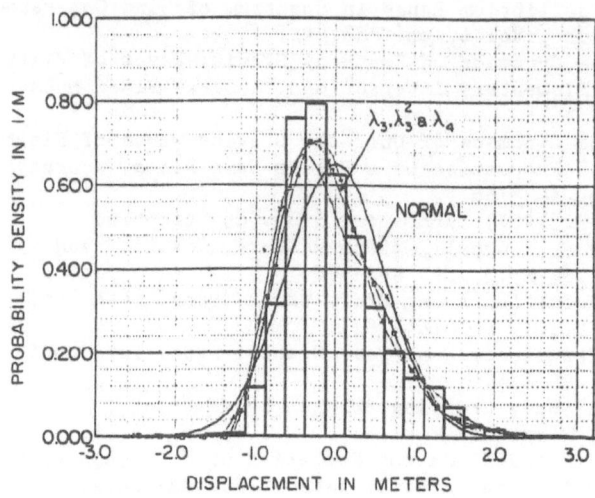

Figure 1
Comparison between
histogram constructed
from measured data,
Gaussian distribution,
and non-Gaussian dis-
tribution (Ochi 1986).

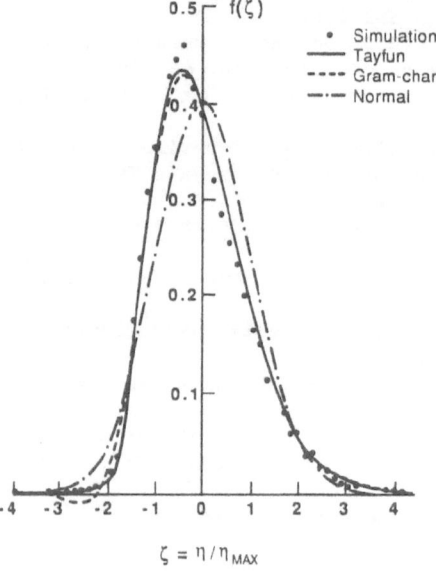

Figure 2
Comparison of probability
density function with data
obtained from simulation study
of wave profile (Tayfun 1980).

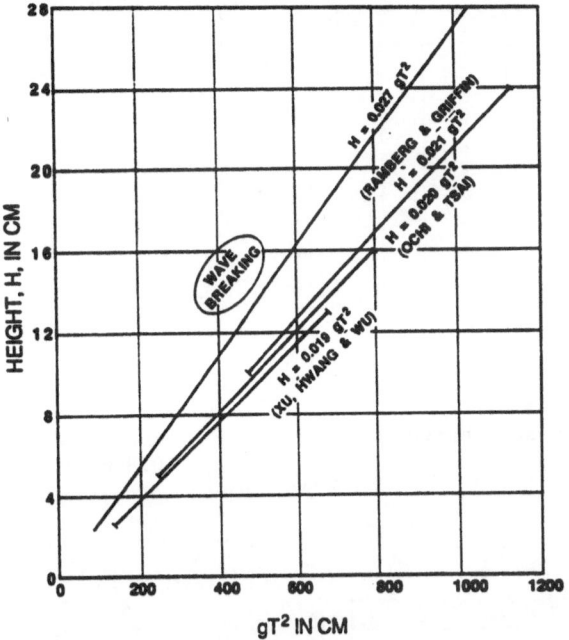

Figure 3 Relationship between wave height and periof for breaking.

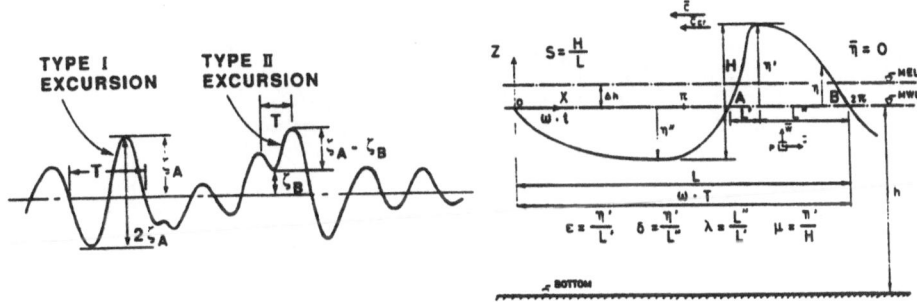

Figure 4 Explanatory sketch of excursions of non-narrow-band random process.

Figure 5 Definition of crest front steepness ε, vertical asymmetry factor λ , and horizontal asymmetry factor μ (Kjeldsen and Myrhaug 1978).

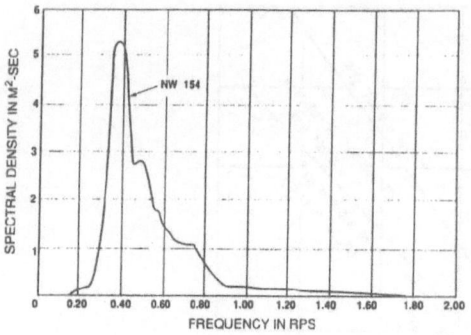

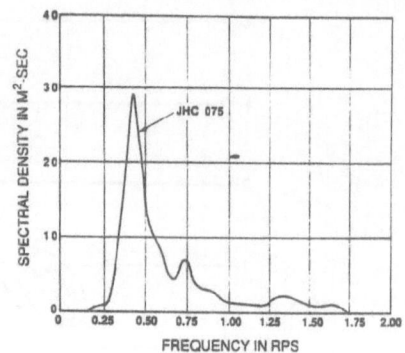

Figure 6
(a) Wave spectrum A:
    significant wave height
    4.6 m.

(b) Wave spectrum B:
    significant wave height
    10.8 m.

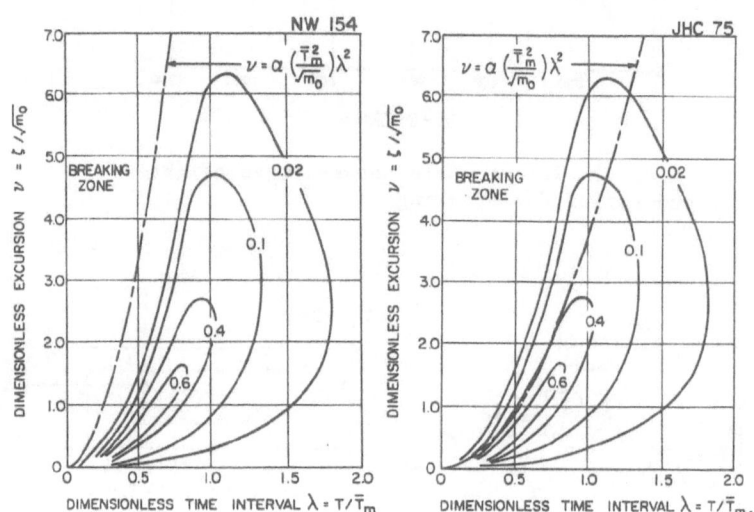

Spectrum A   Spectrum B

Figure 7   Joint probability density function of Type I
excursion and its associated time interval, and lines
indicating breaking condition (Ochi and Tsai 1983).

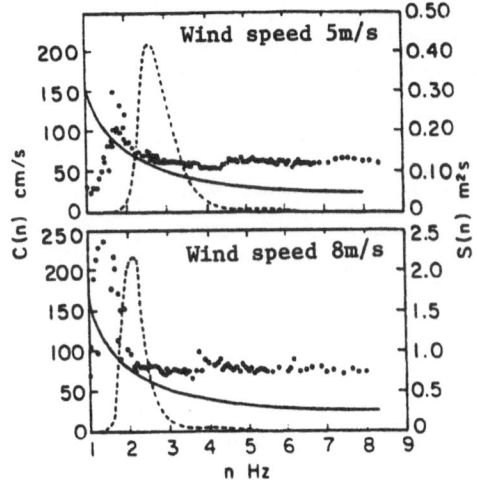

Figure 8
Phase speed of wind waves
measured in the laboratory
plotted as a function of
frequency (circle) and linear
theory (solid line). Spectral
density function (dashed line).
(Mollo-Christensen and
Ramamonjiarisoa 1978).

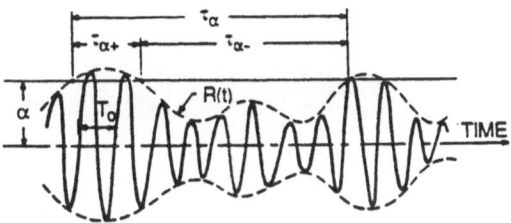

Figure 9   Level crossing of the envelope
of a random process.

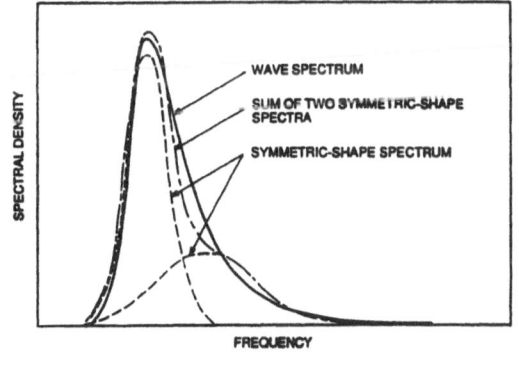

Figure 10
Decomposition of wave spectrum
into symmetric shape spectra.

Table 1 Probability of occurrence of wave group when envelope exceeds a specified level ; two-parameter wave spectrum, H_s = 10.0 m, ω_m =0.42 rps (Ochi and Sahinoglou 1989a).

Level	Wave group	No wave group	
		For $\tau_{\alpha+} < \overline{T}_o$	For $\overline{T}_o \leq \tau_{\alpha+} \leq 2\overline{T}_o$
5.0 m	0.15	0.71	0.14
6.0	0.12	0.76	0.12
7.0	0.08	0.83	0.09
8.0	0.05	0.88	0.07

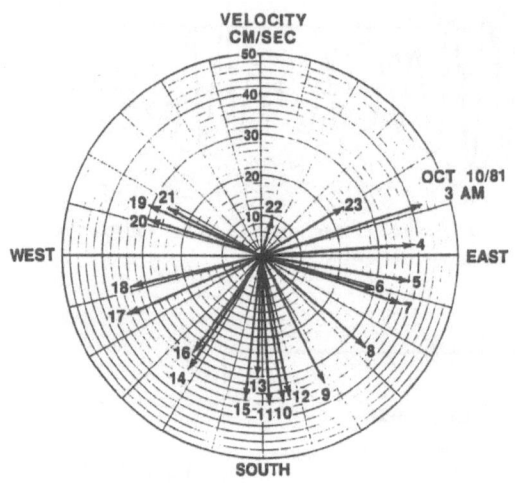

Figure 11 Magnitude of current velocity and its direction as a function of time.

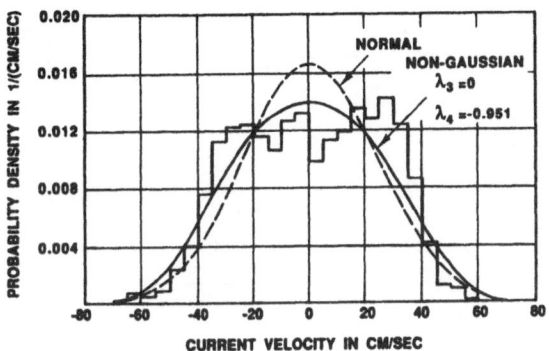

Figure 12 Histogram of deviations from the mean of East-West current velocity.

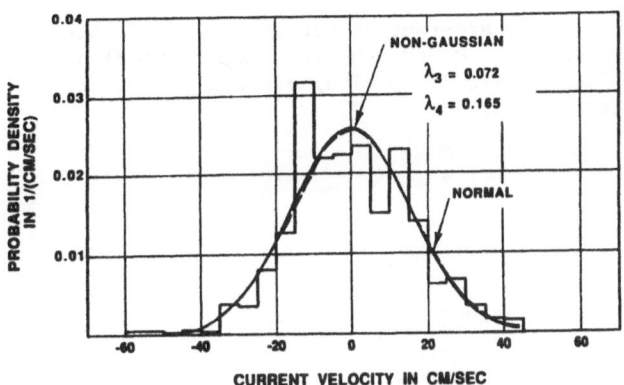

Figure 13 Histogram of deviations from the mean of high frequency (tidal) currents.

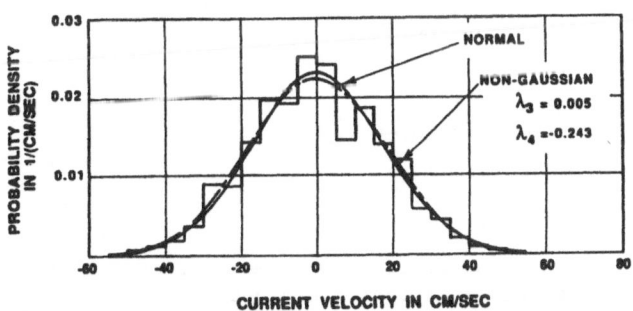

Figure 14 Histogram of deviations from the mean of low frequency (residual) currents.

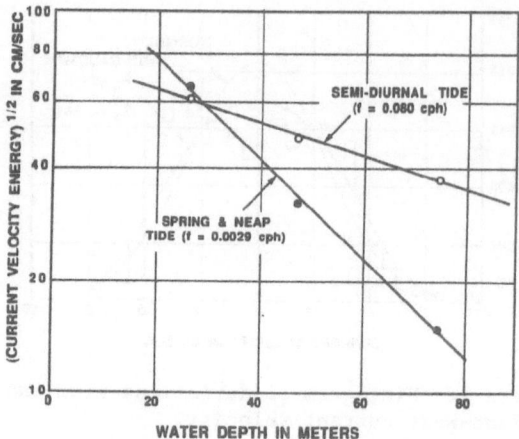

Figure 15   Square-root of current energy density
for frequencies 0.080 cph (12.5 hour period) and
0.0029 cph (14.2 day period) plotted as a function
of water depth.

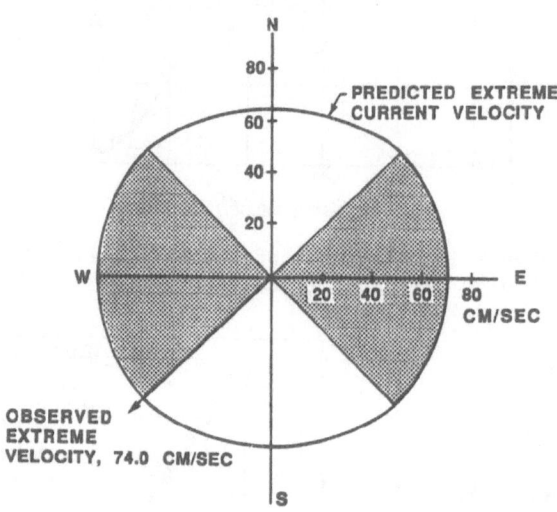

Figure 16   Computed extreme current velocities
expected in a four month observation for
various directions.

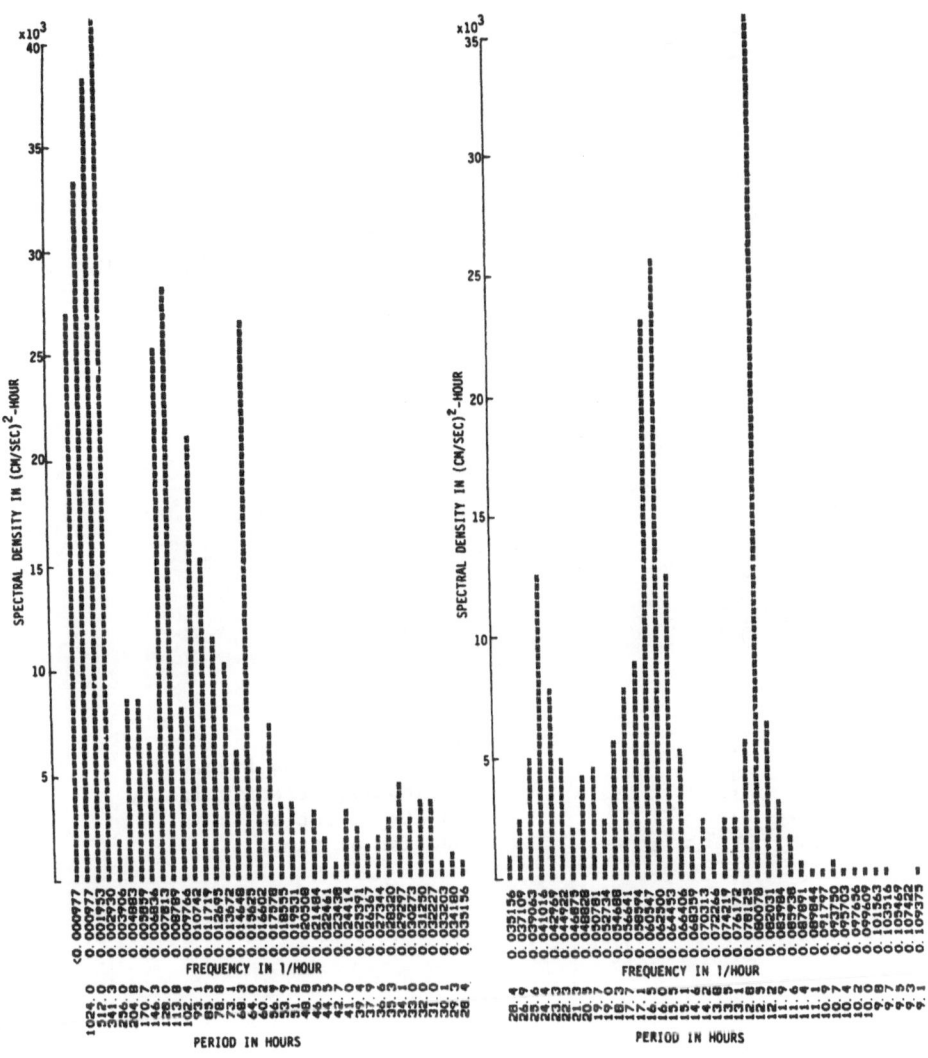

Low frequency (residual) currents   High frequency (tidal) currents

Figure 17   Vector sum of East-West and North-South
energy spectrum.

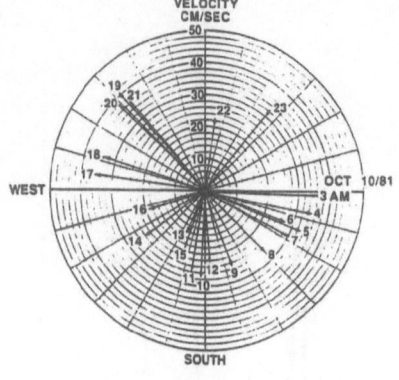

Figure 18
Magnitude of high frequency
(tidal) current velocity and
its direction as a function
of time.

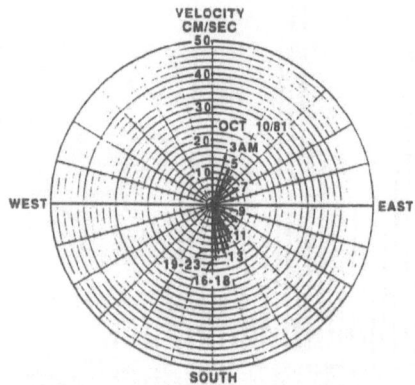

Figure 19
Magnitude of low frequency
(residual) current velocity
and its direction as a
function of time.

DEEP WATER WAVE KINEMATICS MODELS FOR DETERMINISTIC AND STOCHASTIC ANALYSIS OF DRAG DOMINATED STRUCTURES

OVE T. GUDMESTAD, dr. scient,Statoil,
P.O.Box 300,4001 Stavanger, Norway

and

NILS SPIDSØE, dr. ing, SINTEF,7034 Trondheim,NTH,Norway

1   ABSTRACT

A review and comparison of deterministic and   stochastic
methods   for   analysis   of   deepwater   drag   dominated
offshore structures is presented. Particular emphasis is
made   on   the   importance   of   selection of correct wave
kinematics model for calculation of forces on structures
and   the   dynamic   structural   response induced by these
forces. Several wave kinematic models for deep water are
considered   and   the   practical use and consequences for
design of drag   dominated   structures   are   examplified.
Furthermore,recommendations   for   practical analysis are
being   made   and   needed   for   further   research   and
development is outlined.

CONTENTS:

2   INTRODUCTION

During the design of an offshore structure the   designer
has   to   make   a   selection   of   appropriate   models,
parameters and design procedures:

a)   environmental design criteria   to   be   based   on
      available statistics of measured data.

b)   design checks of the structure including checks   of
      all failure modes in the structure to be planned.

A. Tørum and O. T. Gudmestad (eds.), Water Wave Kinematics, 57–87.

c)   wave and current kinematics (model) to be generated from theoretical models based on measurements of the flow below regular waves or in more realistic random wave climate.

d)   force model and related hydrodynamic coefficents to be based on forces measured in realistic wave conditions in the offshore environment or in wave tank.

e)   structural response analysis procedures, including structural modelling and methods for all defined design checks. Dynamics to be included when nonlinearities and structural resonances due to the wave and current actions are significant, through appropriate selection of dynamic model including choice of realistic damping values.

For a careful discussion of the different steps involved in the design of offshore structures see Vugts (1984).

Re: a)

Offshore structures shall be designed to resist forces caused by waves, currents and winds.

For traditional design structural analysis, the offshore wave environment is often described by regular waves. It is normally assumed that the waves are travelling in one selected direction and a two dimensional description of the wave climate is used for regular wave analysis. More advanced structural analysis, however, require a stochastic description of the sea in order to model the true irregularity of the sea surface. This is done through appropriate selection of wave spectra. Although three dimensional models are very important to fully describe the sea surface, two dimensional spectra are normally employed in the design of offshore structures as neglection of 3 dimensional properties in general is proven conservative. For certain shallow water phenomena, however, 3 dimensional effects must be incorporated (Sand et al., 1989).

A structure placed in the sea is to be designed to be safe under the extreme loading situation (Ultimate Limit State Condition, ULS) and to have sufficient fatigue strength to tolerate day-to-day loading over the lifetime of the structure (Fatigue Limit State Condition, FLS),(Ref: NPD, 1989). Furthermore, NPD requires design checks in the Serviceability Limit State condition (SLS) and in the Progressive Limit State condition (PLS).

For the ULS condition a description of the extreme environmental climate is required while a description of the relative occurence of the different wave conditions (wave height/wave period or spectral parameters) is to be given for calculation of fatigue damage.

The associated environmental conditions are given as follows:

* Serviceability limit state (SLS)

    - normal operating environmental conditions apply.

* Ulitmate limit state (ULS)

    - environmental conditions having a probability of occurence of 10^{-2} per year apply.

* Fatigue limit state (FLS)

    - wave scatter diagram to be identified

* Progressive limit state (PLS)

    - check for environmental conditions having a probability of occurence of 10^{-4} per year. Also check of platform in damaged condition.

Re: b)

Structural response and design calculations incorporating the appropriate structural characteristics are to be performed according to recognized codes. For design of jackets reference is made to NPD (1989) or to API RP 2A (1987).

With reference to above discussion, the objective of the paper is two-fold;

1) to discuss deterministic and stochastic analysis of offshore structures

2) to recommend wave kinematics models for deep water applications.

No further discussion of current velocity nor the appropriateness of vectorially adding velocity and current velocity in the forcing Eq (2.2) will be made. For a general discussion of these aspects see Sarpkaya and Isaacson (1981).

It must be noted that the term "deepwater" in the context of this paper is defined in relation to wave height and associated wave period, e.g. by

$$\frac{d}{gT^2} \gtrsim 0.08$$

$$(2.1)$$

where

d = water depth
T = wave period
g = acceleration of gravity

Ref. Fig. 3.3 from Le Mehaute (1976).

A "dynamic sensitive structure" is not necessarity placed in deep waters as the relative dimensions of the structure compared to the water depth determines a structure's dynamic characteristics. A dynamic sensitive structure has a first natural period larger than say, 3 (to 4) seconds whereby it is in resonance with the (more) energetic part of the storm wave spectrum .

Re: c)

The importance of applying correct wave kinematics in a Morison type description of wave loading is apparant from the forcing Eq. (2.2). Several research programs have therefore been designed to measure wave kinematics correctly, conf. MaTS (1982), Vugts (1984) and Gudmestad et al. (1988). To generalize the test results, the measured kinematics values must be compared to results obtained from applying recognized wave theories satisfying the laws of hydrodynamics. The major part of this paper is related to a discussion of wave kinematics models for deep water, conf. Chapters 3 and 4.

Re: d)

Due to the importance of the wave and current action, it is of major importance in the design of offshore structures to :

1. determine the relation between wave particle kinematics and loading on the structures

2. determine the kinematics of the wave particles.

Considerable research into selection of force model has been ongoing over the last decades. Much of this research is documented by Sarpkaya and Isaacson (1981). Although alternative models have been suggested, the

model proposed by Morison et al. (1950), represents
state of art for calculation of forces on framed
structures as jackets, jack-ups and compliant truss
towers. They suggested that the force acting on a
section of a cylinder with diameter D due to wave motion
is made up of two components:

- an inertia force analogous to that on a body
  subjected to a uniformly accelerated flow of an
  ideal fluid

- a drag force analogous to the drag on a body
  subjected to a steady flow of a real fluid
  associated with wake formation behind the body:

$$F_{\text{unit length}} = 0.25 \rho \pi D^2 \, C_m \, \frac{du}{dt} + 0.5 \quad DC_d u |u| \quad (2.2)$$

where

ρ = Density of water

$\frac{du}{dt} = \frac{dv}{dt}$ = water wave acceleration

C_m, C_d = inertia and drag coefficients, respectively

u = summation of water wave velocity v and current
velocity U.

It must be noted that the inertia and drag coefficients,
C_m and C_d, can no longer retain the values found
independently in the uniformly accelerated flow
condition and in the steady flow condition, respectively
when using Equation (2.2). For a summary of Morison type
wave loading, conf. Moe (1989).

For a Morison type wave loading, the drag load dominates
for small diameter structures or for cases with large
wave + current velocity. For slender jacket steel
structures and jackups, the drag load normally dominates
over the inertia load in the ULS condition. The inertia
load however, dominates for the overall ULS design of
large diameter selffloating jackets, gravity structures
and floating platforms. For the fatigue situation the
inertia load is generally dominating for all types of
platforms as the main contribution to the fatigue damage
comes from reduced environmental conditons as compared
to the ULS situation, i.e. from conditions where the
wave and current velocity are considerable reduced
compared with the ULS design velocity.

For large column structures like the concrete structures installed in the North Sea, the _inertia_ loading dominates very much over the drag load. For this type of structures and load conditions, the Morison equation is no longer a sufficient model and more sophisicated methods based on the sink-source principle must be used.

Re: e)

For accurate calculation of the response of offshore structures, a careful inclusion of the structures' dynamic properties is necessary. This is in particular required for structures responding in resonance with the wave climate, i.e. structures with eigenperiode higher than, say 3 to 4 seconds. Discussions regarding dynamic effects will be included in Chapters 3 and 4.

3   DETERMINISTIC STRUCTURAL ANALYSIS

The term "deterministic" means that a well defined regular single wave or a series of regular waves is the basis for the wave load and the response calculations.

In a deterministic quasistatic design wave analysis the dynamic load (wave load) is applied as a static load. The maximum wave load is found by investigating the load for different phase angles and directions of the wave relatively to the structure. This approach has traditionally been adapted by the oil industry. For statically responding structures, the deterministic quasistatic design wave analysis method gives reliable results provided that proper wave kinematics load models are applied. For dynamic sensitive structures, dynamic effets can be accounted for by applying dynamic amplification factors (DAF's) which must be found from other methods of analysis like e.g. a deterministic dynamic analysis or from simplified models accounting for the dynamics of an associated one degree of freedom system, see Biggs (1964). This method should, however, be used with care and only for structures with small or moderate dynamic effets (first natural period below 3-4s)

Statoil has recently installed a slender jacket structure on the Veslefrikk field in the North Sea in 174 m of water (Bærheim and Fossan , 1989) The platform's first natural period is estimated to 3.3s. Table 3.1 shows that the dynamic amplification factor found from a deterministic analysis is very much underestimated compared with the corect DAF found from a time domain stochastic dynamic response analysis conf, Ch 4, method b.

Alternatively to a pseudostatic design wave analysis, a
deterministic time-dependent method can be utilized
whereby a time record of the regular free surface
elevation is used to calculate time dependent loads
taking into account the dynamic response of the
structure, conf. Fig. 3.1. from Bea and Lai (1978). No
attempt is being made to handle the response
statistically, however, ref Chapter 4.

In a pseudostatic deterministic design wave analysis
nonlinearities due to wave forcing, finite amplitude
wave height and selected wave theory are incorporated
through:

-   application of full nonlinar Morison equation

-   integration of waveload to free surface

-   introduction of nonlinear terms in wave kinematics
    model (e.g. by use of higher order Stokes models).

The disadvantages of the pseudostatic deterministic
approach are

-   difficulties to incorporate dynamic effects
    currently.

-   lack of realistic description of the irregular sea.

The wave theory selected for pseudostatic deterministic
analysis is normally chosen from diagrams like those
presented in Fig. 3.2 (Dean, 1970) and in Fig. 3.3 (Le
Mehaute, 1976). Gudmestad and Poumbouras (1986 and 1988)
have however, concluded that the drag force on a single
pile located in deep water varies considerably with the
wave theory (and the wave kinematic model) chosen (Fig.
3.5) .Of particular importance is the varying value of
the higher order harmonics in the drag force caused by
the nonlinear terms of the loading (Fig. 3.4). Similar
variations in the inertia load has not been found
whereby the choice of wave kinematics model for
calculation of inertia load is not as critical as the
calculation of wave velocity for estimation of drag
load.

Although Fig. 3.2 shows that Stream Function or Stokes V
order theory should be used in pseudostatic
deterministic analysis of structures in deep water,
measurements carried out at Caltech, Gudmestad et al.
(1988) show that simple linear theory also gives
reasonable agreement with the measurements if an
apparant return flow in the wave flume is accounted for.

In order to directly account for the measurements, Gudmestad and Connor's theory (1986) should be utilized for the calculation of forces in the crests and the troughs of deep water regular waves. (Fig. 3.6). This theory does also directly account for field observations showing that the wave velocity in the crest is overpredicted by the regular wave theories while the absolute value of the wave velocity under the trough is underpredicted by these theories.(Lloyd's,1984). The measurements at Caltech do furthermore show that the Wheeler method (Wheeler,1970) is not appropriate in deterministic regular wave analysis as the method underpredicts wave kinematics in the crest zone.

Through recent research (e.g. Forrisdal (1985) and Lambrakos et al. (1987)), new and more complicated methods to calculate wave velocity in a wave field have been introduced. In engineering practice these methods are, however, only considered relevant for single regular waves or single 2 dimensional freak wave situations.

The importance of selecting correct wave kinematics model is examplified in Table 3.2 showing that the forces are calcualted to be about 30% higher on the Veslefrikk jacket using Stokes V wave kinematics rather than Wheeler stretching method (Wheeler 1970) for deterministic structural analysis of the jacket.

In order to estimate the Dynamic Amplification Factor, a dynamic transferfunction for the appropriate part of the structure can be constructed by calculating the dynamic response of the structure for a series of waves being stepped through the structure:

$$M\ddot{r} + C\dot{r} + Kr = Q\,(t).\tag{3.1}$$

where

M, C and K are mass, damping and stiffness matrices, respectively. r is the displacement vector and Q(t) is the load vector. Fig. 3.4, however, clearly shows that the response to frequencies being considerably smaller than the first natural frequency of the structure causes higher order harmonics. The energy at these multiples is therefore normally "lumped" to the base frequency of the wave whereby a somewhat inconsistent tranfer function is being constructed.

To calculate the DAF for the selected location in the structure, the static responsefunction is to be determined through:

$$Kr' = Q(t) \qquad (3.2)$$

where

r' is the static displacement.

The DAF is therafter determined as the ratio between the dynamic and the static responsefunctions, Fig. 3.7.

$$DAF = \frac{R_{dynamic}(\omega)}{R_{static}(\omega)} \qquad (3.3)$$

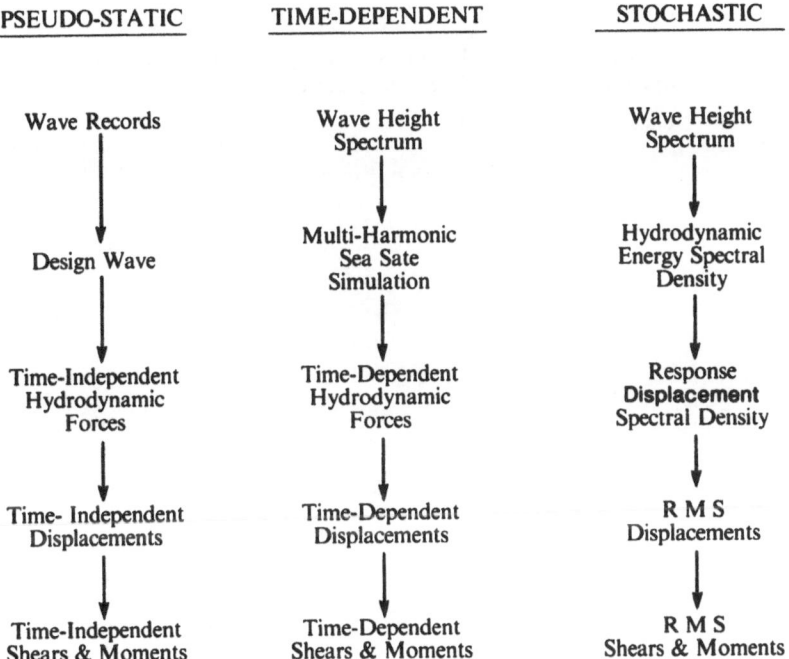

Fig. 3.1 Alternative procedures to wave loading analysis
(Ref. Bea and Lai, 1978)

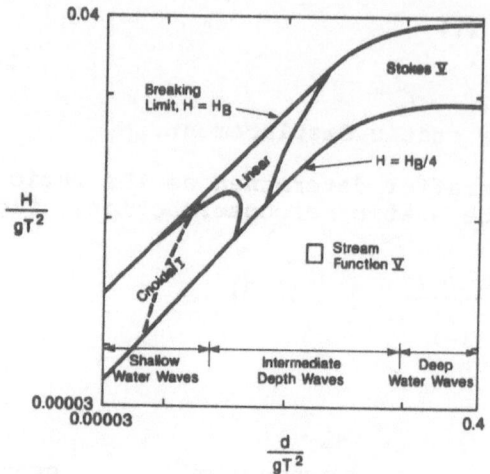

Fig. 3.2 Ranges of wave theories giving the best fit to the dynamic free surface boundary condition

(Ref. Dean, 1970)

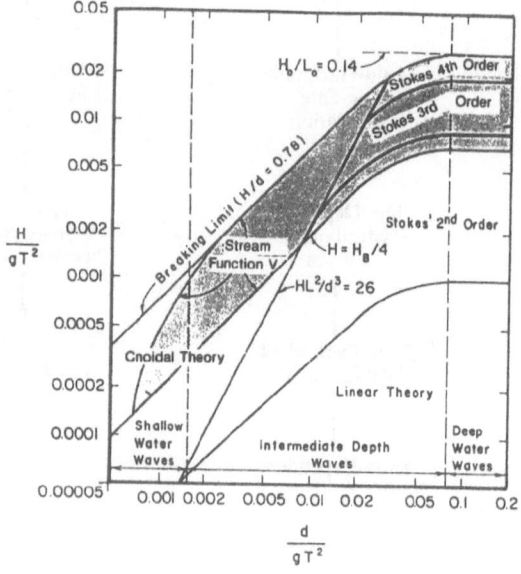

Fig. 3.3 Ranges of suitability for various wave theories as suggested by Le Mehaute (1976)

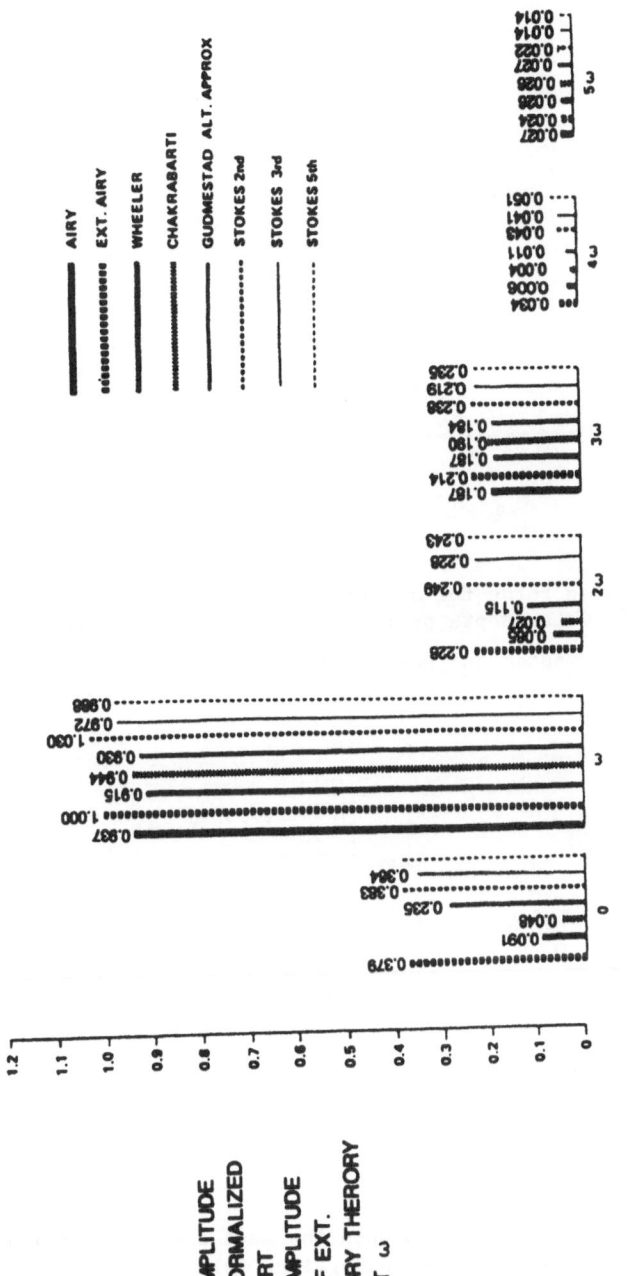

Fig. 3.4 Fourier spectrum for base drag force vs. frequency.
Wave condition 1.d = 150m, T = 17s, H = 30m.

(Ref. O.T. Gudmestad and G.A. Pumbouras, 1986)

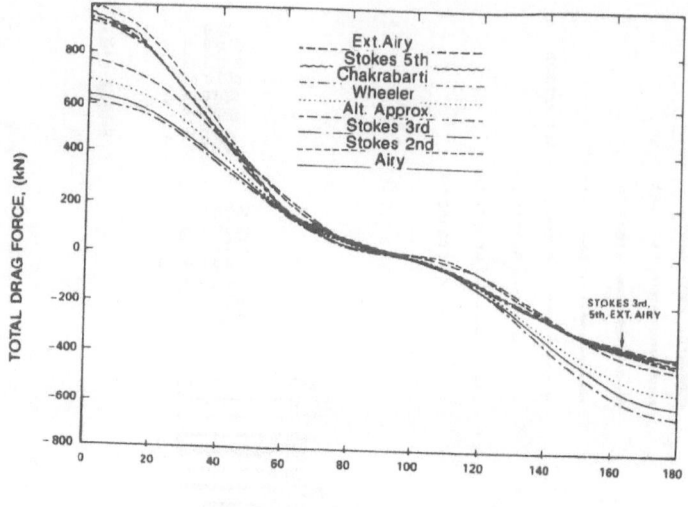

PHASE ANGLE, 0=kx-wt (Degrees)

Fig. 3.5 Total drag force for 30m high wave with period 17s
in 300m water depth as function of phase angle Cd=Cm=1

(Ref. Gudmestad and Poumpouras, 1986)

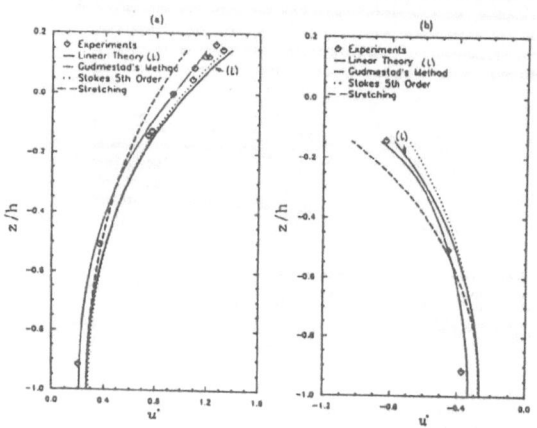

Fig. 3.6.a Horizontal velocities when the wave crest passes (a)
and when the wave through passes (b). H=11.5cm,
T=0.9sec. Caltech tests, Case D.

(Ref. Gudmestad et al., 1988)

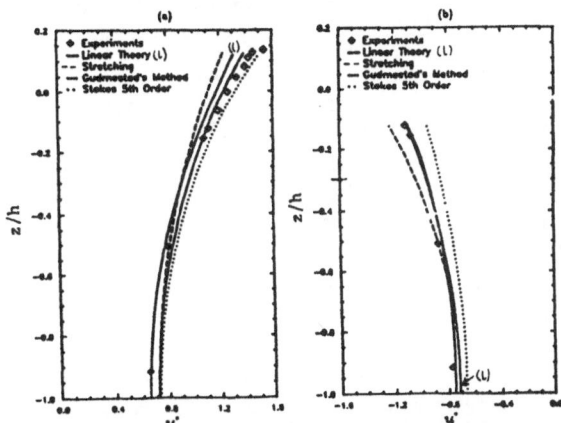

Fig. 3.6.b Horizontal velocities when the wave crest passes
(a) and the wave through passes (b). H=9.9 cm.
T=1.31sec. Caltech tests, Case C.

(Ref. Gudmestad et al., 1988)

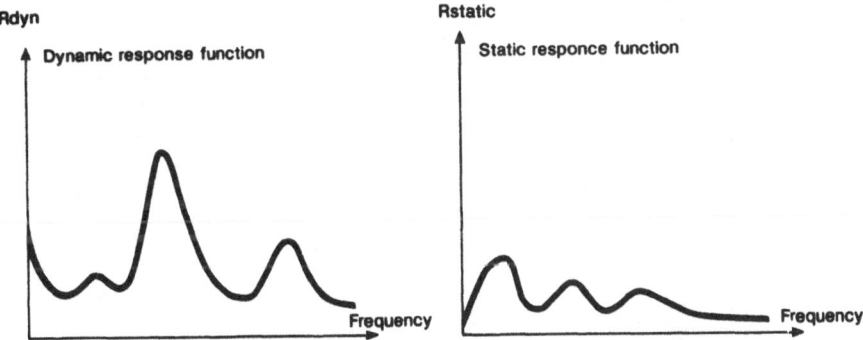

Fig. 3.7 Calculation of DAF in Deterministic Analysis.

RESPONSE QUANTITY	DYNAMIC AMPLIFICATION FACTOR		
	DWA	FDA	TDA
Base shear	1,05	1,03	1,18
Overturning moment	1,05	1,06	1,18
Brace forces			
EL-13m (Brace 75)	1,05	1,05	1,14
EL-36m /Brace 83)	1,05	1,04	1,06
EL-90m (Brace 99)	1,05	1,03	1,05
EL-149m (Brace 115)	1,03	1,03	1,07
Leg forces			
EL-13m (Leg 130)	1,05	1,27	1,38
EL-36m (leg 140)	1,05	1,15	1,33
EL-90m (Leg 148)	1,05	1,08	1,28
EL-149m (Leg 156)	1,05	1,07	1,25

Table 3.1 Comparison of 3D dynamic response of Veslefrikk jacket predicted by the Deterministic quasistatic design wave analysis (DWA), Stochastic linearization in the frequency domain (FDA) and Time domain stochastic dynamic response analysis (TDA) with the vertically extrapolated Airy wave model.

RESPONSE QUANTITY	WAVE THEORY		
	Stokes V 28 m	Wheeler stretching 30m	Stokes / Wheeler
Base shear	24,0	18,1	1,33
Overturning moment	3586	2609	1,37
Brace forces			
EL-13m (Brace 75)	4,37	3,30	1,32
EL-36m (Brace 83)	4,53	4,43	1,32
EL-90m (Brace 99)	3,99	3,08	1,30
EL-149m (Brace 115)	3,32	2,58	1,29
Leg forces			
EL-13m (Leg 136	7,22	3,29	1,15
EL-36m (leg 140)	13,10	10,90	1,20
EL-90m (Leg 148)	23,85	19,27	1,24
EL-149m (Leg 156)	30,90	24,87	1,24

Table 3.2. 3D Response of Veslefrikk jacket calculated using Stokes V order wave theory and Wheeler stretching method and deterministic quasistatic design wave analysis.

Notes 1) Current at Veslefrikk field not included
     2) 28 m high Stokes V order wave equivalent to 30 m high 1. order Airy wave.
     3) Axial force and Base shear in MN, OTM in MNm

4   **STOCHASTIC ANALYSIS**

A full stochastic analysis is based on superposition of
waves of different frequencies, directions' and
amplitudes composing a irregular sea state. It's ability
to describe realistic sea states makes it superior to
the deterministic analysis. Using a Gaussian random wave
model, each sea state is completely described by its
direction wave spectrum. Directional aspects are
however, often described by a spreading function whereby
the wave spectrum reduces to a frequency spectrum $S_\eta(\omega)$
of which various theoretical formulations exists. For
the North Sea the Jonswap spectrum is generally used
while the Pierson - Moskowitz spectrum is widely used in
open sea (Sarpkaya and Isaacson, 1981).

Several methods are available for stochastic analysis.
Two of them will be discussed in terms of their
capability to describe the true loading on the offshore
structure.

a <u>Stochastic linearization in the frequency domain.</u>

The linear stochastic dynamic analysis is based on the
assumption that the wave surface can be regarded as a
stationary Gaussian and ergodic process. The method
involves linearization of the load vector in terms of
the statistical properties of the velocity spectrum
(Borgman, 1969).

The analysis is performed by transforming the
equilibrium equation into the frequency domain. The
structural response is found as

$$S_x \ (\omega) = H \ (\omega) \cdot F \ (\omega) \cdot H^{*^T} \ (\omega) \cdot S_\eta \ (\omega) \qquad (4.1)$$

where

S_x (ω) is the response spectrum
H^x (ω) is the frequency response function
         (mechanical response function)
F (ω) is a matrix relating the wave forces to the wave
         spectrum (hydrodynamic response function).F (ω)
         could include directional spreading if required.

This method restricts selection of wave kinematics model
to the <u>linear Airy wave theory</u>, whereby it is <u>not</u>
possible to include free surface effects as the
contributions of the high frequency wave components
increase unrestricted above the MSL. (Gudmestad, 1989).

Nonlinearities in the wave force model can however, in principle be accounted for through calculation of convolutions of the spectral density function for the velocity (Gudmestad and Connor, 1983).

For an evaluation of the goodness of using linear Airy wave theory, it should be noted that Vugts (1984) concludes that the results from the MaTS (1982) experiments show that the <u>linear random wave model</u> provides a valid theoretical description of the wave kinematics in an irregular sea although the model has the following limitations:.

-   the model and any linear model cannot predict a crest/trough distortion which is also reflected in the measured horizontal velocities: The absolute magnitude of the horizontal velocities is greater under deep wave troughs than under high wave crests.(Lloyd's 1984).

-   the linear random wave theory fails to predict the most extreme events

-   linear theory cannot predict fluid velocity above mean still water level for the high frequency components (see Gudmestad, 1989)

The same conclusions are drawn by Sintef (Sintef, 1987) analyzing full scale measurements.

Following the determination of the response spectrum the design load is calculated. In the case of Gaussion response the extreme value $\hat{x}$ is eq.

$$\hat{x} = \sigma \sqrt{2 L_n N} \qquad (4.2)$$

where σ is the standard deviation of the response spectrum and N is the expected number of maxima for a given duration. If nonlinear forcing terms are included the response is non-Gaussion and extrapolation to maxima at any probability level can be performed from fitted Weibull parameters.

Due to the method's limitations w.r.t.

-   use of linear wave therory
-   negligence of crest effects and associated non linear loading terms
-   difficulty to include nonlinear terms in the forcing function,

the method can not be recommended for __extreme__ response
estimation of dynamic sensitive deep water structures.
This is clearly shown in Table 4.1 giving forces on the
Veslefrikk jacket calculated by use of different
analysis methods.

For __fatigue__ analysis, on the other hand, stochastic
linearization in the freqency domain is appropriate.
Previous analysis (Syvertsen, Thuestad and Remseth,
1983), (Syvertsen and Karunakaran, 1985) shows that the
largest uncertainties in wave loading model for fatigue
analysis are associated with
- selection of wave spectra (the Jonswap and the
  Pierson Moskowitz spectra overestimate the fatigue
  damage with 10-25% compared to measured spectra)
- use of short crested vs long crested waves as long
  crested waves overestimate fatigue damage by a
  factor of 2-3 compared to short crested waves.
- introduction of appropriate spreading function $(\cos^2\theta)$
  where varies between 2 and 6 according to NPD's
  recommendation, NPD 1989) whereby the fatigue
  damage could vary with 0-20%.

b) __Time domain stochastic dynamic response analysis.__

A short term irregular sea state is as for the
stochastic linearization in the frequency domain method
(method a) represented by a wave spectrum $S_\eta(\omega)$.

The nonlinear dynamic analysis is based on a time domain
solution of the dynamic equation of motion (Eq. 3.1.).
In order to perform this analysis, the wave description
must be presented in the time domain by transforming
from the frequency domain to the time domain. Time
series of

$\eta(t)$ sea elevation
$u(t)$ wave particle velocity
$a(t)$ wave particle acceleration

eg. as $\quad u(t) = \sum Re\left\{\sqrt{2\,S_\eta(\omega_i)\,\Delta\omega}\; D_i\; exp\left[j\left(\omega_i t - \theta_i\right)\right]\right\}$ (4.3)

must be constructed;

where

ω_i = spectral frequency, evenly spaced with intervals DW

θ_i = phase angle generated by Monte Carlo simulation
from a uniform distribution between 0 and 2π

$S_\eta(\omega_i)$ = sea elevation spectral density at frequency i.

D_i = depth attenuation function dependent on wave kinematics model chosen (see Appendix).

A step-by step time integration of Eq 3.1 by eg. a Newmark β method produces a continuous response time history. Following this analysis, statistical properties of the response parameter can be calculated and the extreme response during the design storm can be estimated.

Through this analysis the following advantages are included:

- the full nonlinear force equation can be handled

- nonlinearities due to the structures' dynamics and the submergence can be included

- the description of the sea surface is realistic

Using the method, there are, however, certain disadvantages which need further research

- The choice of wave kinematics model is restricted, see discussion below

- Assumptions related to calculation of maximum of a non Gaussion process (nonlinear response history) must be made. A fitted Weibull curve is often assumed.

- The simulation must be performed for a rather long time series in order to obtain realistic estimate of the extreme response value . A time series of 20 minutes is considered minimum.

Through the discussion above, the choice of wave kinematics model appears to be a key element for utilization of this method.

PMB in their Hedop-Project (Hydrodynamic Effects on Design of Offshore Platforms) (Bea, Litton and Pawsey, 1988) have studied forces in extreme irregular hurricane waves. The max forces are predicted deterministically as the max. forces caclulated from the time history.

Three kinematics modes are being used (conf. Appendix):

- Modified Airy theory (also called vertical stretching) represented by Airy wave theory up to the mean water level (MWL) and constant velocity above MWL.

– Wheeler stretching method (Wheeler, 1970) whereby the Airy velocity profile between seabottom and MWL is stretched to the varying water level

– Gudmestad method to calculate velocity in irregular sea states, (Gudmestad, 1989).

The results from Hedop are briefly summarized in Figure 4.1 showing that the vertical stretching method gives very much larger deck displacement than the Wheeler stretching method for a 600 m (2000 ft) Compliant Tower Jacket.

These results represent displacements found in extreme waves in a hurrican sea state, Fig. 4.2.They are not based on a statistical description of the maxima of the response. Despite this, the results give good indication of the relative behaviour of the three wave kinematics models studied by PMB. It is probable that this general trend also would be found in a full time domain stochastic dynamic response analysis.

Steele et al, (1988) compare the measured displacement spectrum of a compliant tower with displacement spectra calculated by using Vertical stretching and Wheeeler stretching. They show that considerable non-measured low frequency higher harmonics are generated by the use of Vertical stretching method, conf. Fig. 4.3. They do also show that the vertical stretching model introduces very large low frequency components in the displacement spectrum of a drag dominated compliant tower.

In a study for Statoil, Sintef (Spidsøe et al, 1989) have calculated the extreme forces on the Veslefrikk jacket using time domain stochastic dynamic response analysis and different wave kinematics models. Although further investigations are required, Sintef's conclusions are as follows:

Non of the models studied in Hedop may be regarded as a true or correct model for time domain stochastic dynamic response analysis. The modified vertically stretched Airy model was found to be conservative as wave model to be used in dynamic response analysis. This model was, however, also shown to have non-physical properties, in particular due to unphysical description of the wave kinematics for intermediate water surface positions, Fig. 4.4. This might have biased the dynamic response predictions. The Gudmestad model **and** the Wheeler stretching model were found to be physically reasonable and slightly non-conservative with respect to calculation of wave forces in the splash zone and with respect to prediction of dynamic response. Fig. 4.4 shows the kinematics profile for selected situations.

Above conclusions were based on comparison between measured forces and simulation of wave forces on a vertical pile using different water particle kinematics models (Brathaug and Karunakaran, 1989):

- Simulation of water particle kinematics from measured surface elevation

- Estimation of C_d and C_m by fitting Morison equation to measured local forces using the simulated water particle kinematics

- Calculation of forces using the estimated force coefficients and the simulated kinematics

- Comparison of measured and simulated local and integrated forces.

Although Sintef's conclusions are based on a study having limited time series, the following general conclusions can be drawn:

- The modified vertically stretched Airy model appears to give conservative results

- Wheeler stretching and Gudmestad kinematics models give physically consistent response values which appears somewhat unconservative.

Table 4.2 gives a comparison of characteristic values of dynamic response of the Veslefrikk jacket using different wave kinematics models.

Results from other studies also show that the Wheeler method appears realistic, Conf. Fig. 4.2 (from Steele et al., 1988).

The three methods discussed above do however, all have serious problems for use in a consistent calculation:

- Non of them satisfy the Laplace equation in the fluid

- The analysis is based on super-position of waves with linear wave heights (Eq 4.3) although the kinematics models all are nonlinear whereby the superpositon principle for adding wave components is violated.

Further model tests are therefore necessary to shed more light on wave kinematics models.

A study sponsored by Statoil, Exxon, Amoco and NTNF is presently being undertaken by NHL with the objective of identifying the most correct wave kinematics model in irregular sea states, Tørum (1989). Preliminary results seems to confirm use of Wheeler stretching and Gudmestad kinematics models.

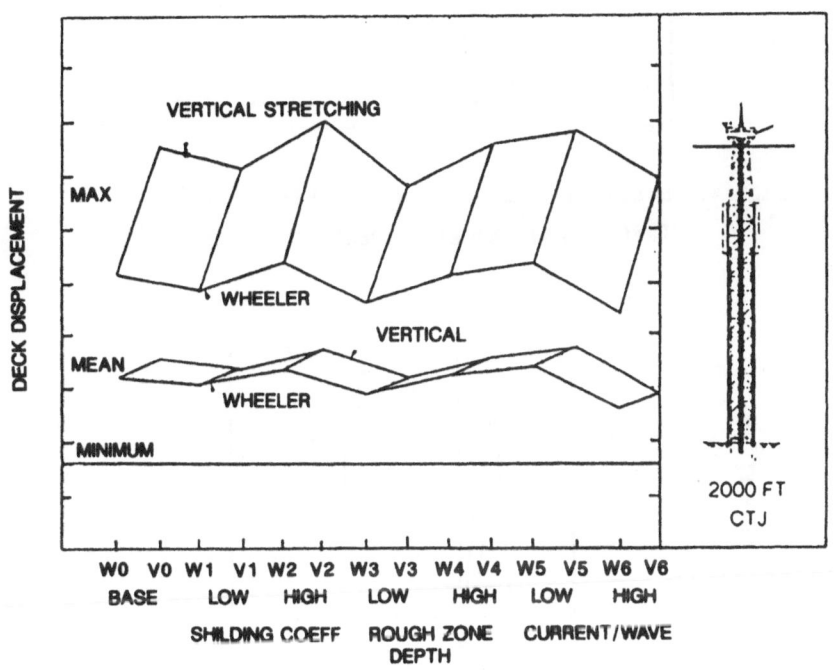

Fig. 4.1 Deck displacement of a 2000 ft CTJ

(Bea et al., 1988)

78

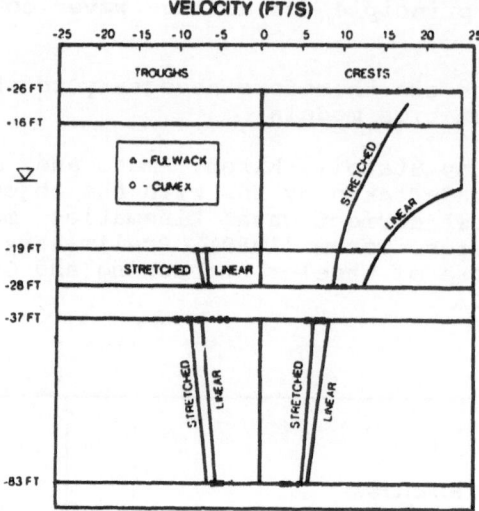

Fig. 4.2 Measured vs predicted wave kinematics
(Ref. Steele et al., 1988)

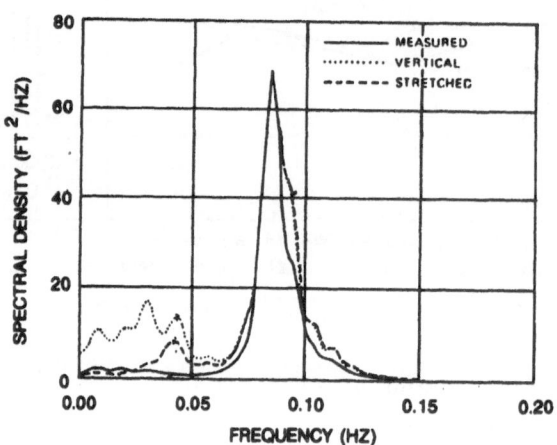

Fig. 4.3 "Lena" Guyed Tower Dynamic Response: Lena
Displacement Spectra, Hurricane Juan
(Ref. Steele et al., 1988)

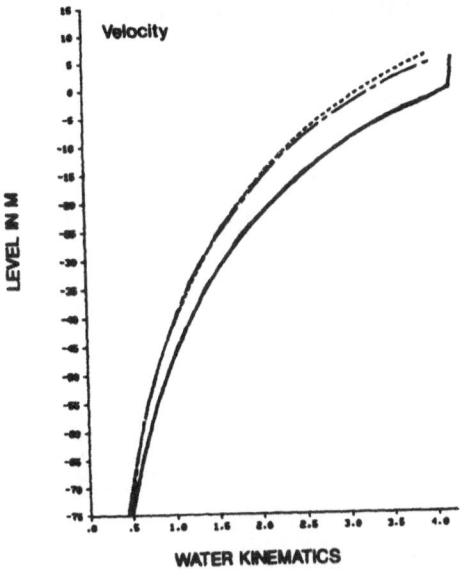

Velocity

LEVEL IN M

WATER KINEMATICS

Example of simulated water particle kinematics at crest position.

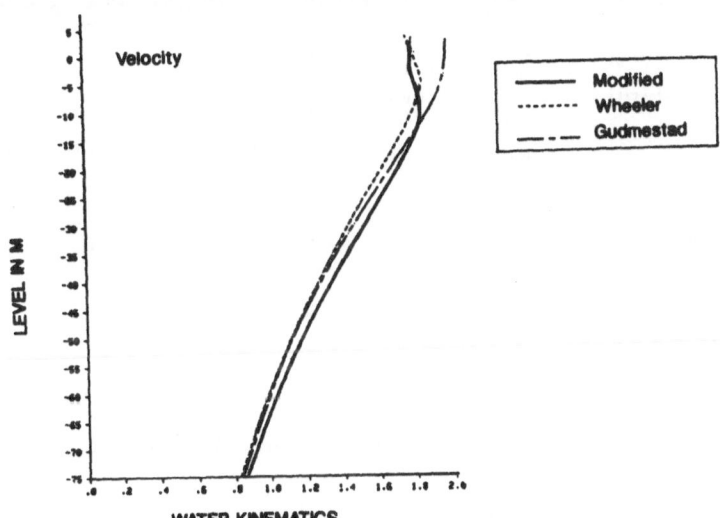

Velocity

	Modified
	Wheeler
	Gudmestad

LEVEL IN M

WATER KINEMATICS

Example of simulated water particle kinematics in position between MWL and crest.

Fig. 4.4 Wave kinematics profile using different water wave
kinematics models

(Ref. Spidsøe et al., 1989)

Response quantity	Extreme response Stochastic linearization in frequency domain			Nonlinear time domain stochastic dynamic response analysis			Design wave analysis	
	Dynamic	Static	DAF	Dynamic	Static	DAF	Wheeler stretching	Stokes V
Base shear	16.4	16.0	1.025	21.82	18.51	1.179	18.1	24.0
Overturning moment	1901	1791	1.061	3778	3213	1.176	2609	3586
Brace forces								
El-13m (Brace 75)	1.82	1.73	1.052	5.34	4.68	1.141	3.30	4.37
El-36m (Brace 83)	2.28	2.19	1.041	4.80	4.54	1.057	3.43	4.53
El-90m (Brace 99)	2.60	2.52	1.032	3.51	3.34	1.051	3.08	3.99
El-149m(Brace 115)	2.61	2.54	1.028	2.02	1.88	1.074	2.58	3.32
Leg.forces								
El-13m (Leg 136)	3.44	2.70	1.274	9.47	1.86	1.380	6.29	7.22
El-36m (Leg 140)	6.06	5.26	1.152	16.61	12.52	1.327	10.90	13.10
El-90m (Leg 148)	12.07	11.15	1.083	27.9	21.81	1.279	19.27	23.85
El-149m (Leg 156)	16.96	15.93	1.065	33.5	26.72	1.254	24.87	30.90

Table 4.1 Forces on the Veslefrikk jacket calculated by using different analysis methods (3D analysis).
Notes:
1. Current not included
2. Vertically stretched Airy theory for Nonlinear time domain analyses
3. Axial force and Base shear in MN and OTM in MNm
4. Dynamic Amplification Factor (DAF) = $\dfrac{\text{Dynamic Response}}{\text{Static Response}}$

Response quantity	Extreme values x_e and Dynamic amplification factors DAF							
	Airy model (integration to MWL only)		Vertically stretched Airy model		Wheeler stretcing		Gudmestad model	
	x_e	DAF	x_e	DAF	x_e	DAF	x_e	DAF
Base shear	17.03	1.29	21.03	1.04	18.31	1.11	17.90	1.13
Overturning moment	1908	1.08	3245	1.04	2723	1.12	2624	1.09
Base forces								
El-13 m (Brace 75)	2.73	1.15	4.62	1.01	3.59	1.13	3.64	1.14
El-36 m (Brace 83)	2.92	1.09	4.31	1.00	3.55	1.11	3.48	1.11
El-90 m (Brace 99)	2.42	1.06	3.30	1.06	2.80	1.06	2.71	1.06
El-149m (Brace 115)	1.80	1.14	2.34	1.24	1.90	1.14	1.81	1.16
Leg forces								
El-13 m(Leg 136)	6.00	2.63	8.14	1.10	6.80	1.27	6.98	1.26
El-36 m(Leg 140)	9.32	1.66	14.09	1.06	11.49	1.20	11.70	1.21
El-90 m(Leg 148)	13.03	1.06	24.19	1.04	20.11	1.16	11.70	1.17
El-149 m(Leg 156)	17,39	1.07	29.22	1.05	25.10	1.12	24.15	1.10

Table 4.2 Comparison of characteristic values for dynamic response of the Veslefrikk jacket calculated by use of different wave kinematics modes (2 D analysis).
Note:
1. Current not included.

5   RECOMMENDATIONS

The present paper has consentrated on discussion of analysis of dynamic sensitive drag dominated structures with particular emphasis on the importance of wave kinematics.

With respect to wave kinematics it is recommended to carry out further research:

-     to fully understand all aspects of measured regular water wave kinematics and compare to predictions of theoretical models. This will also increase our knowledge about transient phenomena in wave tanks.

-     to increase our understanding of wave-current kinematics interactions. This aspect has been outside the scope of the present paper but is considered very important. It is believed that further measurements and analysis of regular wave-current interactions will improve the general knowledge about the phenomena prior to studying the irregular wave-current situation.

-     to develop appropriate and fully consistent wave kinematics models for time domain nonlinear stochastic dynamic response analysis (the underline preferred method of analysis). This should include publication of measurements (Tørum et al. 1989) and further theoretical work (Spidsøe et al. 1989).

As measurements (Tørum 1989) indicate that the stretched Wheeler method and the Gudmestad method (Gudmestad,1989) give sufficient conservative design, these methods might be acceptable. In practice the Wheeler method is the simpler and should be used in engineering applications.

Prior to full confirmation of these measurements, the modified Vertically stretched Airy theory may be used as it is conservative and results in a safe design.

It should however, be noted that Gudmestad's irregular wave kinematics model, which seems interesting for time dynamic nonlinear stochastic dynamic respone analysis, is consistent with Gudmestad and Connor's method for regular waves and that the latter regular wave theory was found to describe the kinematics in regular waves particularly well (Gudmestad et al, 1988)

- to determine the extreme value(s) in time domain nonlinear stochastic dynamic response analysis most accurately.

Although our knowledge about wave processes increases as our models are being continuously refined, we will not be able to to fully understand all aspects of the wave. Our theoretical models will always represent idealizations which hopefully however, will be sufficient for engineering purposes.

The author will therefore end this paper with an old saying from Jæren, Norway describing the wave::

"Here comes a barrel forceful roaring
with no bands
with no bottom
No man in our land
this barrel bonding can".

6 ACKNOWLEDGEMENTS

The authors will express thanks to Statoil and to Sintef for permission to publish the paper. Furthermore, thanks are due to colleagues S.Haver and H.Nordal in Statoil, and to A. Tørum of NHL for valuable discussions. The interests of J.Vugts, Shell in the field of wave forces on structures in irregular waves has also provided inspiration to perform the work.

7 REFERENCES

API RP2A (1987)   : "Recommended Practice for Planning, Designing and Constructing Fixed Offshore Platforms". American Petroleum Institute, 17 edition.

Bea, R.G.and Lai, N.W.(1978)   : "Hydrodynamic Loadings on Offshore Structures". Offshore Tech. Conf., Houston. Paper No OTC 8064 pp 155-168.

Bea, R.G., Litton, R.G.
and Pawsey, S.F.(1988)   : "Hydrodynamic Effects on
                               Design   of   Offshore
                               Platforms (Hedop)." In
                               OCJ Study MMS 88-0057:
                               "Technology   Assessment
                               and Research Program for
                               Offshore   Minerals
                               Operations", US Dept. of
                               the Interior MMS, pp. 85
                               - 90.

Biggs, J.M. (1964)   : "Introduction   to
                               Structural   Dynamics".
                               McGraw Hill, Inc.

Borgman, L.E. (1969)   :"Ocean   Wave Simulation
                              for Engineering Design".
                              J.Waterways and Harbors
                              Division, ASCE Vol. 95 No
                              WW4 pp 557-583.

Brathaug, H.P.
and Karunakaran, D. (1989) :"Measured Irregular Wave
                              Forces on a Pile". Sintef
                              Report to be issued 1989.

Bærheim, M. and
Fossan,T.I.(1989)   :"Weight Optimization of
                              the Veslefrikk Jacket".
                              OTC 6189 Proc. Offshore
                              Technology   Conference
                              1989,Vol. 1, pp 689-700.

Dean, R.G. (1970)   : "Relative validities of
                               Water   Wave   Theories".
                               J.Waterways Harbors and
                               Coastal Engr. Division,
                               ASCE, Vol. 96, No WW1 pp
                               105-119.

Forrisdal, G.J. (1985)   : "Irregular   Wave
                               Kinematics   from   a
                               Kinematic   Boundary
                               Condition Fit (KBCF)".
                               Journal of Applied Ocean
                               Research, Vol. 7, No 4
                               pp 202 - 212.

Gudmestad, O.T. (1989)   :"A   new   Approach   for
                              Estimating Irregular Deep
                              water Wave Kinematics".

84

Accepted for publication in Applied Ocean Research.

Gudmestad, O.T.and
Connor, J.J. (1983)
: "Linearization Methods and the Influence of Current on the Nonlinear Hydrodynamic Drag Force". Applied Ocean Research,Vol. 5, No 4 pp 184-194.

Gudmestad, O.T. and
Connor, J.J. (1986)
: "Engineering Approximations to Nonlinear Deepwater Waves". Applied Ocean Research, Vol. 8, No. 2 pp 76 - 88.

Gudmestad, O.T.,Johnsen, J.M.,
Skjelbreia, J.
and Tørum A. (1988)
:"Regular Water Wave Kinematics". Proc. Int. Conf. on Behaviour of Offshore Structures,(Boss) Trondheim 1988, Vol. 2.

Gudmestad, O.T.and
Poumbouras, G.(1986)
:"Nonlinear Effects in Hydrodynamic Waves with some Applications to Offshore Structures". Proc. 1986 Offshore Mechanics and Arctic Engineering (OMAE), Tokyo pp 155 - 165.

Gudmestad, O.T. and
Poumbouras, G. (1988)
:"Time and Frequency Domain Wave Forces on Offshore Structures". Applied Ocean Research, Vol.10, No 1 pp 43 - 46.

Lambrakos, K.F., Chao,
J.C., Beckman, H.
and Brannon,H.R. (1987)
: "Wave Model of Hydrodynamic Forces on Pipelines". Ocean

Engineering Vol. 14 No 2.

Le Mehaute, B. (1976)

: "An Introduction to Hydrodynamics and Water Waves". Springer-Verlag, Dusseldorf.

Lloyd's (1984)

: "Monitoring and Assessment of Data from Offshore Platforms -Summary Resport". Report OT-08356,Lloyd's Register of Shipping.

MaTS (1982)

: "Wave Kinematics in Irregular Waves". Report M/628/MaTS VM-1-4,Delft Hydraulic Laboratory, Delft.

Moe, G. (1989):

:"Morison type Wave loading" Proc. NATO Advanced Research Workshop, Molde, May.

Morison, J.R.,O.Brian, M.P., Johnson, J.W. and Schaaf, S.A., (1950)

:"The Force Exerted by Surface Waves on Piles". Petrol Trans., AIME, Vol. 189, pp 149-154.

Norwegian Petroleum Directorate (NPD) (1989)

:"Acts, Regulations and Provisions for the Petroleum Activity" NPD Stavanger, Norway.

Sand, S.E., Ottesen Hansen, N.E., Jacobsen, V., Gudmestad, O.T. and Sterndorff, M.J. (1989)

:"Freak Wave Kinematics". Proc. NATO Advanced Research Workshop, Molde, May.

Sarpkaya T. and Isaacson, M.(1981)

:"Mechanics of Wave Forces on Offshore Structures", van Nostrand Reinhold Company.

Sintef (1987)

:"Statistical Analysis of Wave and Current Data Measured in Extreme Sea States". Sintef report STF 71 F87018. Authors: D.Karunakaran and N.Spidsøe.

Spidsøe, N., Karunakaran, D. and Skjåstad, O. (1989)

:"Modelling of Irregular Waves for Stochastic Dynamic Response Analysis of Drag Dominated Offshore Platforms". Sintef Report STF 71 F89021.

Steele, K.M., Finn, L.D. and Lambrakos, K.F. (1988)

:"Compliant Tower Response Prediction Procedures". Proc. Offshore Technology Conference (OTC), 5783.

Syvertsen, K. and Karunakaran, D. (1985)

:"Uncertainties in Fatigue Life Calculations and Extreme Response predictions". Sintef report STF 71 F.85048.

Syvertsen, K., Thuestad T. and Remseth S. (1983)

:"The effect of Short Crested Waves to Fatigue Damage Calculation". ASCE Structures Congress '83. Houston, Oct. 1983.

Tørum, A. (1989)

:"Irregular Water Waves". Proc. NATO Advanced Research Workshop, Molde, May.

Vugts, J. (1984)

:"Offshore Structures Engineering." Lectures presented to the Norwegian Institute of Technology at Trondheim in October 1983. SIPM report EP 60690, August 1984.

Wheeler, J.D.   (1970)   :"Method for Calculating Forces produced by Irregular Waves" Journal of Petroleum Technology, March, pp 359-367.

8   APPENDIX

Depth decay function for wave models used in the time domain nonlinear stochastic dynamic response analysis (Chapter 4):

-    Modified vertically extrapolated Airy model

$$D_i = \begin{cases} \dfrac{\cosh k_i(z+d)}{\cosh k_i} & \text{for } z \leqslant 0 \\[2mm] 1 & \text{for } z > 0 \end{cases} \qquad \text{(Airy model)}$$

-    Vertically stretched model (Wheeler)

$$D_i = \frac{\cosh k_i(z'+d)}{\cosh k_i} \qquad \text{where } z' = (z-\eta)\frac{d}{d+\eta}$$

i.e. water depth corresponds to the level of the instantaneous surface elevation.

-    Gudmestad model,

D_i is a relatively complicated combination of several terms, Gudmestad (1989).

Theory versus Measurement

D.H. PEREGRINE
Dept.of Mathematics
Bristol University
Bristol BS8 1TW
England

ABSTRACT The respective roles of theory and measurement are discussed. After an introduction to the common assumptions invoked to obtain theoretical results some of the interactions between theory and experiment are described. Problems do arise when comparisons are made between theory and experiment, often to the ultimate benefit of both sides. Some particular examples are mentioned. The final discussion concludes that there is no real competition between theory and measurement but that they are complementary.

1. Theoretical background

There is no doubt that the Navier–Stokes equations are adequate to describe air and water motion, but it is impractical to use them directly for any length scale greater than a few centimetres. For the foreseeable future the best we can do is to use the inviscid Euler equations of motion with or without the inclusion of some modelling of turbulence where appropriate. Turbulence modelling is a large, and not entirely successful, area of research on its own.

Fluid boundaries also need description. Rigid impermeable boundaries present no intrinsic difficulty unless there are intricate details of geometry. On the other hand the sea bed may be permeable, movable and/or covered in flexible marine growth. In some of these cases considerable effort is still needed to provide satisfactory mathematical models. Similarly, the free surface between air and water is easily modelled if it is clean and has only gentle slopes. In more severe circumstances, which might include wave breaking, strong winds, surface pollution and ice, much development of mathematical models is still required for the free surface and associated two–phase flow.

Even when suitable physical approximations have been made and a satisfactory mathematical model is obtained, there is usually a need for further mathematical and physical approximations to enable numerical or analytical solutions to be found. In a large majority of cases this results in theoretical studies of irrotational flow, with rigid boundaries and a linearized free surface. For prototype scales, greater than one metre in length scale, and for model scales greater than 10 to 20cm, this is usually a good approximation as long as waves do not become steep or travel a long distance.

A. Tørum and O. T. Gudmestad (eds.), Water Wave Kinematics, 89–101.
© 1990 Kluwer Academic Publishers.

In many cases modelling can be most readily improved by including nonlinear terms. The "state of the art" for irrotational flows is that almost any wave motion short of breaking can be computed for two dimensional examples, whereas in three-dimensions even weakly nonlinear theories are only well developed for the case of slowly-varying waves over a slowly-varying bed.

Of the many other phenomena neglected in these models the most important are breaking and the various effects of turbulence. The dissipative effects of both breaking and turbulence can sometimes be adequately dealt with by semi-empirical terms. An important effect of turbulence is the mean vorticity which is established, for example the variation of currents with depth due to wind shear or bottom friction. Progress is being made in the inviscid modelling of waves with vorticity in two dimensions.

2. Theory and measurement

The classical scientific interactions between theory and measurement are

(a) observations of a phenomenon lead to theory being developed to explain it.

(b) new phenomena predicted by theory lead to experiments and/or observations being made to verify the prediction.

(c) discrepancies between theory and measurement stimulate further development of both.

In the context of water waves, technological and environmental considerations are also relevant. Field measurements are made to provide data for establishing wave climate, though frequently semi-empirical methods must be used in the absence of such data. In both laboratory and prototype studies it is difficult to measure all desired quantities. Theory then complements experiment by permitting estimates of these quantities to be made from the measurements. Perhaps the commonest measurement is a point measure of height as a function of time. Such time series are frequently extrapolated with various assumptions to give energy spectra, velocity and pressure distributions.

As well as the classical use of measurements to provide a data-base or empirical rules in the absence of theory, experiments are compared with approximate solutions in order to judge the range of applicability of theoretical results. For example, if a parameter such as wave steepness must be small, we like to ascertain at what steepness theory and experiment diverge significantly.

3. Comparisons between theory and experiment

In addition to the many approximations made in obtaining theoretical solutions there are often discrepancies, or gaps in measured data which cause considerable uncertainties when comparisons are made with laboratory experiments. Examples include:

water depth: variations of water level might be measured to fractions
of a millimeter but often the mean water level, or even the
still water level, may not be known to the nearest millimeter.

currents are caused by any spatial or temporal variation in wave
amplitudes. These are only rarely measured or even estimated.
They are especially significant in field measurements and in
three–dimensional wave experiments.

geometry: is this as specified ? In my first attempt at serious
experiments I found that the nominally plane horizontal bed
of a flume could hold puddles over 1cm deep when it was
drained. Considerable effort was needed to bring the discrepancy
below 2mm. One of the common difficulties of modelling wave
generation by a paddle in a wave tank is how to model leakage
past the paddle.

waves: since these are the main feature of wave experiments these
are usually measured to verify that the correct waves are
generated, but reflections, free harmonics, and long waves are
usually created at the same time and need monitoring.
e.g. measurements of a "mean" force exerted by a wave group
could be seriously in error if a free long wave is generated
rather than the appropriate bound long wave e.g. see Bowers (1980).

Field experiments and observations bring in numerous further problems such as
bed mobility, time varying winds and currents.

A majority of experimenters are probably aware of these problems, in addition to
the specific difficulties and errors associated with the measuring instruments used.
Theoreticians may be less aware of these difficulties.

4. Some examples

The accuracy of linear theory for gentle waves can be impressive. Early in my career
I had the opportunity to create waves in a large circular tank, approximately 3 metres
in diameter. The two standing wave modes of longest period, those with $J_1(kr)$ and
$J_0(kr)$ surface shapes, could be easily excited and had such small dissipation that their
period could be readily measured and was found to be in agreement with linear theory
to 3 significant figures.

In experiments on solitary waves in a trapezoidal channel (Peregrine, 1969) I
found satisfying agreement with theory (Peregrine 1968) for variation of crest height
across the channel. See figure 1 for a typical experimental photograph. In retrospect,
it may be more scientifically stimulating to stress that I could not generate satisfactory
solitary waves when the channel cross section was of triangular form with one vertical
side and the other at 30° to the horizontal. In addition although most of my wave
velocity measurements were within 2% of the theoretical value, hardly any were
greater than that value.

Figure 1. Head on view of a solitary wave in a trapezoidal channel, showing the variation of crest height across the channel. The diagonal lines on the screen behind the wave help to make the profile more distinct.

Negative or biased results such as these are not often reported since they may be due to some oversight in experimental technique. Examples where discrepancies have been followed up are

(i)   In comparisons of wave transmissions through a small hole in a thin plate. Guiney, Noye and Tuck (1972) found that agreement with theory was much improved when the theory was extended to allow for plate thickness.

(ii) The modelling of wave reflection and transmission at a thin vertical plate immersed to a finite depth to act as a breakwater was improved by Stiassnie, Naheer and Boguslavsky (1984) when a simple model of flow separation from the plate was introduced.

A comparison between numerical solutions of the shallow–water equation with bores (Hibberd and Peregrine, 1979) and experiments on beaches (Packwood, 1980, gave sufficiently good results that we were able to tell the experimenters that their beach was not long enough to avoid overtopping by waves, see figure 2. The moral for experimenters is: look at your experiment don't leave it entirely to computers.

On the other hand, in a different experiment, where we had included a friction term, our run–up comparisons were awful. The last thin film of water scarcely flowed down the beach. However, fair agreement was obtained once we found out that the run–up gauge measured water only when it was more than 2mm deep, figure 3.

Experiments compared with the same numerical method for shallow–water waves, outside the expected validity of the approximations used, show that it may still be a useful tool in the absence of any more effective method (Kobayashi, da Silva and Watson, 1989).

Field observations (Packwood, 1980) for the same type of condition showed that even with predominantly normal incidence for the waves on a beach a considerable part of the records of bore positions were affected by three dimensional motions, figure 4.   The bore positions were taken from a movie which permitted easy observation of these effects.

Our most important comparison with experiment for these bores on beaches is shown in figure 5. Input to the numerical program came from height measurements at the offshore edge of the surf zone in a flume.   Comparison is made between experiments and computations close to the shore. Good agreement is seen, as the first few waves arrive and the set up develops. Thereafter a long period oscillation in the computation disturbs the otherwise good results. Numerical tests showed this is due to relying solely on height measurements at an open boundary since shoreward reflection of reflected waves is possible. Even so such poor boundary conditions are often used because of shortage of velocity measurements.

Another example where reflection interfered is in experiments performed by D. Skyner in the Edinburgh Wide Tank on wave focussing (Peregrine et al,1988). In these, focussed wave patterns were measured and compared with the wave field computed using a nonlinear Schrödinger equation. Agreement was poor until an estimated 5% reflection was included, see figure 6.

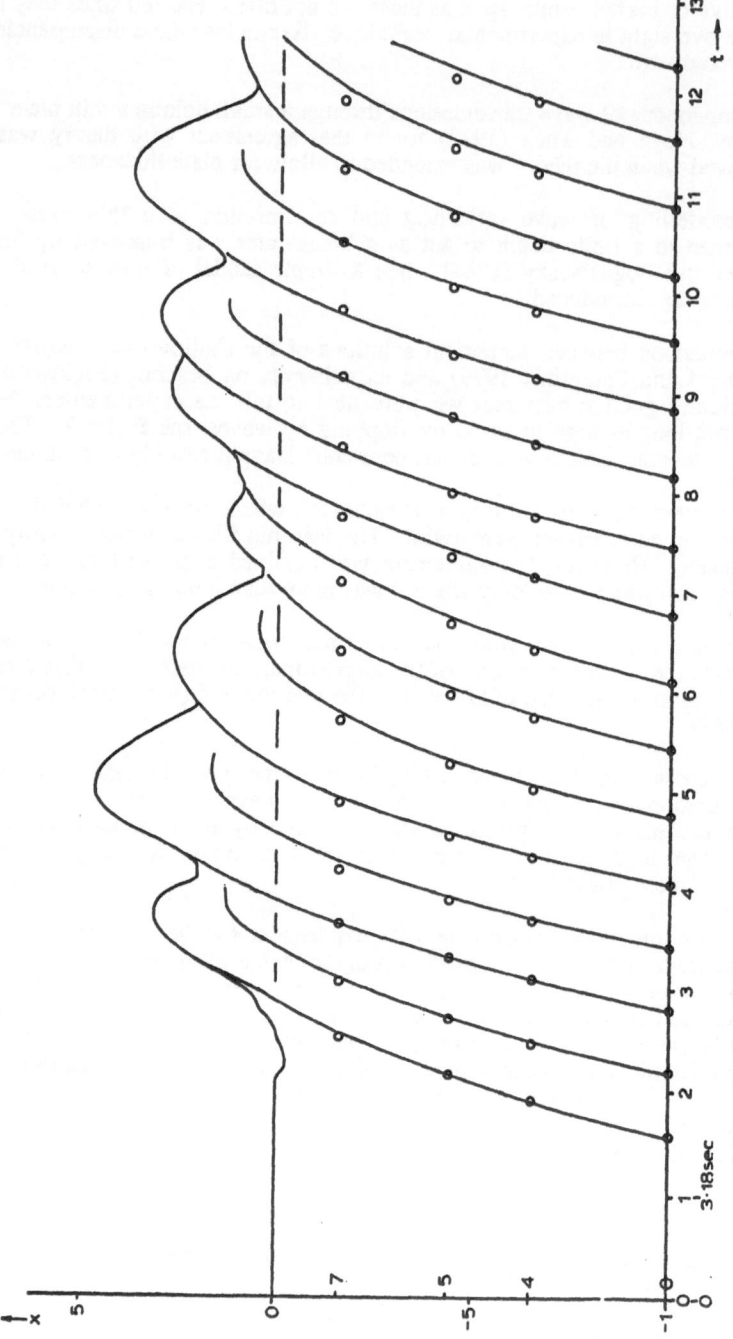

Figure 2. Space-time plot of bore trajectories computed for an experimental beach. The computation uses information from the experiment at the outer (lower) boundary in shallow water equations. Comparisons with measurement are shown by open circles. (Courtesy of A.R.Packwood).

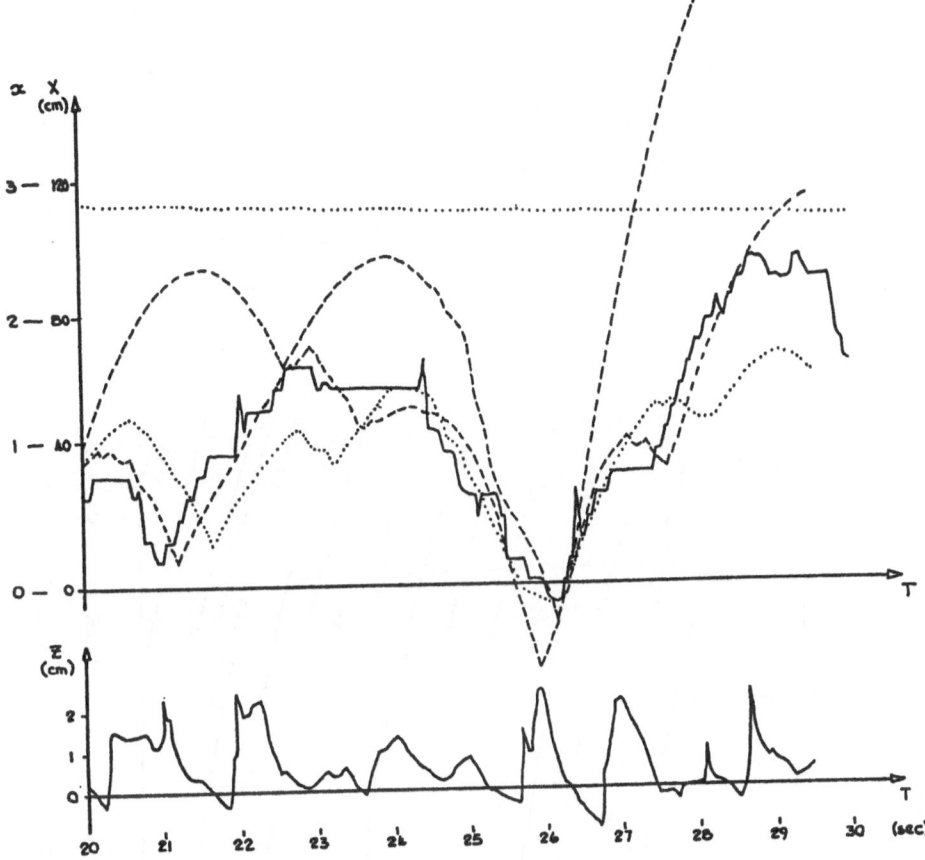

Figure 3. Comparison between theory and experiment for shoreline run–up. Solid line denotes experiment. Dashed lines inviscid shallow–water computation, shoreline and 2mm depth contour. Dotted lines shoreline and 2mm contour with a friction term added to equations. The offshore signal which drives the computations is shown below. (Courtesy of A.R.Packwood).

96

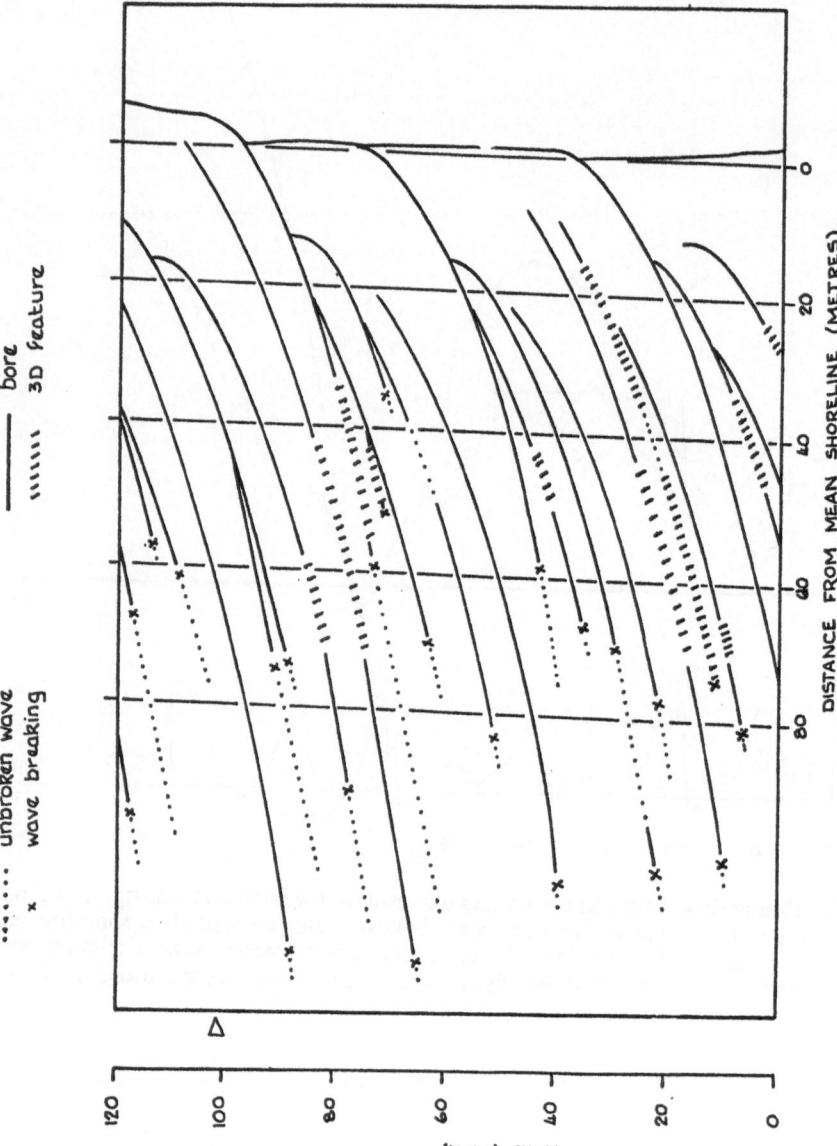

Figure 4. Space–time plot of bores on Putborough Sands, Devon. (Courtesy of A.R.Packwood).

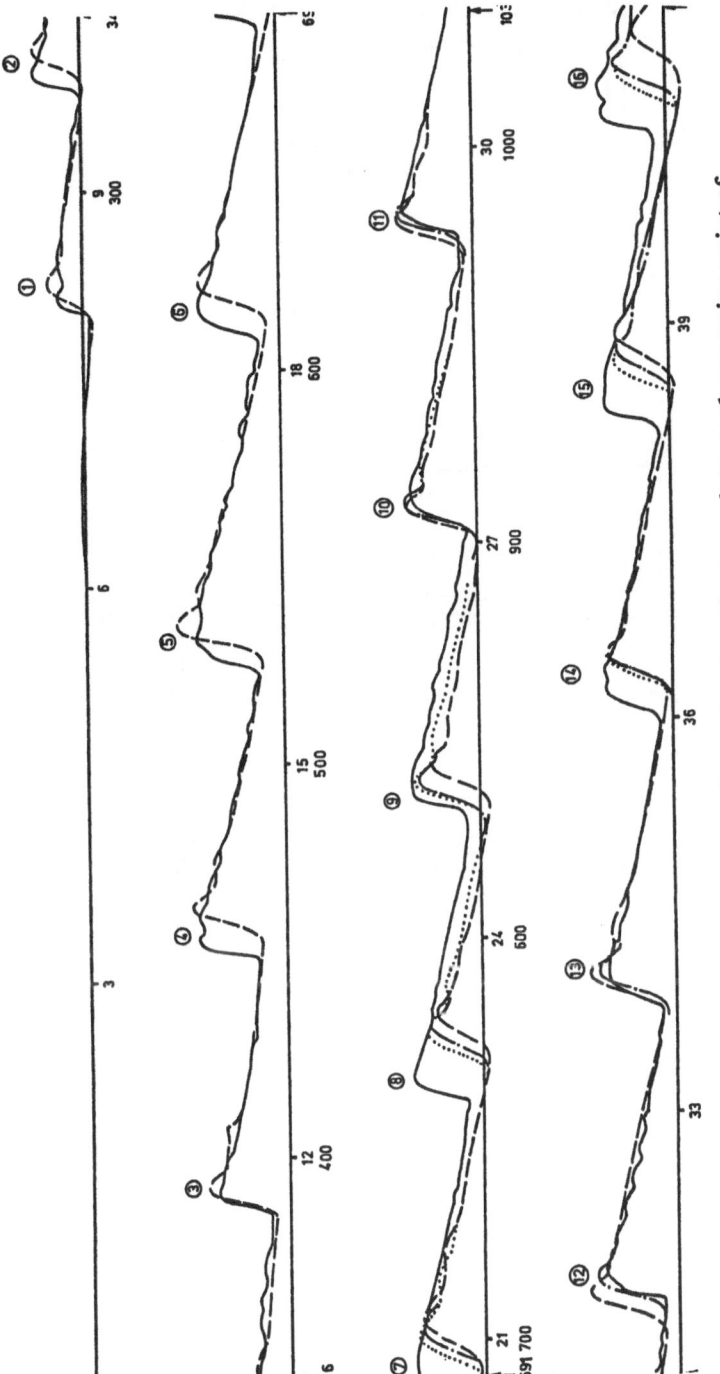

Figure 5. Comparison of theory and experiment at the most shoreward measuring point of figure 2. Solid line shows experimental measurement, other lines are different computations with shallow water equations. (Courtesy of A.R.Packwood).

97

98

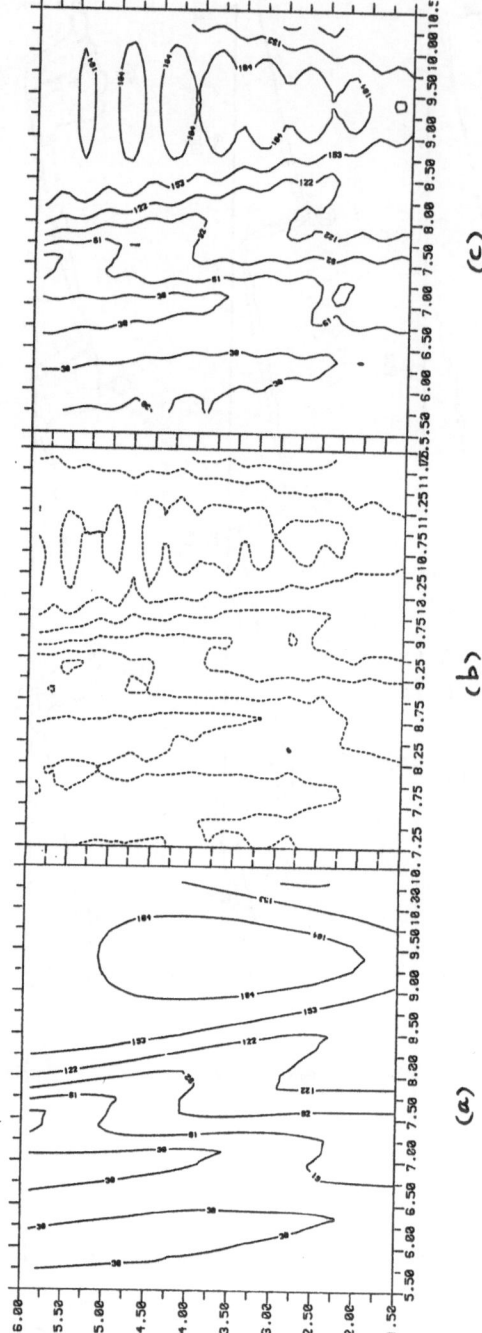

Figure 6. Comparison of surface contours in a focussing experiment with
(a) theory (without reflection), (b) measurements, (c) theory with 5% reflection.

For two dimensional, non breaking waves, numerical experiments with full irrotational flow can be compared with results of simpler theoretical models. We have carried out some examples for deep–water–wave modulation, where the appropriate nonlinear Schrödinger equation is surprisingly good and for undular bore development where Boussinesq's equations were no better than might be expected from the approximations made in their derivation. In the former case there were the typical difficulties of theory/experiment comparison when we needed to deduce how the derive a "slowly–varying amplitude" from the "exact" numerical solutions.

5. Theory versus measurement ?

The title of this paper, and some exchanges between theoreticians and experimenters, imply a contest between theory and experiment. In reality, in any study or application of waves, there must be the best practical combination of both. Indeed very little use could be made of measurements without theory and theory on its own is generally only able to satisfactorily predict wave motion in some laboratory examples. Measurements will always be important to fill the gaps in the theory. Indeed this should be one aim of measurements.

As well as the really difficult areas of severe wave conditions, where some recent work is described by Chan and Melville (1988), there are three important topics which can be pursued with expectation of reasonable progress both experimentally and theoretically.

(i) The irrotational flow assumption needs to be relaxed more often, flows with vorticity should be studied. These might have shear in the vertical or horizontal., Progress has been made since the major review, Peregrine (1976), but more coordination of measurements and theory would help. For example, can theoreticians or experimenters devise an effective way of measuring vorticity ?

(ii) Reflection from coastal regions is often ignored, yet the background noise measured by seismometers, was explained in terms of the pressure on the ocean floor due to standing waves by Longuet–Higgins (1950). This implies significant coastal reflection. Such reflection can be seen from steep coasts and coastal protection works are known to cause severe problems when they reflect waves into sensitive areas. In addition there are the recently discovered effects of "localization" in which waves are reflected by irregular bottom topography (Devillard et al, 1988, Belzons et al, 1989) and reflection by regular undulations of the bottom (see Davies and Heathershaw, 1984). Awareness of these reflection mechanisms has arisen from studies of other types of wave motion and their transfer to water waves is essentially theoretical.

(iii) Three–dimensional effects are a difficult aspect to interpret in most field measurements other than the occasional large–scale field experiment where enough measurements can be made to clarify the three–dimensional nature of wave fields. This is an area where remote sensing can be of particular value. Again, more theory and measurement are needed here.

Research support from the U.K. Science and Engineering Research Council is gratefully acknowledged.

References

Belzons, M., Guazzelli, E. & Parodi, O. (1988) Gravity waves on a rough bottom: experimental evidence of one dimensional localization. J.Fluid Mech. 186, 539–558.

Bowers, E.C. (1980) Long period disturbances due to wave groups. Proc.17th Coastal Engng.Conf. ASCE, Sydney, 1, 610–623.

Chan, E.S. & Melville, W.K. (1988) Deep–water plunging wave pressures on a verical plane wall. Proc.Roy.Soc.London A 417, 95–131.

Davies, A.G. & Heathershaw, A.D. (1984) Surface wave propagation over sinosoidally varying topography. J.Fluid Mech. 144, 419–443.

Devillard, P., Dunlop, F. & Souillard, B. (1988) Localization of gravity waves on a channel with a random bottom. J.Fluid Mech. 186, 521–538.

Guiney, D.C., Noye, B.J. & Tuck, E.O. (1972) Transmission of water waves through small apertures J.Fluid Mech. 55, 149–161.

Hibberd, S. & Peregrine, D.H. (1979) Surf and run–up on a beach. J.Fluid Mech. 95, 323–345.

Kobayashi, D., DeSilva, G.S. & Watson, K.D. (1989) Wave transformation and swash oscillation on gentle and steep slopes. J.Geophys.Res. 94, C1, 951–966.

Longuet–Higgins, M.S. (1950) A theory of the origin of microseisms. Phil.Trans.A. 243, 1–35.

Packwood, A.R. (1980) Surf and run–up on beaches. Ph.D. Thesis, Bristol Univ.

Peregrine, D.H. (1968) Long waves in a uniform channel of arbitrary cross–section. J.Fluid Mech. 32, 353–365.

Peregrine, D.H. (1969) Solitary waves in trapezoidal channels. J.Fluid Mech. 35, 1–6.

Peregrine, D.H. (1976) Interactions of water waves and currents. Adv. in Appl.Mech. 16, 9–117.

Peregrine, D.H., Skyner, D., Stiassnie, M. & Dodd, N. (1988) Nonlinear effects on weakly focussed waves. Proc.21st Conf.Coastal Engng. ASCE.

Stiassnie, M., Naheer, E. & Boguslavsky (1984) Energy losses due to vortex shedding from the lower edge of a vertical plate attacked by surface waves. Proc.Roy.Soc.A 396, 131–142.

EFFECTS OF WATER WAVE KINEMATICS UNCERTAINTY ON DESIGN CRITERIA FOR OFFSHORE STRUCTURES

Robert G. Bea
University of California, Berkeley
Departments of Civil Engineering and
Naval Architecture & Offshore Engineering
202 Naval Architecture Building
Berkeley, CA 94720

ABSTRACT

The effects of water wave kinematics on design criteria for offshore structures is examined in the framework of a comprehensive reliability analysis. The major components contributing to uncertainties in kinematics and hydrodynamic forces are discussed. Basic engineering approaches to evaluations of uncertainties in the components are discussed and illustrated with development of design criteria for an offshore oil and gas drilling and production platform located in a tropical cyclone region.

Introduction

Reliability methods have been developed to assist the ocean engineer in managing uncertainties associated with design, construction, and operation of structures. A major source of uncertainties is that from extreme condition wave loadings, and a major contributor to this source is the uncertainty associated with wave kinematics.

Extreme condition hydrodynamic loadings can be contrasted with nominal condition hydrodynamic loadings that are important to fatigue performance. Evaluation of uncertainties in nominal condition hydrodynamic loadings can be developed using approaches similar to those developed here. However, the characteristics and uncertainties associated with nominal condition hydrodynamic loadings can be very different from extreme condition loadings.

This paper will focus on maximum static forces acting on fixed structures located in deep water (water depths in excess of half the wave length). This focus can be contrasted with a time history of static forces, and with dynamic forces developed in flexible structures.

Note: This paper was prepared to be given as an invited lecture at the NATO ARW "Water Wave Kinematics."

A. Tørum and O. T. Gudmestad (eds.), Water Wave Kinematics, 103–159.

The characterization of uncertainties can be very different for static force time histories, and for dynamic load effects developed in flexible structures. In addition, due to a variety of complex wave-current-sea floor interactions, the characterization of uncertainties associated with shallow water wave forces can be very different from those in deep water.

In addition, this paper will concentrate on global forces developed on structures. Such global forces are reflected in parameters such as "base shear" and "overturning moment." These forces can be contrasted with with local forces acting on various elements or parts of the structure. These local forces can be particularity important near the free-surface (e.g. slamming from waves). There is an important force uncertainty "integrating effect" that takes place along a given member and in a spatially distributed assembly of members.

This paper will primarily consider structures comprised of slender members that can be characterized with hydrodynamic viscous drag forces that are related to the square of the water particle velocities. Such force characterizations will be appropriate for structures comprised of members that have diameters or widths that are less than about two percent of the wave length, or ten percent of the wave height.

Also, this paper will consider structures comprised of large members that can be characterized with hydrodynamic inertia forces that are related linearly to the water particle acceleration. Such a force characterization will be appropriate for members that have diameters than are greater than about ten percent of the wave length, or fifty percent of the wave height.

Perhaps most important, this paper will concentrate on an engineering assessment of the uncertainties associated with extreme condition hydrodynamic loadings for the purpose of developing design criteria for offshore structures. Determining the uncertainties is not an end it itself. Characterization of the uncertainties is a means to the end of evaluating the implications of uncertainties on design criteria. The fundamental objective is to make sound engineering judgements and decisions regarding design criteria for offshore structures.

Reliability Background

In this section, a summary is given of key aspects of a reliability based approach to development of design criteria [1].

LOADINGS AND CAPACITIES

Loadings, S, will be defined as a global load effect developed in the structure by storm winds, waves, and currents. Gravity and/or operational loadings that act on the structure can be included in this load effect.

Capacities, R, will be defined as the ultimate limit state resistance or capacity of the structure for a given type of demand. The capacity of the structure must be expressed in the same terms as S.

PROBABILITIES OF SUCCESS AND FAILURE

The probability or likelihood that the structure will survive the loadings can be expressed as:

$$Ps = P[S < R]$$ (1)

Where P [.] is read as the probability that the demand, S, is equal to or less than the capacity, R. Ps is the probability of success, or reliability.

The probability of failure, Pf, is the compliment of the reliability:

$$Pf = 1 - Ps$$ (2)

The reliability can be computed as follows:

$$Ps = \Phi[\beta]$$ (3)

where $\Phi[\beta]$ is the standard Normal distribution cumulative probability of the variate, β, from - infinity to β. β is commonly termed the "safety index."

In this development, it will be assumed that the probability of the demand and capacity variables can be adequately described with Lognormal distributions (Table 1). In this distribution, the logarithms of the variables are Normally distributed.

Lognormal distributions to describe the probability characteristics of storm loadings and structure capacities have proven to be robust in providing reasonable characterizations for a wide variety of problems. For other types of distributions, the reader should consult references [2,3].

Given Lognormally distributed demands and capacities, β is computed as follows:

$$\beta = \frac{\ln R/S}{\sqrt{\sigma_R^2 + \sigma_S^2}}$$ (4)

or,

$$\beta = \frac{\ln R/S}{\sigma}$$ (5)

With good accuracy, the probability of failure can be computed from ($1 < \beta < 3$):

$$Pf = 0.475 \exp(-\beta^{1.6})$$ (6)

CENTRAL TENDENCY AND UNCERTAINTY MEASURES

$\underline{R}$ is the median ultimate limit state capacity of the structure. $\underline{S}$ is the median force effect imposed on or induced in the structure.

The median capacity or force effect ($\underline{X}$) refers to the values of these parameters for which 50 percent are greater than and 50 percent are less than the given value. The median value is one of three alternative measures of the central tendency characteristics of the probability distributions (Table 1). The other two are the mean or average value (X, also called the expected value), and the mode or most probable value (Xp).

UNCERTAINTY MEASURES

σ_R is the standard deviation of the logarithm of the capacities. σ_S is the standard deviation of the logarithm of the forces. σ is the standard deviation of the logarithms of the demands and capacities. The standard deviation is a measure of the dispersion or variability of the distribution.

The ratio of the standard deviation to the mean is known as the coefficient of variation (V $= \sigma_X / X$). The coefficient of variation can be thought of as a normalized measure of the variability of the parameter of concern.

For example, steel yield strengths commonly have V = 0.05 to 0.10; soil shear strengths have V = 0.10 to 0.50 (higher range to very variable soils such as calcareous soils), dead loadings have V = 0.05 to 0.15, and annual maximum loadings from storms and earthquakes have V = 0.30 to 2.00 (higher range to very uncertain conditions and loadings such as from earthquakes).

TIME CONSIDERATIONS

The time period that often is used to define the probability characteristics of the loadings is one year; thus, the loadings need to be expressed as the likelihoods, P_{Sa} , of the annual maximum loading, S, being equal to or less than some value, s.

Other time periods can be used. For example, the loading characterization could be based on the lifetime (exposure period) of the structure, L (expressed in years). In this case, given that the loadings were independent from year to year (this is another way of saying that the occurrence of a loading in one year does not influence the occurrence of another loading in another year), the lifetime likelihood would be related to the annual likelihood by:

$$P_{SI} = (P_{Sa})^L \qquad\qquad (7)$$

The effect of the exposure period is to increase the median value of the loading and decrease the standard deviation of the loading.

If the capacity were changing as a function of time, for example, due to fatigue degradation of the strength, then Pf could be determined for discrete time intervals recognizing the change in the capacity, and the Pf's summed over the total exposure period.

Relating the annual risk, Pfa, to the lifetime risk, Pf_L is simple if each year is considered a statistically independent event (no correlation of trials from year to year). In this case, for a lifetime of I years:

$$Pf_L = 1 - (1 - Pfa)^L \tag{8}$$

For small Pfa, this gives:

$$Pf_I = L (Pfa) \tag{9}$$

There is correlation of risk from year to year due to statistical dependence through several important variables in Pf including the structure resistance, some of its loadings (e.g dead loadings), and some of the sources of uncertainty (e.g. methods of analysis). Many of the variables are independent of the natural randomness associated with such occurrences as storms and earthquakes, and may be considered constant during the lifetime.

If we take the other extreme assumption, and consider perfect dependence or correlation from year to year, then:

$$Pf_L = Pfa \tag{10}$$

In this paper, the probabilities will be based on a period of one-year (annual), unless otherwise stated.

AVERAGE RETURN PERIOD AND ENCOUNTER PROBABILITY

Another time related quantity of interest is the "average return period" (ARP). For quantities such as wave height and hydrodynamic loading (referred to with the parameter X), the ARP can be expressed as:

$$ARP = \lambda [1 - P_X]^{-1} \tag{11}$$

where λ is the average rate of occurrence of the events of interest (number of events per time unit) measured by X. The mean waiting time between events is λ^{-1}.

P_X is the likelihood that the variable X is equal to or less than a given value, x; or, $P_X = P$ (X < x). This is known as the cumulative distribution function for the variable, X. When P_X is expressed on an annual basis, $\lambda = 1.0$.

For example, if the average number of storms exceeding a given intensity level per year is 0.01, the mean waiting time between storms will be 100 years. This is called the ARP for that storm intensity.

For Lognormally distributed parameters, X, the average return period associated with a particular occurrence of the event, Xe, can be estimated from:

$$ARP_{Xe} = 2.1 \exp \left(\frac{\ln Xe - \ln X}{\sigma_X} \right)^{1.6} \tag{12}$$

The median waiting time is the value of L, denoted by L_{50}. For an exponential distribution of the waiting times:

$$L_{50} = \frac{\ln 2}{\lambda} (ARP) = 0.693 \, ARP \qquad (13)$$

Another way to understand the meaning of the ARP is in terms of encounter probabilities, E. The probability of experiencing at least one event during an exposure period, L, is:

$$E = 1 - \exp(-L / ARP) \qquad (14)$$

When the exposure period is equal to the ARP, E = 63.2 percent.

Hydrodynamic Loading Effects

We will now direct our attention to the definition of the design cyclone conditions, and specifically to determination of the design wave height (H_D) and associated average return period (ARP_D) to define S_D.

TOTAL AND WIND LOADINGS

It will be assumed that the design loading condition, S_D, is composed of a cyclone wind load, Sw, and a cyclone hydrodynamic wave and current load, SH:

$$S_D = S_W + S_H \qquad (15)$$

The cyclone wind load will be defined as:

$$S_W = K_W \, V_{WD}^2 \qquad (16)$$

where K_W is a structure dependent (effective area, drag coefficient) design wind loading parameter, and V_{WD} is the design wind speed (that occurs at the same time as the maximum wave height).

WAVE AND CURRENT LOADINGS

For purposes of this development, calculation of the hydrodynamic load will be based on the traditional Morison force formulation. The global loading developed on the structure will be expressed as:

$$S_H = K_H \, H_D^{\,\alpha} \qquad (17)$$

where K_H is a structure dependent hydrodynamic loading coefficient (effective volume and area, drag and inertia coefficients), H_D is the design wave height, and α is the exponent relating the wave height to the total hydrodynamic force. The effects of current are included implicitly in the hydrodynamic loading, S_H.

For drag force dominated structures, α has a value of about 2; for inertia dominated structures, α has a value of about 1.

To simplify this development, the design wind loading will be assumed as a proportion of the design hydrodynamic loading. For fixed offshore platforms the maximum wind loading (that occurs at the time of the maximum hydrodynamic loading) is generally about 10 to 15 percent of the maximum hydrodynamic loading. Thus it will be assumed that $S_D = 1.15\ S_H$.

STRUCTURE CAPACITY

The ultimate capacity of the structure (R) will be expressed with a Reserve Strength Ratio (RSR) where:

$$RSR = \underline{R} / S_D \tag{18}$$

S_D is the nominal design load, and $\underline{R}$ is the median ultimate capacity. S_D will be expressed by the nominal design global loading (e.g. base shear, overturning moment). Note that the nominal design load, S_D, generally is not the same as the median loading, $\underline{S}$

$\underline{R}$ is the best-estimate or expected ultimate capacity of the structure. All sources of implicit and explicit bias ("bias" = true median value/predicted or true median value /nominal) have been assumed to have been recognized in the analyses so that the bias in the estimate of $\underline{R}$ is unity. Similar statements pertain to the estimates of the loadings.

The RSR is equal to what we traditionally have termed the "factor-of-safety", or:

$$FS = RSR = \underline{R} / S_D \tag{19}$$

Note that if the design analyses were performed for the ultimate capacity of the structure or element in the structure (e.g. a foundation element), that the introduction of the RSR would be unnecessary. However, because the majority of design engineering analyses are performed in the serviceability capacity regime using linear elastic methods, the transformation to R_D is required.

NOMINAL DESIGN VALUES

Sometimes, it is preferable to refer the design capacity or loading to nominal design values such as R_D (or RSR_D) and S_D. When this is done, $\underline{R}$ and $\underline{S}$ must be replaced by the nominal value times the bias factor, B, for the particular nominal value or:

$$\underline{R} = (B_R)\ R_D \tag{20}$$

and,

$$\underline{S} = (B_S)\ S_D \tag{21}$$

DESIGN LOADING

Using the basic expression for the safety index (eq. 5), the nominal design load effect can be expressed as:

$$S_D = [\ \underline{S} / RSR\ ] \exp (\beta\ \sigma) \tag{22}$$

The design load will change directly with an increase in the median loading, and inversely with the design RSR. Also, the design load will increase as an exponential function of the required safety and the combined uncertainties in the loadings and capacities.

The reader is referred to the cited references for additional background on definition of the required safety expressed through the safety index, β [1,4], and on evaluation of the structure capacity characteristics, RSR and σ_R [5].

DESIGN WAVE HEIGHT

Given these developments, the design wave height can be expressed as:

$$H_D = \left(\frac{H^\alpha}{RSR} \exp (\beta \sigma) \right)^{-\alpha} \tag{23}$$

The key hydrodynamic parameters that will influence the determination of the design wave height are the median wave height, $\underline{H}$, the wave loading uncertainty, σ_S, and the wave loading exponent, α.

DESIGN RSR

It is frequently desirable to use the 100-year wave height (H_{100}) in design. In this case, the reliability assessment could be used to define the structure capacity, R, or equivalently the design reserve strength ratio or factor of safety. For Lognormally distributed expected annual maximum wave heights the 100-year wave height is related to the median (2-year) wave height as follows (eq.12):

$$H_{100} = \underline{H} \exp (2.33 \, \sigma_H) \tag{24}$$

The required design RSR is:

$$\underline{RSR} = \exp [(\beta \, \sigma) - (2.33 \, \sigma_H \, \alpha)] \tag{25}$$

It is apparent that the hydrodynamic loading uncertainties will have a dramatic influence on the design capacity expressed through the RSR.

LOAD AND RESISTANCE FACTORS

A design format that is being used in design of offshore structures is known as the Load and Resistance Factor Design (LRFD) format. This format utilizes a load factor (ϕ, generally greater than unity), and a resistance factor (γ, generally less than unity) as follows:

$$\gamma \underline{R} > \phi \underline{S} \tag{26}$$

Thus, the loading, $\underline{S}$, is factored up, and the resistance, $\underline{R}$, is factored down. Generally, the factoring is done so that the design engineer is still able to use linear elastic analysis methods in design computations.

Using an approximation to split the loading and resistance effects it can be shown [1-3] that:

$$\phi = \exp (0.75\ \beta\ \sigma_S) \tag{27}$$

and,

$$\gamma = \exp (-0.75\ \beta\ \sigma_R) \tag{28}$$

If the loading, S, were composed of two components: Sd (for dead loading) and Ss (for storm loading), then:

$$\phi\ \underline{R} > \gamma_d\ Sd + \gamma_s\ Ss \tag{29}$$

and,

$$\gamma_d = \exp (0.75\ \beta\ \varepsilon\ \sigma_d) \tag{30}$$

$$\gamma_s = \exp (0.75\ \beta\ \varepsilon\ \sigma_s) \tag{31}$$

where the "splitting coefficient", ε, can be approximated from:

$$\varepsilon = \frac{\sqrt{\sigma_d^2 + \sigma_s^2}}{(\sigma_d + \sigma_s)} \tag{32}$$

Analysis of Uncertainties

Uncertainty: doubt, lack of sureness, indefinite, not having certain knowledge. Uncertainty leads to risk. Realistic evaluations of uncertainties are one of the critical parts of a reliability based assessment.

In these analyses, uncertainties will be organized into two general categories. The first category will be identified as natural or inherent randomness (Type I = "randomness").

Examples of Type I uncertainty are the maximum annual wave height experienced by a structure at a given location during a given period of time, and the strengths that will be observed from mill tension tests on samples of steel.

The second category will be identified as un-natural, cognitive, or modeling uncertainty (Type II = "modeling"). This type of uncertainty applies to fixed or deterministic, but unknown values of parameters (parameter uncertainty); to modeling uncertainty (imperfect understanding of problems, simplified theories or analytical models used in practice); and to the actual state of the system or element (imprecise knowledge of properties and characteristics).

An example of Type II uncertainty would be to contend that the maximum annual wave height follows a certain type of probability distribution (e.g. Weibull), and has a certain mean, X, and standard deviation, σ_x. Without an infinite amount of data one cannot estimate the values of X and σ_x precisely. Given limited information, one has estimates of X and σ_x, and a quantifiable amount of Type II uncertainty about X and σ_x.

Another way to distinguish Type I and Type II uncertainties is that Type II uncertainty can be reduced by additional information; Type II uncertainty is information sensitive. Research, development, engineering, inspection, quality control/assurance and similar efforts can reduce Type II uncertainties.

There are difficulties associated with some parameters in making unambiguous identifications and quantifications of Type I and Type II uncertainties. For example, in the case of processes used to generate long-term and short-term sea state characterizations, Type I and Type II uncertainties are interwoven in a very complex way in the data and analytical models. It would be very difficult to distinguish clearly Type I and Type II uncertainties in this case.

There can be other types of uncertainties that lead to risk. For example; individual, group, and/or society errors. Evaluation and management of this type of uncertainty is the subject of current reliability engineering research [29, 30].

In this development, the scope of evaluation of uncertainties will be restricted to Type I and Type II for design criteria for new structures. The results will be portrayed with and without Type II uncertainties to allow the design criteria decision process to reflect potential effects on decisions caused by these two types of uncertainty. This approach also allows differentiations to be made between uncertainties that are independent from year to year (Type I), and those that are dependent or correlated from year to year (Type II).

The evaluation of variabilities of the demands and capacities from the components of the demands and capacities that contribute uncertainties will be based on the the Normal function algebraic expressions summarized in Table 2 [31].

Evaluation of "biases" is a particularly important part of an analysis and evaluation of uncertainties. Bias , Bx, in a parameter, x, can be expressed as the ratio of the true value to the predicted or nominal value of the parameter:

$$Bx = X_{true} / X_{predicted} = X_{true} / X_{nominal} \tag{33}$$

This is one means of expressing Type II uncertainties. This evaluation provides an important opportunity to clearly identify differences between values of parameters that are derived from simplified models and the true values, or values that have been defined based on implicit "conservatisms" or "unconservatisms." Subjective "judgement factors" need to be clearly identified, incorporated into the analyses as appropriate, and their effects on the results analyzed. It is here that qualified and informed judgment can play a vital role in reliability assessments.

In the following example application Type II uncertainties will be characterized with the mean and standard deviation: $B_{II}x$ and σ_{IIx}.

Hydrodynamic Forces

The calculation of hydrodynamic forces acting on a structure can be organized into four major components (Figure 1): 1) environmental conditions, 2) structure conditions, 3) kinematics, and 4) forces.

In this section, primary considerations for each of these components will be discussed, and the fundamental reliability related aspects highlighted.

ENVIRONMENTAL CONDITIONS

The environmental hydrodynamic conditions consist chiefly of the short term (few hours) and long term (years) characterizations of the waves and currents that can act on the structure.

In the case of a narrow wave spectrum (component frequencies concentrated in small range) and an assumption of independence of the wave heights, the short term distribution of wave heights can be expressed by a Rayleigh distribution:

$$P(H) = 1 - \exp -[\ 2\ (H/Hs)^2\ ] \tag{34}$$

where P(H) is the probability of not exceeding a given wave height, H, and Hs is the significant wave height (average of the highest one-third wave heights). In most cases, the significant wave height is the height used by the oceanographer in characterizing sea state intensities from either measurements or storm wind and wave numerical models.

The wave height corresponding to the peak of the extreme Rayleigh density distribution of wave heights is Hm, which is the most probable highest wave in a given period. An estimate of Hm is given by:

$$Hm = Hs \sqrt{\frac{\ln N}{2}} \tag{35}$$

where N is the number of waves in a given time period that the sea has a significant wave height Hs. For waves having a 10 sec. average period, and a sea state duration of 3 hours, one could expect about N = 1000 waves, and Hm = 1.86 Hs.

Many extreme sea state conditions are not truly narrow banded, and the waves are not truly independent (there can be correlation between successive wave heights) and consequently, this relationship tends to overestimate the most probable maximum wave heights. Studies of wave records from extreme sea conditions indicates that this relationship can overestimate the wave heights by 10 to 20 percent [6,7]. This can be an important source of bias (true value / predicted or nominal value) in the short-term estimates of the maximum wave heights.

LONG-TERM CHARACTERISTICS

A long-term distribution of Hs has not been derived by theoretical means. There are several distributions that can provide acceptable fits to long-term wave height measurements and long-term wave statistics developed from hindcast analytical models

[8,9]. The Lognormal distribution frequently provides an acceptable fit to the data. For a wide variety of applications, the Weibull distribution can provide acceptable fits to the data. This distribution is expressed by:

$$P(Hs) = 1 \exp -(Hs/Hc)^{\gamma} \tag{36}$$

where Hc and γ are parameters that are determined based on the long-term Hs data. If it is desired to determine the Hs associated with a given ARP:

$$Hs_{ARP} = Hc [\ln (\lambda \ ARP)]^{-\gamma} \tag{37}$$

where λ is the average rate of occurrence of events (in same units as ARP) implied in the distribution P(Hs). If P(Hs) is based on a time period of one year, $\lambda = 1.0$.

DEVELOPMENT OF TROPICAL CYCLONE CHARACTERISTICS

A long term-distribution of the expected maximum wave heights could be developed for tropical cyclones as follows. A probability distribution could be developed to express the annual likelihoods of experiencing various minimum differential barometric pressures, ΔPm, from tropical cyclone tracks crossing within a distance, R, of the proposed structure location. The distance R could be represented by the typical radius to maximum winds developed in intense tropical cyclones (e.g. R = 20 to 25 km) [10]. Weather records and analytical storm models could be used to develop the annual probability distribution, P[ΔPm].

The maximum wind associated with the minimum differential barometric pressures could be expressed as [10]:

$$Vm = C (\Delta Pm)^{1/2} \tag{38}$$

Where Vm is the 10 minute average wind speed in meters per second at an elevation of 10 meters, and ΔPm is the differential pressure in millibars. Available data indicates that the wind speed factor, C, is in the range of 4.0 to 4.5 ($\pm 1\sigma$), with an average value, $\underline{C} = 4.25$ ($V_c = 6 \%$).

The maximum significant wave height (meters), Hsm, could be estimated as:

$$Hsm = \Psi \ Vm \tag{39}$$

The factor Ψ can be based on traditional storm wind - wave hindcasting techniques [11], and/or on measured data. Analyses of tropical cyclones indicate that the value of Ψ can range from 0.27 to 0.33 ($\pm 1\sigma$) with an average value, $\underline{\Psi} = 0.3$ ($V_{\Psi} = 10\%$)

The expected maximum wave height, Hm, could be estimated from the short term wave height distribution based on 1000 waves (expressing a 3-hour duration of the maximum sea state intensity at the location):

$$Hm = \zeta \ Hsm \sqrt{\frac{\ln N}{2}} \qquad (40)$$

Recognizing that the spectral widths of the extreme sea state will not be truly narrow, nor the wave heights independent, a correction factor, ζ, in the range ($\pm 1\sigma$) of 0.85 to 1.0 could be introduced [6,7], with an average value, $\zeta = 0.93$ ($V_z = 8$ %). Thus, the expected maximum wave height could be estimated as $Hm = 1.73$ Hs.

The period associated with the maximum wave could be developed in several ways [6,7]. One approach could be based on measured maximum wave heights and the associated periods of these waves recorded during intense tropical cyclone sea states. This approach could indicate that the maximum wave heights are at or near limiting steepness; height to length ratios (H/L) would lie in the range ($\pm 1\sigma$) of 1/10 to 1/15, with a best estimate of 1/12 ($V_{H/L} = 20\%$).

In deep water, the wave length, L, can be related to the wave period, T, as follows:

$$L = (g / 2\pi) \ T^2 \qquad (41)$$

where g is the acceleration due to gravity.

Using a best estimate value of wave steepness of H/L = 1/12, the expected wave period (Tm in sec.) associated with the maximum wave height could be expressed in terms of the maximum wave height (Hm in meters) as:

$$Tm = 2.77\sqrt{H} \qquad (42)$$

The expected maximum current, conditional on the time of occurrence of the maximum wave could be estimated as:

$$Um = G \ Vm \qquad (43)$$

where G represents the combined effect of the wind shear stress, wind and water densities, eddy viscosity coefficients, and Coriolis parameter. Based on available measurements and numerical model results [12], a reasonable range ($\pm 1\sigma$) for G is 0.030 to 0.045, with a best estimate value of $\underline{G} = 0.04$ ($V_G = 12.5\%$).

Um is a depth averaged current developed in the upper mixed layer of the water column (thickness = ϑ). For intense cyclones and deep water conditions, this thickness could range from $\vartheta = 50$ to 150 meters [12]. This thickness appears to be a function of the wave heights generated by the storm (e.g. $\vartheta = 5$ to 10 times Hsm).

Thus, given the probability distribution of annual ΔPm, we are now able to develop directly the probability distributions of Vm, Hs, Hm, Tm conditional on the occurrence of Hm, and Um conditional on the occurrence of Hm.

Information can also be provided on storm wave angles of approach to the structure, and on directional spreading in the waves. If the structure can be reliably oriented or is very asymmetrical, then sometimes advantage can be taken the angles of wave approach.

Wave directional spreading effects can be taken into account by making corrections to the two-dimensional, long crested, uni-directional wave kinematics.

The characterization of the design storm current speed, directions, and profile through the water column is made conditional on the time of occurrence of the maximum waves [12]. This strategy is adopted to prevent unrealistic combinations of extreme conditions. In formulating this problem, the expected maximum wave height has been identified as the primary hydrodynamic force measure of the intensity of a tropical cyclone.

Other types of environmental conditions and other types of structures could indicate other force intensity measures to be appropriate. A reasonable guideline is to choose the environmental condition intensity measures that exert the greatest influences on the forces produced in the critical parts of the type of structure of concern, and preserve the magnitude, spatial, and temporal characteristics of the parameters when parameter combinations are chosen. Formal probabilistic methods are available to assist with spe cialized problems of loading combinations [2,3].

Storm tides and surges can also play an important role in characterizations of design conditions, particularly for shallow water structures, and structures whose design is sensitive to wave crest elevations. This consideration will be illustrated in a later section of this paper.

STRUCTURE CONDITIONS

The structure conditions consist of member sizes, locations, and inter-connectivity; or the geometry and topology of the structure.

For the purposes of this development, the members comprising the structure will be assumed to be prismatic cylinders having diameters, D, and lengths, L.

Of particular importance for many offshore platforms, is the recognition of the close spaced groups of elements (e.g. well conductors and pipeline risers) located in these structures. In addition, the large projected areas represented by appurtenances such as boat landings, barge bumpers, and cathodic protection elements must be included. This recognition can be introduced through corrections to the characterizations of the member areas and volumes, through corrections to the kinematics, or through corrections to the hydrodynamic force coefficients.

Marine fouling effects also must be described. Marine fouling not only can lead to a dramatic increase in the viscous drag forces and decrease in inertia forces, but as well, dramatically increase the projected area of the below-water structure elements and the weight imposed on these elements. Marine fouling is a location, time and depth below mean seal level consideration. Also, marine fouling can be controlled by anti-fouling coatings/claddings, and cleaning programs.

Frequently, there is as much or more uncertainty associated with the marine fouling geometry effects, as there is with its effects on the force coefficients. Generally, marine fouling effects on the member areas and volumes are introduced into the structure characterization, and in characterization of the hydrodynamic force coefficients.

KINEMATICS

Calculation of the kinematics (water particle velocities and accelerations) associated with the extreme condition waves is generally accomplished using traditional two-dimensional wave theories [13]. These kinematics are referred to as the "free-field" kinematics (unaffected by the presence of the structure).

It has been recognized that the free-field kinematics are only one part of a complex picture of water flow around structural elements [14]. Of particular importance are the wakes (local flow intensifications) that are generated by the structure-water interactions. These wakes have proven to be important not only for dense assemblages of elements, but as well, for single elements.

Wave kinematics, free-field and local, are dramatically influenced by currents. A wide variety of types of currents can act on or in the water column during the passage of a tropical cyclone. The problems of estimating extreme condition storm associated currents is perhaps even more difficult than that of estimating the wave conditions.

If strong enough, the currents can have important effects on the wave characteristics. In addition, the currents can act to sweep away high velocity vortices generated by the wave crest passing the member before the vortices can collide with the member during later phases of the wave motion. Since significant currents are present in most intense storms, this is an important consideration in interpreting wave flume and U-tube test results in which the vorticity is trapped or not swept away from the members.

The wave kinematics are influenced by the directional characteristics of the waves. For extreme wave conditions associated with the central portions of intense tropical cyclones, the wave energy is highly spread (frequently over angles in excess of 90 degrees) [11]. Thus, design wave kinematics determined for such conditions using long-crested, two-dimensional wave kinematics must be corrected to recognize these effects (or alternatively, to use multiple directional waves to simulate the directional spreading effects).

Kinematics well below the free-surface seem to be reasonably described by traditional wave theories. Kinematics in the free-surface zone (wave crest-trough region) are another matter. Measurements in this zone are very difficult, and hence, the measurements that have been made have many questions regarding their proper interpretation and reliability. Deductions based on near-surface kinematics measurements regarding the suitability or unsuitability of a theory to compute kinematics are also in question. This is an important area of present research [13].

LINEAR WAVE THEORY KINEMATICS

In deep water (water depths , d > L / 2), for the purposes of this development an acceptable characterization of uni-directional (long-crested) wave water particle horizontal velocities, u_w, and accelerations, a_w, can be developed from linear or Airy wave theory:

$$u_w = (\pi H / T) e^{kz} \cos (\theta) \qquad (44)$$

and,

$$a_w = (2 \pi^2 H / T^2) e^{kz} \sin (\theta) \tag{45}$$

where w is the wave circular frequency ($\omega = 2 \pi / T$), k is the wave number ($k = 2 \pi / L$), z is the vertical coordinate measured up (+) and down (-) from the still water level, and θ is the wave phase angle ($\theta = kx - \omega t$), x is the horizontal coordinate measured from the wave crest, and t is the time coordinate).

CURRENTS

The water particle velocity is generally assumed to be reasonably represented by the vectorial superposition of the wave velocity, u_w, and the current velocity, u_c, or:

$$u = u_w + u_c \tag{46}$$

For waves propagating on a uniform current, a modified wave period, Te, is required to allow recognition of the wave-current interactions on the wave kinematics:

$$Te = T / (1 - u_c / C) \tag{47}$$

where C is the wave celerity ($C = L/T$). For adverse currents ($-u_c$), the wave period will be shortened.

DIRECTIONAL SPREADING EFFECTS ON KINEMATICS

Extreme storm condition maximum waves often are short-crested and associated with significant directional spreading of the wave energy. A correction to uni-directional, long-crested wave velocities acting on along a single vertical surface piercing cylinder has been developed based on a $\cos^{2n}$ directional spreading function [15]:

$$R_{uw} = \left(\frac{C(n)}{C(n+1)} \right)^{1/2} \tag{48}$$

and for accelerations:

$$R_{aw} = \left(\frac{C(n)}{C(n+1)} \right)^{1/4} \tag{49}$$

where:

$$C(n) = \frac{\Gamma (n+1)}{\sqrt{\pi} \; \Gamma (n + 1/2)} \tag{50}$$

Γ is the Gamma function for the quantity (.).

The directional spreading coefficient, n, is frequently in the range of 0.5 to 5 near the center of intense tropical cyclones. For $n = 1$, $R_{uw} = 0.87$, and $R_{aw} = 0.93$.

For multiple arrays of vertical cylinders, the kinematics corrections are somewhat greater than for a single cylinder (function of specific geometry and spacings of cylinders). This is a very complex directional wave component kinematics spatial and temporal coherence problem[15,16]. For this development, the directional spreading effects will be introduced by the kinematics correction factors R_{uw} and R_{aw}.

KINEMATICS DATA

In recent years, reliable instrumentation has been developed to permit accurate measurements of wave kinematics in the ocean. Several recent in-ocean measurement programs have provided value insights into wave kinematics associated with large waves generated by tropical cyclones [17,18]. Figure 2 summarizes kinematics results from one such program, the Conoco Test Structure (CTS) [17]. Force data from the CTS will be discussed later in this paper.

The CTS was instrumented with multiple wave staffs and electromagnetic kinematics transducers were used to measure wave conditions and water particle velocities and accelerations. The kinematics transducers were located at elevations -1.53 m. and -6.10 m. (referenced to mean water level) in a water depth of 54 meters. The resultant kinematics (x and y horizontal components), including any currents that were present were evaluated.

The predicted kinematics were based on a variety of wave theories; the one shown here is based on the Stokes Fifth Order Theory. No currents were added to the wave kinematics. The wave heights used in the analyses were based on the recorded average heights and periods for the preceding and following troughs. The peak recorded and predicted velocities during the passage of the wave crest were evaluated.

The data summarized in Figure 2 are reported in terms of an error ratio (or bias = measured value / predicted value), for three different ranges of wave heights. While the coefficients of variation for the three data sets are comparable ($V = 0.18$), the median or mean values tend to decrease as the wave heights increase. The Stokes Fifth Order Theory median prediction bias is about 0.85 for the largest range of wave heights, implying a 15 percent overprediction of the peak velocities. It is important to note that if the storm associated currents were included in the wave kinematics, the bias would have been even smaller (implying a greater tendency to overpredict the kinematics).

LOCAL FORCES

The method used in this development for hydrodynamic force calculation is based on the Morison formulation. The total hydrodynamic force per unit length, F, is comprised of a drag force, Fd, and an inertial force, Fi:

$$F = Fd + Fi \tag{51}$$

where,

$$Fd = C_d (\rho / 2) (D) u |u| \tag{52}$$

$$Fd = Kd \, u |u| \tag{53}$$

and,

$$Fi = C_m \, (\rho \, \pi \, D^2 / 4) \, (a) \tag{54}$$

$$Fi = Km \, (a) \tag{55}$$

F is the force normal to the member axis, u is the velocity of the fluid particles normal to the member axis, a is the acceleration of fluid particles normal to the member axis, r is the mass density of the water, D is the member diameter, C_d is the drag coefficient, and C_m is the inertia coefficient.

The drag and inertia coefficients are time varying functions of the flow conditions around the cylinder [19]. The time-averaged or constant Cd and Cm are a functions of the Reynold's number, Re, Keulegan-Carpenter number, KC (= u_m T/D, where u_m is the maximum water particle velocity and T is the period of the oscillatory motion), and relative roughness, ε/D (ε is the roughness height). If the free-field kinematics have not been corrected for member spacings and wake encounters, C_d and C_m also may be characterized as functions of these effects [20].

Based on evaluations of a large number of laboratory tests involving rough, slender, cylindrical members subjected to a wide variety of wave and current flow conditions, Nath [21] has suggested that a total force coefficient, C_υ (Figure 3),can be used to compute the total wave and current force:

$$F = C_\upsilon \, \rho / 2 \, (D \, L) \, u \, | \, u \, | \tag{56}$$

$$F = K_\upsilon \, u \, | \, u \, | \tag{57}$$

The total force coefficient is expressed as a function of KC for KC < 20 as:

$$C_\upsilon = 2 \, \pi^2 / KC \tag{58}$$

For KC >20, $C_\upsilon = C_{ds} \cong 1.0$, where C_{ds} is the steady flow drag coefficient.

INTEGRATED FORCES

Based on linear wave theory and $u_c = 0$, the total lateral maximum force acting on a prismatic, vertical, cylinder extending from the sea floor through the free-surface in a water depth, d, can be expressed as:

$$F_{dm} = Kd \, Ku \, H^2 \tag{59}$$

$$F_{im} = Ki \, Ka \, H \tag{60}$$

Ku is a function that integrates the velocities over the length of the cylinder [19]. It has the dimensions of acceleration and is a function of d/T^2 and H/T^2. For linear waves (H/L = 0), Ku = g/8. For a wave steepness H/L = 1/12, and Stokes Fifth Order wave theory,

$Ku \cong g/4.5$; for a wave steepness $H/L = 1/7$, $Ku \cong g/3$. In shallow water breaking wave conditions, $Ku \cong g$.

Thus, in addition to the wave height, the wave steepness can have a very important effect on the integrated maximum drag force. The maximum horizontal drag force component will occur at the wave crest for all surface waves.

Ka is a function that integrates the accelerations over the length of the cylinder. It also has dimensions of acceleration, and is a function of d/T^2 and H/T^2. For linear waves, $Ka = g/2$. For a Stokes Fifth Order wave theory and a wave steepness of $H/L = 1/7$, $Ka \cong g/2.25$. In shallow water breaking wave conditions, $Ka \cong g/3$.

The maximum inertia force component will occur at the $\theta = 90$ degrees (still water level intersection with the wave profile) position for a linear wave. For finite values of H/T^2, the position of the maximum inertia force approaches $\theta = 0$ (in phase with the drag force) for very small values of H/T^2. This explains how and why the total force coefficient based approach suggested by Nath is workable. Also, it is apparent that wave steepness can have a dramatic influence on the combination of drag and inertia forces on structures that have significant force contributions from both components.

Similar expressions can be developed to express the overturning moments developed on a surface piercing vertical cylinder. These expressions have similar relationships as for the lateral forces; thus, they will be omitted here. For deep water wave conditions, the effective lever arms of the total lateral force components are located at or slightly below the still water depth, d. Thus, a reasonable approximation for the overturning moments at the sea floor is:

$$M_{dm} = d (F_{dm}) \tag{61}$$

and,

$$M_{im} = d (F_{im}) \tag{62}$$

WAVE AND CURRENT FORCES DATA

Several sets of laboratory and field test data will be discussed to provide insights into the variability of the force coefficients and the forces themselves.

Figure 4 summarizes results from a recent series of laboratory towing tank tests reported by Roddenbush [22]. The drag and inertia coefficients are summarized in Figure 4 for a towed, vertical cylinder (towing allows development of large Re's and KC's and good knowledge of the kinematics acting on the cylinder). All of the test data shown are for the cylinders in a roughened condition ($\varepsilon / D > 1 / 100$) and towed to produce high Re (up to 2×10^6) flow conditions. The data are reported in terms of the amplitude of motion of the cylinder to the cylinder diameter ($A/D = KC / 2\pi$). The cylinder was moved in a wide variety of paths intended to simulate steady, harmonic two dimensional, and irregular - three dimensional flow around the cylinder. At high A/D numbers associated with extreme sea state conditions, $\underline{Cd} = 1.1$, $\sigma_{Cd} = 0.10$, $\underline{Cm} = 1.3$, and $\sigma_{Cm} = 0.23$.

Figure 5 summarizes results from the Ocean Test Structure (OTS) [23]. The OTS was a 4-leg structure designed as a scaled model of a platform and built specifically for the purpose of gathering kinematics and wave load data. The structure was located in a water depth of 20 meters in the Gulf of Mexico.

The data shown are based on measured kinematics and measured local forces [23]. The data indicate that for smooth cylinders and high KC flow conditions, $\underline{Cd}$ = 0.68, and σ_{Cd} = 0.18. For rough, marine fouled cylinders and high KC flow conditions, $\underline{Cd}$ = 1.0, and σ_{Cd} = 0.25.

The OTS also was able to record global forces. The measured forces were compared with computed values developed using the EXVP -D wave theory [19] (reported to give kinematics within one percent of the Stokes Fifth Order theory), Cm = 1.5, and Cd varied over the structure depending on the KC number and roughness of the elements. Typically, Cd was of the order of 0.7 for clean members and 1.0 for fouled members at high KC numbers.

Figure 6 summarizes the results. The composite bias (calculated force / measured force, the reciprocal of the error ratio used in this paper) was 11 percent, and the coefficient of variation 28 percent. For the winter storm period that produced the largest waves (winter storm long-crested wave conditions), the bias increased to 19 percent, and the coefficient of variation to about 30 percent.

Local wave and current force data from an earlier wave in-ocean measurement program is summarized in Figure 7. The experiment was identified as Wave Force Project II (WFP II), and consisted of a 0.76 meter diameter vertical cylinder equipped with force measurement transducers at various elevations located in a water depth of 30 meters. The project measured wave heights and local wave forces. Thus, the force coefficients that are reported involve uncertainties associated with the analyses of wave and current kinematics. The wave kinematics were computed with a Stokes Fifth Order theory. No currents were added to the computed kinematics.

The data indicate that the flow coefficients are functions of elevation along the test cylinder. The transducers located closer to the free surface tend to have lower Cd's and higher Cm's. This could be due to either the wave kinematics being over predicted at these elevations, or near surface effects on the flow coefficients. Recent laboratory experiments indicate that there are important changes in the flow coefficients near the free surface [24].

These data indicate that $\underline{Cd}$ = 0.6 to 1.0, and σ_{Cd} =0.65 to 0.68; $\underline{Cm}$ = 1.0 to 2.3, and σ_{Cm} = 0.60 to 1.02. The much greater scatter in the coefficients is due to the calculated kinematics as contrasted with the coefficients based on measured kinematics.

Results from integration of the local forces along the test cylinder to determine the global resultant forces for the twenty highest recorded waves in summarized in Figure 8 [25]. The results are shown for kinematics computed with both Stream Function and Stokes Fifth Order theories [13,19]. For both predictions, a constant drag coefficient was used. In the case of the Stokes based kinematics, a constant Cd = 0.5 was used. In the case of the Stream Function based kinematics, Cd's that varied with the Re's were used (Cd for Re > 10^6 = 0.55) [26].

These data indicate that the bias in predicted forces has a mean value close to unity, and a coefficient of variation of about 17 percent. A important force variability integrating effect has taken place along the length of the cylinder; the resultant integrated force has a much lower variance than the local forces.

The last data set that will be discussed is that from the Conoco Test Structure (CTS). The CTS was a full-scale, 6-leg drilling and production platform located in 54 meters of water in the Gulf of Mexico [27]. In addition to the kinematics and wave height measurements discussed earlier, the CTS was equipped with local wave force transducers and a large number of braces, legs, and piles instrumented to record global forces.

Results from analyses of the maximum lateral hydrodynamic forces developed on the platform (global forces) are summarized in Figure 9 [28]. The predicted total lateral forces were based on the measured wave heights and periods, Stokes Fifth Order predictions of the kinematics, and $Cd = 0.7$ and $Cm = 1.7$. It is important to note that this platform was marine fouled, and that currents were not included in the kinematics calculations. The data indicate an unbiased prediction at the 50-th percentile (Fd measured / Fd predicted = 1.0), and $\sigma_{Fd} = 0.33$.

This same data set was re-analyzed when it was discovered that there was a discernable relationship between the wave heights (or sea state intensities) and the global force biases. Figures 10 and 11 summarize results from these analyses [28].

Figure 10 summarizes the results for the data set (Block 12) in which the waves were essentially uni-directional and regular. As before, Stokes Fifth Order theory was used to evaluate the kinematics and no currents were added. A $Cd = 0.7$ and $Cm = 2.0$ was used to calculate the global forces. The results indicate a biased estimate at the 50-th percentile of 1.2, and $\sigma_{Fd} = 0.19$.

In the case of the data set in which the waves were short-crested (Figure 11, Block 6), and the sea state very directionally spread, the results indicated a biased estimate at the 50-th percentile of 0.85, and $\sigma_{Fd} = 0.34$. Wave directional spreading ($n \approx 1$) had a marked effect on the wave kinematics and resultant forces. This effect was not recognized in the prediction model.

When currents were added to the kinematics, the Block 12 data set indicated a median bias of 1.0; the Block 6 data set indicated a median bias of 0.64. Increases in the drag coefficient has comparable effects. Using a drag coefficient of 0.8 (without current), reduced the median error in Block 12 to 1.05, and in Block 6 to 0.73.

In the case of the highly directional spread sea condition force data, an unbiased estimator could be developed by the addition of current (based on hindcast results) to the computed kinematics, and the introduction of a directional spreading correction to the wave kinematics [28].

Example Application

An example problem will be used to illustrate how the foregoing background can be applied to development of extreme condition wave height design criteria for an offshore oil drilling and production platform. The platform will be a conventional, steel, template-type, pile supported platform sited in a water depth of 130 meters off the Northwest Shelf of Western Australia.

LONG-TERM WAVE AND CURRENT CHARACTERISTICS

As described earlier, a meteorologic-oceanographic study has been performed for the platform location, and a long-term probability distribution of tropical cyclone significant wave heights generated (Hs, Figure 12). In addition, an evaluation has been made of short-term sea state characteristics associated with intense cyclones, and a long-term probability distribution developed for the expected maximum wave heights (Hm, Figure 12). A mean directional spreading coefficient of $n = 1$ ($\pm 1\sigma$ range of 0.5 to 5) has been estimated for the extreme sea states.

In a similar manner, the meteorologic-oceanographic study defined a long-term probability distribution of tropical cyclone associated currents that would be acting on the platform at the same time as the expected maximum wave heights (u_c, Figure 13). The study indicates that the current has a profile that is essentially uniform with depth, and has a principal direction in line with the principal direction of the maximum waves.

At the 100-year return period, Hs = 11.6 meters, Hm = 20.0 meters, and u_c = 1.2 meters per second. The probability distribution of the logs of the annual Hm's has a standard deviation, $\sigma_H = 0.30$. The probability distribution of the logs of the annual u_c's (conditional on time of occurrence of Hm) has a standard deviation, $\sigma_c = 0.30$.

GLOBAL WAVE AND CURRENT FORCES

Analyses of hydrodynamic forces developed on the platform by various combinations of wave heights, periods, and forces (Figure 14) indicates that the global forces (base shear, overturning moment) vary with the square of the wave height; thus, $S_{Hm} = Kd\ Ku\ H^2$.

The forces were computed using traditional long-crested, uni-directional waves that had a steepness of 1/12, uniform current speeds consistent with the given wave height, and the Morison force formulation with a drag coefficient, Cd = 0.6 and inertia coefficient = 2.0. The currents accounted for 25 to 30 percent of the total maximum force.

The maximum lateral forces, S_m, defined as a function of Hm, and the probability distribution of annual Hm have been used to define the probability distribution of the expected annual maximum total lateral force acting on the proposed platform (Figure 15). The probability distribution of the logs of the annual S_m has a standard deviation, $\sigma_{Sm} = 0.56$ and coefficient of variation, $V_{Sm} = 61$ percent (reflects only uncertainties in the expected annual maximum wave heights).

WAVE HEIGHT UNCERTAINTIES

In the next step, a Type II uncertainty will be introduced to recognize prediction errors associated with the expected maximum wave heights (Figure 16). This evaluation was developed by comparing hindcast and measured maximum wave heights in severe tropical cyclones. The hindcast wave heights were based on the process outlined earlier that was initiated with the storm's minimum barometric pressure.

The data indicates a median bias (measured/predicted) of $B_{IIHm} = 1.1$ and a $V_{IIHm} = 0.13$. Note that the uncertainty that would be deduced from the compounded uncertainties in the hindcast coefficients (eqns.37 - 39) is $V_{Hm} = 0.14$.

The evaluation of this uncertainty illustrates one of the difficulties of separating Class I and Class II uncertainties. The comparison of the measured maximum and expected maximum wave heights to define the Class II uncertainties in the predicted wave heights has implicitly incorporated a Class I uncertainty, the natural variability in the maximum wave height measured at a given location in a given storm. Thus, a portion of $Vm = 0.13$ could be attributed to Class I variability.

WAVE AND CURRENT FORCE UNCERTAINTIES

Turning to the hydrodynamic forces, there are two paths that could be followed to evaluate uncertainties. One would be to evaluate each of the components contributing to forces uncertainties; kinematics and force calculations, conditional on specification of the cyclone waves and currents. A second approach would be to use measured global wave force data measured on prototype platforms in tropical cyclones, avoiding the explicit evaluation of kinematics uncertainties. Both approaches will be illustrated.

The first approach will be developed by evaluating the uncertainties in the key components that determine the integrated hydrodynamic forces acting on a vertical cylinder (eqn. 59).

UNCERTAINTIES IN KINEMATICS COEFFICIENT, Ku

A principal component of uncertainty in Ku is contributed by the calculation of the wave kinematics, conditional on a given wave height and period. This is illustrated by the differences in Ku caused by different wave theories (e.g. Ku Linear = $g/8$, Ku Stokes Fifth = $g/3$ to $g/4.5$). The major part of this difference is concentrated in the kinematics in the crest region of the wave [13,17].

For the assumption of long-crested waves, and based on the CTS kinematics test data, the computed kinematics source of uncertainty in Ku will be evaluated as a Type II uncertainty with $B_{IIKu} = 0.72$, and $V_{IIKu} = 0.36$.

An additional contributor to uncertainty in Ku is current. As noted earlier, currents account for 25 to 30 percent of the computed total maximum lateral loadings developed on the example platform. Yet, the kinematics data from the CTS, and the force data from OTS, WFII, and CTS do not show the anticipated effects of currents. This paradox seems to be the result of a series of compensating effects in the calculations of wave forces (e.g. neglect of shielding and near-surface reductions in the force coefficients).

To recognize the effects of the uncertainties in the current effects on Ku, and based on the measured wave force data cited, a Type II uncertainty will be introduced with $B_{IIKu} = 0.75$, and $V_{IIKu} = 0.15$.

Another principal contributor to the uncertainty in Ku is the wave steepness. This is a natural or inherent source of uncertainty. Based on the tropical cyclone wave steepness background developed earlier in this paper, and on the background on the influence of wave steepness on Ku, this source of uncertainty will be characterized with $B_{IKu} = 1.0$, and $V_{IKu} = 0.10$.

The last contributor to the uncertainty in Ku that will be introduced is directional spreading effects on the wave kinematics. This factor will be accounted for using the single cylinder directional spreading kinematics corrections given in eqn. and eqn. This source of uncertainty will be evaluated as Type II. A best estimate directional spreading coefficient, $n = 3$ will be used based on the meteorologic-oceanographic studies. This would indicate a kinematics correction, $R_{uw} = 0.87$. Since the correction is being applied to the force, $B_{IIKu} = 0.76$. Given a $\pm 1\sigma$ range for n of 0.5 to 5, $V_{IIKu} = 0.26$.

UNCERTAINTIES IN FORCE COEFFICIENT, Kd

The principal component of uncertainty in Kd is Cd. Some of the variability in Cd is Type I (inherent) and some is Type II (modeling). Based on the laboratory tests reported by Roddenbush [22], on vertical rough (marine fouled) cylinders subjected to high KC wave flow conditions (Figure 4), Type I variability in Cd (and in Kd) is estimated as $V_{IKd} = 0.10$.

Based on the OTS data, there is a Type II bias in the Cd used to compute the hydrodynamic forces. The data indicate at high KC and Re flow conditions, and for marine fouled cylinders, the average $Cd \cong 1.0$. Thus, $B_{IIKd} = 1.0 / 0.6 = 1.67$. Also, $V_{cd} = 0.25$ (total); thus, $V_{IIcd} = 0.23$.

INTEGRATION OF UNCERTAINTIES - DRAG FORCE DOMINATED PLATFORM

Given these results, assuming no correlation of the parameters, and using the Normal function algebraic expressions (Table 2), the result of the first approach to the evaluation of Type I and II uncertainties in the annual maximum lateral forces can be summarized as follows (summary of uncertainties given in Table 3):

$$B_{ISm} = B_{IHm}^2 \, B_{IKd} \, B_{IKu} = 1.0 \tag{63}$$

$$V_{ISm} = \sqrt{(\, 2 \, V_{IHm} \,)^2 + V_{IKd}^2 + V_{IKu}^2} \; = 0.62 \tag{64}$$

$$B_{IISm} = B_{IIHm}^2 \, B_{IIKd} \, B_{IIKu} = 0.83 \tag{65}$$

$$V_{IISm} = \sqrt{(\, 2 \, V_{IIHm} \,)^2 + V_{IIKd}^2 + V_{IIKu}^2} \; = 0.58 \tag{66}$$

The combined Type I and Type II uncertainties would give $B_{Sm} = 0.83$, and $V_{Sm} = 0.85$.

Note that the single largest source of Class II uncertainty for the drag force dominated platform is attributed to the wave kinematics influences on the hydrodynamic forces (Table 3, $\sigma_{Ku} = 0.47$). As might be expected for this particular region, the tropical cyclone wave heights constitute the single largest source of Class I uncertainty ($\sigma_{Hm} = 0.30$).

The second approach that could be used is based on measured wave force data. This approach would evaluate the wave force uncertainties conditional on the measured wave and current characteristics, thereby avoiding explicit consideration of the kinematics uncertainties.

The evaluation will be based on the wave force measurements from the CTS. The data (Block 6, characteristic of the wave conditions close to the center of tropical cyclones) indicate a $B_{IIFm} = 0.64$ ($Cd = 0.7$) and $V_{IIFm} = 0.34$. Correcting for the drag coefficient used in the calculation of wave forces on the example platform($B_{IIFm} = 0.64 \times 1.1 = 0.70$),and introducing the uncertainty due to the hindcast wave heights:

$$B_{IISm} = B_{IIHm}^2 \, B_{IIFm} = 0.85 \tag{67}$$

$$V_{IISm} = \sqrt{(2 \, V_{IIHm})^2 + V_{IIFm}^2} = 0.43 \tag{68}$$

Using the results developed earlier for the Class I uncertainties (Table 3), $B_{ISm} = 1.0$, and $V_{ISm} = 0.62$.

Thus the combined Type I and Type uncertainties would give $B_{Sm} = 0.85$, and $V_{Sm} = 0.75$.

The two approaches to assessment of uncertainties associated with hydrodynamic loadings from extreme waves yield the same mean bias ($B_{Sm} = 0.85$), and coefficients of variation in the range $V_{Sm} = 0.75$ to 0.85.

DESIGN WAVE HEIGHTS

If one wanted to derive the appropriate design wave heights based on these results, one could use the following quantities: $B_{ISm} = 1.0$, $V_{ISm} = 0.60$ ($\sigma_{ISm} = 0.55$), $B_{IISm} = 0.85$, $V_{IISm} = 0.50$ ($\sigma_{IISm} = 0.47$), $B_{Sm} = 0.85$, $V_{Sm} = 0.78$ ($\sigma_{Sm} = 0.69$), $\alpha = 2.0$, $H = 10$ meters, $RSR = 2.0$, $\sigma_R = 0.25$, $s = 0.60$ (Type I) to 0.73 (Types I and II),and $\beta = 3.5$.

Based on the Type I uncertainties, $H_D = 20$ meters (eqn.23). Based on the combined Type I and Type II uncertainties, $H_D = 23$ meters. The Type II uncertainties add 3 meters of wave height to the minimum H_D, and the kinematics uncertainties contribute the majority of this difference (Table 3). The 3 meters increase in design wave height would equate to a 32 percent increase in the maximum total lateral wave force.

Note that the method to compute the hydrodynamic loadings is implicitly integrated into the design wave heights (Stokes Fifth Order Wave Theory, $Cd = 0.6$, $Cm = 2.0$, inclusion of current velocities).

DESIGN CAPACITY BASED ON 100-YEAR WAVE HEIGHT

If one wanted to use the 100-year design wave height of 20 meters, then based on the combined Type I and II uncertainties, $RSR = 2.7$.

If the evaluation of the platform capacity were based on nominal capacities of the elements, and $B_{RSR} = 1.25$, then the nominal design $RSR = 2.2$.

LOAD AND RESISTANCE FACTORS

If one wanted to determine the loading factor for an LRFD format and elements dominated or controlled by the cyclone loadings, then the load factor for the median loading would be:

$$\phi = \exp (0.75 \times 3.5 \times 0.69) = 6.12 \tag{69}$$

If the 100-year return period 20 meter wave height were used in design, then $F_{100} / F = 20^2 / 10^2 = 4$, and $\phi_{100} = 6.12 / 4 = 1.53$. For elements that were loaded to about 50 percent of their nominal value by dead loading with $\sigma_d = 0.10$, then $\phi_{100} = 1.3$ (eqns.31 and 32).

SAFETY INDICES AND THE PROBABILITY OF FAILURE

If one wanted to determine the safety index and probability of failure of the platform with RSR = 2 and H_D = 20 meters (100-year value), for the Type I and II uncertainties as postulated, then:

$$\beta = \ln [(RSR / B) \exp (2.33 \, \sigma_H \, \alpha)] / \sigma \tag{70}$$

and β = 3.3 to 3.5, and Pf = 0.03 to 0.06 percent per year.

One could evaluate Pf for a variety of exposure periods, I, (P_{fI}) using several different approaches. The assumption of independence of all platform trials from year to year (failing to recognize correlation of Type II uncertainties) would yield the results shown in Figure 17. Also, shown in this figure are the results for the approximation $P_{fI} = I \, Pfa$. In addition, the results are shown for a consistent treatment of independent (Type I) and dependent (Type II) uncertainties. For this particular problem, there are no substantial differences in results.

INERTIA FORCE CONTROLLED PLATFORM

If the hydrodynamic loadings developed on the platform were controlled by the inertia force component, the maximum lateral force would be related linearly to the wave height (eqn. 60).

Based on the data that has been previously cited, Table 4 summarizes the uncertainties in the oceanographic conditions, kinematics, and forces for a platform whose hydrodynamic forces are controlled by the inertia force component. The Type I $B_{ISm} = 1.0$, and $V_{ISm} = 0.33$; and the Type II $B_{IISm} = 0.66$ (due to use of Cm = 2.0 in force calculations as compared with expected value Cm = 1.3 based on laboratory and field test results), and $V_{IISm} = 0.33$. The combined Type I and II $B_{Sm} = 0.66$, and $V_{Sm} = 0.47$.

Note that in the case of the inertia force dominated platform, the magnitude of the kinematics uncertainties is considerably smaller. This is due to the muted influence of the crest region accelerations on the total integrated forces, and the lack of substantial influences from the current kinematics.

Based on these uncertainties, an RSR = 2.0, and a safety index β = 3.5, the design wave height H_D = 17 meters. This height could be compared with the H_D = 23 meters for the platform whose forces were drag force dominated.

Stated in another way, if the inertia force controlled platform were designed with the same wave height as for the drag force controlled platform, the safety index β = 4.1 (compared with β = 3.3 for the drag force controlled platform). The inertia force controlled platform would inherently be much safer than the drag force controlled platform.

Conclusions

Hydrodynamic force criteria for design of offshore platforms are not only a function of the oceanographic environment, the wave kinematics, and the hydrodynamics, but as well of the type of analyses employed in the design, the structure characteristics, and the level of reliability required of that structure.

In the case of drag force dominated structures, it has been demonstrated that the primary contributor of Class II uncertainty (modeling) is due to wave kinematics. In the case of inertia force dominated structures, uncertainties in wave kinematics are still a substantial contributor to the resultant uncertainties in hydrodynamic forces.

Reliability based evaluations of hydrodynamic loadings on offshore structures provides a logical and consistent basis with which engineers can examine and quantify the effects of uncertainties on design criteria for such structures.

Perhaps the most important aspect of such evaluations is communications between the wide range of areas of technical expertise and engineering disciplines involved in design of marine structures. The reliability approach provides a framework for the conduct of these communications. It allows one to clearly demonstrate how the different aspects that characterize hydrodynamic loadings influence each other and the overall structure design process.

The quality of reliability based evaluations of hydrodynamic loadings on coastal and ocean structures should not be judged by the numerical results produced by such an approach; the quality should be judged by the quality of the decisions and engineering that it produces.

Acknowledgements

The author would like to express his appreciation to the organizers and sponsors of the NATO Advanced Research Workshop on "Water Wave Kinematics" for the encouragement to write this paper.

References

[1] Bea, R.G. (December 1989) 'Reliability Based Design Criteria for Coastal and Ocean Structures,' Proceedings, Coastal and Ocean Engineering Conference, Institution of Engineers, Australia, Adelaide, South Australia.

[2] Melchers, R.E. (1988) Structural Reliability Analysis and Prediction, Jacaranda Wiley Limited, Brisbane, Queensland, Australia.

[3] Madsen, H.O, Krenk, S., and Lond, N.C. (1986) Methods of Structural Safety, Prentice-Hall of Australia Pty. Ltd., Sydney, Australia.

[4] Bea, R.G. (1989) 'Acceptable Reliability of Coastal and Ocean Structures,' Institution of Engineers, Australia, Civil Engineering College Eminent Overseas Speaker Program.

[5] Bea, R.G. (1989) 'Reliability Based Evaluations of Coastal and Ocean Structure Capacities,' Institution of Engineers, Australia, Civil Engineering College Eminent Overseas Speaker Program.

[6] Haring, R.E., Osborne, A.R., and Spencer, L.P.(1976) 'Extreme Wave Parameters Based on Continental Shelf Storm Wave Records, Proceedings, Coastal Engineering Conference, American Society of Civil Engineers, Cpt. 10.

[7] Forristall, G.Z. (May 1978) 'On the Statistical Distribution of Wave Heights in a Storm,' Journal of Geophysical Research, Vol. 83, No. C5.

[8] Carter, D.J.T., Challenor, P.G., Ewing. J.A., Pitt, E.G., Srokosz, M.A., and Tucker, M.J. (1986) 'Estimating Wave Climate Parameters for Engineering Applications, Offshore Technology Report OTH 86 228, Department of Energy, London.

[9] Nolte, K.G. (1974) 'Statistical Methods for Determining Extreme Sea States,' Proceedings, Second Int. Conference on Port and Ocean Engineering Under Arctic Conditions.

[10] Holland, G.J. (1980) 'An Analytic Model of the Wind and Pressure Profiles in Hurricanes, 'Monthly Weather Review, American Meteorological Society, Vol. 108.

[11] Young, I.R. (Sept. 1988) 'Parametric Hurricane Wave Prediction Model,' Journal of Waterway, Port, Coastal and Ocean Engineering, American Society of Civil Engineers, Vol. 114, NO. 5.

[12] Cooper, C.K. (May 1988) 'Parametric Models of Hurricane Generated Winds, Waves, and Currents in Deep Water,' Proceedings, Offshore Technology Conference, Paper 5738, Houston, Texas.

[13] Sobey, R. J. (Sept. 1989) 'A Review of Recent Developments in Wave Kinematics,' Proceedings, Int. Conf. on Hydraulic and Environmental Modelling of Coastal Estuarine and River Waters, University of Bradford, England, Gower Publishing Co.

[14] Lambrakos, K.F., Chjao, J.C., Beckmann, H., Bannon, H.R. (1987) 'Wake Model of Hydrodynamic Forces on Pipelines,' Ocean Engineering, Vol. 14, No. 2.

[15] Dean, R.G.(May 1977) 'Hybrid Method of Computing Wave Loading, Proceedings, Offshore Technology Conference, OTC No. 3029, Houston, Texas.

[16] Mitwally, H., and Novak, M. (March 1989) 'Wave Forces on Fixed Offshore Structures in Short-Crested Seas,' Journal of Engineering Mechanics, American Society of Civil engineers, Vol. 115, No. 3.

[17] Ohmart, R.D., and Gratz, R.L. (May 1978) 'A Comparison of Measured and Predicted Ocean Wave Kinematics,' Proceedings, Offshore Technology Conference, OTC No. 3276, Houston, Texas.

[18] Haring, R.E., Olsen, O.A., and Johansson, P.I. (1979) 'Total Wave Force and Moment Vs Design Practice,' Proceedings of the Specialty Conference, Civil Engineering in the Oceans IV, American Society of Civil Engineers, Vol. II.

[19] Sarpkaya, T., and Isaacson, M (1981).Mechanics of Wave Forces on Offshore Structures, Van Nostrand Reinhold Company, Melbourne, Victoria, Australia.

[20] Heideman, J., and Sarpkaya, T. (May 1985) 'Hydrodynamic Forces on Dense Arrays of Cylinders,' Proceedings, Offshore Technology Conference, OTC No. 5008, Houston, Texas.

[21] Nath, J.H. (Nov. 1987) 'On Wave Force Coefficient Variability,' Journal of Offshore Mechanics and Arctic Engineering, Transactions of American Society of Mechanical Engineers, Vol. 109.

[22] Roddenbusch, G., Kallstrom, C. (May 1986) 'Forces on a Large Cylinder in Random 2-Dimensional Flow,' Proceedings, Offshore Technology Conference, OTC No. 5096.

[23] Heideman, J.C., Olsen, O.A., and Johansson, P.I. (1979) 'Local Wave Force Coefficients,' Proceedings of the Specialty Conference, Civil Engineering in the Oceans IV, American Society of Civil Engineers, Vol. II.

[24] Kjeldsen, S.P. and Akre, A.B. (1985).'Wave Forces in Vertical Piles Near the Free Surface Caused by 2-Dimensional and 3-Dimensional Breaking Waves,' Report No. 1.10, Programme for Marine Structures, Marintek, Trondheim, Norway.

[25] Bea, R.G., and Lai, N. W. (May 1978) 'Hydrodynamic Loadings on Offshore Platforms,' Proceedings, Offshore Technology Conference, OTC No. 3064.

[26] Ohmart, R.D., Gratz, R.L., and Mallery, G.O. (May 1970) 'Instrumentation Program for Evaluation of Offshore Structure Design,' Proceedings, Offshore Technology Conference, OTC No. 1264, Houston, Texas.

[27] Ohmart, R.D (Nov. 1984) 'Hydrodynamic Loading of Steel Structures,' Journal of the Waterway, Prot, Coastal and Ocean Engineering, American Society of Civil Engineers, Vol. 110, No. 4.

[28] Bea, R.G., Pawsey, S.F., and Litton, R.W. (May 1988) 'Measured and Predicted Wave Forces on Offshore Platforms,' Proceedings, Offshore Technology Conference, OTC No. 5787.

132

[29] Bea, R.G. (1989) 'Organizational Aspects of Reliability Management in Design, Construction, and Operation of Coastal and Ocean Structures,' Institution of Engineers, Australia, Civil Engineering College Eminent Overseas Speaker Program.

[30] Pate'-Cornell, E., and Bea, R.G. (April 1989) 'Organizational Aspects of Reliability Management: Design, Construction, and Operation of Offshore Platforms,' Research Report No. 89-1, Dept. of Industrial Eng. and Eng. Management, Stanford Univ.

[31] Haugen, E.B. (1968) Probabilistic Approaches to Design, John Wiley & Sons, Inc., Sydney, Australia.

Table 1. Lognormal Distribution Formulas

Median of the parameter, $X = \underline{X} = X_{50} = $ 50-th percentile value (half of values are larger than $\underline{X}$ and half are smaller than $\underline{X}$.

The mean of the parameter, X, can be computed from n values of the variable X as:

$$X = \frac{\sum\limits^{n} X}{n}$$

The standard deviation of a variable, σ_X, can be computed from the mean of the variable, X, and n values of the variable X, as:

$$\sigma_X{}^2 = \sum \frac{(X - X)^2}{n}$$

Standard Deviation of ln X can be determined as:

$$\sigma_{\ln X} = 0.39 \, [\, \ln X_{90} - X_{10} \,]$$

where X_{90} and X_{10} are the 90th and 10th percentile values, respectively.

The standard deviation of X is:

$$\sigma_X = X \sqrt{\exp (\sigma^2{}_{\ln X}) - 1}$$

Table 1. (continued)

The ratio of the standard deviation to the mean is known as the coefficient of variation (V):

$$V_X = \frac{\sigma_X}{X}$$

For Lognormal variates, σ is related to V_X as:

$$\sigma_{\ln X} = \sqrt{\ln (1 + V_X{}^2)}$$

For $V_X < 0.5$, $\sigma_{\ln X} \cong V_X$. Also, V is related to σ as:

$$V_X = \sqrt{\exp (\sigma^2{}_{\ln X}) - 1}$$

The mean of X can be determined as:

$$X = \underline{X} \exp (0.5 \, \sigma^2{}_{\ln X})$$

The probability density function (probability that X is a given value) is:

$$p_x(X) = \frac{1}{X \sqrt{2\pi} \, \sigma_X} \exp \left\{ -\frac{1}{2} \left(\frac{1}{\sigma_X} \ln \frac{X}{\underline{X}} \right)^2 \right\}$$

The cumulative probability function (probability that X < x) is:

$$P_X(X) = \int_0^X \frac{1}{\sqrt{2\pi}} \exp \left\{ - \frac{(\ln X - \underline{X})^2}{2\sigma^2} \right\}$$

Table 2. Algebra of Normal Functions

For the addition or subtraction of two random variables, $(X \pm Y) = Z$, the mean (same as mode and median) of the resultant distribution can be calculated as:

$$Z = X \pm Y$$

The standard deviation of the resultant distribution can be calculated as:

$$\sigma_Z = \sqrt{\sigma_X{}^2 + \sigma_Y{}^2 \pm 2\rho\,\sigma_X\,\sigma_Y}$$

For the multiplication of two random variables, $(X\,Y) = Z$, the mean of the resultant distribution can be calculated as:

$$Z = X\,Y$$

The standard deviation of the resultant distribution can be calculated as:

$$\sigma_Z = X\,Y\sqrt{(1 + \rho^2)\,[V^2{}_X + V^2{}_Y + (V^2{}_X\,V^2{}_Y)]}$$

For the division of two random variables, $(X / Y) = Z$, the mean of the resultant distribution can be calculated as:

$$Z = X / Y$$

The standard deviation of the resultant distribution can be calculated as:

$$\sigma_Z = (X^2 / Y^2)\sqrt{V^2{}_X + V^2{}_Y - 2\rho\,(V_X\,V_Y)}$$

When the random variables X and Y can be considered independent ($\rho = 0$), and V_X and V_Y are small ($V \ll 1$), then for either the mulitiplication or quotient of two random variables:

$$V_Z \cong \sqrt{V_X{}^2 + V_Y{}^2}$$

For the square of X ($Z = X^2$):

$$Z = X^2 + \sigma_X{}^2$$

$$\sigma_Z = \sqrt{4\,X^2\,\sigma_Z{}^2 + 2\sigma_Z{}^4}$$

Table 2. (continued)

The "correlation coefficient", ρ, expresses how strongly two variables, X and Y, are related to each other. It measures the strength of association between two variables.

If $\rho = \pm 1$, they are perfectly correlated, so that knowing X allows one to make perfect predictions of Y ($-\rho$ implies X increases as Y decreases). If $\rho = 0$, they have no correlation, or are independent, so that the occurrence of X has no affect on the occurrence of Y.

The correlation coefficient can be computed from data in which the results of n samples of X and Y are developed:

$$\rho = \frac{\Sigma X Y - n \underline{X} \underline{Y}}{\sqrt{(\Sigma X^2 - n \underline{X}^2)(\Sigma Y^2 - n \underline{Y}^2)}}$$

Table 3. Summary of Uncertainties in Hydrodynamic Loadings For Drag Force Dominated Platform

Uncertainty Component	Type I		Type II	
	B	σ	B	σ
Conditions (Hm)	1.0	0.30	1.1	0.13
Kinematics (Ku)	1.0	0.10	0.41	0.47
Force Coef. (Kd)	1.0	0.10	1.67	0.23
Resultant (first approach)	1.0	0.62	0.83	0.58
Total Forces (Fm)	1.0	0.16	0.70	0.34
Resultant (second approach)	1.0	0.62	0.85	0.43

Table 4. Summary of Uncertainties in Hydrodynamic Loadings For Inertia Force Dominated Platform

Uncertainty Component	Type I		Type II	
	B	σ	B	σ
Conditions (Hm)	1.0	0.30	1.1	0.13
Kinematics (Ku)	1.0	0.10	0.93	0.20
Force Coef (Kd)	1.0	0.10	0.65	0.23
Resultant	1.0	0.33	0.66	0.38

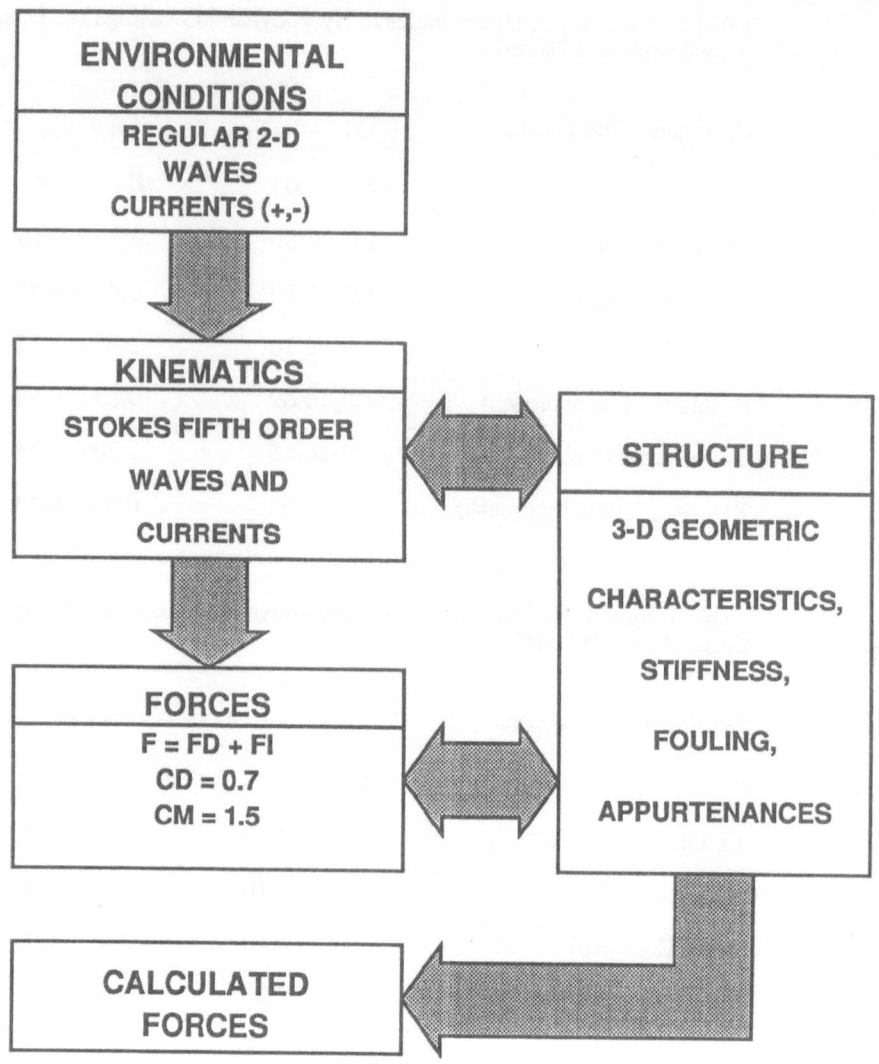

Figure 1. Procedure to Calculate Hydrodynamic Forces

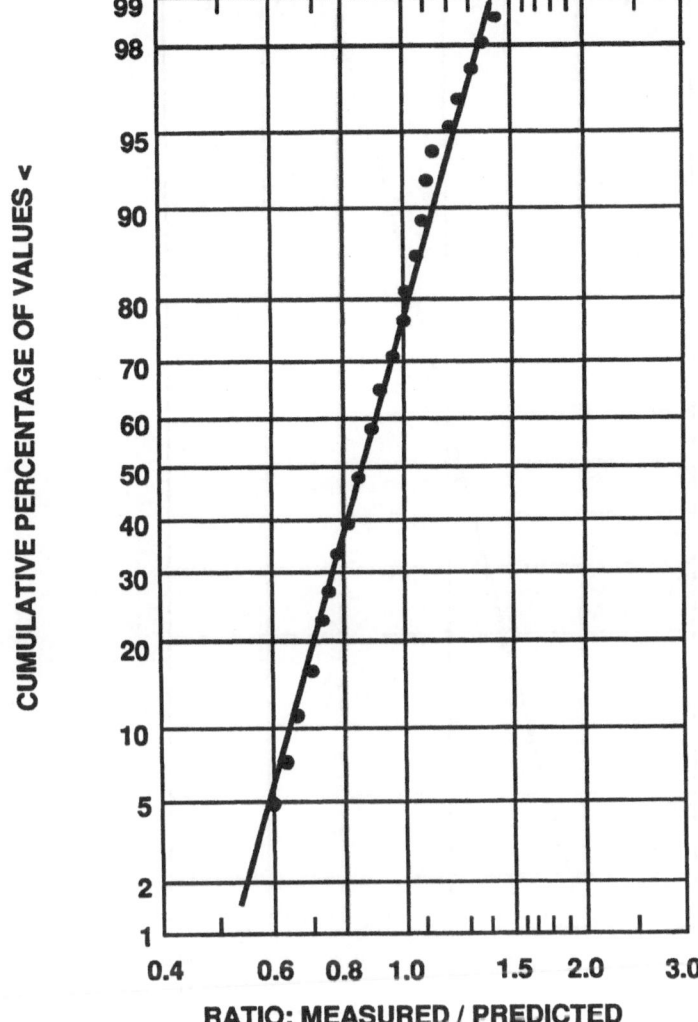

Figure 2(a). Distribution Using Stokes Theory for 38 Waves
          Height Range 24-35 feet

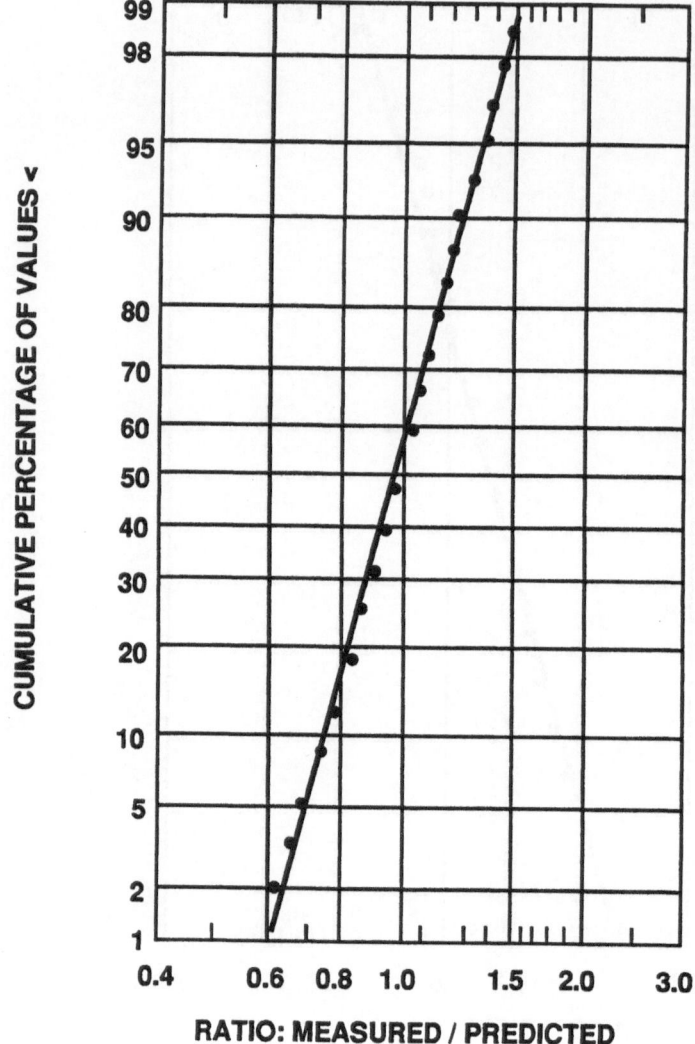

**Figure 2(b). Distribution Using Stokes Theory for 125 Waves
Height Range 20-24 feet**

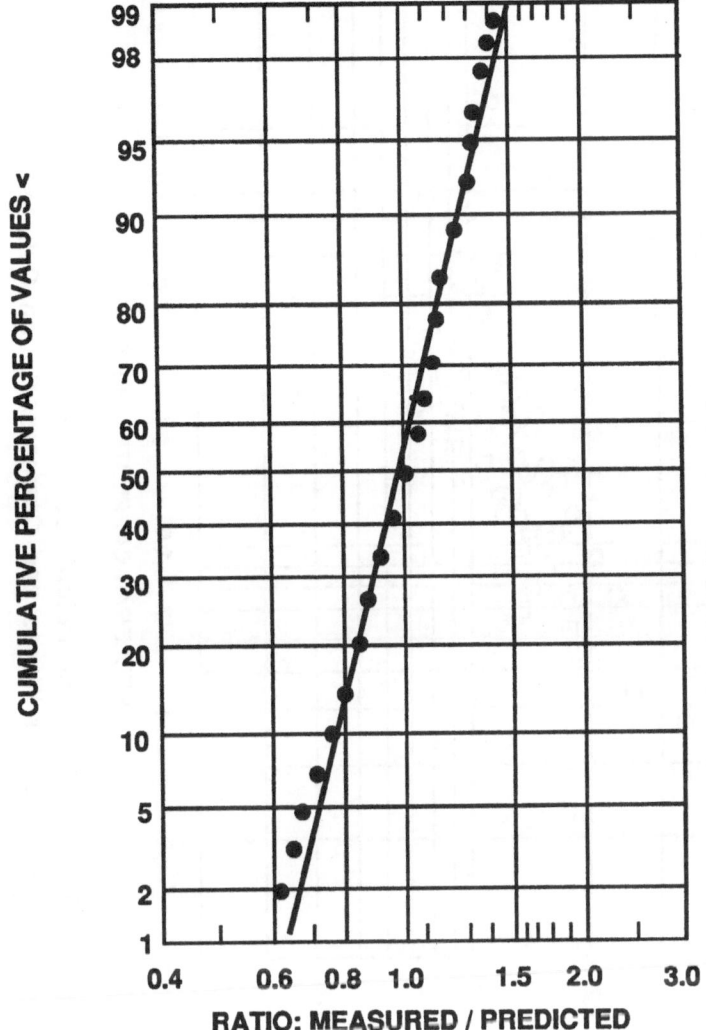

**Figure 2(c). Distribution Using Stokes Theory for 284 Waves
Height Range 16-20 feet**

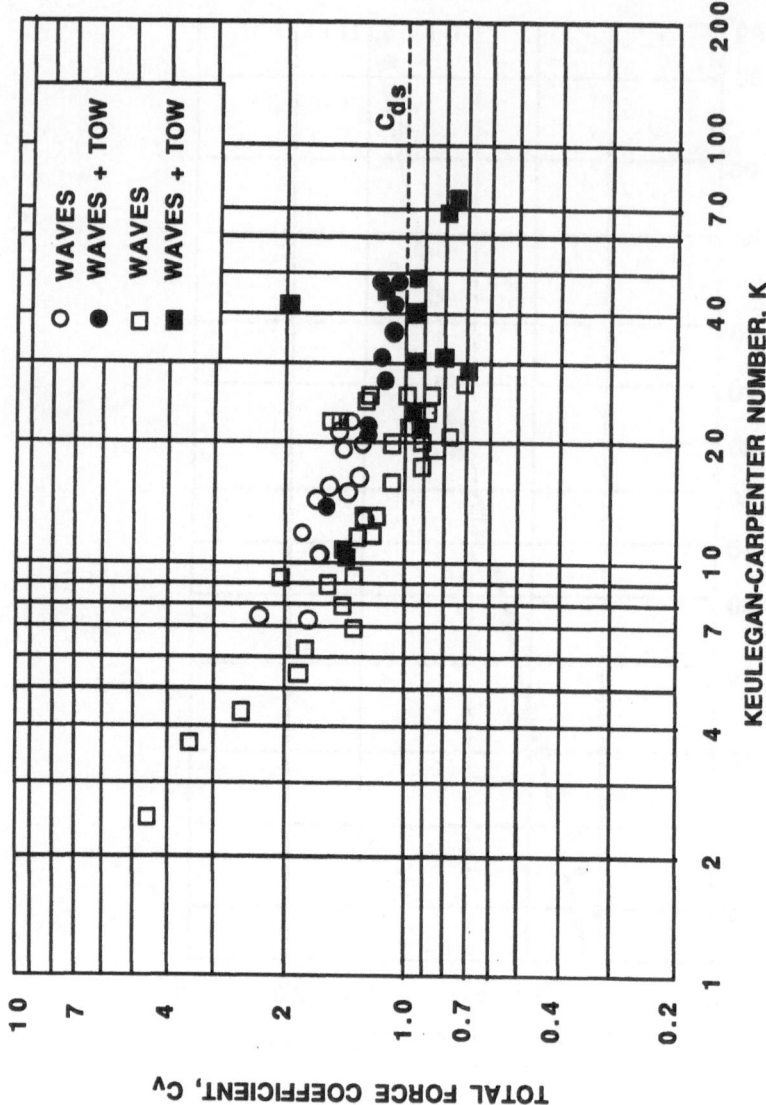

Figure 3. Total Force Coefficient as Function of Keulegan-Carpenter Number for Waves and Currents, Rough Cylinder

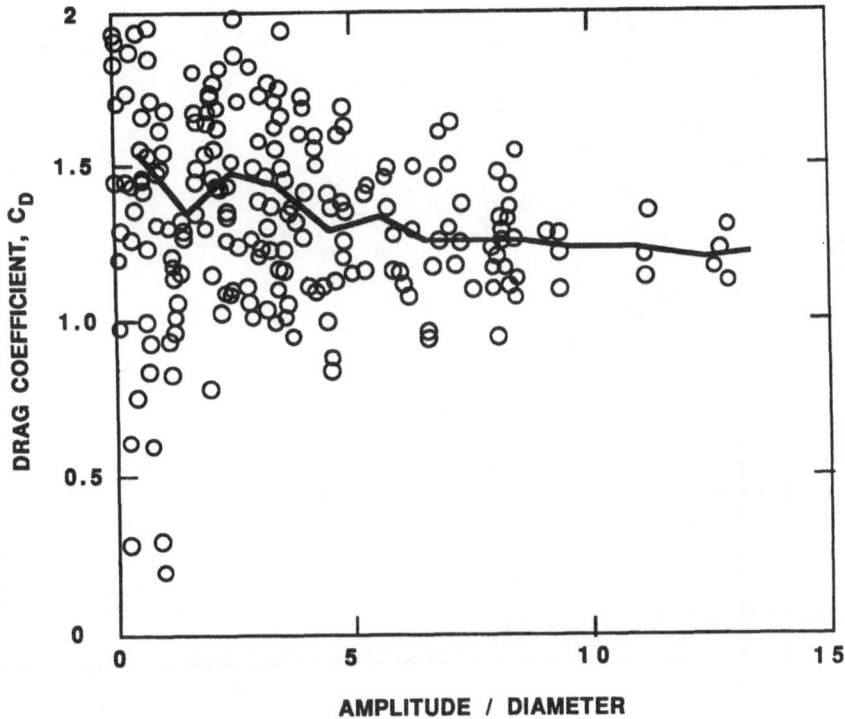

Figure 4(a). **Drag Coefficient Scatter Plot for the Rough Cylinder: SSPA Data**

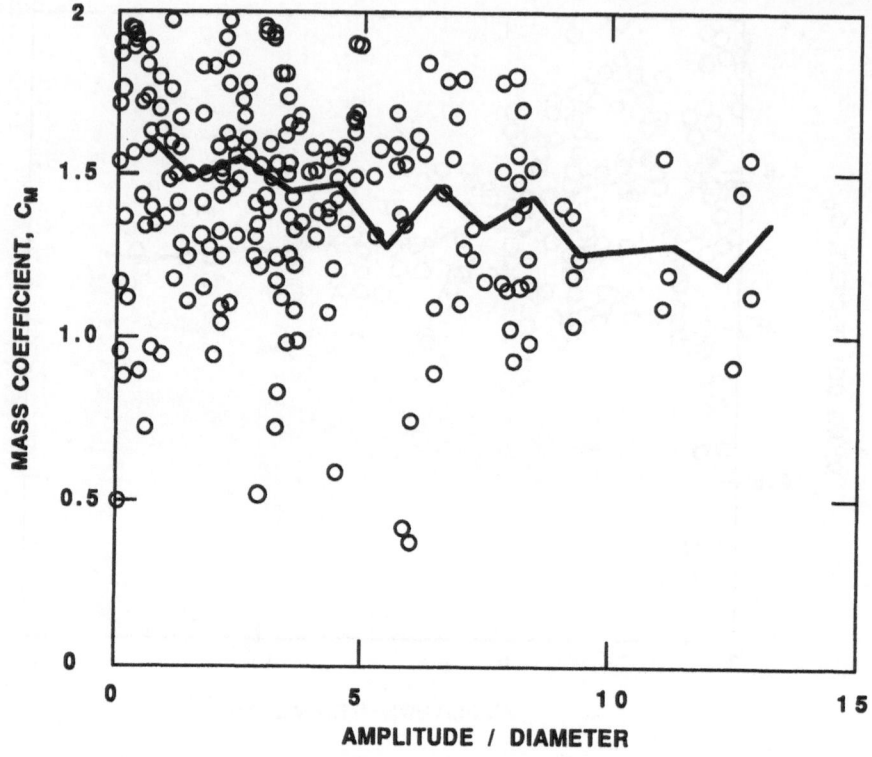

Figure 4(b). Mass Coefficient Scatter Plot for the
Rough Cylinder: SSPA Data

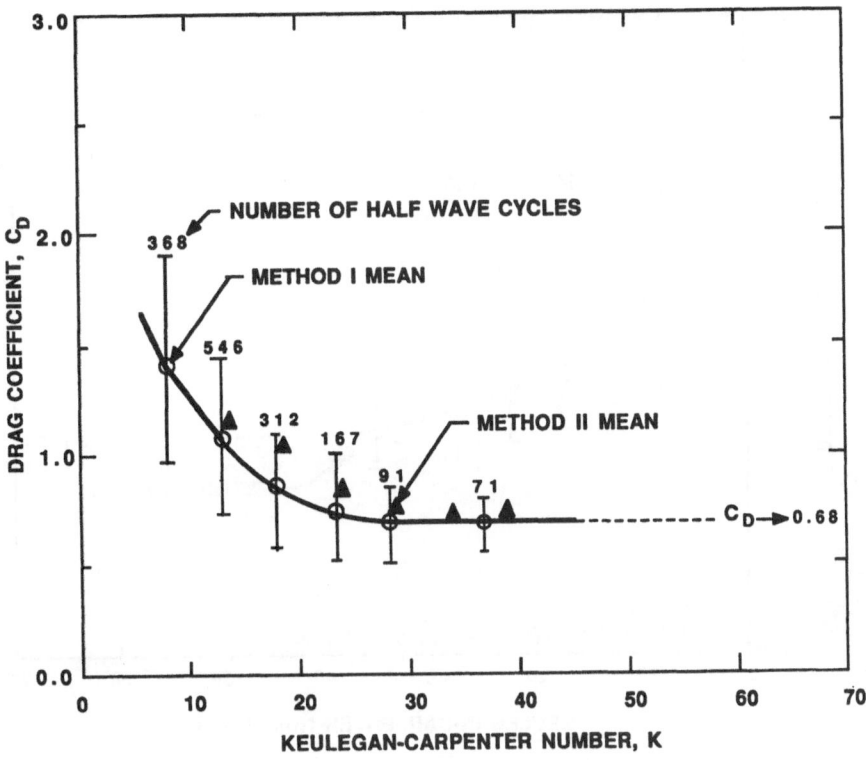

Figure 5(a). Drag Coefficient as a Function of Keulegan-Carpenter Number for Smooth Wave Force Transducers

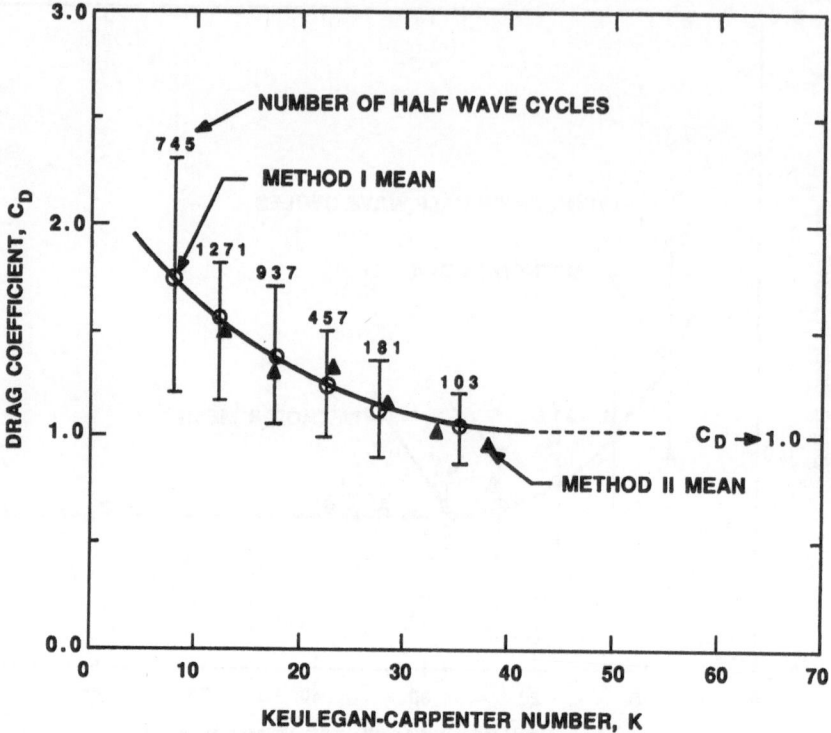

Figure 5(b). Drag Coefficient as Function of Keulegan-Carpenter Number for Barnacle Encrusted Wave Force Transducers

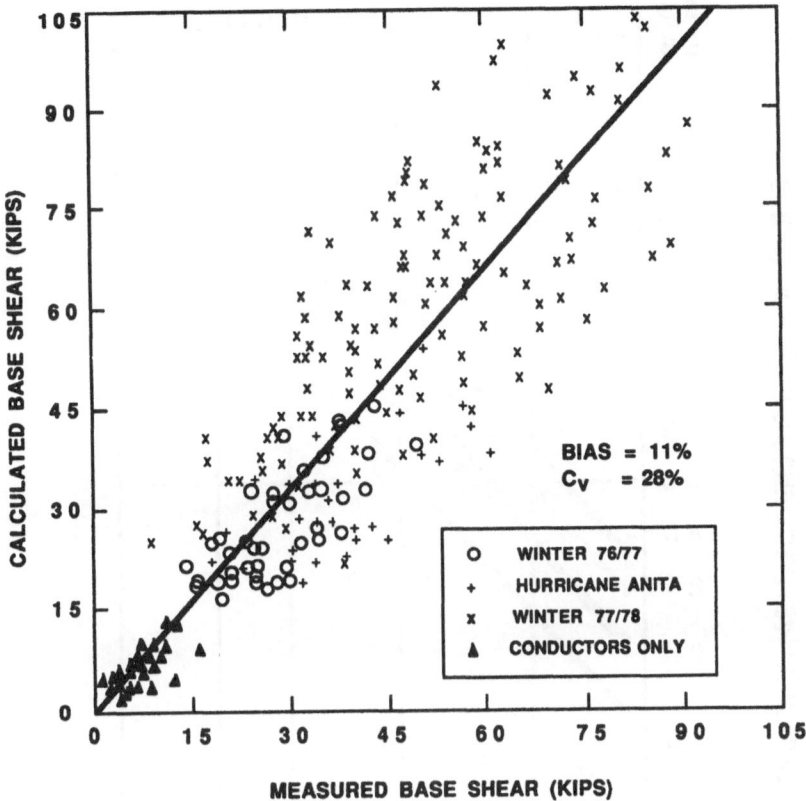

Figure 6. Measured and Calculated Base Shears on
Ocean Test Structure

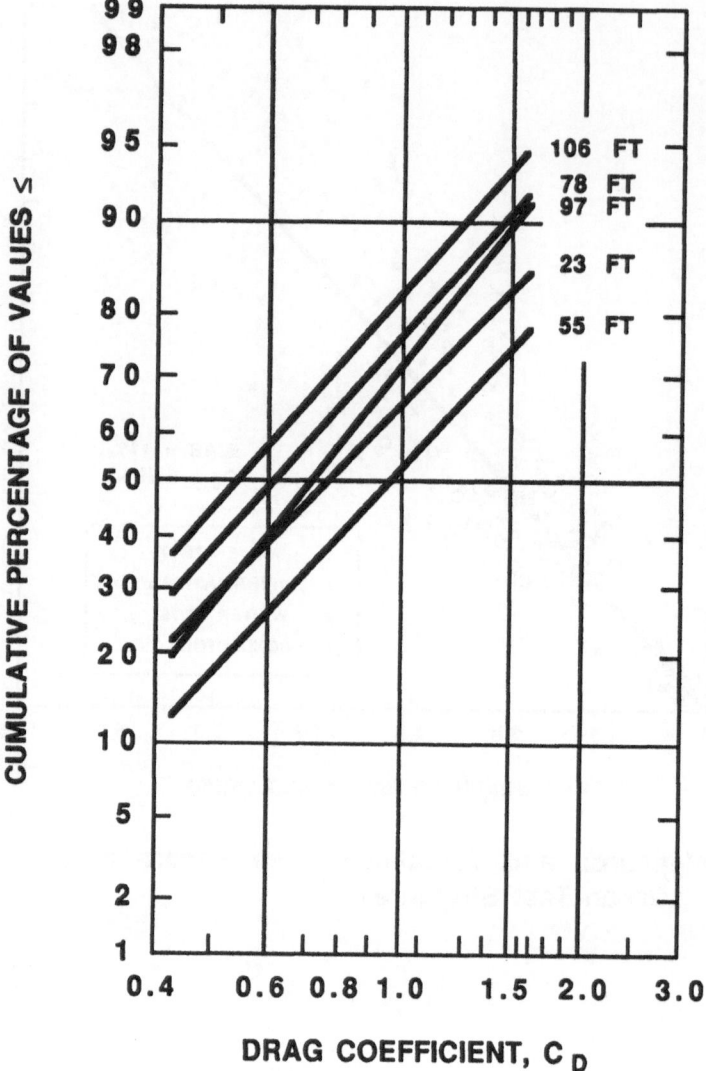

Figure 7(a). Probability Distributions of Drag Coefficients for Different Depths Below Mean Water Level - Wave Force Project II

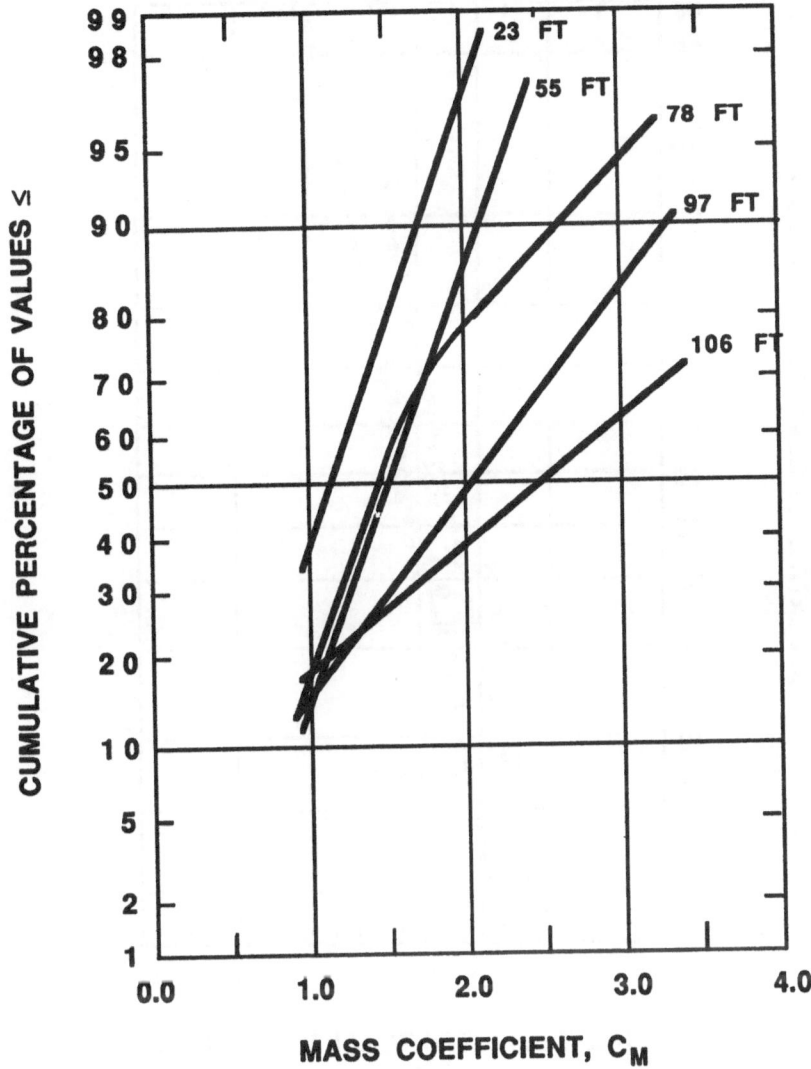

Figure 7(b). Probability Distributions of Mass Coefficients for Different Depths Below Mean Water Level - Wave Force Project II

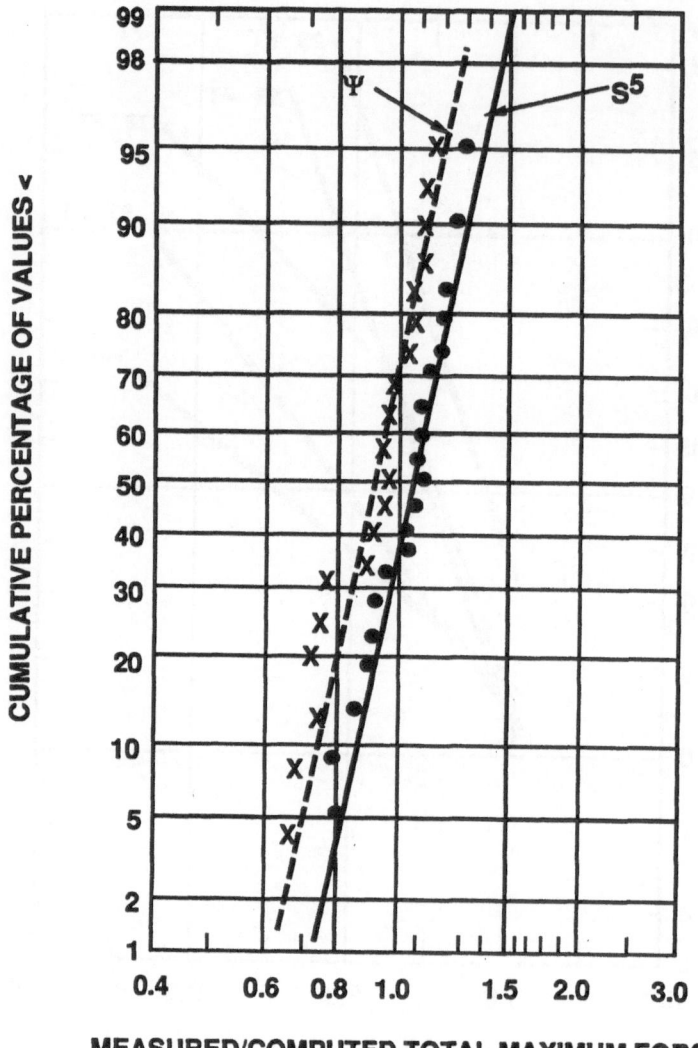

Figure 8. Probability Distributions of Ratios of Measured to Computed (Stokes Fifth Order, S^5, and Stream Function, Ψ) Maximum Total Wave Forces for 20 Highest Waves in Wave Force Project II

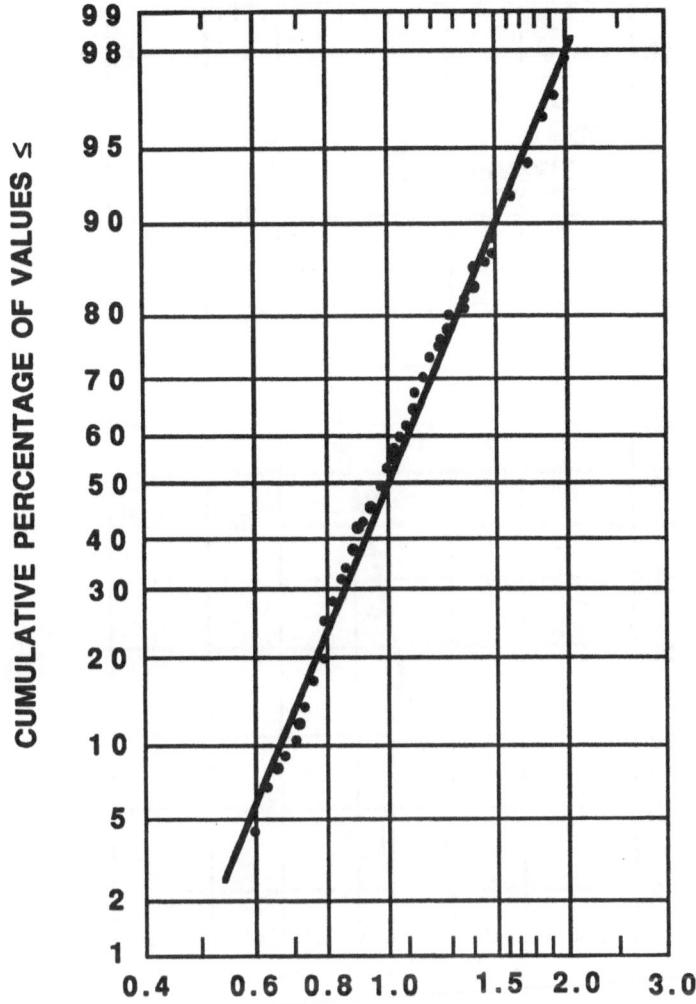

MEASURED/ COMPUTED TOTAL MAXIMUM FORCE

Figure 9. Probability Distributions of Ratio of Measured
to Computed (Stoke's 5th Order) Total Maximum
Wave Forces on Conoco Test Platform

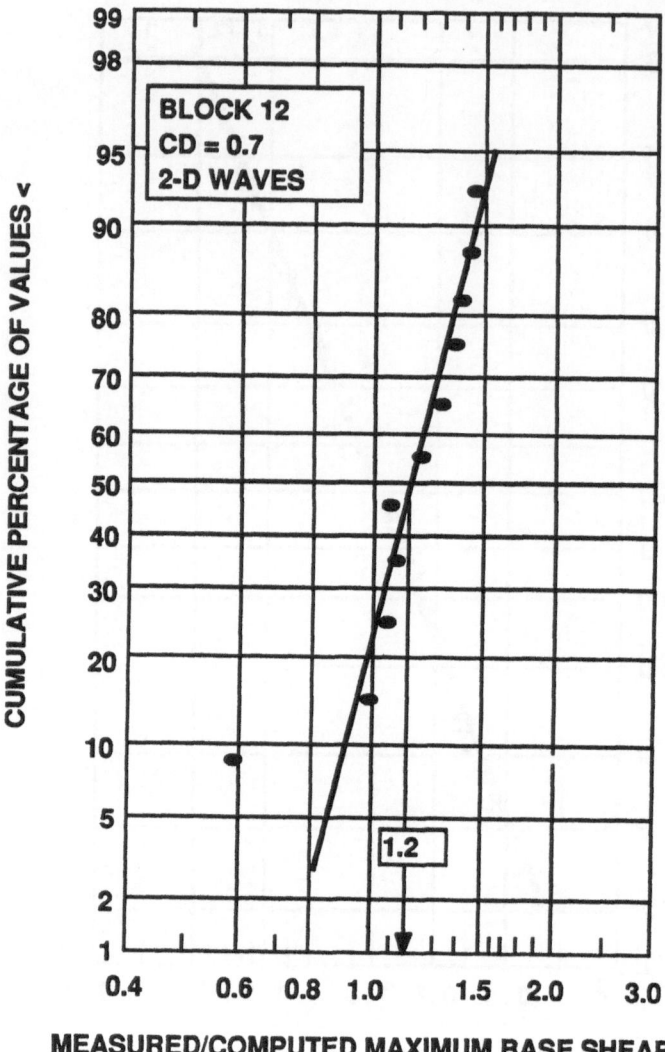

Figure 10. Probability Distributions of Ratios of Measured to Computed
Total Maximum Wave Forces on Conoco Test Platform For
Low Directional Spreading Conditions

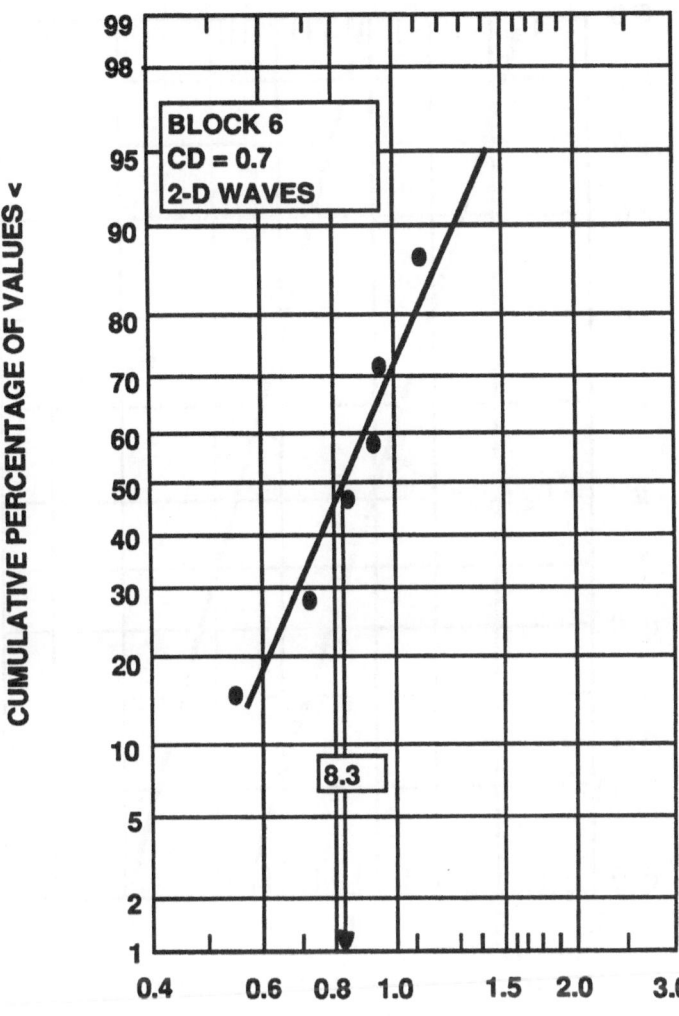

MEASURED/COMPUTED MAXIMUM BASE SHEAR

Figure 11. Probability Distributions of Ratios of Measured to Computed Total Maximum Wave Forces on Conoco Test Platform For High Directional Spreading Conditions

154

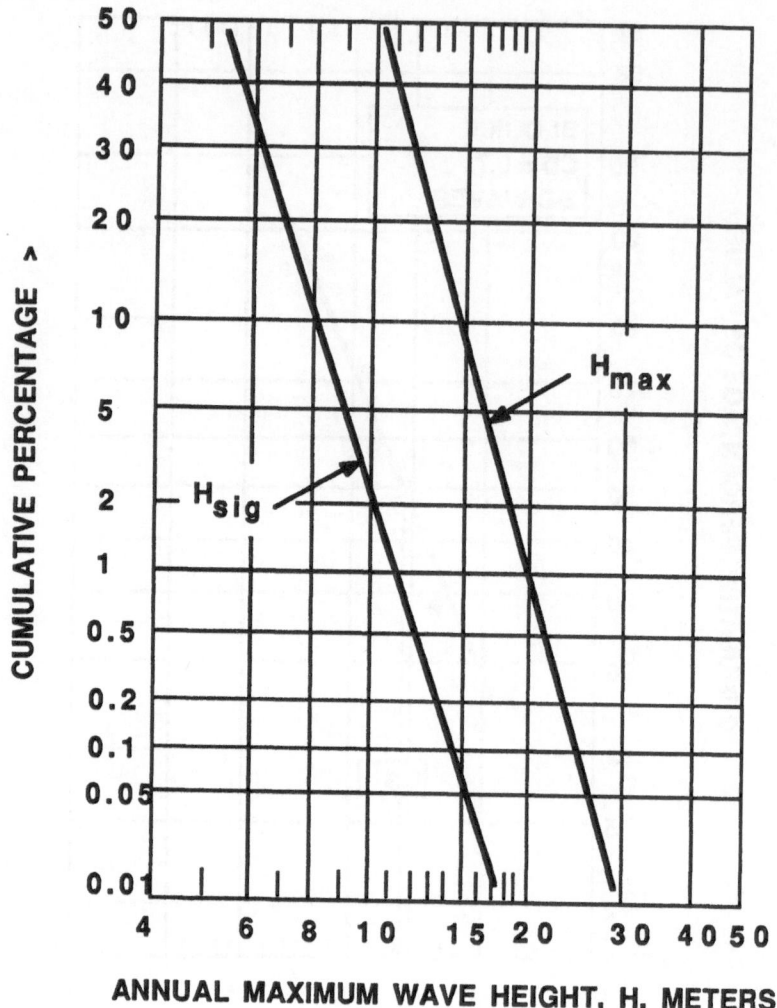

Figure 12. Probability Distribution of Annual Expected Maximu
(Hmax) and Significant (Hsig) Wave Heights

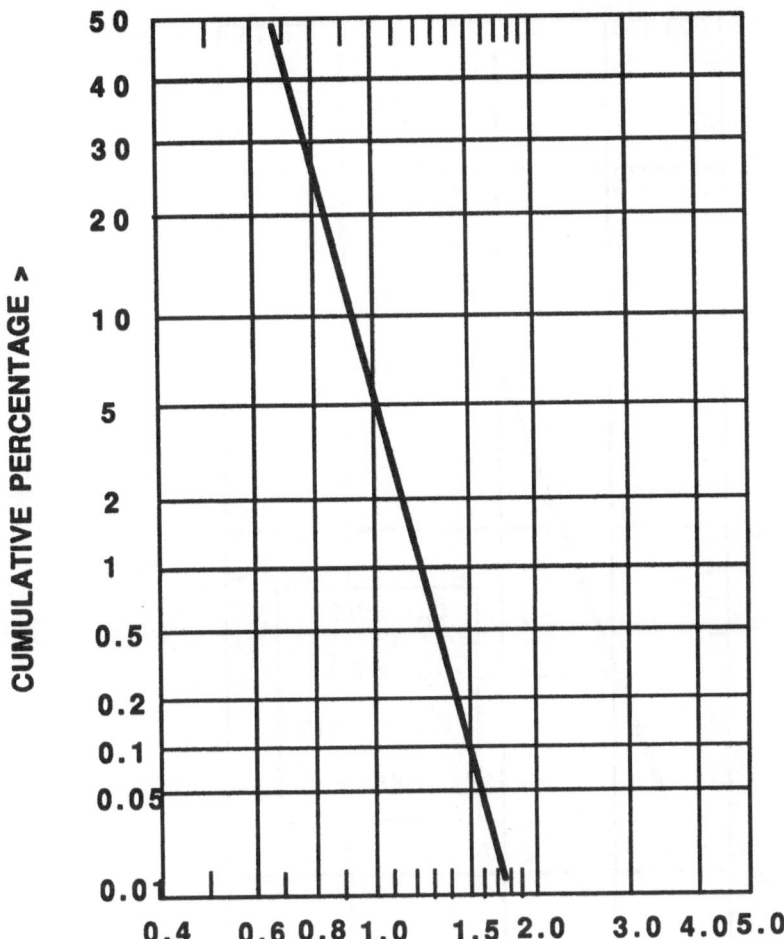

ANNUAL MAXIMUM SURFACE CURRENT VELOCITY -
U_C (METERS PER SEC)

Figure 13. Probability Distribution of Annual Expected
Maximum Surface Current Velocity

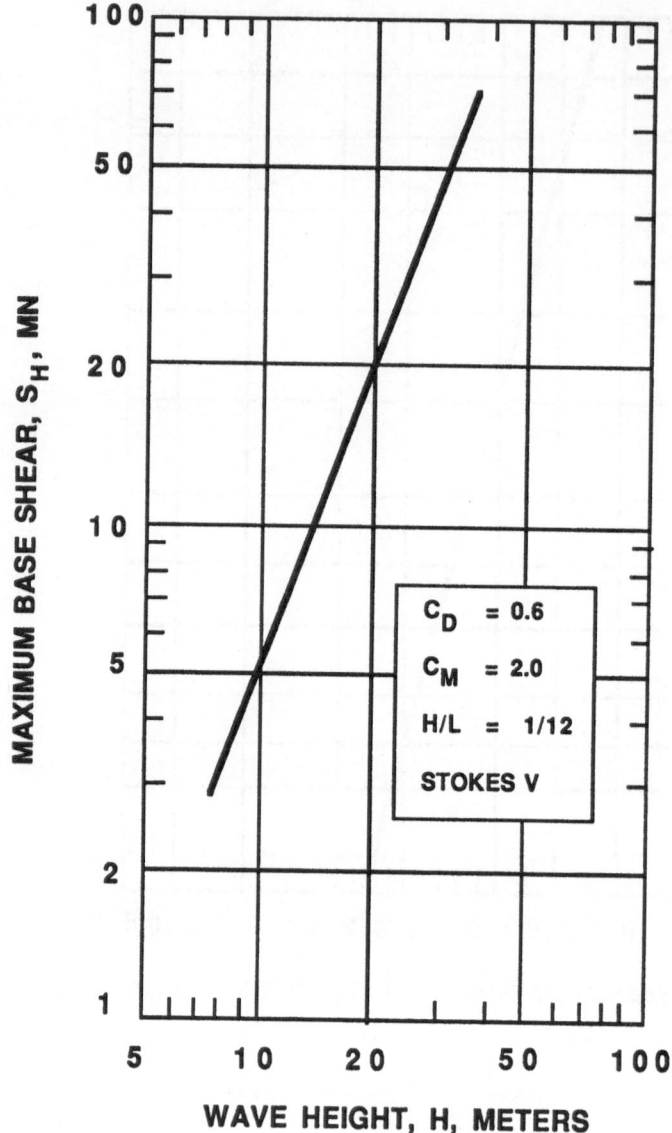

Figure 14. Calculated Maximum Total Lateral Force on Example Platform as Function of Wave Height

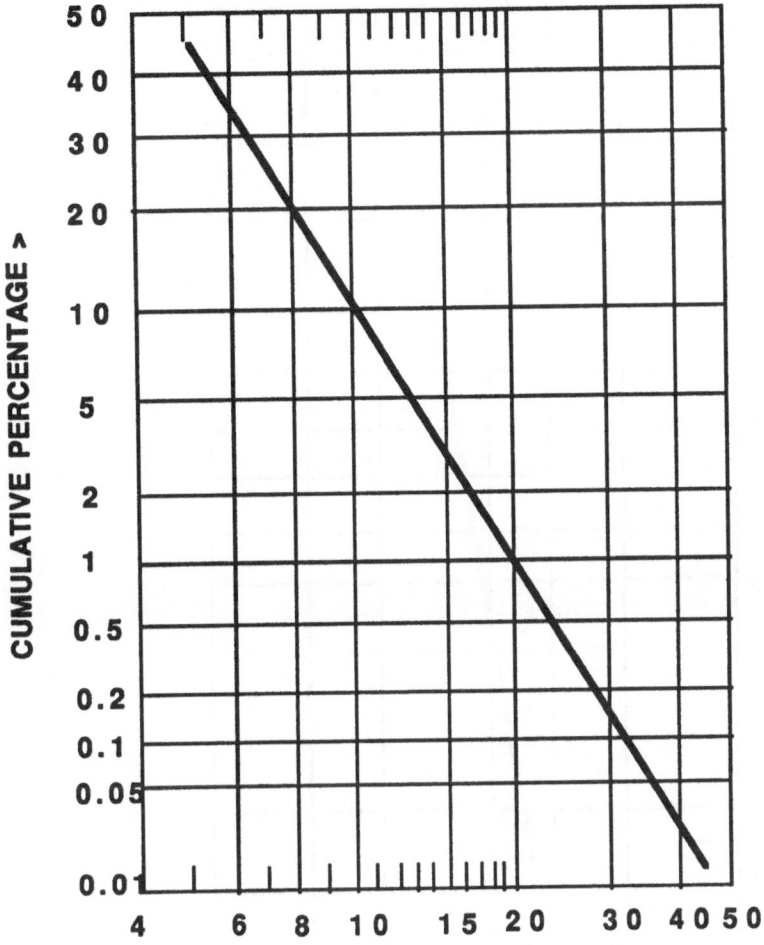

ANNUAL EXPECTED TOTAL MAXIMUM BASE SHEAR - S_H (MN)

Figure 15. Probability Distribution of Annual Expected
Maximum Base Shear on Example Platform

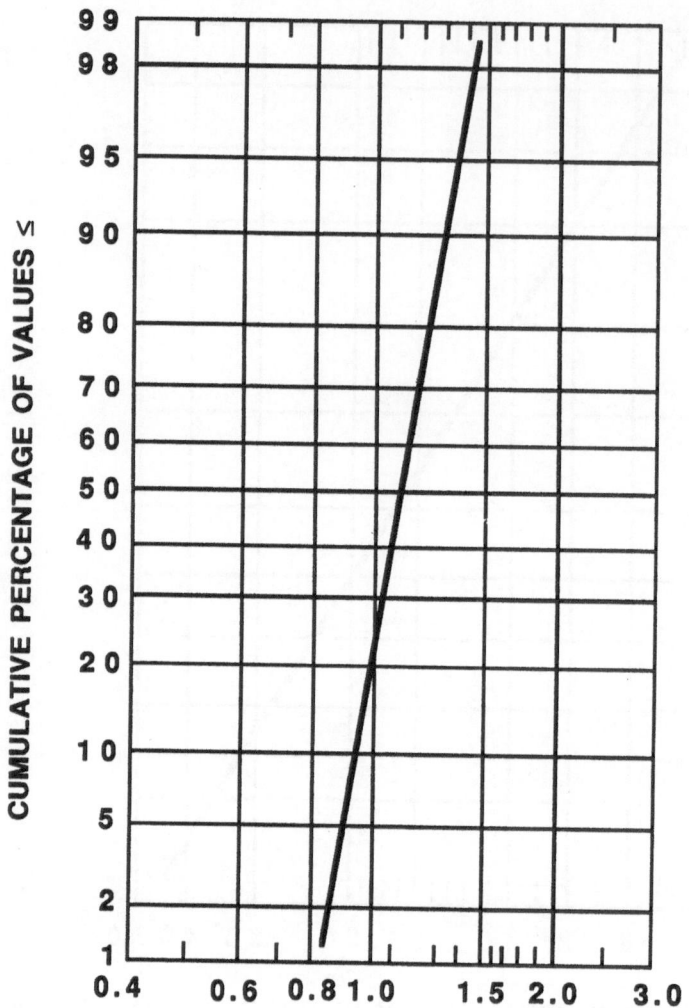

B$_H$ = MEASURED/ PREDICTED MAXIMUM WAVE HEIGHTS

Figure 16. Probability Distribution of Wave Height Prediction Bias, B$_H$

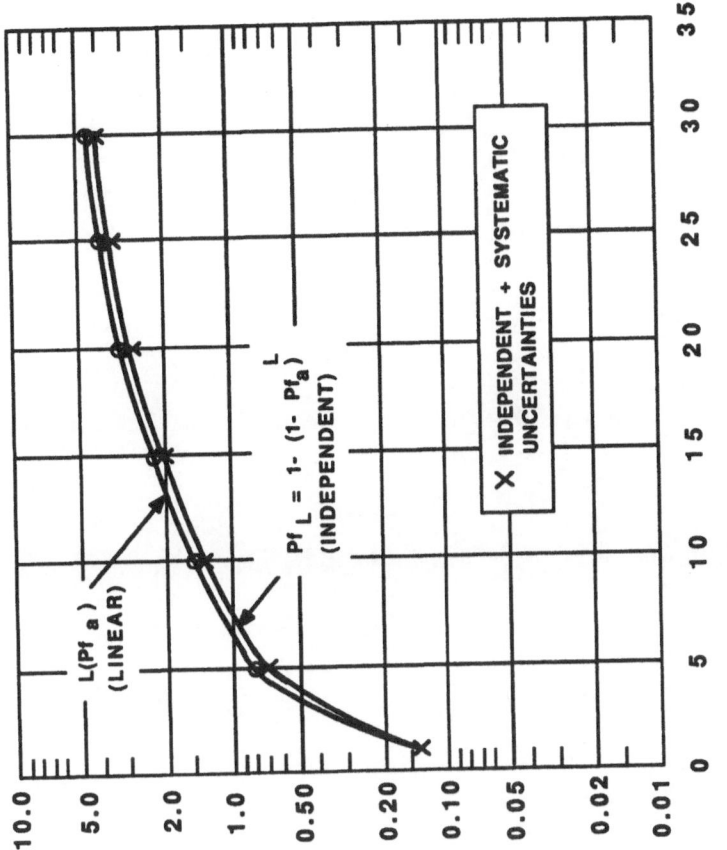

Figure 17. Probability of Failure as Function of Exposure Period, L, for Independent and Systematic Uncertainties Compared with Perfectly Independent and Linear Approximations

Deep Water Wave Kinematics

SLOW MODULATION OF WEAKLY NONLINEAR WAVES

CHIANG C. MEI
Massachusetts Institute of Technology
Cambridge, MA USA

EDMOND LO
Physical Sciences Co.
Andover, MA USA

YEHUDA AGNON
Technion
Haifa, Israel

ABSTRACT. In oceans, surface waves are not only irregular and of finite amplitude, but also nonstationary, in that their average features change with space and time. The nonstationarity is inherently due to nonlinear interactions among waves of different frequencies. Here we discuss two examples. The first is the evolution of unidirectional waves in deep water. Group splitting is shown to develop in both two and three dimensional cases. Individual crests in the transient groups may exceed the threshold of breaking. The second example is the excitation of trapped long waves on a submarine ridge by incident short wave groups. Such long waves may be important to moored ships of semisubmercibles.

1. Introduction

In many engineering problems, the sea waves are modelled as a stationary process. For small amplitudes, linearity is often assumed; the theory of stationary random process can be applied to irregular waves with many frequencies. For finite amplitudes, much progress has been made on the calculation of steady waves that propagate without change of form (Stokes waves, cnoidal waves, etc.). The complex causes of generation and nonlinear effects inherent in the physical process, however, make irregular sea waves transient. Therefore for engineering purposes, a truly realistic theory must include nonstationarity, in addition to irregularity and finite amplitudes. Indeed these three aspects are causes and effects of one another because of nonlinearity.

In this paper we review some past works on the effects of nonstationarity. To facilitate analysis and understanding we only consider cases when nonlinearity and irregularity are both weak. Specifically, we assume that waves have small slope and narrow frequency band initially. Nonlinear exchange of energy exists among waves of different frequencies within the spectrum and among waves within and outside the spectrum. This makes the subsequent process nonstationary with

A. Tørum and O. T. Gudmestad (eds.), Water Wave Kinematics, 163–183.
© 1990 Kluwer Academic Publishers.

wavelengths and frequencies much broader than the original.

Two examples will be discussed here. In the first (Section 2) we examine the evolution of deep water waves propagating in one direction only. The initial instability and the following transient modulations are shown. In the second (Section 3) diffraction of waves by a submerged ridge creates waves in several directions. Nonlinearity enhances long waves which can be resonated on the ridge.

Throughout this paper, the term deep water refers to kh > O(1) where finite depth to kh = O(1). For the North Sea, the typical design wave period is about 12 sec. From the dispersion relation ω^2 = gk tanh kh it can be shown that kh = 1.4 if h = 50 m but kh = 9 if h = 360 m. Therefore both deep water and finite depth are of practical interest.

2. Evolution of Deep Water Waves in One Direction

Dispersive waves with slightly different frequencies have slowly modulated envelopes which propagate at the group velocity: Cg = dω/dk. Let Δk be the measure of the narrow-bandedness. The group has the length scale 1/Δk, and Δk/k is the small ratio of wave length to group length. If nonlinearity is omitted, it is known that (i) the envelope propagates at the group velocity without change of form within a time range 1/$\Delta\omega$ or a space range 1/Δk. (ii) the envelope flattens slowly by dispersion after a much longer time t > O(1/$\Delta\omega$) or distance x > O(1/Δk). Thus there is no stationarity for sufficiently long time or distance.

If nonlinearity is small but not negligible (ϵ = kA << 1), then dispersion and nonlinearity compete with each other, so that waves can be unstable and groups do not simply flatten, in the time range t = O(1/$\Delta\omega$) or space range x = O(1/Δk). In any case, nonstationarity is the rule rather than the exception when one considers long time or long distances.

Most of the interesting physics occurs when Δk/k is of the same order as kA where A is the typical amplitude. Under this condition, the envelope is known to be governed by the nonlinear Schrödinger equation if $\epsilon^2\omega t$ = O(1) and $\epsilon^2 kx$ = O(1). Since the seminal work of Benjamin and Feir (1964) much has been learned from the solution of the nonlinear Schrödinger equation (see Yuen & Lake (1980) or Mei (1983) for reviews).

When kA increases or the propagation time or distance is sufficiently large, the Schrödinger equation is inadequate. For example, Feir (1967) has observed that initially symmetric and bell-shaped envelopes became forward-leaning after a short distance from the wavemaker. Further downstream two distinct groups emerged, with the smaller one trailing. More extensive experiments on wave packets of various durations have been performed by Su (1982), who revealed further information about group formation and splitting. In contrast, the Schrödinger equation, which is symmetric with respect to the spatial coordinate in the coordinate system moving with the group velocity, only predicts splitting into symmetric envelopes if they are symmetrical initially (Chu & Mei, 1971; Yuen & Ferguson, 1978). Furthermore, the groups that split off do not separate. In experiments involving initially uniform wavetrains (Lake et al., 1977; Melville, 1982), initially equal sideband disturbances are known to grow at equal rates only for a short time, after which the lower sideband grows faster, attaining a larger maximum than the upper sideband. Again, the Schrödinger equation fails to predict this unequal growth.

A theoretical limitation of the Schrödinger equation is that it is accurate only to the third order in kA. Dysthe (1979) has extended the time range by including the fourth order terms for deep water waves. In addition to introducing asymmetric terms, there is now also a term representing the feedback from the long waves. With these, Dysthe obtained a much better agreement in the linear stability analysis when compared with the exact results of Longuet-Higgins (1978). In light of this, it is natural to see how Dysthe's extension of the cubic Schrödinger equation affects the nonlinear stage of evolution.

In this section, we first outline the derivation of the governing equations. After assigning the relevant length and time scales, multiple scale analysis is introduced. The governing equations are then expanded to the fourth order in kA. Evolution equations governing the wave envelope and generated long waves are obtained. The derived equations are solved numerically, first for one-dimensional propagation. Finally, two-dimensional propagation in infinite depth is studied.

2.1. GOVERNING EQUATIONS

We assign (x,y,z) to be the usual rectangular coordinates with z being vertical. No atmospheric forcing is assumed. For the irrotational flow of an inviscid and incompressible fluid, the surface wave is described by a velocity potential $\phi(x,y,z,t)$ and a free surface displacement $\zeta(x,y,t)$

$$\nabla^2 \phi = 0 \qquad\qquad -h \leq z \leq \zeta \qquad (1)$$

$$\phi_z + \phi_x h_x + \phi_y h_y = 0 \qquad\qquad z = -h \qquad (2)$$

$$\phi_{tt} + g\phi_z + [|\nabla\phi|^2]_t + \frac{1}{2}\nabla\phi \cdot \nabla[|\nabla\phi|^2] = 0 \qquad z = \zeta \qquad (3)$$

where g is the gravitational acceleration and h is the depth of the sea bottom. The free surface displacement ζ is related to ϕ by the Bernoulli's equation

$$\phi_t + \frac{1}{2}|\nabla\phi|^2 + g\zeta = 0 \qquad\qquad z = \zeta \qquad (4)$$

It is well known that the case of weak nonlinearity and slow modulation (narrow bandedness) can be effectively treated by the method of multiple scales. Thus we let x be the direction of propagation of the waves, and define the following cascade of multiple variables in terms of ϵ.

$$\begin{aligned}
&x, x_1 = \epsilon x, x_2 = \epsilon^2 x, x_3 = \epsilon^3 x \\
&y, y_1 = \epsilon y, y_2 = \epsilon^2 y \\
&z, z_1 = \epsilon z \\
&t, t_1 = \epsilon t, t_2 = \epsilon^2 t, t_3 = \epsilon^3 t
\end{aligned} \qquad (5)$$

Note that (x,y,z,t) are the fast variables describing the short waves, while the rest are slow variables. The vertical scale z_1 is needed if the depth considered is large, and is of the order $(1/\epsilon)$.

We assume a topography that is essentially flat, and the waves essentially unidirectional. Let

$$S = kx - \omega t \tag{6}$$

be the phase function. The unknowns ϕ and ζ will be expanded in a perturbation series in ϵ and in harmonics of S:

$$\phi = \sum_{n=0}^{\infty} \epsilon^n \phi_n = \sum_{n=0}^{\infty} \epsilon^n \sum_{m=-n}^{m=n} \phi_{nm} e^{imS} \tag{7}$$

$$\zeta = \sum_{n=1}^{\infty} \epsilon^n \zeta_n = \sum_{n=1}^{\infty} \epsilon^n \sum_{m=-n}^{m=n} \zeta_{nm} e^{imS}$$

where ϕ_{nm} and ζ_{nm} are functions of the slow variables and z. For ϕ and ζ to be real, we require that

$$\phi_{nm} = \phi_{n,-m}^* , \quad \zeta_{nm} = \zeta_{n,-m}^* \tag{8}$$

where * denotes complex conjugates.

Note that the expansion for ϕ starts at n = 2. It can be shown that ϕ_{00} and ϕ_{10}, if kept, are functions only of the slow variables, and would represent the ambient weak currents not originated by waves. But such a weak current does not affect waves in important ways and is ignored here. The short waves are represented at leading order by ϕ_{11} which contributes at first order to the free surface through ζ_{11}. As is known (see e.g., Mei, 1983), the wave-induced currents correspond to ϕ_{n0}, $n \geq 1$ for $kh \sim O(1)$ and $n \geq 2$ for $kh \geq O(1/\epsilon)$.

The expansions (7) are substituted into the governing equations for ϕ (1)–(3), and into the Bernoulli Equation (4). The remaining procedure is lengthy but can be found in many sources (e.g., Mei, 1983). We only summarize the results for a bathemetry such that h is of order $(1/k\epsilon)$ throughout. Then the kinematic boundary condition (2) should be simplified to

$$\frac{\partial \phi_{nm}}{\partial z} = 0 \qquad\qquad z = -h \equiv -h_1/\epsilon \tag{9}$$

In fact, ϕ_{nm} for $m \geq 1$ will always be exponentially small.

At the first order n = 1, the potential is

$$\phi = \epsilon \phi_{11} e^{iS} + * \tag{10}$$

where

$$\omega^2 = gk \tag{11}$$

This is just the usual linear solution for the leading order short waves. Note that ϕ_{11} is a function of the slow variables x_1, x_2 as well as y and t, i.e., it varies over long propagation distances. The free surface is given by

$$\zeta_1 = \frac{1}{2}\left(A\, e^{iS} + *\right) \tag{12}$$

The wave-induced current is represented to leading order by ϕ_{20} which is governed by

$$\nabla_1^2 \phi_{20} = 0 \qquad\qquad -h_1 \leq z_1 \leq 0 \tag{13}$$

with the boundary conditions

$$g\frac{\partial \phi_{20}}{\partial z_1} = \frac{\omega^3}{2k}\frac{\partial}{\partial x_1}|A|^2 \qquad\qquad z_1 = 0 \tag{14}$$

$$\frac{\partial \phi_{20}}{\partial z_1} = 0 \qquad\qquad z_1 = -h_1 \tag{15}$$

After invoking the solvability of $O(\epsilon) \rightarrow O(\epsilon^4)$ for the first harmonic (m=1, n=1,2,3,4) we obtain the evolution equation for A at various orders. The results can be combined to give

$$\frac{\partial A}{\partial t_1} + \frac{\omega}{2k}\frac{\partial A}{\partial x_1} + \epsilon\left\{ i\frac{\omega}{8k^2}\frac{\partial^2 A}{\partial x_1^2} - i\frac{\omega}{4k^2}\frac{\partial^2 A}{\partial y_1^2} + \frac{\omega k^2}{2}i|A|^2 A \right\}$$

$$+ \epsilon^2\left\{ \frac{\omega}{8k^3}\left(3\frac{\partial^2}{\partial y_1^2} - \frac{1}{2}\frac{\partial^2}{\partial x_1^2}\right)\frac{\partial A}{\partial x_1} + \omega k|A|^2\frac{\partial A}{\partial x_1} - \frac{3}{4}\omega k A^2\frac{\partial A^*}{\partial x_1}\right.$$

$$\left. + k(i\frac{\partial \phi_{20}}{\partial x_1} + \frac{\partial \phi_{20}}{\partial z_1})A\right\} = O(\epsilon^3) \qquad\qquad z_1 = 0 \tag{16}$$

The terms proportional to ϵ are associated with the usual cubic Schrödinger equation, while the terms proportional to ϵ^2 correspond to Dysthe's extension. The latter group includes two linear dispersive terms, two nonlinear terms involving A, and two terms accounting for the effects of the wave-induced current ϕ_{20}. Having served its purpose ϵ, which can be regarded as an ordering parameter in (2.43), can be set to unity and the original coordinates restored.

For normalization, we use the typical properties of the first harmonic, the amplitude scale a_0 where ϵa_0 is the characteristic amplitude, the wave number k and the wave frequency ω, as the scales to define the dimensionless variables as follows:

$$A' = A/a_o \qquad (\phi'_{20}, \psi') = (\phi_{20}, \psi)/a_o^2$$

$$\xi = \epsilon\omega\left(\frac{x}{C_g} - t\right) \qquad \eta = \epsilon^2 kx \tag{17}$$

$$y', z' = \epsilon k(y, z) \qquad H = \epsilon kh$$

After normalization, we omit primes for brevity and obtain for the wave induced current

$$\left(4\frac{\partial^2}{\partial\xi^2} + \frac{\partial^2}{\partial y^2} + \frac{\partial^2}{\partial z^2}\right)\psi = 0 \qquad\qquad -H < z < 0 \tag{18}$$

$$\frac{\partial\psi}{\partial z} = \frac{\partial}{\partial\xi}|A|^2 \qquad\qquad z = 0 \tag{19}$$

$$\frac{\partial\psi}{\partial z} = 0 \qquad\qquad z = -H \tag{20}$$

while the wave amplitude A is governed by

$$\frac{\partial A}{\partial\eta} + i\frac{\partial^2 A}{\partial\xi^2} - \frac{i}{2}\frac{\partial^2 A}{\partial y^2} + i|A|^2 A$$
$$+ \epsilon\left[\frac{\partial^2 A}{\partial y^2 \partial\xi} + 8|A|^2\frac{\partial A}{\partial\xi} + 4iA\frac{\partial\psi}{\partial\xi}\bigg|_{z=0}\right] = 0 \tag{21}$$

Note that the uniform Stokes wave

$$A = e^{-i\eta}, \quad \psi = 0 \tag{22}$$

is a special solution.

2.2. ONE-DIMENSIONAL EVOLUTION OF A WAVE PACKET IN DEEP WATER

For one-dimensional propagation, the set of equations can be solved by a split-step pseudo-spectral method. Details of the scheme are reported in Lo & Mei (1985). Suffice it to mention that a domain of ξ is chosen according to the problem and periodicity in ξ over the computational domain is required.

Lo & Mei (1985) studied several examples of unidirectional waves. Among them is the evolution of a wave packet for which the following physical features are known experimentally (Feir (1967) and Su (1982)). In particular, a single wave packet, originally symmetrical fore and aft, may split into several wave packets in an asymmetrical way. New groups emerge on both sides of the original peak and are of different sizes and shapes. They also tend to separate from one another. Based on a set of equations which are equivalent to the cubic Schrödinger equation, Chu & Mei (1970) found theoretically the splitting of groups. However, for an originally symmetric group the side groups remain symmetric fore and aft. The analytical theory by Zakharov and Shabat (1972) of the cubic Schrödinger equation shows further that the evolution of these groups is cyclic in time and the groups do not separate. With the new terms due to Dysthe, we have simulated some of Su's experiments for wave packets of initially rectangular envelope. The following sets of initial inputs were considered: wave slope $\epsilon = 0.09$, number of waves in the packet $= 5, 10$. For each case we calculated A and ψ from which the free surface displacement can be found:

$$\zeta = \epsilon^2 \frac{\partial \phi}{\partial \xi} + \left(\frac{1}{2} A - i\epsilon \frac{\partial A}{\partial \xi} - \frac{3}{8} \epsilon^2 |A|^2 A \right) e^{iS}$$

$$+ - (A^2 - iA \frac{\partial A}{\partial \xi}) e^{2iS} + \frac{3}{16} \epsilon^2 A^2 e^{3iS} + * \tag{23}$$

We remark that the last term in Eq (21), which represents the Doppler effect of wave-induced current on waves, is critical to the initial side-band instability, but it is the new nonlinear term $8\epsilon |A|^2 \frac{\partial A}{\partial \xi}$ that is essential in the nonlinear stage of evolution. In particular, if A is symmetric in ξ at $\eta = 0$, this term will render A asymmetric at large time. This is clearly shown in our results (Fig 1). The average trend of frequency change decreases in the front group and increases in the rear group (see Fig 2). This implies that the group at the front travels faster than that at the rear, leading to eventual separation. Moreover, the calculated height of each group is close to the measured values. However, the predicted rate of spreading is not in good agreement with the measured values.

In the case of side-band instability of uniform Stokes waves this new nonlinear term in question leads to faster growth of the lower side-band at first, causing an apparent downshift of frequency. But for sufficiently small ϵ this trend can be reversed and quasi recurrence may follow (Lo & Mei, 1985). In available experiments breaking follows the fast initial growth of the lower side-band and obliterates the recurrence.

2.3. TWO-DIMENSIONAL MODULATIONAL INSTABILITY AND WAVE PACKET EVOLUTION

From equations (18)-(21), a linearized sideband instability analysis of the Stokes waves may be made by assuming

$$A = 1 + \tilde{b} \exp(-i\eta + i\delta), \quad \phi = \tilde{\phi} \tag{24}$$

with the disturbance

$$(\tilde{b}, \delta, \tilde{\phi}) = (b_0, \delta_0, \phi_0) \exp i(\alpha\xi + \beta y + \theta y) << 1 \tag{25}$$

where α and β are the wave number components of the disturbance along x and y directions, then it can be shown that

$$\theta = -\epsilon P_1 \pm \left\{ P_2 \left[P_2 - 2\left(1 - \frac{4\epsilon\alpha^2}{k} \coth kH \right)\right]\right\}^{\frac{1}{2}} \tag{26}$$

$$P_1 = \alpha\beta^2 - 8\alpha, \quad P_2 = \alpha^2 - \beta^2/2, \quad k = (4\alpha^2 + \beta^2)^{\frac{1}{2}} \tag{27}$$

If θ is complex, instability may occur. This is Dysthe's result (modified slightly for finite kH). The result according to the lower order Schrödinger theory (Benney & Roskes, 1969) is obtained by omitting the $O(\epsilon)$ term above; the region of instability in the α-β plane is then an open strip (see Fig 4). Martin & Yuen have shown that initial disturbances may grow to induce higher harmonics which may fall in this strip. This leads to many unstable modes and to chaos. With the new $O(\epsilon^2)$ terms, the region of instability is narrower (see Fig 4). According to an accurate numerical theory for arbitrary kA (McLean, 1982) or to the more complicated Zakharov's equation (1968), the instability region is curved also shown in Fig 3. McLean calls this the instability of Class I since there is another three-dimensional instability (Class II) for finite kA. Qualitatively this also has the effect of inhibiting the higher harmonics from entering the zone of instability. By an approximate solution of Zakharov's equation, Stiassnie & Shemer (1987) show that for low enough $\epsilon < 0.15$, Class I (two-dimensional) dominates over Class II for a long time.

Litvak et al. (1983) made a brief study of the asymptotic fate of an initially two-dimensional wave packet ($A(\xi,y,0)$ = given). By considering the second moments of an isolated wave packet

$$\bar{a}^2 = \iint \xi^2 |A|^2 \, d\xi \, dy \qquad \bar{b}^2 = \iint y^2 |A|^2 \, d\xi \, dy \tag{28}$$

they conclude analytically that an initial pulse always flattens. No information on the intermediate stage of development was given.

We now examine the evolution of a two-dimensional wave envelope. By the following rescaling

$$\xi = \frac{\xi'}{\gamma}, \quad y = \frac{y'}{\delta} \tag{29}$$

we fix the following two-dimensional pulse: the computational domain to be $\xi' \epsilon [0,2\pi]$, $y' \epsilon [0,2\pi]$. The initial envelope is described by

$$A(\xi',y',0) = \begin{cases} -[1+\cos(8\sqrt{(\xi'-\pi)^2+(y'-\pi)^2})] & \text{if} \quad 8\sqrt{(\xi'-\pi)^2+(y'-\pi)^2} < \pi \\ \\ 0 & \text{otherwise} \end{cases} \tag{30}$$

The center of the pulse is at $\xi' = \pi$ and $y' = \pi$ and A vanishes outside the circle

$$(\xi' - \pi)^2 + (y' - \pi)^2 = \left(\frac{\pi}{8}\right)^2 \tag{31}$$

which is well within the boundaries of the computational domain. Mapping back to ξ and η, the edge of the wave envelope is an ellipse

$$(\gamma\xi - \pi)^2 + (\delta y - \pi)^2 = \left(\frac{\pi}{8}\right)^2 \tag{32}$$

Thus the lengths of the axes are $\frac{\pi}{4\gamma}$ and $\frac{\pi}{4\delta}$ in ξ and y directions or $\frac{\pi}{4\gamma\epsilon k} = \frac{\lambda}{8\gamma\epsilon}$ and $\frac{\pi}{4\delta\epsilon k} = \frac{\lambda}{8\delta\epsilon}$ in physical coordinates. The ellipse contains $m_\xi = \frac{1}{8\gamma\epsilon}$ waves in the ξ direction and $m_y = \frac{1}{8\delta\epsilon}$ waves in the y direction. Our computations were made by specifying ϵ, γ and δ. In both cases discussed below, $\epsilon = 0.15$. In Fig 4 the initial wave packet is longer in the y' direction ($m_y = 30$) than in the ξ direction ($m_\xi = 20$). Therefore, much of the center part evolves essentially two-dimensionally. But fore and aft asymmetry in the ξ direction develop so that crescent-shaped wave packets appear. In comparison the less accurate Schrödinger equation gives only a symmetric evolution as shown in Fig 5.

Other examples of this type of study can be found in Lo & Mei (1985b).

3. Long Wave Generation by Short Waves Scattered by a Submarine Ridge

So far we have only considered unidirectional waves. However, when there are reflected and incident short waves, their envelopes will also propagate in different directions. Over a long time or distance ($\epsilon^2\omega t = O(1)$ or $\epsilon^2 kx = O(1)$) the nonlinear, transient coupling of these envelopes can be important. However, little is known on these couplings when there is scattering. Nevertheless, considerable work exists on the physics near and within $\epsilon kx = O(1)$ from the scatter. Of particular engineering interest (in slow drift oscillations, harbor oscillations due to wind waves, etc.) are the long waves which can be induced by nonlinear interaction of the short waves. As an example, Agnon & Mei (1988) examined the possible resonance of long waves on a ridge when a train of short waves arrives obliquely from one side. We shall explain the ideas of their analysis for the simpler case of a step shelf.

We first recall that for a step shelf with a depth discontinuity along the y-axis ($h = h'$, $x < 0$; $h = h'$, $x > 0$). The wave field generated by an obliquely incident long wave with wave number vector $\vec{k} = (\alpha, \beta)$ from $x \sim -\infty$ is

$$\eta_1 = A(e^{i\alpha x} + R\, e^{-i\alpha x})\, e^{i\beta y} \qquad\qquad x < 0$$

$$\tag{33}$$

$$\eta_2 = A\, T\, e^{i\alpha' x + i\beta y} \qquad\qquad x > 0$$

where

$$\alpha^2 + \beta^2 = \frac{\omega^2}{gh} \qquad \alpha'^2 + \beta^2 = \frac{\omega^2}{gh'} \qquad (34)$$

The reflection coefficient is

$$R = \frac{\alpha h - \alpha'h'}{\alpha h + \alpha'h'} \qquad (35)$$

The inclinations of the incident and transmitted waves are respectively

$$\tan\theta = \beta/\alpha \quad \text{and} \quad \tan\theta' = \beta/\alpha' \qquad (36)$$

If

$$k^2 > \beta^2 > k'^2 \quad \text{or} \quad \frac{\omega^2}{gh} > \beta^2 > \frac{\omega^2}{gh'} \qquad (37)$$

θ is real while θ' is imaginary, there are then no transmitted waves, i.e., the reflection is complete. This happens only if $h < h'$. As a corollary, if there is a ridge with depth h, $|x| > L$ and depth h', $|x| < L$, then long waves can be trapped on the ridge. These long waves may be excited by wind or by transient inputs, but not by long waves incident from outside the ridge according to a linearized theory.

The nonlinear theory with short wave diffraction can be outlined as follows. First we examine two trains of monochromatic short waves incident on the shelf. When their frequencies and wave numbers are slightly different, they form periodic groups with very long modulational periods and wavelength. The scattering coefficients for each wave train are known from the linearized theory. By nonlinear interactions these short waves force time-periodic long waves which are bound to the short wave groups. In addition free long waves are radiated from

the step and propagate at the speed $\sqrt{gh}$ or $\sqrt{gh'}$. The amplitudes of these free long waves are determined by continuity conditions at the step. Being periodic in time, this long wave solution can be used to construct slowly varying transient solution by Fourier superposition.

The mathematics is facilitated by the combined use of the methods of multiple scales and matched asymptotics. Referring to Fig 6, we define the near field of a step as the region with a length scale $O(k^{-1})$. Within the near field the potential depends on x, z, t, y and t_1, but is independent of τ_1, i.e.,

$$\Phi = \psi(x,y,z,t/y_1,t_1) \qquad (38)$$

In this near field both propagating and evanescent modes of the short waves are present. In the far field the potential depends in addition on x_1

$$\Phi = \phi(x,y,z,t/x_1,y_1,t_1) \qquad (39)$$

Only the propagating modes of the short waves are effective and the ray theory can be used. Finally ϕ and ψ are matched such that

$$\phi(kx_1 << 1) = \psi(kx >> 1) \tag{40}$$

We let the first order complex amplitude of two trains of sinusoidal short waves be

$$A^{\pm} e^{\mp i\Omega(t_1 - \mu x_1 - \nu y_1)} \tag{41}$$

where $A^{\pm}$ are the amplitudes of potentials, $2\epsilon\Omega$ is the frequency separation and $2\epsilon\Omega\mu$ and $2\epsilon\Omega\nu$ are the separation of wave number components. Then the incident and reflected short waves far to the left of the shelf have the total potential

$$\psi_{11} = [A\, e^{i(\alpha x + \gamma y - \omega t)} + B\, e^{i(-\alpha x + \gamma y - \omega t)}]\, f_o(z) \tag{42}$$

with

$$f_o(z) = \sqrt{2} \cosh k(z+h) \,/\, [1 + \sinh^2 kh]^{\frac{1}{2}} \tag{43}$$

$$
\begin{aligned}
A &= A^+ e^{-i(\Omega t_1 - \mu x_1 - \nu y_1)} + A^- e^{i(\Omega t_1 - \mu x_1 - \nu y_1)}\\[4pt]
B &= B^+ e^{-i(\Omega t_1 + \mu x_1 - \nu y_1)} + B^- e^{i(\Omega t_1 + \mu x_1 - \nu y_1)}
\end{aligned} \tag{44}
$$

The potential of the transmitted short wave far to the right is

$$\psi_{11} = D\, e^{i(\alpha' x + \gamma y - \omega t)}\, f_o(z) \tag{45}$$

with

$$D = D^+ e^{-i(\Omega t_1 - \mu' x_1 - \nu y_1)} + D^- e^{i(\Omega t_1 - \mu' x_1 - \nu y_1)} \tag{46}$$

If the envelope and the short waves are in the same direction, then

$$\mu Cg = \alpha/k \quad\text{and}\quad \nu Cg = \beta/k \tag{47}$$

The reflection R and transmission T coefficients are known from the linearized theory so that

$$B = AR, \quad D = AT \tag{48}$$

By quadratic interaction, the long wave potential $\phi_{10}(x_1, y_1, t_1)$ is governed by

$$\frac{\partial^2 \phi_{10}}{\partial t_1^2} - gh\, \nabla_1^2\, \phi_{10} = -f_o(0)\,(k^2 - \sigma^2 + \frac{2\omega}{Cg})\,\frac{\partial}{\partial t_1}\,[|A|^2 + |B|^2] \tag{49}$$

far to the left of the step and by

$$\frac{\partial^2 \phi_{10}}{\partial t_1^2} - gh' \, \nabla_1^2 \, \phi_{10} = -f_o'(0) \, (k'^2 - \sigma^2 + \frac{2\omega}{Cg}) \, \frac{\partial}{\partial t_1} \, [|D|^2] \tag{50}$$

far to the right of the step. The forcing terms on the right are related to the amplitudes of the short waves. The forcing terms on the right hand sides of (49) and (50) have steady mean and transient parts which oscillate at the frequency $2\epsilon\Omega$. Sorting out the periodic part we get the set-down long wave bound to the envelope of the short waves. Thus we have on the left of the step the bound wave potentials:

$$\phi_{10}^A + \phi_{10}^B = \text{Re} \; \mathcal{A} \; e^{-2i\Omega(t_1 - \mu x_1 - \nu y_1)} + \text{Re} \; \mathcal{B} \; e^{-2i\Omega(t_1 + \mu x_1 - \nu y_1)} \tag{51}$$

where

$$\mathcal{A} = (\frac{m}{-2i\Omega}) \, A^+ A^{-*} \qquad \mathcal{B} = (\frac{m}{-2i\Omega}) \, B^+ B^{-*}, \tag{52}$$

and on the right of the step

$$\phi_{10}^D = \text{Re} \; \mathcal{D} \; e^{-2i(\Omega t_1 - \mu' x_1 - \nu y_1)} \tag{53}$$

with

$$\mathcal{D} = (\frac{m}{-2i\Omega}) \, D^+ D^{-*} \tag{54}$$

In addition, we must add free waves, which are the homogeneous solutions of (49) and (50), on the left of the step

$$\phi_{10}^G = \text{Re} \; \mathcal{G} \; e^{-2i\Omega(t_1 - \nu y_1)} \cos 2\Omega \beta x_1 \qquad\qquad x_1 < 0 \tag{55}$$

where

$$\beta^2 + \nu^2 = (gh)^{-1} \tag{56}$$

and on the right of the step

$$\phi_{10}^H = \text{Re} \; \mathcal{H} \; e^{-2i\Omega(t_1 - \nu y_1)} \, e^{-2\Omega \beta' x_1} \qquad\qquad x_1 > 0 \tag{57}$$

with

$$-\beta'^2 + \nu^2 = (gh')^{-1} \tag{58}$$

If β and β' are both real then ϕ^H decays exponentially in x_1 as $x_1 \sim \infty$ and ϕ_{10}^G is a free wave reflected to the left of the step. Therefore this free wave is trapped on the shelf. Now the sum of free and bound waves must match across the step through the near field. We first match the potential. It can be shown that

$$\psi_{10} = \psi_1(y_1, t_1) \tag{59}$$

so that $(\phi_{10})_{left}$ and $(\phi_{10})_{right}$ can be matched directly to the leading order. To match the fluxes, we have to use the fact that the net flux across from the near field is:

$$\int_{S_\infty^-} \frac{\partial \psi_{20}}{\partial x}\, dz - \int_{S_\infty^+} \frac{\partial \psi_{20}}{\partial x}\, dz = hU_- - h'U_+ \tag{60}$$

where U_- and U_+ are the Stokes drift, easily calculated for each short wave train. Each term on the left-hand-side above is then matched to $\frac{\partial \phi_{10}}{\partial x_1}$. The coefficients $\mathscr{G}$ and $\mathscr{H}$ are thus determined. Details are omitted but can be found in Agnon & Mei (1988).

Based on this approach, Agnon & Mei examined sinusoidal long waves on a submarine ridge which has the depth h' within $|x| < L$ in a sea of constant depth h. The natural frequencies $\epsilon \Omega_n$ of trapped modes on a shelf of width 2L is

$$2\Omega_n \beta' \epsilon L = \tan^{-1} \frac{\beta h}{\beta' L} + \frac{n\pi}{2} \qquad\qquad n = 0, 1, 2, \dots \tag{61}$$

The results for sinusoidal modulation and sinusoidal long waves can be used to give transient response by Fourier transform. Fig 7a shows the incident envelope which is itself a modulated sinusoidal signal with a Gaussian envelope. It is interesting that as the incident wave passes, there are persistent oscillations in the wake much like the effect of a passing ship in a channel (Fig 7b). These resonances should be of interest in platform design on a ridge.

ACKNOWLEDGMENT The authors have been supported by the U.S. Office of Naval Research and the U.S. National Science Foundation.

REFERENCES

Agnon, Y. & Mei, C.C. (1988) "Trapping and resonance of long shelf waves due to groups of short waves", *J. Fluid Mech.* 195, 201-221.

Benney, D.J. & Roskes, G.J. (1969) "Wave instabilities", *Studies Appl. Math.* 48, 377-385.

Chu, V.H. & Mei, C.C. (1971) "The nonlinear evolution of Stokes waves in deep water", *J. Fluid Mech.* 47, 337-351.

Dysthe, K.B. (1979) "Note on a modification to the nonlinear Schrödinger equation for application to deep water waves", *Proc. Roy. Soc. Lond.* A 369, 105-114.

Feir, J.E. (1967) "Discussion: Some results from wave pulse experiments", *Proc. Roy. Soc. Lond.* A **299**, 54–58.

Lake, B.M., Yuen, H.C., Rungaldier, H. & Ferguson, W.E. (1977) "Nonlinear deep water waves: Theory and experiment. Part 2: Evolution of a continuous wave train", *J. Fluid Mech.* **83**, 49–74.

Litvak, A.G., Petrova, T.A., Sergeev, A.M. & Yunakovskii, A.D. (1983) "On the self-effect of two-dimensional gravity wave packets on the deep water surface", in R.Z. Sagdeev (ed.), *Nonlinear and Turbulent Processes in Physics*, vol. 2. G&B/Harwood.

Lo, E. & Mei, C.C. (1985a) "A numerical study of water-wave modulation based on a higher order nonlinear Schrödinger equation", *J. Fluid Mech.* **150**, 395–416.

Lo, E. & Mei, C.C. (1985b), "Long-time evolution of surface waves in coastal waters", Technical Report No. 303: MIT Department of Civil Engineering.

Longuet-Higgins, M.S. (1978) "The instabilities of gravity waves of finite amplitude in deep water. Part II: Subharmonics", *Proc. Roy. Soc. Lond.* A **360**, 489–505.

Martin, D.U. & Yuen, H.C. (1980) "Quasi-recurring energy leakage in the two-dimensional nonlinear Schrödinger equation", *Phys. Fluids* **23**, 881–883.

McLean, J.W. (1982a) "Instabilities of finite-amplitude water waves", *J. Fluid Mech.* **114**, 315–330.

Mei, C.C. (1983) *The Applied Dynamics of Ocean Surface Waves*, Wiley Interscience.

Melville, W.K. (1982) "The instability and breaking of deep-water waves", *J. Fluid Mech.* **115**, 165–191.

Stiassnie, M. & Shemer, L. (1987) "Energy computations for evolution of Class I and II instabilities of Stokes waves", *J. Fluid Mech.* **174**, 299–312.

Su, M.Y. (1982) "Evolution of groups of gravity waves with moderate to high steepness", *Phys. Fluids* **27**, 2595–97.

Yuen, H.C. & Ferguson, W.E. (1978a) "Relationship between Benjamin-Fier instability and recurrence in the nonlinear Schrödinger equation", *Phys. Fluids* **21**, 1275–77.

Yuen, H.C. & Lake, B.M. (1980) "Instabilities of waves on deep water", *Ann. Rev. Fluid Mech.*

Zakharov, V.E. (1968) "Stability of periodic waves of finite amplitude on the surface of a deep fluid", *J. Appl. Mech. Tech. Phys.* (Engl. Trans.) **2**, 190–194.

Zakharov, V.E. & Shabat, A.B. (1972) "Exact theory of two-dimensional self-focusing and one-dimensional self-modulation of waves in nonlinear media", *Sov. Phys.* JETP **34**, 62-69.

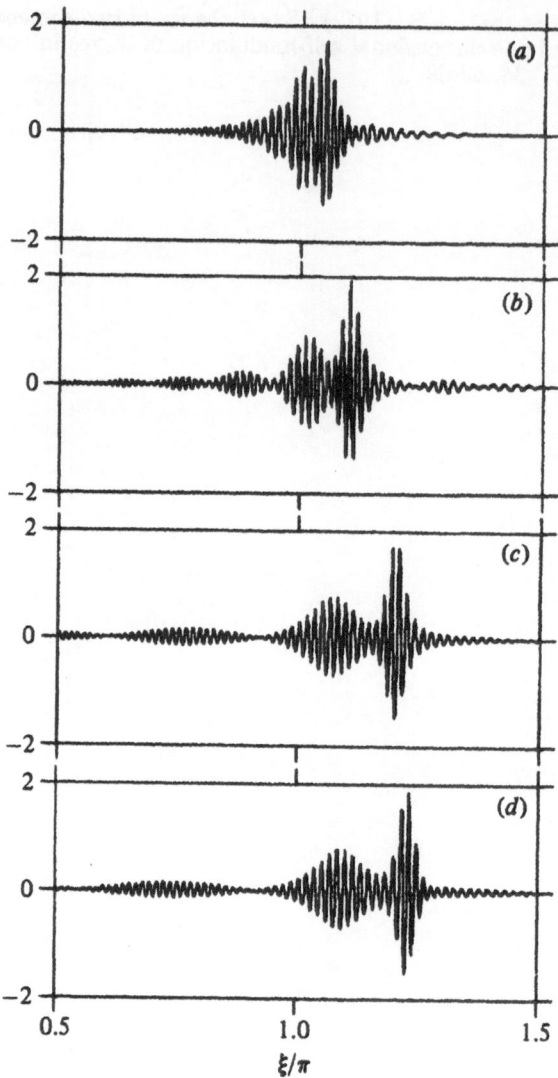

Figure 1 Computed ζ of short wave packet at the measuring stations of Su's experiments. The probe stations x_0 are (a) 18.3m; (b) 42.7m; (c) 91.5m; (d) 106.7m, from the wavemaker. There are 10 waves in the original rectangular packet with central frequency $\omega/2\pi = 0.96$ Hz and wave steepness $\epsilon = 0.09$.

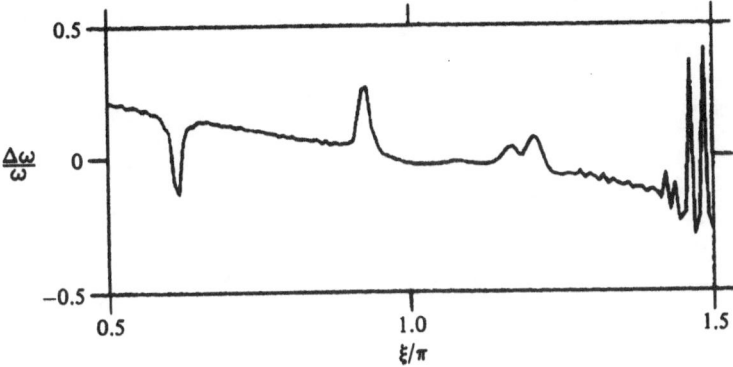

Figure 2 Frequency shift at $x_0 = 106.7$m.

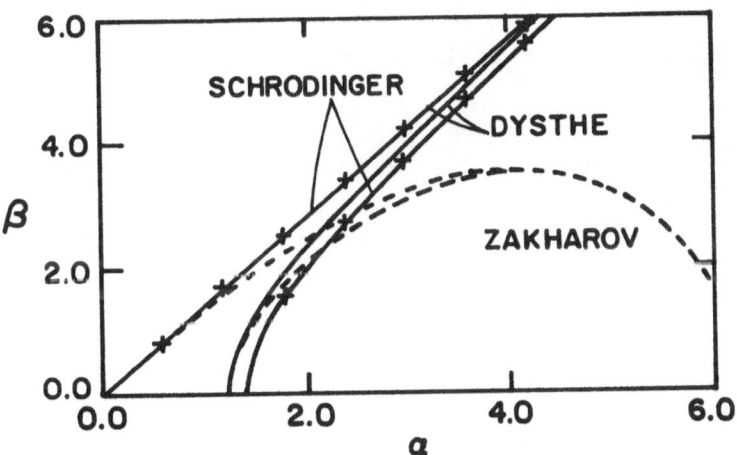

Figure 3 Instability region according to 3 theories.

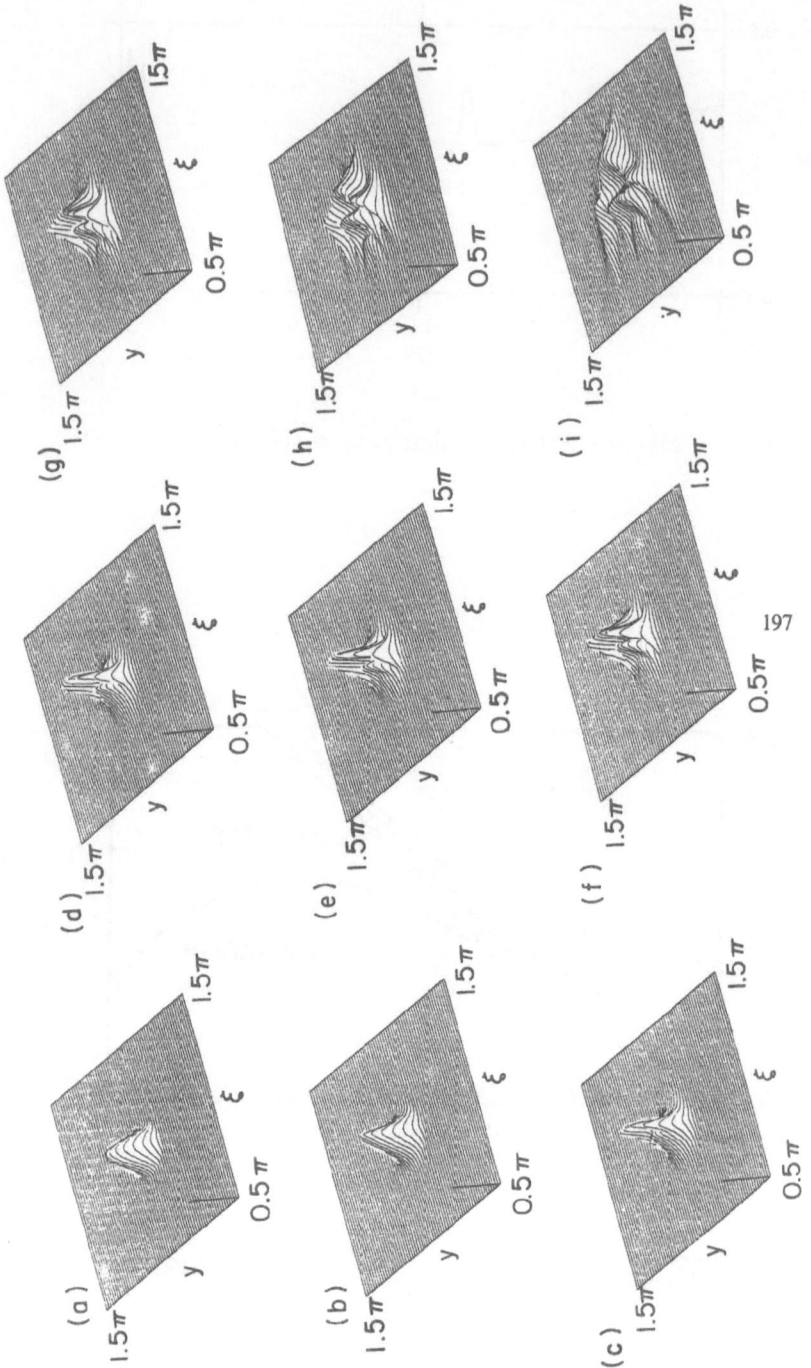

Figure 4 Evolution of a 2-D solitary envelope according to Eq (2.1).

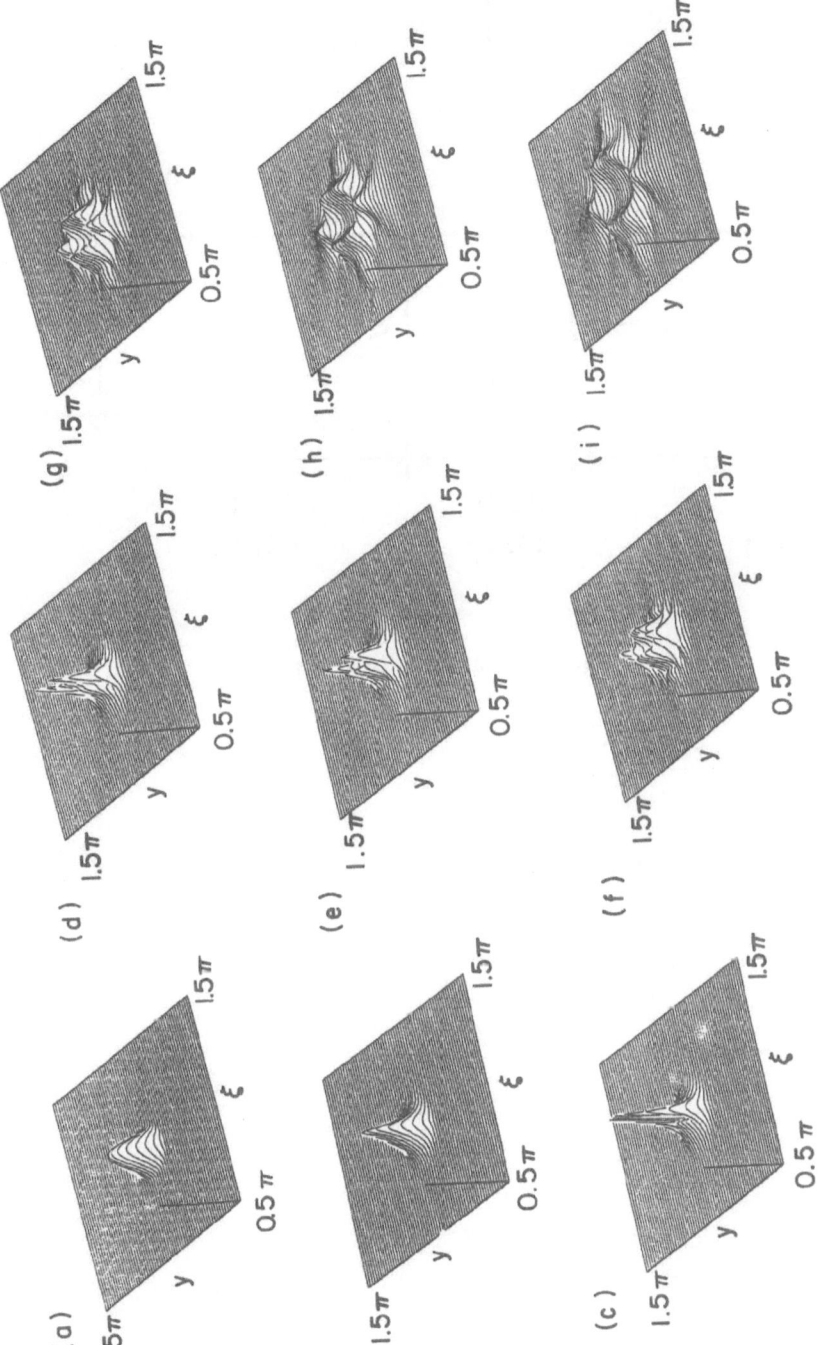

Figure 5 Evolution of a 2-D solitary envelope according to Schrödinger equation.

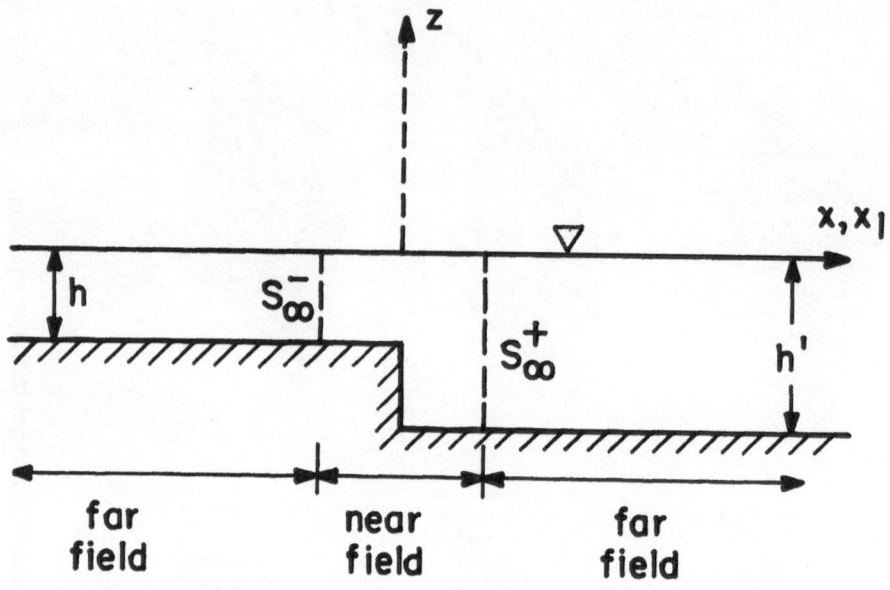

Figure 6 Scattering by a step.

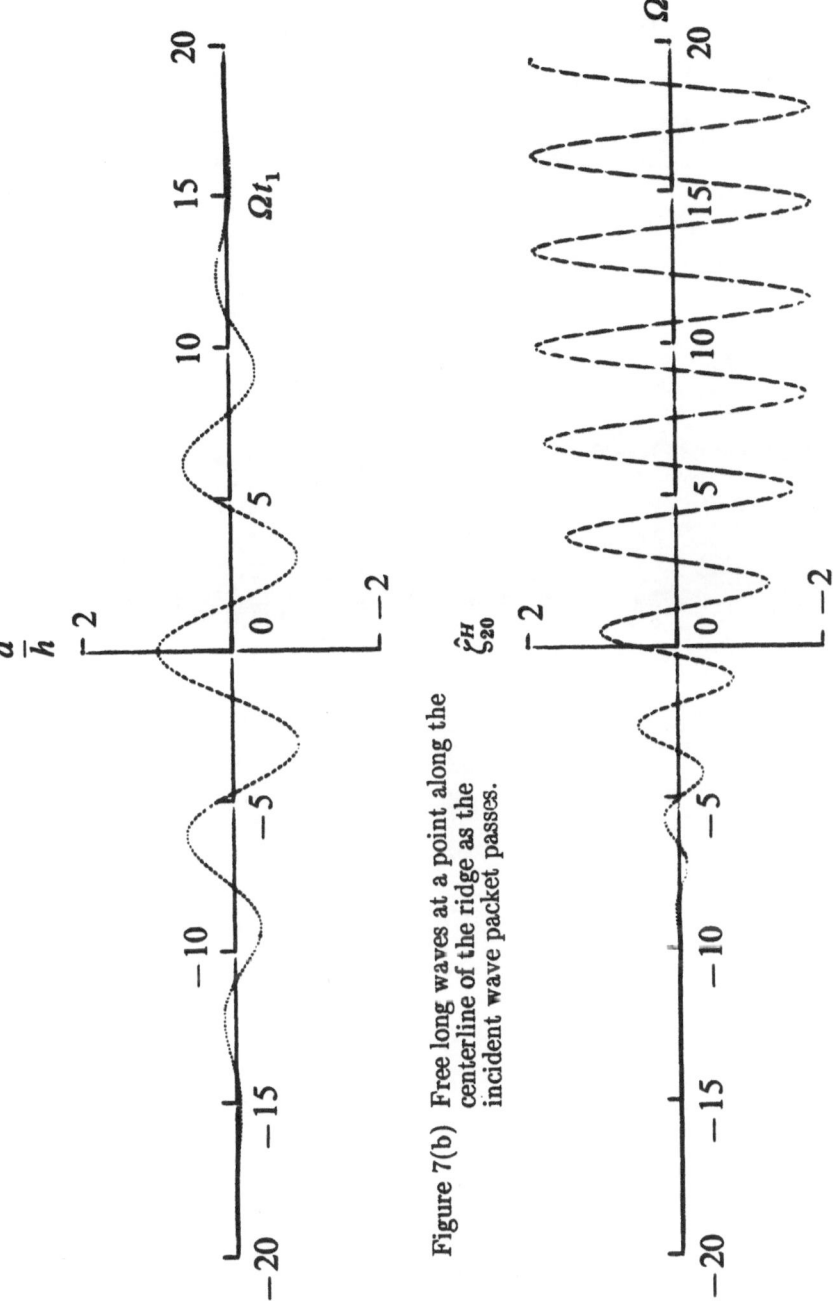

Figure 7(a) Envelope of incident short waves.

Figure 7(b) Free long waves at a point along the
centerline of the ridge as the
incident wave packet passes.

NEAR SURFACE IRREGULAR WAVE KINEMATICS

P. KLINTING and V. JACOBSEN
Danish Hydraulic Institute
Offshore Structures Division
Agern Allé 5
2970 Hørsholm
Denmark

ABSTRACT. Accurate assessments of the wave kinematics in the near surface zone are extremely important for obtaining reliable estimates of
hydrodynamic loads and response of offshore structures. Measurements
to verify kinematics models in this area are, however, very rare due
to the inherent problems. A model test program including measurements
and analyses of near surface kinematics in irregular waves is presented. The model set-up provided continuous time series of wave kinematics which have been compared with predictions from 2nd order irregular wave theory.

1. INTRODUCTION

Knowledge about wave kinematics can be obtained from different sources
such as field measurements and model tests. Since the early seventies,
several offshore structures have been equipped with environmental monitoring systems which in general focus on surface elevation and current velocities in the upper region of the water column. Analyses of
records taken during storm conditions have been applied to evaluate
the applicability of commonly used wave theories. Field measurements
of kinematics are, however, in general limited to levels beneath the
lowest expected trough as most instruments need to be fully submerged
to provide accurate recordings. This also applies to most model basin
measurements.

For estimating design loads and response of offshore structures,
accurate assessment of the kinematics in the region from wave trough
to wave crest is a key issue, since a major part of the impact results
from this region.

For the static, maximum single wave design, various wave theories
may be used for estimating the near surface kinematics, such as
Stokes' fifth order theory and Dean's stream function theory.

When time series of waves are applied, near surface kinematics
may be found using linear theory as a basis with various stretching
methods to achieve estimates of kinematics above mean sea level. Vertical, Delta, Gudmestad and Wheeler stretchings have been applied,

185

A. Tørum and O. T. Gudmestad (eds.), Water Wave Kinematics, 185–200.
© 1990 *Kluwer Academic Publishers.*

(Rodenbusch and Forristall, Gudmestad, Wheeler) and these stretchings may be used for 2-D as well as 3-D waves.

Stretching methods appear reasonably sound from an engineering point of view. Few data are, however, available to substantiate their applicability in the region above mean water level.

It is well known that real waves possess bound harmonics, (Dean and Sharma, Mansard et.al., Sand and Mansard). The superharmonics tend to add to the wave crests, but as they travel with the celerity of the basic waves, the associated velocities cannot be correctly derived from a linear wave approach. This fact is not considered when stretching methods are applied.

This paper presents measurements and analysis of near surface kinematics for two irregular sea states generated in the offshore basin at Danish Hydraulic Institute. The measurements have been compared with predictions based on 2nd order theory for irregular waves, i.e. the total elevation has been separated into the basic linear part and the superharmonics for the calculation of velocities. Comparison with a traditional Wheeler stretching is also included.

2. THE MODEL SET-UP

2.1. Principle of Operation

Most current meter probes need to be fully and continuously submerged. In order to measure the kinematics near an oscillating surface, the current probe may be moved in a pattern similar to the elevation pattern.

This principle has been applied in the test programme. The set-up is shown in Fig. 1. It consists of: Current meter, movable frame, hydraulic piston, wave gauges and amplifiers. The current meter is mounted on the movable frame, the movement of which is controlled by the hydraulic piston. The piston movement is controlled by the amplifier in a servoloop, which ensures that the movement is identical to the surface level recorded by the wave gauge held in a fixed position just upstream the current meter.

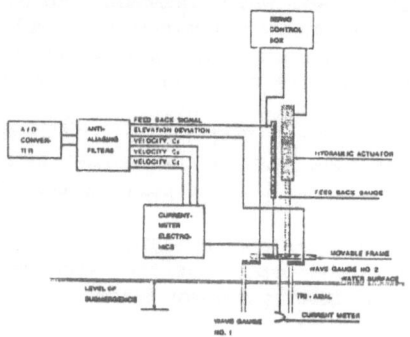

Figure 1. Sketch of set-up with movable current meter.

The current meter is thus kept at a fixed level below the instantaneous sea surface giving continuous velocity recordings. By varying the vertical position of the current meter in subsequent runs with identical wave time series, the velocity variation can be determined as function of depth as well as time.

2.2. Current Meter:

The triaxial ultrasonic current meter was of the type Minilab SD 12.

It was calibrated dynamically, using the movable frame in a horizontal position. The set-up was subjected to forced irregular movements in calm water where the relative velocity could easily be determined. The calibration included the determination of frequency dependent gain and phase functions which are influenced by the time constants inherent in the probe and in the current meter electronics.The measured signals were corrected for the influence of time constants, using specially designed digital filters.

2.3. Test Basin:

The measurements were made in DHI's deep water test facility. This basin has the dimensions 20 x 30 x 3 m. One of the long sides is equipped with 60 individually controlled wave flaps, see Fig. 2. The wave generator is capable of producing uni-directional as well as multi-directional irregular seas. The maximum significant wave height is of the order of 0.25 m and the peak periods are in the range from 1 to 2.3 secs.

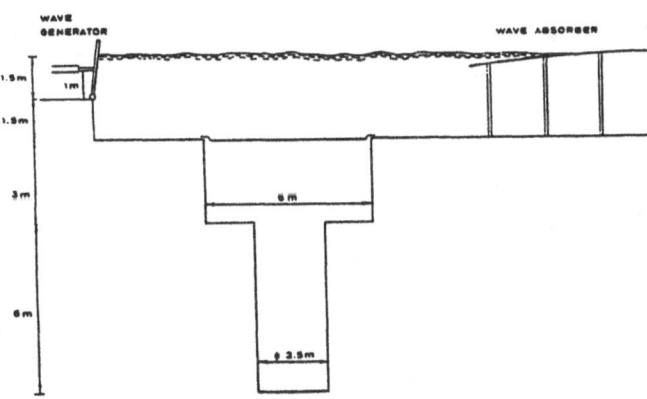

Figure 2. The deep water wave basin.

2.4. Wave Conditions

The wave conditions were calibrated prior to the kinematic measurements, using wave gauges in the position where the current meter was located during the measurements of the kinematics. Reuse of the digital control signals for the wave generator ensured a high degree of repeatability.

The control signals were calculated, using linear wave theory. Uni-directional as well as directionally spread seas were applied in the program.

This paper discusses the uni-directional wave time series characterized by:

Jonswap spectra with peak enhancement factor $\gamma = 3.3$; significant wave heights of $H_{mo} = 0.25$ m and 0.20 m with peak period of $T_p = 2.2$ s and 1.5 s, respectively.

The two wave trains have different wave steepness, namely: 3.3% and 5.7%, respectively. Accordingly, the contents of higher harmonic contribution will differ significantly in the two wave trains.

The duration of the irregular sea tests was 15 minutes.

3. DATA COLLECTION and ANALYSES

3.1. Data Collection

The wave induced velocities in different levels were obtained through repetition of the tests, and placing the probe in different vertical positions. The recording levels were 6, 9, 12 and 18 cm below the sea surface. The measured velocity time series were joined into one file containing apparently simultaneous velocity readings in different depths. Due to small disturbances the paths of the current meter movements in the repeated test were not exactly identical. However, the actual positions were recorded by the feedback on the hydraulic piston and the exact levels are used when comparing the measured velocities with the theoretically derived series. All signals were logged at a scanning rate of 20 Hz.

3.2. Theoretical Series

3.2.1. Surface elevation

The anticipated linear wave elevation time series were used for the generation of the control signals for the wave actuators. However, wave-wave interaction produces 2nd order bound harmonics and the wave generators constitute a boundary which generates free harmonics. The measured surface elevations thus contain the linear waves plus the bound and free harmonics. In the analysis, the bound superharmonics have been separated from the total elevation, and the two parts have been applied in the velocity calculations.

3.2.2. Kinematics

From the measured surface elevation, the wave kinematics in different fixed levels can be found including the contributions from the bound superharmonic waves. After having calculated the wave kinematics in closely spaced constant levels ($\Delta Z = 0.04$ m), the theoretical horizontal velocities in the varying levels were found applying interpolation along the vertical axis for each time step. The actual elevation time series for each run was used in this calculation.

In the calculation to 2nd order of the elevation series and the velocity series the frequency range has been limited to the interval between 0.5 f_p and 2.0 f_p for the linear waves and up to 2.5 f_p for the bound harmonics. This limitation has almost no effect on the calculated elevation series, but it reduced the effect of noise in the calculation of the velocities which are sensitive to high frequency components.

Short-periodic wavelets propagating on top of much longer waves may, so to say, feel the long-periodic surface elevation as the mean water level, see Gudmestad (1989). According to this, the contribution of the short-periodic wave to the near surface kinematics should be proportional to the amplitute of the wavelet itself, independent of the magnitude of the long waves. When representing mathematically the waves as series expanding from the mean water level, the contribution from short-periodic waves will be extrapolated to levels beyond their own amplitude, thus producing unrealistically high velocities. The applied frequency range for the velocity calculations tends to avoid this situation, which is estimated to occur for wave lengths shorter than one fourth of the long wave lengths.

3.3. DATA ANALYSES

3.3.1. Crossing Analyses of Velocities

Crossing analyses have been made simultaneously on the theoretical and measured velocity series. The maximum velocities beneath the crests and the troughs have been extracted, and the measured velocities are compared with the theoretical velocities. The full test duration (15 minutes) is included corresponding to approximately 500 individual waves. A level cut-off has been introduced excluding the smallest waves from the analyses.

3.3.2. Spectral Analyses of Velocities

Cross-spectra between the measured and the calculated velocities have been prepared for the four different recording levels. The cross-spectra form the basis for calculation of gain and phase functions.

4.   PRESENTATION AND DISCUSSION OF THE RESULTS

4.1. Time Series

Fig. 3 shows an extract of recorded elevation and velocity time series, and it is observed that the form of the two compares very well.

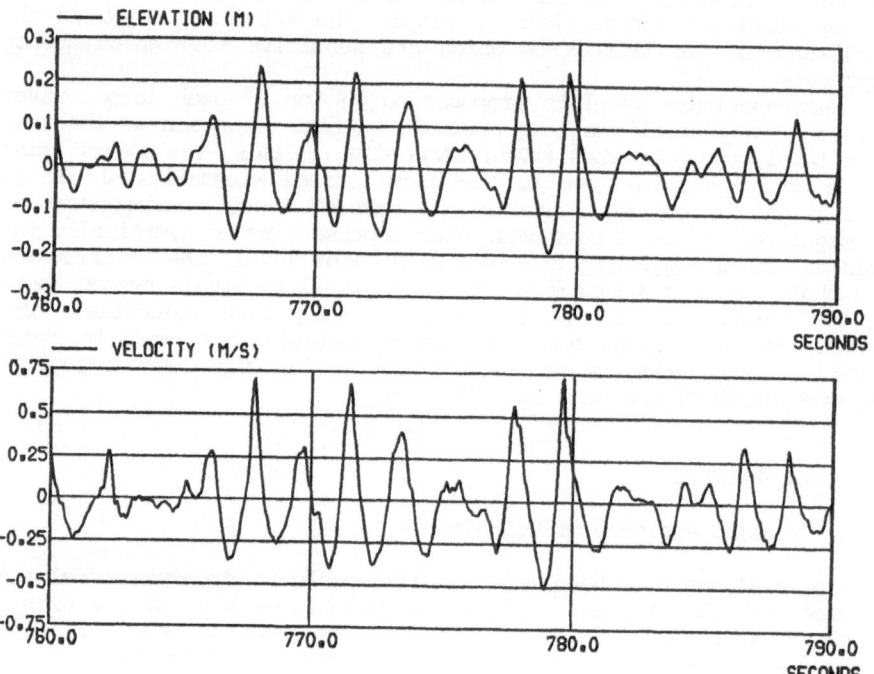

Figure 3. Recorded surface elevation and velocity (at a level of 12 cm below surface).

Measured and predicted velocities in two levels (6 and 18 cm below sea surface) are shown in Figs. 4 and 5.

The predicted and measured time series follow each other quite well, with some minor discrepancies. The recorded velocity amplitudes under the crests (positive velocities) appear to be slightly smaller than the predicted ones. For the troughs, the measured and predicted velocity amplitudes seem to be of the same magnitude although with some variation.

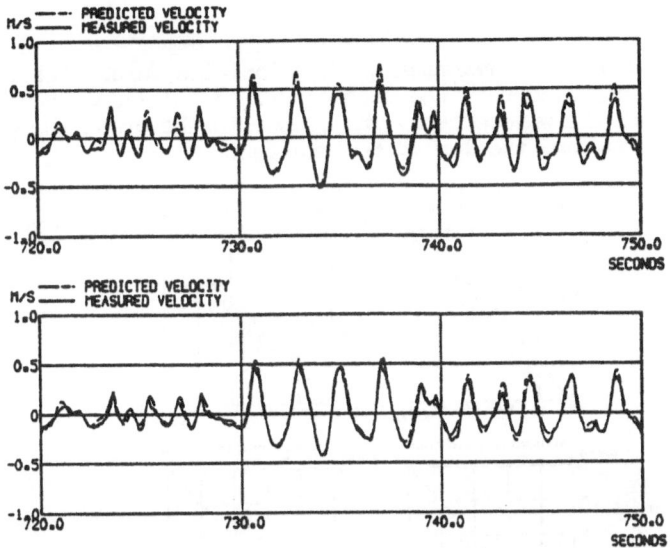

Figure 4. Measured and predicted velocities 6 and 18 cm below sea surface. H_s = 0.25 m, T_p = 2.2 s.

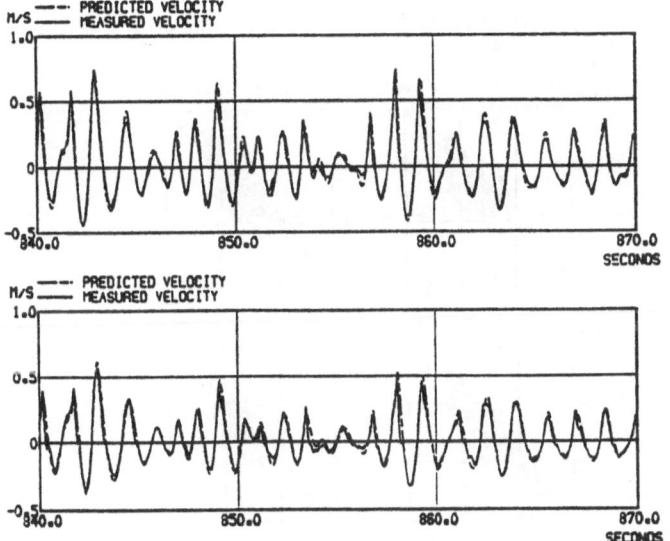

Figure 5. Measured and predicted velocities 6 and 18 cm below sea surface. H_s = 0.2 m, T_p = 1.5 s.

4.2. Velocity Amplitudes

Figs. 6 to 9 show the measured peak velocities versus the theoretical ones. It is observed that the theoretical calculations overpredict the velocities at the wave crests. The closer to the crest the greater the overprediction, and the larger the scatter.

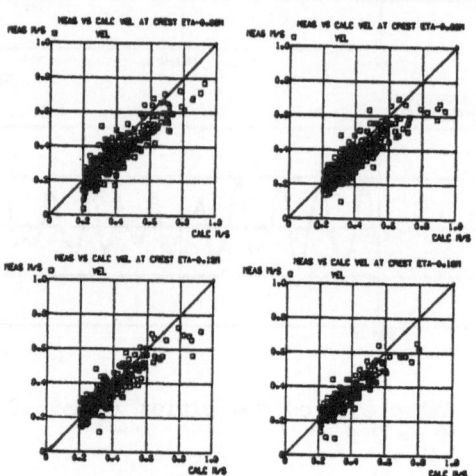

Figure 6. Peak velocities under crests. Measured vs calculated. $H_s = 0.25$ m, $T_p = 2.2$ s.

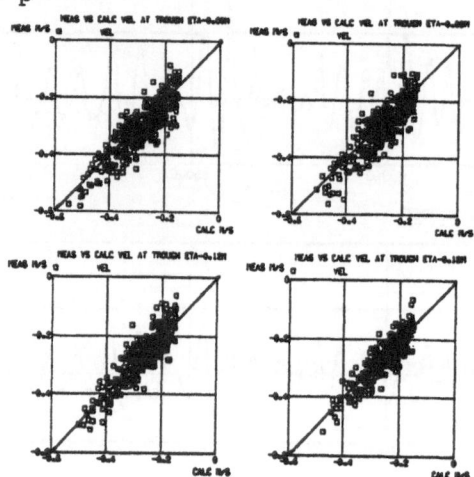

Figure 7. Peak velocities under troughs. Measured vs calculated. $H_s = 0.25$ m, $T_p = 2.2$ s.

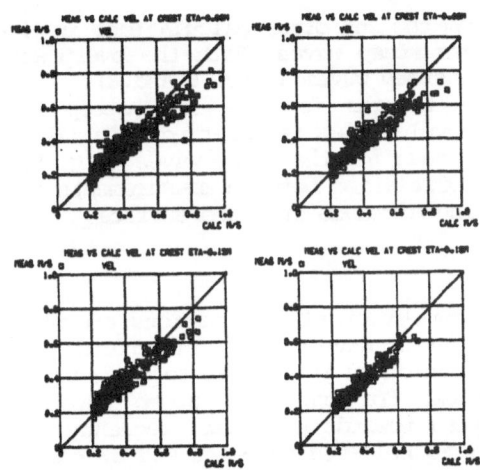

Figure 8. Peak velocities under crests. Measured vs calculated.
$H_s = 0.2$ m, $T_p = 1.5$ s.

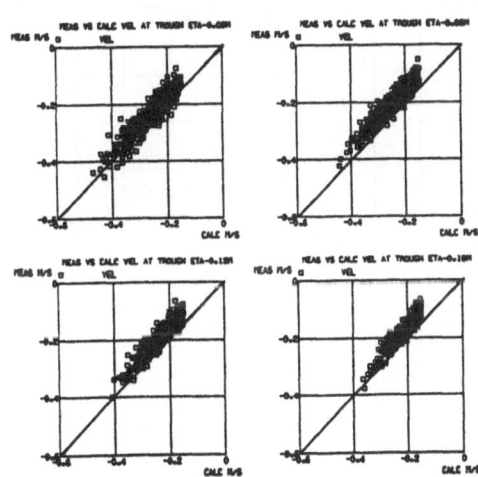

Figure 9. Peak velocities under troughs. Measured vs calculated.
$H_s = 0.2$ m, $T_p = 1.5$ s.

The overprediction from the 2nd order calculations in the crest region is greater for the less steep waves. Both the overprediction and the scatter reduce for the steep waves (Figs. 6 and 8).

Beneath the wave troughs for the less steep waves, a slight underprediction of the velocities is observed (Fig. 7) while an almost perfect correlation is seen for the steep waves (Fig. 9).

In general, the correlation between measurements and calculations increase with increasing distance from the surface.

4.3. Cross spectra

The results for the cross-spectral computations are shown in Figs. 10 and 11 for the two different wave series. In general, the shape of the spectra for calculated velocity is similar to the shape of the spectra measured for velocities. For the less steep waves (Fig. 10) the spectral peak is shifted towards a higher frequency for the measured velocities. This shift is believed to be caused by reflected waves.

Within frequencies containing a significant amount of energy, the phase correspondance between the measured and the predicted velocities is documented. The phase lag observed corresponds to less than half the time step of the data logging discretization. The coherency function shows a high degree of correlations within the frequency range containing wave energy. Finally, the gain function confirms the over-estimation of the velocities obtained from the 2nd order calculations.

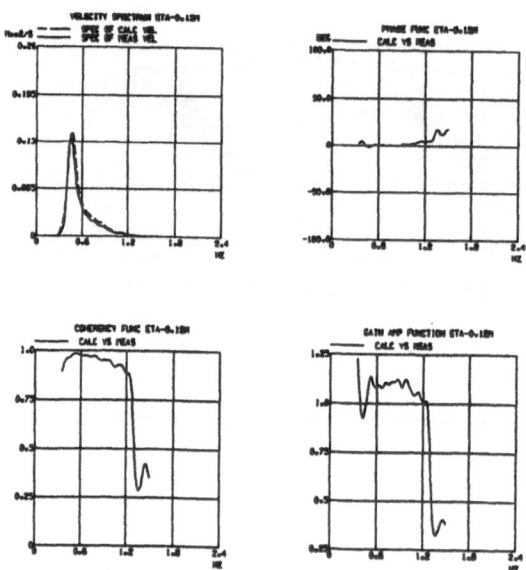

Figure 10. Cross-spectral analysis of measured and calculated velocities at a level of 12 cm below sea surface. H_s = 0.25 m, T_p = 2.2 s.

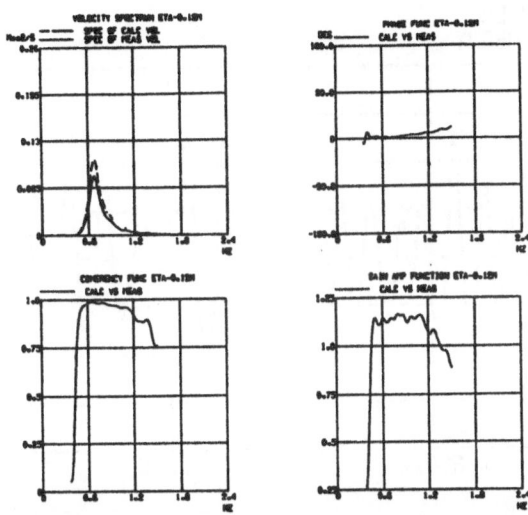

Figure 11. Cross-spectral analysis of measured and calculated velocities at a level of 12 cm below sea surface. H_s = 0.20 m, T_p = 1.5 s.

4.4. Discussion of Results

4.4.1. Linear vs 2nd Order Approach

Traditional analysis of a recorded wave elevation time series regards
the total signal as being made up of linear wavelets each travelling
with its own celerity according to the dispersion relation. However,
non-linear wave-wave interaction in irregular waves produces bound
harmonics (see e.g. Sand et al., Mansard et al., and Dean and Sharma).
The amplitudes of the superharmonics tend to add to the crest
elevation of the larger waves. If linear theory is applied for the
total elevation, an exaggerated increase in velocities under the
crests will result.

This effect is demonstrated by the plot in Fig. 12, where velocities at mean water level have been calculated using linear theory
and the 2nd order approach applied in this paper. The overestimation
of crest velocities for the high waves is noticed.

4.4.2. Velocities Above Mean Water Level

In the analyses presented so far, the velocities at each level have
been calculated by simply inserting the value of the vertical coordinate into the basic formulas. This means that extrapolations are used
above mean water level. The use of the extrapolation technique is
known to produce too large estimates in connection with linear analysis, and may also be responsible for the overestimation of peak crest
velocities, cf. Figs. 6 and 8.

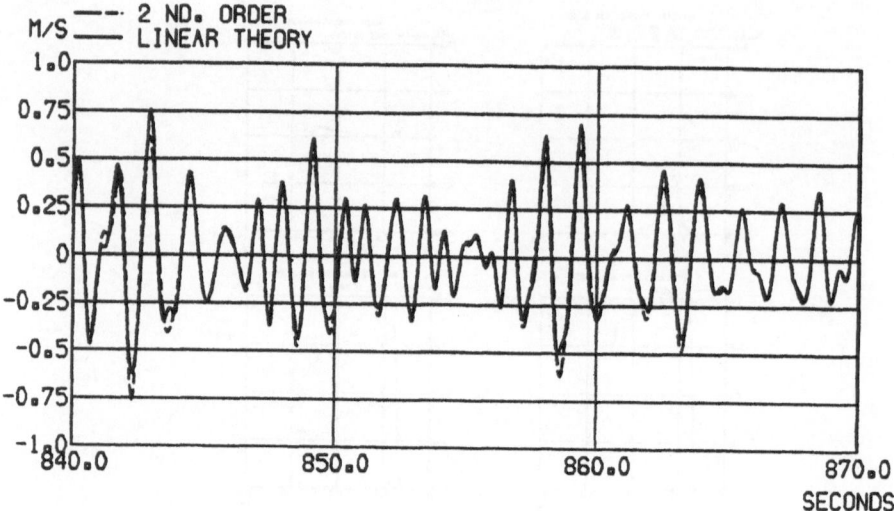

Figure 12. Time series of velocities at mean water level calculated by linear and 2nd order theory.

Various stretching methods have been used in connection with linear theory to reduce the overestimation. The coordinate transformation introduced by Wheeler is frequently applied. A similar technique can be used for the present data, i.e. the calculated velocities from the 2nd order approach in constant levels are stretched up to the actual sea surface.

The result of this exercise is illustrated in Fig. 13. The reduction in calculated peak crest velocities is significant, and the measured velocities are in general larger than the calculated ones in contrast to the results from the extrapolation technique (cf. Fig. 8a).

The results of using traditional Wheeler stretching on the time series are shown in Fig. 14, where the peak velocities under the crests are presented. An almost perfect correlation between measured and calculated velocity amplitudes is found for the recording level closest to the sea surface. Below this level Wheeler stretching underpredicts the velocity amplitudes.

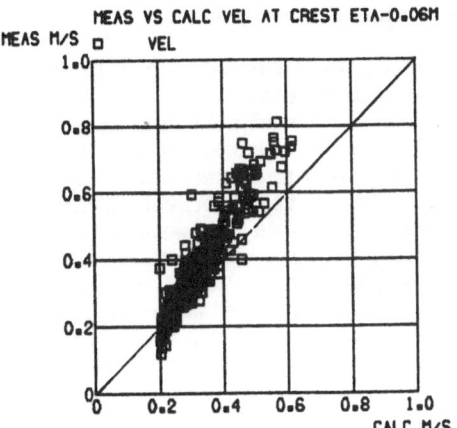

Figure 13. Peak velocities under crests. Measured vs calculated. Co-ordinate stretching applied. H_s = 0.2 m, T_p = 1.5 s.

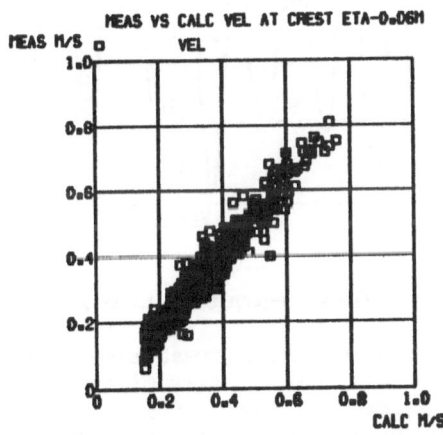

Figure 14. Peak velocities under crests. Measured vs calculated. Traditional linear Wheeler stretching. H_s = 0.2 m, T_p = 1.5 s.

4.4.3. Measured and Calculated Elevations

The surface elevation time series to be expected in the point of velocity recordings have been calculated. The calculations are based on
2nd order theory, i.e. including bound sub- and superharmonics as well
as the parasitic or free harmonics created by the wave generator boundary. An example of calculated and measured wave elevation time series
is shown in Fig. 15 below. The overall correspondence between the two
series is quite good. Discrepancies may, however, also be observed. It
is believed that these discrepancies are mainly caused by reflected
waves. The amplitude of the reflected waves are reduced compared to
the generated waves, but as a reflected wave which e.g. adds to the
crest elevation will decrease the velocity, the effect on the measured
kinematics can be noticeable especially for the small amplitude waves.

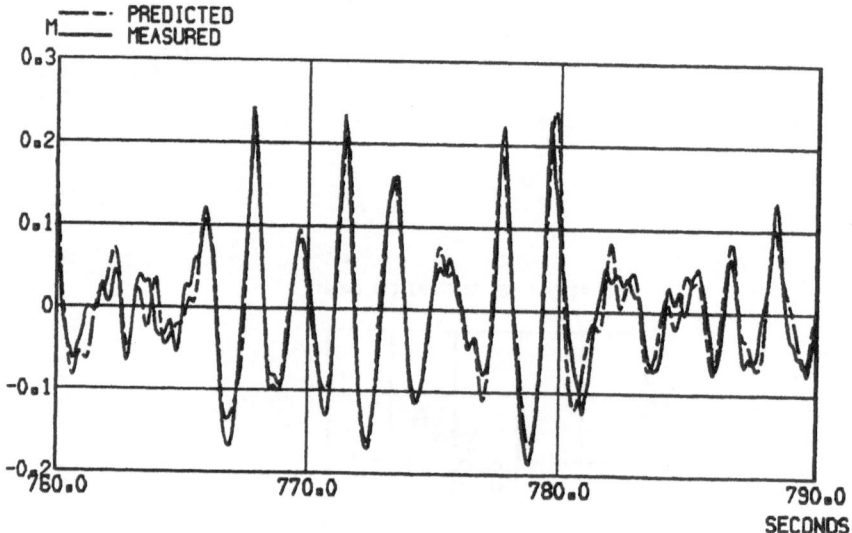

Figure 15.   Measured and calculated elevation time series.

A comparison of measured and predicted crest elevations is shown in
Fig. 16. The correlation between the two sets of data is high, and the
scatter reflects the influence from wave reflection and other inaccuracies in the numerical/physical wave generation. The scatter in the
plot is similar to the scatter in Figs. 6 to 9 for velocity amplitudes. It may thus be concluded that the scatter in the velocity data
is mainly due to non-perfect physical wave generation, a problem that
we have to face in model tests.

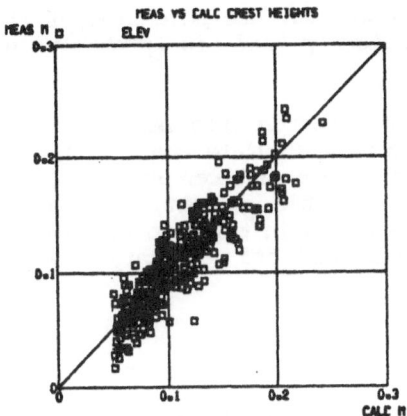

Fig. 16   Comparison of measured and predicted crest elevations.

5.   SUMMARY AND CONCLUSIONS

Time series of near surface deep water irregular wave kinematics have been recorded and compared with calculated velocities using 2nd order irregular wave theory including bound superharmonics.
    From the analyses, the following can be summarized:

1. For a real sea with bound harmonics, linear theory predicts higher velocities near the sea surface than 2nd order irregular theory.

2. 2nd order irregular wave theory provides good estimates of measured velocities beneath the wave troughs.

3. The crest velocities are overestimated when extrapolation is used.

4. When substituting the extrapolation with coordinate stretching (similar to Wheeler stretching), the crest velocities are underestimated.

5. Traditional linear Wheeler stretching provided quite good estimates of velocity amplitudes under the crests. For the wave series with the largest steepness the Wheeler stretching yielded almost perfect estimates close to the wave crests but provided smaller velocity amplitudes than measured for the lower recording levels.

From 3 and 4 it would appear that a procedure where extrapolations are made to a certain level above mean water level and coordinate stretching applied from this level could provide better agreement with measured velocity amplitudes. Such a procedure has been exercised and excellent agreement was obtained. The selection of transition level, is, however, quite arbitrary and is probably not universally applicable.

In conclusion, still more research into near surface irregular wave kinematics is needed to obtain accurate and reliable estimates of hydrodynamic loads, which for most offshore structures are proportional to the velocity squared. In this context, continuous recordings of velocities, like those presented in this paper, from either laboratory or field tests, will form a vital and necessary data base for analysis and developments.

ACKNOWLEDGEMENTS

Part of the study programme reported here has been funded by the Danish Ministry of Energy under the Energy Research Programme.

6. LIST OF REFERENCES

Dean, R.G. and Sharma, J.N.:
'Simulation of Wave Systems Due to Nonlinear Directional Spectra'
Preprints. Intl. Symp. on Hydrodynamics in Ocean Engineering. Trondheim, August 1981.

Gudmestad O.T:
'A New Approach for Estimating Irregular Deep Water Wave Kinematics'. Accepted for publication in Applied Ocean Research.

Mansard E.P.D., Sand S.E., Klinting P.:
'Sub- and Superharmonics in Natural Waves'. Proc. Sixth Int. Conference on Offshore Mechanics and Arctic Engineering, Houston, Texas. March 1987.

Rodenbusch, G. and Forristall, G.Z.:
'An Empirical Model for Random Directional Wave Kinematics Near the Free Surface'. Proc. 18 Annual Offshore Technology Conference, Houston. May 1986.

Sand S.E., and Mansard E.P.D.:
'Reproduction of Higher Harmonics in Irregular Waves'. J. Ocean Engng. Vol 13, No 1. (1986), pp.57-83.

Wheeler, J.D.:
'Method for Calculated Forces Produced by Irregular Waves'. Journal of Petroleum Technology, March 1970.

PRACTICAL WAVE MODELLING

B. BARNOUIN
IFREMER
Engineering and Technology Division
P.O. BOX 70
29263 Plouzané
FRANCE

ABSTRACT. The wave modelling activity at IFREMER is currently emphasizing a largest incorporation of real sea data analysis into the development of engineers-oriented formulations. This approach offers an opportunity to bring together user's needs and theoretical knowledge. Three examples are given about wave groups, wave dissymetry and seastate stationnarity ; preliminary results are given and the ongoing developments are mentionned.

1. Introduction

Wave modelization and wave statistics have provided subjects for theoretical studies for many years, at IFREMER just as in other research institutes.

Following the needs expressed by engineers using these results, a recent tendancy is to adjust - or to complement - this knowledge through a more empirical approach, based upon extensive use of real sea data.

To illustrate this point, three studies recently completed at IFREMER are exposed :
- . About wave groups effects : a study devoted to the modelization of the envelope of the low frequency part of the wave spectrum [1] ;
- . About crests and through statistics : a study devoted to the wave dissymetry importance [2] ;
- . About seastate durations : a study devoted to the determination of the transition between two "stationnary" seastates. [3].

A. Tørum and O. T. Gudmestad (eds.), Water Wave Kinematics, 201–214.

2. Spectral Models for Wave Induced Drift Forces

2.1. DEFINITIONS

This analysis is concerned with low frequency structural excitation due to wave action, i.e. the sort of forces that are proportionnal to the square of the wave envelope.

The so-called "wave grouping" phenomenon may have dramatic influence on these forces as long as it may be considered as periodic, (with long regular intervals between the groups) for such structures as articulated columns, moored tankers, deep compliant towers, etc..

Two different sources of excitation may happen simultaneously : the low frequency component of the wave process itself (due to non-linear propagation effects) and the low frequency excitation component due to a non-linear interaction between the wave field and the structure.

The first one has been analysed, and the litterature (see [4]) provides the following model :

$$\eta(t) = \sum_{n,m} \xi_{n,m}(t)$$

$$\xi_{n,m}(t) = G_{nm}(f, f) [a_n a_m + b_n b_m \cos(\Delta w_{mn} t) +$$

$$(a_m b_n - a_n b_m) \sin \Delta w_{mn} t]$$

where :

- a_n and b_n are the Fourier coefficients of the wave process

- $W_{mn} = w_m - w_n$

- $G_{mn}(f, f)$ is a transfer function given in [4]

This expression (experimentally verified in [5]) allows for a simple relationship between the low frequency components of $\eta(t)$ and of $\eta^2(t)$, in the case of a narrow-banded spectrum.

The second source of excitation is usually computed from the squared envelope of the wave height time history $\eta(t)$. We show hereunder that it is possible to use the squared elevation directly :

The Fourier transform of $\eta^2(t)$ is : $F(f) * F(f)$

where $F(f)$ is the Fourier transform of the wave process $\eta(t)$, and is supposed to be null when f is lower than f_o.

[env(t)], the envelope of $\eta(t)$, is the modulus of the analytical signal $\eta_a(t)$ associated to $\eta(t)$, and, by definition :

$$F_a(f) = 2 \quad F(f) \text{ for positive frequencies}$$
$$= 0 \qquad \text{for negative} \qquad "$$

Therefore, the Fourier transform of $[env(t)]^2$ is $:2[F(f)*F(f)]$

and is thus equivalent to the Fourier transform of $\eta^2(t)$
for all $f < 2 \ f_0$

In the following we shall concentrate on the process $\eta^2(t)$.

2.2. WAVE GROUPS AND GAUSSIAN WAVE MODEL

Before going further, it must be reminded that the existence of wave groups should not be opposed to the random phase model (i.e. the gaussian model).
   It is shown in [6], for example, that the grouping factor (correlation between successive wave heights) is not influenced by a randomisation of the phase.
   Nevertheless, it is easy to exemplify the phenomenon by performing a simple numerical simulation of a signal with a narrow rectangular bandwith Δf.
   It may be observed that the empirical distribution of the duration between successive maxima of its envelope exhibits a strong modal value (at a period which depends upon Δf, for example around 70 sec.for $\Delta f = 0.04$ Hertz) (fig. 1).

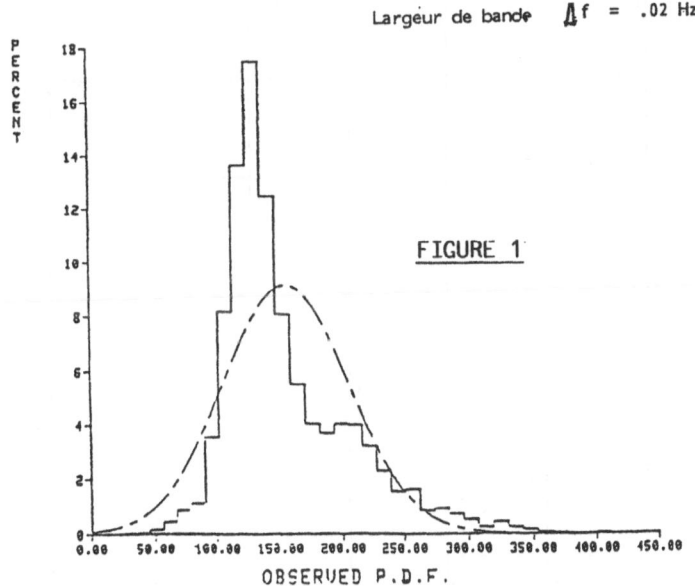

Largeur de bande $\Delta f = .02$ Hz

FIGURE 1

OBSERVED P.D.F.

On the other hand, it is known that the spectrum of its squared envelope is monotonously decreasing (because of the random phase hypothesis) and, therefore, does not exhibit any spectral peak corresponding to this modal value.

Then, the spectrum, which conveys information about the average energy distribution, may not be significant in terms of time-history maxima distribution.

In [7] the spectra of the squared elevation have been plotted, together with the ones deduced by direct convolution, for several records taken during the "CAMILLE" hurricane. (see fig. 2).

FIGURE 2 (from (7))

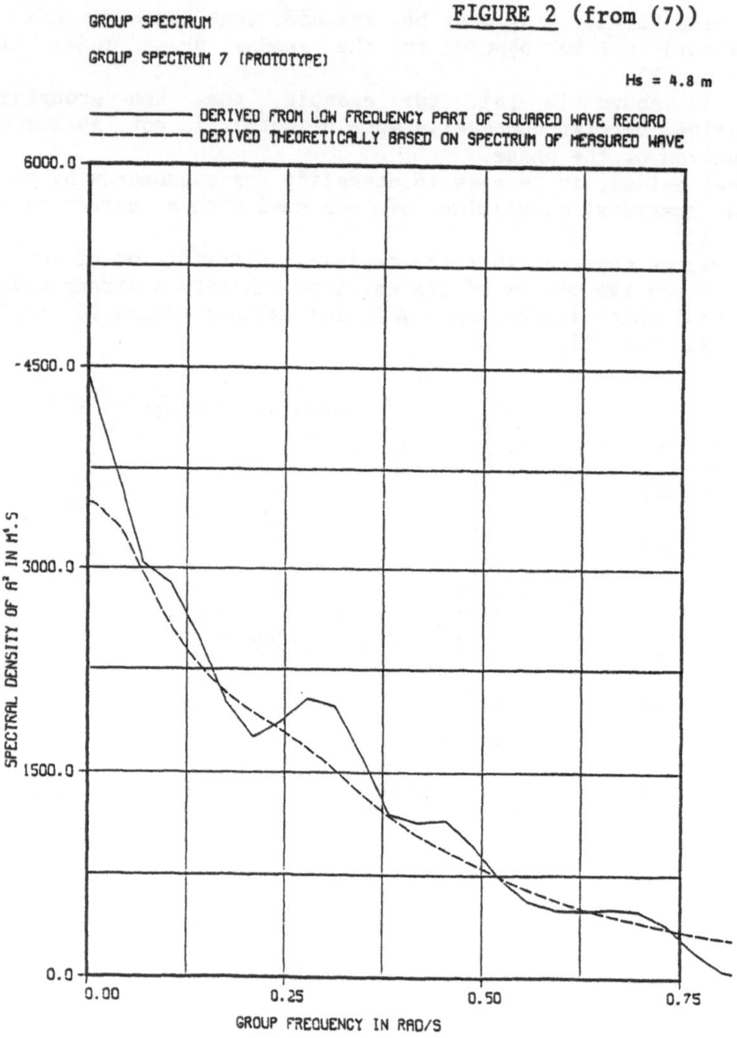
GROUP SPECTRUM
GROUP SPECTRUM 7 (PROTOTYPE)
Hs = 4.8 m
DERIVED FROM LOW FREQUENCY PART OF SQUARED WAVE RECORD
DERIVED THEORETICALLY BASED ON SPECTRUM OF MEASURED WAVE
6000.0
4500.0
3000.0
1500.0
0.0
SPECTRAL DENSITY OF A² IN M⁴.S
0.00
0.25
0.50
0.75
GROUP FREQUENCY IN RAD/S

They show that both spectra agree satisfactorily and don't have any statiscally relevant peak. This experimental result confirm the theoretical results mentionned hereabove.

2.3. PARAMETRIZATION OF DRIFT-LIKE FORCES SPECTRA

Similar calculations and plots have been conducted for about one hundred North Sea 20 mn records.

Besides the spectra of $\eta^2(t)$ and of $[env(t)]^2$, another spectrum of $[\tilde{\eta}(t)]^2$ was drawn in order to verify that there is not significant effect of phases :

$$\tilde{\eta}(t) = F^{-1} \{| F (f) | .\exp i \emptyset(f)\}$$

is obtained from $\eta(t)$ by randomisation of the phase.

Here again, same conclusions as above were reached. (see fig. 3).

Thus, it was attempted to find a parametric model for this monotonous decreasing spectrum that fits rougly a linear-logarithmic function with slope $1/\beta$ (in the low frequency range) :

$$S_{\eta^2}^{L.F.}(f) = \frac{k^2}{2\beta} \exp (- \frac{f}{\beta})$$

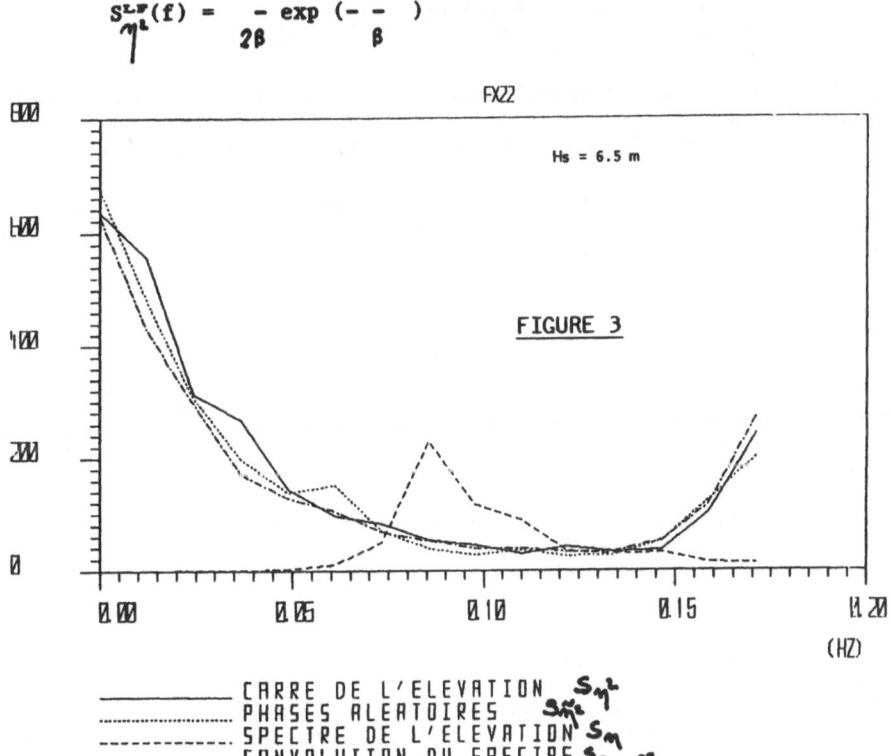

FIGURE 3

It may be noticed that this is the convolution product of :

$$F(f) = \frac{k}{\beta} \exp - (\frac{f-\alpha}{\beta}) \quad \text{(for } f \geqslant \alpha, 0 \text{ elsewhere)}$$

Therefore, if one defines a bandwidth parameter

$$bw = \frac{4}{m_0^2} \int_0^\infty S_\eta^2(f) (f - \frac{m_0}{m_1})^2 df$$

it may be shown that, m_0 and m_1 having simple expressions in k, β, α, it comes :

$$bw = \beta \sqrt{\frac{m_2 m_0 - m_1^2}{m_0^2}}$$

It is thus suggested that the following model can be used :

$$S_\eta^{LF}(f) = 4 [\int_0^\infty S_\eta^2(f) df] \exp (- \frac{f}{bw})$$

where $S_\eta(f)$ is the original wave spectrum (in the normal wave frequency range).

This parametrization has yet to be tested experimentally on records from other sites.

The question which is now being adressed is how to simulate forces which have such spectra with respect to a possible existence of wave-group pseudo-periodicity.

3. Statistics of Successive Wave Heights and Throughs

Recent studies have shown that some compliant structures can be extremely sensitive to the design wave crest absolute elevation.

The wave profile is known to be dissymetrical and the maximum forces are induced by the passage of the crest.

The objective of this study was to quantify from real sea data the statistical differences between crests heights and throughs amplitudes, and to compare these to the values obtained in the simulation of the same seastate (normal engineering practice).

They are referred to the commonly used "zéro-up crossing" values (HUZC).

The "zéro-down crossing" values (HDZC) are also given for reference.

Definitions (and notations) of these parameters (and of their calculation from records) comply with the I.A.R.H. recommandations. [8]

TABLE 1.

SAMPLE 1. 40 records of 1122 secs each.

	HC	HT	HUZC	HDZC	HS	TZ
H1/10	8.67	7.33	7.53	7.50		
H1/3	6.69	5.87	6.04	6.03	6.28	8.62
HMax	17.77	12.52	13.10	12.50	(5208	waves)

SAMPLE 2. Simulation of 32.768 secs of wave elevation
from a jonswap spectrum.
γ = 2.2, H_s = 6.30 m, T_p = 13 secs.

	HC	HT	HUZC	HDZC	HS	TZ
H1/10	7.93	7.90	7.50	7.51		
H1/3	6.28	6.25	6.04	6.03	6.30	9.86
HMax	12.53	14.21	12.62	12.45	(3325	waves)

HC = Twice crest elevation
HT = Twice trough elevation

The results of table 1 show that although the simulation provides very good HUZC's (and HDZC's), it fails naturally to represent non-linearity effects, i.e. the dissymetry between the HC's and the HT's.

The second step of the analysis was to fit p.d.f. laws to these data and to compare extrapolated values deduced from these fitted distributions :

TABLE 2. COMPARISON OF DISTRIBUTIONS.

- WHOLE SAMPLE
SAMPLE 1 : WEIBULL PARAMETERS

	ORIGIN	SCALE	SHAPE
HC	0	4.875	2.065
HT	0.2	4.113	2.073
HUZC	0.2	4.117	2.053
HDZC	0.2	4.194	2.094

- WITH THRESHOLD = H1/3
SAMPLE 1 : WEIBULL PARAMETERS

	ORIGIN	SCALE	SHAPE
HC	4.88	2.036	1.247
HUZC	4.60	1.600	1.242

SAMPLE 2 :

	ORIGIN	SCALE	SHAPE
HC	4.62	1.869	1.309
HUZC	4.60	1.677	1.334

The results of table 2 show that :
- the best fits are obtained with Weibull laws very close to Rayleigh (shape factors close to 2)
- the Weibull parameters are strongly dependant of the use of a threshold value, if one choose to adjust the higher values (the upper third in the present case) in order to improve the extrapolation ;
- in both cases, the HC and HUZC follow Weibull laws nearly identical in shapes.

The most probable extreme values predicted for a duration equal to eight times the sample length are as follows :

Real HUZC   15.02
Real 2*HC   18.63

Simulated HUZC   14.10
Simulated 2*HC   15.31

The maximum crest elevation is therefore underestimated (about 20%) by the distribution derived from the up-zero-crossing value.

As a third step, in order to refine the differences between crest and ZUP values, a similar statistical analysis was performed on the ratios HC/HUZC and HT/HTZC (ratios of values obtained on the same "wave"). The statistics of these ratios were computed for different classes of wave heights (quoted in the left column of table 3).

TABLE 3. STATISTICS OF RATIO's HC/HUZC.

	m	σ	min	max
3.5 m	1.034	0.25	0.10	2.0
5-7 m	1.067	0.18	0.58	1.71
7-9 m	1.098	0.16	0.60	1.42
9-11 m	1.092	0.10	0.93	1.35

This table shows that the wave non-linearity increases with wave height ; complementary calculations show that the difference between crest and ZUP heights, (unpredicted by the simulation) are in the real case about 11% for H1/3, 16% for H1/10 and 36% for Hmax.

As a conclusion, it seems necessary to introduce the wave non-linearity in time domain simulation if one wants to reproduce crests statistics, which may have significant effects for some type of structures.

4. Stationnary and Transient States of Random Sea

Stationarity is the main hypothesis needed to derive realistic theoretical models of random seas. Besides, when performing a simulation in order to extrapolate the computed structural behaviour, the designer needs some information about the most probable duration of the sea state he has simulated. This provided the motivation for evaluating the duration of stationarity of a wave process.

Let us define a sea-state as a stationary state of the piecewise stationary (in the broad sense) stochastic random wave process X_t.

The severity of a sea-state is usually described in term of the significant wave height H_s. A realistic method to discuss variations in sea-state, is therefore to consider changes in H_s.

Data coming from radar distance meter measurements made in the Frigg Field from QP structure, are used extensively. The wave data were acquired permanently at a 2 Hz sampling frequency. Starting in 1984, on board calculations of the usual reduced parameters (H_s, T_s,...) are recorded every 20 minutes.

Let $H_s(i)$ be the empirical estimator of H_s over the consecutive record number i. The following model was calibrated for detecting abrupt changes in the discrete process H_s. As long as stationarity is verified, there exists a constant V such that :

$$H_s(i) = V + \theta_i \qquad (1)$$

The θ_i are independent normal random noises N (0,σ), with zero mean and standard deviation σ (hypothesis H_o). Morever a sea-state changing is found to be due mainly to a significant change in mean of θ_i. So, after this change, θ_i is N (μ,σ) distributed (hypothesis H_1).

A sequential detection method, based on Hinkley's test [9] is then implemented. We consider the statistical serie S_i indexed by the record number, and defined recursively by,

$$S_o = 0 \text{ and } S_i = [S_{i-1} + H (\theta_i)]^+$$

where $[f]^+ = $ Sup $(f,0)$. The function H is the log-likelihood of θ under the two simple hypothesis H_o and H_1 :

$$H (\theta) = \frac{\mu}{\sigma^2} (\theta - \frac{\mu}{2})$$

A rupture is then detected after a crossing of the discrete process S_i, at a fixed level S. A stable threshold level S is evaluated, from a sensitivity study.

For example (see figure 4) :
- 43 is the index of the first exceedance of S threshold
- 38 is the effective index marking the end of the stationary state
- 43-38 = 5 is a recorder for the delay of detection.

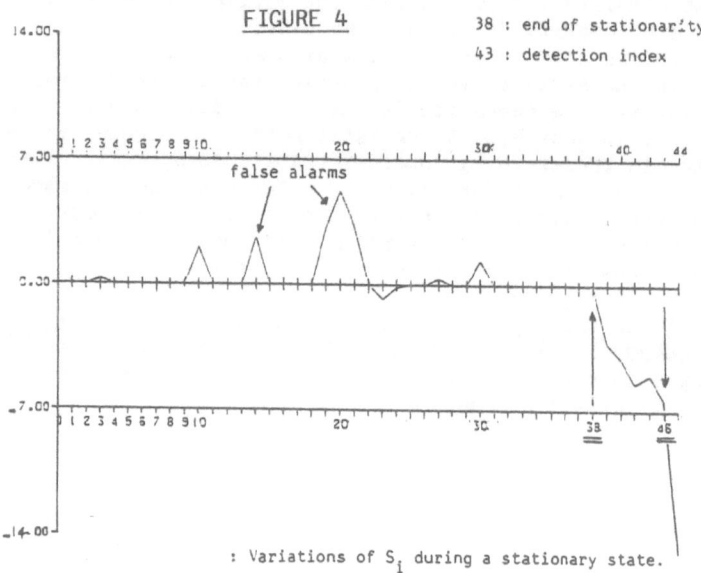

FIGURE 4

38 : end of stationarity

43 : detection index

false alarms

: Variations of S_i during a stationary state.

This modelling allows to identify a very large spectrum of sea-states (see figure 5).

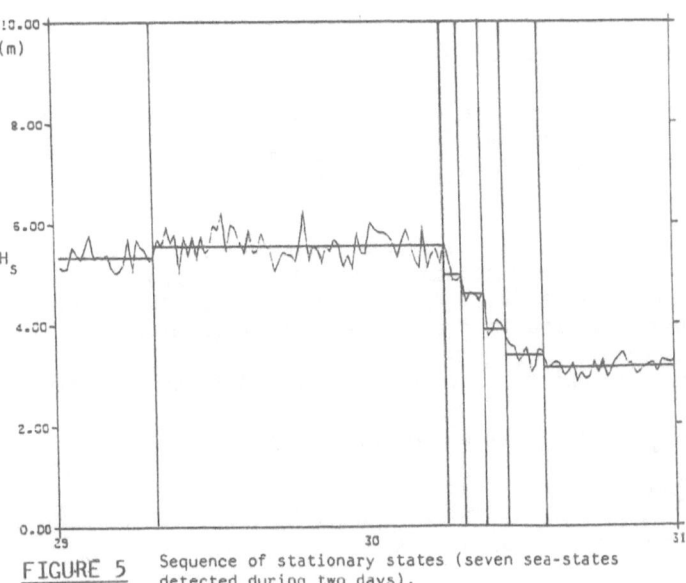

FIGURE 5   Sequence of stationary states (seven sea-states detected during two days).

It seems complementary to understand the development of a sequence of sea-states.

We define a transient state as a state where the property of stationarity in the broad sense, fails.

The problem comes to study the transition between two stationary states. We start from an initial state $(H_o = V_o)$ in view of establishing another one $(H_\infty = V_\infty)$. The values V_o and V_∞ are unknown.

Let $\Delta = V_\infty - V_o$ be the jump amplitude of the transition.

The term V is now a series which satisfies the condition,

$$V_o, \quad V_i = f(V_{i-1})$$

The function f is a contraction such that $f(V_\infty) = V_\infty$.

The simplest approach is to consider convex hull modelling. The serie V is then defined by,

$$V_i = (1-C) V_\infty + C V_{i-1}$$

where C is a damping coefficient. The coefficient C gives some measure of the rate of convergence of the transition, as we have,

$$V_i = V_\infty - \Delta C^i$$

In the present study, a stable value is found to be $C = .8$. The equation (1) becomes,

$$H_\infty(i) = V_\infty - \Delta C^i + \theta_i$$

A sequential detection method is then implemented as it follows :

- we have first to define some initial values for V_∞ and . The estimators $\hat{V}_\infty$ and $\hat{\Delta}$ are the coefficients of the linear least square fitting to the k (ex: k=5) first points, $(C^i, H_\infty(i))$, $i = 0,\ldots,k-1$;

- the variables θ_i are independent normal random noises $N(0,\sigma_i)$ with zero mean and standard deviation σ_i (hypothesis H_0). After changing θ_i is $N(\mu_i, \sigma_i)$ distributed (hypothesis H_1). The parameters μ_i and σ_i depend linearly of $V_\infty - \Delta C^i$;

- we consider the statistical serie S_i, defined recursively by,

$$S_k = 0, \quad S_i = [S_{i-1} + H(\theta_i)]^+ \quad \text{if } i > k$$

where,

$$H(\theta_i) = \frac{\mu_i}{\sigma^2_i} \left(\theta_i - \frac{\mu_i}{2} \right)$$

The end of the transient state is then detected after a crossing of the discrete process S_i, at a fixed level S. As long as the hypothesis H_0 doesn't fail, the estimators $\hat{V}_\infty$ and $\hat{\Delta}$ are sequentially updated.

The figure 6 indicates that the non-linear dynamics of the transitions are well approximated.

Several fluctuations on the average energy content are better described by fitting sequences of transient states, than stationary ones. This is a support for considering H_∞ as a continuous stochastic process, in storm modelling. The assumption of stationarity, otherwise so tractable, fails in such conditions.

For each sea-state, let T_{stat} be its duration. The variable T_{stat} was found to be distributed following an exponential type, with consequently a very large dispersion.

Note that 20 minutes is a lower bound for T_{stat} due to the time length of recording. As a result, the duration of st ationarity of some sea-states may be shorter than this lower bound. This is especially true for large values of H_∞.

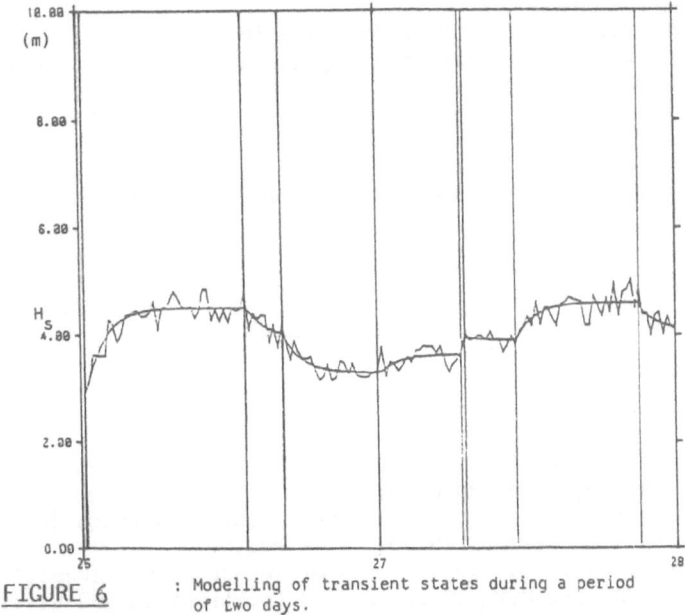

FIGURE 6 : Modelling of transient states during a period of two days.

The conditional expectation of the duration of stationarity (given H_s) was found approximately constant (= 3 hrs), in the range of observed values. More generally, the duration of stationarity and the rms level are stochastically independent. These investigations confirm that three hours may be considered as the mean duration of a sea-state in this specific site.

Data selected here, provide a strong support to further study aiming to apply the Markov chain theory, in describing the sequences of sea-states along with the estimation of some transition probabilities.

5. Further Developments and Conclusion

The three studies cited hereabove examplify the need to rely on large sets of real sea data to verify, transform - or, even, suggest - theoretical developments in order to get useful engineering results.

All three of them deserve complementary studies in order to reach this goal; which are currently undertaken, within the French Research Group CLAROM.

More specifically :

. Sea state stationnarity durations are involved into :
"Times scales and statistical uncertainties in the prediction of extreme environmental conditions", to implement practically the derivation of the extreme design waves confidence intervals (to be published in the Reliability journal)
and into :
"Characterization of sea-states for fatigue purposes" published at OMAE 89.

. Wave grouping and dissymetry effects are taken into account in :
"Numerical simulation of actual wave properties" to implement cheap methods that provide elevation time histories representing more properties than the mere spectral shapes. (submitted to OMAE 90).

6. REFERENCES

[1] "Groupage de vagues et excitation basse fréquence", Rapport IFREMER DIT/SOM 88-198, Marc PREVOSTO.

[2] "Statistiques expérimentales des différentes hauteurs de vagues", Rapport IFREMER DIT/SOM 89-90, Agnès ROBIN.

[3] "Stationnary and transient states of random seas", Submitted to the Journal of Marine Structures, Design Construction and Safety in Nov. 88, Jacques LABEYRIE (at IFREMER).

[4] "Long period waves in natural waves trains", Prog.Rep. 46, Inst. Hydrodyn. and Hydraulic Eng., Techn. Univ. Denmark, N.E. OTTESEN HANSEN.

[5] "Wave grouping described by bounded long waves", Ocean Engineering, Vol. 9, n° 6, 1982, S.E. SAND.

[6] "Wave grouping studies by means of correlation techniques", Norwegian Maritime Research n° 4, 1981, H. RYE, E. LERVIK.

[7] N.S.M.B. report Z50058, J. PINKSTER.

[8] "List of seastate parameters". Supplement to bulletin n° 52, 1986, by: International Association for Hydraulic Research.

[9] "A survey of design methods for failure detection in dynamics systems", Automatica, Vol. 12, 1976, by A.S. WILLSKY.

WAVE THEORY PREDICTIONS OF CREST KINEMATICS

Rodney J. Sobey[1]
Department of Civil Engineering
University of California at Berkeley
Berkeley, CA 94720, U.S.A.

1 INTRODUCTION

In much of the practice of coastal and ocean engineering, the question of which wave theory to use has been largely of academic interest. Consideration of shallow water and/or higher order wave theories has been complicated by the relative inaccessibility of these theories. The literature is extensive and frequently confusing, which has often led to the adoption of linear wave theory, whether or not it is appropriate. In much of the practical parameter range, mathematical predictions for steady, progressive waves are available from a range of orders from Stokes, Cnoidal and Fourier theories. The differences are often considerable.

Further complications are introduced by efforts to generalize steady progressive wave theory to real sea states, where spatial and temporal irregularity are characteristic features of the water surface. In both cases however, the crucial aspects of the mathematical physics are the nonlinear free surface boundary conditions, so that there remains a close relationship between regular (i.e. steady) and irregular wave theory.

Crest kinematics in particular have been the subject of considerable recent interest. The details of crest profiles and especially velocities and accelerations in the neighborhood of the crest are critical in considerations of the breaking wave process and in wave loading predictions from the O'Brien-Morison equation. The entire trough-crest region is equally dominant in forcing the nearshore circulation.

This review will focus on the prediction of crest kinematics in regular and irregular waves. Initial consideration is given to the theoretical background and predictive capability of steady progressive wave theory. This leads to a discussion of crest kinematics in regular and irregular waves. The review is completed by consideration of theories for the prediction of kinematics in irregular waves.

2 STEADY PROGRESSIVE WAVE THEORY

Mathematical Background. Progressive waves of permanent form are steady in a frame of reference moving at the phase speed C. Accordingly, it is convenient to adopt a steady or moving x, z reference frame that is located at the mean water level (MWL), or alternatively at the sea bed, and moves at speed C with the wave

---

1 Prepared while on leave at Department of Civil Engineering, Delft University of Technology, 2600 GA Delft, The Netherlands.

A. Tørum and O. T. Gudmestad (eds.), Water Wave Kinematics, 215–231.

crest, rather than an unsteady or fixed X, Z, t reference frame. Assuming that the flow is incompressible and irrotational, the mathematical formulation may be presented in terms of the Euler equations, in terms of the velocity potential function $\phi(x,z)$ or $\Phi(X,Z,t)$ or in terms of the stream function, $\psi(x,z)$.

Unnecessary complications are avoided by the choice of the stream function, where the field equation is the Laplace equation

$$\frac{\partial^2 \psi}{\partial x^2} + \frac{\partial^2 \psi}{\partial z^2} = 0 \tag{2.1}$$

where the velocity components (u,w) are $(\partial\psi/\partial z, -\partial\psi/\partial x)$.

This field equation is subject to the following boundary conditions:
(i) Bottom boundary condition, representing no flow through the horizontal bed, is
$$\psi(x,-h) = 0 \qquad \text{at } z = -h \tag{2.2}$$
(ii) Kinematic free surface boundary conditions, representing no flow through the free surface, is
$$\psi(x,\eta) = -Q \qquad \text{at } z = \eta(x) \tag{2.3}$$
where $\eta(x)$ is the free surface and $-Q$ is the constant volume flow rate per unit width under the steady wave.
(iii) Dynamic free surface boundary condition, representing constant atmospheric pressure on the free surface, is

$$\frac{1}{2}\left[\left(\frac{\partial\psi}{\partial x}\right)^2 + \left(\frac{\partial\psi}{\partial z}\right)^2\right] + g\eta = R \qquad \text{at } z = \eta(x) \tag{2.4}$$

where g is the gravitational acceleration and R is the Bernoulli constant in the steady frame.
(iv) Wave is periodic in space and time and maintains a permanent form (i.e. a stable profile shape), which in the steady frame requires the wave profile to be symmetric in x about the crest.

The adopted solution methodology depends primarily on the dimensionless depth kh, where $k = 2\pi/L$ is the wave number and h is the mean water depth. For larger values (deep water), a Stokes wave theory would be used. For smaller values (shallow water), a Cnoidal wave theory would be used. Fenton (1989) has given the following demarcation between fifth order theories:

$$kh = 0.292\exp\left(1.87\frac{H}{h}\right) \tag{2.5}$$

which is dependent directly on the wave height and indirectly, through the wave number, on the order of the theory and any coflowing current. Alternatively, a Fourier approximation theory may be used, regardless of the value of kh or H/h.

The theoretical development of rational and consistent theories from first to fifth order for Stokes and Cnoidal waves, and from first to (theoretically) any truncation order for Fourier waves have much in common. Sobey et al (1987) have emphasized the common aspects in establishing a coordinated presentation of these theories, closely following a sequence of papers by Fenton (1979, 1985) and Rienecker and Fenton (1981). The detail of these wave theories is crucial but this is not the context in which to dwell on such detail. Each theory is analytically complicated, becoming also numerically complicated at higher orders and for all orders with Fourier wave theory. Sobey et al (1987) have presented tabular summaries of the predictive equations for Stokes, Cnoidal and Fourier theories. Microcomputer implementations of these theories are available.

The general applicability of Fourier wave theory is achieved at some computational cost but it is nonetheless a computational task that can be accommodated on a microcomputer. An analysis of alternative formulations of Fourier wave theory (Sobey 1989) identifies truncation order as the crucial parameter and shows that published solutions differ only in the approach to the limit wave. Sobey (1988) gives general recommendations on the selection of truncation order.

Measures of Validity. One measure of appropriateness of a wave theory is a comparison of theoretical predictions with the imposed free surface boundary conditions, the approach adopted by Dean (1970). This may seem unnecessarily selective in considering only Equations 2.3 and 2.4 but it is quite appropriate. The analytical foundation of Stokes, Cnoidal and Fourier wave theories all ensure that the field equation (Equation 2.1) and the bottom boundary condition (Equation 2.2) are satisfied identically. Dean adopted the mean square errors in the kinematic and dynamic free surface boundary conditions as measures of validity. His results are predictable. Stokes waves do best in deep water, Cnoidal waves do best in shallow water, the transition region between Stokes and Cnoidal waves is somewhat uncertain, and Fourier waves do very well throughout, except in very shallow water.

The restriction of attention to the free surface however may not provide a sufficiently comprehensive evaluation to identify potentially spurious solutions. The complexity of the analytical detail almost guarantees that potential problems are deeply embedded in the predictive equations. To establish without question the appropriateness of a wave prediction is perhaps an unreasonable goal, but to establish that there are no questionable features of the solution field is a reasonable objective. From experience, it can prove most revealing (Sobey and Goodwin 1986). Perusal of the complete solution field for all field variables (horizontal and vertical velocity, horizontal and vertical acceleration and pressure) as well as the water surface profile is essential. Computer graphics is the appropriate tool, specifically surface plots of each field variable in the steady reference frame between adjacent wave crests.

The ultimate test of the validity of any theory is a comparison of theoretical predictions with laboratory and field measurements. This was the approach initially adopted by Le Méhauté et al (1968) but their detailed conclusions in 1968 may no longer be appropriate. There have been significant refinements in Stokes, Cnoidal and Fourier wave theory since that time. The well-known Skjelbreia and Hendrickson (1962) Stokes V theory has been shown to be incorrect at fifth order. The impact of a coflowing current on higher order wave theories has been identified by Fenton (1985). There is certainly a case for repeating these comparisons.

Of course, none of these theories can describe the limit wave with its theoretical slope discontinuity at the crest. Higher-order theory can approach the limit wave but will not converge for wave heights above the limit wave.

Predicted Kinematics. A successful steady wave theory provides predictions of the water surface profile $\eta(X, t)$ and a range of field variables including the horizontal velocity $U(X, Z, t)$, the vertical velocity $W(X, Z, t)$, the horizontal acceleration $dU(X, Z, t)/dt$, the vertical acceleration $dW(X, Z, t)/dt$ and the dynamic pressure $P_{dyn}(X, Z, t)$. Their variation throughout the solution is considerable and the response to changes in wave height, water depth, period and current is also significant.

Details of two predicted solution fields are included to illustrate the range of conditions. The wave theory in both cases was Fourier XVIII, as implemented by Sobey (1989). The given wave parameters for Figure 2.1 are $H= 1$ m, $h= 100$ m, period $T= 10$ s and Eulerian current $C_E = 0$; these are typical deep water conditions. The solutions are presented as surface plots in the steady reference frame where $x = X - Ct$ and $z = Z$. Part (a) is the horizontal velocity, part (b) the vertical velocity, part (c) the dynamic pressure and part (d) the total horizontal acceleration. The response patterns here are not significantly different from Airy theory which does a tolerable job for small to moderate wave steepness in deep water. The water surface profile is almost symmetric with a zero crossing close to the quarter wave length position. The horizontal velocity is large and positive at the crest and large and negative at the trough, zero velocity at the water surface almost corresponding with the zero crossing of the MWL. Velocity magnitudes decay towards zero at a depth of order $- L/2$. The vertical velocity is zero under both the crest and the trough, and at the bed by virtue of the bottom boundary condition (Equation 2.2). It peaks at the water surface near the zero crossing of the MWL. The horizontal acceleration follows much the same pattern as the vertical velocity, expect that it is small but non-zero at the bed. The dynamic pressure also follows the same pattern as the horizontal velocity, being large and positive under the crest, large and negative under the trough but quite small within the body of the fluid.

The given wave parameters for Figure 2.2 are again $H=1$ m, period $T= 10$ s and current $C_E=0$, but $h=2$ m; these are typical shallow water conditions. The only change is the water depth but the solution fields are very different. The water surface profile has a sharp crest and a long flat trough, with the zero crossing position being shifted very much closer to the crest. The horizontal velocity changes little with depth, being small, uniform and negative under the long trough but relatively intense under the sharp crest. This horizontal asymmetry in the almost depth-uniform horizontal velocity has important implications to sediment transport under waves. In situations where the threshold velocity for initiation of sediment motion falls in magnitude between the trough and crest velocities, it suggests a mechanism for a net sediment transport in the direction of wave motion. Note that Airy theory does not predict an inequality between the magnitudes of the crest and trough velocities. The vertical velocity has become very sharply concentrated towards at maximum that remains at the water surface, but has been shifted much closer to the crest than even the zero crossing of the water surface profile. The horizontal acceleration is similarly concentrated towards a maximum at essentially the same water surface location, but the vertical structure remains much more uniform. The dynamic pressure again follows much the same pattern as the horizontal velocity.

3 CREST KINEMATICS IN REGULAR WAVES

It is apparent from the horizontal velocity and acceleration surfaces in Figures 2.1 and 2.2 that the crest region provides a very significant part of the total load on a vertical surface-piercing cylindrical structure. This is especially true for the drag force component (proportional to $U|U|$) but there is also an increasingly significant

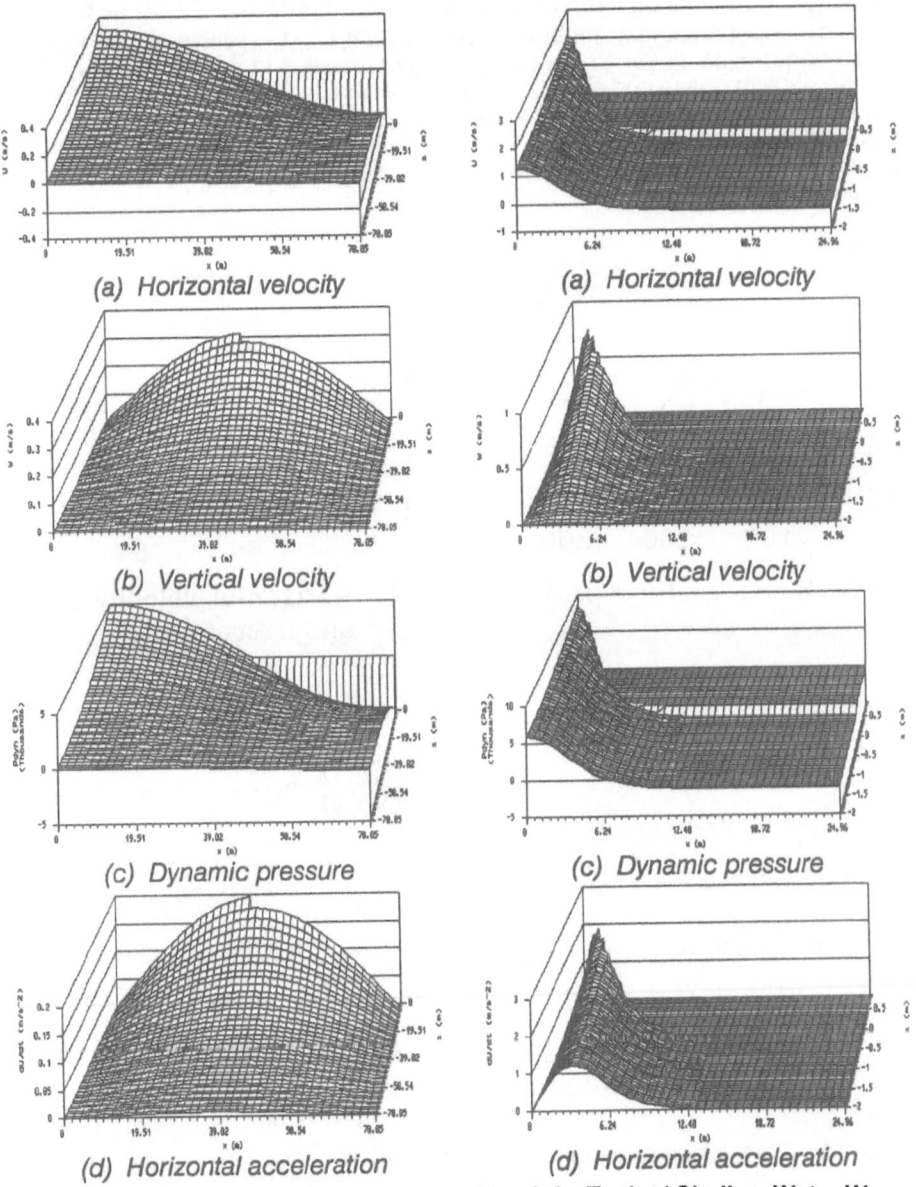

(a) *Horizontal velocity*

(a) *Horizontal velocity*

(b) *Vertical velocity*

(b) *Vertical velocity*

(c) *Dynamic pressure*

(c) *Dynamic pressure*

(d) *Horizontal acceleration*

(d) *Horizontal acceleration*

Fig. 2.1 *Typical Deep Water Wave Predictions from Fourier XVIII Theory.* [H=1 m, h=100 m, T=10 s and C_E=0 m/s]

Fig. 2.2 *Typical Shallow Water Wave Predictions from Fourier XVIII Theory.* [H=1 m, h=2 m, T=10 s and C_E=0 m/s]

contribution to the inertial force component (proportional to dU/dt) as the wave becomes steeper and the water becomes shallower and the phase of the maximum acceleration moves closer to the crest. The kinematics in this region vary much more rapidly with phase and elevation than elsewhere in the solution field. The crest kinematics are also quite sensitive to the choice of wave theory. Attention is quite naturally directed towards a region that has such a disproportionate influence on the wave loading. It is also clearly a crucial region in any attempt at wave theory validation.

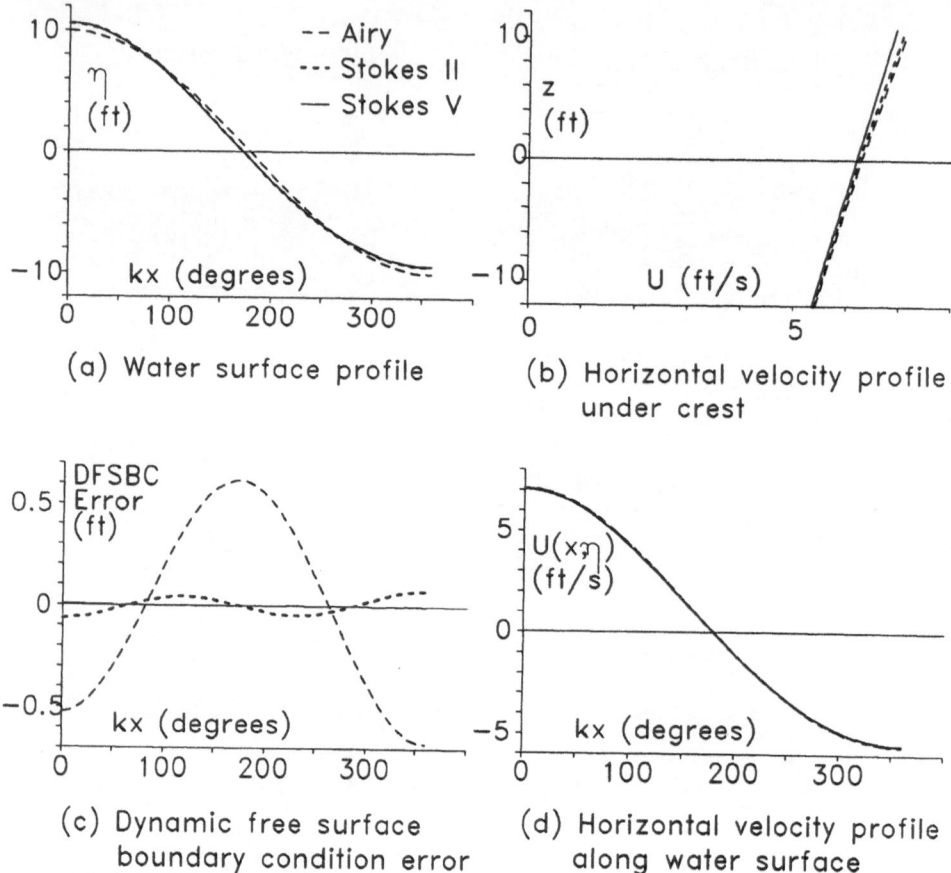

(a) Water surface profile

(b) Horizontal velocity profile under crest

(c) Dynamic free surface boundary condition error

(d) Horizontal velocity profile along water surface

Figure 3.1 *Crest Kinematics for a Deep Water Wave*

The predictive capability of steady wave theory in the crest region has been the subject of some concern in recent years, especially among designers of marine structures. From a theoretical viewpoint, it is hardly surprising that problems are identified in the crest-trough region. Stokes, Cnoidal and Fourier wave theories automatically satisfy the field equation (Equation 2.1) throughout the solution domain and also the bottom boundary condition (Equation 2.2), but the nonlinear free surface boundary conditions (Equations 2.3 and 2.4) are satisfied only approximately. Errors are thus concentrated at the free surface.

Figure 3.1, expanding on an example used by Forristall (1985), illustrates the level of approximation at the free surface. The wave height H is 20 ft., the water is deep, the wave period T is 10 s and the current C_E is zero. The figure includes Airy (or linear or Stokes I) theory, together with the Fenton (1985) Stokes II and Stokes V theories using the Stokes' first definition of phase speed. Part (a) shows the predicted water surface profile, part (b) the vertical profile of horizontal velocity under the crest, part (c) the residual error in the dynamic boundary condition at the predicted free surface (in length units) and part (d) the profile of horizontal velocity along the predicted water surface. The residual error in the kinematic boundary condition at the predicted free surface is similar to part (c). While the free surface boundary conditions errors are not insignificant for Airy theory, they are relatively small in comparison to the wave height of 20 ft. Note also that the free surface boundary condition errors rapidly decline for the higher order approximations, being quite insignificant at fifth order.

Background to Comparisons of Theory and Measurement. Measurements of crest kinematics are difficult and rare. A number of poor comparisons of theory and measurement have been reported (e.g. Le Méhauté et al 1968, Gudmestad and Connor 1986). The evidence from the few available measurements however is not conclusive and it would seem premature to reject the predictive potential of steady wave theory. Errors both in measurement and in application of theory can be substantial and more than sufficient to negate any hasty conclusions.

Laboratory measurements of "regular" waves are notoriously irregular. They are influenced by harmonic contamination from the wave maker, reflection from the beach and the wave maker, resonant modes within the flume and bound long wave motions, all of which violate the basic assumptions of steady wave theory. More detailed comparisons of regular wave kinematics with laboratory experiments must acknowledge these realities of the flow field in a flume. Such problems are essentially laboratory effects and should not distract attention from the real problem, namely whether gravity wave theory is generally capable of predicting crest kinematics. The core of gravity wave theory is the field equation (Laplace), the two free surface boundary conditions and the bottom boundary condition. It is clear from both profile asymmetry and the variability of neighboring waves that symmetry about the crest and the periodic lateral boundary conditions are frequently violated in a wave flume and it is unreasonable to insist on enforcing them in making comparisons with theory and experiment. An alternative approach might be some local approximation (see Section 5) that predicts the kinematics from a known water surface record using the field equation and the bottom and free surface boundary conditions but not profile symmetry and the lateral boundary conditions. Such an approach incorporates the essential physics of gravity wave theory and should provide a valuable measure of the predictive potential. It is nonetheless common and not entirely inappropriate to compare laboratory waves with steady wave theory.

The choice of a steady wave theory to approximate individual laboratory (or field) waves can also be troublesome. Some common theories have known mathematical errors. The Stokes V theory of Skjelbreia and Hendrickson (1962) has a known error at fifth order (Fenton 1985) and the tabulated Dean (1974) solutions for near-limit waves (Case D) are potentially spurious (Chaplin 1980, Sobey 1989). Published details of many analytical theories (Cnoidal theory in particular)

222

are very complicated (and prone to typographical errors). Many are also incomplete, perhaps excluding predictive equations for some dependent variables, or lack flexibility in the consideration of current.

Influence of Current. The transformation from the theoretical reference frame (that moves with the wave crest such that the motion is steady) to the physical (fixed in space) frame in which the motion is unsteady is often not given the attention it deserves. It involves the specification of the appropriate definition (Stokes' first or second) of phase speed and the associated current. The differences are potentially significant (Sobey et al 1987) and compounded by the fact that most published wave theories have automatically assumed Stokes' first definition of phase speed (and often also zero current) in the problem formulation. The current is assumed to be depth-uniform and any vertical structure is ignored. A Stokes theory with a linear velocity profile (i.e. constant vorticity) by Kishida and Sobey (1988) has demonstrated that vorticity has only a very minor influence on the wave-induced kinematics. This may not however be a general observation.

The adoption of a higher order wave theory may not be realistic where due attention has not been given to the influence of tidal or other ocean currents on the wave kinematics. Alternatively, if the current is not known, higher order precision can not and should not be expected.

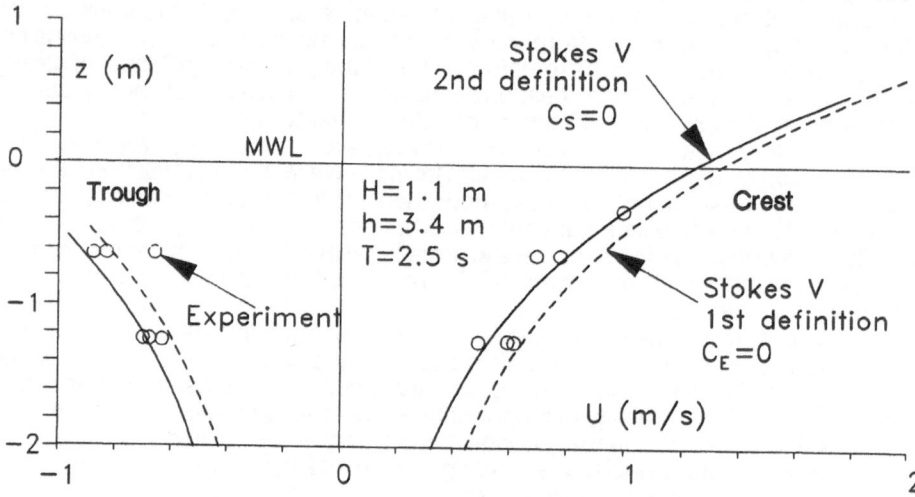

Figure 3.2 *Influence of Phase Speed Definition on Prediction of Crest and Trough Kinematics for Deep Water Laboratory Wave of Nath and Kobune (1978)*

Comparisons of theory and experiment by Nath and Kobune (1978) in deep water conditions and Le Méhauté et al (1968) in shallow water conditions serve to illustrate the potential for misinterpretation. The Nath and Kobune experiments were conducted in a wave flume and compared with Stokes V wave theory with Stokes' first definition of phase speed and zero current (i.e $C_E = 0$). The theoretical predictions of horizontal velocity under the crest and the trough were offset from the measurements, as shown by the dashed line in Figure 3.2. Stokes' second definition of phase speed with zero mass transport velocity (i.e C_s = 0) however is appropriate in a wave flume. The corrected predictions are also shown as the solid line in Figure 3.2 and give little cause for doubting the predictive capability of an appropriate steady wave theory.

Figure 3.3 is a rather more subtle illustration of the significance of current. It shows the experimental measurements of Le Méhauté et al (1968) in shallow water conditions compared to Fourier XVIII wave theory with zero Eulerian current (C_E = 0) and Fourier XVIII wave theory with zero mass transport or Stokes drift (C_s = 0). Neither of these assumptions regarding the coflowing current at the measurement section appears to be adequate, both overpredicting the horizontal velocity profile under the crest. The return current was not measured but it appears from Figure 3.3 that C_E is negative and that it exceeds C_s in magnitude. Such a laboratory current is indeed possible at the measurement section, as shallow water waves of significant amplitude were achieved by contracting both the width and the depth of the flume at the measurement section. The measurement section will act like a venturi on the small return current in the flume and an increase in magnitude of order C_s, sufficient to closely match the experiments, is plausible.

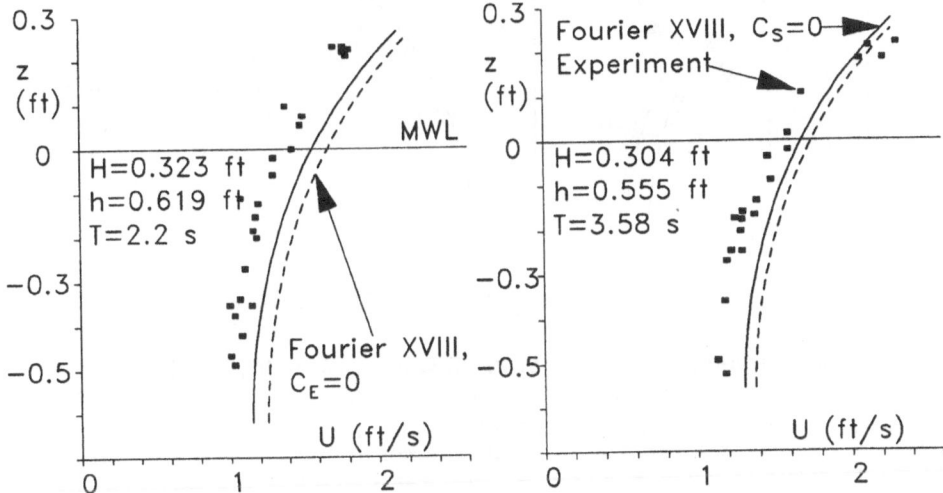

Figure 3.3 *Influence of Current on Predictions of Crest Kinematics for Shallow Water Laboratory Waves of Le Méhauté et al (1968)*

Empirical Adjustments to Steady Wave Theory in Deep Water. Reportedly poor theoretical predictions of crest kinematics in deep water have led to the introduction of a number of empirical adjustments to Airy wave theory (e.g. Wheeler 1970, Chakrabarti 1971, Gudmestad and Connor 1986, Lo and Dean 1986) in an effort to match the measurements. These empirical adjustments generally seek to reduce the errors in the free surface boundary conditions but, inevitably, these theories no longer satisfy the field equation.

Much is made of a perceived overprediction of the horizontal velocity under the crest by higher order steady wave theory in the presentation and development of these empirical adjustments. Gudmestad and Connor (1986), for example, cite the Nath and Kobune (1978) comparison (see Figure 3.2) of laboratory measurements with Stokes V theory (and Stokes' first definition of phase speed and zero current) and further laboratory measurements by Delft Hydraulics (1982). Figure 3.4 shows a sample comparison of the Delft measurements and the Gudmestad and Connor empirical approximation (both listed in Gudmestad and Connor 1986, Table 10b), together with the Fenton (1985) Stokes V prediction for both the Stokes' first and second definitions of phase speed and zero current.

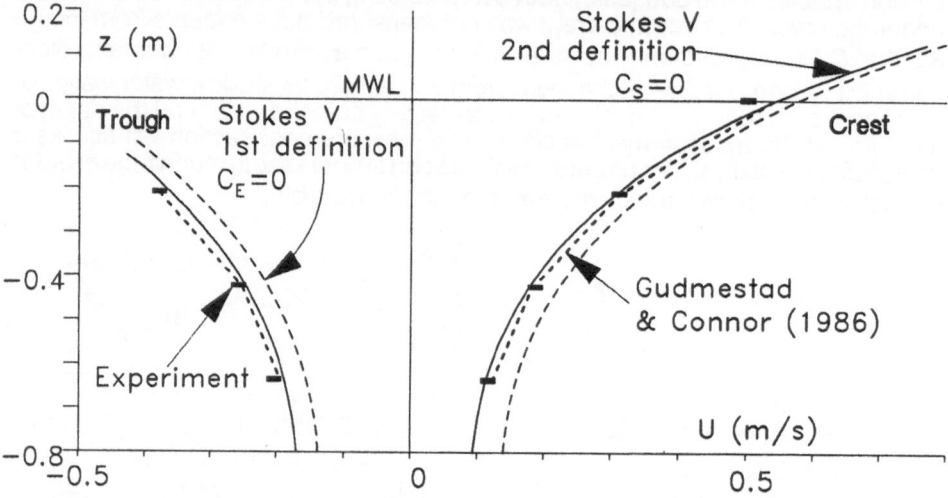

Figure 3.4 *Crest and Trough Kinematics for a Laboratory Wave of Delft Hydraulics (1982)*

Again, an appropriate higher order steady wave theory (here Stokes V or Fourier to a suitable truncation order) with $C_s = 0$ provides a credible prediction of the horizontal velocity under the crest and trough respectively. Gudmestad and Connor argue that their approximation is superior to those of Wheeler (1970) and Chakrabarti (1971); it would appear from Figures 3.2 and 3.4 however that there is little reason to doubt the predictive capability of the appropriate steady wave theory in deep water, or indeed even in shallow water (Figure 3.3).

Perceived inadequacies of steady wave theory appear to be almost entirely attributable to errors in the application of steady wave theory. On the basis of published data, there is no evidence to suggest that steady wave theory, when used correctly, is at all inadequate. The need for empirical adjustments to steady wave theory is not supported by published measurements. Accordingly, these empirical theories are not recommended and should be used with caution.

4 CREST KINEMATICS IN LINEAR RANDOM WAVES

The prediction of crest kinematics in real sea states defines a somewhat different problem. The pragmatic model of directional real sea states remains the Gaussian random wave model, which involves linear superposition of very many Airy waves of different amplitudes a_n, wave numbers k_n, wave frequencies ω_n and random phases ϕ_n. Each component is a free wave mode, such that each k_n, ω_n pair satisfies the Airy dispersion relationship. In discrete form, the water surface is

$$\eta(\underline{x}, t) = \sum_n a_n(\underline{x}) \exp(i(\underline{k}_n \cdot \underline{x} - \omega_n t + \phi_n)) \qquad (4.1)$$

Such a model inherits all of the advantages and disadvantages of Airy wave theory. It will simulate the directionality and random phasing of real sea states but is unlikely to perform well in the prediction of crest kinematics. Below the MWL, it has proved tolerably satisfactory under both laboratory (Delft Hydraulics 1982) and field (Forristall 1981) conditions.

The reliance on Airy theory is at the same time an advantage and a disadvantage. Linear superposition is a prerequisite of the powerful spectral description of a real sea state but Airy theory simultaneously maximizes the free surface boundary condition errors. Recourse to higher order approaches is certainly possible (see Section 5), but at the expense of loosing the convenience of the spectral description.

High frequency contamination of linear crest kinematics. High frequency contamination is often cited (Forristall 1985, Lo and Dean 1986) as perhaps the major error source in the prediction of crest kinematics from linear superposition. This problem arises from attempts to extrapolate Airy theory into the crest region. Strictly, Airy wave theory limits wave heights to infinitesimally small and restricts the vertical extent of the solution domain to the horizontal MWL.

Within the solution domain ($-h \le z \le 0$), the horizontal velocity, for example, is

$$u(x, z, t) = a\omega \frac{\cosh k(h + z)}{\sinh kh} \cos(kx - \omega t) \qquad (4.2)$$

Following the general Stokes theory, velocity predictions in the crest region are available from a Taylor series expansion about the MWL (which is within the solution domain). In practice, the result differs little from simply using Equation 4.2 with kz small and positive. Equation 4.2 is invariably used for extrapolation into the crest region, but note that $z \le \eta$ is implicit.

For a real sea state, the horizontal velocity field within the strict solution domain ($-h \le z \le 0$) is

$$\underline{u}(\underline{x}, z, t) = \sum_n a_n \omega_n \frac{\cosh k_n(h + z)}{\sinh k_n h} \cos(\underline{k}_n \cdot \underline{x} - \omega_n t + \phi_n) \qquad (4.3)$$

The extension of Equation 4.3 into the crest region appears automatic. This does lead however to some unrealistically large oscillatory velocities near the crest at the high frequency or high wave number end of the spectrum, where $k_n z$ values are no longer small.

This high frequency contamination has been illustrated by Forristall (1985) through superposition of a small high frequency component (H_2=2 ft., T_2=3 s) on the Figure 3.1 wave (H_1=20 ft., h=deep, T_1=10 s) considered previously. The water surface profile is

$$\eta(x,t) = \eta_1 + \eta_2 = \frac{H_1}{2}\cos(k_1 x - \omega_1 t) + \frac{H_2}{2}\cos(k_2 x - \omega_2 t) \quad (4.4)$$

and the horizontal velocity is

$$u(x,z,t) = \omega_1\eta_1 \frac{\cosh k_1(h+z)}{\sinh k_1 h} + \omega_2\eta_2 \frac{\cosh k_2(h+z)}{\sinh k_2 h} \quad (4.5)$$

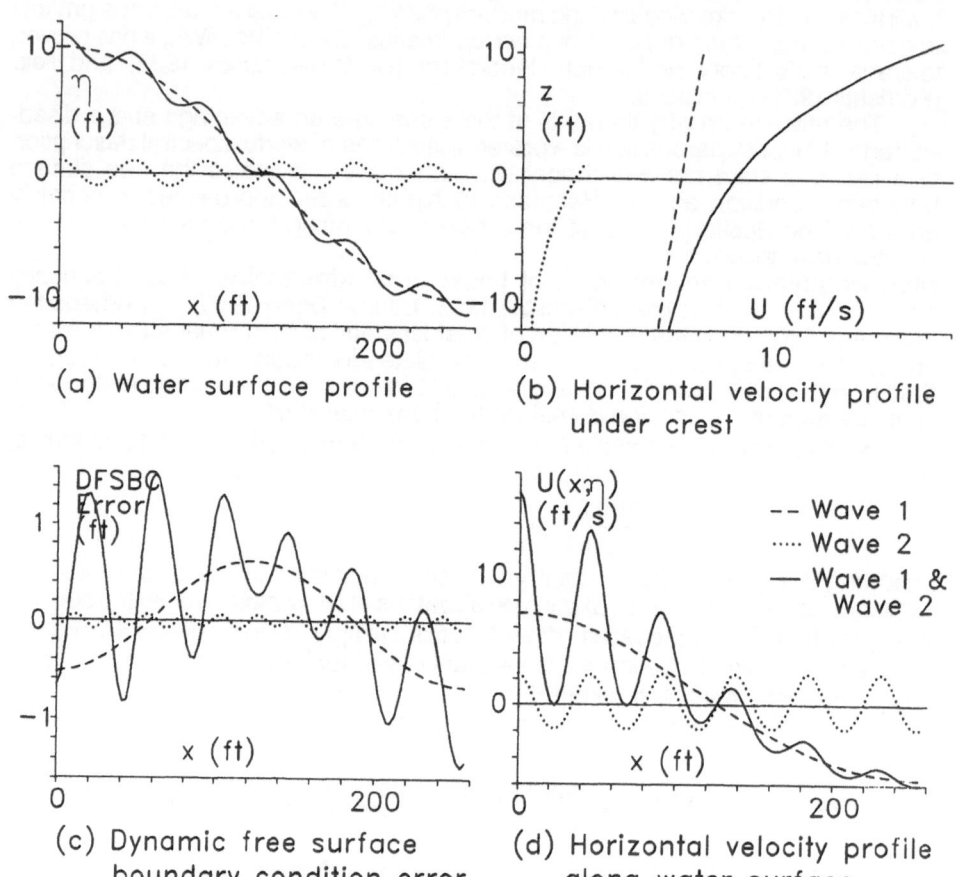

(a) Water surface profile

(b) Horizontal velocity profile under crest

(c) Dynamic free surface boundary condition error

(d) Horizontal velocity profile along water surface

Figure 4.1 *Apparent High Frequency Contamination of Crest Kinematics*

Figure 4.1 shows the predicted water surface, the predicted horizontal velocity profile under the crest and along the predicted water surface, together with the dynamic free surface boundary condition errors. Also shown are the predictions for waves 1 and 2 alone. Note that the free surface boundary condition errors are now substantially larger than those in Figure 3.1 and that there are large high frequency oscillations near the crest. These boundary condition errors are responsible for the unrealistically large high frequency oscillations in the horizontal velocity profile along the water surface. This unrealistic high frequency response is cited by both Forristall (1985) and Lo and Dean (1986) as a rationale for rejecting the predictive potential of crest kinematics from the Gaussian random wave model. As shown in part (b), the problems quite obviously originate from using Equation 4.5 with $z > \eta_1$ and $z > \eta_2$ Such points are above the water surface of the respective components and certainly beyond the predictive capability of Equation 4.2. Most of this so-called high frequency contamination results from an irrational extrapolation of linear wave theory above the MWL.

Predictions remain rational within the Airy solution domain $(-h \leq z \leq 0)$ but the Gaussian random wave model should not be extrapolated beyond the MWL. Recall also that the free surface boundary condition errors are not especially small with Airy theory (see Figure 3.1). It must be concluded that the linear superposition approximation has severely limited capabilities in predicting the crest kinematics above the MWL.

5 KINEMATICS IN IRREGULAR WAVES

In field and laboratory measurement programs, it is rare that every kinematic variable is measured and there is frequently a need to infer the time histories of those that were not measured. The water surface time history is often available and it is desired say to estimate the velocities, accelerations and pressures. Alternatively, a dynamic pressure time history from a submerged transducer might be available and it is desired to infer the water surface, velocities and accelerations.

In principle, the problem formulation retains much of the classical steady wave formulation, except that the waves are irregular and not steady in any reference frame, so that the unsteady forms of the kinematic and dynamic free surface boundary conditions, respectively

$$W = \frac{\partial \eta}{\partial t} + U \frac{\partial \eta}{\partial x} \qquad \text{at } Z = \eta \qquad (5.1)$$

$$\frac{\partial \Phi}{\partial t} + \frac{1}{2}(U^2 + W^2) + g\eta = \overline{B} \qquad \text{at } Z = \eta \qquad (5.2)$$

must be used, where $\Phi(X, Z, t)$ is the velocity potential function and $\overline{B}$ the Bernoulli constant, both in the fixed frame. Further, the periodic lateral boundary conditions are not appropriate, together with any assumptions that impose symmetry about a vertical axis through a wave crest. The field equation (Laplace) and the bottom boundary condition remain unchanged, although the latter may be generalized to accommodate a sloping bed.

Methodologies that have been introduced fall somewhat short of this goal and can generally be categorized as local approximations. Applicability in a global sense is compromised in an effort to achieve fidelity in a local sense. Note that this contrasts with the general approach of steady wave theory where local fidelity (especially near the wave crest) is perhaps sacrificed in the global interest. Given however that significant problems such as crest kinematics are strongly related to local errors in the free surface boundary conditions, there is intrinsic value in pursuing this approach.

Locally Steady Approximation. One apparent compromise (Dean 1965, Fenton 1986) is to assume that the wave field is locally steady with local phase speed C, such that variations with X and t can be locally combined as $x = X - Ct$, as in steady wave theory. This is a key assumption as it infers the spatial x variation from a time history of a dependent variable at a fixed X position. It also implies a single dominant mode that alone (as in steady wave theory) satisfies the dispersion relationship, all the higher harmonics being bound wave modes that do not satisfy the dispersion relationship.

Dean (1965) introduced a variation on steady Fourier wave theory for situations where the water surface time history $\eta(t)$ is known for a single wave. He assumed that the local stream function in the locally steady reference frame could be represented as a truncated Fourier series

$$\psi(x, z) = -\bar{u}(h + z) + \sum_{j=1}^{N} \sinh jk(h + z)(A_j \cos jkx + B_j \sin jkx) \quad (5.3)$$

The irregularity is introduced through the sine terms in the Fourier series, which do not appear in the related Fourier steady wave theory. The field equation (Equation 2.1) and bottom boundary condition (Equation 2.2) remain satisfied exactly and the Fourier coefficients are determined numerically to best fit the kinematic and dynamic free surface boundary conditions (Equations 2.3 and 2.4) at the known water surface nodes. Dean applied this methodology to a complete measured wave from trough to following trough, but it could equally be applied to a shorter segment of a wave record.

Fenton (1986) assumed that the local solution could be represented by a truncated polynomial series for the complex potential function

$$\phi(x, z) + i\psi(x, z) = \sum_{j=0}^{M} \frac{a_j}{j + 1}[x + i(h + z)]^{j+1} \quad (5.4)$$

where the a_j coefficients are real. Again, the field equation and bottom boundary condition are satisfied exactly and the polynomial coefficients could be determined numerically to best fit the kinematic and dynamic free surface boundary conditions at the known water surface nodes. Fenton actually considered the more complicated problem of estimating the water surface as well as the balance of the kinematics from a pressure time history at a submerged location. This is a familiar problem associated with bottom pressure wave recorders, for which Airy wave theory has been consistently used - even in shallow water. This polynomial procedure was shown to cope reasonably well for longer waves but not so well for shorter waves, where the vertical variation tends to exponential and where presumably the hyperbolic sine variation in Equation 5.3 is more suitable. The reverse comment would seem to be equally appropriate, such that these two approaches are complementary rather than competitive.

Field Solutions under Simulated Water Surface. Forristall (1985) noted that the Equation 5.1 form of the kinematic free surface boundary condition could be written in terms of the normal gradient of the velocity potential at the water surface

$$\frac{\partial \Phi}{\partial n} = \frac{-\frac{\partial \eta}{\partial t}}{\sqrt{1 + \left(\frac{\partial \eta}{\partial x}\right)^2}} \tag{5.5}$$

and that both the gradient and the location of the boundary are fixed if the $\eta(X, t)$ surface is known. The bottom boundary condition similarly defines a gradient condition at the bed. Forristall chose to undertake a complete field solution of the Laplace equation for a single irregular wave at a particular time. The solution domain is closed by the location of vertical lateral boundaries at adjacent troughs; zero potential is specified along these lateral boundaries, which is in principle a periodic lateral boundary condition. The solution domain at a particular time is thus specified without using the dynamic free surface boundary condition. A full field solution of the Laplace equation under such conditions is numerically straightforward but time consuming. The boundary integral method would perhaps be more efficient, especially as interest was focussed on the rather limited region of the solution domain near the crest.

This is again an incomplete definition of an irregular wave. The dynamic free surface boundary condition is not imposed and periodic lateral boundary conditions are retained. The need for a locally steady assumption is avoided by the specification of additional information, namely the space and time history of the water surface. Unfortunately, the water surface is generally not known in the spatial detail required and measured data is typically the water surface time history at a single X position.

Forristall considered several strategies for the estimation of the $\eta(X, Y, t)$ surface, including the linear Gaussian random wave model and a second order correction (in the deep water Stokes sense) to this linear surface, following Sharma and Dean (1979). Forristall (1986) has extended this methodology to two horizontal spatial dimensions with the $\eta(X, Y, t)$ surface estimated from a directional spectral description provided by the Gaussian random wave model (superposition of free modes) with second order (bound wave modes) corrections. This procedure becomes computationally intensive (Cray + vector processing).

6 CONCLUSIONS

Fidelity in representation of the nonlinear free surface boundary conditions is confirmed as the dominant consideration in efforts to achieve a satisfactory predictive capability for crest kinematics. It is this aspect of the mathematical physics that is compromised, by definition, in the lowest order wave theories, namely Airy or Stokes I or linear in deep water and Cnoidal I in shallow water. It is hardly surprising that the predictive capability of these lowest order theories is least satisfactory in the trough to crest region. Higher order theories much more faithfully represent the free surface boundary conditions and can be expected to be much more satisfactory in the trough-crest region.

This predictive capability for regular waves can only be achieved however if proper attention is given to any coflowing current. This is especially important in comparisons of wave theory with laboratory measurements where some inappropriate theoretical predictions have led to a lack of confidence in existing theoretical capabilities. A re-examination of these comparisons suggests that the predictive capability of higher order steady wave theory is indeed sound. It is the application of this theory that has not always been satisfactory. A range of empirical ('stretching') approximations have been introduced to compensate for a perceived inadequacy of steady wave theory but the rationale for these approximations is suspect and they are arguably unnecessary.

The popular Gaussian random wave model for real sea states is reasonably successful in deep water for the prediction of kinematics below the MWL. It should not be used however for extrapolation into the crest region. This is outside the theoretical solution domain for the separate Airy wave components and frequently even above the water surface of the separate Airy waves; it results in large amplitude but clearly spurious high frequency oscillations of predicted field variables in the trough-crest region.

Consideration of irregular waves is incomplete and requires further attention. Rational methodologies should focus on high fidelity representation of the free surface boundary conditions and refrain from imposing constraints that are appropriate only for a steady wave. Several local approximations have been introduced specifically for situations where the kinematics are to be predicted from a known water surface profile. The predictive potential and comparative performance of these methods needs detailed evaluation.

Nevertheless, these approaches fall short of the ultimate goal of a predictive capability for kinematics in an irregular sea state. Mathematical problems introduced by the nonlinear free surface boundary conditions are maximized at precisely the same region where physical interest is concentrated and where the magnitudes of the dependent variables are most extreme. A continuing interest in crest kinematics is certain to remain a valid concern among practicing engineers and researchers in coastal and ocean engineering for the foreseeable future.

REFERENCES

Chakrabarti, S.K. (1971). "Discussion on 'Dynamics of single point mooring in deep water'." Journal of Waterways, Harbors and Coastal Engineering Division, ASCE, 97, 588-590.

Chaplin, J.R. (1980). "Developments of stream-function wave theory." Coastal Engineering, 3, 179-205.

Dean, R.G. (1965). "Stream function representation of nonlinear ocean waves." Journal of Geophysical Research, 70, 4561-4572.

Dean, R.G. (1970). "Relative validity of water wave theories." Journal of Waterways, Harbors and Coastal Engineering Division, ASCE, 96, 105-119.

Dean, R.G. (1974). "Evaluation and development of water wave theories for engineering application." Coastal Engineering Research Center, Special Report No 1, 2 Volumes.

Delft Hydraulics Laboratory, (1982). "Wave kinematics in irregular waves." MaTS Report VM-1-4.

Fenton, J.D. (1979). "A high-order cnoidal wave theory." Journal of Fluid Mechanics, 94, 129-161.

Fenton, J.D. (1985). "A fifth-order Stokes theory for steady waves." Journal of Waterway, Port, Coastal and Ocean Engineering, ASCE, 111, 216-234.

Fenton, J.A. (1986). "Polynomial approximation and water waves." Procs., 20th International Conference on Coastal Engineering, Taipei, ASCE, 1, 193-207.

Fenton, J.D. (1989). "Nonlinear wave theories" In: Le Méhauté, B. and Hanes, D.M, eds. The Sea, Vol.9: Ocean Engineering Science. Wiley, New York, 1989, in press.

Forristall, G.Z. (1981). "Kinematics of directionally spread waves." Procs., Conference on Directional Wave Spectra Applications, Berkeley, ASCE, 129-146.

Forristall, G.Z. (1985). "Irregular wave kinematics from a kinematic boundary condition fit (KBCF)." Applied Ocean Research, 7, 202-212.

Forristall, G.Z. (1986). "Kinematics in the crests of storm waves." Procs., 20th International Conference on Coastal Engineering, Taipei, ASCE, 1, 208-222.

Gudmestad, O.T. and Connor, J.J. (1986). "Engineering approximations to nonlinear deepwater waves." Applied Ocean Research, 8, 76-88.

Gudmestad, O.T., Johnsen, J.M., Skjelbreia, J. and Torum, A. (1988). "Regular water wave kinematics." Procs., International Conference on Behaviour of Offshore Structures, Trondheim, Tapir Publishers, 2, 789-803.

Kishida, N. and Sobey, R.J. (1988). "Stokes theory for waves on linear shear current." Journal of Engineering Mechanics, ASCE, 114, 1317-1334.

Le Méhauté, B., Divoky, D. and Lin, A. (1968). "Shallow water waves: A comparison of theories and experiments." Procs., 11th International Conference on Coastal Engineering, London, ASCE, 1, 86-107.

Lo, J. and Dean, R.G. (1986). "Evaluation of a modified stretched linear wave theory." Procs., 20th International Conference on Coastal Engineering, Taipei, ASCE, 1, 522-536.

Nath, J.H. and Kobune, K. (1978). "Periodic theory velocity prediction in random wave." Procs., 16th International Conference on Coastal Engineering, Hamburg, ASCE, 1, 340-359.

Rienecker, M.M. and Fenton, J.D. (1981). "A Fourier approximation method for steady water waves." Journal of Fluid Mechanics, 104, 119-137.

Sharma, J.N. and Dean, R.G. (1979). "Development and evaluation of a procedure for simulating a random directional second order sea surface and associated wave forces." Ocean Engineering Report 20, University of Delaware.

Skjelbreia, L. and Hendrickson, J. (1962). "Fifth order gravity wave theory." Procs., 7th ICCE, The Hague, ASCE, 184-196.

Sobey, R.J. and Goodwin, P. (1986). "Which wave theory?" Procs., IBM University AEP Conference, San Diego.

Sobey, R.J., Goodwin, P., Thieke, R.J. and Westberg, R.J.jr. (1987). "Application of Stokes, Cnoidal and fourier wave theories." Journal of Waterway, Port, Coastal and Ocean Engineering, ASCE, 113, 565-587.

Sobey, R.J. (1988). "Truncation order of Fourier wave theory." Procs., 21st International Conference on Coastal Engineering, Malaga, ASCE, 1, 307-321.

Sobey, R.J. (1989). "Variations on Fourier wave theory." International Journal for Numerical Methods in Fluids, in press.

Wheeler, J.D. (1970). "Method for calculating forces produced by irregular waves." Procs., 1st Annual Offshore Technology Conference, Houston, 1, 71-82.

REVISION TO THE UK DEn GUIDANCE NOTES

N.D.P. BARLTROP
Atkins Oil & Gas Engineering Limited,
Woodcote Grove,
Ashley Road,
EPSOM,
Surrey KT18 5BW.

1. INTRODUCTION

The UK Department of Energy is revising the Guidance Notes that it issues as a basis for the Certification of Offshore Structures. WS Atkins were contracted to prepare a new section on fluid loading and to produce a supporting background document. A working group was set up with representatives from designers, oil companies, certifying authorities and research institutions to provide advice and guidance as the work progressed. The working group also provided a forum for preliminary consultation with interested parties with respect to the guidance notes themselves. British Maritime Technology and Lloyds Register of Shipping, who were represented on the steering committee, also produced papers that formed the basis of some of the material in the background document.

2. CONTENT

The documents discuss the following subjects:-

Particle kinematics - regular waves *
- irregular waves
- wave current interaction *
- breaking waves

Fluid loading - Morison's equation *
- vortex shedding
- diffraction
- slam & slap
- pressure
- interference

Extreme and fatigue analysis
- deterministic/spectral methods
- sensitivity to assumptions.

In this paper discussion is limited to the subjects marked with a * in the list above.

A. Tørum and O. T. Gudmestad (eds.), Water Wave Kinematics, 233–246.
© 1990 Kluwer Academic Publishers.

3. REGULAR WAVES

Regular waves are commonly used as part of the design/analysis process yet there is still uncertainty over what wave theories are appropriate in given water depths for particular height, period combinations.

Dean (1970) and Le Mehaute (1976) proposed wave theory selection diagrams (Figures 1 and 2) that are commonly used, however neither diagram was based on a study of the accuracy of the different theories for the calculation of particle kinematics, wave lengths and crest elevations which are of primary concern to the offshore structural designer.

We therefore conducted a small study using our program: ASAS-WAVE and Dean's stream function tables. We compared linear, Stokes 5 and stream function theories 3, 5, 7, 9 and 11 and estimated the range of applicability of each theory on the assumption that if two theories gave the same result the lower order theory was still valid. The diagram we produced is shown in Figure 3.

Since it is often convenient to use linear theory insead of a higher order theory, figures were also produced showing the error involved in using linear theory. An example, showing the error in crest velocity, is shown as Figure 4.

4. WAVE-CURRENT COMBINATION

Various methods of approximately combining wave and current are in common use, eg:-

1) Current profile stretching to the instantaneous surface
2) Mass continuity with profile stretching
3) Cut-off in trough, uniform current addition in crest

These methods are shown in Figure 5.

The various options can be tested for the simple case of a uniform current (ie. constant over the water depth), which is consistent with the irrotational flow assumption of the commonly used wave theories. This shows that effectively the wave should be considered to be riding on the current and that the kinematics of a current plus wave of given length can be obtained by simply adding the wave and current velocities.

This demonstrates that the frequently used mass continuity method (Figure 5, Method 2) will underestimate the horizontal particle velocity in the crest and underestimate the loading on a jacket structure.

Methods 1 and 3 are both correct for the case of uniform current. Of these, Method 1 is the most plausible since it approximately represents the convection of particles from eg. crest to trough in the wave when that particle has its horizontal velocity increased by the presence of the current. Method 3 implies that a given particle may be significantly accelerated and decelerated during the wave cycle by the presence of the current and this seems unlikely.

Therefore the method of current profile stretching, but without imposing wave continuity, has been recommended for modelling wave-current interaction.

For cases where the wave period is specified, instead of the wave length, then it is in principle necessary to modify the no current wave period. This is because the wave will travel faster on a co-linear current, and therefore the wave period is dependent on the current:

$$\frac{L}{T^u} = \frac{L}{T^0} + u$$

where L = wave length

u = current velocity in line with the wave

T^u = wave period with current

T^0 = wave period for wave of length L without current

$$\therefore T^u = \frac{L}{\frac{L}{T^0} + u}$$

5. MORISON'S EQUATION

There is still considerable uncertainty in the values of C_d and C_m that should be used with Morison's Equation. Figures 6 and 7 show drag coefficients measured for smooth and rough vertical cylinders in various experiments as a function of Keulegan-Carpenter Number $-K_c$.

At high K_c there is some evidence that C_d tends to the steady flow value as shown by Figure 8. Measured inertia coefficients are shown in Figures 9 and 10. At low Kc higher Cds are commonly measured although it is not clear to the author whether C_d is changing or whether the phase of the inertia force changes so that some of the dominant inertia force is interpreted as drag force.

Commonly designers take no account of either the effect of K_c or surface roughness on C_d or C_m. In order to determine the significance of ignoring the effect of K_c we estimated C_d and C_m

values as functions of Kc and calculated a total force coefficient C_f. We then repeated the exercise with a steady flow roughness and C_m which dropped from 2.0 to about 1.9 as roughness increased from $^1/10000$ to $^1/50$.

The results shown in Figures 9 and 10 demonstrate that constant coefficients can be used to obtain the correct total force on an individual member over a range of Kc so justifying the principle of using constant coefficients.

However, designers usually use values which are more readily associated with experimentally determined smooth cylinder coefficients, even when the cylinders are expected to be considerably roughened by marine growth. From the available data this would be expected to result in a considerable underestimation of the loading on rough cylinders. However, it seems likely that the use of the lower than realistic value of C_d can be justified by arguing that it contains a load reduction factor to account for a number of conservative assumptions also made by designers. These conservative assumptions include:

a) Separate extreme values of wave height, wave period, current and mean water level are combined.

b) Regular wave theories are used.

c) The seas are assumed uni-directional.

d) Possible benefits from shielding effects are excluded.

Some studies suggest that, at least for the UK sector of the North Sea, these factors may justify the load reduction factor implied by the use of the smooth cylinder drag coefficients. A study on this was performed by Lloyds and is included as an Appendix to the background document.

Acknowledgement

The authors thank the UK Department of Energy for permitting the publication of this paper.

References

Dean R.G., 1965, 'Relative Validity of Water Wave Theories', Journal Waterways, Harbours and Coastal Engineering Div., ASCE, Vol. 96, No. 1, pp. 105-119.

Le Mehaute B., 1976, 'An Introduction to Hydrodynamics and Water Waves', Springer-Verlag, Dusseldorf.

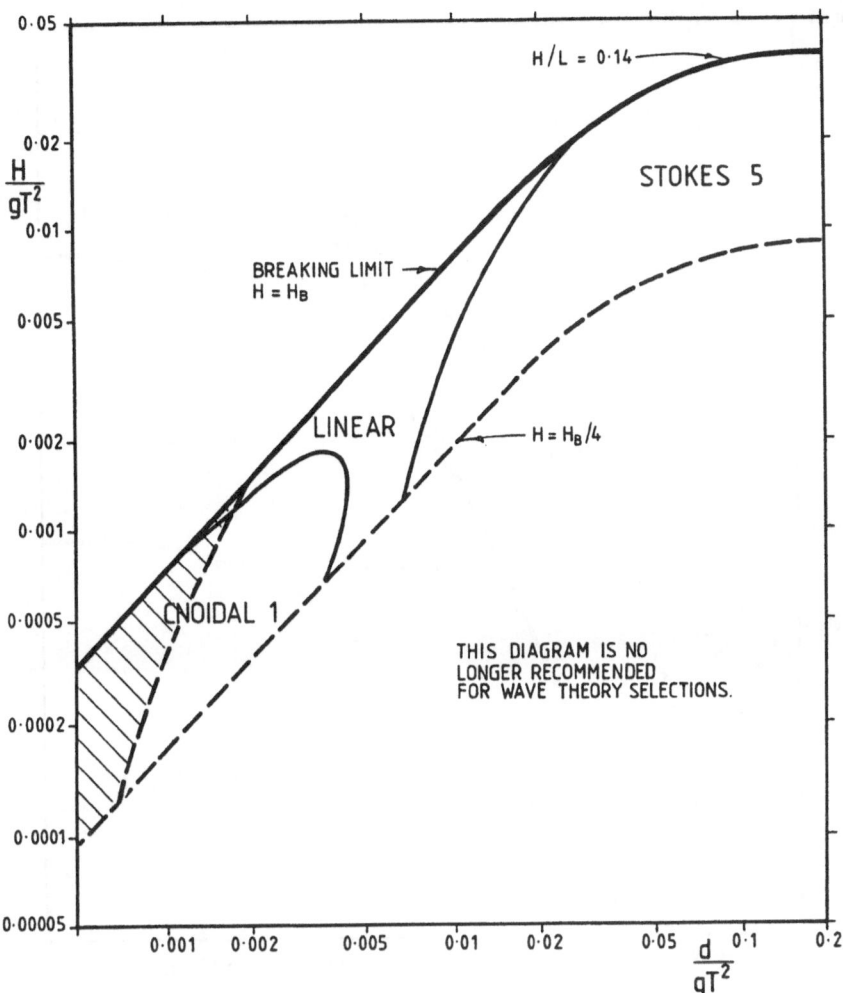

FIGURE 1 RANGES OF WAVE THEORIES GIVING THE BEST FIT TO THE DYNAMIC FREE SURFACE BOUNDARY CONDITION (DEAN, 1970)

238

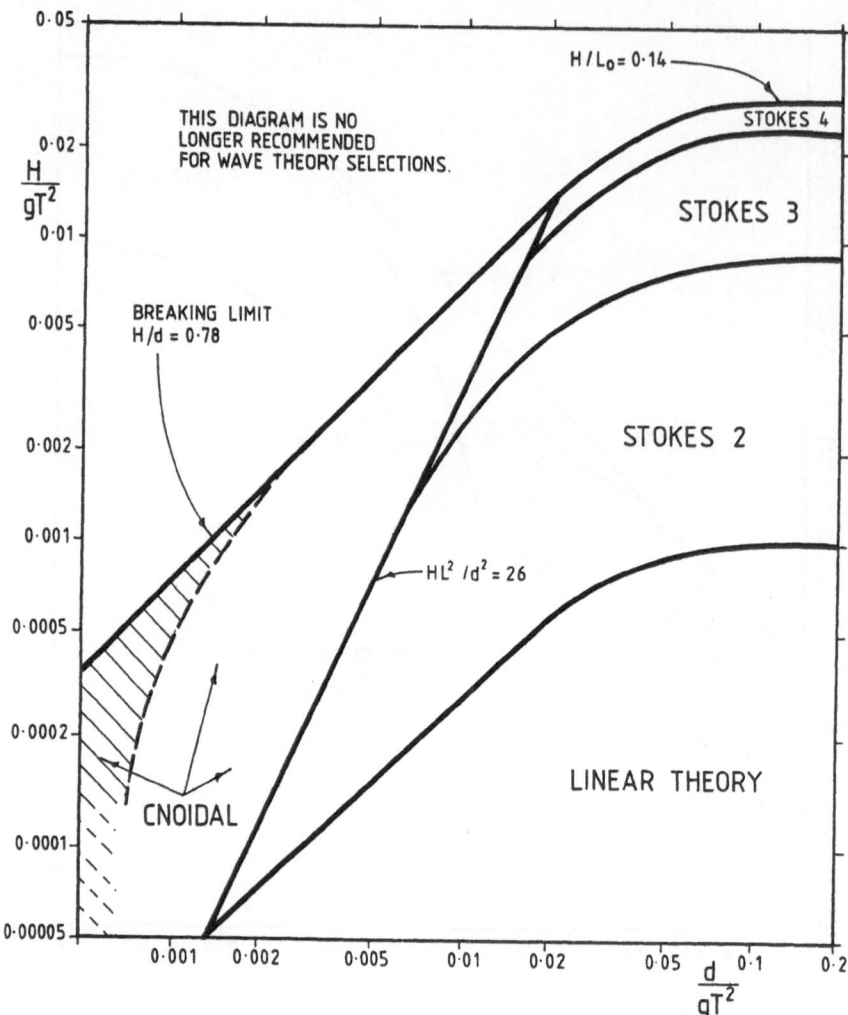

FIGURE 2   RANGES OF SUITABILITY FOR VARIOUS WAVE THEORIES AS SUGGESTED BY LE MEHAUTE (1976)

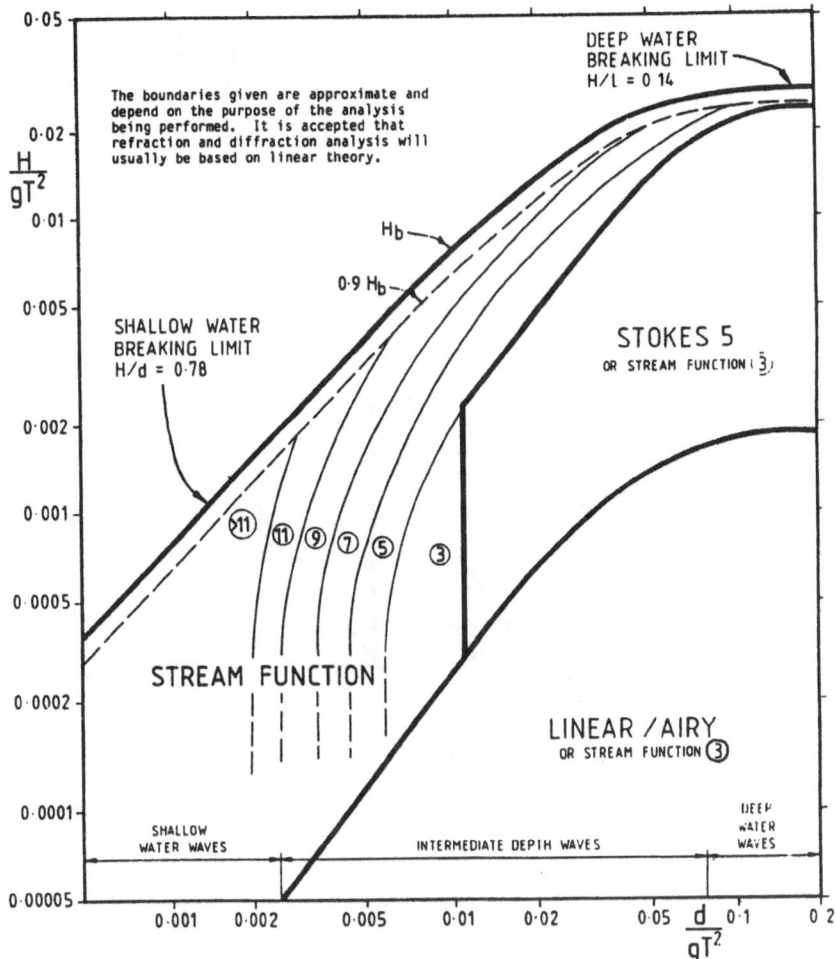

The boundaries given are approximate and depend on the purpose of the analysis being performed. It is accepted that refraction and diffraction analysis will usually be based on linear theory.

DEEP WATER BREAKING LIMIT
H/L = 0 14

Hb

0·9 Hb

SHALLOW WATER BREAKING LIMIT
H/d = 0·78

STOKES 5
OR STREAM FUNCTION (3)

STREAM FUNCTION

LINEAR / AIRY
OR STREAM FUNCTION (3)

SHALLOW WATER WAVES

INTERMEDIATE DEPTH WAVES

DEEP WATER WAVES

$\frac{H}{gT^2}$

$\frac{d}{gT^2}$

FIGURE 3 REGULAR WAVE THEORY SELECTION DIAGRAM (LOG SCALES)

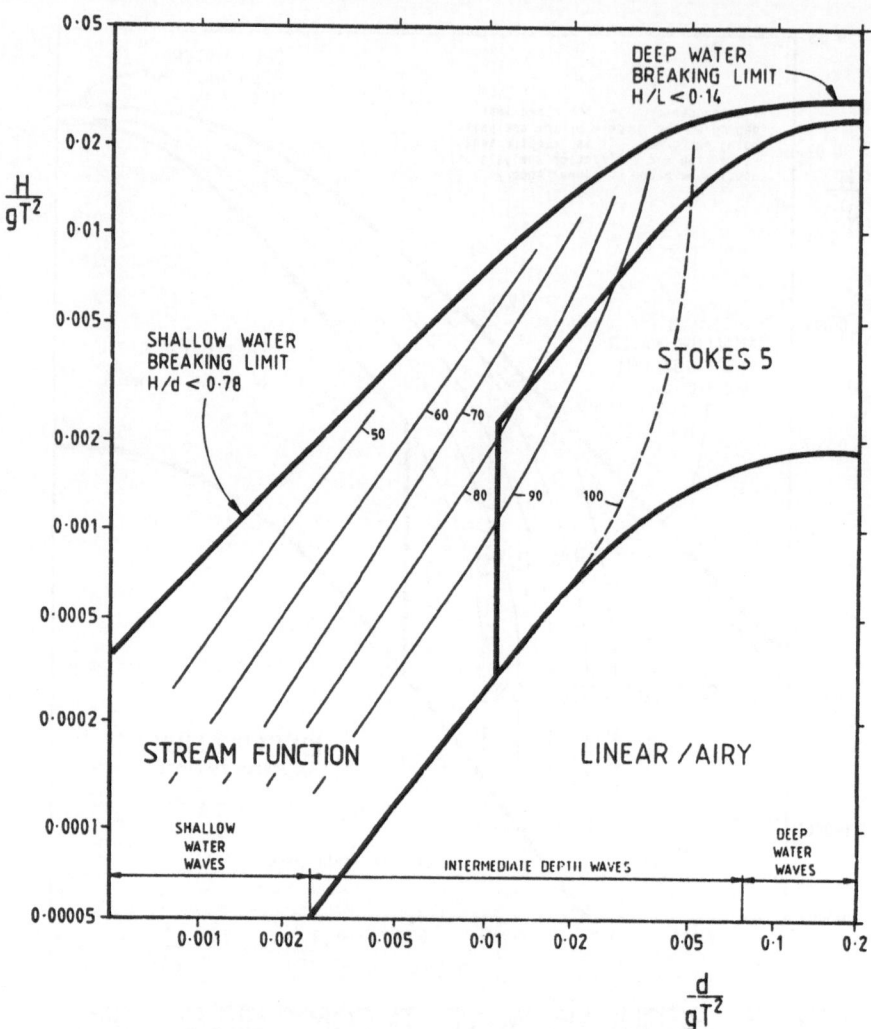

FIGURE 4 HORIZONTAL PARTICLE VELOCITY AT THE WAVE CREST AIRY THEORY AS % OF THEORY SHOWN IN FIGURE 3.

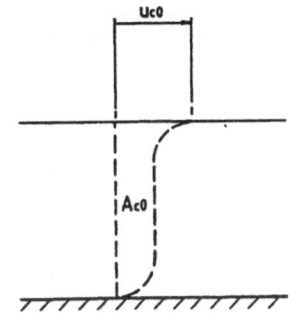

Current Profile with no Waves

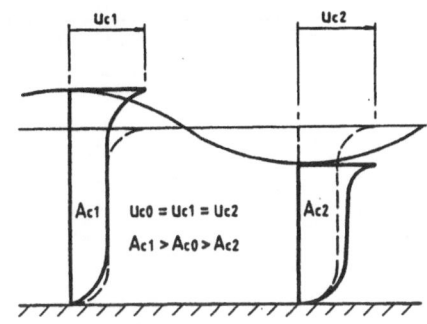

Method 1 - Current Profile Stretching

Key
--- current profile without wave

—— current profile in the wave

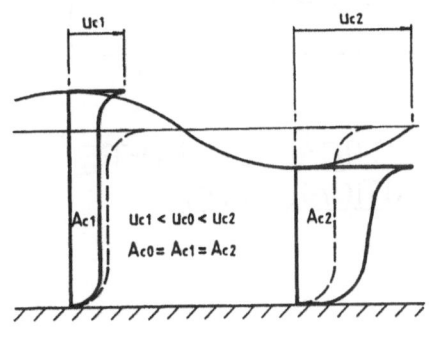

Method 2 - Mass Continuity with Profile Stretching

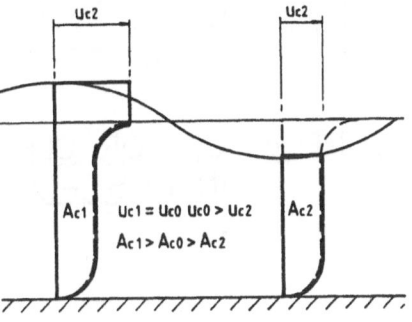

Method 3 - Cut-Off at Surface with Uniform Current Addition

FIGURE 5 VARIOUS METHODS OF COMBINING A CURRENT PROFILE WITH THE VARIATION IN INSTANTANEOUS WATER DEPTH DUE TO WAVE ACTION

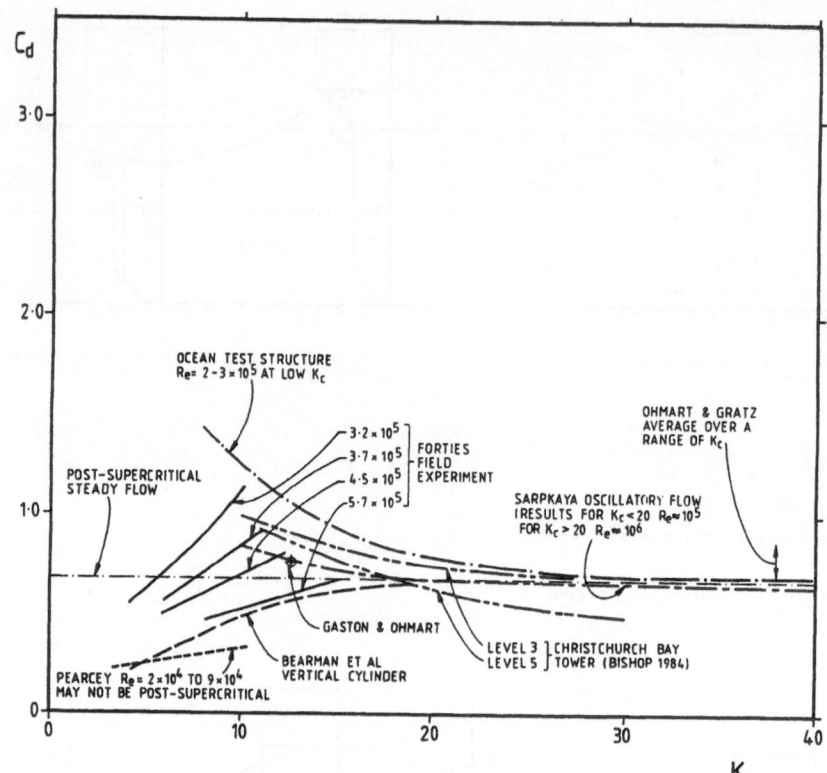

FIGURE 6   Cd FOR CLEAN VERTICAL CYLINDERS
IN POST−SUPERCRITICAL FLOW

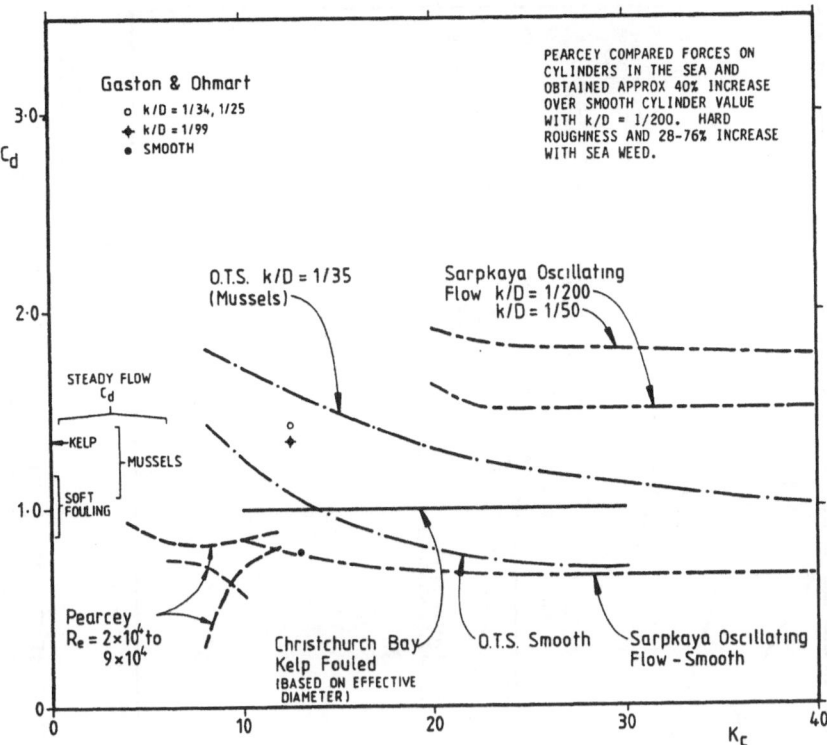

PEARCEY COMPARED FORCES ON
CYLINDERS IN THE SEA AND
OBTAINED APPROX 40% INCREASE
OVER SMOOTH CYLINDER VALUE
WITH k/D = 1/200. HARD
ROUGHNESS AND 28-76% INCREASE
WITH SEA WEED.

Gaston & Ohmart

○ k/D = 1/34, 1/25
✦ k/D = 1/99
● SMOOTH

O.T.S. k/D = 1/35
(Mussels)

Sarpkaya Oscillating
Flow k/D = 1/200
k/D = 1/50

STEADY FLOW
C_d

KELP

MUSSELS

SOFT
FOULING

Pearcey
$R_e = 2 \times 10^4$ to
9×10^4

Christchurch Bay
Kelp Fouled
(BASED ON EFFECTIVE
DIAMETER)

O.T.S. Smooth

Sarpkaya Oscillating
Flow - Smooth

C_d

3·0

2·0

1·0

0

0   10   20   30   40

K_c

FIGURE 7 C_d FOR ROUGH VERTICAL CYLINDERS IN POST—SUPERCRITICAL FLOW

244

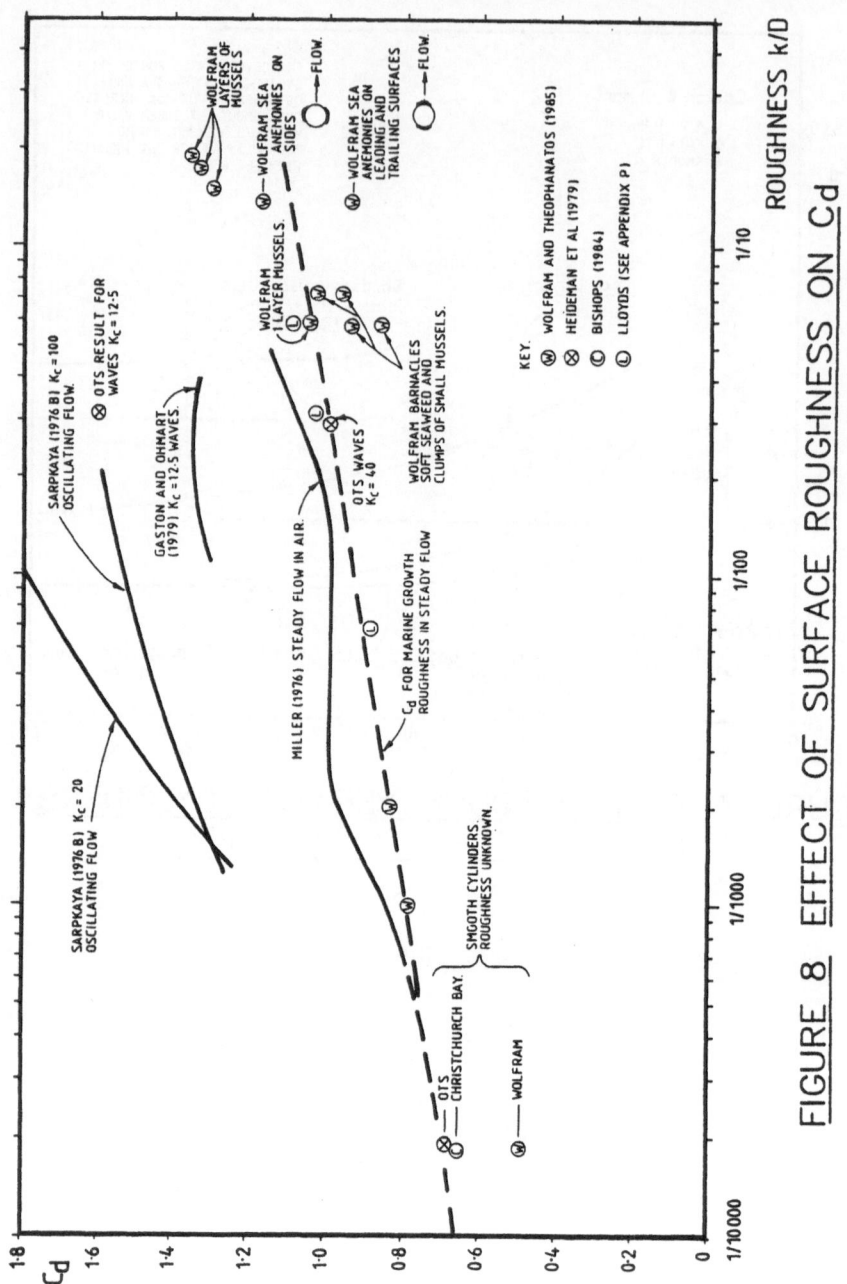

FIGURE 8   EFFECT OF SURFACE ROUGHNESS ON C_d

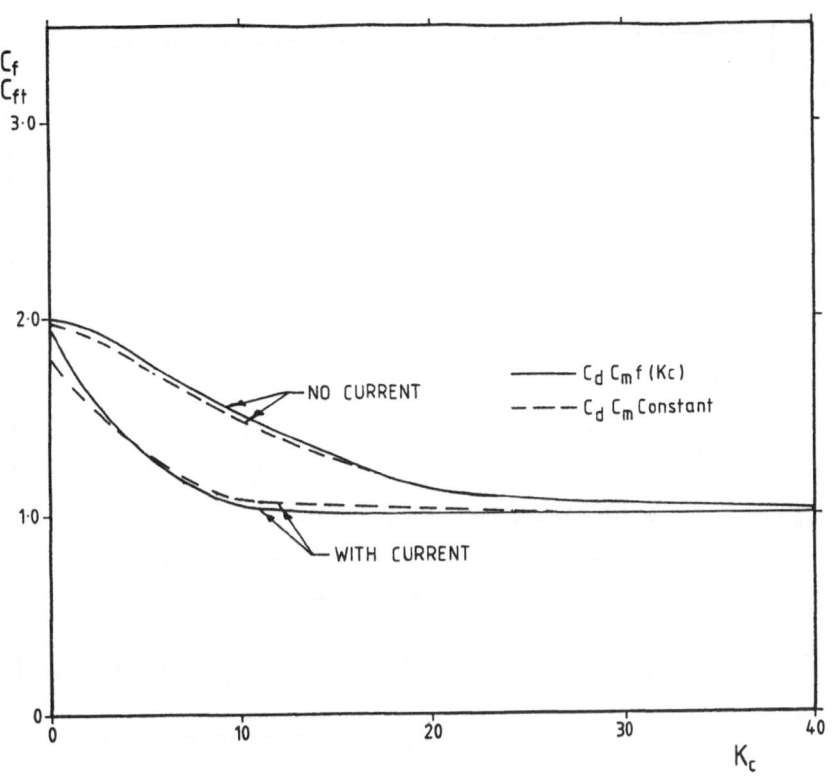

FIGURE 9   COMPARISON OF TOTAL FORCES USING SIMPLIFIED AND BEST ESTIMATE ROUGH CYLINDER FORCE COEFFICIENTS

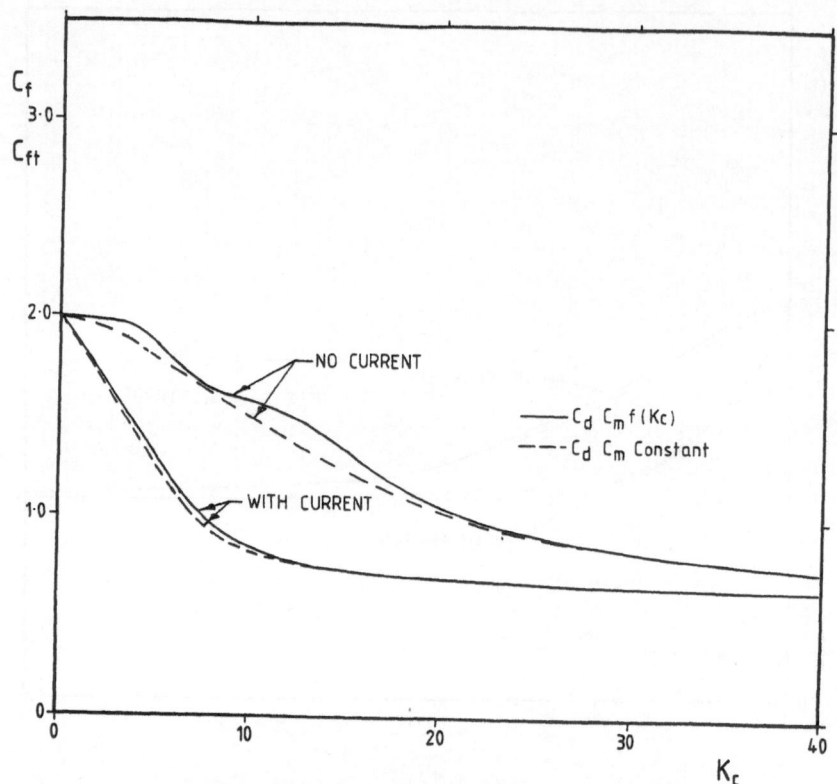

<u>FIGURE 10</u> <u>COMPARISON OF TOTAL FORCES:</u>
<u>SIMPLIFIED AND BEST ESTIMATE SMOOTH</u>
<u>CYLINDERS FORCE COEFFICIENTS</u>

CONDITIONAL SIMULATION OF OCEAN WAVE KINEMATICS AND COMPARISONS WITH STORM FIELD MEASUREMENTS

by

Leon E. Borgman
James Allender
Herald Krogstad
Stephen Barstow
Tore Audunson

ABSTRACT

Estimates of the directional wave spectrum and measurements of sea surface elevation time series are combined to produce conditional simulations of the subsurface water particle velocity time series. These simulations are compared with water particle velocities measured in North Sea storms. The conditionally simulated kinematics agree reasonably well with the data.

INTRODUCTION-SOME PHILOSOPHY

A very significant and revealing new perspective from which to examine interpolation and prediction has arisen in a number of different fields of science, sometimes associated with the new fields of chaos and fractal theory. This new perspective has variously been referred to as conditional simulation (Borgman, 1982; Borgman, Taheri, and Hagan, 1983), stochastic interpolation (Hewett, 1986), and fractal interpolation (Barnsley, 1988, p. 207-247). The terminology is interpreted to include extrapolation and prediction, as well as interpolation. The phrase "stochastic interpolation" will be used in the subsequent discussion to refer to this collection of techniques.

The underlying question motivating these developments may be stated as, 'If you have measurements in space and time, how do you best fill in the phenomena between or beyond the measurements or predict the space time continuum field of values for some co-related phenomena?'. The classical answer to this, based on "Occam's Razor" or "Rule of Simplicity," is that you use the most simple curve which passes through the data. The new rule of stochastic interpolation claims that you should fill in the region with simulated values which have the same statistical properties as the neighboring data, which pass through the measurements, and which trick the eye into believing it is seeing real high resolution measurements."

Stochastic interpolation has many advantages and several disadvantages and dangers. The purpose of interpolation, predic-

A. Tørum and O. T. Gudmestad (eds.), Water Wave Kinematics, 247–263.
© 1990 Kluwer Academic Publishers.

tion, and simulation is to provide reasonable versions of the real world behavior which are consistent with the measurements and physical structure of the process. This latter physical structure is usually violated by interpolation with Occam's rule since the interpolation will almost always show much more "smoothness" than would be present in real high resolution data. Also such subsidiary physical properties as spectral and covariance structure and higher order statistical moments are not preserved. Thus, Occam's Rule" interpolation/prediction procedures may immediately offend the eye as being unnatural and artificial.

Stochastic interpolation, prediction and simulation tries to be true to the measurements and to maintain the structure believed to exist in the space-time process. Unfortunately, this leads usually to a loss of uniqueness. That is, there will be many curves which pass through the data and have the desired statical structural relations. Stochastic interpolation, therefore, produces a suite or collection of possibilities that are representative of those interpolations and predictions that appear the most realistic in some sense relative to the application. Uniqueness is traded for realism and it seems reasonable that most ocean engineers would prefer realism.

As an illustration, Fig. 1 shows two stochastic interpolations which pass through the dots. Figure 2 gives an analogous stochastic interpolation in two-dimensions which goes through the prespecified circled specified values.

The imposition of additional physical structure, not contained in the actual measurements, when interpolating is common practice in many fields. For example, in finite element modeling, it is standard to use splines which impose continuity of higher order derivatives. In ocean wave studies, one often plots the logarithm of the spectral density versus the logarithm of frequency to seek a straight line relation at higher frequencies and then extends the straight line past the frequencies estimable from the data. This latter practice is really just a way to estimate the structure which has come to be called the "similarity rule" in fractal analysis (Feder, 1988, p. 217).

All of the above indicates that there are ways to impose other types of continuity than just that in the prosaic real space-time continuum. One can seek continuity in frequency domain, continuity in various derivatives, or continuity in other ways such as covariance relations and higher order polyspectra. Various of these, as appropriate in the application, can be imposed in the interpolations and prediction schemes. Incidentally, in this perspective, the usual conventional simulation procedures may be considered as predictions when you have no data.

The statistical procedure of considering measurements as being representable by a deterministic signal plus noise has, rightly so, been condemned as archaic and incorrect recently by some authors (See Casti, 1989, pp. 466-481). A more acceptable view would seem to be to take the process as evolving deterministically in such a way as to be unpredictable in long term behavioral aspects. This is not randomness in the sense of a chance mechanism, but in the sense of subjective uncertainty (De Finetti, 198) or chaos theory (Gleick, 1987; Devaney, 1987). However, probability as a model of personal uncertainty or likelihood of outcome, can still be used in a consistent way (De Finetti, op. cit.).

The greatest danger in stochastic interpolation and prediction is the possibility that one may lose sight of what is the imposed subjective structure and what is an immediate consequence of the actual data. To a large extent this danger can be eliminated by never considering just one interpolation or prediction of the space-time assemblage, but always working with a suite or collection of alternative versions of the reality consistent with the data and the hypothesized structure. The similarities between versions from the ensemble exhibit the commonality imposed by the data. At the same time, the differences indicate the inadequacy of the current data to really uniquely define the space-time continuum of the process.

INTRODUCTION - WAVE KINEMATICS
The stochastic interpolation, extrapolation, and prediction of ocean wave kinematics has been used in one way or another by many ocean engineers. Simulations of wave fields have been applied, particularly for studies in nonlinear dynamic compliant motion. It is difficult to avoid working in time domain for such problems, and pure simulations, not conditioned to agree with data, are a natural tool. Such stochastic predictions will be called "unconditional simulations". In contrast to this, simulations which are forced to agree with data will be called "conditional simulations" or "stochastic interpolations".

The technique of conditional simulation arose in mining engineering (Journel, 1974) and was extended to a variety of geostatistics applications (Borgman, 1983). It has been introduced as a technique for wave kinematics (Borgman, 1982), and more recently has been intensively studied in research funded by the U. S. Minerals Management Services and administered by the U. S. Navy Civil Engineering Laboratory, Port Hueneme (Borgman, 1990, in press). In a parallel way, the EPA Environmental Monitoring Systems Laboratory, Las Vegas, has sponsored research at the University of Wyoming into conditional simulation of random fields of contaminant data. This effort has produced a number of Ph.D. thesis studies at U. W. into various aspects of the procedure. A good presentation of the statistical background for the procedures is given by Brillinger, (1981, pp.286-290).

Conditional simulation of ocean wave kinematics has been used by Oceanor (Trondheim, Norway) in studies of data from the Wadic Project (Allender, et. al., 1989, in press; Vartdal,

Krogstad, and Barstow, 1989). It is being investigated in studies at Chevron's La Habra laboratories, and in moored platform dynamics by the U. S. Navy Civil Engineering Laboratory in Port Hueneme, California. As the result of the research directed by NCEL, a public domain software package for conditional simulation of wave kinematics has been prepared by the senior author of this paper. After a suitable shakedown to eliminate "bugs", the computer software, called SIMBAT, will be made available freely to the ocean engineering profession.

BASIC THEORY

Full details of the theory will be available in a chapter entitled "Irregular Ocean Waves: Kinematics and Forces", by Borgman in <u>The Seas, vol. 9, Ocean Engineering Science</u> now in press and scheduled for publication in 1990. Only a brief and limited summary of the theory will be given here.

The Fundamental Theorem of Conditioning: Consider a vector, partitioned into two parts,

$$\underline{V} = \begin{bmatrix} \underline{V}_1 \\ \underline{V}_2 \end{bmatrix} \tag{1}$$

Conceptually, $\underline{V}_1$ can be thought of as the part of the random vector $\underline{V}$ for which a measured realization, $\underline{v}_1$, is available. $\underline{V}_2$ is that part of the vector which is unknown and is to be simulated. Also let the covariance matrix of $\underline{V}$ be similarly partitioned as

$$C = \begin{bmatrix} C_{11} & C_{12} \\ C_{12}{}^T & C_{22} \end{bmatrix} \tag{2}$$

Theorem: A multivariate normal conditional simulation of $\underline{V}_2$, given $\underline{V}_1 = \underline{v}_1$, can be achieved in two steps
   (1) An unconditional simulation of $\underline{V}$ is performed to obtain $\underline{V}_1$ and $\underline{V}_2$.
   (2) Then, a conditional simulation is produced by ("T" denotes transpose)

$$\overset{\sim}{\underline{V}}_2 = C_{12}{}^T C_{11}{}^{-1} (\underline{v}_1 - \underline{V}_1) + \underline{V}_2 \tag{3}$$

The unconditional simulation of $\underline{V}$ in step (1) can be developed by any method which represents C as

$$C = A A^T \tag{4}$$

Then an unconditional simulation is obtained by

$$\underline{V} = A \underline{Z} + \underline{\mu} \tag{5}$$

where $\underline{Z}$ is a vector of independent standard normal random variables and $\underline{\mu}$ is the vector mean to be imposed. Techniques based on a Choleski decomposition of C or an eigenvector analysis of C (Borgman, 1982) can be introduced.

The required input for this theorem is the covariance matrix and the mean of $\underline{V}$. For ocean waves, these follow from the wave spectral density and the linear theory of waves. Only the sea surface elevation at (x_0, y_0) and the horizontal components of velocity at (x, y, z) will be treated here. Theory for all components of velocity and acceleration, the pressure anomaly, and the sea surface elevation at any set of space locations are given in the "Seas" reference mentioned earlier.

The three properties can be represented as

$$\eta(x_0, y_0, t) = \sum_{m=1}^{M} \sum_{j=0}^{J-1} a_{mj} \cos[k_m x_0 \cos(\theta_j) + k_m y_0 \sin(\theta_j) - 2\pi f_m t + \Phi_{mj}] \tag{6}$$

$$\begin{bmatrix} v_x(x,y,z,t) \\ v_y(x,y,z,t) \end{bmatrix} = \sum_{m=1}^{M} \sum_{j=0}^{J-1} a_{mj} (2\pi f_m) \frac{\cosh[k_m(d+z)]}{\sinh[k_m d]}$$

$$\cdot \begin{bmatrix} \cos(\theta_j) \\ \sin(\theta_j) \end{bmatrix} \cdot \cos[k_m x \cos(\theta_j) + k_m y \sin(\theta_j) - 2\pi f_m t + \Phi] \tag{7}$$

with

$$f_m = m/(N\Delta t) \quad \text{and} \quad (2\pi f_m)^2 = g k_m \tanh(k_m d)$$

where a_{mj} are independent Rayleigh-distributed random variables, and Φ_{mj} are independent random phases uniformly distributed on $(0, 2\pi)$ radians. These formulas can be rearranged into complex form by combining a_{mj} and Φ_{mj} into a single complex-valued amplitude as

$$\eta(x_0, y_0, t) = \sum_{m=0}^{N-1} \sum_{j=0}^{J-1} A_{mj} (2\pi f_m) e^{-k_m[x_0 \cos(\theta_j) + y_0 \sin(\theta_j)]}$$

$$\cdot e^{i2\pi f_m t} \tag{8}$$

$$
\begin{bmatrix} V_x(x,y,z,t) \\ V_y(x,y,z,t) \end{bmatrix} = \sum_{m=0}^{N-1} \sum_{j=0}^{J-1} A_{mj} \ (2\pi f_m) \ \frac{\cosh[k_m(d+z)]}{\sinh[k_m d]}
$$

$$
\cdot \begin{bmatrix} \cos(\theta_j) \\ \sin(\theta_j) \end{bmatrix} e^{-k_m[x \cos(\theta_j) + y \sin(\theta_j)]} \ e^{i2\pi f_m t} \tag{9}
$$

The three wave properties have mean zero in this model. The A_{mj} are related to the ocean wave spectral density by

$$
E[|A_{mj}|^2] = S(f_m, \theta_j) \ \Delta f \ \Delta \theta \tag{10}
$$

for $1 \le m \le M \le N/2$ and where $A_{mj} = A_{N-m,j}^*$ with the superscript denoting complex conjugation.

TWO PROCEDURES FOR CONDITIONAL SIMULATION

Attention will be limited here to the simulation of V_x (x, y, z, t) or V_y (x, y, z, t) given η (x_o, y_o, t) for $0<t<T$ over the full interval. Other procedures in the SIMBAT software can be used to conditionally simulate if only a partial record of η (x_o, y_o, t) are available, say for $t \epsilon S$ where S is a subset of (0, T).

The conditioning on a wave record which extends over the full time interval of interest, can best proceed in frequency domain. Two procedures are available. One works with the cross spectra between η and (v_x, v_y) at each frequency. The second proceeds with an initial conditional simulation of the complex matrix of A_{mj} , given $\eta(x_o, y_o, t)$ for $0<t<T$, and then uses the simulated A_{mj} in (9) to get v_x and v_y. The second procedure has many advantages and is the basic technique used in SIM4 of the SIMBAT software. However, the data comparisons which will be given below were made with the older cross-spectral matrix approach developed by Borgman (1973, 1980, 1982). The reader is referred to those sources and the new SEAS paper in press for further theoretical details.

SOME DATA COMPARISONS

How well does conditional simulation perform in following subsurface wave velocities. Data comparisons from the WADIC Project in the North Sea will be given. This project had the primary objective to evaluate operational, commercially available directional wave measurement systems under severe open ocean conditions. We refer to Allender et al.(1989) for a more complete description of WADIC which took place at the Edda platform in the North Sea. Current measurements were carried out at 8 different depths from a specially designed instrument tower attached to the NE corner of the platfom. In addition, 5 EMI laser altimeters recorded the surface height at the side of the tower. The original plans were for two current meters above the free surface, but the sea floor subsidence at Ekofisk caused the tower to be mounted about 1.5m lower than expected. The tidal level variations are generally less than .5m in this area. The highest current meter was a Marsh McBirney (MMB) electromagnetic current meter, whereas the remainder were Simrad Ultrasonic Meters (UCM-10). The sampling interval was 2 Hz, normally for a 40 minutes period every 3 hours, but hourly during storms.

Figure 3 shows the directional spectrum from 5 Nov., 1985, 1940hr. and measured vs. conditionally simulated time series. The thin lines are measurements and the thick lines simulations conditional Laser 3 about 7m to the side of the tower holding the current meters. The depth and the horizontal current components are marked on the graph. In this case, the coherence between the conditional and the simulated series is not one, and hence, a perfect agreement is not really expected.

On the subsequent plot (Fig. 4), the waves are almost exactly along the local y-axis and the agreement for the y-component is very good. There are, however occasional offsets. A statistical comparisons is given in Fig.5 where the distribution functions are compared for wave crests (more than 2m above MSL) and wave troughs (less than 2m below MSL). In general, these comparisons come out very close. Figure 6) shows a simulation involving current meters partly in free air. The simulation is first carried out without regard to the free surface, and portions of the surface in free air are discarded afterwards. The comparison is still good, but a closer look at the uppermost current meter reveals a slight overprediction compared to the measurements. By cutting the velocity transfer functions at an approproate frequency (.2 Hz) for the uppermost current meter (1m above MWL), the measurements and the simulations match quite well (Figure 7).

A conditionally simulated (and predicted) series will only show a perfect match to an actually measured series if these series have perfect coherence, i.e. if for every frequency, the cross spectrum Sxy satisfies abs(Sxy) = sqrt(abs(Sxx)^2 * abs(Syy)^2). (The coherence may be computed from the transfer functions and the directional spectrum. For example, it is easy to show that the surface elevation and one component of the horisontal current will not have perfect coherence unless the directional distribution is unidirectional in the direction of the current.) In some of the examples from the WADIC experiment, this explains why the predictions/simulations for the y-channels show better correspondence than the x-channels.

CONCLUSIONS
Stochastic interpolation, extrapolation, and prediction through conditional simulation appear to be useful tools for the ocean engineer in treating wave kinematics. The very recent availability of computer software which will make these tech-

niques more readily available to the engineer without requiring extensive statistical training, should allow much more extensive experience to be accumulated relative to their effectiveness

It should be pointed out that this paper has only examined one application of conditional simulation. The technique is useful in many more scenarios. For example, a short measured section of wave record with an exceptionally high wave or with a group of successive severe waves can be embedded in a much longer wave record through a conditional simulation of the rest of the time series. Then all the kinematics can be conditionally simulated based on this full time series.

The chapter in The Seas, vol. 9 (Borgman, in press) goes into the development of a procedure based on (x,y,z) expansion of conditionally simulated wave kinematics with Legendreorthogonal polynomials. This technique appears quite promising as a way to apply conditional simulations to vibrating or moored structures.

The full evaluation and implementation of these various techniques will probably extend over a number of years into the future. However, it seems likely that these methods, or techniques very similar, will find more and more use in the study and design of ocean structures under attack by complex directional seas and currents.

ACKNOWLEDGEMENTS

The opinions expressed in the paper are primarily those of the senior author. The associate authors have participated in many spirited discussions with him on these topics, and had the major role in the analysis of the WADIC Project data results reported and discussed in the paper.

All of the authors wish to express their most sincere appreciation to the following oil companies for their sponsorship and support of the WADIC Project and their permission to present data from the project in this paper: Chevron Oil Field Research Company, Conoco Norway, Inc., A/S Norske Shell, BP Petroleum Co. Norway, Unocal Norge A/S, Arco Norway Inc., Esso Norge A/S, Mobil Exploration Norway Inc., Phillips Petroleum Co. Norway A/S and Co-Venturers (including Fina Exploration Norway Inc., Norsk Agip A/S, Elf Aquitaine Norge A/S, Norsk Hydro A/S, Den Norske Stats Oljeselskap A/S, and Total Marine Norsk A/S), UK Department of Energy and the Continental Shelf, and Petroleum Technology Research Institute Ltd.

REFERENCES

Allender, J.; Audunson, T.; Barstow, S. F.; Bjerken, S.; Krogstad, H.; Steinbakke, P.; Vartdal, L.; Borgman, L. E.; and Graham, C. (1989, accepted for publication), "The WADIC Project: A Comprehensive Field Evaluation of Directional Wave Instrumentation", Coastal Engineering

Barnsley, Michael (1988) *Fractals Everywhere*, Academic Press, Inc., Boston, 394 pp.

Borgman, L. E. (1972), "Statistical Properties of Fast Fourier Transform Coefficients", *Tech. Paper No. 76-9*, U. S. Army Corps Engr. Coastal Eng. Res. Center, Fort Belvoir, VA. (Also. *Res. Paper No. 23*, College of Commerce and Industry, Univ. of Wyo., Laramie.

Borgman, L. E. (1980) "Conditional Simulation of Ocean Wave Properties", *Proc. 19th Conf. on Coastal and Ocean Engineering*, Sydney, Australia, 15 pp.

Borgman, L. E. (1982), "Techniques for Computer Simulation of Ocean Waves", *Topics in Ocean Physics* (Osborne, A. and Rizzoli, Ed) pp. 387-417, North Holland Pub. Co., Amsterdam.

Borgman, L. E.; Taheri, M; and Hagan, R. (1983), "Three-Dimensional, Frequency-Domain, Simulation of Geological Variables", *Geostatistics for Natural Resource Characterization*,NATO ASI Series (Verly, David, Journel, and Marechal, Ed.) Part I, pp. 517-541, D. Reidel Pub. Co., Dordrecht.

Borgman, L. E. (1990, in press), "Irregular Ocean Waves: Kinematics and Forces", *The Seas, vol. 9, Ocean Engineering Science*.

Brillinger, David (1981), *Time Series, Data Analysis And Theory* Holden-Day, Inc., San Francisco, CA., 540pp.

Casti, John L. (1989), *Alternate Realities, Mathematical Models of Nature and Man*, John Wiley and Sons, Interscience Publications, New York, 493 pp.

Devaney, Robert L. (1987), *An Introduction to Chaotic Dynamical Systems*, Addison-Wesley Pub. Co., Inc., Redwood City, CA, 320 pp.

DeFinetti (1974), *Theory of Probability*, two volumes, John Wiley amd Sons, New York, vol. I: 300 pp, vol.II: 375 pp.

Feder, Jens (1988) *Fractals*, Plenum Press, New York, 283 pp.

Gleick, James (1987), *Chaos, Making A New Science*, Viking Press, New York, 352 pp.

Hewett, T. A. (1986), "Fractal Distributions of Reservoir Heterogeneity and their influence on Fluid Transport", *Proc. 61st Annual Tech. Conf., Amer. Soc. Petroleum Eng.*, New Orleans, LA, Oct. 5-8, SPE 15386.

Journel, Andre (1974), "Geostatistics for Conditional Simulation of Ore Bodies" _Economic Geology_, vol. 69, pp. 673-687.

Vartdal, L.; Krogstad, H. E.; and Barstow, S. F. (1989) "Measurement of Wave Properties in Extreme Seas during the WADIC Experiment", _Proc. 1989 Offshore Tech. Conf._, Houston, Texas, May 1-4, OTC paper 5964, pp. 73-82.

Conditional Simulation

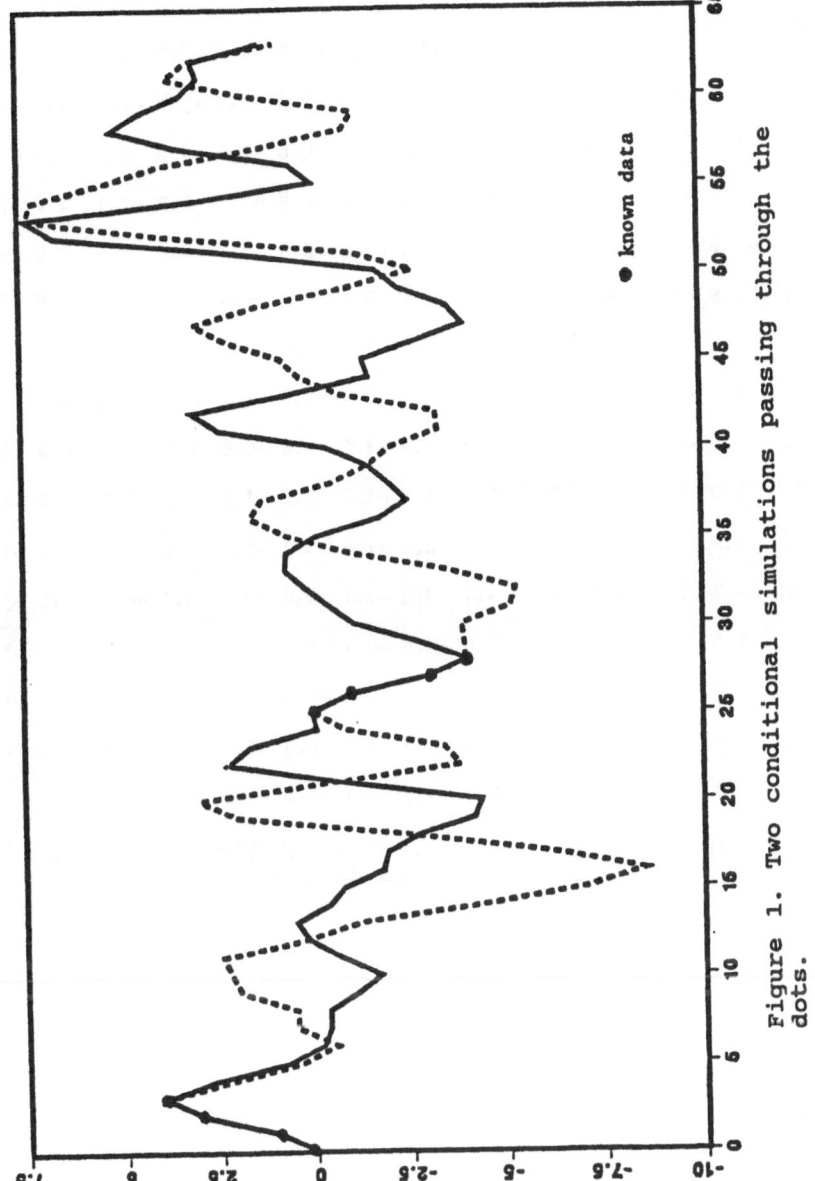

Figure 1. Two conditional simulations passing through the dots.

Realization 1

(1.0)	3.4	4.5	1.3	0.7	0.4	-1.8	1.1	-0.7	-1.9	-1.5	-0.3	3.3	0.5	-3.9	0.2
-1.2	1.6	(3.0)	4.5	3.4	2.1	-1.2	0.7	-2.6	-5.3	-3.8	-2.4	1.7	1.0	-3.9	-1.2
-1.3	-2.3	(0.0)	0.8	2.1	0.9	-2.9	1.0	0.6	-2.5	-1.8	-0.5	3.1	3.5	-1.4	1.2
-0.4	1.0	-0.5	1.3	3.9	3.9	-2.0	0.0	0.9	-1.2	-1.3	-1.1	1.2	2.1	-2.7	0.3
-1.4	-0.6	-1.2	2.3	4.1	5.5	-0.1	-0.2	-0.5	-2.4	-0.9	0.6	2.9	5.1	-0.1	-0.1
1.4	0.9	-1.2	3.7	4.7	6.1	2.0	2.4	2.3	-0.4	1.3	2.2	2.0	4.6	2.5	2.1
0.1	0.9	-1.8	3.2	3.6	3.6	0.8	1.8	1.9	-3.1	-1.4	2.6	3.4	4.7	3.3	0.8
0.8	0.9	-2.0	3.0	4.5	2.1	-1.8	-0.1	3.9	-0.2	-1.4	0.0	0.5	2.3	4.1	2.2
0.6	1.8	-0.5	1.9	4.9	3.8	-0.5	-1.6	1.6	-0.6	0.5	1.8	0.6	0.5	2.0	0.1
-0.4	1.0	0.7	0.3	2.5	2.0	-1.3	-1.6	1.5	-1.6	-0.4	2.2	1.0	1.2	3.7	0.1
-1.1	-0.2	1.7	0.9	2.9	3.0	-0.6	-1.6	1.7	-0.5	(-1.0)	3.1	0.9	-1.0	2.7	-0.2
0.6	0.7	2.3	0.4	2.2	3.7	1.3	-0.1	0.6	-0.8	-3.1	-0.8	-1.6	-3.7	0.7	0.4
0.7	1.0	1.7	-2.0	-0.7	1.0	-0.8	-0.1	-1.3	-0.1	-0.2	0.4	-0.7	(-3.0)	-1.3	-0.2
0.4	2.3	3.7	-1.6	-1.0	0.9	-2.2	0.0	-3.0	-3.3	-1.3	1.2	0.9	-3.0	-2.6	-0.8
1.2	2.8	4.7	-0.8	-1.1	1.6	-1.5	1.6	0.3	-0.9	-0.5	1.5	1.3	-2.7	-2.8	0.7
0.4	1.7	3.6	-0.7	-1.7	0.7	-1.6	-0.1	-2.8	-3.9	-2.8	0.6	2.9	-1.6	-4.4	-0.1

Figure 2. A 2-D conditional simulation constrained to agree with the circled data points.

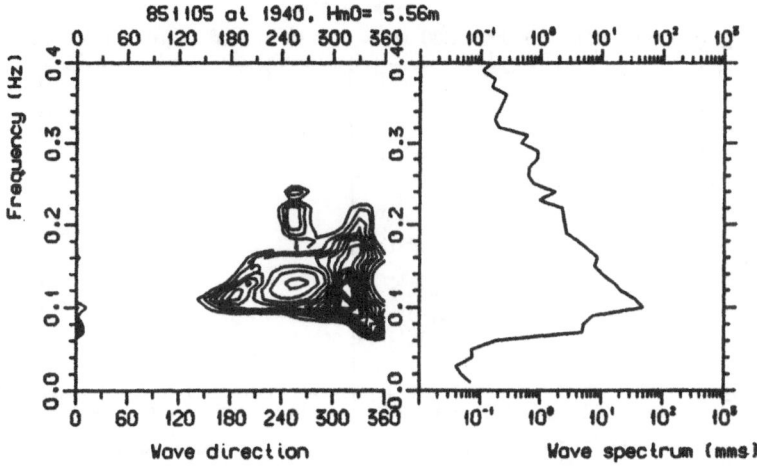

851105 at 1940, HmO= 5.56m

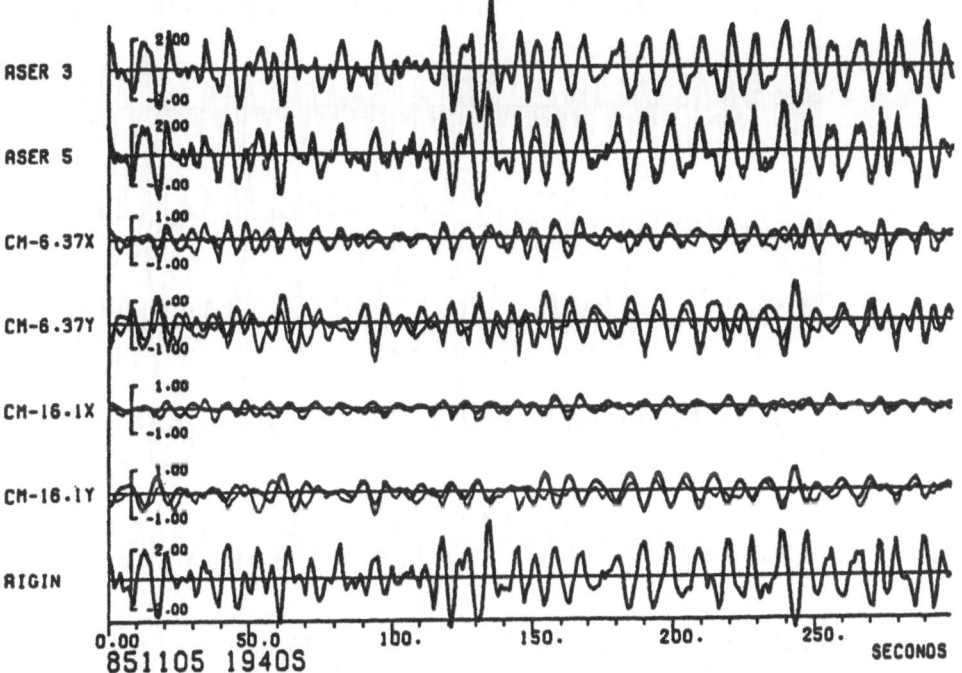

Figure 3. Conditional simulations (thick lines) versus
measurements (thin lines) for totally submerged
current meters.

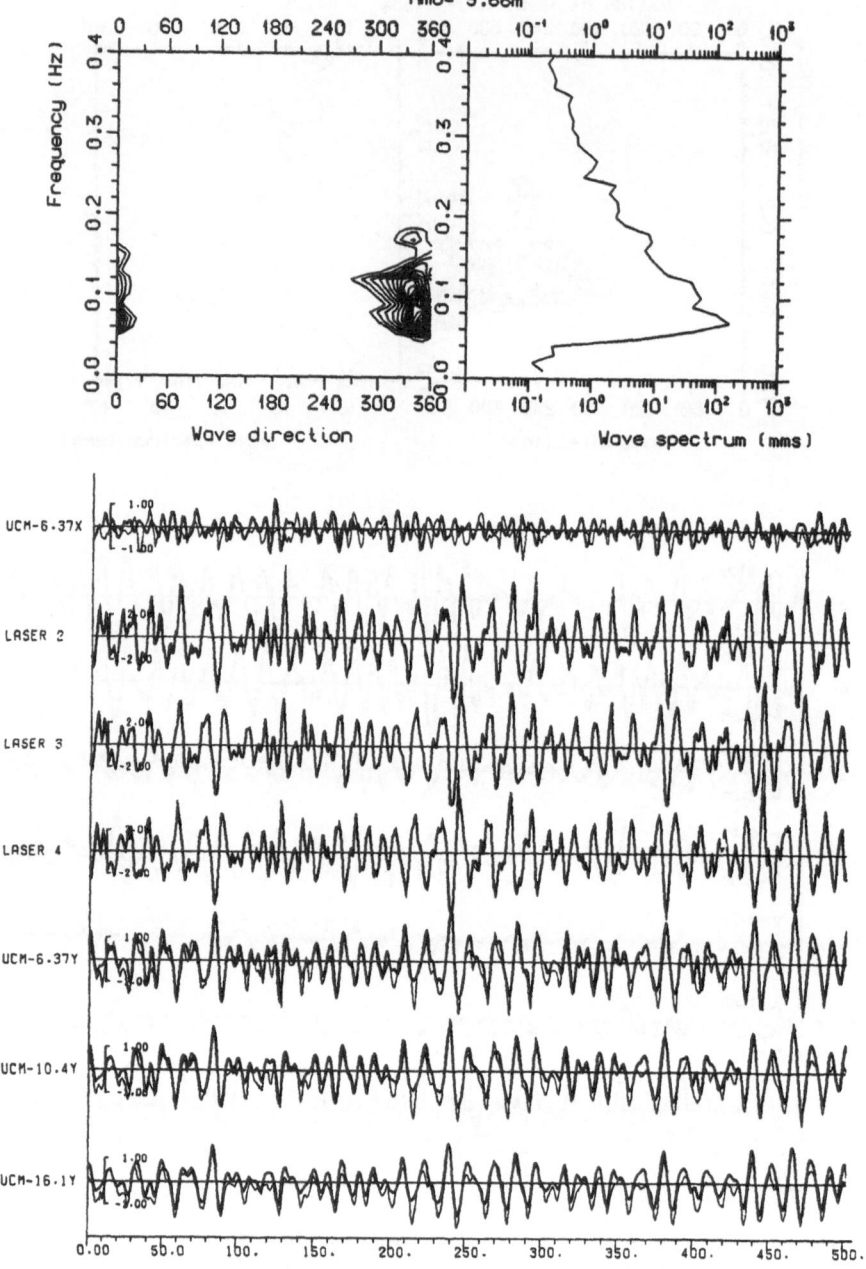

Figure 4.   Conditional simulation (thick lines) versus
            measurements (thin lines) for Nov. 6, 1985, 0420
            hrs. for totally submerged current meters.

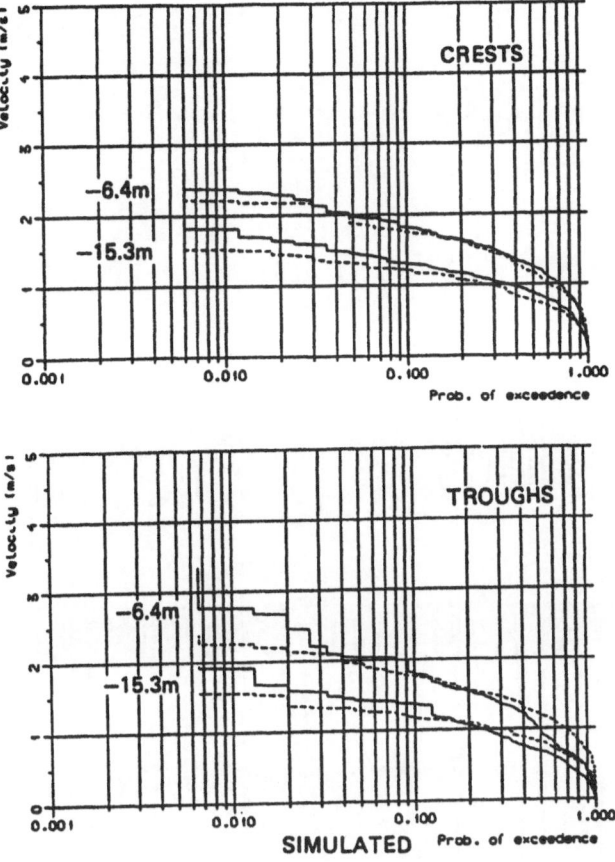

Figure 5. Cumulative plots for conditional simulations and measurements at crest and trough for November 6, 1985, 0430 hours.

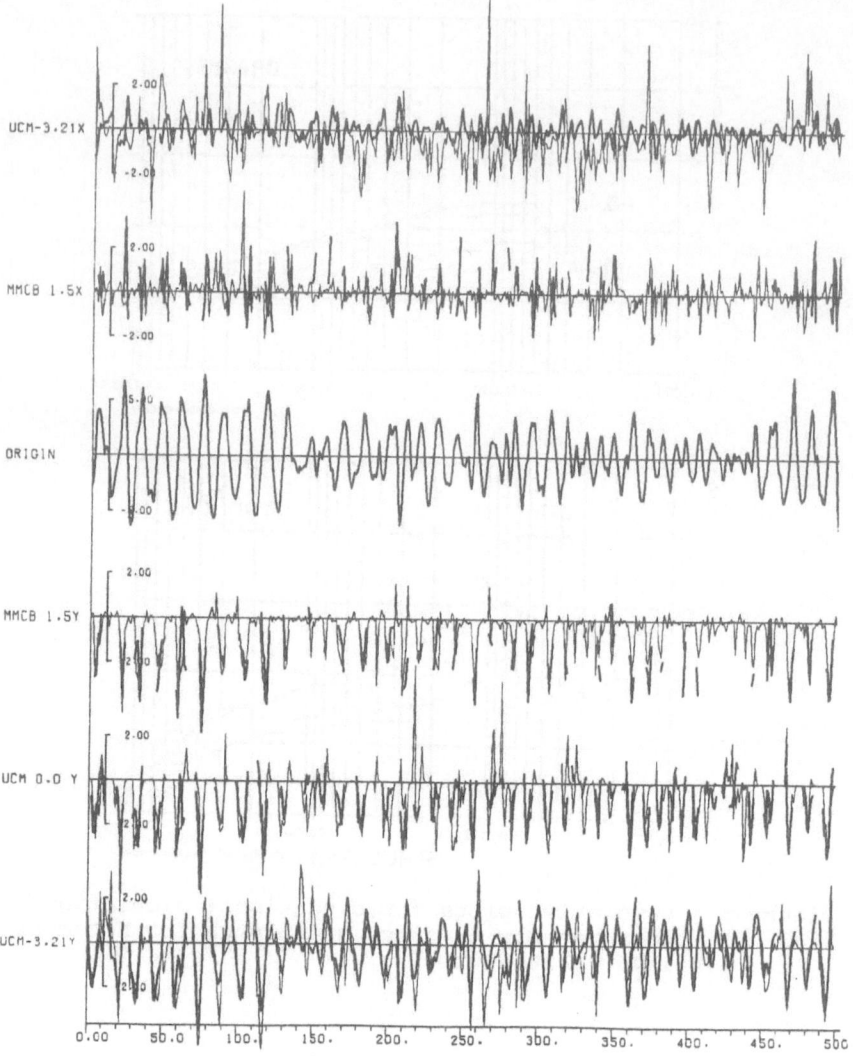

Figure 6.   Conditional simulations (thick lines) versus
            measurements (thin lines) for Nov. 6, 1985,
            0020 hrs. in the splash zone.

EXCEEDENCE PROBABILITY DISTRIBUTIONS FOR LWT AND MEASUREMENTS AT THE STORM PEAK +1m ABOVE MSL

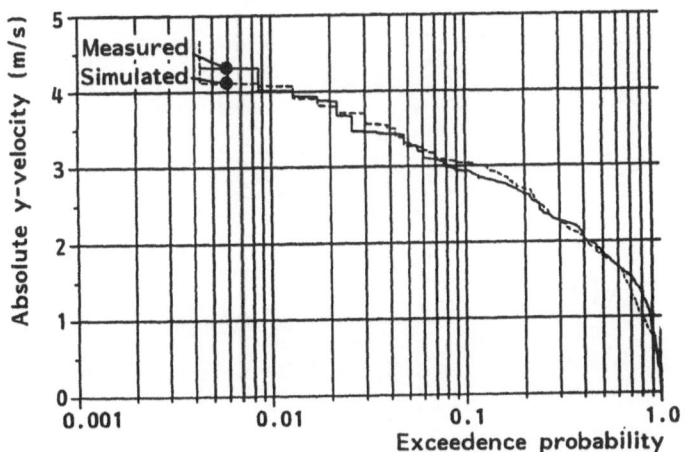

Figure 7. Measured vs. conditionally simulated velocities for the y-component of the Marsh McBirney current meter after the velocity transfer function has been truncated at 0.2 Hz.

SOME ASPECTS OF THE KINEMATICS OF SHORT WAVES OVER LONGER GRAVITY WAVES
ON DEEP WATER

C. KHARIF
Institut de Mécanique Statistique de la Turbulence
12, Avenue du Général Leclerc
13003 MARSEILLE, France

ABSTRACT. Modulations in wave number and in amplitude of short waves propagating over the surface of longer gravity waves on deep water is known to be of main importance for various fundamental and applied problems. They can be studied either by integrating the ray equations coupled with the action conservation principle established by Bretherton and Garrett (1969) or by treating the stability of the long waves to superharmonic perturbations. The aim of this note is to check the validity of the former approach when the nonlinearity of the longer wave is taken into account together with the capillarity effect on the short waves. A system of first order differential equations is derived from approximate formulation and solved numerically along the characteristics. A numerical investigation of superharmonic normal mode perturbations of finite amplitude Stokes wave has been developed by using the QZ algorithm. The variations of wave number and amplitude of the short waves over the profile of the long wave, obtained by the two different methods are similar, showing that ray equations and action conservation principle can be applied for strongly nonlinear inhomogeneous moving media.

1. INTRODUCTION

The energy and the wavelength of short waves are known to be affected by longer waves. How their amplitude and their wave vector are distributed along the phase of longer waves is a crucial question in fundamental and applied hydrodynamic fields. In this study the short waves are considered as linear waves and the longer wave is assumed to be fully nonlinear. The back reaction of the small waves on the larger wave is ignored.
          Henyey et al. (1988) have studied the dynamics of small waves riding on longer waves using a canonical formulation. They extended the calculation of Longuet-Higgins (1987) to include gravity-capillary waves and to allow three dimensional wave field. In this note the results given by the classical ray equations and the Bretherton and Garrett principle are compared to those derived from the stability computations.

A. Tørum and O. T. Gudmestad (eds.), Water Wave Kinematics, 265–279.
© 1990 Kluwer Academic Publishers.

Firstly the method of calculation of the surface velocities and orbital accelerations in the longer wave is briefly presented. Then, the approximate formulation and the stability study are developed. The last section reports on the results for pure gravity waves and gravity-capillary waves.

2. FINITE AMPLITUDE GRAVITY WAVES

The long gravity surface wave considered herein is a Stokes wave on deep water. In a frame of reference moving with constant speed C, the unperturbed surface defined for a wavenumber K = 1 is given by the usual parametric representation :

$$x = - \frac{\varphi}{C} - \sum_1^\infty H_n \sin n \frac{\varphi}{C}$$

$$z = \frac{H_0}{2} + \sum_1^\infty H_n \cos n \frac{\varphi}{C}$$

(2.1)

where x and z are respectively the horizontal and the vertical coordinates and φ is the potential velocity. The unknown coeffficients H_n and the phase velocity C are calculated by an iterative scheme developed by Longuet-Higgins (1985). Then the wave amplitude writes :

$$A = H_1 + H_3 + H_5 + \ldots$$

The orbital velocity and the orbital acceleration are given by :

$$U^2 = - 2z$$

(2.2)

$$a = - U^6 (x_\varphi + i z_\varphi) (x_{\varphi\varphi} + i z_{\varphi\varphi})$$

(2.3)

3. FORMULATION OF THE PROBLEM

3.1. Hamilton equations and action conservation

In this section we consider the dynamics of the small waves in a coordinate system s (s_1, s_2) tied to the longer wave. Wavelets superimposed upon and interacting with a much longer wave are considered as slowly varying wave trains of small amplitude propagating in an inhomogeneous moving medium as shown in figure 1. The vector wavenumber k (k_1, k_2) of the short waves is parallel to the long wave surface. The amplitude a of the small waves is defined in terms of normal distance from the long wave surface. We assume :

$$ak \ll 1 \qquad\qquad K \ll k$$

where ak is the wave steepness of the short waves and K the wavenumber
of the finite amplitude long waves.

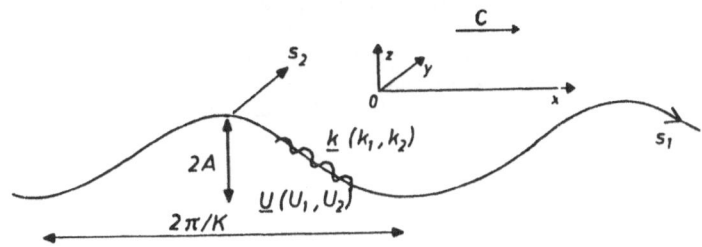

Figure 1. Definition sketch

The ray trajectories are given by :

$$\frac{dk_1}{dt} = - \frac{\partial W}{\partial s_1} \qquad ; \qquad \frac{ds_1}{dt} = \frac{\partial W}{\partial k_1} \qquad \text{(a)}$$

$$\text{(3.1)}$$

$$\frac{dk_2}{dt} = - \frac{\partial W}{\partial s_2} \qquad ; \qquad \frac{ds_2}{dt} = \frac{\partial W}{\partial k_2} \qquad \text{(b)}$$

$$W(k, \ s, \ t) = \sigma + k \cdot U \qquad \text{(3.2)}$$

$$\sigma^2 = g' \ (k_1{}^2 + k_2{}^2)^{\frac{1}{2}} + \frac{T}{\rho} \ (k_1{}^2 + k_2{}^2)^{3/2} \qquad \text{(3.3)}$$

$$g' = g - a \qquad \text{(3.4)}$$

$$g' = |g'| \qquad \text{(3.5)}$$

where s_1, s_2 and t are the space variables and the time variable,
W(k,s,t) is the local dispersion relation, σ is the intrinsic frequency,
U is the local orbital velocity in the long wave as seen by an observer
travelling with the long wave speed C, g' is the effective vector gravity, a is the orbital acceleration in the long wave and T is the surface
tension coefficient. Since the pressure gradient of the long wave has no
component tengential to the surface, g' is always normal to the free
surface.

In the frame of reference chosen, the long gravity wave is steady and the free surface is a streamline. In consequence the velocity **U** is always tangent to the free surface. In the case of a Stokes wave the component U_2 is zero. The component U_1 and the acceleration g' are determined from equations (2.2), (2.3), (3.4) and (3.5). In order to integrate equations (3.1) we write U_1 and g' as functions of the variable x by using Fourier transforms. Then the derivatives dU_1/dx and dg'/dx can be easily obtained.

Let $z = \eta(x)$ be the equation of the long wave surface. Then we can write :

$$ds_1 = (1 + \eta^2_x)^{\frac{1}{2}} \, dx$$

$$ds_2 = dy$$

and use these expressions to define x and y as independent variables instead of s_1 and s_2. Equations (3.1) are integrated numericaly with x, y and t as variables. The initial conditions are taken at the mean surface level of the long wave. The solutions give the change in wavenumber along the rays.

In order to determine the amplitude modulation we use the Bretherton and Garrett (1969) formulation. These authors established the action conservation principle as :

$$\frac{\partial}{\partial t} \left(\frac{E}{\sigma} \right) + \nabla \cdot \left[(\mathbf{U} + \mathbf{C}_g) \, \frac{E}{\sigma} \right] = 0 \tag{3.6}$$

where

$$E = \frac{1}{2} \rho \, g' \, a^2 \left[1 + \frac{T(k_1^2 + k_2^2)}{\rho g'} \right]$$

is the wave energy density in a frame of reference moving with the local current velocity and $\mathbf{C}_g$ is the group velocity of the short waves defined as :

$$\mathbf{C}_g = \frac{\partial \sigma}{\partial \mathbf{k}}$$

In the present case the medium is time independant, so that the frequency is conserved along the trajectories and the equation (3.6) may be written as :

$$\nabla \cdot \left[(\mathbf{U} + \mathbf{C}_g) \, \frac{E}{\sigma} \right] = 0 \tag{3.7}$$

3.2. Stability

A more extensive calculation relative to the modulation of short wave is derived using the numerical investigations of Kharif and Ramamonjiarisoa (1988). The motion of deep water waves on an inviscid irrotational incompressible fluid obeys a well-known set of a linear equation and nonlinear boundary conditions. In first place we ignore the effect of the capillarity.

In a frame of reference moving at some constant speed C, the basic equations are :

$$\nabla^2 \psi = 0 , \quad -\infty < z < \eta , \quad \lim_{z \to -\infty} \psi = - Cx \tag{3.8}$$

$$\psi_t + \eta + \frac{1}{2} (\psi_x^2 + \psi_y^2 + \psi_z^2) = \frac{1}{2} C^2 \tag{a}$$

on $z = \eta(x,y,t)$ (3.9)

$$\eta_t + \psi_x \eta_x + \psi_y \eta_y - \psi_z = 0, \tag{b}$$

where x, y are the horizontal coordinates, z is the vertical coordinate, t is the time, $\psi(x,y,z,t)$ is the velocity potential, and $z = \eta(x,y,t)$ is the free surface. As usual, the gravitational acceleration and the wavelength are taken, respectively, to be unity and 2π without loss of generality.

The system above is known to admit two-dimensional steady solutions (Stokes waves with phase speed C). This study deals with the stability of these solutions to two and three dimensional perturbations. Let :

$$\eta = \overline{\eta} + \eta' \qquad \qquad \psi = \overline{\psi} + \psi' \tag{3.10}$$

where $(\overline{\eta}, \overline{\psi})$ and (η', ψ') correspond, respectively, to the unperturbed and infinitesimal perturbative motions ($\eta' << \overline{\eta}$, $\psi' << \overline{\psi}$).

The first order pertubation equations can be written as :

$$\nabla^2 \psi' = 0 , \quad \quad -\infty < z < \overline{\eta}$$

$$\psi'_t = - \eta' - \overline{\psi}_x \psi'_x - \overline{\psi}_z \psi'_z - (\overline{\psi}_x \overline{\psi}_{xz} + \overline{\psi}_z \overline{\psi}_{zz}) \eta' \tag{a}$$

on $z = \overline{\eta}(z)$ (3.11)

$$\eta'_t = - \overline{\psi}_x \eta'_x - \overline{\eta}_x \psi'_x - (\overline{\psi}_{xz} \overline{\eta}_x - \overline{\psi}_{zz}) \eta' + \psi'_z \tag{b}$$

with solutions of the following form :

$$\eta' = e^{-i\sigma t} e^{i(px + qy)} \sum_{-\infty}^{+\infty} a_j e^{ijx} \qquad (a)$$

(3.12)

$$\varphi' = e^{-i\sigma t} e^{i(px + qy)} \sum_{-\infty}^{+\infty} b_j e^{ijx} e^{\sqrt{(p+j)^2 + q^2}\, z} \qquad (b)$$

An eigenvalue problem for σ with eigenvector $\mathbf{u} = [a_j, b_j]$ is derived from (3.11) :

$$(\mathbf{A} - i\sigma \mathbf{B})\, \mathbf{u} = 0 \qquad (3.13)$$

where A and B are complex matrices depending on the unperturbed wave steepness AK and the arbitrary real numbers p and q. The eigenvalue satisfies :

$$\left| i\sigma \mathbf{B} - \mathbf{A} \right| = 0$$

The unperturbed surface is given by the usual parametric representation (2.1).

The purpose is to study the modulations of perturbations with wavelengths smaller than that of the umperturbed wave, so we will only considered the case of superharmonic stability (p = 0). The deflection due to the perturbations may be written as :

$$\eta' = e^{-i\sigma t} e^{iqy} \alpha(x) e^{i\phi(x)}$$

where $\alpha(x)$ is the envelope and $\phi(x)$ the phase of the short waves.

Then, with simple transformations, it is easy to calculate the wavenumber and the amplitude of the short wave as defined in section 3.1, along the surface coordinates of the long wave.

If the effect of capillarity is taken into account only for the short waves, we have to add supplementary terms to the equation (3.11.a) :

$$\varphi'_t = -\eta' - \overline{\psi}_x \varphi'_x - \overline{\psi}_z \varphi'_z - (\overline{\psi}_x \overline{\psi}_{xz} + \overline{\psi}_z \overline{\psi}_{zz})\eta'$$
$$+ \kappa\, [(1 + \overline{\eta}_x{}^2)^{-\frac{1}{2}}\, \eta'_{yy} + (1 + \overline{\eta}_x{}^2)^{-3/2}\, \eta'_{xx} - 3(1 + \overline{\eta}_x{}^2)^{-5/2}\, \overline{\eta}_{xx}\overline{\eta}_x \eta'_x] = 0$$
$$, z = \overline{\eta}(x)$$

where $\kappa = K^2 T/\rho g$ is the non dimensional surface tension coefficient. Then the procedure is similar to that presented previously.

Let us consider the special case in which the unperturbed wave has zero amplitude. The dispersion relation may be written as :

$$\sigma_n = -n \pm [n^2 + q^2]^{1/4} [1 + \kappa(n^2 + q^2)]^{\frac{1}{2}}$$

where n is the wave number of the perturbation in the x direction. When n and q become significantly large we cannot neglect the term κ (n^2 + q^2), even if $\kappa \ll 1$.

4. RESULTS

Modulations in wave number and in amplitude of superharmonic modes travelling in the same sense (positive sense) or in opposite sense (negative sense) with respect to the long basic gravity wave, have been computed for various values of the steepness AK using two different methods. The results presented in this paper deal with the two dimensional mode n = 10 and the three dimensional mode (n,q) = (10,10).

Figures 2 and 3 display respectively the evolution of the normalized amplitude a/$\bar{a}$ ($\bar{a}$ is the amplitude of the short waves at the mean level of the long wave) and wave number k of short gravity waves (two dimensional mode n = 10) propagating in the positive sense, as a function of the horizontal coordinate x between the crest and the trough of the long wave.

With the linear dispersion relation (3.2) and the Bretherton and Garrett equation (3.7) the amplitude and wave number evolutions are found to be very close to the evolutions obtained from the stability calculations of the long wave to superharmonic perturbations.

Similarly, figures 4 and 5 show the evolution of the relative amplitude and wave number of gravity capillary waves (two dimensional mode n = 10) travelling in the positive sense. In this case small difference appears at the crest for AK = 0.40. Figures 6 and 7 display the curves corresponding to gravity capillary waves (two dimensional mode n = 10) travelling in the negative sense. In figures 8 and 9 are plotted the curves corresponding to the three dimensional mode (p,q) = (10,10).

As a main conclusion an equivalence exists between the ray equations based on the linear dispersion relation coupled with the action conservation principle and the eigenvalue problem derived from the exact equations.

Generally, instabilities arise when frequencies of two modes of opposite signatures collide (McKay and Saffman, 1986). According to the previous conclusion, it seems possible to predict the instability of wave of permanent form by using the linear dispersion relation with the signature added, for any value of the wave steepness AK.

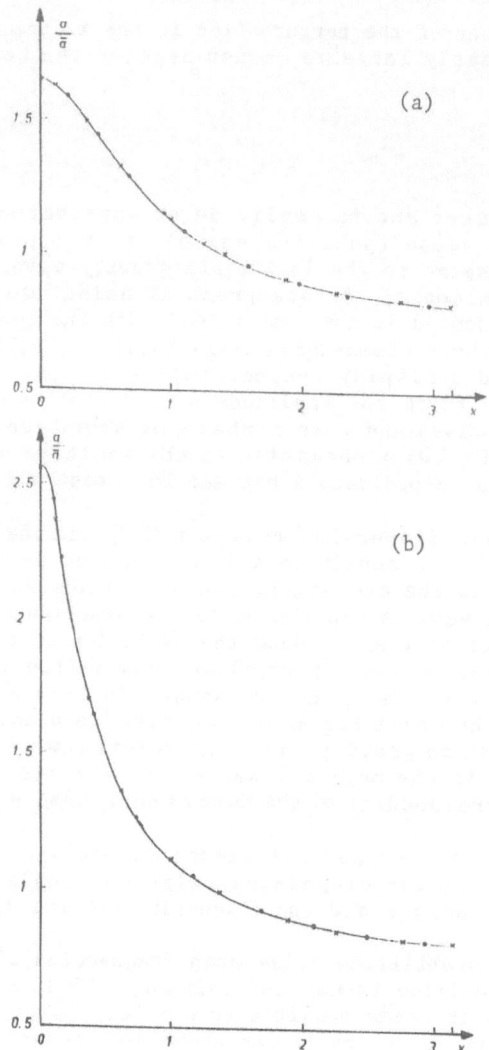

Figure 2. Evolution of the normalized amplitude a/ā of two dimensional short gravity waves propagating in the <u>positive</u> sense as a function of the horizontal coordinate x between the crest and the trough of the long wave. K = 1 ; x : Stability (n = 10) ; • : Action conservation ; (a) AK = 0.3 ; (b) AK = 0.4.

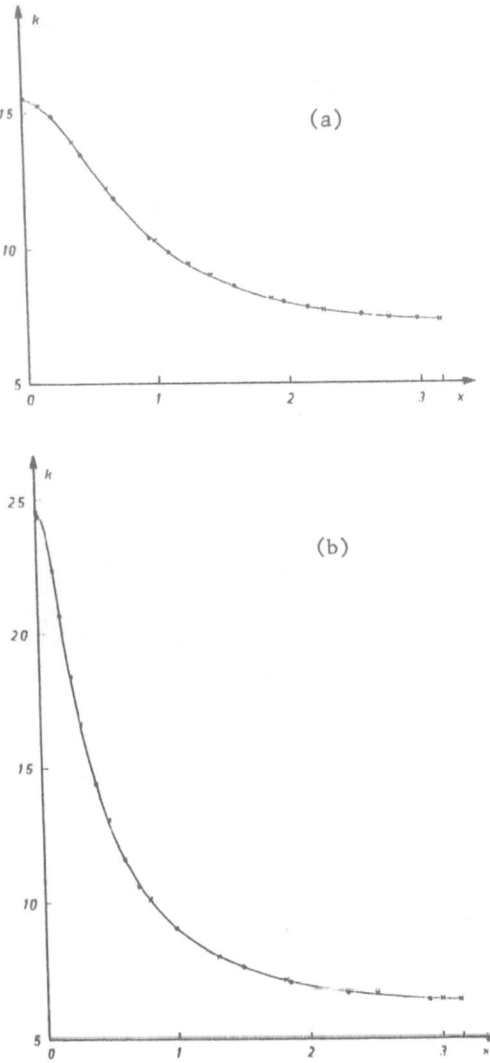

Figure 3. Evolution of the wave number k of two dimensional short gravity waves propagating in the <u>positive</u> sense as a function of the horizontal coordinate x between the crest and the trough of the long wave. K = 1 ; x : Stability (n = 10) ; ● : Ray equations ; (a) AK = 0.3 ; (b) AK = 0.4.

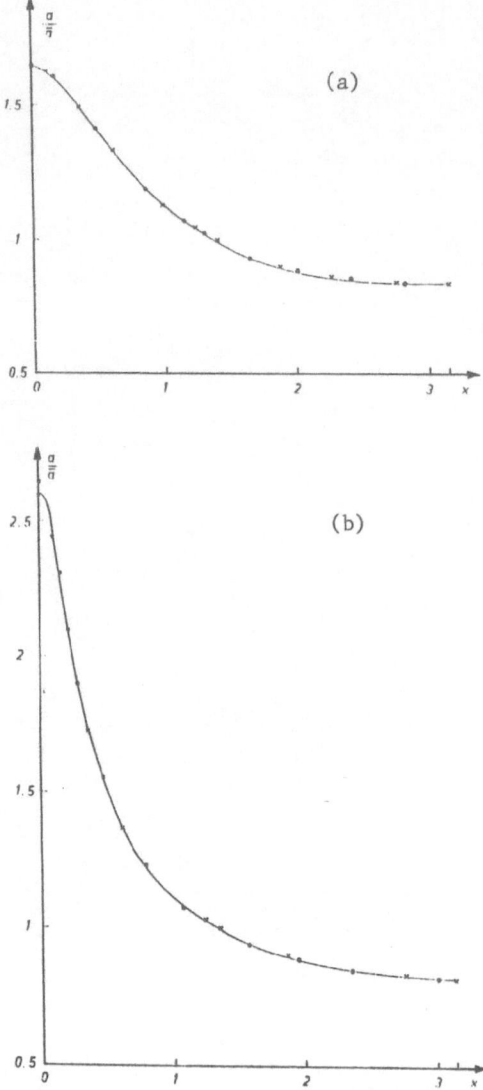

Figure 4. As Fig. 2 but for two dimensional short gravity capillary waves propagating in the <u>positive</u> sense.
K = 1 ; κ = 0.0003 ; x : Stability (n = 10) ; ● : Ray equations ;
(a) AK = 0.3 ; (b) AK = 0.4.

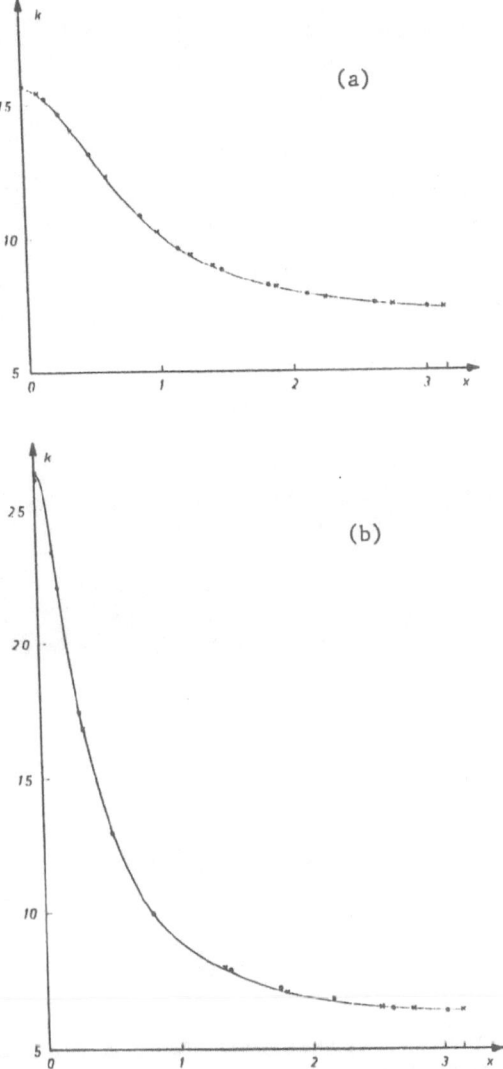

Figure 5. As Fig. 3 but for two dimensional short gravity capillary waves propagating in the <u>positive</u> sense.
K = 1 ; κ = 0.0003 ; x : Stability (n = 10) ; ● : Action conservation ;
(a) AK = 0.3 ; (b) AK = 0.4.

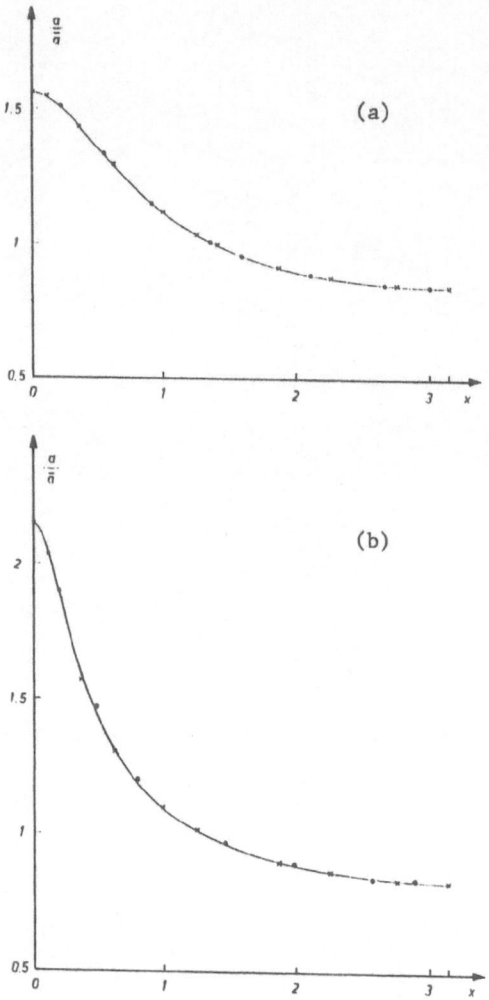

Figure 6. As Fig. 2 but for two dimensional short gravity capillary waves propagating in the <u>negative</u> sense.
K = 1 ; κ = 0.0003 ; x : Stability (n = 10) ; ● : Action conservation ; (a) AK = 0.3 ; (b) AK = 0.4.

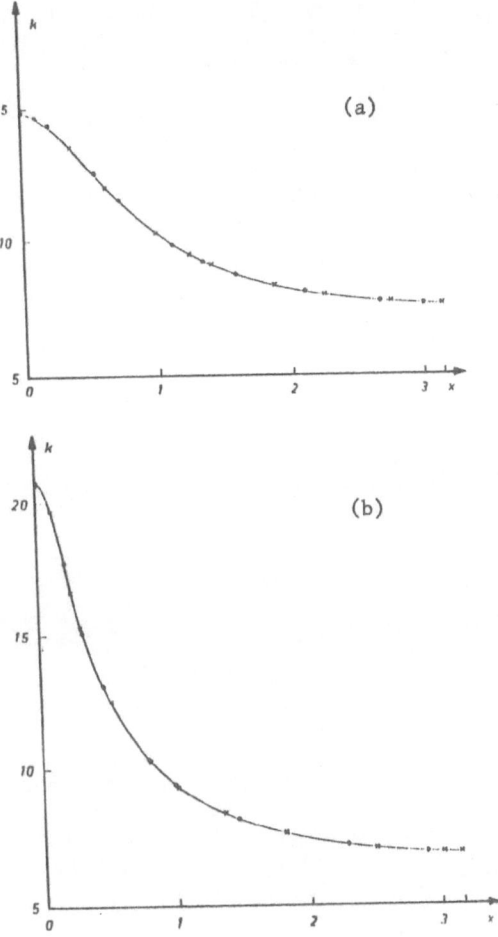

Figure 7. As Fig. 3 but for two dimensional short gravity capillary waves propagating in the <u>negative</u> sense.
K = 1 ; κ = 0.0003 ; x : Stability (n = 10) ; ● : Ray equations ;
(a) AK = 0.3 ; (b) AK = 0.4.

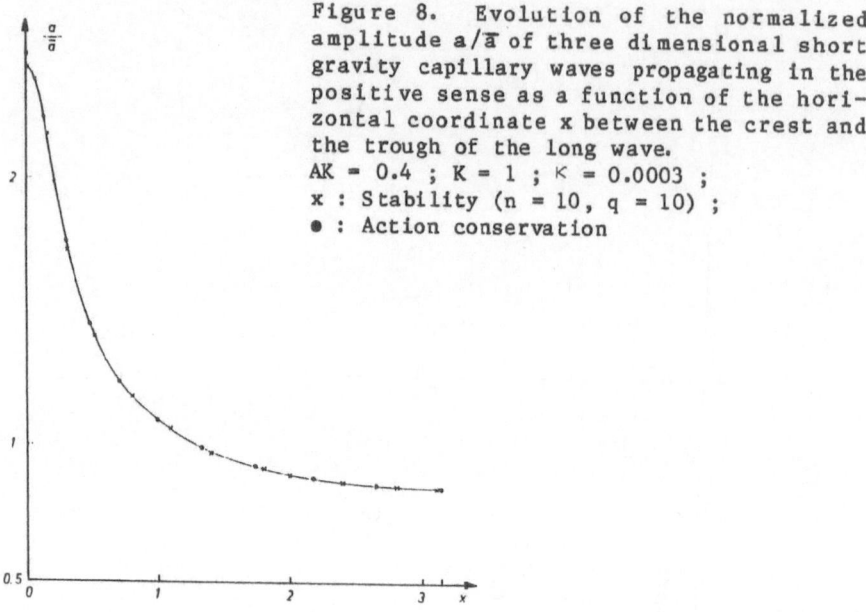

Figure 8. Evolution of the normalized amplitude $a/\bar{a}$ of three dimensional short gravity capillary waves propagating in the positive sense as a function of the horizontal coordinate x between the crest and the trough of the long wave.
$AK = 0.4$; $K = 1$; $\kappa = 0.0003$;
x : Stability (n = 10, q = 10) ;
● : Action conservation

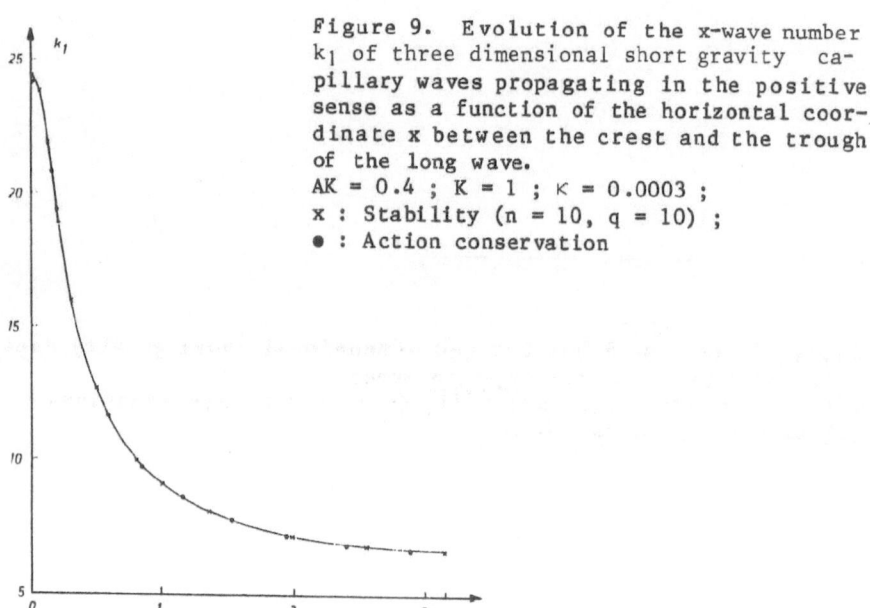

Figure 9. Evolution of the x-wave number k_1 of three dimensional short gravity capillary waves propagating in the positive sense as a function of the horizontal coordinate x between the crest and the trough of the long wave.
$AK = 0.4$; $K = 1$; $\kappa = 0.0003$;
x : Stability (n = 10, q = 10) ;
● : Action conservation

5. ACKNOWLEDGEMENT

I would like to thank Dr. A. Ramamonjiarisoa for the many helpful discussions during the course of this work.

6. REFERENCES

Bretherton, F. and Garrett, C. (1969) 'Wave trains in inhomogeneous moving media' Proc. Roy. Soc. Lond., A302, 529–554.

Henyey, F., Creamer, D., Dysthe, K.B., Schult, R.L. and Wright, J.A. (1988) 'The energy and action of small waves riding on large waves' J. Fluid Mech., 189, 443–462.

Kharif, C. and Ramamonjiarisoa, A. (1988) 'Deep-water gravity wave instabilities at large steepness' The Physics of Fluids, vol. 31, n° 5, pp.1286–1288.

Longuet-Higgins, M.S. (1985) 'Bifurcation in gravity waves' J. Fluid Mech., 151, 457–475.

Longuet-Higgins, M.S. (1987) 'The propagation of short surface waves on longer gravity waves' J. Fluid Mech., 177, 293–306.

McKay, R.S. and Saffman, P.G. (1986) 'Stability of water waves' Proc. Roy. Soc. Lond. A406, 115–125.

IRREGULAR WATER WAVE KINEMATICS

Alf Tørum
Norwegian Hydrotechnical Laboratory - SINTEF Group
Norwegian Institute of Technology
N-7034 Trondheim
Norway

James E. Skjelbreia
Norwegian Marine Technology Research Institute A/S (MARINTEK)
SINTEF Group
P.O. Box 4125 - Valentinlyst
N-7002 Trondheim
Norway

1.   INTRODUCTION

There are several theories and methods to calculate the kinematics of water waves. Most of the theories and methods have been developed for regular waves, like the linear theory, Stokes higher order theories, stream function theory, etc. The stream function theory can also be applied for an irregular surface profile, using the assumption that the profile travel without change of form.

Probably the most used concept for irregular waves is that the waves are composed of a sum of regular linear wave components, each component travels at its own speed, direction and phase. This is a concept that can be applied easily to many engineering applications, like for example response calculations of offshore structures. However, it is difficult to apply this concept in the wave crest region because the high frequency components will introduce erroneously high water particle velocities.

Forristal (1985) developed a method that also can be used for irregular waves with a time and space varying surface profile. However, the required computer time is so large that the method can not be used so easily for response calculations of an offshore structure.

Different engineering methods have been developed to circumvent the limitation with the linear theory for irregular waves, e.g. Wheeler (1979), Chakrabarti (1971), Gudmestad and Connor (1986), Lambrakos (1981), Gudmestad (1988) and others. However, such methods should be checked against measurements of the water particle velocities.

There has been different water particle velocity measurements programs, both in laboratories and in the field. Field measurements have the advantage that measurements are then done in "true" waves,

A. Tørum and O. T. Gudmestad (eds.), Water Wave Kinematics, 281–295.

nevertheless, the hydrodynamic conditions may not then be so well controlled as in a laboratory test set up and the measurements much more difficult to obtain, to analyse and interpret.

The most extensive laboratory investigation on water wave kinematics is probably the one carried out in the Netherlands, Bosma and Vugts (1981). This measurement program was carried out in a laboratory flume for longcrested waves. The velocity measurements were obtained with a Laser Doppler Velocity Meter (LDV), mainly below the still water line, but also some measurements were carried out slightly above the still water level. The results of the measurements were compared with the linear wave superposition concept. The main conclusion was that there was a good agreement between the measurements and the calculated water particle velocities below the still water level. It seemed to be a fair agreement between the measurements above the still water line and the calculated velocities. However, since the measurements were made at a location only one standard deviation of the surface elevation above the still water level, there remained still uncertainties on the water particle velocities closer to the wave crest.

Gudmestad et al (1988) presented results of measurements of regular water wave kinematics and compared the results with different wave theories and methods (linear, Stokes 2nd and 5th order, stream function theory, Gudmestad's method, Wheeler's method and Chakrabarti's method). The Eulerian type measurements showed an apparent return current which for regular waves compensate for the mass transport (Langragian). When taking this apparent return current into account the best agreement between calculations and measurements was obtained by the Stokes 5th order and the stream function theory. If the return current was not accounted for, Gudmestad's method gave the best agreement between calculated and measured wave kinematics under the wave crest and the wave through. Gudmestad's method, however, gives a flatter velocity-time curve at the wave crest and a steeper velocity-time curve at the wave through, contrary to what higher order wave theories and measurements show. The Wheeler method gave a poor agreement when assuming only one wave frequency. However, results obtained by others, Lambrakos (1989), indicate that Wheeler's method gives better agreement when the regular wave profile is decomposed into different Fourier components.

In the project for long crested irregular water wave kinematics we have emphasized the region between the wave crest and the wave through, although measurements at lower locations have also been carried out.

Only a limited amount of results will be presented in this paper.

2. TEST SET UP AND TEST PROCEDURES

The measurements were carried out in a laboratory wave tank as shown in Figure 1, approximately 30 m long and 1 m wide. The water depth during the irregular wave tests was 1.30 m.

The wave absorber at the end of the flume was made of expanded steel metal plates with different distances between the plates and different perforation of each plate. This is a design tested out by the Hydraulics Laboratory of the National Research Council of Canada, Jamieson and Mansard (1987). The reflection coefficient for 1st order waves is approximately 5 % over a wide range of frequencies.

The surface elevations were measured with conductivity type wave gauges. The vertical "rod" of the supporting frame was "streamlined" to give minimum water surface disturbance.

The water surface elevation were measured at 7 locations along the flume as shown in Figure 1. Since the velocities were measured at one point at a time, it was necessary to repeat the wave conditions several times by the same control signal to the wave paddle. The motion of the wave paddle repeated quite well. However, the waves at each gauge did not repeat exactly from run to run. This was mainly because there were small cross oscillations in the flume. A floating plastic mat in front of the wave absorber, Figure 1, helped damping out the cross waves, but not completely.

It was decided to measure the waves at the location of the velocity measurements with 5 wave gauges evenly spaced across the wave flume and use the average of surface elevation measured at these five gauges as the nominal surface elevation at the location of the water velocity measurements. Wave calibration with these five gauges were carried out before any velocity measurements were carried out.

The wave paddle control signal was generated by starting out with a target wave spectrum (JONSWAP, $\gamma = 3.0$). This spectrum was then simulated by 1000 equally spaced frequency component with randomly picked phases.

The water particle velocity measurements were carried out with a Laser Doppler Velocimeter (LDV). The measurements were taken at points located halfway between the glass walls on each side of the flume.

Each run of irregular waves were carried out for approximately 13 min duration or for approximately 650 - 1300 waves, depending on the peak period of the wave spectra. There was a waiting period between each run of approximately 30 minutes. The sampling frequency was 40 Hz.

3. TEST PROGRAM

The velocity measurements were done for the following test program, covering intermediate and deep water.

RANDOM WAVE CASES

T_p sec	H_s m	d m	H_s/T_p^2	d/T_p^2	H_s/d
1.2	0.10	1.30	0.069	0.87	0.08
1.4	0.15	1.30	0.077	0.63	0.12
1.8	0.20	1.30	0.062	0.38	0.16
2.4	0.25	1.30	0.043	0.41	0.19
1.67	0.156	1.30	0.056	0.44	0.13
1.69	0.20	0.6	0.069	0.21	0.33

REGULAR WAVE CASES

1.45	0.275	1.30	0.13	0.59	0.22
1.48	0.36	0.6	0.16	0.27	0.60

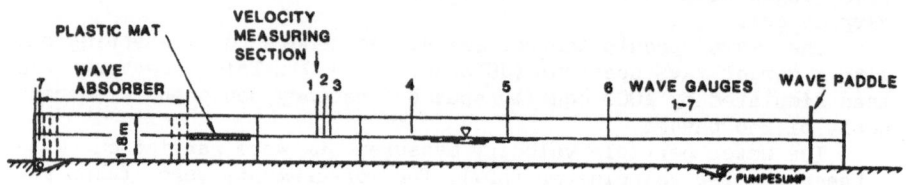

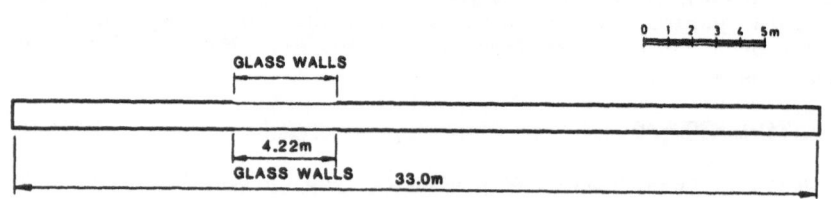

Figure 1.

In addition tests with a two component wave spectrum was carried out, ω_1 = 2.6 sec^{-1} and ω_2 = 2.99 sec^{-1} (T_1 = 2.4 sec, T_2 = 2.1 sec) for d = 1.30 m.

The measurements were done for several points as shown in Figure 2.

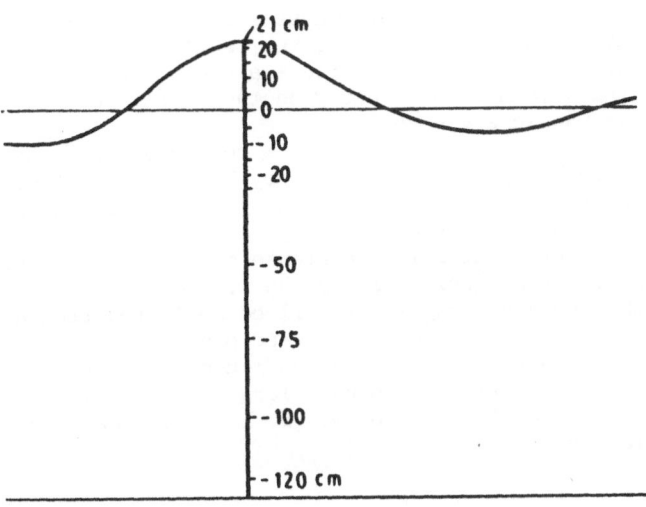

Figure 2.

4.   ANALYSIS

In a wave flume there will be generated 1st order waves and 2nd order waves. The bound long waves will be "true" waves. However, the wave generator had no compensation for the spurious long waves generated by the 1st order motion of the wave flap. Thus spurious long waves were generated also.

Thus the horizontal water particle velocities are composed of different components:

$$u = u_{1i} + u_{1r} + u_{ib} + u_{rb} + u_{if} + u_{rf}$$

where

u_{1i} = "first" order incoming wave
u_{1r} = "first" order reflected waves
u_{ib} = incoming bound wave

u_{rb} = reflected bound wave
u_{if} = incoming free long wave
u_{rf} = reflected free long wave

The major contribution to the horizontal water particle velocities for the larger waves is the "first" order incoming wave. The other waves may contribute with some few per cent of the first order wave. However, the "other" waves should be considered, except the reflected bound waves which are very small.

The spurious waves in the flume are expected to be reflected fully from the wave absorber. Later on they will be re-reflected from the wave paddle and so on.

The bound long waves will also be reflected from the wave absorber. However, they will not necessary be fully reflected. The work of Symonds et al (1982) indicated that the reflection of bound long waves as long free waves will depend on the bottom slope, the location of the breaking point. Kostense's (1985) test results supported to some extent qualitatively the work of Symonds et al.

How the bounded long waves will be reflected from a wave absorber consisting of a row of perforated plates is not known. We have therefore to live with the bound incoming waves, the incoming free waves and the reflected bound long waves (small and negligible) and the reflected free waves. They will have to be sorted out analytically.

4.1. Reflected 1st order waves

The incoming and reflected waves were sorted out by a general least square method for an arbitrary number of wave gauges along the flume, Funke and Mansard (1980). For two gauges the method is basically the same as derived by Goda and Suzuki (1976).

Figure 3 shows an example of analysed incoming and reflected wave spectra as well as the ratio between the amplitudes of the reflected and the incoming wave components. The results are for peak period T_p = 2.4 sec and significant wave height H_s = 0.25 m. The analysis is based on three wave gauges located 0.74 m and 1.16 m apart.

The diagram of Figure 3 shows that the reflections are low for waves with frequencies around the peak frequency, reflection ratios of approximately 5 %. For lower frequencies the estimate of the reflections is very uncertain because the distance between the gauges are then very small compared to the wave length. For the higher frequencies the reflection increase. This is believed not to be "true" reflections, but is partly a result of the gauge spacing is too large, and phase errors become a problem.

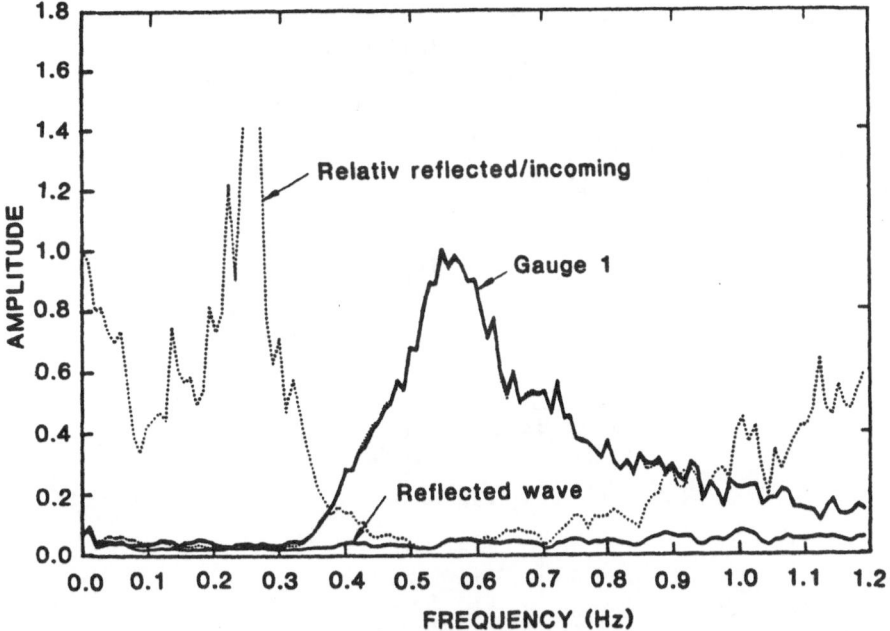

Figure 3.

4.2. Incoming and reflected long waves

Figures 4 and 5 shows wave spectra from two wave gauges, gauge 4 and gauge 7, along the flume, Figure 1, for a wave spectrum with T_p = 2.4 sec and significant wave height of 0.26 m. The central portion of the spectrum does not vary significantly from gauge to gauge, while the low frequency portion varies significantly.

The natural frequency of long wave oscillations of the flume has been calculated according to the following formulae:

$$T_n = \frac{2L}{(n + 1)\sqrt{gd}}$$

where

    L = length of flume = 29 m
    g = acceleration of gravity = 9.81 m/sec²

288

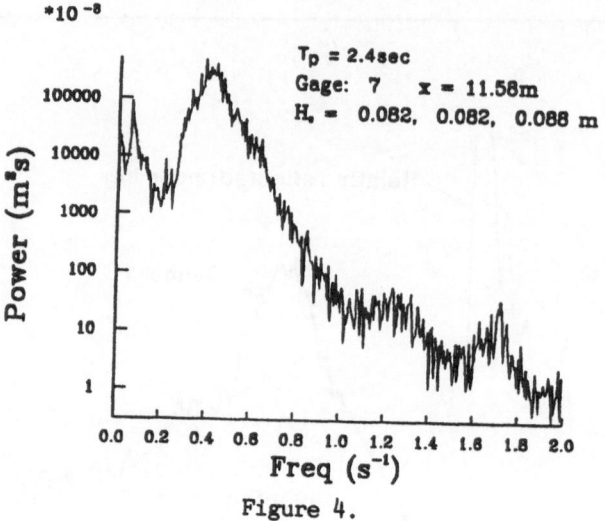

Figure 4.

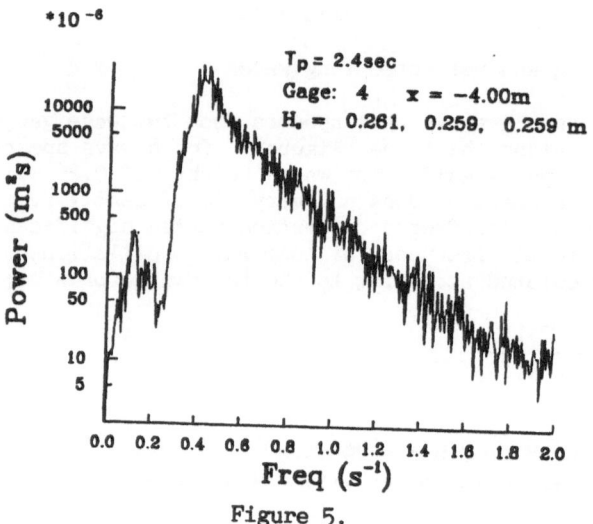

Figure 5.

d = water depth = 1.3 m
n = mode number

T_0 = 16.24 sec	f_0 = 0.0615 Hz	
T_1 = 8.12 "	f_1 = 0.123 "	
T_2 = 5.41 "	f_2 = 0.185 "	
T_3 = 4.06 "	f_3 = 0.246 "	
T_4 = 3.24 "	f_4 = 0.309 "	
T_5 = 2.70 "	f_5 = 0.370 "	

The spectrum of gauge no 7 (at the end of the flume) shows clearly a peak for the fundamental mode, while the spectrum from gauge no 4 closer to the middle of the flume shows a peak at a higher mode.

4.3. Velocity measurements

Figure 6 shows example of time traces of surface elevation z, horizontal and vertical velocities at the still water line for T_p = 2.4 sec and H_s = 0.25 m. The measurements have in this case been compared with velocities calculated according to the concept of superposition of linear waves. The calculations have been based on the measured profile at gauge 1 and no corrections have been made for reflected waves and long waves. Figure 7 shows time traces where the measurements have been compared to the Wheeler (1970) method and the linear wave super-position method under the same conditions. The calculated elevation in both figures were obtained from a spectral decomposition of the free surface elevation, energy above 8 Hz was set to zero.

Figures 8 and 9 show example of time traces of surface elevation z, horizontal and vertical velocities at an elevation +0.2 m (z/h = 0.154). Figure 8 is a detail of part of the time trace of Figure 7. In this case the measurements have also been compared to the Wheeler method.

Figures 10 and 11 show time traces of horizontal and vertical velocities respectively for different elevations. The measurements have been compared to the Wheeler method.

The velocity measurement results will further be compared with several theories and methods to calculate the water particle velocities:

Linear theory
Stream function
5th order
Wheeler method
Gudmestad method

290

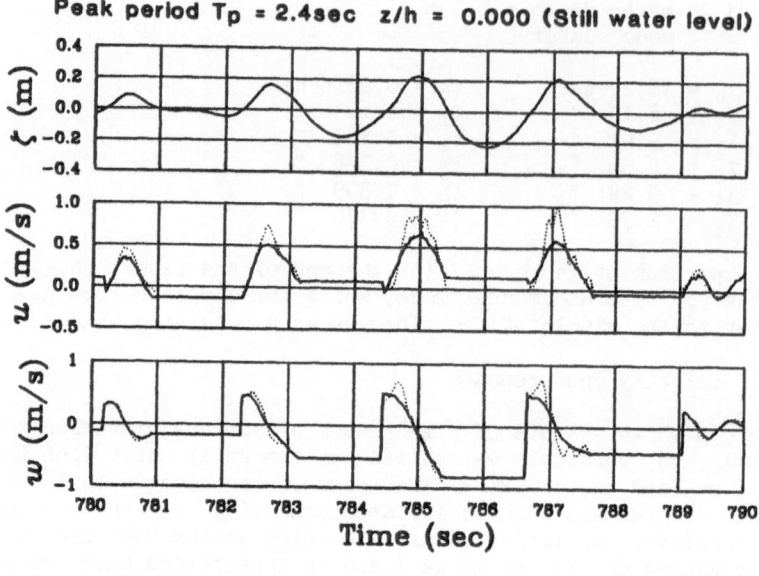

Figure 6.

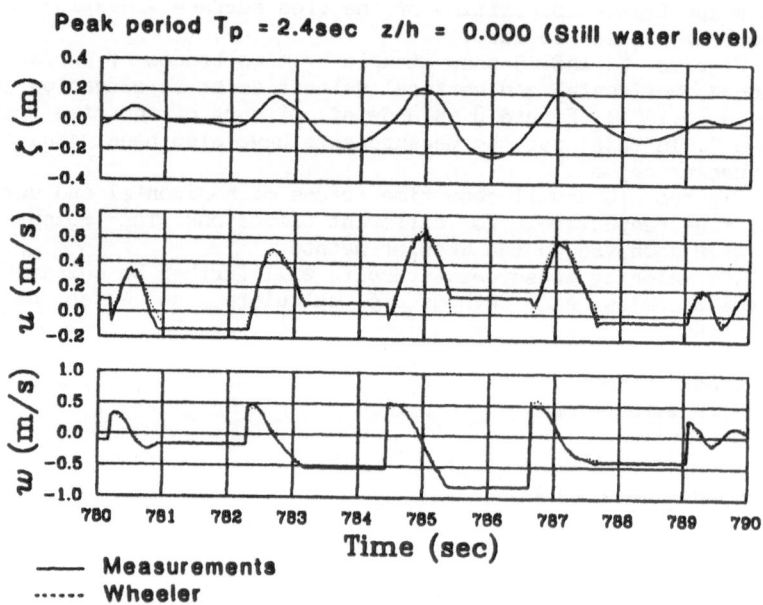

Figure 7.

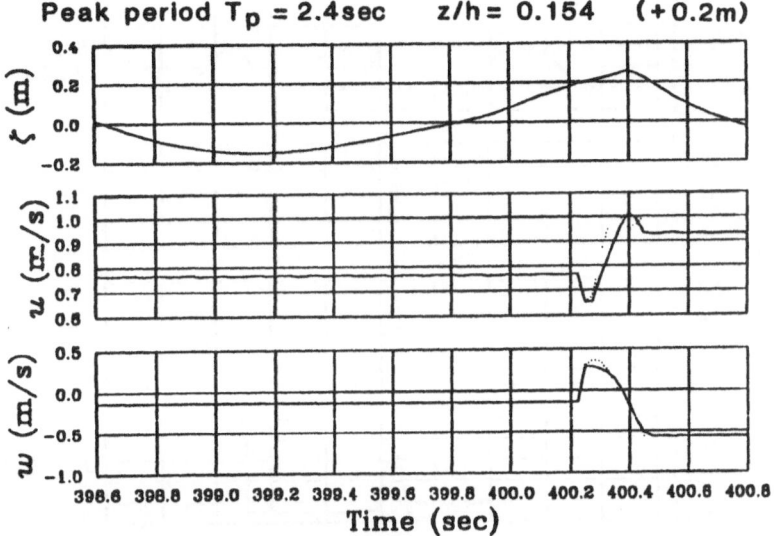

Figure 8.

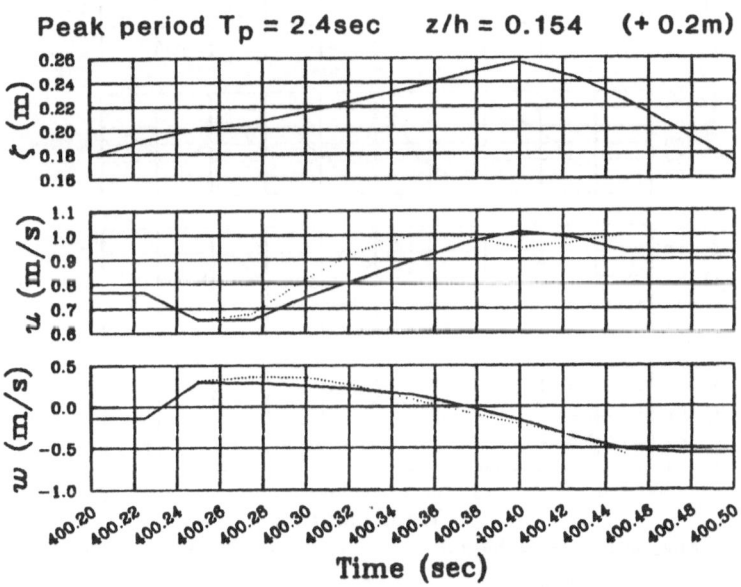

Figure 9.

292

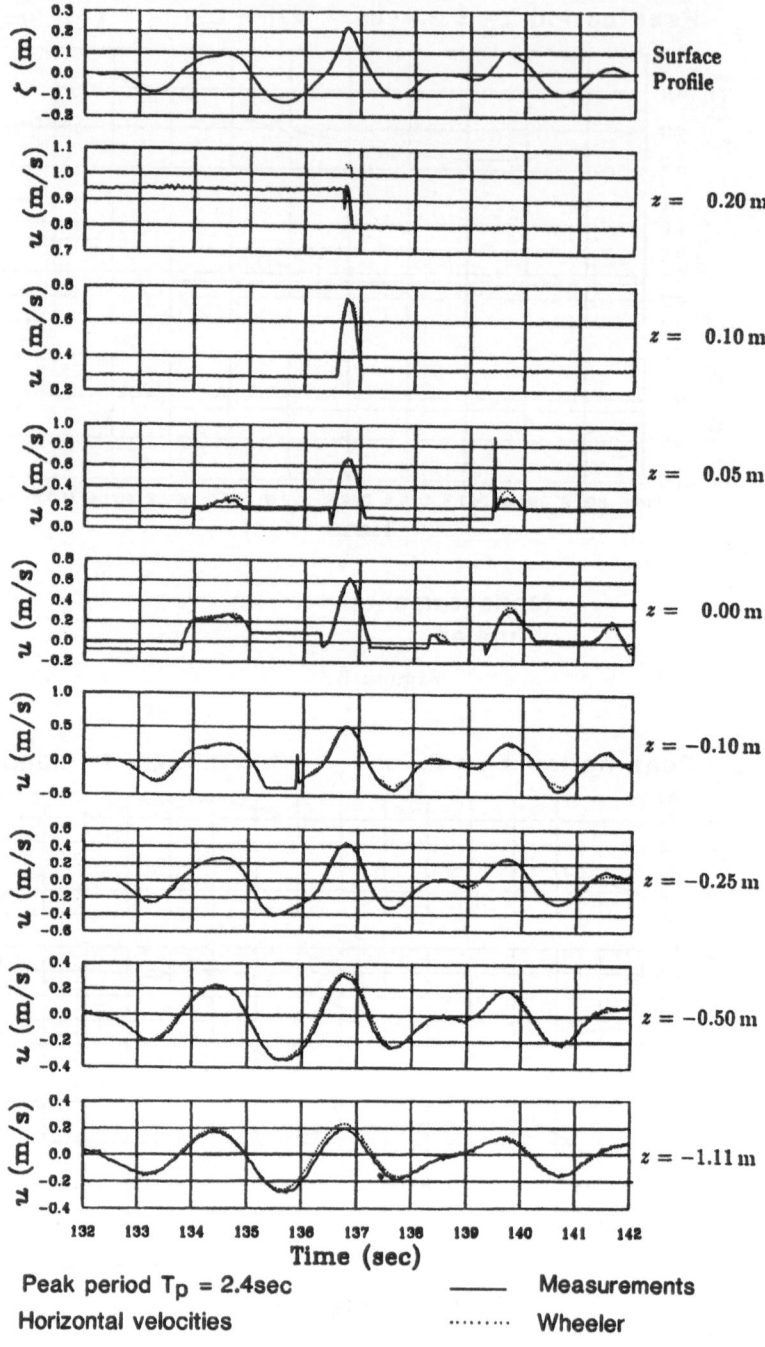

Figure 10.

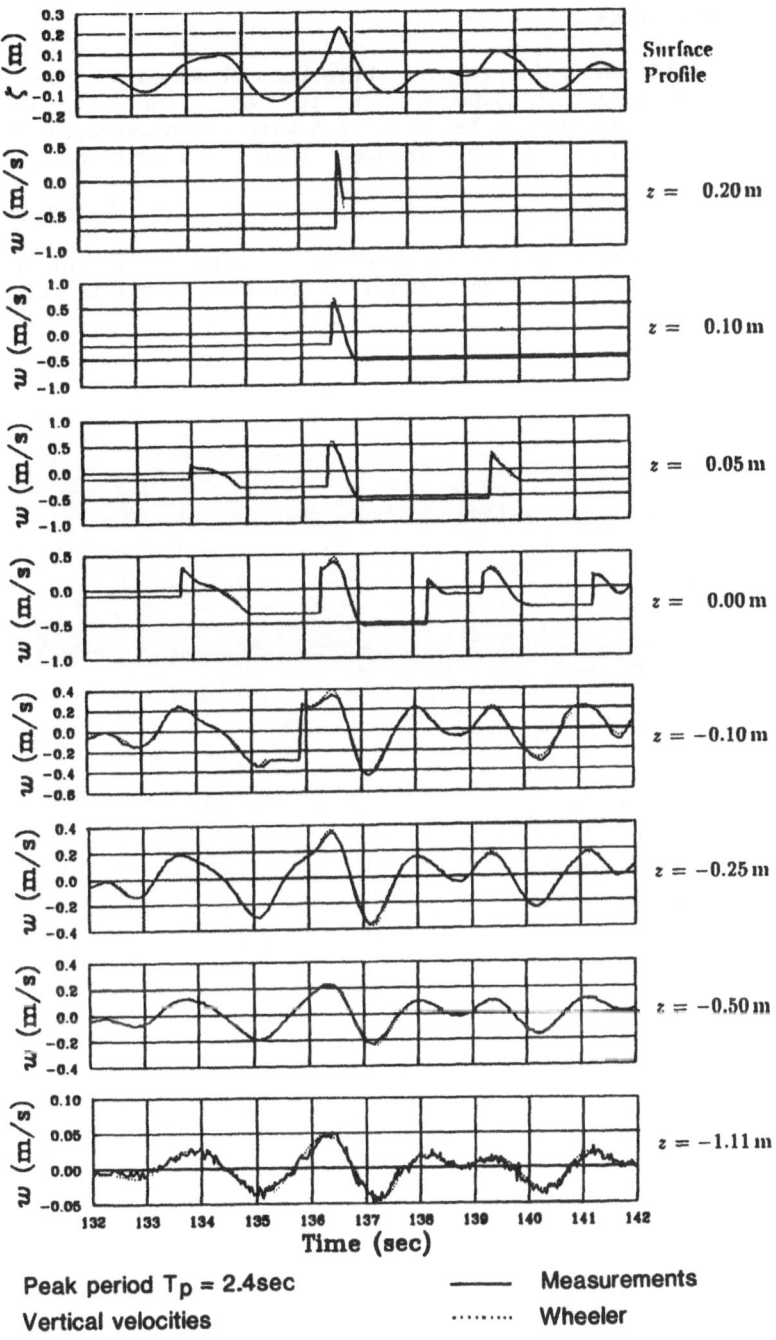

Figure 11.

5. CONCLUSIONS

Only limited analysis results have been obtained and presented. Based on the results obtained so far it seems that the Wheeler method gives a reasonable agreement with the measurement in the surface zone and much better agreement then the linear theory.

6. ACKNOWLEDGEMENT

We appreciate the financial support and professional interaction with personnel from NTNF, Amoco, Exxon and Statoil to carry out this study.

REFERENCES

BOSMA, J. and VUGTS, J.H. (1981): Wave kinematics and fluid loading in irregular waves. International Symposium on Hydrodynamics in Ocean Engineering, The Norwegian Institute of Technology

CHAKRABARTI, S.K. (1971): Discussion on "Dynamics of Single Point Mooring in Deep Water". Journal of Waterways, Harbour and Coastal Engr. Division, ASCE 1971, 97 (WW3) pp 588-590.

FORRISTAL, G.Z. (1985): Irregular wave kinematics from a kinematic boundary condition fit (KBCF). Applied Ocean Research, Vol 7, No 4.

GODA, Y. and SUZUKI, Y. (1976): Estimation of incident and reflected waves in random wave experiments. Proc. 15th International Conference on Coastal Engineering Conference, June 11-17, Honolulu, Hawaii.

GUDMESTAD, O.T. and CONNOR, J.J. (1986): Engineering Approximation to Non-Linear Deep Water Waves. Applied Ocean Research, Vol 8, No 2.

GUDMESTAD, O.T. (1988): A New Approach for Estimation of Irregular Deep Water Wave Kinematics. Submitted Ocean Research, August.

GUDMESTAD, O.T., JOHNSEN, J.M., SKJELBREIA, J. and TØRUM, A. (1988): Regular Water Wave Kinematics. Proc. International Conference on Behaviour of Offshore Structures (BOSS'88), Norwegian Institute of Technology.

JAMIESON, W.W. and MANSARD, E.P.D. (1987): An efficient upright wave absorber. Paper presented at the ASCE Speciality Conference on Coastal Hydrodynamics, University of Delaware, June 29-July 1.

KOSTENSE, J.K. (1985): Measurements of surf beat and set down beneath wave groups. Publication No 338, April. Delft Hydraulics Laboratory.

LAMBRAKOS, K. (1981): Extended Velocity Potential Wave Kinematics Journal of Waterway, Port, Coastal and Ocean Division, Proc. ASCE Vol 107, No WW 3, August.

LAMBRAKOS, K. (1989), Exxon. Personal communication.

MANSARD, E.P.D. and FUNKE, E.R. (1980): The measurement of incident and reflected spectra using a least squares method. Seventeenth International Conference on Coastal Engineering.

SYMONDS, G., HUNTLEY, D.A. and BOWEN, A.J. (1982): Two-dimensional surf beat long wave generation by a time-varying breakpoint. Journal of Geophysical Res., Vol 87, No C1, pp 492-498.

WHEELER, J.D. (1970): Method for calculating forces produced by irregular waves. J. Petroleum Tech, March, pp 119-137.

COMPUTATIONAL MODELLING OF VELOCITIES AND ACCELERATIONS IN STEEP WAVES

J. SKOURUP & I.G. JONSSON
Inst. of Hydrodynamics and Hydraulic Engineering
Tech. Univ. of Denmark, DK-2800 Lyngby, Denmark
S.T. GRILLI & I.A. SVENDSEN
Dept. of Civil Engineering
Univ. of Delaware, Newark DE 19716, USA

ABSTRACT. The Boundary Integral Equation Method is used for the spatial solution of a non-linear, periodic water wave problem. The temporal updating of the free water surface is performed using the exact non-linear free surface boundary conditions. A method for updating water particles in the interior of the fluid domain enables one to follow the particle motion as the wave propagates in time. A structure is introduced in the computational domain, and its influence on the particle trajectories is visualized. The computational model presented here is two-dimensional, but there are (in principle) no limitations for the model to be extended to three spatial dimensions.

INTRODUCTION

The numerical modelling of the interaction between steep water waves and large off-shore structures is by no means a trivial task. If the steepness of the waves is small, and the waves are time-harmonic, then the time-dependency can be separated from the wave equation, and a solution for the velocity potential can be found. Then forces and moments on structures can be evaluated.

However, when the wave steepness becomes large (and hence the non-linear terms become dominant), or when the wave is not time-periodic (irregular), we must use a more general formulation which takes these factors into account. One of the most efficient methods for this purpose is the Boundary Integral Equation Method (BIEM).

The first contribution using the BIEM for the modelling of steep waves and the overturning of waves was given by Longuet-Higgins & Cokelet (1976). Cokelet (1979) used this model for computations of velocities and accelerations within the fluid volume of a breaking wave, and the same computational model was used by Peregrine et al. (1980) for a study of the interior velocity and acceleration fields around the tip of breaking waves. These computations were all performed in a conformably mapped space, and since only waves on infinite depth were considered, the computations were restricted to the free surface. An extension to finite-depth modelling was made by New et al. (1985).

A. Tørum and O. T. Gudmestad (eds.), Water Wave Kinematics, 297–312.
© 1990 Kluwer Academic Publishers.

Vinje & Brevig (1981) developed a model based on Cauchy's integral theorem. They performed the computations in the physical space with finite depth, and velocities and accelerations within a breaking wave were computed. In a later paper (Brevig et al., 1981) the forces due to breaking waves on a fixed or moving submerged circular cylinder (pipeline) were calculated.

An efficient method for the temporal updating of the free water surface was developed by Dold & Peregrine (1984). In this method the influence of the higher order derivatives along the free water surface was taken into account, and this permitted the use of large time-steps with a good accuracy, and the method became very efficient for the modelling of e.g. overturning waves. In this model a conformably mapped space and Cauchy's integral theorem were used, and hence their computations (as in the previous models) were restricted to two-dimensional (2-D) problems.

Examples of three-dimensional (3-D) models are still rare. One of the first was presented by Isaacson (1982), where the interaction of steep waves with fixed or floating structures of arbitrary shape was computed. Due to closure problems of the computational domain, only transient problems were modelled. In a recent paper (Dommermuth & Yue, 1987) an axisymmetric heaving problem was studied, and satisfactory results were obtained, but the far-field closure of the computational domain is yet not fully developed.

In the present paper a 2-D physical-space, non-linear BIEM is used for modelling steep water waves and wave-structure interaction, and, contrary to the 2-D models described above, there are in principle no restrictions for this model to be extended to 3-D problems. Thus, even though the present model is only 2-D, the formulation and ideas can all be extended to 3-D problems.

The spatial solution of the governing differential equation is performed by using Green's 2nd identity, and the temporal updating of the free water surface is obtained from the fully non-linear free surface boundary conditions in a way similar to that outlined by Dold & Peregrine (1984).

Computational results, showing the high degree of non-linearity in the present model, are given. A simple method (based on the BIEM) for a time-stepping of water particles within the fluid domain is developed and presented for the first time. The method makes use of the velocities and accelerations of these particles. A fixed structure with a simple shape is incorporated in the computational model, and the flow field around it is evaluated by use of the time-stepping of interior points.

MATHEMATICAL FORMULATION

An inviscid and incompressible fluid, and an irrotational flow described by a velocity potential ϕ is considered, with ϕ defined by:

$$\nabla\phi = \vec{u} = (u,w) \tag{1}$$

where $\vec{u}$ is the velocity vector, and u and w are the horizontal and vertical velocity components, respectively.

Using the continuity equation in a fluid domain $\Omega(t)$ (where t denotes time) surrounded by the boundary $\Gamma(t)$ (see Fig. 1) gives us a Laplace equation for the velocity potential ϕ

$$\nabla^2\phi = 0 \qquad (2)$$

Using Green's 2nd identity

$$\phi(\vec{x}) = \int_\Gamma [\phi(\vec{\xi}) \, G_n(\vec{x},\vec{\xi}) - G(\vec{x},\vec{\xi}) \, \phi_n(\vec{\xi})] \, d\Gamma \qquad (3)$$

(where subscript n denotes differentiation with respect to the outwards normal vector defined in Fig. 1) permits us to solve (2) provided that either ϕ or ϕ_n is given on all parts of the boundary Γ.

The integration point $\vec{\xi}$ belongs to the boundary Γ, while the observation point $\vec{x}$ can be situated anywhere inside, outside, or at the boundary of the domain; however, in the solution of the Laplace equation (2), $\vec{x}$ will be a boundary point. G is the free space Green's function given in 2-D as

$$G = \frac{1}{\alpha} \ln | \vec{x} - \vec{\xi} | \qquad (4)$$

where α depends on the boundary geometry ($\alpha = \pi$ at a smooth part of the boundary, and $\alpha = 2\pi$ at a point inside Ω).

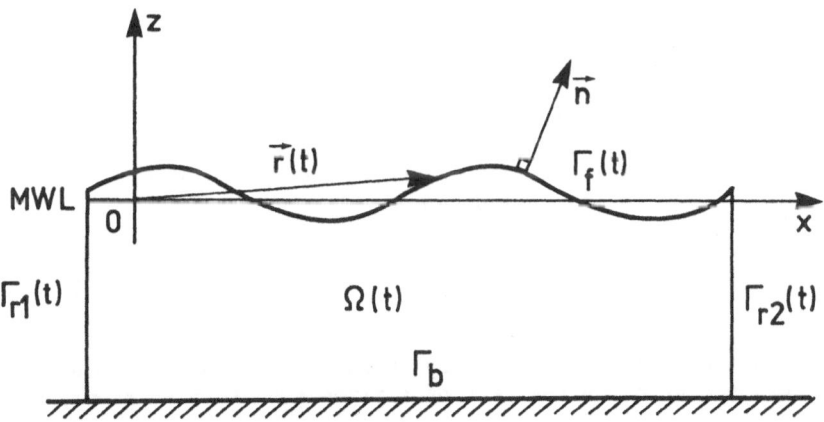

Fig. 1 Integration domain and definition of boundaries. The x-axis at mean water level (MWL).

The boundary conditions at the free water surface $\Gamma_f(t)$ are the kinematic condition

$$\frac{D\vec{r}}{Dt} = \left(\frac{\partial}{\partial t} + \vec{u} \cdot \nabla\right) \vec{r} = \vec{u} = \nabla\phi, \qquad z = \eta \qquad (5)$$

where $\vec{r}$ is a position vector of a particle at the free water surface, and the dynamic condition (Bernoulli's equation)

$$\frac{D\phi}{Dt} = -gz + \frac{1}{2} \mid \nabla\phi \mid^2 - \frac{P_a}{\rho}, \qquad z = \eta \qquad (6)$$

since $D\phi/Dt$ equals $\partial\phi/\partial t + \mid \nabla\phi \mid^2$. In (6), g is the acceleration due to gravity, P_a is the atmospheric pressure, and ρ is the specific mass of water.

At the bottom Γ_b (which is considered horizontal and impermeable) the boundary condition is

$$\frac{\partial\phi}{\partial n} = 0 \qquad (7)$$

The waves modelled in the following are space-periodic (but not necessarily time-periodic), and without a net current below wave trough level. Therefore, the following periodicity conditions can be imposed on the vertical boundaries $\Gamma_{r1}(t)$ and $\Gamma_{r2}(t)$:

$$\frac{\partial\phi}{\partial n} [\Gamma_{r1}(t)] = -\frac{\partial\phi}{\partial n} [\Gamma_{r2}(t)] \qquad (8a)$$

$$\phi [\Gamma_{r2}(t)] = \phi [\Gamma_{r1}(t)] \qquad (8b)$$

for the same z-values at the two boundaries. The horizontal distance between $\Gamma_{r1}(t)$ and $\Gamma_{r2}(t)$ has to be an integer number of wave lengths.

When a structure is present in the computational domain, the boundary condition on its surface is

$$\frac{\partial\phi}{\partial n} = \nabla\phi \cdot \vec{n} = V_n \qquad (9)$$

where V_n is the velocity of the body surface in the direction of the unit normal vector $\vec{n}$. In the case of a fixed structure we simply have $V_n = 0$.

In the temporal updating of the free surface to the next time-step, the non-linear free surface boundary conditions (5) and (6) are used in connection with truncated Taylor expansions in time for the position vector $\vec{r}$ of the free surface particles and for the velocity potential ϕ, following the ideas given by Dold & Peregrine (1984),

$$\vec{r}(t+\Delta t) = \vec{r}(t) + \sum_{k=1}^{n} \frac{(\Delta t)^k}{k!} \frac{D^k \vec{r}(t)}{Dt^k} + O[(\Delta t)^{n+1}] \tag{10}$$

$$\phi(\vec{r}(t+\Delta t), t+\Delta t) = \phi(\vec{r}(t), t) + \sum_{k=1}^{n} \frac{(\Delta t)^k}{k!} \frac{D^k \phi(\vec{r}(t), t)}{Dt^k}$$

$$+ O[(\Delta t)^{n+1}] \tag{11}$$

The coefficients in (10) and (11) are obtained by successive solutions of the Laplace equation for the velocity potential ϕ and its time derivatives, where a solution of one Laplace problem provides the non-linear boundary conditions for the next.

Here the series (10) and (11) are truncated at n=2, but it should be mentioned that the terms in them are obtained from the fully non-linear boundary conditions (5) and (6).

Thus, to obtain the coefficients in the series (10) and (11), two Laplace problems must be solved; the first providing ϕ and ϕ_n at the whole boundary, and the second giving ϕ_t and ϕ_{tn} at the whole boundary.

The boundary conditions at the free water surface for the second Laplace problem are found from the Eulerian form of the dynamic free surface boundary condition (cf. (6))

$$\frac{\partial \phi}{\partial t} = - gz - \frac{1}{2} | \nabla \phi |^2 - \frac{P_a}{\rho} \tag{12}$$

The periodicity conditions (8a) and (8b) are still used as boundary conditions at the vertical boundaries $\Gamma_{r1}(t)$ and $\Gamma_{r2}(t)$, but they are now differentiated with respect to t. At the bottom Γ_b the impermeability condition (7) differentiated with respect to t is used as a boundary condition.

The two Laplace problems are solved at the same time level, and hence in the same geometry. This makes the solution of the second problem very fast compared to the first Laplace solution, since changes only are made in the right-hand-side vector (load vector) of the system of equations. The coefficient matrix (stiffness matrix) only depends on the geometry and remains the same for all Laplace problems at the same time level.

After the solutions of the Laplace problems for ϕ and ϕ_t have been found by the BIEM, the quantities ϕ, ϕ_n, ϕ_t and ϕ_{tn} are known at the boundary. Then, by use of (3), we are able to determine the values of the velocity potential ϕ and its time derivative ϕ_t at any point $\vec{x}$ within the computational domain $\Omega(t)$. Furthermore, analytical differentiations of (3) provide integrals to determine the values of ϕ_x,

ϕ_z, ϕ_{xt}, ϕ_{zt}, ϕ_{xx} and ϕ_{xz} at the point $\vec{x}$. These unknowns only appear outside the relevant integrals (since G and its derivatives are only functions of the geometry, which is known), and their numerical evaluation is therefore very fast. From the results we may deduce the particle velocity components u and w, the acceleration components a_x and a_z, and the excess pressure p^+ ($= p + \rho gz$) at the interior point given by the position vector $\vec{x}$ as:

$$u(\vec{x}) = \phi_x(\vec{x}) \tag{13}$$

$$w(\vec{x}) = \phi_z(\vec{x}) \tag{14}$$

$$a_x(\vec{x}) = \phi_{xt}(\vec{x}) + \phi_{xx}(\vec{x})\phi_x(\vec{x}) + \phi_{xz}(\vec{x})\phi_z(\vec{x}) \tag{15}$$

$$a_z(\vec{x}) = \phi_{zt}(\vec{x}) + \phi_{xz}(\vec{x})\phi_x(\vec{x}) - \phi_{xx}(\vec{x})\phi_z(\vec{x}) \tag{16}$$

$$p^+(\vec{x}) = - \rho \left[\phi_t(\vec{x}) + \frac{1}{2}\{ [\phi_x(\vec{x})]^2 + [\phi_z(\vec{x})]^2 \} \right] \tag{17}$$

where in (16) $\phi_{zz}(\vec{x})$ has been replaced by $-\phi_{xx}(\vec{x})$. The equations (13) - (17) are all written in the exact non-linear form ((17) is obtained from the Bernoulli equation), and the only approximation introduced is the numerical evaluation of the integrals to determine ϕ_x, ϕ_z, ϕ_{xt}, ϕ_{zt}, ϕ_{xx} and ϕ_{xz}.

A procedure for a time-stepping of points in the fluid domain $\Omega(t)$ is developed in a similar way as the one for the updating of points at the free water surface (i.e. based on truncated Taylor series), and it reads

$$x_i(t+\Delta t) = x_i(t) + u(\vec{x})\Delta t + a_x(\vec{x})\frac{(\Delta t)^2}{2} + O[(\Delta t)^3] \tag{18}$$

$$z_i(t+\Delta t) = z_i(t) + w(\vec{x})\Delta t + a_z(\vec{x})\frac{(\Delta t)^2}{2} + O[(\Delta t)^3] \tag{19}$$

with the interior point $\vec{x}$ denoted by (x_i,z_i). Hence a visualization technique that enables to follow the traces of water particles in the fluid domain is established.

NUMERICAL SOLUTION

The input to the model is the initial geometry of the domain $\Omega(t=0)$, and the initial free surface velocity potential $\phi(\vec{r}(t=0), t=0)$. Then the simulation is performed by successive solutions of Laplace problems (2) by (3) and time-stepping by (10) and (11). The time-stepping of interior points is performed by use of (18) and (19), if required.

The collocation method is chosen for the numerical solution of (2) by (3), so the whole boundary is discretized by inserting N nodal points on it. In all the corners double points are inserted due to the discontinuity of the normal vector, and appropriate compatibility conditions are introduced here.

The variation of the geometry over each boundary interval (i.e. between two adjacent nodal points) is modelled either by linear interpolation functions or by Hermite cubic splines, while the variation of the fields related to the fluid is modelled by linear interpolation functions. A Gauss-Legendre quadrature formula is used for the numerical integration over the regular intervals (i.e. intervals where there is no confluence between the node of observation and one of the nodes of integration). Special methods are used for the numerical integration over singular intervals, and a detailed description of these can be found in Grilli et al. (1989). When an interior point is close to the boundary curve, the evaluation of velocities, accelerations, and pressure becomes inaccurate due to the weak singularity there. Improved results (for a slightly increased computational cost) are found when a large number of integration points (e.g. 16) in the Gauss-Legendre quadrature formula is used at boundary intervals close to the relevant interior point.

VERIFICATION AND NON-LINEAR FEATURES OF THE NUMERICAL SCHEME

A demanding test of the numerical scheme is made when a wave with exact shape and of large steepness is used as input to the computational model, and computations covering several wave periods are performed. The initial data for a periodic, steep, non-linear wave of exact shape are generated by using the stream function wave model from Rienecker & Fenton (1981). A wave, with the (Eulerian) wave period $T = 6.14$ s, wave height $H = 5.00$ m (and a spatial resolution by 24 Fourier components) at the water depth $h = 10.0$ m, is considered. This gives a wave length $L = 55.2$ m, and hence a wave steepness of $H/L = 9.06\%$, which is close to the maximum wave steepness at this particular water depth. The boundary is discretized by inserting 92 nodes with 49 of these at the free water surface which is approximated by Hermite cubic splines. In the numerical integration a 4 point Gauss-Legendre quadrature formula is used for each interval at the boundary.

A horizontal dimension of one wave length was followed over more than 9 wave periods by using time-steps of $T/120$. Shortly before 10 wave periods were reached, the computations were terminated due to growing oscillations in the highest derivatives of the velocity potential ϕ. These could possibly be suppressed by smoothing, but it has been chosen

304

to avoid smoothing in these computations, since it is believed that smoothing could destroy information about the wave development, just as well as it could improve the results. By using a reduced time-step and more quadrature points in the numerical integration over the boundary, the computations can be followed further. The error on the total energy in the computational domain is kept within 0.03% during the computations, except in the final steps just before termination where it grows to about 0.9%. The mean water level (MWL) stays within 0.0002m from the original MWL during the computations. In Fig. 2 the free surface profile is depicted after each (Eulerian) wave period during nine wave periods.

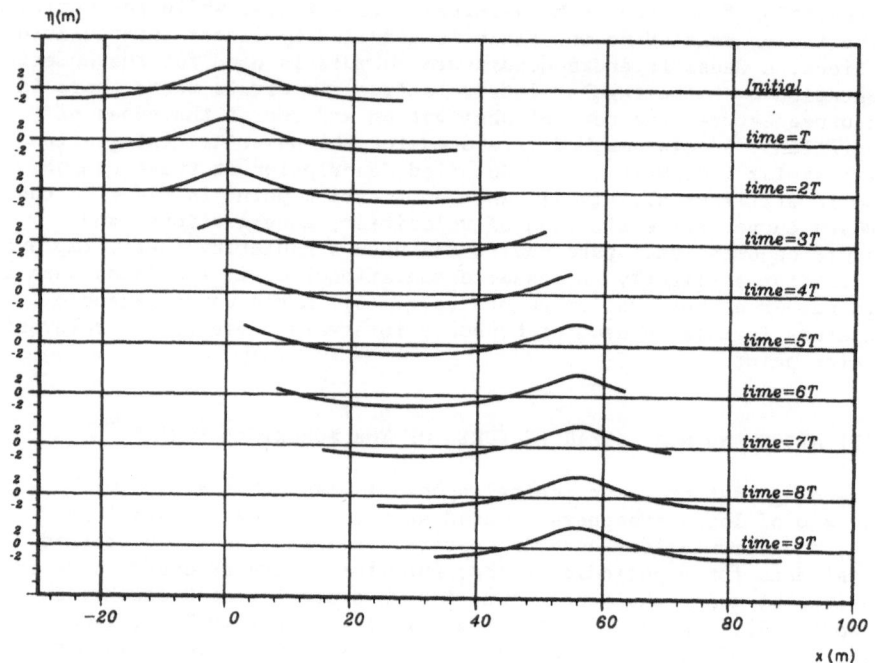

Fig. 2 Free surface profiles for a steep stream function wave. One profile is shown each (Eulerian) period. There is no vertical exaggeration.

A verification of the kinematics of particles within the fluid is performed by computing the velocity and acceleration components, and the excess pressure at some points within the fluid domain, and comparing them with those obtained from the stream function wave (SFW) theory. In Table 1 a comparison is made for particles situated along the vertical line through x = 0.0 m (line through the wave crest) and the vertical line through x = 4.6 m, at time t = 0.

x — 0.0 m, η — 3.34 m (wave crest)

z (m)	u (m/s) BIEM	SFW	w (m/s) BIEM	SFW	a_x (m/s^2) BIEM	SFW	a_z (m/s^2) BIEM	SFW
3.0	4.627	4.697	0.000	0.000	0.000	0.000	-4.282	-3.070
2.5	4.337	4.358	0.000	0.000	0.000	0.000	-3.057	-2.980
2.0	4.039	4.052	0.000	0.000	0.000	0.000	-2.868	-2.866
1.0	3.519	3.527	0.000	0.000	0.000	0.000	-2.597	-2.596
0.0	3.089	3.095	0.000	0.000	0.000	0.000	-2.309	-2.305
-1.0	2.735	2.740	0.000	0.000	0.000	0.000	-2.019	-2.015
-2.0	2.443	2.447	0.000	0.000	0.000	0.000	-1.740	-1.736
-3.0	2.203	2.206	0.000	0.000	0.000	0.000	-1.477	-1.473
-4.0	2.007	2.011	0.000	0.000	0.000	0.000	-1.230	-1.226
-5.0	1.850	1.853	0.000	0.000	0.000	0.000	-1.000	-0.996
-6.0	1.727	1.730	0.000	0.000	0.000	0.000	-0.784	-0.779
-7.0	1.634	1.638	0.000	0.000	0.000	0.000	-0.579	-0.574
-8.0	1.569	1.573	0.000	0.000	0.000	0.000	-0.382	-0.378
-9.0	1.530	1.535	0.000	0.000	0.000	0.000	-0.185	-0.188

x — 4.6 m, η — 2.06 m

z (m)	u (m/s) BIEM	SFW	w (m/s) BIEM	SFW	a_x (m/s^2) BIEM	SFW	a_z (m/s^2) BIEM	SFW
1.5	2.793	2.798	2.071	2.068	2.883	2.998	-1.152	-1.192
1.0	2.643	2.644	1.885	1.888	2.738	2.740	-1.246	-1.230
0.0	2.368	2.369	1.572	1.574	2.303	2.299	-1.240	-1.232
-1.0	2.133	2.134	1.309	1.311	1.944	1.942	-1.175	-1.169
-2.0	1.933	1.935	1.087	1.088	1.657	1.655	-1.073	-1 068
-3.0	1.765	1.767	0.895	0.896	1.427	1.426	-0.951	-0.947
-4.0	1.625	1.628	0.729	0.729	1.245	1.244	-0.818	-0.815
-5.0	1.512	1.514	0.582	0.581	1.101	1.102	-0.682	-0.679
-6.0	1.421	1.424	0.450	0.449	0.992	0.992	-0.545	-0.541
-7.0	1.353	1.355	0.329	0.327	0.910	0.911	-0.408	-0.404
-8.0	1.304	1.308	0.216	0.214	0.854	0.856	-0.272	-0.269
-9.0	1.275	1.279	0.106	0.106	0.811	0.823	-0.131	-0.134

Table 1 Velocities and accelerations at interior points of a steep wave.
Comparison of results from present model and results from the
stream function wave theory.

An excellent agreement between the BIEM and the SFW results is
readily seen. The differences between the two results are less than 1%,
except in points very close to the free surface, where it is slightly
larger (the difference on a_z at (x,z) — (0.0,3.0) is considered to be
caused by an inaccurate integration). By observing the vertical
acceleration at x — 4.6 m, an interesting result appears both from the
BIEM and from the SFW computations; namely that the magnitude of the
vertical acceleration has a maximum not at the water surface, but close
to it. The same variation is found for the horizontal acceleration at

other vertical lines through the fluid. The cause for this variation of
the accelerations with depth is unknown, and no references about this
particular phenomenon have been found in the literature.

The motion of water particles at the free surface is readily
obtained from the free surface nodes during the computation, since the
time-stepping procedure is Lagrangian (particle following). Thus,
following a computational node provides the orbit of a water particle at
the free surface during the computation. In Fig. 3 the orbit of the
water particle (situated at the wave crest initially) is depicted during
more than nine wave periods. The position of the particle after each
Eulerian wave period is marked by an □ in Fig. 3.

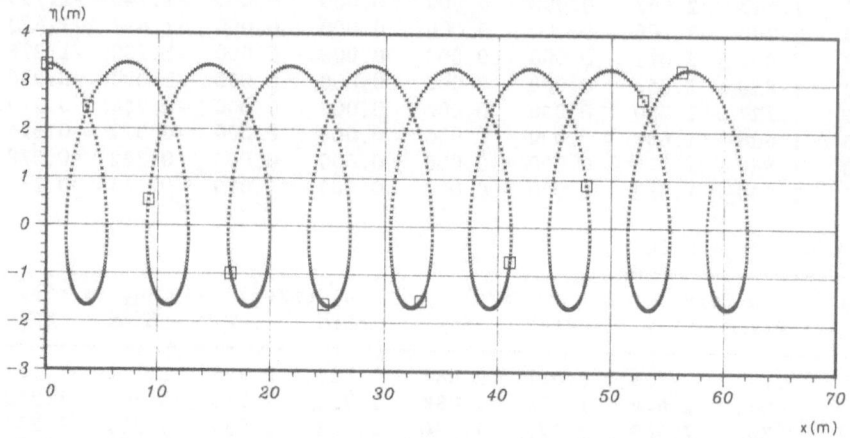

Fig. 3 Particle orbit during 9.92 Eulerian wave periods for a steep
stream function wave. $T_E = 6.14$ s, $T_L = 6.94$ s.

From Fig. 3 it is clearly seen that there is a drift of the free
surface particles in the direction of wave propagation, as the
computations proceed in time (Stokes drift). The drift of one particle
over one Lagrangian period T_L (the time it takes a water particle to go
from one wave crest to the next) is found to be as large as 13% of the
wave length. Furthermore it is found that the ratio between the
Lagrangian period and the Eulerian period T is about 1.13 for particles
at the free water surface for this particular wave. By performing time-
stepping of interior points (by (18) and (19)) the Lagrangian period for
these is readily obtained, and it is found that it is highly dependent
on the distance below the water surface (at the bottom it equals T_E).
When model tests (or field measurements) involving steep water waves are
made, this variation in the Lagrangian period should be taken into
account, if the wave period is determined from observation of particles.

PARTICLE KINEMATICS AROUND HORIZONTAL CYLINDER

The computation of the interaction between waves and a submerged
horizontal cylinder with its axis parallel to the wave crests has been

treated widely in the literature, beginning with Dean (1948), who showed that linear deep water waves undergo a phase shift as they pass over a cylinder, and further that there is no reflection from the cylinder. Ursell (1950) used a multipole method, and derived expressions for the first-order forces on the cylinder. Ogilvie (1963) extended Ursell's method and provided expressions for the mean second-order ("drift") force at the cylinder. Chaplin (1984) performed studies on the non-linear forces and mass transport around a horizontal submerged cylinder, and experimentally verified that there is a mass-transport around the cylinder, as it could be predicted by Milne-Thompson's (1968) circle theorem.

Structures are simple to incorporate in the BIEM model, and results from computations of the flow field around these, and the resulting wave force vector $\vec{F}$ on these is evaluated by the present model.

The flow field around a fixed horizontal circular cylinder due to waves may be visualized by following a number of water particles in the fluid volume around the cylinder by use of (18) and (19) for the time-stepping of interior points. An example with a wave of exact shape with the data H = 5.00 m, T = 8.78 s, at a depth h = 100 m, providing L = 122 m, and a submerged circular cylinder of diameter D = 10.0 m, and the cylinder axis (which is parallel with the wave fronts) situated 20.0 m below MWL, is depicted in Fig. 4, where the computations cover one (Eulerian) wave period (the wave moves from left to right).

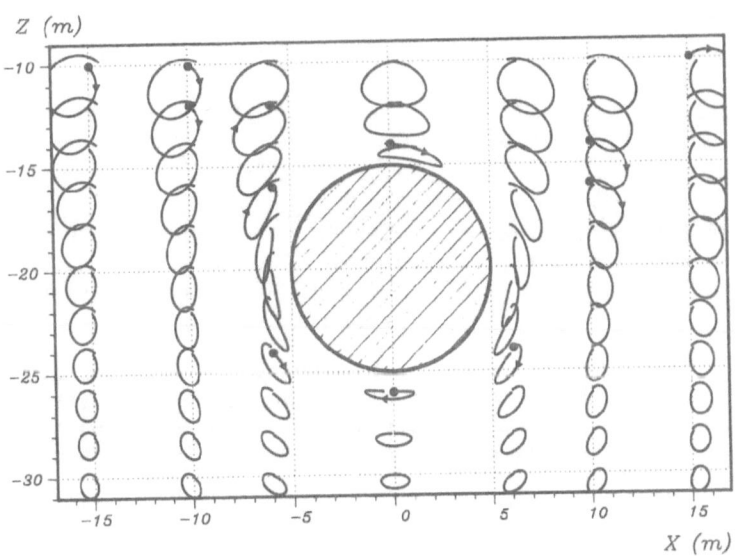

Fig. 4 Particle motion around a cylinder of diameter D = 10.0 m during one Eulerian wave period. The wave data are: T_E = 8.78 s, H = 5.00 m, and L = 122 m, at h = 100 m. ●: Initial position, ▶ : Orbital direction.

The particle orbits (which are in the clockwise direction since the wave is moving from left to right) are clearly seen to be influenced by the cylinder, even at distances which equal the diameter of the cylinder.

By increasing the wave height to H - 10.0 m (providing a wave length of L - 128 m for the same wave period and water depth as before), we find a larger influence of the particle paths from to the cylinder, and the results from the computations covering one (Eulerian) wave period are depicted in Fig. 5.

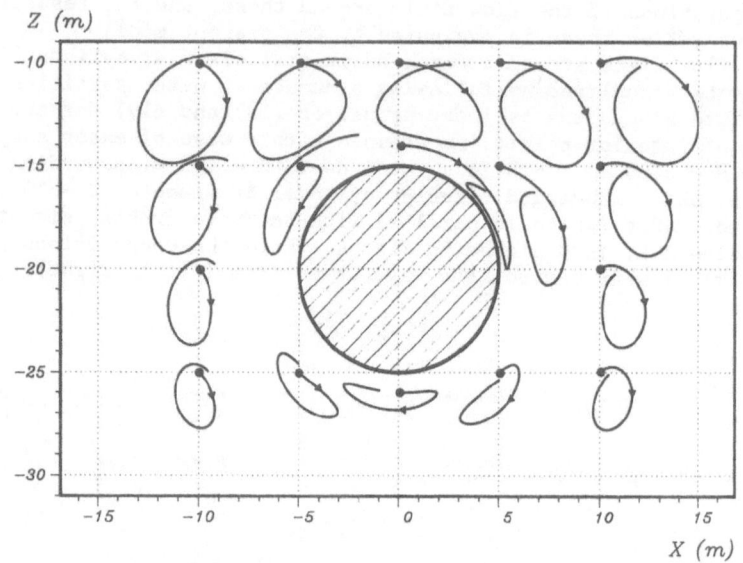

Fig. 5 Particle motion around a cylinder of diameter D - 10.0 m during one Eulerian wave period. The wave data are: T_E - 8.78 s, H - 10.0 m, and L - 128 m, at h - 100 m. ●: Initial position, ▶ : Orbital direction.

It is clearly seen in Fig. 5 that the water particles are actually "transported" around the cylinder as the wave propagates (from left to right). By extending the computations to involve seven (Eulerian) wave periods this is even more clearly seen, and this is depicted in Figs. 6a and 6b.

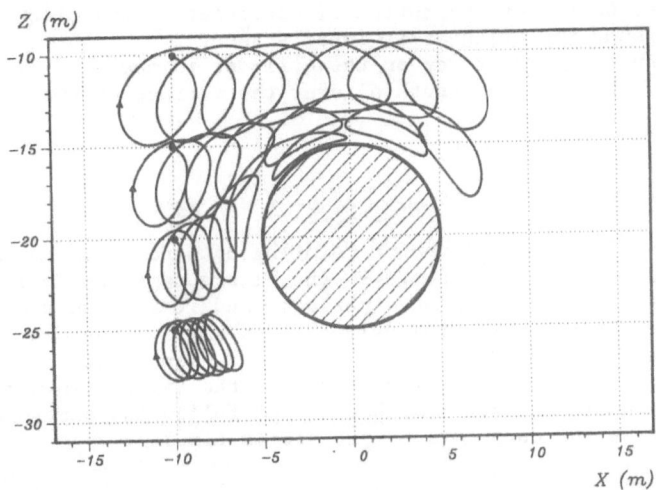

Fig. 6a Particle motion around a cylinder of diameter D = 10.0 m during
seven Eulerian wave periods. The wave data are: T_E = 8.78 s,
H = 10.0 m, and L = 128 m at h = 100 m. ● : Initial position,
▶ : Orbital direction.

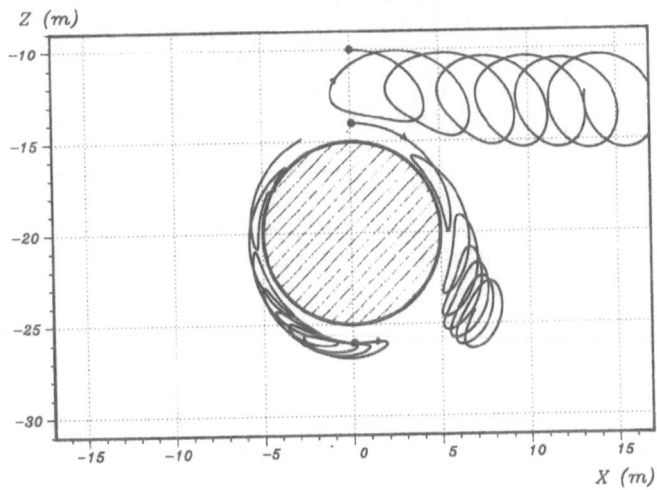

Fig. 6b Particle motion around a cylinder of diameter D = 10.0 m during
seven Eulerian wave periods. The wave data are: T_E = 8.78 s,
H = 10.0 m, and L = 128 m at h = 100 m. ● : Initial position,
▶ : Orbital direction.

310

In Figs. 6a and 6b the motion of particles in the clockwise direction around the cylinder is readily seen.

The resulting <u>wave</u> force on the cylinder in computed by integrating the excess pressure p^+ over the surface of the structure

$$\vec{F} = - \int_{cyl} p^+ \, \vec{n} \, ds \qquad (20)$$

where $\vec{n}$ is a (unit) normal vector pointing outwards from the body. In Fig. 7 the wave force components F_x and Fz are depicted from computations covering seven wave periods, and it is seen that the effects of reflection from the cylinder are small. The computations were terminated after seven Eulerian wave periods, since the vertical boundary then reached the cylinder (due to the Lagrangian updating of the boundary where the vertical boundaries follow the particles at the free surface as the computations proceed in time).

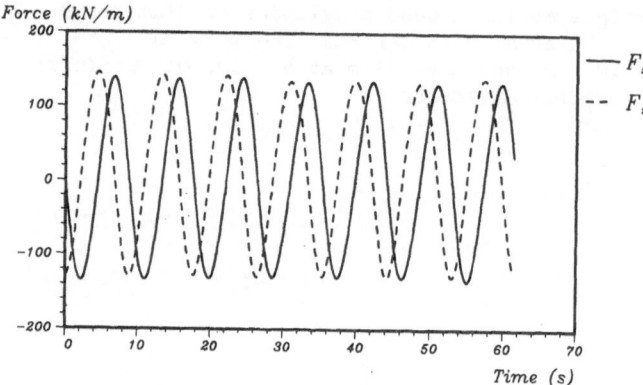

Fig. 7 Force vector components F_x and F_z at circular cylinder during seven wave periods.

SUMMARY AND CONCLUDING REMARKS

A 2-D Boundary Integral Equation Method combined with a non-linear time-stepping procedure is presented for the modelling of non-linear water waves, and an application to the modelling of a steep wave during several wave periods is given. A method for updating interior points of the fluid domain is developed, and computational results are given, showing the influence on the particle paths from a fixed submerged horizontal cylinder. This method does only give reliable results when there is no flow separation around the cylinder, but since the Keulegan-Carpenter numbers for the computational examples are low, and the structure does not have any sharp corners, the effects due to flow separation are insignificant. Wave forces on submerged, fixed structures are simple to evaluate by the present computational model.

The computational model presented here is a physical space model in 2-D, and there are no restrictions in the formulation that prevent the model from being extended to 3-D; prevailing models in the literature are not able to be extended to 3-D due to their mathematical formulation. However, new conditions at the lateral boundaries of the computational domain must be introduced if irregular waves in 3-D is modelled, and work still remains to be done there if a 3-D computational scheme should be developed.

ACKNOWLEDGEMENT

This research project, the span of which included the first author's stay at the University of Delaware, USA during Sep. 1987 - June 1988, is supported by The Danish Technical Research Council (STVF J. No. 5.17.3.6.03). This support is gratefully acknowledged.

REFERENCES

Brevig, P., Greenhow, M. & Vinje, T. (1981). Extreme wave forces on submerged cylinders. Second International Symposium on Wave and Tidal Energy. Cambridge, England. Paper E2, pp 143-166.

Chaplin, J.R. (1984). Mass transport around a horizontal cylinder beneath waves. Journal of Fluid Mechanics, 140, pp 175-187.

Cokelet E.D. (1979). Breaking waves - The plunging jet and interior flow field. Mechanics of Wave-induced Forces on Cylinders. Ed. by T.L. Shaw. San Francisco: Pitman, pp 287-301.

Dean, W.R. (1948). On the reflection of surface waves by a submerged cylinder. Proc. Camb. Phil. Soc., 44, pp 483-491.

Dold, J.W. & Peregrine, D.H. (1984). Steep unsteady water waves. An efficient computational scheme. 19th Coastal Engineering Conference, Houston, USA. Vol. 1, pp 955-967.

Dommermuth, D.G. & Yue, D.K.P. (1987). Numerical simulation of non-linear axisymmetric flows with a free surface. Journal of Fluid Mechanics, 178, pp 195-219.

Grilli, S.T., Skourup, J. & Svendsen, I.A. (1989). An efficient boundary element method for non-linear water waves. Engineering Analysis, vol. 6, no. 2.

Isaacson, M. de St. Q. (1982). Nonlinear-wave effects of fixed and floating bodies. Journal of Fluid Mechanics, 120, pp 267-281.

Longuet-Higgins, M.S. & Cokelet, E.D. (1976). The deformation of steep surface waves on water. I. A numerical method of computation. Proc. Roy. Soc. of London. A. 350, pp 1-26.

Milne-Thompson, L.M. (1968). Theoretical Hydrodynamics. 4th ed. Macmillan.

New, A.L., McIver, P. & Peregrine, D.H. (1985). Computation of overturning waves. Journal of Fluid Mechanics, 150, pp 233-251.

Ogilvie, T.F. (1963). First and second order forces on a cylinder
    submerged under a free surface. Journal of Fluid Mechanics, 16,
    pp 451-472.

Peregrine, D.H., Cokelet, E.D. & McIver, P. (1980). The fluid mechanics
    of waves approaching breaking. 17th Coastal Engineering Conference,
    Sydney, Australia. Vol. 1, pp 512-528.

Rienecker, M.M., & Fenton, J.D. (1981). A Fourier approximation method
    for steady water waves. Journal of Fluid Mechanics, 104,
    pp 119-137.

Ursell, F. (1950). Surface waves in the presence of a submerged circular
    cylinder. Proc. Camb. Phil. Soc., 46, pp 141-158.

Vinje, T. & Brevig, P. (1981). Numerical simulation of breaking waves.
    Adv. Water Resources, 4, pp 77-82.

A VISCOUS MODIFICATION TO THE OSCILLATORY MOTION BENEATH A SERIES OF PROGRESSIVE GRAVITY WAVES

C. Swan
Department of Engineering
University of Cambridge
Cambridge CB2 IPZ
ENGLAND.

ABSTRACT. A perturbation analysis is presented for steady monochromatic waves propagating without change of form in water of constant depth. The analysis allows for the development of a fully diffused vorticity profile, and thereby incorporates the influence of the viscous boundary layers. At a third order of wave steepness it is found that the proposed solution differs significantly from the existing irrotational theories. A new series of experimental measurements using laser Doppler anemometry are presented. It was found that an irrotational solution overestimates the amplitude of the oscillatory motion in the upper half of the flow field, and underestimates the amplitude in the lower half. In each case the present viscous solution provides a better description of the wave kinematics, compared with the traditional irrotational solution.

1. Introduction.

There have, of course, been many investigations of the oscillatory motion beneath a series of progressive gravity waves. The first realistic description was presented by Stokes in 1847. He neglected the effects of viscosity and assumed that the motion, having been generated from rest through the action of "ordinary forces", was necessarily irrotational. More recently, a number of higher order solutions have been obtained. For example, De (1954), Skjelbreia and Hendrickson (1961), and Fenton (1985) each produced analytical solutions to a fifth order of wave steepness. In addition, the numerical work by Dean (1965 and 1974), Schwartz (1974), Cokelet (1977), Chaplin (1980), and Riendecker and Fenton (1981) have generated a number of very high order solutions.

There has also been a number of experimental studies in which the wave characteristics have been measured within a laboratory wave flume. In many cases, the agreement between the theoretically predicted wave kinematics and the experimental observations is reasonably good. This is particularly true when the velocity measurements are taken directly after the wave has been generated. This is a common experimental

A. Tørum and O. T. Gudmestad (eds.), Water Wave Kinematics, 313–329.
© 1990 *Kluwer Academic Publishers.*

technique since it eliminates the problems associated with wave reflection. However, in eliminating one difficulty, the experimentalist has created an artificial environment in which the wave form has only existed for perhaps two or three wave periods. The wave kinematics may therefore be very different from those waves which have been allowed to develop in both space and time.

There are a number of experimental observations which tend to support this view. For example, the observations due to Morison and Crooke (1953) indicate that the present irrotational solutions over-estimate the magnitude of the oscillatory motion in the upper layers of the flow field, and underestimate the magnitude in the lower layers. More recently, Beech (1978) and Simons (1980) observed the velocity field within the near bed region and also found that the velocities were larger than the predicted irrotational motion; while Anastasiou et. al. (1982) concluded that the large positive velocities in the near surface region are overestimated by an irrotational solution.

The theoretical descriptions noted above have one important point in common: they all assume that the wave motion is irrotational, and therefore that no vorticity exists throughout the depth of the flow field. However, it is clear from the conduction solution proposed by Longuet-Higgins (1953) that a vorticity profile can exist within the interior of the flow field. Indeed, he showed that given sufficient time, a fully diffused vorticity profile would develop and that such a distribution would have a significant effect upon the time-mean component of the water particle velocity.

The object of the present paper is thus to consider the effects of a fully diffused vorticity profile. The final solution will be valid throughout the depth of the flow field and must be compatible with the development of the viscous boundary layers at both the bed and the oscillating free surface. The derivation of the analytical solution will not be described in detail (this information has already been presented by Swan 1989a). A brief description will be given, and the final solution quoted without proof. The paper will be primarily concerned with the experimental observations, and whether the apparent departures from the existing irrotational solutions can be explained by the diffusion of vorticity from the viscous boundary layers.

2. Theoretical work.

The difficulty with any analysis which incorporates an oscillating free surface is the application of the boundary conditions. Both the kinematic and the dynamic free surface constraints are strongly non-linear, and must be applied on a surface for which there can be no prior knowledge of its position. To overcome this difficulty, a transformation was adopted in which the whole of the Cartesian co-ordinate framework (x,z) was mapped onto an orthogonal curvilinear framework (ξ,η). Within the final co-ordinate system the line $\eta=0$ describes the free surface, and the co-ordinate axes move with the wave speed c, thereby creating a steady frame of reference.

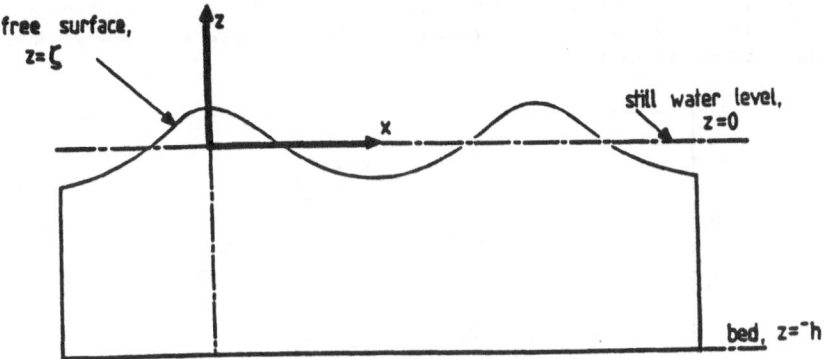

Figure 1a. The Cartesian co-ordinate system (x,z).

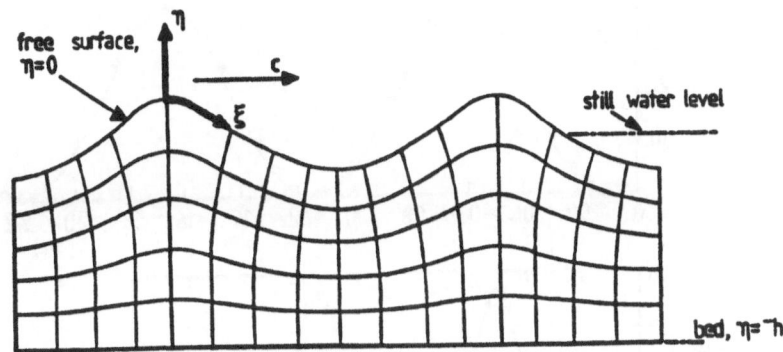

Figure 1b. The curvilinear co-ordinate system (ξ, η).

To achieve this type of transformation an initial estimate of the surface profile was required. The experimental observations shown in figure 2. suggest that Stokes' third order elevation provides a reasonably good description of the surface profile over a wide range of wave forms. The following solution was therefore taken as an initial estimate:

$$\zeta = b\cos(kx-\sigma t) + \frac{b^2 k\cosh(kh)[\cosh(2kh)+2]\cos(2(kx-\sigma t))}{4\sinh^3(kh)}$$
$$+ \frac{b^3 k^2[1+8\cosh^6(kh)]\cos(3(kx-\sigma t))}{64\sinh^6(kh)} \qquad (1)$$

316

where ζ is the height above mean water level, b is the amplitude of the first harmonic, k is the wave number, σ is the wave frequency, and h the undisturbed water depth.

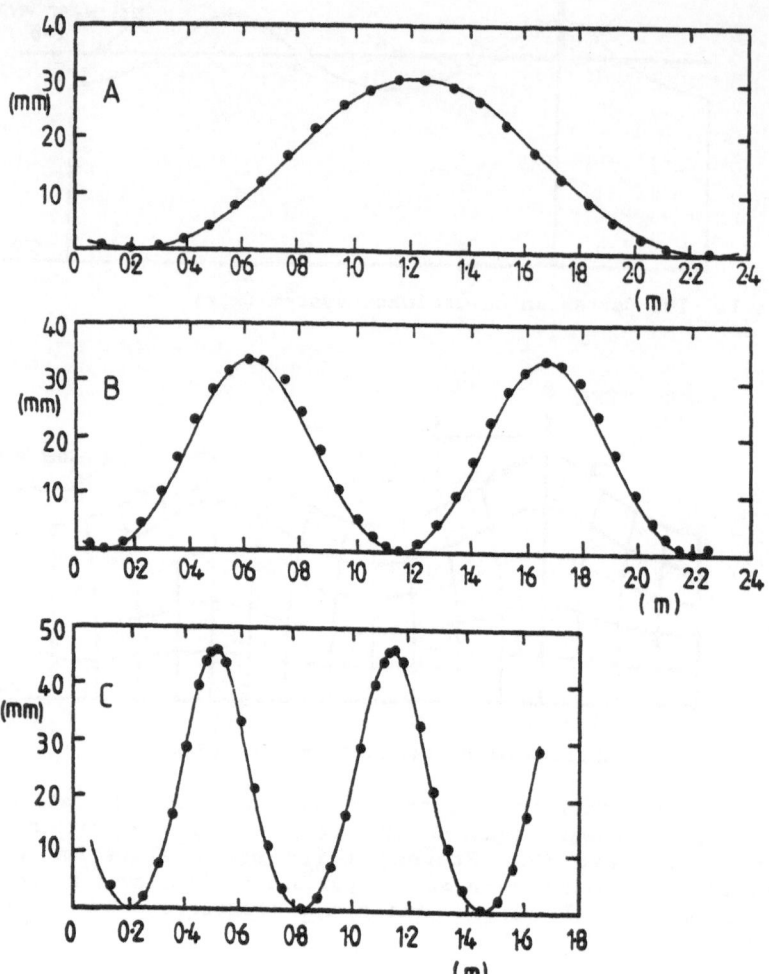

Figure 2. The surface elevation. (A) kh=1.11, (B) kh=2.16, (C) kh=3.86.
———————————— 3rd. order irrotational solution, ● data points.

The (ξ, η) co-ordinate system may now be established via a series of successive transformations. These are similar to the transformation adopted by Benjamin (1959). The only difference is that they will be applied at each order of the perturbation (see below), and the constant coefficients A_n and B_n will be defined such that they eliminate all the terms at the largest existing order. In this way the line $\eta=0$ will define the free surface profile to any order of approximation.

$$\eta = z - \sum_{n=0}^{n} \frac{A_n b^n \cosh(nk(h+\eta))\cos(nk\xi) - B_n}{\sinh(kh)}$$

$$\xi = x - \sum_{n=0}^{n} \frac{A_n b^n \sinh(nk(h+\eta))\sin(nk\xi)}{\sinh(kh)}$$

$$(2)$$

The Jacobian of this transformation is given by:

$$J = \frac{\partial(\xi,\eta)}{\partial(x,z)} \tag{3}$$

If we consider a two-dimensional monochromatic wave train propagating in water of constant depth, then the vorticity equation in general orthogonal co-ordinates is given by:

$$\nu D^4 \psi + J \left(\frac{\partial \psi}{\partial \xi} \frac{\partial(D^2\psi)}{\partial \eta} - \frac{\partial \psi}{\partial \eta} \frac{\partial(D^2\psi)}{\partial \xi} \right) = 0 \tag{4}$$

where ν is the kinematic viscosity. The differential operator, D, is defined by:

$$D = J \left(\frac{\partial^2}{\partial \xi^2} + \frac{\partial^2}{\partial \eta^2} \right) \tag{5}$$

and the stream function ψ is related to the velocity components u_ξ and w_η in the usual manner:

$$u_\xi = \sqrt{J} \frac{\partial \psi}{\partial \eta} \qquad \text{and} \qquad w_\eta = -\sqrt{J} \frac{\partial \psi}{\partial \xi} \tag{6}$$

It is not possible to find an exact analytical solution to equation 4. However, if a small perturbation approach is adopted, then a solution can be identified at successive orders of the wave steepness. This requires that the stream function ψ, the Jacobian J, and the two velocity components u and w, each be expanded in terms of some small parameter ϵ, which in this case is taken to represent the wave steepness. The perturbation expansion will be of the form:

$$f = f_0 + \epsilon f_1 + \epsilon^2 f_2 + \epsilon^3 f_3 + \epsilon^4 f_4 + \ldots\ldots\ldots \tag{7}$$

where f represents ψ, J, u, or w. Having obtained the general solution to equation 4, the only remaining step is to satisfy the boundary conditions. These are relatively easy to apply within the (ξ,η) co-ordinate system:

(a) The kinematic free surface condition:

$$\left(\frac{\partial \psi}{\partial \xi}\right)_{\eta=0} = 0 \tag{8a}$$

(b) The zero tangential stress condition:

$$\left[J\nabla^2\psi + \left(\frac{\partial J}{\partial \eta}\frac{\partial \psi}{\partial \eta}\right) - \left(\frac{\partial J}{\partial \xi}\frac{\partial \psi}{\partial \xi}\right)\right]_{\eta=0} = 0 \tag{8b}$$

(c) The constant normal stress constraint:

$$\left(\frac{\partial P}{\partial \xi}\right)_{\eta=0} = 0$$

which on substituting for (i) and (ii) in the Navier-Stokes equation and defining F as the gravitational body force in the η direction yields:

$$\left[FJ^{-0.5} - 0.5\frac{\partial}{\partial \xi}\left(J\frac{\partial \psi}{\partial \eta}^2\right)\right]_{\eta=0} = 0 \tag{8c}$$

(d) The vertical velocity component must be zero at the bed:

$$\frac{\partial \psi}{\partial \xi} = 0 \quad \text{at} \quad \eta = -h \tag{8d}$$

(e) The horizontal velocity component just outside the bottom boundary layer must match the existing higher order solutions within the near bed region. Hence, at $\eta = -h$ we obtain:

$$\sqrt{J}\,\frac{\partial \psi}{\partial \eta} = \text{Eulerian velocity solution} \atop \text{Sleath (1972)} \tag{8e}$$

(f) Finally, for the purpose of experimental comparison the horizontal drift through any vertical cross-section must be zero:

$$\int_{z=-h}^{z=0} U_L \, dz = 0 \tag{8f}$$

where U_L is the time-mean Lagrangian velocity component in the horizontal direction.

At the first and second orders of wave steepness the boundary conditions can be satisfied exactly without any modification of the surface profile. Physically, this should perhaps have been expected since at this order of approximation the solution merely represents the superposition of Stokes' (1880) irrotational flow field and the time independent terms identified by Longuet-Higgins' (1953) conduction solution. Since both these solutions are well known, they will not be further discussed.

At the third order of wave steepness the solution indicates a significant departure from the traditional irrotational solution. The third order modification to the fundamental components of the velocity in the horizontal and vertical directions are given by:

$$u_3 = \frac{b^3 k^2 \sigma \{L1(kh)\sinh[k(h+z)](\frac{z}{h}+1)+L2(kh)\cosh[k(h+z)]\}\cos(\sigma t - kx)}{h}$$

(9)

$$w_3 = \frac{b^3 k^2 \sigma \{L1(kh)\cosh[k(h+z)](\frac{z}{h}+1)-L3(kh)\sinh[k(h+z)]\}\sin(\sigma t - kx)}{h}$$

where the subscript "3" indicates the order of the velocity components, and the functions L1(kh), L2(kh), and L3(kh) are constant for any particular wave form. These are shown on figure 3.

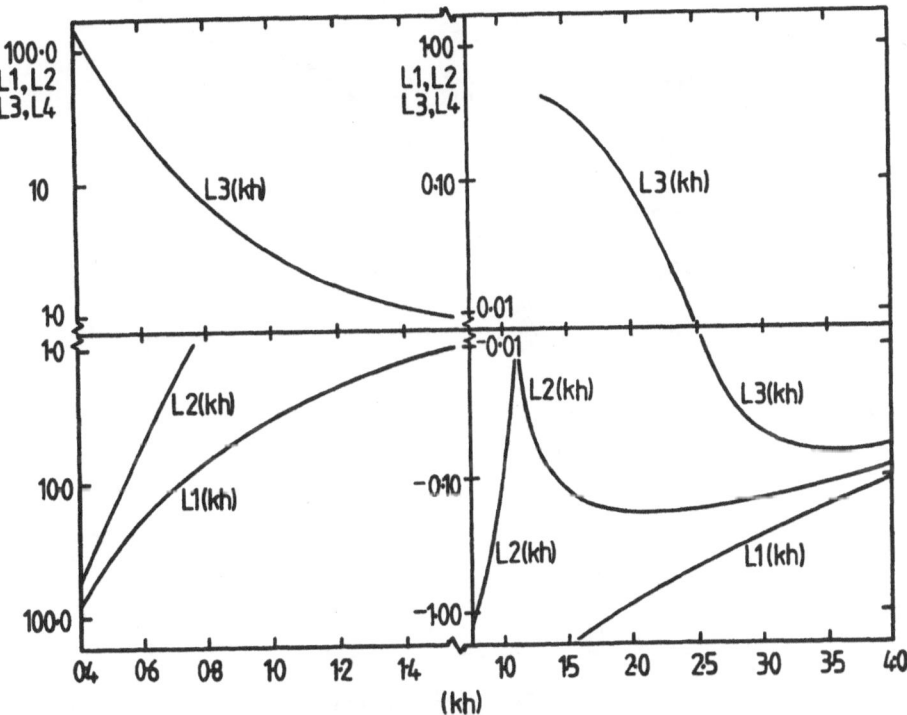

Figure 3. The functions L1(kh), L2(kh), and L3(kh).

Finally, it should be noted that the above analysis is not dependent upon the initial estimate of the surface profile. It represents an exact third order solution which satisfies all the boundary conditions (8), and the governing vorticity equation (4). If a detailed derivation of this solution is sought, then the reader is directed to Swan (1989a).

3. Experimental work.

The experimental observations due to Morison and Crooke (1953) show a number of important departures from the predicted irrotational motion. At the mean water level a third order irrotational solution over-estimates the velocity amplitude by 16%; while in the near bed region an irrotational solution underestimates the velocity amplitude by 28%. The present solution appears to account for much of this behaviour. Figure 4. shows a typical set of results in which the measured amplitude of the horizontal velocity component is compared with the predicted irrotational motion and the present viscous solution given in (9).

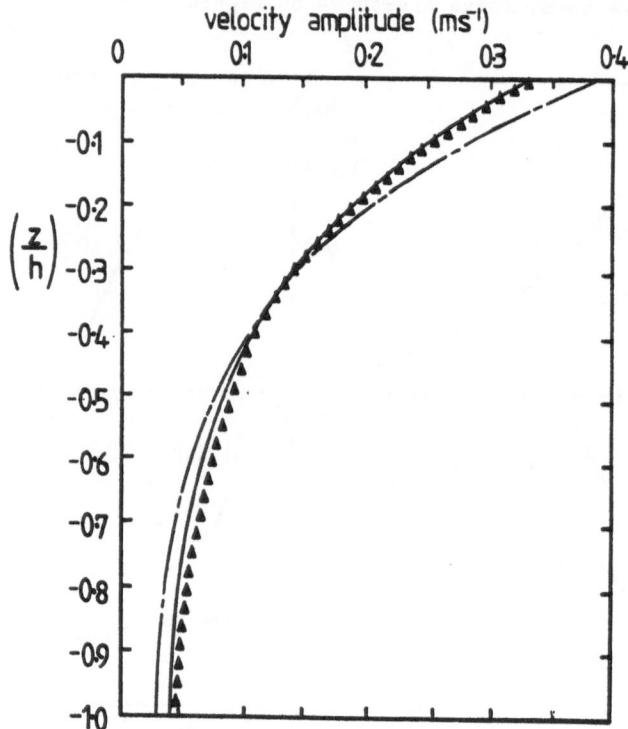

Figure 4. The horizontal velocity component. ———— - ———— 3rd. order irrotational solution, ———————— 3rd. order viscous solution, ▲▲▲▲▲▲▲▲▲ Mean of experimental data (Morison and Crooke, 1953).

For the same test conditions Morison and Crooke (1953) also provided an experimental description of a particle orbit in the upper layers of the flow field. In this instance, they eliminated the effect of the mass transport velocity, thereby creating a closed particle loop. This was achieved by measuring the apparent particle drift over one wave cycle, and then apportioning the negative of this motion over the corresponding length of the particle path. This data is given on Figure 5.

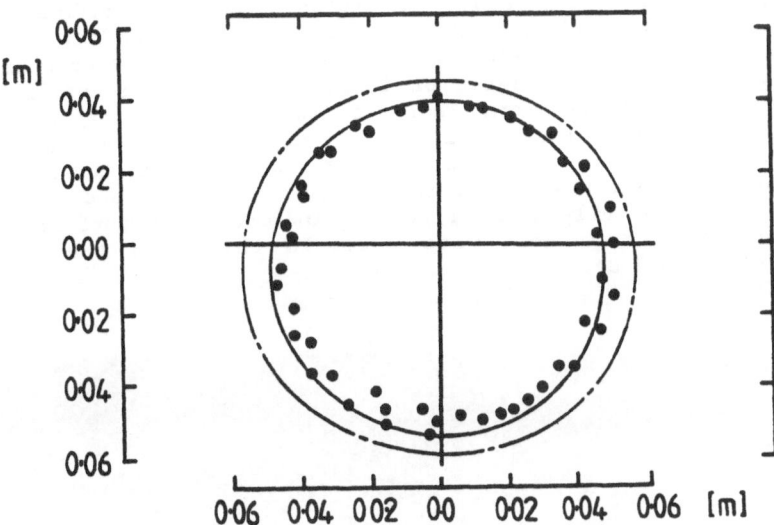

Figure 5. A particle orbit. ———— – ———— 3rd. order irrotational solution. ———————— 3rd. order viscous solution.
● data points (Morison and Crooke, 1953)

More recently, Beech (1978) investigated the velocity field beneath a series of progressive gravity waves using a laser Doppler anemometer. His observations within the lower regions of the flow field show a similar trend to those presented by Morison and Crooke (1953). However, many of the wave forms which Beech investigated were too long for his experimental flume. In these cases it is doubtful whether stable conditions actually developed. Furthermore, he did not report any measurements of the reflection coefficient associated with these long waves. Without this information, it is very difficult to draw any conclusions from these experimental observations. A similar study was completed by Simons (1980). In this case the measurements appear to be of a much higher standard, but they were concentrated within the lower layers of the flow field, and the investigation was limited to one wave form (kh=1.0).

In view of this shortage of reliable data, the author sought to confirm the characteristics of the oscillatory motion under strictly controlled conditions for a range of wave forms. The observations were

322

conducted within the Cambridge University Engineering Department's wave flume. This has an overall length of 17.5m, a width of 0.59m, and a maximum depth of 0.45m. The paddle mechanism consisted of a hinged flap located in a deepened section at one end of the wave flume. Further details of the experimental arrangement are provided by Swan (1987).

To obtain experimental data which will allow a valid comparison with the theoretically predicted behaviour, we must generate and maintain a regular wave train within the finite length of an experimental flume. This is, of course, extremely difficult. In the first case, wave reflection must be reduced to an absolute minimum. To achieve this, the beach slope was set at 1:20, and covered with pebbles having an average diameter of 25mm. The reflection coefficient was measured for each wave form. This was determined by moving a wave gauge along the length of the wave flume, and monitoring the variation in the wave height. Even in the worst cases, corresponding to the largest wave lengths, the reflection coefficient never exceeded 4%, and in most cases it was significantly less. Figure 6 indicates the surface elevation record for the longest wave form. In this case the reflection coefficient is 3.8%.

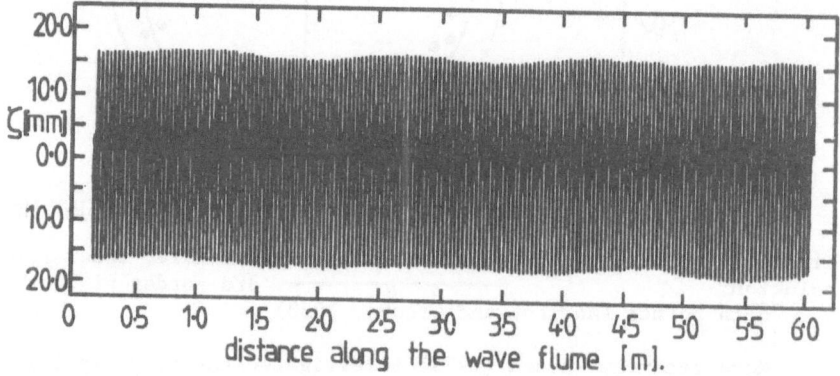

Figure 6. The variation in the surface elevation along the wave flume.

Another problem which frequently arises within a laboratory flume is wave modulation. This has been the subject of several investigations. It has been found to be particularly prevalent where the wave profile has been generated by a simple hinged paddle, as is the case in the present experiments. However, wave modulation is closely associated with highly non-linear interactions occurring within the steeper wave forms. As a result, its effects can be neglected within the present investigation provided the observations are restricted to those waves in which the steepness (H/L) is considerably less than its limiting value. The wave profile shown in figure 7 represents the steepest wave form considered within the present investigation (H/L=0.075). The measurements were taken at one point and show the variation in the surface elevation with time. There is no evidence of wave modulation, and the wave train appears to be extremely regular.

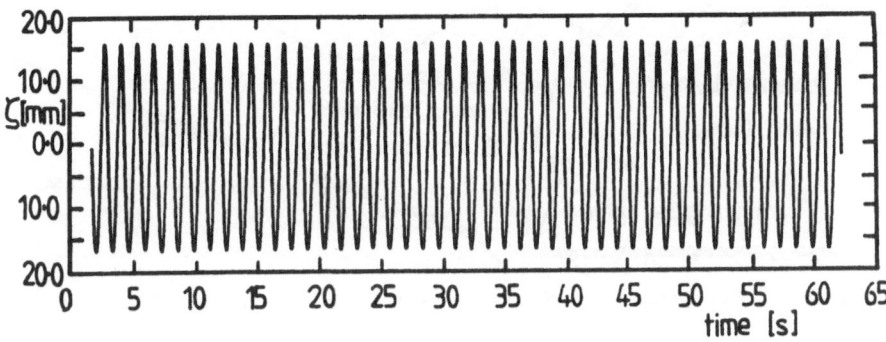

Figure 7   The variation in the surface elevation with time (H/L=0.075).

Difficulties can also arise due to seiching within a laboratory flume. This is generated when the paddle is initially switched on, and corresponds to a large scale longitudinal motion within the wave flume. Fortunately, this type of motion will decay with time provided there are no resonant interactions to counteract the viscous damping. Since the present investigation is concerned with the oscillatory motion some time after the wave motion was first generated, it may be assumed that the decay time was sufficient to eliminate the effects of seiching. In any case, the surface profile records, similar to that indicated in figure 7, would reveal any long term variation in the surface elevation. This was not observed in any of the test data.

Finally, there are a number of problems associated with the finite length of the wave flume. If the velocity profile is allowed to develop in both space and time, then we must ensure that the end conditions have a minimal effect on the velocity field in the working sections of the wave flume. The near shore region is, of course, associated with wave steepening and eventual breaking. This leads to the generation of vorticity within this area which may be convected backwards by the return flow established within a wave flume. The end conditions thus represent a source of vorticity which has not been accounted for in the present analysis. Indeed, when Longuet-Higgins (1953) considered this point, he concluded that the governing equation was indeterminate without further knowledge of the end conditions. However, from an experimental point of view there is substantial evidence to suggest that the addition of a flexible curtain at the toe of the beach is very effective at reducing the influence of the near shore region. The curtain shown in figure 8 was first suggested by Russell and Osorio (1957). In a recent series of observations Swan (1989b) found that in the absence of a protective curtain the whole wave flume could eventually be dominated by instabilities arising within the near shore region. Alternatively, with a curtain in place, no such instabilities could be identified, and the working sections appeared largely unaffected by the near shore region.

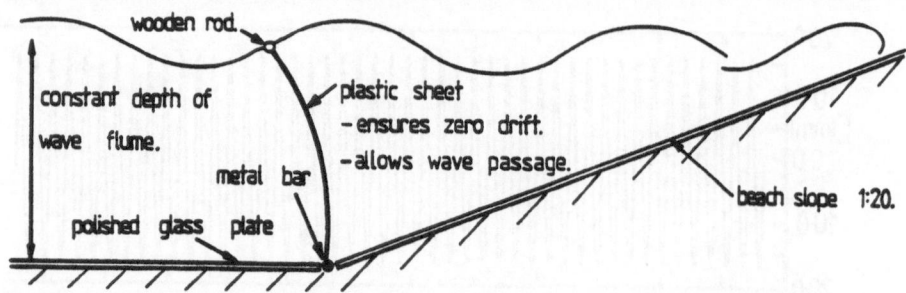

Figure 8.   The plastic curtain installed at the toe of the beach.

Having established a regular wave train the underlying velocity field was measured using a laser Doppler anemometer. The final system consisted of a 35mW helium-neon laser together with a system of plane polarizing lens. This allowed the two components of the velocity field to be measured from a three beam system, and avoided the need for two different frequencies of laser light. The measuring volume was estimated to be of the order of $1mm^3$.

Taking into account the experimental difficulties mentioned above, and the quality of the measuring system, it was estimated that the underlying velocity field could be determined to ±4%. The experimental data presented in figures 4. and 5. show that the oscillatory velocity differs from the predicted irrotational solution by as much as +28% and -16%. Consequently, the present measurements should be sufficiently accurate to identify whether a consistent pattern of discrepancies do arise over a range of wave forms.

The experimental data is given in figures 9. and 10. Figures 9a-9d show the amplitudes of both the horizontal and the vertical velocity components for four different wave types. To supplement this information the steepest wave form (case 9c.) is further investigated. In figures 10a and 10b, two particular depths are considered (z=-0.04m, and z=-0.32m) and the variation in the horizontal component of the oscillatory velocity is shown throughout a complete wave cycle.

4.   Discussion of results.

The experimental data presented in figures 9 and 10 were measured under strictly controlled conditions. The importance of wave reflection, wave modulation, wave breaking, three dimensionality, and longitudinal seiching were all considered. After a number of modifications to both the wave flume, and the range of wave forms to be considered, it was concluded that these effects could effectively be eliminated (figures 6 and 7). They did not have an appreciable effect upon the characteristics of the flow field. Despite all these precautions, the experimental observations clearly indicate that under certain circumstances the velocity field beneath a series of progressive gravity waves can be significantly different from the predicted irrotational motion.

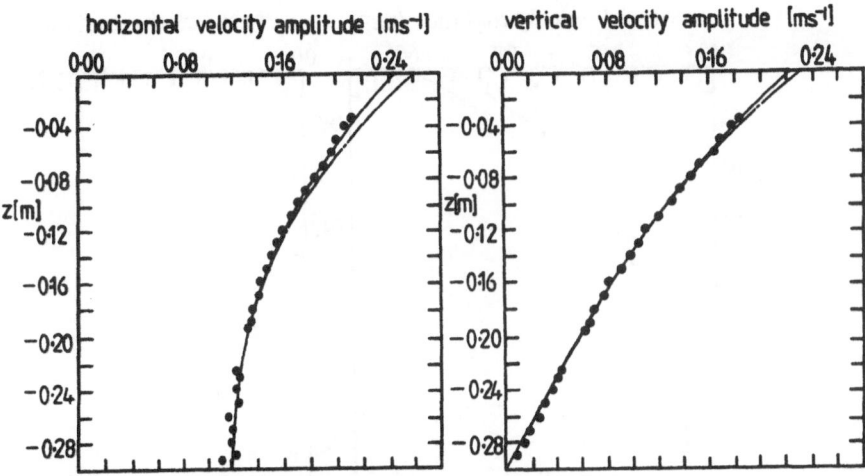

Figure 9a   The amplitudes of the orbital velocity. (a=36.2mm, h=0.30m, σ=6.45s^{-1}, $k_1 h$=1.43)————————— 3rd. order irrotational solution. ——————————— 3rd. order viscous solution.

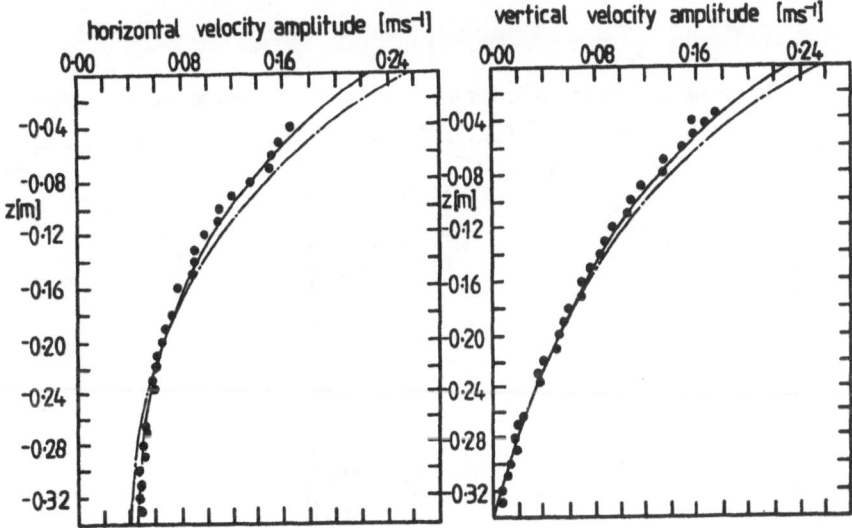

Figure 9b.   The amplitudes of the orbital velocity. (a=30.3mm,   h=0.34m, σ=8.57s^{-1}, $k_1 h$=2.58).

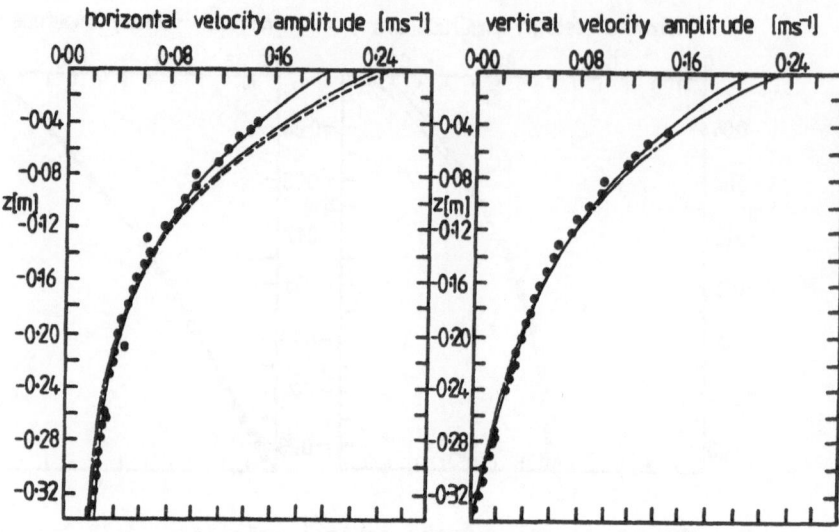

Figure 9c. The amplitudes of the orbital velocity. (a=25.0mm, h=0.34m, σ=9.61s⁻¹, k₁h=3.21). ———————— Numerical solution (Williams, 1985).

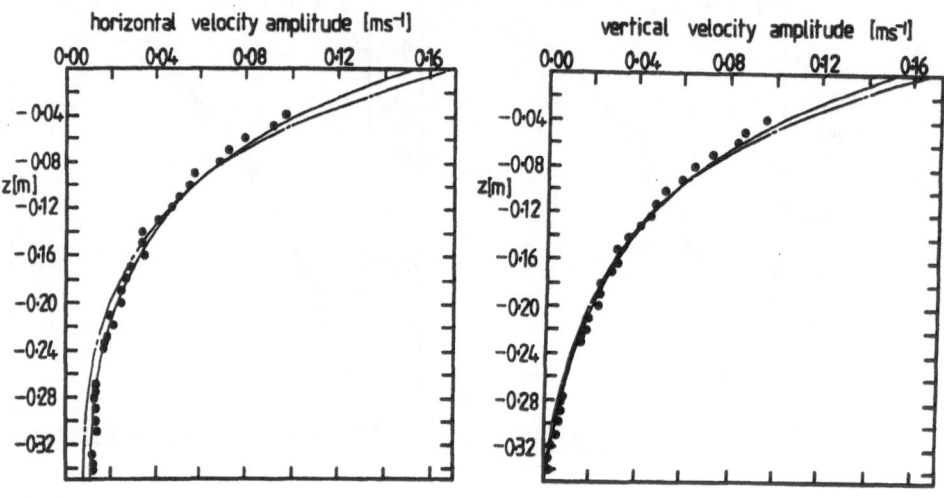

Figure 9d. The amplitudes of the orbital velocity. (a=16.3mm, h=0.35, σ=10.37s⁻¹, k₁h=3.84).

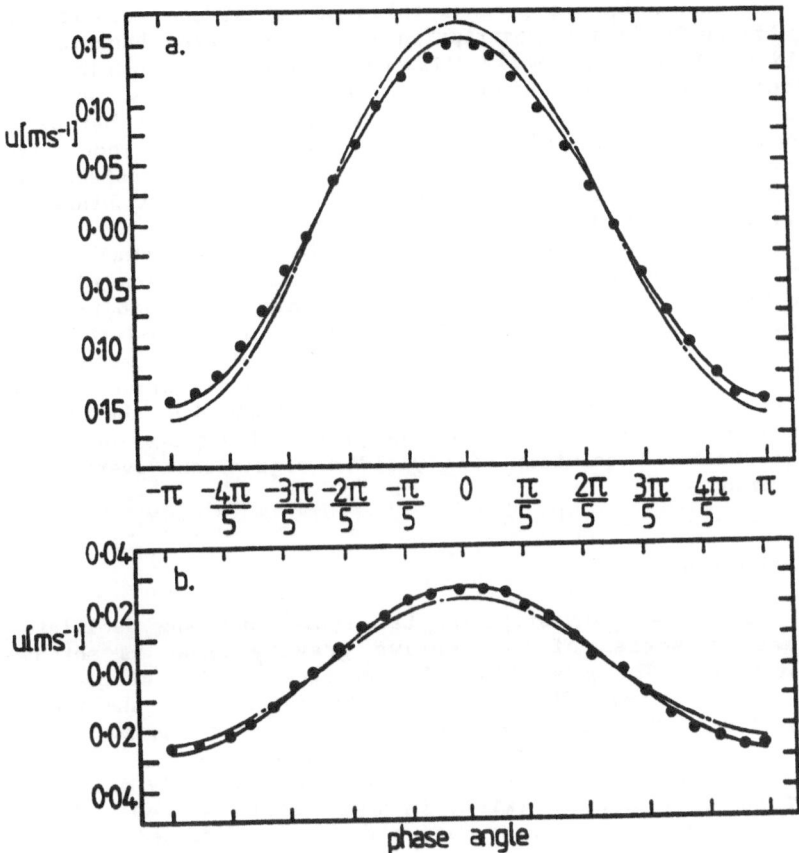

Figure 10. Cyclic variation in the orbital velocity. Case a: z=-0.04, Case b: z=-0.32. (a=25.0mm, h=0.34m, σ=9.61s^{-1}, k_1h=3.21). ————·——— 3rd. order irrotational solution, ————————— 3rd. order viscous solution.

According to the classification code proposed by Dean (1970) each of the wave forms considered above are third order waves. However, a third order irrotational solution consistently overestimates the amplitude of the oscillatory motion in the upper layers of the flow field; and underestimates the amplitude in the lower layers. In figure 9c, which corresponds to the steepest wave conditions (H/L=0.075), the numerical solution proposed by Williams (1985) is also shown. This solution is very similar to Cokelet's (1977) exact solution, and sums up a very large number of terms within an irrotational expansion. It is clear from this comparison that the apparent discrepancies cannot be explained by the neglected terms within a higher order irrotational solution.

In contrast, the present third order viscous solution provides a significantly better description of the observed behaviour. The flow field is at least partially diffused and therefore contains a rotational velocity component which cannot be predicted by any irrotational solution. The supposition that the vorticity profile has predominately arisen due to vertical diffusion appears to be correct. Although the present solution is only valid for waves of moderate height, in water of intermediate depth, there is every indication that other wave forms would be equally effected.

The observed departures from the irrotational motion are of significant practical importance. The wave kinematics within the near bed region are crucial in determining the sediment transport characteristics. If an irrotational solution is used then the present experimental observations indicate that the effectiveness of the transport processes are likely to be underestimated. Equally, the extremes of the wave loading are often determined by the wave kinematics within the near surface region. The present investigation indicates that an irrotational solution will yield a significant overestimate within this area.

5. Conclusions.

The present investigation has shown that the oscillatory motion beneath a series of progressive gravity waves is not necessarily irrotational. The generation of vorticity within the boundary layers, and the subsequent diffusion into the interior of the flow field not only affects the time independent terms (Longuet-Higgins, 1953) but also the oscillatory component of the wave motion.

It should not, however, be assumed that a fully diffused vorticity profile will always exist. In many circumstances there will be insufficient time or space, and a partially diffused vorticity profile will result. In this case, an irrotational solution will provide a lower bound to the characteristics of the wave motion; and the present viscous solution will provide an upper bound.

Perhaps the most important aspect of the present solution is the size of the viscous modification. Although the solution has only been developed to a third order of wave steepness, it seems likely that the higher order terms will be equally affected. If this is indeed the case, then the direct numerical solution of the irrotational equations may perhaps be of less interest than was previously assumed.

6. References.

Anastasiou, K. Tickell, R.G. and Chaplin, J.R. (1982) "Measurements of particle velocities in laboratory-scale random waves." Coastal Eng. 6:233-254.

Beech, N.W. (1978) "Boundary layers due to gravity waves." Ph.D. dissertation, Univ. of Nottingham.

Benjamin, T.B. (1959) "Shearing flow over a wavy boundary."
J. Fluid Mech. 6:161-205.

Chaplin, J.R. (1980) "Developments of stream function theory."
Coastal Eng. 3:179-205.

Cokelet, E.D. (1977) "Steep gravity waves in water of arbitrary
uniform depth." Philos. Trans. Roy. Soc. Lond. A 286:183-230.

De, S.C. (1954) "Contributions to the theory of Stokes' waves."
Proc. Cambridge Phil. Soc. 51:713-36.

Dean, R.G. (1965) "Stream function representation of non-linear
ocean waves." J. Geophys. Res. 70:4561-72.

Dean, R.G. (1970) "Relative validities of water wave theories."
Proc. A.S.C.E. J. Waterw. Harbours Coastal Eng. Div. 96:105-119.

Dean, R.G. (1974) "Evolution and development of water wave theories
for engineering application." Vols. i and ii. Special rep. no. 1,
U.S. Army Coastal Engin. Res. Center, Virginia.

Fenton, J.D. (1985) "A fifth order Stokes' theory for steady waves."
J. Waterw. Port Coastal Ocean Engng. 111(2):216-34.

Longuet-Higgins, M.S. (1953) "Mass transport in water waves."
Philos. Trans. Roy. Soc. A 245:535-81.

Morison, J.R., & Crooke, R.C. (1953) "The mechanics of deep water,
shallow water, and breaking waves." U.S. Army, Corps. of
Engineers, B.E.B. Tech. Memo. No. 40. pp 1-14.

Riendecker, M.M., & Fenton, J.D. (1981) "A Fourier approximation
method for steady water waves." J. Fluid Mech. 104:184-96.

Russell, R.C.H. & Osorio, (1957) "An experimental investigation of
drift profiles in a closed channel." Proc. 6th. conf. Coastal
Eng. Miami, pp587-603.

Simons, R.R. (1980) "The interaction between waves and currents."
Ph.D. dissertation, London University.

Schwartz, L.W. (1974) "Computer extension and analytical
continuation of Stokes' expansion for gravity waves." J. Fluid
Mech. 62:553-78.

Sleath, J.F.A. (1972) "A second approximation to mass transport by
water waves." J. Mar. Res. 30(3):295-304.

Skjelbreia, L., & Hendrickson, J.D. (1961) "Fifth order gravity wave
theory." Proc. 7th. Conf. Coastal Engng. 1:184-96.

Stokes, G.G. (1847) "On the theory of oscillatory waves." Trans.
Cambridge Philos. Soc. 8:441-55.

Stokes, G.G. (1880) "Supplement to a paper on the theory of
oscillatory waves." Mathematical and physical papers, vol. 1
Cambridge University Press, pp 314-26.

Swan, C. (1987) "The higher order dynamics of progressive waves."
Ph.D. dissertation, Cambridge University.

Swan, C. (1989a) "A rotational wave theory." Submitted to Proc.
A.S.C.E. J. Waterw. Harbours Coastal Eng. Div.

Swan, C. (1989b) "Convection within an experimental wave flume."
To appear in: J. Hyd. Res.

Williams, J.E. (1985) "Tables of progressive gravity waves." Pitman
London.

APPLICATION OF GAUSSIAN WAVE PACKETS FOR SEAKEEPING TESTS OF OFFSHORE STRUCTURES

G. F. CLAUSS, Prof. Dr.-Ing.
Institut für Schiffs- und Meerestechnik
Technische Universität Berlin
Salzufer 17-19, 1000 Berlin 10, F.R.G.

ABSTRACT. For seakeeping tests of offshore structures transient wave techniques are widely used. The method presented here is based on a special Gauss-modulated amplitude spectrum. These wave groups of limited length can be superimposed, the actual surface elevation being a function of packet characteristics and initial time lag.

The application of this technique is demonstrated in two cases presenting a semisubmersible and a crane vessel. It is shown, that a particular problem needs a tailored wave packet containing sufficient energy all over the relevant frequency range. These packets are designed by superimposing individual Gaussian wave groups. The new seakeeping test techniques has the following advantages

- The wave train is well defined at any location of the tank.
- The wave elevation and spectrum can be standardized or easily adapted to any specific problem.
- The duration of the test is very short; reflections (beach) do not interfere the results. At the culmination point the length of the wave train is small. This facilitates short duration seakeeping and manoeuvring tests.
- The results show high resolution and are in good agreement with regular wave test data.

Summarising, the Gaussian wave packet method is a versatile technique yielding precise and highly resoluted results in a short time.

1. Introduction

In offshore engineering a great variety of new structures has been designed during the past decade. They are fixed,

A. Tørum and O. T. Gudmestad (eds.), Water Wave Kinematics, 331–344.

articulated or floating and show a large diversity of shapes. For their analysis model tests are carried out in wave
tanks or basins using regular waves or irregular sea spectra. Both methods are very time consuming.

With regular waves, even with a great number of individual runs, sharp peaks of transfer functions may be lost.
With irregular sea tests the statistical analysis requires
at least 20 minutes to limit excessive scattering of the results. As an alternative this paper presents an experimental
technique working with wave groups, i.e. transient wave
trains of limited length. Validity and limitations of this
method are compared to other techniques. Its flexibility and
efficiency is demonstrated at some applications in offshore
engineering. It should be noted that the scope of this investigation is not focused on the simulation of the natural
seaway for oceanographic research but concentrates on opti-
mum techniques for model testing. The method, however, is
universally applicable.

2. Background

An excellent classification of different approaches to laboratory wave generation has been published by Funke and Mansard /1/ including

- mechanical and electronic, harmonic and non-harmonic
  synthesizers,
- filtered white noise and pseudo random noise generators,
- reproduction of natural sea waves,
- Fourier transform generation and analysis techniques
  including transient wave methods.

Against the background of the historical development the review shows that advanced numerical systems are required to
achieve a realistic physical simulation of wave conditions
which

- best represent nature and
- serve the experimental analysis of marine structures.

The general distinction between deterministic and non-
deterministic spectral amplitude models appears to be less
significant, particularly, since - even at random amplitude
and phase - specific conditions and constraints are preset
to satisfy the clients requirements. In view of this argument most wave simulation techniques are conditional or
partly deterministic methods /1/. It should be pointed out
that even at deterministic simulation techniques uncertain-

ties and irregularities are experienced in connection with

- the partly unknown dynamic transfer functions of the electronic-hydraulic-mechanical wave generating system,
- reflections from basin walls, wave absorbers and wave board as well as distortions of the initial wave related to the kinematics of the rigid wave generator,
- standing wave oscillations as well as recirculating currents in the test flume or basin of limited dimensions, resulting from the mass transport of waves,
- numerical inaccuracies according to the limited length of the registration or cut-out sections of it.

In this context even model tests in regular waves may become confusing under certain conditions /2/.

3. Gaussian wave packets

Gaussian wave trains are special types of transient waves /3-6/. They are composed of an infinite number of superimposed harmonic components, i.e.

$$a(x,t) = \int_{-\infty}^{\infty} a(k) \cdot \exp[i(kx - \omega t)] \, dk \; ,$$

with a Gauss-shaped amplitude spectrum

$$a(k) = \frac{a_0}{s\sqrt{2\pi}} \cdot \exp[-(k-k_0)^2/2s^2] \; .$$

The standard deviation s of the amplitude spectrum is called the form factor of the Gaussian wave packet. This yields the integral

$$a(x,t) = \int_{k=-\infty}^{\infty} \frac{a_0}{s\sqrt{2\pi}} \cdot \exp[-(k-k_0)^2/2s^2] \exp[i(kx-\omega t)] \, dk$$

which can be integrated in closed form if the wave frequency

$$\omega(k) = \sqrt{gk\,th(kd)}$$

is expanded into a three term Taylor series

$$\omega(k) = \omega_0 + A(k-k_0) + \frac{B}{2}(k-k_0)^2$$

with

$$\omega_0 = \omega(k=k_0) \qquad A = \frac{d\omega}{dk}\bigg|_{k=k_0} \qquad B = \frac{d^2\omega}{dk^2}\bigg|_{k=k_0}$$

Substitution and some algebraic manipulations yields the function of the Gaussian wave train profile

$$a(x,t) = a_0 \sqrt{\frac{1}{1+is^2Bt}} \cdot \exp[-\frac{1}{2}\frac{s^2}{1+s^4B^2t^2}(x-At)^2] \exp[i(k_0x-\omega_0t+\frac{1}{2}\frac{s^4Bt}{1+s^4B^2t^2}(x-At)^2)]$$

Note that at s=0 the expression degenerates to a simple harmonic wave. A typical Gaussian wave group at three different locations is shown in Fig. 1.

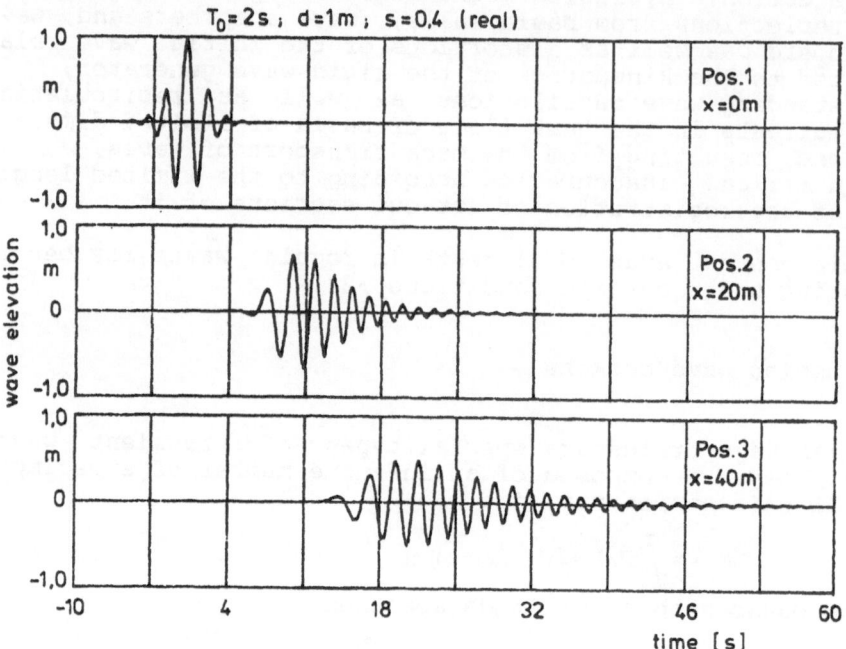

Fig. 1   Gaussian wave group at three different locations

The above expression for the Gaussian wave train contains real and imaginary components. For the discussion of the propagation characteristics of such packets it is convenient to separate the complex function into its real and imaginary parts. Lengthy algebraic operations /7/ result in

$$\underset{\substack{Re\\Im}}{}\{a(x,t)\} = a_0 \underbrace{\sqrt[4]{\frac{1}{1+s^4 B^2 t^2}}}_{X1} \cdot \underbrace{\exp\left[-\frac{1}{2}\frac{s^2}{1+s^4 B^2 t^2}(x-At)^2\right]}_{X2} \times \underbrace{\substack{\cos\\ \sin}\left\{k_0 x - \omega_0 t + \frac{arctg(-s^2 Bt)}{2} + \frac{1}{2}\frac{s^4 Bt}{1+s^4 B^2 t^2}(x-At)^2\right\}}_{X3}.$$

This shows that the Gaussian wave packet is defined by the damping term X1, the modulation function X2 and the oscillation term X3. The factor B of the Taylor series plays an important role in the damping term. For this reason B is called a damping factor. The modulation function travels with the velocity A in positive x-direction. Therefore, the factor A of the Taylor expansion can be interpreted as the group velocity of the Gaussian wave packet. A and B follow from the differentiation of the dispersion relation

$$A = \frac{d\omega}{dk}\Big|_{k=k_0} = \frac{C_0}{2}\left[1 + \frac{2k_0d}{sh(2k_0d)}\right]$$

with

$$B = \frac{d^2\omega}{dk^2}\Big|_{k=k_0} = C_0 \cdot d\left[\frac{1 - 2k_0d \cdot coth(2k_0d)}{sh(2k_0d)} - \frac{1}{2sh(2k_0d)}\left(\frac{sh(2k_0d)}{2k_0d} - \frac{2k_0d}{sh(2k_0d)}\right)\right]$$

$$C_0 = \sqrt{\frac{g}{k_0} th(k_0d)}$$

The use of the central wave number k_0 is inconvenient. It is more transparent to introduce the dominant period T_0, which is related to the central wave number k_0 by the dispersion relation. Thus the Gaussian wave packet is characterized by this period T_0, the water depth d and the form factor s. The three terms X1, X2, X3 are now discussed in detail.

The damping term X1 has its maximum (X1=1) at the concentration point x=t=0. The amplitudes increase up to this location as the long wave components overtake the shorter ones. Beyond the culmination point the amplitudes decrease again as the long waves are now travelling away due to their higher celerity. With increasing values of the form factor s the damping term deteriorates rapidly with time. As the form factor s is also the standard deviation of the amplitude spectrum, the total wave energy is distributed over a larger frequency range. Thus a large form factor implies a wide spectrum with many wave components of different lengths resulting in a quick dispersion of the packet.

Multiplying the damping term X1 with the modulation function X2 yields the envelope of the Gaussian wave packet. The modulation function governs the shape of the wave packet and has the characteristic of a distorted Gauss-distribution. Its skewness and variation in the time domain depend on form factor s, group velocity A, damping factor B as well as on time and location respectively.

The expression for the Gaussian wave packet is completed by multiplying the envelope function X1 * X2 with the oscillating term X3 which shows a characteristic phase shift compared with a simple harmonic wave of the same frequency ω_0. The phase shift has two components. The arctg-term is nearly constant as it fades away rapidly from zero at t=0 to sign(t) * $\pi/4$. The second term decreases at the maximum of the modulation function. Thus the wave component which coincides with the maximum of the modulation function has a wave period close to the dominant period.

3.1 GENERATION AND PROPAGATION OF GAUSSIAN WAVE PACKETS

In the following sections the Gaussian wave packets are characterised by the amplitude function f(t) at various locations and by the accompanying Fourier spectra F(ω), i.e.

$$F(\omega) = \frac{1}{2\pi}\int_{-\infty}^{+\infty} f(t)\, e^{-i\omega t}\, dt,$$

336

The spectra are calculated using Fast Fourier transformation technique (FFT). The area under the Fourier spectrum yields the half of the maximum amplitude of the Gaussian wave packet at x=t=0, which is also a direct measure of the total energy contained in the wave train.

As discussed earlier, the generation of the actual wave group requires modification of the input signal using the transfer function of the wave generator, which is the combination of its electrical/mechanical and hydrodynamic transfer functions, both in magnitude and phase. These relationships have been determined by using superimposed Guassian wave packets and regular waves.

As an illustration of the propagation behaviour of Gaussian wave packets Fig. 2 shows a 0.8-s-wave train followed by a faster 2-s-wave group. At the culmination point x=t=0 both wave groups coincide and penetrate each other. At later positions the long wave group has overtaken the short one. Both theoretically as well as experimentally the Fourier spectra of each individual wave packet are clearly identified even at the culmination position.

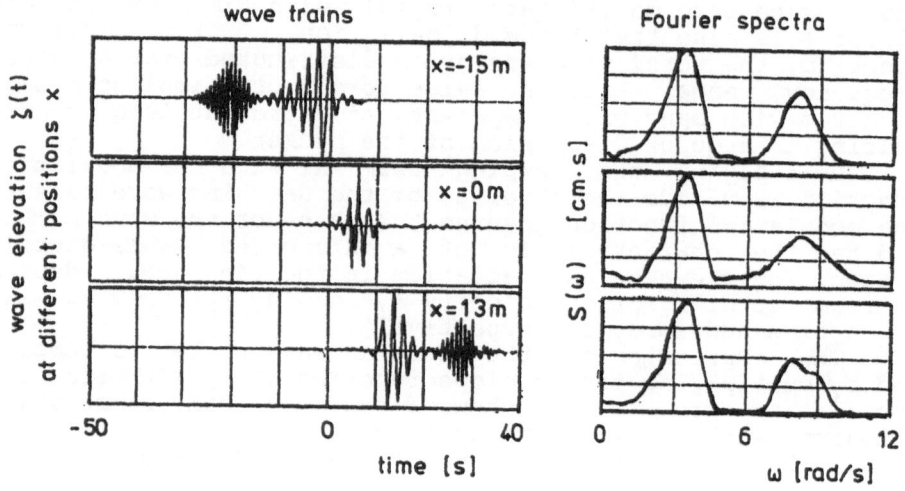

Fig.2 Superposition of two Gaussian wave packets (T_0=0.8 s, T_0=2.0 s) - experimental results

At all positions the Fourier spectrum is invariant according to its shape and area. The agreement between theory and experiment is quite good.

3.2 APPLICATION OF WAVE PACKETS FOR SEAKEEPING ANALYSIS

The practical application of wave packets takes into account that sufficient wave energy should be available in a frequency range relevant to model characteristics. As individual wave groups act independently, they can be arbitrarely combined for the design of broad energy spectra. According to the fact that complicated electrohydraulic transfer functions have to be considered, the theoretical relations between wave train and Fourier spectrum are just one minor step for evaluating the electrical input signal. For this reason the whole process is performed numerically. We define the wave spectrum and the wave packet at the concentration point, i.e. at zero phase x=t=0, and convert this tailored signal to the upstream position at the wave board by using an average group velocity. Introducing the appropriate electrohydraulic and hydrodynamic transfer functions we finally yield the relevant electrical control signal. With this procedure the actual time history of the wave packet is known at any position of the wave tank. For experimental tests it is recommended to select a model position shortly before the concentration point. As the wave train and the model response are of limited length the numerical integration is well defined suppressing any statistical scatter and yielding smooth spectra.

The transfer function results from the ratio of the Fourier transforms of the wave group and the model response, $F_\zeta(\omega)$ and $F_R(\omega)$ respectively, i.e.

$$H_{\zeta R}(\omega) = |H_{\zeta R}(\omega)| e^{-i\varepsilon_{\zeta R}(\omega)} = \frac{F_R(\omega)}{F_\zeta(\omega)}$$

The complex division directly yields the frequency dependent transfer function by magnitude and phase. Specified wave groups are repeatable without variation. The time window of analysis is selected interactively to avoid errors due to beach, wave board or model reflections. Problems arise if slightly damped resonance motions are observed which are not fading out within the time window frame. In this case sidelobe-leakage effects may distort the resulting spectra and time history tapering by special numerical window techniques is required /8/. As resonant motions depend on system damping, the response of the structure is nonlinear within this frequency range. In this case model tests in regular waves of different heights are recommended.

The application of the wave-packet technique is examplified by two case studies, discussing model tests with a semisubmersible and a crane vessel during operation. The results are compared with regular wave tests.

3.2.1 Case 1: Semisubmersible RS 35

The RS 35 is a 37,000 tons deepwater drilling and production system with a variable load capacity of nearly 15,000 tons. The basic construction - a toroidal double-wall hull with four columns carrying an integrated deck of modular design - shows excellent motion characteristics and a high safety standard /9/.

Seakeeping tests have been performed at a scale of 1:53. Fig. 3 shows the time histories of typical experiments with a wave train of superimposed Gaussian wave packets, the semisubmersible being positioned at three different locations. The same driving signal and consequently the same wave train is used in all three tests. In the first run the semisubmersible is positioned in the converging range of the wave train (left column). In the second run the structure is located at the concentration point (middle column) and in the third run the platform is exposed to the wave train in its diverging range (right column). The upper set of records shows the time histories of the wave train in front and on the beam of the semisubmersible as well as its heave, pitch and surge response. The middle set of plots represents the Fourier spectra of the above records. Finally, the lower set of diagrams shows the heave, pitch and surge transfer functions compared with regular wave test data. Generally it can be stated that the agreement of the different test results is excellent. However, it is recommended to test the structures in the converging phase of Gaussian wave packets to avoid wave breaking and other nonlinear effects that may occur near the concentration point /5/.

Investigating hydrodynamic weakly damped structures - like semisubmersibles - one is confronted with the problem, that the persistant resonance motions exceed the available data aquisition time interval. The abrupt cutoff of the time histories yields the so called "leakage"-effect when the Fourier transformation is applied. The Fourier spectra and, in consequence, the transfer functions may be distorted. The leakage effect may affect the complete spectrum. Especially frequency ranges with low values of the transfer function are sensitive. This problem can be overcome, using the high flexibility of the Gaussian wave train technique. As described earlier, it is easy to generate wave packets with different spectral properties. Using wave trains with limited bandwidth it is possible to produce spectra which contain only a small amount of energy near the resonance frequencies of the structure. Thus all motions are decreasing during the data recording interval and the leakage effect does not occur. With this technique the transfer functions are not distorted.

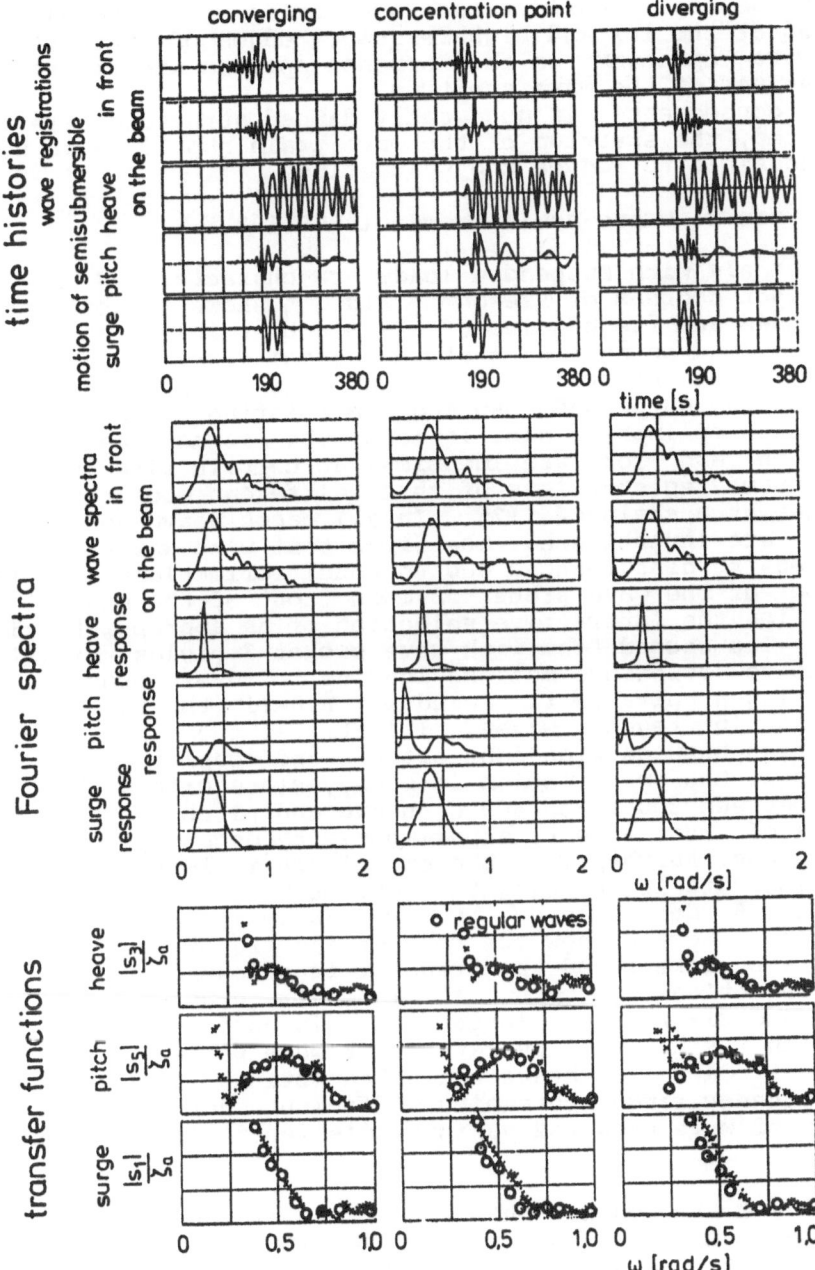

Fig. 3   Seakeeping   tests of the semisubmersible RS 35 at three different locations

In general considerable scattering of the results is observed at the low and high frequency edges of the narrow band wave spectrum as the response spectra are related to small values of the wave spectrum.

In consequence, if not weakly damped structures are investigated, the spectrum of the wave train should be as broad as possible to reduce the scattering.

3.2.2 Case 2: Crane semisubmersible DB 102

Crane vessels with suspended loads are characterized by at least eight degrees of freedom. Fig. 4 shows typical resonance modes and frequencies of a semisubmersible vessel, the suspended load being 4.65 % of the vessel displacement. Apparently, the natural frequencies are influenced by the weight of the suspended load and the hoisting wire length. In the case of inphase oscillations, the frequencies of heave, pitch and roll are identical at certain wire lengths. Fig. 5 shows registrations of three different model tests in head seas at a scale of 1:75. In all cases heave and pitch motions are investigated: The first test with wave packet 1 indicates the disturbances due to model reflections at the position of the wave probe, 8 m in front of the model. It examplifies the interactive selection of an appropriate time window. The second test with wave packet 2 yields significant heave resonance motions leading to severe problems in designating a suitable time window. Finally the third test, performed in regular waves, illustrates the difficulty to distinguish between transient effects and coupled oscillations. While the surge motion is steady-state within the interval between 200 s and 400 s, heave and pitch motions show complicated beat effects superimposed by decaying resonance oscillations. Further, at the end of the registration beach reflections are influencing the motions.

Fig. 6 demonstrates the FFT-analysis of the above tests. The diagrams show the Fourier spectra of the two different wave packets and the transfer functions of the heave and pitch motions. Note that at higher frequencies - see the right hand diagrams - a larger scale is used to improve the resolution. Fig. 6 illustrates that wave packet 1 yields consistent data for heave and pitch motions within the entire frequency range containing sufficient energy. Analyzing the data of wave packet 2 various discrepancies are experienced:

- at low frequencies the heave transfer function depends significantly on the size of the time window,
- at higher frequencies the effects of side-lobe-leakage result in severe scattering /8/.

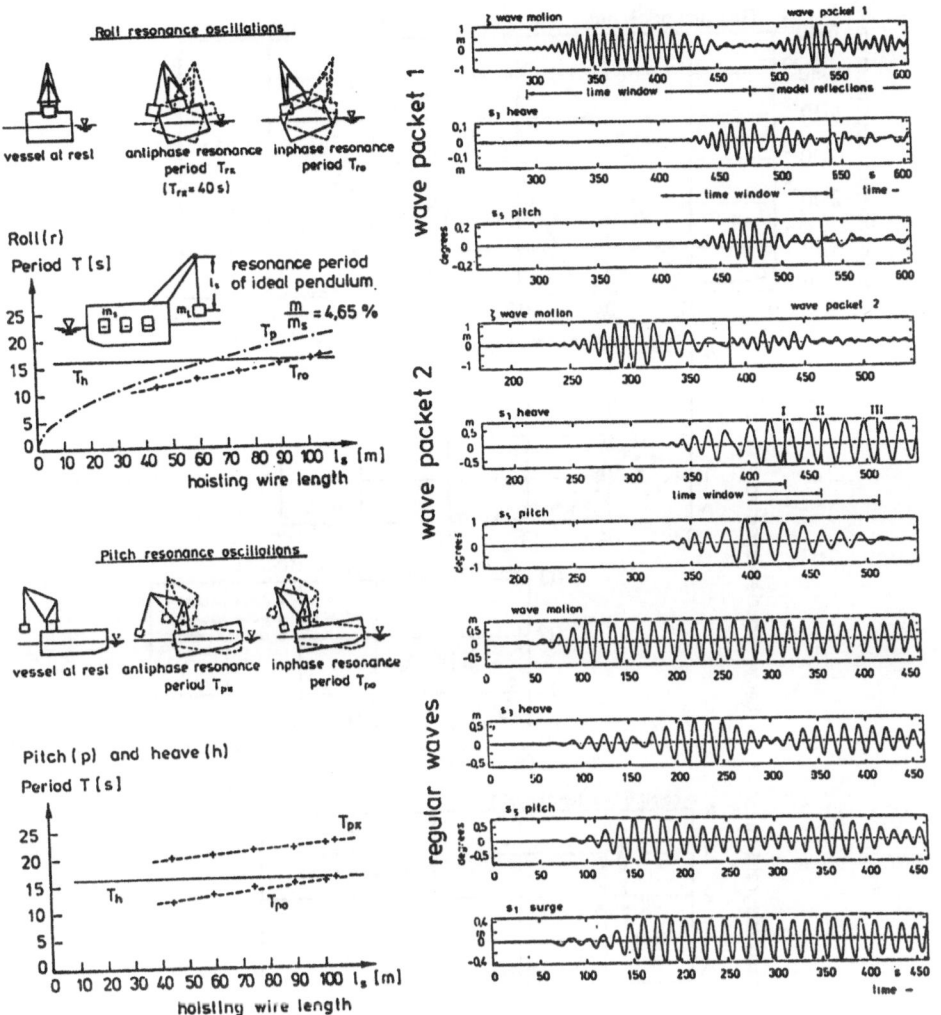

Fig. 4 The resonance modes of the crane semisubmersible DB 102 and their frequencies against hoisting wire length

Fig. 5 Registration of three different model tests with the crane semisubmersible DB 102 (scale 1:75)

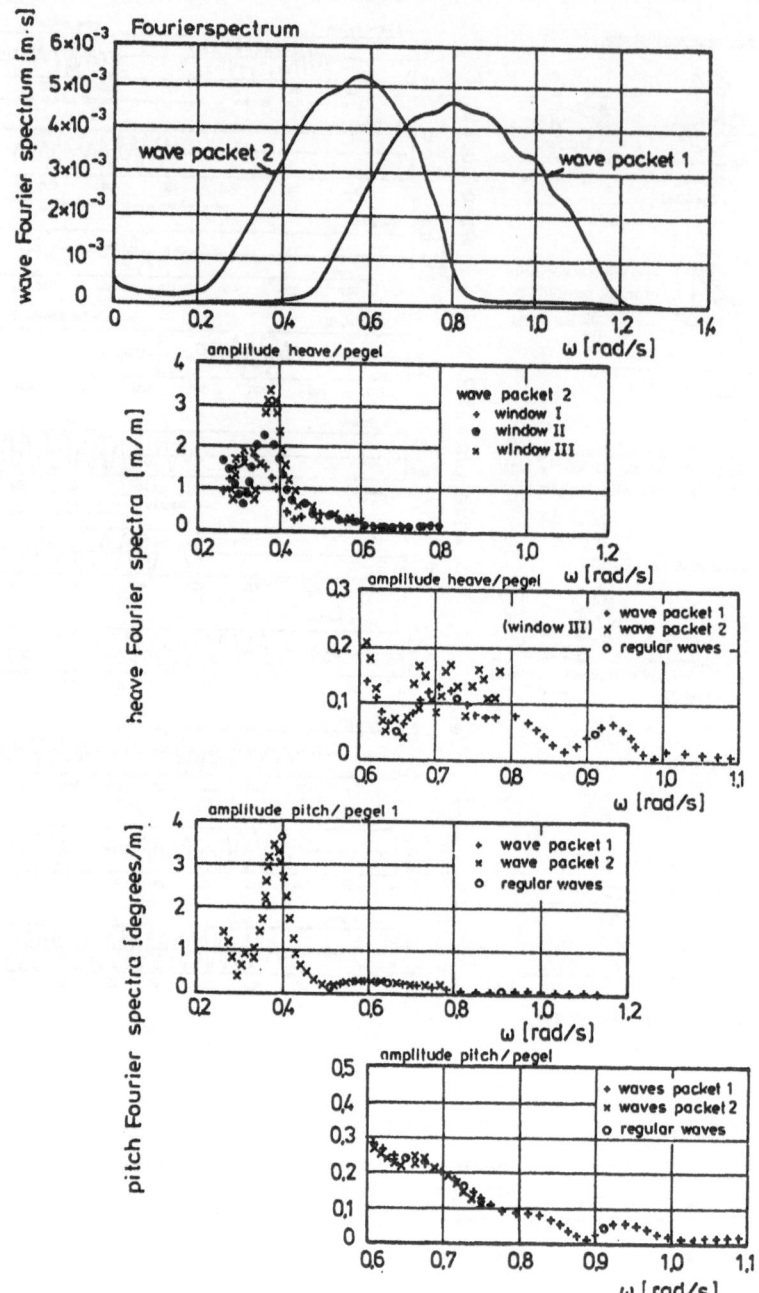

Fig. 6   Fourier spectra of wave packets and the resulting transfer functions of heave and pitch motions

This example shows that seakeeping tests with transient waves yield excellent data, if it is provided that appropriate wave packets are selected:

- they should be short enough to avoid disturbances by beach or wave board reflections,
- they must contain sufficient energy within the relevant frequency range,
- finally, in the vicinity of natural frequencies, the spectral energy should be small enough to limit the time window to an acceptable duration.

4. Conclusions

Seakeeping tests with Gaussian wave packets show considerable advantages

- The wave train is well defined at any location of the tank. Any test is exactly repeatable.
- The wave elevation and spectrum can be standardized or easily adapted to any specific problem.
- The duration of the test is very short; reflections from beach or wave board do not interfere the results. At the culmination point the length of the wave train is small. This facilitates short duration seakeeping and manoeuvring tests.
- the results show high resolution and are in good agreement with regular wave test data.

In comparison to the random wave technique (see /2/), the Fourier transform of the wave train is unique, the spectra are deduced without any statistical scattering. This also results in smooth transfer functions. The controlled repeatability of arbitrary wave groups allows the definition of standard wave packets which could be applied at any towing tank, world wide, for comparative tests.

Problems, however, arise at nonlinear behaviour of the investigated models. In the case of resonance motions detailed tests with regular waves are recommended. If statistical informations are required, e.g. with the analysis of slow drift motions of moored structures, random sea model tests are inevitable. In most cases, however, the Gaussian wave packet method is a superior technique yielding precise and highly resolved results in a short time.

344

5. Acknowledgements

This paper is a continuation of the research work of Dr. Jan Bergmann which was financed by the Federal Ministry of Research and Technology, BMFT, Bonn. On the basis of his excellent results the Gaussian wave technique was further developed and became a standard method for seakeeping tests. The financial support is gratefully acknowledged. The authors are also indebted to the "Institut für Wasserbau und Wasserwirtschaft" for their generous support of the experiments.

6. References

/1/ Funke, E.R.; Mansard, E.P.D.: A Rationale for the Use of the Deterministic Approach to Laboratory Wave Generation. IAHR-SEMINAR, Lausanne 1987

/2/ Clauss, G.F.; Riekert, T.: Generation of Wave Groups for Model Testing and Some Applications in Coastal and Offshore Engineering. 2nd Int. Symp. on Wave Research and Coastal Engineering, University of Hannover, F.R.G., Oct. 12-14, 1988

/3/ Coulson, C.A.: Waves, a Mathematical Account of the Common Types of Wave Motion. Oliver and Boyd, Interscience Publ. Inc., NY, 1949

/4/ Kinsman, B.: Wind Waves, their Generation and Propagation on the Ocean Surface. Prentice-Hall Inc., Englewood Cliffs, NJ, 1965

/5/ Clauss, G.; Bergmann, J.: Gaussian Wave Packets - A New Approach to Seakeeping Tests of Ocean Structures. Applied Ocean Research, Vol. 8, No. 4, 1986

/6/ Chakrabarti, S.K.; Libby, A.R.: Further Verification of Gaussian Wave Packets. Applied Ocean Research, Vol. 10, No. 2, pp. 106 - 108, 1988

/7/ Bergmann, J.: Gaußsche Wellenpakete - Ein Verfahren zur Analyse des Seegangsverhaltens meerestechnischer Konstruktionen. PhD Thesis, published at the Institute of Naval Architecture and Ocean Engineering, Technical University Berlin, 1985

/8/ Bendat, J.S.; Piersol, A.G.: Random Data. 2nd Edition, New York 1986

/9/ Clauss, G.F.: Stability and Dynamics of Semisubmersible After Accidental Damage, OTC 4729, Houston 1984

DETERMINATION OF THE SURFACE ELEVATION PROBABILITY DISTRIBUTION OF WIND WAVES USING MAXIMUM ENTROPY PRINCIPLE*

WITOLD CIEŚLIKIEWICZ
Polish Academy of Sciences
Institute of Hydroengineering
ul. Kościerska 7, 80-953 Gdańsk
Poland

ABSTRACT. Probability density function of the surface elevation of a non-Gaussian random wave field is obtained. The method of determination of probability distribution is based on the maximum entropy principle. The results compare well with field experiment data.

1. Introduction

There are two proposals of the non-Gaussian density functions of the surface in the existing literature. The first one has been given by Longuet-Higgins (1963) who uses the Gramm-Charlier series in the Edgeworth's form. The second one is due to Tayfun (1980), essentially improved by Huang et al. (1983).

The Longuet-Higgins approach, however, has some drawbacks. The density function proposed by Longuet-Higgins may achieve negative values in some domain of argument. Moreover, by the integrating it from minus to plus infinity we obtain the value different from one. And, at last, the variance, asymmetry and flatness coefficients calculated using this density function differ from the initial parameters of this distribution.

The authors of the second proposal have worked out a formula for the probability density based on the representation of free surface elevation in the Stokes expansion of the 2nd and 3rd order. The limitation of such an approach is the fact that in some wave conditions, especially for coastal waves, the Stokes expansion is not sufficiently precise.

The aim of the present study is to analyse the possibility of the probability density determination for a free surface elevation, on the basis of the maximum entropy principle. The method proposed is an alternative to the above mentioned approaches, and does not contain their drawbacks. We use no special representation of the random field of wind waves, so we hope that our method can find applications in a wide range of wave conditions.

At the final part of this paper the theoretical results with experimental data of Ochi and Wang (1984) is compared. These data, obtained during severe storms for a wide range of water depth, also for steep waves, prove the correctness of the present approach. A good consistency of calculated probability functions with experimental values is noticed.

---

* Presented by S. R. Massel

A. Tørum and O. T. Gudmestad (eds.), Water Wave Kinematics, 345–348.

2. Maximum entropy probability density for a random wind wave field

Assume that the mean value of free surface elevation ξ is zero. Further, let the variance σ^2 and the moments μ_3 and μ_4 of the third and fourth order of this random variable be given. Let $\Phi(\sigma^2, \mu_3, \mu_4)$ denote the following set of probability density functions $\rho(\xi)$:

$$\Phi(\sigma^2, \mu_3, \mu_4) = \left\{ \rho(\xi): \quad \rho(\xi) \geq 0, \quad \int_{-\infty}^{+\infty} \rho(\xi)\, d\xi = 1, \quad \int_{-\infty}^{+\infty} \xi\, \rho(\xi)\, d\xi = 0, \right.$$
$$\left. \int_{-\infty}^{+\infty} \xi^2 \rho(\xi)\, d\xi = \sigma^2, \quad \int_{-\infty}^{+\infty} \xi^3 \rho(\xi)\, d\xi = \mu_3, \quad \int_{-\infty}^{+\infty} \xi^4 \rho(\xi)\, d\xi = \mu_4 \right\} \quad (1)$$

We shall determine the representative distribution for this set, i.e. the distribution which maximizes entropy on Φ. Due to the maximum entropy principle (Jaynes 1957), the representative distribution gives the most objective and full description of the "statistical knowledge" contained in the moments σ^2, μ_3 and μ_4. The entropy S connected with the distribution ρ is defined as a functional of the following form:

$$S[\rho] = - \int_{-\infty}^{+\infty} \rho(\xi) \ln \rho(\xi)\, d\xi \quad (2)$$

We shall find the representative distribution for the set Φ (1), i.e. such that

$$S[\rho^*] = \max_{\rho \in \Phi} S[\rho] \quad (3)$$

by the use of Lagrangian multipliers method.

Consider the functional

$$L[\rho] = - \int_{-\infty}^{+\infty} d\xi\, \rho(\xi) \left[\ln \rho(\xi) + \sum_{p=0}^{4} \alpha_p \xi^p \right] \quad (4)$$

From the condition of the extremum existence

$$\left. \frac{\delta L[\rho]}{\delta \rho(\xi)} \right|_{\rho = \rho^*} = 0 \quad (5)$$

where $\delta/\delta\rho(\xi)$ denotes a functional derivative, we obtain that

$$\rho^*(\xi) = A^{-1} \exp\left\{ -\sum_{p=1}^{4} \alpha_p \xi^p \right\} \quad (6)$$

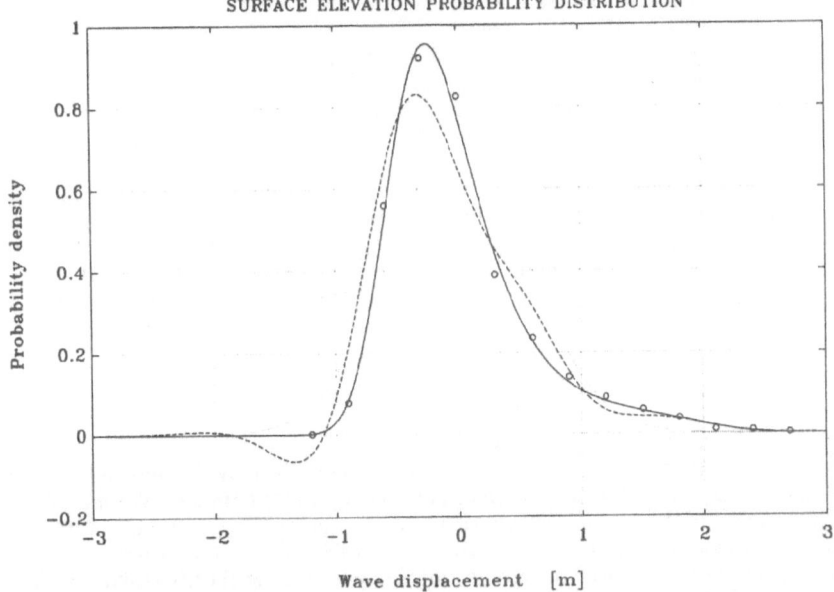

Fig. 1. Comparison of theoretical probability density functions and field experiment data (open circle) of Ochi and Wang (1984) for $\sigma_\zeta = 0.56$, $\lambda_3 = 1.26$, $\lambda_4 = 4.98$. Dotted, Gaussian; solid, maximum entropy distribution; dashed, result of Longuet-Higgins (1963).

In the above formula there is $A^{-1} = \exp\{-1 - \alpha_0\}$. To guarantee the finiteness of entropy (2) we have to assume that

$$\alpha_4 > 0 \tag{7}$$

Using definition (1) of the set of probability density functions Φ, we can determine the unknown Lagrangian multipliers α_p and the normalization constant A. Then, we have

$$\begin{cases} I_0 = A, & I_1 = 0, & I_2 = A\sigma^2 \\ I_3 = A\mu_3, & I_4 = A\mu_4 \end{cases} \tag{8}$$

where

$$I_n = \int_{-\infty}^{+\infty} \xi^n \exp\left\{-\sum_{p=1}^{4} \alpha_p \xi^p\right\} d\xi \tag{9}$$

The system (8) consists of five nonlinear equations with respect to five unknown quantities A, α_1, ..., α_4. One can easily prove that only three equations of this systems are independent. In fact, we can reduce (8) to the following three equations:

$$\begin{cases} I_0 = A, & I_1 = 0 \\ I_2 - \dfrac{\sigma^2}{\mu_3} I_3 = 0 \end{cases} \tag{10}$$

So there should hold two relations between the unknown quantities A, α_1, ..., α_4. Indeed, one can check easily the following identities:

$$\begin{cases} \alpha_1 I_0 + 2\alpha_2 I_1 + 3\alpha_3 I_2 + 4\alpha_4 I_3 \equiv 0 \\ I_0 - \alpha_1 I_1 - 2\alpha_2 I_2 - 3\alpha_3 I_3 - 4\alpha_4 I_4 \equiv 0 \end{cases} \tag{11}$$

which allow, by use of (8), to express e.g. α_1 and α_2 in terms of α_3 and α_4 :

$$
\begin{cases}
\alpha_1 = -\sigma^3 \left(\dfrac{3}{\sigma} \alpha_3 + 4\lambda_3 \alpha_4 \right) \\[2ex]
\alpha_2 = \dfrac{1}{2\sigma^2} - \dfrac{\sigma^2}{2} \left(\dfrac{3\lambda_3}{\sigma} \alpha_3 + \lambda_4 \alpha_4 \right)
\end{cases}
\tag{12}
$$

where $\lambda_3 = \mu_3/\sigma^3$ and $\lambda_4 = \mu_4/\sigma^4$.

Our task reduces finally to solving the system of three nonlinear equations (10) with unknown quantites A, α_3 and α_4. Note that the unknown Lagrangian multipliers α_3 and α_4 appear in (10) as parameters of integrals of the type (9). We are not able to calculate such integrals analytically, so numerical calculations are needed. Note that the first of equations (10) allows us to calculate the normalization constant A provided we know the multipliers α_3 and α_4. Thus we have to solve the following system of equations

$$
\begin{cases}
I_1 = 0 \\
I_2 - \sigma \lambda_3 I_3 = 0
\end{cases}
\tag{13}
$$

with respect to the unknowns α_3 and α_4.

In Fig. 1 an example of comparison of density function obtained by the maximum entropy principle and the experimental data of Ochi and Wang (1984) is made. We marked also the density function of the normal distribution and the one of Longuet-Higgins (1963). As one can see, the theoretical results of this work are consistent with the experimental data.

This short contribution is based on more detailed paper of the author (Cieślikiewicz 1988) in which one can find more comparisons of the maximum entropy probability distributions with field experiment data of Ochi and Wang (1984).

3. Conclusions

The method of determination of free surface elevation probability distribution by using the maximum entropy principle proved to be effective in the case of wind waves of given variance, asymmetry and flatness coefficients. We are able to describe the shape of a probability density function. The hypothesis that a wind wave random field is a subject of maximum entropy principle has been confirmed by comparison to experimental results in nature.

4. References

Cieślikiewicz, W., (1988), "Determination of free surface elevation probability distribution of wind waves on the basis of the maximum information principle," *Inter. Rep. Inst. Hydroeng. P.A.S.*, Gdańsk (in Polish).

Huang, N. E., Long, S. R., Tung, C. C., Yuan, Y., Bliven, L. F., (1983), "A non-gaussian statistical model for surface elevation of nonlinear random wave fields," *Journal of Geophysical Research*, **88**, No. C12, 7597–7606.

Jaynes, E. T., (1957), "Information theory and statistical mechanics," *Phys. Rev.*, **106**, 620–630.

Longuet-Higgins, M. S., (1963), "The effect of non-linearities on statistical distributions in the theory of sea waves," *Journal of Fluid Mechanics*, **17**, 459–480.

Ochi, M. K., Wang, W. C., (1984), "Non-gaussian characteristics of coastal waves," *Proceedings of the Nineteenth Coastal Engineering Conference*, Vol. 1, 516–531.

Tayfun, M. A., (1980), "Narrow-band nonlinear sea waves," *Journal of Geophysical Research*, **85**, No. C3, 1548–1552.

Shallow Water Wave Kinematics

MATHEMATICAL MODELLING OF SHORT WAVES IN SURF ZONE

Th. KARAMBAS and Chr. KOUTITAS
Division of Hydraulics and Environmental Engineering
Department of Civil Engineering
ARISTOTLE UNIVERSITY OF THESSALONIKI 54006
GREECE

ABSTRACT.
A numerical model for the propagation of breaking waves is developed. Using an apropriate F.D. scheme in the solution of BOUSSINESQ type of equations, a third-order accuracy is obtained, without the need of including the additional SERRE terms.
By providing the above equations with a suitable dissipative mechanism by introducing a dispersion term (using the Boussinesq eddy viscosity concept), we are able to simulate breaking waves. In this way it is possible to compute both the dissipation of the wave height and set-up, and in the 2-D case, the longshore currents. In the above cases the radiation stresses are genarated automatically.

1) INTRODUTION

Assuming that the vertical velocity linearly increases from the bottom to the sea surface and averaging the Eulerian equation over the depth we obtain the BOUSSINESQ type of equations.

$$\zeta_t + [(d+\zeta).U]x = 0 \qquad \text{Continuity}$$

$$Ut + U.Ux + g\zeta_x = \tfrac{1}{2}d.(d.U)xxt - \tfrac{1}{6}d^2 Uxxt - g\frac{U|U|}{C^2.(d+\zeta)} + Vx.Uxx$$

$$\text{x-Momentum}$$

$$(1.1)$$

d is the water depth, U is the mean, over the depth, horizontal velocity, ζ is the surface elevation, C=Chezy resistance ($m^{\frac{1}{2}}/sec$) and Vx=eddy viscocity coefficient (m^2/sec).

351

A. Tørum and O. T. Gudmestad (eds.), Water Wave Kinematics, 351–365.
© 1990 Kluwer Academic Publishers.

The third derivative dispersive term in the above equation includes the effects of the finite vertical acceleration on the pressure distribution, which is no longer hydrostatic.

In this way it is possible to simulate non-linear finite amplitude water waves so long as d/L is of the same order or smaller than H/d.

$$O(d/L) \sim O(H/d) < 1$$

The numerical solution of Boussinesq equations can describe the effects of most of the wave phenomena such as shoaling, refraction, diffraction. In addition, the automatic appearance of radiation stresses in the computations leads to the production of non-linear irregular waves calculation, when interactions between frequencies occurs.

By introducing a dispersion term (Vx.Uxx) we are able to simulate the wave deformation in the surf zone, i.e. wave breaking with dissipation (predicting set-up and surf beats) and, in the 2-D case, the longshore current induced by this, including wave-current the interaction. In this case also the radiation stesses are generated automatically.

In the derivation of equations (1.1) the two parameters, H/d are assumed to be small. Near the breaking point and shoreward H/d takes values around 1. In this case the equations (1.1) seem to be invalid. T. Yasuda et all (1982,1982) solving numerically the K.d.V equation with the same order as (1.1), concluded that accurate results can be expected using appropriate F.D. schemes, even for H/d-O(1), without the need of including additional terms.

2) NUMERICAL SOLUTION

Using central finite-differences in the approximation of partial derivatives both in space and time, we have:

$$Ft \Big|_i^n = \frac{F_i^{n+1} - F_i^{n-1}}{2.\Delta t} \qquad\qquad Fx \Big|_i^n = \frac{F_{i+1}^n - F_{i-1}^n}{2.\Delta x}$$

centered in point i Δx at time n Δt.

Expansion using Taylor series of the space derivatives gives:

$$\zeta_x \Big|_i^n = \frac{\zeta_{i+1}^n - \zeta_{i-1}^n}{2.\Delta x} - \frac{(\Delta x)^2}{6} \zeta_{xxx}^n$$

$$Ux \Big|_i^n = \frac{U_{i+1}^n - U_{i-1}^n}{2.\Delta x} - \frac{(\Delta x)^2}{6} U_{xxx}^n$$

or using the linearized equations:

$$Uxxx = -1/d \cdot \zeta_{xxt}$$

$$\zeta_{xxx} = -1/g \cdot Uxxt$$

Replacing the above expresions in the significant linear terms of (1.1) $g \cdot \zeta_x$ and $d \cdot Ux$:

$$g \cdot \zeta_x = [g \cdot \zeta_x]^{F.D.} + \frac{(\Delta x)^2}{6} \cdot Uxxt$$

$$d \cdot Ux = [d \cdot Ux]^{F.D.} + \frac{(\Delta x)^2}{6} \cdot \zeta_{xxt}$$

In comparison to the Boussinesq term $d^2/3$ Uxxt the truncation error becomes important when d and Δx are of the same order.
That shows the importance of including the two corrections on the F.D. approximation of eq. (1.1).
In the same way for the non-linear terms:

$$U \ Ux \Big|_i^n = U_i^n \frac{(U_{i+1}^n - U_{i-1}^n)}{2 \cdot \Delta x}$$

$$(U(d+\zeta))x \Big|_i^n = \frac{U_{i+1}^n \cdot (d+\zeta)_{i+1}^n - U_{i-1}^n \cdot (d+\zeta)_{i-1}^n}{2 \cdot \Delta t}$$

and the third derivative term:

$$Fxxt \Big|_i^n = \frac{F_{i+1}^{n+1} - 2 \cdot F_i^{n+1} + F_{i-1}^{n+1} - F_{i+1}^{n-1} + 2 \cdot F_i^{n-1} - F_{i-1}^{n-1}}{2 \cdot \Delta t \cdot \Delta \overset{2}{x}}$$

In order to solve the linear system of equations derived from the above F.D. approximation we use the Bi-tridiagonal method. (Rosenberg, 1969).
In some cases the system becomes unstable. Indroducing a weighting factor α, for the terms Ux, ζ_x we are able to eliminate instability:

$$Fx \Big|_i^n = (1-2\alpha) \frac{F_{i+1}^n - F_{i-1}^n}{2 \cdot \Delta x} + \alpha \cdot \frac{F_{i+1}^{n+1} - F_{i-1}^{n+1}}{2 \cdot \Delta x} + \alpha \cdot \frac{F_{i+1}^{n-1} - F_{i-1}^{n-1}}{2 \cdot \Delta x}$$

354

Keeping α very small (=0.005) we avoid any reduction of accuracy.

Low courant numbers (Cr < 0.5) are required to obtain a stable scheme, and so there is no need to correct truncation errors of order Δt.

Fig. 1 shows the propagation of a Limiting Height solitary wave H=4 m in depth d=5 m with H/d=0.8. The same order of the ratio H/d is expected in the surf zone. In 2,200 m travel there is no reduction in the wave height and no developement of the dispersive tail at its trailing edge.

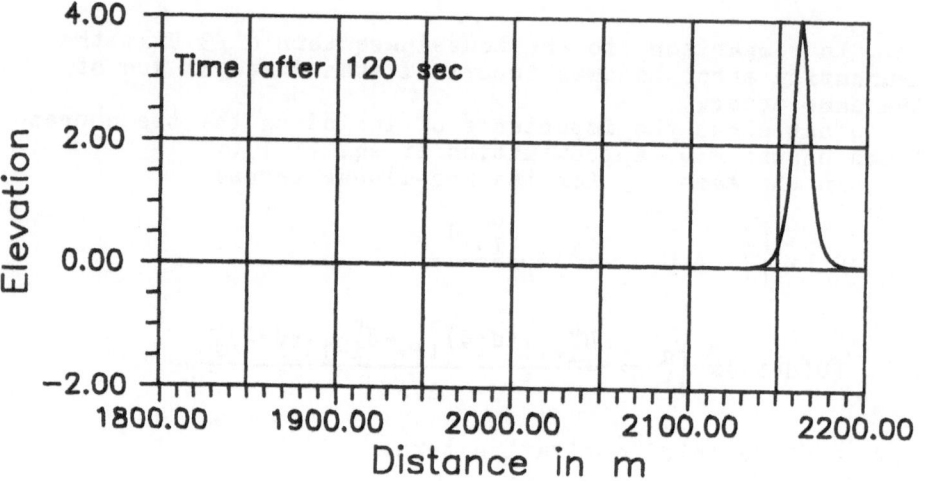

Figure 1. Propagation of a Limiting Height Solitary Wave. Δx=2.5 m, Δt=0.125, H=4 m and d=5 m.

3) WAVE BREAKING

Wave breaking is the transformation of the irrotational wave motion into vortical motions inclunding turbulence.

There exist two dominant breaker types: spillers and plungers. According to Miller the two types genarate similar patterns of motion but at different scales (Basco, 1985).

Svendsen et al. (1978) separated surf zone in three regions:

1. Outer region with rapid transitions of wave shape.
2. Inner region the flow of the front of the breaking wave is very similar to that of a moving bore or a hydrau-

lic jump.
    3. Run-up region.

    Based on experimental data (Goda 1970) we can give an empirical relation which relates the wave elevation at breaking point with the depth, wave length in deep waters and beach slope:

$$\zeta_b/d_b = A - B \ln(d_b/Lo) - E \left(\ln(d_b/Lo)\right)^2$$

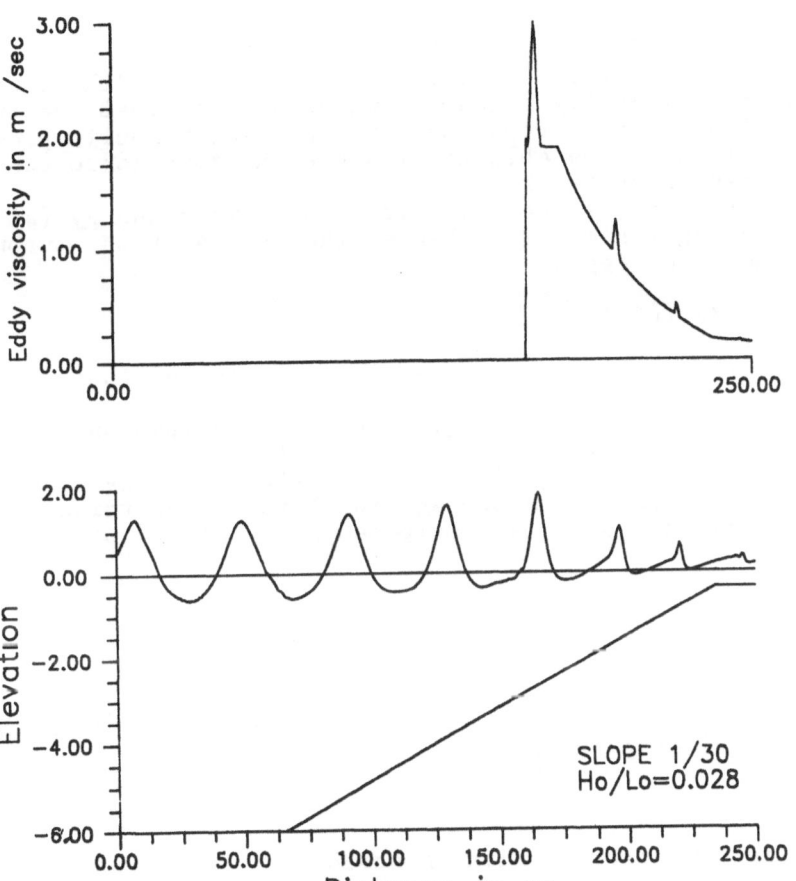

Figure 2. Numerical solution of eg. 1.1: a)Shoaling, breaking and dissipation after breaking of cnoidal waves (slope 1/30), b)Eddy viscosity coefficient (eq. (5.4)).

Where ζ_b, d_b are the elevation and depth respectively, and Lo the wave length in deep waters, and:

Slope	A	B	E
1/30	0.096	0.215	0.017
1/20	0.082	0.230	0.017
1/10	0.077	0.245	0.012

The above relation gives some indications about the position of breaking point and is very usefull for inter-period wave models (No direct calculation of the height H)

4) ZURF ZONE TURBULENCE

In this model the Reynolds stresses $-\rho.\overline{u_i{}'.u_j{}'}$, on the right-hand side of eq. (1.1), are modelled using the Boussinesq eddy viscosity concept, which assumes that the turbulent stresses are propotional to the mean velocity gradients. Thus we are able to simulate the turbulence conditions caused by breaking.

The breaking-induced dissipation of wave energy (at a rate D per unit area) is taken as the source of the turbulent kinetic energy k:

$$k = 1/2 \ (\overline{u}'^2 + \overline{w}'^2)$$

u', w' are the turbulence velocity fluctuations.

Turbulent kinetic energy k and its dissipation rate $(\varepsilon \sim D/(\rho.d))$ are related through: $\varepsilon \sim k^{3/2}$

Usually the energy dissipation after breaking is approximated by that of a propagating bore. According to Lamb energy in a bore is dissipated at a rate D' per unit span:

$$D' = 1/4.\rho.(\frac{h_2 + h_1}{2h_1.h_2})^{\frac{1}{2}} \ (h_2 - h_1)^3 \ . \ g^{3/2}$$

The eddy viscosity coefficient is proportional to the turbulent kinetic energy k.

$$V\tau = 1.\sqrt{k}$$

l is the turbulent length scale

or using the above relationship of k and ε:

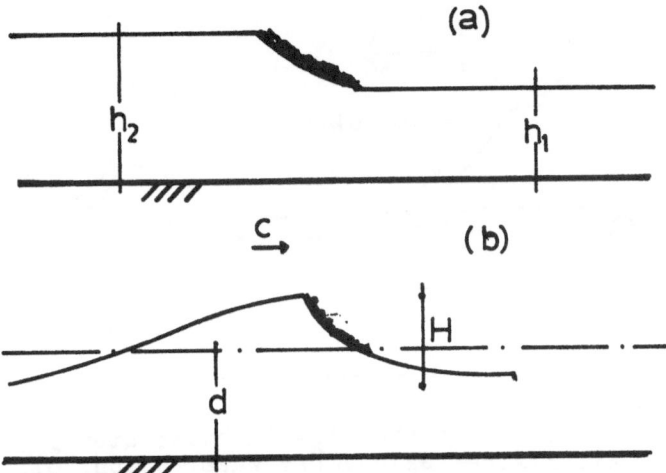

Figuré 3. Similarity between quasi-steady breaking
waves in shallow waters and steady bores.

$$V\tau = \varepsilon^{1/3}.1^{4/3}$$

ε is the rate of turbulent energy dissipation per
unit mass

This method gives the value of $V\tau$, inside the surf
zone, intergrated over the wave period T and so it can be
used only partialy in the inter-period model which is developed here.

5) EDDY VISCOSITY COEFFICIENT

Peregrine and Svendsen (1978), based on visual observations, proposed a qualitative model which describe the
nature of the flow in quasi-steady waves. They concluded
that the turbulent flow is in part like a mixing layer
and a part liko a wake.
This similarity is only applicable in the inner region
where the breaking waves propagate like stable bores.
Assuming that, in the outer region, the unresolved
scales of motion can be treated as turbulence, we can use
the above coefficient in both regions.

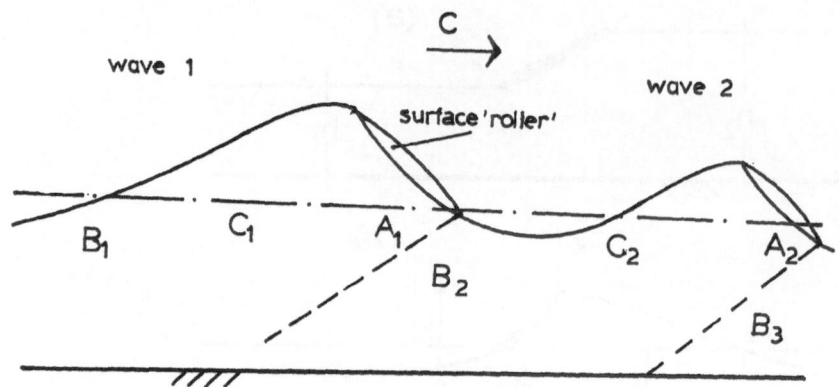

Figure 4. The nature of the flow in quasi-steady
breaking waves. Ai: Mixing region, Bi: Wake region, Ci;
Interaction region.

Battjes and Sakai (1980) concluded that, in the wake
region, the turbulent velocity magnitude (u') and the
velocity defect ($U_{crest}-U_{bottom}$) are proportional to $x^{\frac{1}{2}}$
(x is the downstream distance from the toe) while the turbulent length scale l is proportional to $x^{-\frac{1}{2}}$.

The eddy viscosity in this region may be estimated
as (Tennekes and Lumley, 1972).

$V\text{-wake} = Us.l/R\tau$

with $R\tau = 12.5$

Considering a fixed point in the breaking zone (where
the distance x is replaced by the time t) the value of
$V\text{-wake}$ is constant over a wave period and proportional to
$Usmax$ and l max.($Usmax$ and $lmax$ are related to the crest
phase of the wave).

Alternatively another -constant over the time- value
of V can be evaluated using the formulae refering to models intergrated over the wave period:

$V\text{-bore} \sim \varepsilon^{-1/3}.l^{4/3} \sim (D/\rho)^{1/3}.d$

assuming that

$$\bar{\epsilon} \sim D/(\rho.d) \quad \text{and} \quad 1 \sim d$$

where: $\bar{\epsilon}$ mean over the period rate of dissipation per unit mass
           D dissipation per unit area
           d, ρ depth and density respectively

According to Stive (1984)

$$D = A\epsilon \ (1/4 \ \rho.g.c.H^3)/(d.L)$$

$$A\epsilon = 2 \ \tanh \ (5 \ \xi o)$$
$$(\xi o \ \text{Irribaren number})$$

L = Wave length
C = $\sqrt{gd}$, celerity
H = Height

By replacing L=cT (T=period) we have:

$$D = A\epsilon.1/4 \ \rho.g.H^3/(d.T)$$

Let's consider now that the wave height decay whithin the surf zone is given by the relationship.

$$H = H_b.\Gamma(x)$$

H_b: Height at breaking point
$\Gamma(x)$: an empirical function, where x is hthe distance from the breaking point to the shore.

$$\Gamma(x) = (a.x./d_b+1)^{-2}$$

with   a   slope

                     0.045   1/20
                     0.040   1/30
                     0.028   1/65
                     0.020   1/80
                     0.016   1/100

If we substitute the above relations in the eddy viscosity coefficient, we obtain:

$$V\text{-}bore = \theta.(A\epsilon/T)^{1/3}.\gamma_b.d_b.(d^{2/3}.\Gamma(x)) \qquad (5.1)$$

θ: is a factor to be determinated

$$\gamma_b = H_b/d_b = 1.9(Sb/(1+2\ Sb))^{1/2} \quad \text{(Svendsen, 1987)}$$

$$Sb = 2.3\ \xi o$$

From the relation for γ_b/d_b, given in paragraph 3, we can determine directly the breaking depth beyond which and up to the shore the eddy viscocity coefficient V takes the value of (5.1)

In equation (5.1) dissipation is predicted even when H/d<<1 which is unrealistic. In the case of an horizontal bottom breaking would continue until wave height reach a limiting value, Hs=0.3 + 0.4.d (Horikawa and Kuo, 1966 and Dally et al., 1984).
After this point there is no decay and V-bore should be taken equal to zero.

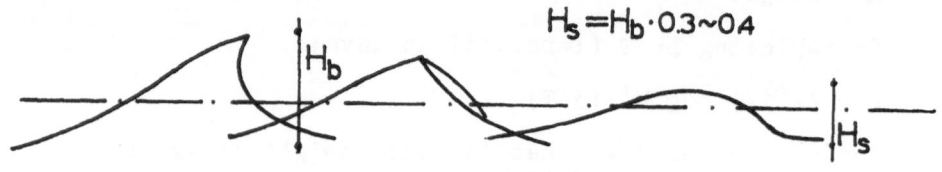

$$H_s = H_b \cdot 0.3 \sim 0.4$$

Figure 5. Shelf beach presentation of the surf zone. H_b wave height at breaking point, Hs stable wave height.

The dissipation rate would be:

$$D = D' - Ds = A\epsilon \cdot 1/4 \cdot \rho \cdot g \cdot H^3/(d \cdot T) \cdot (\Gamma'^3(x) - (0.4/\gamma)^3)$$

$$(Hs = 0.4\ d_b = 0.4/\gamma_b \cdot H_b)$$

and $\quad V\text{-bore} = \theta(A\epsilon/T)^{1/3} \cdot \gamma_b \cdot d_b [d^{2/3}\{\Gamma'^3(x) - (0.4/\gamma_b)^3\}^{1/3}]$

$$(5.2)$$

with

$$\Gamma(x) = .(0.2/d_b;x + 1)^{-1/2}$$

Let's assume now that the two constant values of V-wake and V-bore (=V-mean over the period) are proportional to one another i.e. V-wake ~ V-bore.

We are now going to try to simulate the mixing layer region. Then the eddy viscosity, in mixing region, can be expressed by Prandtl mixing-length hypothesis.

$$Vl = l^2m \left| \frac{\partial u}{\partial z} \right|$$

with lm proportional to local layer width δ

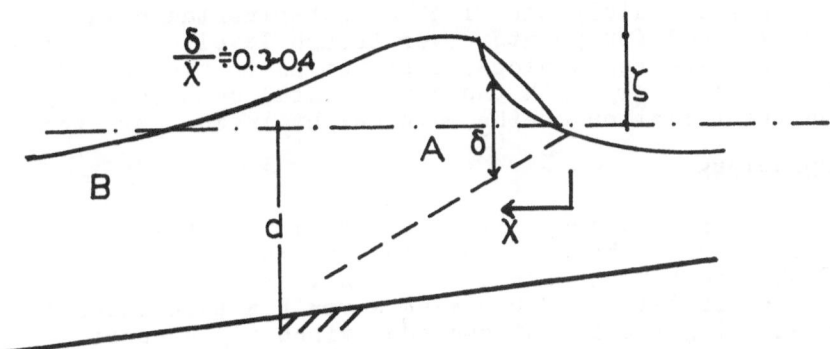

Figure 6. Growth rate of mixing region (Stive, 1980) A: mixing layer region, B:wake region.

and

$u = U-(z^2/2 + d.z + d^2/3).Uxx$ (Peregrine, 1972)

U = mean over the depth horizontal velocity

Assuming that $\delta \sim x \sim \zeta$:

$Vl = \mu.\zeta^2.(d+\zeta) Uxx$

intergrated over the depth.

Figure 4 (based on experiments and observations) makes clear that only a part of the energy of the wave "i+1" has been dissipated when the next wave "i" passes

and there is a superposition of the wake region of "i+1"
wave with the mixing region of "i" wave.

Therefore the total eddy viscosity coefficient, in a
fixed point of the breaking zone (consindering quasi-state
situation) is:

V-total = V-wake + V-layer

or from (5.1) and (5.3)

$$V\tau = \theta \ (A\epsilon/T)^{1/3}\gamma_b d_b(d^{2/3}\Gamma(x)) + \mu.\zeta^2(d+\zeta) \ . \ \left|U_{xx}\right| \ (5.4)$$

In the wake region only the constant over the time
part of V-total (V-wake) remain because of the small va-
lues of ζ and U_{xx} ($U_{xx} \rightarrow 0$) while in mixing region both
part of V-total relation are important because U_{xx} takes
its maximum values in the steep front side of the wave
(see fig. 2).

Unfortunately the lack of experimental data for ϵ
(or k) and l for points, over trough level, with different
steepnes and beach slope, restrict us to a comparison of
our results only with the experimental data refering to
the tranformation of the wave height inside surf zone.

CONLUSIONS

Using an appropriate F.D. implicit scheme, a third
order accuracy has been obtained in the numerical solution
of the Boussinesq equations, even when H/d O(1).

A simulation of breaking waves has been added in this
model using the Boussinesq eddy viscosity concept.

In order to simulate the turbulent flow we have made
the hypothesis that there are two main regions in a breaker: the wake and the mixing-layer region.

In the wake region the eddy visosity cofficient (constant over the time) has been taken proportional to the
dissipation rate D in a moving bore, while in the mixing
layer region Prandl mixing-length hypothesis has been adopted.

The model seems to correctly simulate the wave dissipation after breaking.

REFERENCES

Basco, D.R. (1985). 'A qualitative description of wave
breaking' J.Waterw.Port Coastal Eng. ASCE 111:171-188.

Battjes,J.A. (1975) 'Modeling of turbulence in the
surf zone', Proc. Symp. Modeling Techniques, ASCE pp 1050-
1061.

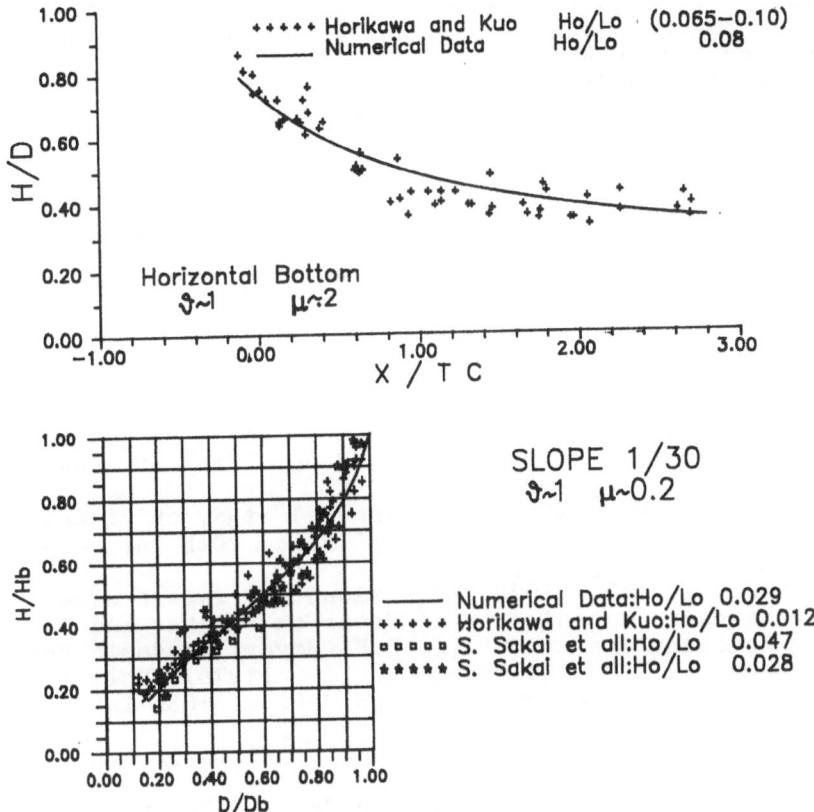

Figure 7. Wave decay on plane beaches. Comparison with experimental data (Horikawa and Kuo, 1966).

Battjes, J.A. and Janssen, J.P.F.M. (1978). 'Energy loss and set-up due to breaking of random waves' Proc. 16th Int. Conf. Coastal Eng., ASCE pp 569-587.

Battjes, J.A. and Sakai T. (1980) 'Velocity field in a steady breaker', Proc. 17th Int. Conf. Coastal Eng., ASCE pp 498-511.

Battjes, J.A. (1986) 'Energy dissipation in breaking solitary and periodic waves, Communications on hydraulic and geotechnical engineering, TU Delft Report nr 86-6.

Battjes, J.A. (1988) 'Surf-zone dynamics', Ann. Rev. Fluid Mech. 20:257-293.

Dally, W.R., Dean R.G. and Dalrymple R.A. (1984) 'A model for breaker decay on beaches' Proc. 19th Int. Conf. Coastal Eng., ASCE pp 82-98.

Goda, U. (1970), A synthesis of breaker indices, Trans. Jap. Soc. Civil Eng., vol 180, pp 39-49 (in Japanese)

Horikawa, K. and Kuo C.T. (1966), 'A study on wave transformation inside surf zone', Proc. 10th Int. Conf. Coastal Eng., ASCE pp 69-81.

Johns, B (1980) 'tThe modelling of the approach of bores to a shoreline', Coastal Engineering, 3:207-219.

Koutitas C., (1988), Mathematical Models in Coastal Engineering Pentech Press.

Madsen, P.A, and Svendsen, I.A. (1983) 'Turbulent bores and hydraulic jumps', J. Fluid Mech. vol. 129, pp 1-25.

Mizugushi, M. (1980) 'An heuristic model of wave height distribution in surf zone'. Proc. 17th Int. Conf. Coastal Eng., ASCE pp 278-289.

Mizugushi, M. (1986) 'Experimental study on kinematics and dynamics of wave breaking' Proc. 20th Int. conf. Coastal Eng., ASCE pp 589-603.

Peregrine, D.H. (1972) 'Equations for waters waves and approximations behind them' Waves on Beaches and Resulting Sediment Transport' (ed R.E. Meyer) Academic Press

Peregrine, D.H. and Svendsen, I.A. (1978) 'Spilling breakers, bores and hydraulic jumps', Proc. 16th Int. conf. Coastal Eng., ASCE pp 540-550.

Rodi W., (1980) Turbulence models and their application in hydraulics, IAHR.

Rosenberg, D.U. (1969) Methods for the Solution of Differential Equations, Elsevier N.Y.

Sakai, S., Hiyamizu, K., Saeki, H. (1986),'wave height decay model within a surf zone', Proc. 20th Int. Conf. Coastal Eng., ASCE pp 686-696

Stive, M.J.F. (1980) 'Velocity and pressure field of spilling breakers', Proc. 17th Int. conf. Coastal Eng., ASCE pp 547-566.

Stive, M.J.F. and Wind, H.G. (1982) 'A study of radiation stress and set-up in the nearshore region', Coastal Engineering, 6:1-25.

Stive, M.J.F. (1984) 'Energy dissipation in waves breaking on gentle slopes', Coastal Engineering, 8:99-127.

Svendsen, I.A., Madsen, P.A. and Hansen J.B. (1978) 'Wave characteristics in the surf zone', Proc. 16th Int. Conf. Coastal Eng., ASCE pp 520-539.

Svendsen, I.A., Madsen, P.A (1984) 'A turbulent bore on a beach' J. Fluid Mech. vol. 148, pp 73-96

Svendsen, I.A. (1984) 'Wave heights and set-up in a surf zone' Coastal Engineering, 8:303-329.

Svendsen, I.A. (1987) 'Analysis of surf zone turbulence', J. of Geophysical Research, vol. 92, no C5, pp
5115-5124.

Tennekes, H. and Lumley, J.L. (1972), A First Course
in Turbulence, The MIT Press, pp 104-145.

Uasuda, T., Goto, Sh. and Tsuchiya, Y. (1982), 'On
the relation between changes in intergral quantities of
shoaling waves and breaking inception', Proc. 18th Int.
Conf. Coastal Eng., ASCE pp 23-37.

Yasuda, T., Yamashita, T., Goto, Sh. and Tsuchiya,
Y. (1982), 'Numerical calculations for wave shoaling on a
sloping bottom by K-dV equation', Coastal Eng. in Japan,
JSCE, Vol. 25.

FREE AND FORCED CROSS-SHORE LONG WAVES

H. A. SCHÄFFER & I. G. JONSSON
Inst. of Hydrodynamics and Hydraulic Engineering
Tech. Univ. of Denmark, DK-2800 Lyngby, Denmark
I. A. SVENDSEN
Dept. of Civil Engineering
Univ. of Delaware, Newark DE 19716, USA

ABSTRACT. Low frequency motion in and outside the surf zone generated by grouping in the incident short waves is considered in a two-dimensional model. This groupiness has two primary effects, which are shown to be of the same order of magnitude. One is the long forced wave bound to the incident short waves, and the other is a time-varying position of the break point, which also contributes to the forcing of low frequency motion (or long waves).
Free, long waves are formed where the short-wave forcing undergoes changes due to shoaling or breaking or by reflection of bound waves. Mathematically this is a consequence of imposed boundary and matching conditions.
The presence of the induced long waves significantly changes the bottom velocities in a way that suggests a close connection with the formation of longshore sand bars.

1. INTRODUCTION

The purpose of the present work is to study the low frequency motion in the surf zone, i.e. the fluctuations of the mean water surface. This slow motion, which has a period of the order of several minutes, is usually termed surf beats or infragravity waves.

Since Longuet-Higgins and Stewart (1962, 1964) showed that short-wave modulation results in a long forced wave (set-down wave) which is phase-locked to the propagation of the short-wave envelope, a number of authors have treated the long wave problem. Longuet-Higgins and Stewart presented a solution which was derived for a constant depth. This solution becomes singular as the forcing becomes resonant in shallow water. They explained some early measurements of surf beats by Munk (1949) and Tucker (1950) as outgoing free waves arising from the release of these bound waves as they are reflected somewhere near the coastline.

After these first recordings, numerous observations have shown that the energy at surf beat frequencies can be substantial and in some cases even exceed that of the short wind waves (Wright et al., 1982). Likewise

A. Tørum and O. T. Gudmestad (eds.), Water Wave Kinematics, 367–385.

the amplitudes arising from surf beats and short waves can be comparable at the shoreline (Guza and Thornton, 1982, 1985).

In a lengthy WKB-expansion with multiple scales in time and off-shore direction Foda and Mei (1981) treated the 3-dimensional case under the assumption that the long waves were of the same magnitude as the short waves. They obtained equations for wave evolution as well as interactions at different orders, and applied their derivations to the problem of a closed coast using empirical relations for the breaking of the short waves. Several other 3-dimensional models of long wave genera-tion at a coast (e.g. edge waves) have been given in the literature.

In the present work we restrict ourselves to two dimensions so that only the so-called leaky modes are modelled, and trapped-mode edge waves cannot be represented. The model is closely related to the pioneer work of Symonds et al. (1982) who, using linearized, depth-integrated conser-vation equations for mass and momentum, focussed on the effect of a time-varying break point position of the short waves concluding this to be a possible source of surf beats. This generating mechanism, which had not been studied previously, is combined with the effect of an incident, long, forced wave, which was neglected by Symonds et al. Furthermore, a more general description of the short-wave breaking allows us to study effects of short-wave modulation being transmitted into the surf zone.

In the open sea Molin (1982), (starting with the Laplace equation for the velocity potential) showed that discontinuities in the bottom slope result in emission of free long waves in addition to the long forced waves that are present. His results are confined to deep water for the short waves and shallow water for the long waves and do not include stationary set-down. Mei and Benmoussa (1984) generalized these results to a 3-dimensional treatment including stationary set-down and allowed for arbitrary depth relative to the short waves, using equations from a WKB-expansion by Chu and Mei (1970), which are equivalent to the conservation equations used by Symonds et. al (1982). Liu (1988) sugges-ted a different method of solution to the governing equations and gave corrections to the boundary conditions used by Mei and Benmoussa. These boundary conditions, however, need minor adjustments in order to satisfy continuity in the flux of mass, momentum and energy as will be shown.

Other related works restricted to constant water depth are Bowers (1977), who showed that long, forced waves are a possible source of harbour resonance; Sand (1982), who analyzed the impact of these long waves on laboratory models; and Ottesen Hansen (1978), and Ottesen Hansen et al. (1980), who analyzed long waves forced by a spectrum of short waves.

2. MATHEMATICAL FORMULATION

2.1. Governing equations

It is convenient to use the depth-integrated and time-averaged conservation equations of mass and momentum, where the time averaging is taken over one short wave period. Following Phillips (1979) we have conservation of mass:

$$\frac{\partial \bar{\zeta}}{\partial t} + \frac{\partial \bar{Q}}{\partial x} = 0 \tag{1}$$

and conservation of momentum:

$$\frac{\partial \bar{Q}}{\partial t} + \frac{\partial}{\partial x}\{ \frac{\bar{Q}^2}{h+\bar{\zeta}} + \frac{S_{xx}}{\rho} \} + g(h+\bar{\zeta}) \frac{\partial \bar{\zeta}}{\partial x} = 0 \tag{2}$$

see Fig. 1 for definitions. Here $\bar{Q}$ is the total volume flux defined as the sum of the volume flux Q_s of the short waves, and the volume flux of the current $Q_c = (h+\bar{\zeta}) \; U_c$

$$\bar{Q} = Q_s + Q_c \tag{3}$$

where $U_c = U_c(x,t)$ is the uniform current under wave trough level assumed in the derivation of (1) and (2), see also Fig. 1. This current is allowed to have a slow variation in time as well as in space, and together with $\bar{\zeta}$ it describes the low frequency motion under investigation. Defining the total mean horizontal velocity U by

$$U = \frac{\bar{Q}}{h+\bar{\zeta}} = \frac{Q_s}{h+\bar{\zeta}} + U_c \tag{4}$$

equation (1) and (2) can be written

$$\frac{\partial \bar{\zeta}}{\partial t} + \frac{\partial}{\partial x}\{ (h+\bar{\zeta}) \; U \} = 0 \tag{5}$$

and

$$(h+\bar{\zeta}) \frac{\partial U}{\partial t} + U \frac{\partial}{\partial t}(h+\bar{\zeta}) + U \frac{\partial \bar{Q}}{\partial x} + \bar{Q} \frac{\partial U}{\partial x} + g(h+\bar{\zeta}) \frac{\partial \bar{\zeta}}{\partial x} = - \frac{1}{\rho} \frac{\partial S_{xx}}{\partial x} \tag{6}$$

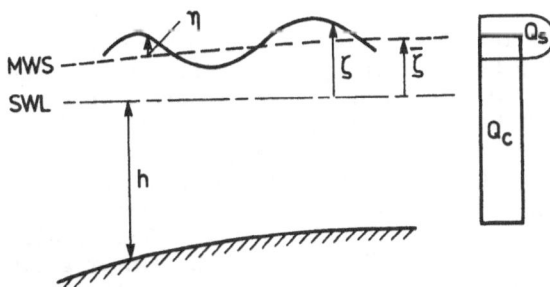

Fig. 1 Definition sketch.

The second and third terms cancel out by continuity, and we get

$$\frac{\partial U}{\partial t} + U \frac{\partial U}{\partial x} + g \frac{\partial \bar{\zeta}}{\partial x} = -\frac{1}{\rho} \frac{1}{h+\bar{\zeta}} \frac{\partial S_{xx}}{\partial x} \qquad (7)$$

which together with (5) are the nonlinear shallow water equations with a forcing term. Thus we can regard the surf beats as a long wave phenomenon, the forcing being analogous to a fictitious surface pressure p_s given by

$$p_s = \int \frac{1}{h+\bar{\zeta}} \frac{\partial S_{xx}}{\partial x} \, dx \qquad (8)$$

For a constant depth the fictitious pressure was introduced by Longuet-Higgins and Stewart (1962). It turns out to be a convenient concept when studying the transfer of energy between the short and the long waves.

Once the forcing term is known from the short waves, the surf beat wave can in principle be found by solving (5) and (7) simultaneously, subject to relevant boundary or initial conditions. However, the forcing term itself depends weakly on the solution through the total water depth $h+\bar{\zeta}$, and the interaction between the short waves and the current corresponding to the surf beat wave. Upon linearizing the equations this dependency vanishes, since we get

$$\frac{\partial \bar{\zeta}}{\partial t} + \frac{\partial}{\partial x} \{ hU \} = 0 \qquad (9), \qquad \frac{\partial U}{\partial t} + g \frac{\partial \bar{\xi}}{\partial x} = -\frac{1}{\rho h} \frac{\partial S_{xx}}{\partial x} \qquad (10)$$

In this formulation the undisturbed depth h is consequently used in the calculation of the radiation stress, where also the current interaction is neglected. Thus the energy feed-back from the long waves to the short waves is not accounted for. Eliminating U from (9) and (10) we get

$$\frac{\partial}{\partial x} \{ gh \frac{\partial \bar{\zeta}}{\partial x} \} - \frac{\partial^2 \bar{\zeta}}{\partial t^2} = -\frac{1}{\rho} \frac{\partial^2 S_{xx}}{\partial x^2} \qquad (11)$$

which is the linear shallow water equation with a forcing term. This linear equation can also be obtained by a WKB-expansion as shown by Chu and Mei (1970), and Mei and Benmoussa (1984). In order to describe the nonlinearity of the surf beat waves as in (1) and (2) or (5) and (7) this expansion would have to be carried to higher order.

Carrier and Greenspan (1958) solved the nonlinear shallow water equations for free long waves (i.e. the homogeneous versions of (5) and (7)) and found that, although the solution for the surface elevation close to the shoreline was very different from the linear solution, the run-up height was the same as could be predicted from the linearized equations (see e.g. p 527 Mei, 1983). This fact gives a little comfort when dealing with the linear theory.

A well known solution to (11) is the forced long wave described by Longuet-Higgins and Stewart (1962) for constant depth (see (54)).

Restricting ourselves to periodic solutions we introduce the Fourier expansions of $\bar{\zeta}$ and S_{xx}

$$\bar{\zeta} = \sum_{n=0}^{\infty} \frac{1}{2} (\xi_n e^{in\omega t} + *) \tag{12}$$

$$S_{xx} = \sum_{n=0}^{\infty} \frac{1}{2} (S_n e^{in\omega t} + *) \tag{13}$$

(where * denotes the complex conjugate of the preceding term) by which (11) is transformed into the ordinary differential equations

$$\frac{d}{dx} (gh \frac{d\xi_n}{dx}) + (n\omega)^2 \xi_n = - \frac{1}{\rho} \frac{d^2 S_n}{dx^2} ; \quad n = 0, 1, 2, \dots \tag{14}$$

The case $n = 0$ determines the stationary set-down or set-up, while the surf beat wave is governed by the rest of (14) which we shall concentrate on in the following.

To solve these equations a description of the radiation stress and appropriate boundary conditions is required.

2.2. The radiation stress

To the lowest order of approximation the radiation stress S_{xx} associated with the short waves can be obtained from linear Stokes theory with appropriate modifications in the surf zone. We have

$$\frac{1}{\rho g} S_{xx} = \frac{1}{2} |A|^2 (\frac{2c_g}{c} - \frac{1}{2}) \tag{15}$$

where A is a complex amplitude which describes the modulation of the short waves, and c_g and c are their group velocity and phase velocity, respectively. The most simple case of short waves forcing long waves is that of a superposition of two regular wave trains with slightly different angular frequencies

$$\omega_1 = \omega_s (1 + \epsilon), \quad \omega_2 = \omega_s (1 - \epsilon) \tag{16}$$

where ω_s is their mean value, and ϵ is a small perturbation parameter. (Alternatively one could have specified a wave number perturbation as done by Mei and Benmoussa (1984), and also used by Liu (1988), and Schäffer and Svendsen (1988), but their approach has the disadvantage that the perturbation parameter then has to refer to some fixed depth. Note that the two approaches result in different definitions of ϵ). The change $\epsilon\omega_s$ in the angular frequency corresponds to a change Δk in the wave number

$$\Delta k = \frac{\Delta k}{\Delta \omega} \epsilon \omega_s \approx \frac{\epsilon \omega_s}{c_g} \qquad (17)$$

Letting the two amplitudes be a and δa ($\delta \ll 1$) we have

$$\eta_1 = \frac{a}{2} \exp\{ i (\int [k + \epsilon \frac{\omega_s}{c_g}] \, dx + \omega_s[1 + \epsilon]) \} + *$$

$$\eta_2 = \frac{\delta a}{2} \exp\{ i (\int [k - \epsilon \frac{\omega_s}{c_g}] \, dx + \omega_s[1 - \epsilon]) \} + * \qquad (18)$$

and their superposition

$$\eta = \eta_1 + \eta_2 = \frac{1}{2} A \exp\{ i (\int k \, dx + \omega_s t) \} + * \qquad (19)$$

where A is

$$A = a (e^{i\theta} + \delta e^{-i\theta}) \quad (20), \qquad 2\theta = \int \frac{\omega}{c_g} \, dx + \omega t \qquad (21)$$

ω being the angular difference frequency $\quad \omega = \omega_1 - \omega_2 \qquad (22)$

Outside the surf zone energy conservation prescribes

$$\frac{\partial |A|^2}{\partial t} + \frac{\partial}{\partial x}\{ c_g |A|^2 \} = 0 \qquad ; \qquad h \geq h_B \qquad (23)$$

where from (20)

$$|A|^2 = a^2 \{1 + \delta^2 + \delta(e^{2i\theta} + *)\}$$
$$\approx a^2 \{1 + \delta(e^{2i\theta} + *)\} \qquad ; \qquad h \geq h_B \qquad (24)$$

Here subscript B refers to the break point. Choosing deep water as the reference depth for the short waves (23) yields

$$a = a_\infty \{ \frac{c_{g\infty}}{c_g} \}^{1/2} \qquad ; \qquad h \geq h_B \qquad (25)$$

where subscript ∞ indicates deep water. Thus to the leading order the shoaling of each wave train is independent of the other.

Within the surf zone the short-wave amplitude can be modelled in a variety of ways, the choice of the breaking criterion being essential. One simple model is obtained by assuming that the short-wave modulation is totally destroyed by breaking, so that the wave-height decay is solely dependent on the local water depth (see Fig. 2, case $\kappa = 1$). This implies a time-varying break point position, the effect of which was studied by Symonds et al. (1982).

Another extreme is to assume that the break point is fixed so that the short-wave modulation is transmitted into the surf zone (see Fig. 2, case $\kappa = 0$). This was done by Schäffer and Svendsen (1988). In the present work we shall introduce a parameter κ, the value of which prescribes the degree of variation of the break point position, and through this the degree of short-wave modulation transmitted into the surf zone. Taking the decay after breaking of the individual waves to be proportional to the water depth we get inside the surf zone

$$|A| = \gamma h \qquad ; \qquad h \leq h_B \qquad (26)$$

where γ is a slowly varying factor of proportionality with a modulation that follows from the modulation of $|A|$. If γ_0 denotes the ratio of short-wave amplitude to water depth in the case of vanishing modulation ($\gamma_0 = 0.4$ in all computations), we have

$$|A|^2 = \gamma^2 h^2 = \gamma_0^2 \{1 + (1-\kappa)\delta(e^{2i\theta} + *)\} h^2 \; ; \quad h \leq h_B \qquad (27)$$

Requiring the short-wave to be continuous at the break point we get

$$a_B^2 \{1 + 2\delta \cos(2\theta_B)\} = \gamma_0^2 \{1 + (1-\kappa)2\delta \cos(2\theta_B)\} h_B^2 \qquad (28)$$

or

$$
h_B = \frac{a_B}{\gamma_0} \left\{ \frac{1 + 2\delta \cos(2\theta_B)}{1 + (1-\kappa) 2\delta \cos(2\theta_B)} \right\}^{1/2}
$$

$$
= \frac{\bar{a}_B}{\gamma_0} \{ 1 + \kappa\delta \cos(\omega t) \} + 0(\delta^2)
$$

$$(29)$$

taking the lower limit in the integral expression for θ (21) to be the mean position of the break point, also choosing $t = 0$ when the short-wave amplitude has a maximum at this point. Here $\bar{a}_B$ denotes the mean amplitude at breaking taken as the amplitude at the break point for vanishing short-wave modulation ($\delta \to 0$). Eq. (29) may be taken as a definition of κ after which (27) follows from (28). For $\kappa = 0$ we have a constant break point position and a full transmission of groupiness into the surf zone (Schäffer and Svendsen, 1988). For $\kappa = 1$ the break point oscillates just so much that all groupiness is destroyed, and the short-wave amplitude is solely a function of depth inside the surf zone (Symonds et al., 1982).

Note that κ is taken as an independent parameter although it ought to be determined for a given topography and given incident wave conditions. It is evident that we always have $\kappa > 0$ since high waves

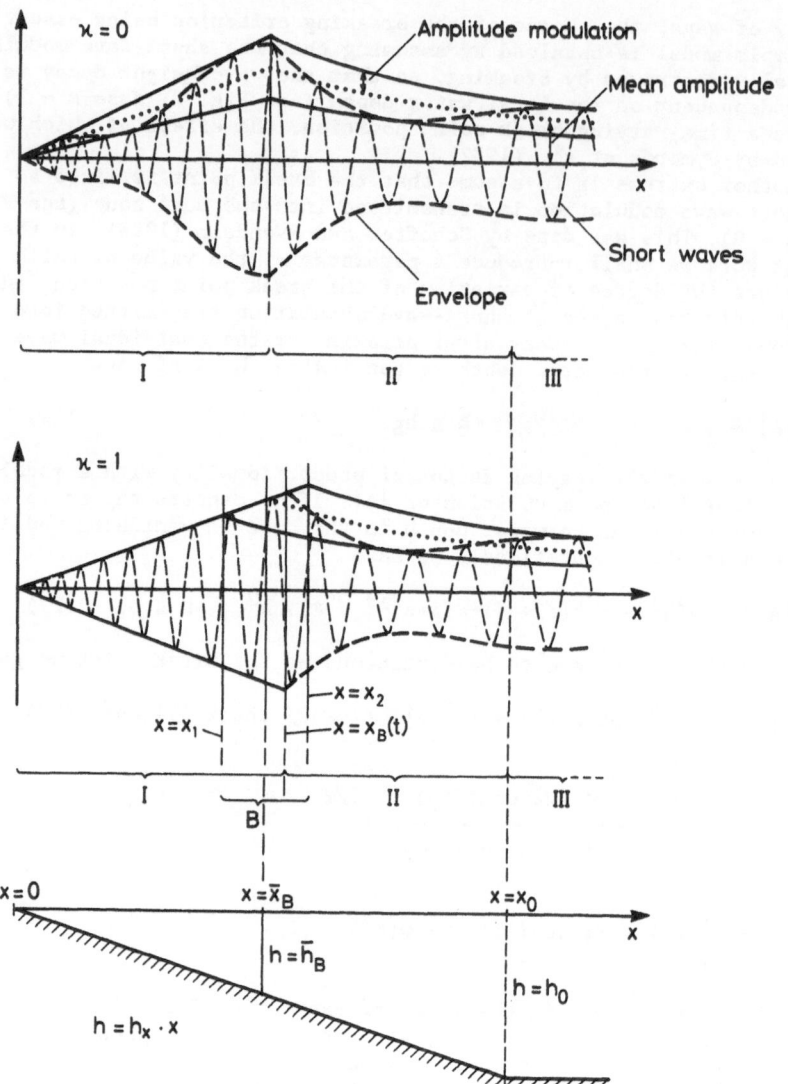

Fig. 2 Sketch of bottom topography and two examples of short-wave modelling with associated definitions of regions I, II, B, and III.

must break at greater depths than low waves. Higher waves might even break so early that they decay sufficiently to end up being the lower waves inside the surf zone. This means that the modulation is reversed when groups pass the zone of initial breaking, and it corresponds to $\kappa > 1$. Since amplification of the modulation is out of the question we may conclude that κ will always be in the range $0 < \kappa < 2$. A closer analysis of the results for monochromatic waves collected by Goda (1970) indicates that κ increases with increasing short-wave steepness ranging from $\kappa \approx 1.0$ to 1.3.

2.3. General solution

If the stationary case of $n = 0$ of which (14) can be integrated directly is excluded, the method of variation of parameters yields

$$\xi_n = - \xi_n^{(1)} \{ -\alpha_n + \int_{x_\ell}^{x} \frac{\xi_n^{(2)} q_n}{W_n} \, dx \}$$

$$+ \xi_n^{(2)} \{ -\beta_n + \int_{x_\ell}^{x} \frac{\xi_n^{(1)} q_n}{W_n} \, dx \}; \quad n = 1, 2, 3, .. \quad (30)$$

where $\xi_n^{(1)}$ and $\xi_n^{(2)}$ are homogeneous solutions (free waves), W_n is their Wronskyan, q_n is the right hand side of the normalized equation (14)

$$q_n = - \frac{1}{\rho gh} \frac{d^2 s_n}{dx^2} \quad ; \quad n = 1, 2, 3, \ldots \quad (31)$$

and α_n and β_n are arbitrary complex constants. Any lower limit of integration x_ℓ can be chosen.

2.4. Matching conditions

It is evident that we shall have to match the solutions along the common boundary of two regions e.g. inside and outside the break point. Since we are dealing with a 2nd order differential equation, two matching conditions are needed. One is obtained by requiring the surface elevation to be continuous

$$[\bar{\zeta}]_{x_m^-}^{x_m^+} = 0 \quad (32)$$

where subscript m denotes the match point and superscripts + and - indicate limits from the right and left respectively. The other arises from specifying continuity in mass and momentum flux. Using (10) this yields

$$
[\frac{\partial \bar{\zeta}}{\partial x}]_{x_m^-}^{x_m^+} = - \frac{1}{\rho g h_m} [\frac{\partial S_{xx}}{\partial x}]_{x_m^-}^{x_m^+} \tag{33}
$$

since $\bar{Q}/h$ must be continuous. This condition can be shown to be consistent with continuity in the energy flux of the long wave. Similar expressions are valid for the individual Fourier components. The condition (33) renders a kink in $\bar{\zeta}$ possible. This result is a generalization of the usual kink in the mean water surface, which exists in the classical calculation of wave set-up corresponding to the stationary case of n = 0 and matching at the break point.

Mei and Benmoussa (1984) and Liu (1988) used the condition that $\partial \bar{\zeta}/\partial x$ was continuous when matching solutions at discontinuities in the bottom slope. Strictly this condition is not appropriate since it is only in case of deep water for the short waves that the right hand side of (33) vanishes. However it is mainly when matching at the break point that this is important, since the kink in the radiation stress is rather large here due to the short-wave breaking.

2.5. Boundary conditions

Assuming that the surf beat wave is not steep enough to break, we specify full reflection at the coastline. In the limit of $h \to 0$ this condition gives a standing wave, but because of the forcing that takes place we cannot expect this to happen away from the coastline. In a complex representation, where $\xi_n^{(1)}$ and $\xi_n^{(2)}$ are free progressive waves travelling in opposite directions, full reflection requires $|\alpha_n| = |\beta_n|$ in (30) when the lower limit of integration x = x_ℓ is taken at h = 0. This condition is sharpened by eliminating the singularity at h = 0 to get a finite amplitude at the shoreline. If $\xi_n^{(1)}$ and $\xi_n^{(2)}$ are specified so that their sum is regular, then the condition is $\alpha_n = \beta_n$.

On the seaward boundary we specify that there are no incident free long waves. In order to identify a pure forced wave, i.e. an expression in which no free wave solutions are hidden, a constant depth is necessary. If $\xi_n^{(1)}$ is the incident free wave, we have $\alpha_n = 0$. Note that this α_n is different from the α_n mentioned above, since they correspond to two different regions.

2.6. Specific solution

So far all quantities and derivations are valid for an arbitrary water depth h(x) with the restrictions that h is 'sufficiently' smooth, and that it has a monotonic variation within the surf zone. The last condition is obviously necessary in view of (26).

In the following we shall confine ourselves to a plane sloping beach. The geometry is sketched in Fig. 2, where also divisions into regions I, II and III are shown. Region I is the surf zone, region II is the shoaling zone and region III is the 'shelf zone' - a constant depth zone inferred to enable us to specify a seaward boundary condition for the solutions in the regions I and II. The governing equations used are shallow water equations for the long wave motion, and the presence of region III also serves the purpose of limiting the water depth in order not to violate the assumptions behind these equations. Note that although we have shallow water in region III for the long waves, the short waves may well be deep water waves.

In regions I and II we have

$$h - h_x \, x \qquad ; \qquad h \leq h_0 \qquad \qquad (34)$$

and the free wave solutions (homogeneous solutions) to (14) are given by the Hankel functions of the first and second kind of order zero

$$
\begin{aligned}
\xi_n^{(1)} &- H_0^{(1)}(ny) \\
\xi_n^{(2)} &- H_0^{(2)}(ny)
\end{aligned}
\qquad ; \qquad n - 1, \, 2, \, 3, \ldots \qquad (35)
$$

where

$$y - \frac{2\omega}{gh_x} \sqrt{(gh)} \qquad ; \qquad \frac{dy}{dx} - \frac{\omega}{\sqrt{(gh)}} \qquad (36)$$

Their Wronskyan is (see 9.1.17, Abramowitz and Stegun, 1972)

$$W_n - - \frac{4i}{\pi ny} \frac{d}{dx}(ny) - - \frac{2i}{\pi x} \qquad (37)$$

and we get

$$\frac{q_n}{W_n} - - \frac{i\pi}{2\rho gh_x} \frac{d^2 s_n}{dx^2} \qquad (38)$$

The assumption of wave groups made up from only two wave components reduces the expression for the radiation stress (13) to (superscript J refers to the three regions, see Fig. 2)

$$S_{xx}^J - S_0^J + \frac{1}{2}(S_1^J \, e^{i\omega t} + *) \qquad ; \qquad J - I, \, II, \, III \qquad (39)$$

where S_0^J and S_1^J can be found from (15) with (24) for $J - II$, III and (27) for $J - I$. However, in the zone of initial breaking, which we shall refer to as region B, limited by $x - x_1$ and $x - x_2$ (see Fig. 2, case $\kappa - 1$) we part of the time have shoaling conditions and part of the time surf zone conditions at each point. Thus in region B, S_0 and S_1 are <u>not</u> the Fourier coefficients of the radiation stress, although they yield the right S_{xx} when inserted in (39). This is due to the implicit time-dependence of S_0 and S_1 given through the variable size of regions I and II. According to (13) we have

$$S_0 = \frac{1}{2\pi} \int_0^{2\pi} S_{xx} \, d(\omega t) \tag{40}$$

and

$$S_n = \frac{1}{\pi} \int_0^{2\pi} S_{xx} \, e^{-in\omega t} \, d(\omega t) \; ; \quad n = 1, 2, 3, \ldots \tag{41}$$

Since the goal is to study the low frequency motion and not the stationary set-down and set-up, S_0 is irrelevant. Using that $t = 0$ was chosen so that S_{xx} is an even function of time we get in region B

$$S_n^B = \frac{2}{\pi} \{ \int_0^\tau S_{xx}^I \cos(n\omega t) \, d(\omega t) + \int_\tau^\pi S_{xx}^{II} \cos(n\omega t) \, d(\omega t) \} \tag{42}$$

where $0 < t < \tau/\omega$ and $\tau/\omega < t < \pi$ are the time intervals of which we have surf conditions and shoaling conditions, respectively. Since these intervals vary from point to point, we have $\tau = \tau(x)$. From (29) and (34) we get (for $\kappa\delta \neq 0$)

$$\tau(x) = \cos^{-1}\{ \frac{\gamma_0 h_x x / \bar{a}_B - 1}{\kappa\delta} \} \quad ; \quad x_1 \leq x \leq x_2 \tag{43}$$

Note that both (29) and (43) are expressions for the curve in the x-t plane on which breaking occurs; in (29) the point of breaking is given with t as the independent variable while in (43) the time of breaking is given with x as the independent variable.

Important for the following derivations is that S_0 is $O(1)$ and S_1 is $O(\delta)$ and that the extent of region B is also $O(\delta)$. This means that when the contributions to (42) from S_1 are integrated over region B to get the surf beat elevation (cf. (30), (31)) only terms of $O(\delta^2)$ will occur. As before we only retain terms of $O(\delta)$ so that only terms of $O(1)$ need be kept when computing S_n^B. Away from region B only S_1 (and not S_0) contributes to the forcing of surf beat oscillations. Altogether this means that the influence of the overall short-wave forcing is of the same order of magnitude as the influence of the time-varying break point position. Physically we may explain the importance of the latter as a result of oscillations in the starting point, the set-up effecting the whole surf zone and not only the small region B. Had S_n^B only been a weighted mean of S_n^I and S_n^{II}, it would have been of no significance.

With the above arguments we get from (42) (to the sufficient accuracy)

$$S_n^B = \frac{2}{\pi} \{ \int_0^\tau S_0^I \cos(n\omega t) \, d(\omega t) + \int_\tau^\pi S_0^{II} \cos(n\omega t) \, d(\omega t) \} \tag{44}$$

from which

$$\frac{dS_n^B}{dx} = \frac{2}{\pi} \{ \int_0^\tau \frac{dS_0^I}{dx} \cos(n\omega t) \, d(\omega t) + \int_\tau^\pi \frac{dS_0^{II}}{dx} \cos(n\omega t) \, d(\omega t) \}$$
(45)

since the terms arising from the variable limits of the two integrals cancel, because S_0 is continuous at the break point. Further

$$\frac{dS_n^B}{dx} = \frac{2}{\pi} \{ \frac{dS_0^I}{dx} - \frac{dS_0^{II}}{dx} \} \sin(n\omega\tau)$$
(46)

Returning to the expression for the surf beat elevation (30) we can write a particular solution in region B as

$$\xi_n^B = \frac{\pi}{\rho g h_x} J_0(ny) \{ [Y_0(ny) \frac{dS_n^B}{dx} \frac{x}{x_1}] + \int_{x_1}^x Y_1(ny) \frac{dS_n^B}{dx} \frac{d(ny)}{dx} dx \}$$

$$- \frac{\pi}{\rho g h_x} Y_0(ny) \{ [J_0(ny) \frac{dS_n^B}{dx} \frac{x}{x_1}] - \int_{x_1}^x J_1(ny) \frac{dS_n^B}{dx} \frac{d(ny)}{dx} dx \}$$
(47)

where integration by parts has been used after rewriting the expression in terms of Bessel and Neumann functions. Now in region B all variables except τ can be taken at $x = \bar{x}_B$ (the position of the break point for $\delta \to 0$) only introducing errors of $O(\delta)$ which after integration over region B will be $O(\delta^2)$ which we accept. Being uninterested in the details of the solution in region B we shall only find the 'jump' in the surf beat elevation from $x = x_1$ to $x = x_2$. After some algebra we obtain

$$[\xi_1^B]_{x_1}^{x_2} = \frac{\kappa\delta}{\rho g h_x} (\frac{dS_0^{II}}{dx} - \frac{dS_0^I}{dx})_{x=\bar{x}_B}$$
(48)

and

$$[\xi_n^B]_{x_1}^{x_2} = 0 \qquad ; \qquad n = 2, 3, 4,\ldots$$
(49)

Further

$$[\frac{d\xi_1^B}{dx}]_{x_1}^{x_2} = - \frac{1}{\rho g \bar{h}_B} (\frac{dS_1^{II}}{dx} - \frac{dS_1^I}{dx})_{x=\bar{x}_B}$$
(50)

and

$$[\frac{d\xi_n^B}{dx}]_{x_1}^{x_2} = 0 \qquad ; \qquad n = 2, 3, 4, \ldots \qquad (51)$$

(Equation (50) was first obtained in terms of S_0 and then rewritten in terms of S_1 so it is not in contradiction to the above remarks about only S_0 being important in region B). Thus to the leading order ($0(\delta)$) there are no higher harmonics present. To this order of accuracy we can let region B vanish, keeping the jump conditions (48) and (50) which may then be written

$$[\xi_1]_{\bar{x}_B^-}^{\bar{x}_B^+} = \frac{\kappa\delta}{\rho g h_x} [\frac{dS_0}{dx}]_{\bar{x}_B^-}^{\bar{x}_B^+} \qquad (52)$$

and

$$[\frac{d\xi_1}{dx}]_{\bar{x}_B^-}^{\bar{x}_B^+} = - \frac{1}{\rho g \bar{h}_B} [\frac{dS_1}{dx}]_{\bar{x}_B^-}^{\bar{x}_B^+} \qquad (53)$$

both of which are $0(\delta)$, the latter implicitly through S_1. Note that (53) is in accordance with (33) which originated from continuity in mass and momentum flux.

To sum up, the surf beat elevation is found from (30) with (31) (or (38) in regions I and II) using (15), (24) and (27) for the radiation stress. The six complex constants (α^J, β^J), $J = I, II, III$ are determined from the boundary conditions (cf. 2.6) and the matching conditions (cf. 2.5) where (52) is used at $x = \bar{x}_B$ instead of (32).

In region III analytical integration of (30) is possible and the result is ($\xi_n = 0$, $n = 2, 3, 4, \ldots$)

$$\xi_1^{III} = - \frac{1}{\rho} \frac{S_1^{III}}{g h_0 - c_{g0}^2} + \beta_1^{III} \exp\{ - i \frac{\omega x}{\sqrt{(g h_0)}} \} \qquad (54)$$

where $\alpha^{III} = 0$ has been used. Here the first term is the bound long wave found by Longuet-Higgins and Stewart (1962, 1964) and the second term is the free, long wave propagating seawards.

Before showing some numerical results we shall demonstrate that the solution of Symonds et al. retaining only terms up to $0(\delta)$ is a special case of the present model. Neglecting the forcing away from the surf zone, using shallow water theory when calculating the radiation stress, and specifying $\kappa = 1$ corresponding to Symonds et al. we see that there is no time-dependent forcing anywhere but in region B and from (52) and (53) we get

$$[\frac{2\xi_1}{3\gamma_0^2 \bar{x}_B h_x}]_{\bar{x}_B^-}^{\bar{x}_B^+} = - \delta \qquad (55), \qquad [\frac{d\xi_1}{dx}]_{\bar{x}_B^-}^{\bar{x}_B^+} = 0 \qquad (56)$$

Here (55) gives the jump in the surf beat elevation over region B corresponding to the non-dimensional form of Symonds et al. Thus to the leading order in δ their solution is found solely by matching free long-wave solutions (a standing wave in the surf zone represented by a Bessel function and a seawards propagating wave outside the surf zone represented by a Hankel function) requiring a jump $-\delta$ over region B.

3. NUMERICAL RESULTS AND DISCUSSION

The following non-dimensional quantities are used in the numerical computations:

$$\frac{\xi_1}{a_\infty \delta} \; ; \quad h \, \frac{\omega_s{}^2}{g} \; ; \quad h_x \, \frac{\omega_s}{\omega} \; ; \quad \varepsilon_s = \frac{\omega_s{}^2}{g} \, a_\infty \; ; \quad \chi = \frac{\overline{h}_B \omega_s{}^2/g}{(h_x \omega_s/\omega)^2}$$

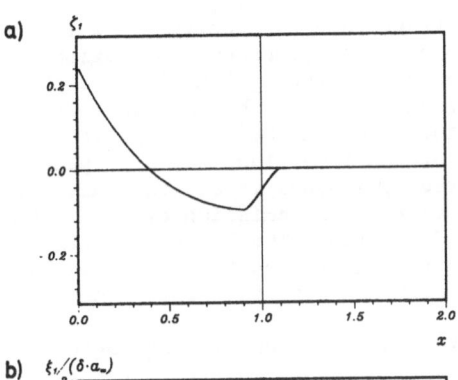

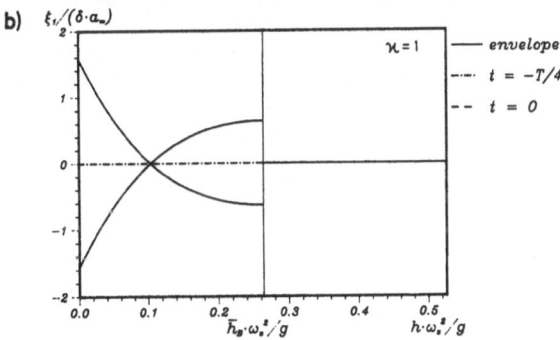

Fig. 3a Reproduction of the first harmonic of the surf beat elevation from Symonds et al. (1982), with their non-dimensional forms of the elevation and the off-shore coordinate. Here $\chi = 3.670$ and $\delta = 0.1$.

Fig. 3b Surf beat elevation and its envelope evaluated from the present model, modified corresponding to Symonds et al. Here $h_x \omega_s/\omega = 0.2677$ and $\varepsilon_s = 0.1$ resulting in $\overline{h}_B \omega_s{}^2/g = 0.263$ and $\chi = 3.670$.

Note that ω_s^2/g is the wave number and ϵ_s is the steepness of the un-modulated short waves in deep water. The parameter χ is introduced since it turns out to be important for the features of the solution (Symonds et al.). Sufficient input parameters are the depth in the shelf zone $h_0\omega_s^2/g$, and any two of the three $h_x\omega_s/\omega$, χ, and (ϵ_s or $\bar{h}_B\omega_s^2/g$).

Fig. 3a,b demonstrates the agreement between the model by Symonds et al. and the present one. Fig. 3a shows their solution for $\chi = 3.670$ and $\delta = 0.1$ in their non-dimensional form of off-shore coordinate and surf beat elevation (cf. (55) and on). To certify their results we have re-derived all their model expressions and Fig. 3a is actually a plot from our computations and not a direct transfer of their Fig. 9a. As predicted above (cf. (55)) the result fits well with a jump $-\delta = -0.1$ over the region of initial breaking (region B), except for a difference in sign, which exists because the figure shows the elevation at $t = T/2$ ($T = 2\pi/\omega$) and not $t = 0$. Fig. 3b shows the surf beat wave and its envelope obtained from the present model modified as to neglect the forcing outside the surf zone, to use shallow water theory when calculating the radiation stress, and to specify $\kappa = 1$ (all modifications corresponding to Symonds et al.). Except for region B the elevations in Fig. 3a and Fig. 3b are indistinguishable. The shallow water assumption makes the elevation in Fig. 3a,b approximately 20% too large.

A special feature of the solution shown in Fig. 3a,b is that for the given value of χ there are no free, long waves propagating seawards from the surf zone. As stated by Symonds et al. this has the following explanation: From region B there is an emission of free, long waves in both seaward and shoreward directions. The latter is reflected at the coastline and proceeds back through region B and radiates seaward in anti-phase (for $\chi = 3.670$) with the directly seaward emitted free, long wave.

When the forcing outside the surf zone is not neglected, the solution looks quite different, as shown in Fig. 4 for $\kappa = 0, 1, 2$. (Note that $\kappa = 0, 2$ are extremes which are probably unrealistic). Here and in Fig. 3b the input parameters were $h_x\omega_s/\omega = 0.2677$ and $\epsilon_s = 0.1$, resulting in $\bar{h}_B\omega_s^2/g = 0.263$ and $\chi = 3.670$. Also $h_0\omega_s^2/g = 1$ was used. Although this corresponds to intermediate water depth rather than deep water for the short waves the solution is almost independent of $h_0\omega_s^2/g$ as long as it is close to unity or greater.

Fig. 5 shows the variation of the surf beat amplitude at the coast-line versus κ. Also a pseudo reflection coefficient R, defined as the ratio between the amplitudes in region III (the shelf zone) of the seaward propagating free wave and the incident long, forced wave, is shown versus κ. The two curves are measures of the magnitude of the surf beat wave inside and outside the surf zone. One observes that inside the surf zone, the surf beat activity decreases with increasing κ, while outside the surf zone the activity has a minimum around $\kappa = 1.25$ (for the particular input parameters).

In Fig. 6 a somewhat more shallow beach is considered for $\kappa = 1$. Other input parameters are $h_x\omega_s/\omega = 0.1$ and as before $\epsilon_s = 0.1$ and $h_0\omega_s^2/g = 1$. This yields $\chi = 26.3$ and R = 5.04. Fig. 6a shows the surf beat elevation and its envelope while Fig. 6b gives the amplitude of the corresponding horizontal velocity relative to the bottom velocity of the

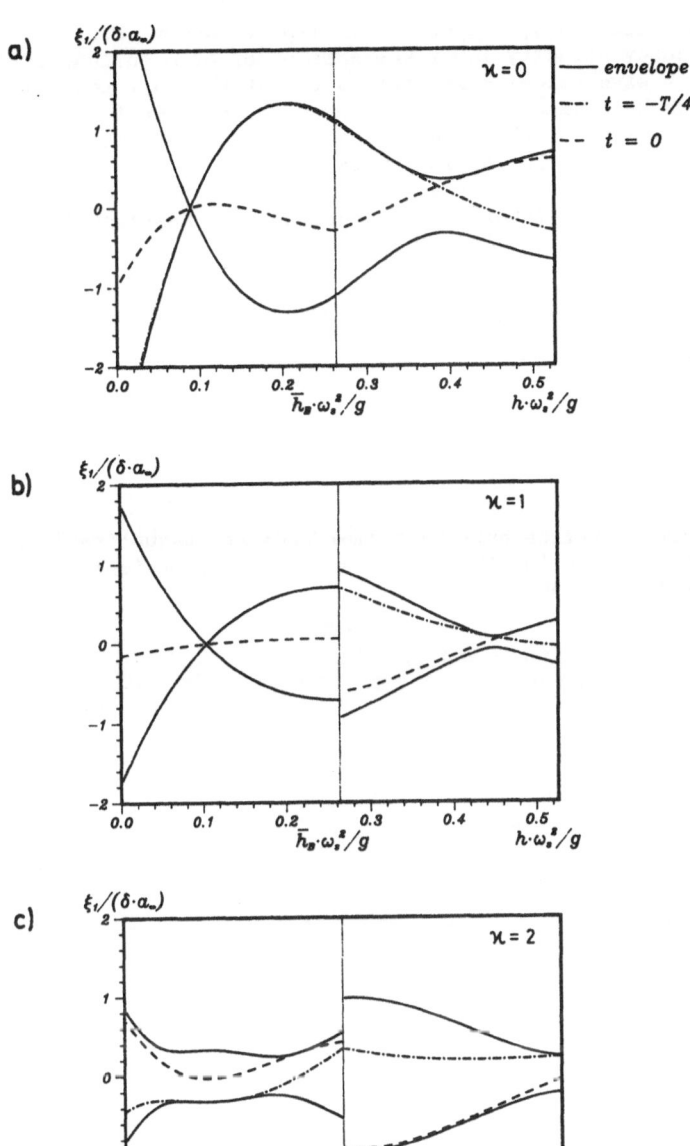

Fig. 4 Surf beat elevation and its envelope for the same topography and incident wave conditions as in Fig. 3b for a) $\kappa = 0$, b) $\kappa = 1$, and c) $\kappa = 2$.

un-modulated short waves, computed from linear theory. For e.g. $\delta = 0.1$ the low-frequency velocity is of the same order of magnitude or larger than the short-wave velocity in large parts of the surf zone. This phenomenon has been widely reported from field observations in recent years e.g. by Wright et al. (1982), and it is evident that it has great influence on the formation of longshore bars.

In a later publication it will be shown that the present model agrees quite well with laboratory measurements by Kostense (1984).

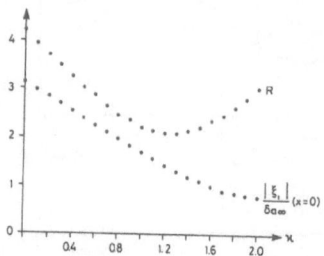

Fig. 5 Maximum shoreline surf beat amplitude and pseudo reflection coefficient R versus κ for the same topography and incident wave conditions as in Fig. 3b and Fig. 4.

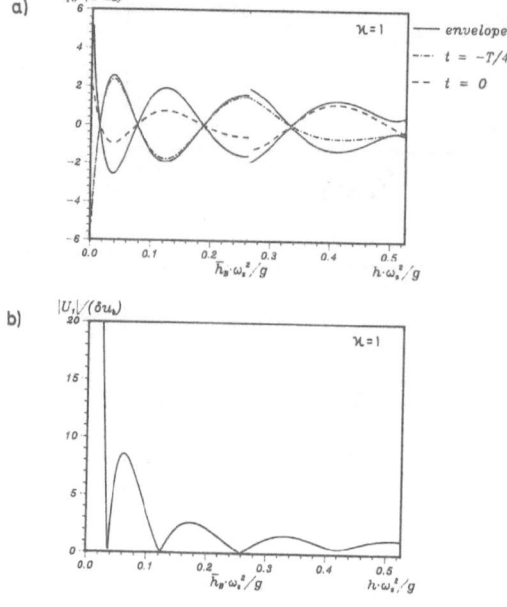

Fig. 6 Shallow beach conditions with $h_x \omega_S / \omega = 0.1$ and $\varepsilon_s = 0.1$ resulting in $\bar{h}_B \omega_S^2 / g = 0.263$ and $\chi = 26.3$. a) Surf beat wave and its envelope. b) Surf beat particle velocity relative to the local unmodulated short-wave bottom velocity found from linear theory.

REFERENCES

Abramowitz , M. and I. A. Stegun (1974). Handbook of Mathematical
    Functions. Dover, New York.

Bowers, E. C. (1977). Harbour resonance due to set-down beneath wave
    groups. J. Fluid Mech. 79, 71-93.

Carrier, G. F. and H. P. Greenspan (1958). Water waves of finite
    amplitude on a sloping beach. J. Fluid Mech., 4, 97-109.

Chu, V. H. and C. C. Mei (1970). On slowly varying Stokes waves. J.
    Fluid Mech., 41, 873-87.

Foda, M. A. and C. C. Mei (1981). Nonlinear excitation of long-trapped
    waves by a group of short swells. J. Fluid Mech., 111, 319-45.

Goda, Y. (1970). A synthesis on breaker indices. Trans. Jap. Soc. Civ.
    Engrs. 2, Part 2.

Guza, R. T. and E. B. Thornton (1982). Swash oscillations on a natural
    beach. J. Geophys. Res., 87, C1, 483-491.

Guza, R. T. and E. B. Thornton (1985). Observations of surf beat. J.
    Geophys. Res., 90, C2, 3161-3171.

Kostense, J. K. (1984). Measurements of surf beat and set-down beneath
    wave groups. Proc. 19th Int.Conf.Coast.Eng.,Houston,Texas, 724-40.

Liu, Philip L.-F. (1988). A note on long waves induced by short-wave
    groups over a shelf. Submitted to J. Fluid Mech. May 1988.

Longuet-Higgins, M. S., and R. W. Stewart (1962). Radiation stress and
    mass transport in gravity waves with application to 'surf beats'.
    J. Fluid Mech., 13, 481-504.

Longuet-Higgins, M. S., and R. W. Stewart (1964). Radiation stresses in
    water waves: A physical discussion with applications. Deep Sea
    Research, 11, 529-562.

Mei, C. C. (1983). The Applied Dynamics of Ocean Surface Waves. Wiley.

Mei, C. C. and C. Benmoussa (1984). Long waves induced by short-wave
    groups over an uneven bottom. J. Fluid Mech. 139, 219-35.

Molin, B. (1982). On the generation of long-period second-order
    free-waves due to changes in the bottom profile. Ship. Res. Inst.
    Papers 68, Tokyo, Japan.

Munk, W. H. (1949). Surf beats. Trans. Am. Geophys. Union 30, 849-854.

Ottesen Hansen, N.-E. et al. (1980). Correct reproduction of group-
    induced long waves. Proc. 17th Int.Conf.Coast.Eng.,Sydney, 784-800.

Ottesen Hansen, N.-E. (1978). Long period waves in natural wave trains.
    Prog. Rep. 46, Inst. Hydrodyn. Hydr. Eng.,Tech.Univ.Denmark,13-24.

Phillips, O. M. (1977). The Dynamics of the Upper Ocean. 2nd Ed.
    Cambridge Univ. Press.

Sand, S. E. (1982). Long wave problems in laboratory models. J.
    Waterway, Port, Coastal and Ocean Div., Proc. ASCE, WW 4, 492-503.

Schäffer, H. A. and I. A. Svendsen (1988). Surf beat generation on a
    mild slope beach. Proc. 21st Int. conf. Coastal Eng.,Malaga, Spain.

Symonds, G., G. A. Huntley and A. J. Bowen (1982). Two dimensional surf
    beat: Long wave generation by a time-varying break point. J.
    Geophys. Res., 87, C1, 492-498.

Tucker, M. J. (1950). Surf beats: Sea waves of 1 to 5 min. period. Proc.
    Roy. Soc. London. A 202, 565-573.

Wright L. D., R. T. Guza and A. D. Short (1982). Surf zone dynamics on a
    high energy dissipative beach. Mar. Geol., 45, 41-62.

COMPUTATION OF NONLINEAR WAVE KINEMATICS DURING PROPAGATION AND RUNUP ON A SLOPE

S. GRILLI and I.A. SVENDSEN
Ocean Engineering Program, Department of Civil Engineering
University of Delaware
Newark-DE 19716, USA

ABSTRACT. An efficient Boundary Element Method (BEM) solving fully nonlinear waterwave problems in the physical space has been developed in our earlier papers. Its detailed numerical features have been presented elsewhere. In the present paper, this method is applied to the computation of the interaction between highly nonlinear solitary waves and plane steep and gentle slopes. Kinematics of the waves is calculated during propagation and runup on a slope. In particular, the internal velocity field above the slope and the pressure force on the slope are computed in detail during runup and rundown of the wave. The results show features of the wave flow such as jet-like up-rush, stagnation point and breaking during backwash. A comparison is made with accurate experimental results (runup, surface elevations), with orther fully nonlinear solutions, and also with the predictions of the Shallow Water Wave Equation. The results show that the BEM used here is capable of accurately describing the wave flow and interaction with a plane slope($2.88°$ to $90°$), in great detail.

1 Introduction

1.1 Numerical modelling of nonlinear water waves

Over the last decade numerical solution of the exact nonlinear equations for water waves, using a Boundary Integral Equation (BIE) description, has become an increasingly successful method. In particular, one of the most impressive achievements of this method was the modelling of overturning waves (plunging breaker). Practically all previous contributions, however, utilized space periodicity of the waves and most of them also were based on a conformal mapping of physical space onto a plane in which the equivalent of the free surface was the only part left of the boundary (the BIE was then solved in the transformed plane). A few contributions used some other complex variable-dependent techniques. There are many advantages of this method which has been explored extensively by many authors. Particularly noteworthy for significant steps to the development are contributions by Longuet-Higgins & Cokelet 1976 [23], Vinje & Brevig 1981 [37] and Dold & Peregrine 1984 [7]. The procedure, however, also has some important limitations which are associated with the use of complex variables and with the assumption of waves that are periodic in space.

In our previous papers (Grilli, Skourup & Svendsen 1988, 1989 [14,15], Grilli & Svendsen 1989a,b [16,17]) we presented an equally efficient and accurate method, without utilizing

A. Tørum and O. T. Gudmestad (eds.), Water Wave Kinematics, 387–412.

those two crucial assumptions and we demonstrated its application to some problems for which the above-mentioned version is not well suited. In this two-dimensional model (2D), we considered the wave motion in the physical space and described it using a free space Green's function to transform the Laplace differential equation into an integral equation that involved only values of the velocity potential and its normal derivative along the physical boundary. Our procedure is a Boundary Element Method (BEM) developed for higher order elements. The time integration is an the explicit scheme correct to second order in Δt using the shape functions of the BEM.

1.2 Interaction of nonlinear waves with maritime structures

An accurate theoretical prediction of the interaction of real waves with maritime structures is essential for the study of coastal protection. These structures have very often their main dimension in the long-shore direction and can be analyzed quite accurately by 2D models in their cross-section (Figure 1a) which can reasonably be represented by a steep plane slope, or a combination of plane slopes and berms. The sea bottom in the shallow water area in front of such a structure can itself be approximated by a gentle plane slope on which the very high waves (e.g. greater than 0.7 times the water depth) will break without any significant reflection, before reaching the structure (Figure 1a, wave (a)).

The numerical modelling of the interaction of waves with a maritime structure aims at determining the maximum and minimum vertical elevations of the waves during their reflection on the structure (runup and rundown), the reflected wave characteristics (phase shift, reflection coefficient), and the pressure distribution on and wave particle velocity close to the slope (essential for the estimation of local and global stability of the structure and of its elements); finally the prediction of the breaking of waves during runup and backwash provides important data for determining local reinforcements of the structure. The numerical modelling of the interaction between very high waves and gentle slopes in front of maritime structures will concentrate on the shoaling and breaking processes. Notice, however, that a gentle slope can also be the approximation of a beach, which makes it interesting to also studying with the model non-breaking runups of smaller waves on gentle slopes.

Although the model can generate and propagate fully nonlinear waves in a two-dimensional physical space, it cannot yet simulate the interaction with a structure of long time series of periodic waves, like they occur in nature. Waves are generated in the model by simulating a wavemaker at one end of a numerical wave tank and are propagated toward a structure modelled at the other end of the tank. After interacting with the structure, partially reflected waves propagate back in the tank, toward the wavemaker, where they reflect again, and so on. From that time onward, these waves interact with other waves generated in the tank, which soon leads to the development in the model of partially standing waves. This reflection from the wavemaker, also essentially occurs in a physical tank and prevents us from running long term numerical experiments with periodic waves.

Moreover, though this kind of model is capable of modelling overturning waves, the computations break down when the tip of the *first* breaking wave hits the free surface. Thus, even if we were able to model long time series of waves, breaking would constitute an important limitation for using such a model for design purposes.

A solitary wave, however, does not pose the same problem. The reflection of a solitary wave on a slope is the reflection of *one* well defined wave only whose detailed study can be made before the wave propagates back to the wavemaker and interacts with it. Solitary wave propagation and interaction with structures can thus well be simulated with the present version of our model. Furthermore solitary waves ressemble the waves generated in a *Tsunami* and, as pointed out by Fenton & Rienecker 1982 [9], they are also the fastest and the most massive waves, and have the greatest impulse. Their pressure forces on maritime structures can thus be used as design criteria.

Hence, in the present paper, we apply the model to the computation of the interaction between solitary waves and plane slopes (both steep and gentle). Generation, propagation, runup, breaking and reflection of the waves are studied in detail and show that our BEM is capable of quite accurately describing the flow in great detail and over extensive periods of time. After briefly describing the model features in Section 2, the generation of a solitary wave by a "numerical wavemaker" is presented in Section 3. The interaction of solitary waves with steep slopes is presented in Section 4 and with gentle slopes in Section 5.

2 General features of the model

In the present form of the two-dimensional model, nonlinear wave problems are solved in the physical space $\Omega(t)$ (Figure 1b). The solution of the Laplace problems (continuity equation for a 2D irrotational inviscid fluid) is based on a non-complex method : a high order BEM, using the free space Green's function $G(\mathbf{x}, \mathbf{x}_l)$ which transforms the Laplace differential equation into an integral equation that involves only values of the velocity potential and its normal derivative along the physical boundary $\Gamma(t)$,

$$\alpha(\mathbf{x}_l)\phi(\mathbf{x}_l) = \int_\Gamma [\frac{\partial\phi}{\partial n}(\mathbf{x})G(\mathbf{x}, \mathbf{x}_l) - \phi(\mathbf{x})\frac{\partial G(\mathbf{x}, \mathbf{x}_l)}{\partial n}] d\Gamma \tag{1}$$

where $\mathbf{x} = (x, z)$ and $\mathbf{x}_l = (x_l, z_l)$ are position vectors for points on the boundary, and $\alpha(\mathbf{x}_l)$ is a geometric coefficient.

In our BEM, collocation nodes are distributed along the entire boundary to describe the variation of boundary geometry as well as boundary conditions and the unknown functions of the problem. Between the collocation points, the variation of all quantities is described by means of shape functions (linear to quartic) or quasi-splines (geometry modelled by cubic splines and field variables by linear shape functions). For this purpose, the boundary is divided into elements each of which contains two or more nodes (in most of the computations reported hereafter, quadratic and quasi-spline elements have been used). Mapping onto isoparametric reference elements is used and the Jacobian of the mapping function is determined analytically. The regular integrals are computed by Gaussian quadrature, and a kernel transformation is applied to the singular integrals which are then computed by a Gauss-like quadrature weighted with a logarithmic singularity.

The nonlinear kinematic and dynamic boundary conditions are imposed on the free surface $\Gamma_f(t)$. Solitary waves are generated by simulating a piston-type wavemaker motion on the boundary $\Gamma_{r1}(t)$ and by imposing the continuity of its normal velocity. Along the stationary bottom Γ_b and the slope Γ_{r2}, a zero normal velocity is imposed.

The time integration of the free surface conditions is the explicit scheme developed by Grilli, *et al.* 1989 [15]. It is based on a truncated Taylor expansion (Dold & Peregrine 1984 [7]) which has been modified to using the shape functions or the quasi-splines of the BEM. A version correct to second order in Δt ($n = 2$) has been developed for the computations presented here. The updating of the free surface position $\mathbf{r}$ from time t to $t + \Delta t$ reads,

$$\mathbf{r}(t + \Delta t) = \mathbf{r}(t) + \sum_{k=1}^{n} \frac{(\Delta t)^k}{k!} \frac{D^k \mathbf{r}(t)}{Dt^k} + O[(\Delta t)^{n+1}] \qquad \text{on } \Gamma_f(t) \tag{2}$$

and the analogous for ϕ.

To get the coefficients in the expansions, we solve a succesion of Laplace problems (two in our case) for the velocity potential and its time derivatives, each problem solution providing the nonlinear free surface boundary conditions for the next one. The expansion coefficients are expressed in terms of $\phi, \frac{\partial \phi}{\partial n}, \frac{\partial \phi}{\partial t}$ and $\frac{\partial^2 \phi}{\partial t \partial n}$ and their derivatives along the free surface. We adopt a coordinate system at the free surface defined by (s, n), the tangential and normal unit vectors to the free surface. This formulation provides simpler expressions for the high order derivatives than a cartesian system. For the evaluation of the derivatives along the free surface we use a sliding 4th order polynomial element independant of the BEM discretization. It provides a local continuity on the boundary of at least the 2nd derivatives.

The overall stability of the method has permitted us so far to avoid smoothing procedures. Notice that all the present features of this 2-D model can in principle be extended to 3-D problems.

A detailed account of the 2D-BEM applied to our water wave problem is given in Grilli, Skourup & Svendsen 1989 [15].

3 Wave generation by a wavemaker

3.1 Generation of a first order solitary wave

The model described above (Section 2) can propagate waves in space and time. Since the computations are made in physical space, however , and only cover a limited region of an infinitely long wave tank, the waves to be studied have to be generated at the outer edge of the computational area. Waves will be generated in this study by simulating a wavemaker movement. Other methods exist, however, for generating waves without moving any body (see e.g. Grilli, *et al.* 1988, 1989 [14,15], Grilli & Svendsen 1989 [16]).

The generation of waves by a numerical wavemaker requires the movement of one of the tank boundaries on a way it creates the desired wave. In many experimental studies (Goring 1978 [11], Losada, *et al.* 1986 [24]) and numerical computations (Kim, *et al.* 1983 [20]), the waves were generated by a piston motion wavemaker (physical or numerical) which was moved to create a *first-order* solitary wave according to Goring's method. This type of generation has been reproduced here as closely as possible (see Appendix A for detail).

Within the frame of Boussinesq approximation, the inverse scattering theory (Gardner, *et al.* 1967 [10], Hammack & Segur 1974 [19]) will predict that, unless it corresponds to an exact solitary wave, a distributed elevation of the free surface will disintegrate into one or more solitons and a tail of disturbances. Thus, a *first-order* solitary wave of appreciable height generated in the model may be expected to exhibit disturbances and modifications of its profile while it propagates down the wave tank. In fact Goring 1978 [11] found that the limit for an accurate solitary wave reproduction by his method was $\frac{H}{d} = 0.2$. In accordance with this, it is found that the higher the waves, the more pronounced is the amplitude of the tail of disturbances they are shedding (see Section 4.2).

3.2 Free surface intersection with moving bodies

If the waves are generated by simulating a wavemaker or a body movement, there will be a corner on the boundary curve where body surface and free surface meet and the same situation appears when dealing with the intersection with other surface piercing structures.

The flow near such intersections between a free surface and a solid body has given rise to substantial concern in the literature since Kravtchenko 1954 [21] showed that for linear waves, the intersection with a moving wall would generally create an incompatibility between the flow requirements of the boundary conditions along the free surface and the wall. Kravtchenko's derivations indicated that a logarithmic singularity would result at the wall. This was in fact further substantiated by Peregrine 1972 [29] for the case of an impulsively accelerated wall, and his findings seemed to be confirmed by the experiments of Greenhow & Lin 1986 [13].

In a more detailed investigation of the linear case, Roberts 1987 [30] found that if the acceleration remains bounded, the amplitude at the wall will also be bounded (although not particularly smoothly behaved) and that over a small but finite time gravity cannot be neglected, as Peregrine does, even with an impulsive start. Roberts found that this does, under certain constraints, limit the free surface elevation to finite values at the moving wall. He also found that away from the wall the free surface variation *is* logarithmic to leading order, as verified by Greenhow & Lin's experiments, even when the amplitude at the wall remains finite (even the dramatic acceleration they produced by a sledge hammer is still not mathematically an impulsive start). This problem has also been studied by Cointe, et al. 1988 [6].

In conclusion the existence at the corner of the logarithmic singularity that has been haunting many investigators using the BIEM or BEM, has still to be proved theoretically for the fully nonlinear case. In our model it does not seem to be a problem provided sufficient care is exercised during the first few time steps starting the computations (bounded acceleration).

It should be noticed that even if the actual mathematical solution of the flow equations is proved to be singular close to the corner, the approximate numerical solution of the problem can only provide results according to the functions used in the numerical model for representing this solution. In the case of the BEM, for instance, if the potential is modelled on the free surface by a piecewise polynomial approximation of unknown magnitude (shape functions within each boundary element), the numerical solution close to the corner will never become anything but a polynomial variation. In particular, a logarithmic variation will never be approximately represented in the results (as observed in our computations). If the corner singularity of the solution can be identified mathematically, however, one can substitute the appropriate singular function (whatever it is) in the numerical model close to the corner. This technique, sometimes named "singular shape function" has been very succesful in such fields as fracture mechanics. To apply here, it would require to determine the leading order of the solution in the corner for the fully nonlinear case.

Care must also be taken to ensure the well-posedness of the flow equations and boundary conditions to be solved by the numerical method. Numerical singularity can occur in the BEM, for instance, if one does not take into account that both the potential and its normal gradient are known on the body part of the corner. Therefore we use double nodes at the corners (providing two different normal directions and unknowns on the two sides of each corner) and we include the relations between the ϕ's and the $\frac{\partial \phi}{\partial n}$'s at the corner points, in the BEM system, as compatibility conditions (for detail see Grilli, $et~al.$ 1988, 1989 [14,15]). Lin, Newman & Yue 1984 [22] handled the corner problem by specifying both the potential and the stream function at the body part of the corner in their Cauchy integral theorem formulation. Doing so, they could generate nonlinear waves by a wavemaker. Using a similar technique, Greenhow & Lin 1985 [12] and Yim 1985 [38] simulated nonlinear wedge entries in water. Dommermuth & Yue 1987 [8], following the same principle as Lin, et al., imposed both the potential and its normal derivative at the body part of the corner in their BEM formulation. Since these methods seem to work, it is assumed that they treat the corner problem in a way similar to the method we use in our model, although the authors do not give details.

4 Interaction of a solitary wave with a steep slope

4.1 Background

Runup and reflection from plane slopes of non-breaking solitary waves have been studied quite extensively in the literature. In fact, the solitary wave runup problem using the exact equations was studied numerically already by Camfield & Street 1967 [3] by means of a marker and cell method. More commonly, however, the runup of solitary waves has been analyzed using long wave theory or Boussinesq theory. The special case of reflection from a vertical wall is equivalent to the problem of collision between two opposing solitary waves of the same height and several studies have been published utilizing that analogy. A first order solution was obtained by the Inverse Scattering Technique developed in the 1960's and 70's (Gardner,$et~al.$ 1967 [10]), and a third order analysis was developed by Su & Gardner 1969 [33] and Su & Mirie 1980 [32] and solved numerically by Mirie & Su 1982 [25]. Higher order analysis were also done recently by Byatt-Smith 1988 [2]. Fenton & Rienecker 1982 [9] used a Fourier Method to obtain very accurate results for the same problem.

The runup on a slope was analyzed numerically on the basis of the Boussinesq equations by Pedersen & Gjevik 1983 [28], by a low order BEM by Kim, $et~al.$ 1983 [20] and Synolakis 1987 [35], using the SW equation, generalized the approach of Carrier 1966 [4] to apply to the situation also considered in this paper with a constant depth region in front of the slope.

Notice that the method we have developed here makes it possible to model any kind of bottom or slope geometry, like obstacles of arbitrary shape, shelf or combination of slopes and berms as long as flow separation is not important.

4.2 Generation and propagation

Since our numerical computations solve the full nonlinear equations, the computed waves should be expected closely to follow what actually happens in a wave flume after the generation of the wave motion. Figure 2 shows profiles of waves with seven different amplitudes $\frac{H}{d}$, generated by a piston wavemaker (according to Goring's first order solution) in a tank of depth $d = 0.3$m, at the (different) time when the crest of each of them passes a point 6 m in front of the wave generator. In accordance with the comments above (Section 3), it is found that the higher the waves, the more pronounced is the amplitude A (Figure 1b) of the tail of disturbances they are shedding. A closer inspection also shows that, although the highest parts of the crests are closely symmetric, there are non-negligeable differences between front and rear for the lowest 25-35 % of each profile.

As the waves propagate further away from the wave generator, the asymmetry decreases and the part of the tail closest to the wave turns into a long, shallow trough. Figure 3 shows this in a comparison between two waves of the same height ($\frac{H}{d} = 0.257$), one having propagated only 6m (20 d) since generated, the other after 26m (86.7 d). The slow transformation of the wave profile as it propagates and adjusts itself to a stable form, also results in a moderate decrease in wave height.

4.3 Reflection from a vertical wall

Before considering the reflection from a slope, we analyze the more simple reflection from a vertical wall. In Figure 4, the maximum elevation ("runup", R_u) at the wall computed for different wave amplitudes is compared to the results of Fenton & Rienecker 1982 [9] (F&R) obtained for that situation using a Fourier Method, and to the third order results by Su & Mirie 1980 [32] (S&M). The general impression is that for amplitudes above 0.3, the results of the BEM fall a little below the other two methods. Part of this can be because the incoming wave is not quite a solitary wave, part could be computational inaccuracies (different number of nodes N_Γ and time steps Δt were used in the computations : N_Γ= 136 or 158, Δt= 0.0075 to 0.020s).

If we analyze the mass and energy balance of the system for a wave of $\frac{H}{d}$=0.456, we find that for most of the propagation, the total energy E_T remains constant to within 0.01 %. In the brief interval of the rapid surface movements at the wall, however, the total energy increases by about 0.06 %. A detailed analysis of the volume error as a function of t shows that, in this case, each change in total energy is matched by a similar change in volume. This suggests that the reason for the energy gain is that during the short interval of the process where the surface is moving really fast up and down along the wall, the (fixed) time step used in the computations is too large, so that the surface movement is not predicted with quite the same accuracy during that part of the process.

Figure 5 shows the kinetic energy E_K during the BEM computations. One sees the kinetic energy has its minimum a fraction of a second before runup and never becomes quite zero. In spite of the increase in total energy during reflection, refered to above, one finds there is a slight decrease in kinetic energy (of 0.288 % in this case). These phenomena are further substantiated by the analysis of the velocity field made in Section 4.5.

F&R also compute total maximum forces and moments of forces acting on the wall, In view of the close agreement described above, it is not surprising that our BEM computations

$\frac{H}{d}$	N_Γ	$N_{\Gamma f}$	$\frac{\Delta x}{d}$	$\Delta t(s)$	CPU (s)
0.269	136	81	0.42	0.015	0.745
0.269	378	241	0.42	0.015	5.041
0.457	158	101	0.33	0.010	0.925

Table 1: Discretization and computation time for the BEM computations. $N_{\Gamma f}$ is the number of nodes on the free surface and $\frac{\Delta x}{d}$ represents the distance between 2 nodes on the free surface

β	$\frac{H_i}{d}$	$\frac{H_r}{d}$	K_r	$\frac{R_u}{d}$	$\frac{R_d}{d}$
90°	0.255	0.245	0.963	0.545	0.012
90°	0.456	0.416	0.913	1.071	0.056
70°	0.259	0.243	0.937	0.580	0.014
70°	0.437	0.378	0.866	1.062	0.096
45°	0.269	0.222	0.825	0.674	0.137
45°	0.457	0.358	0.785	1.257	0.386

Table 2: BEM computations : Solitary wave on a slope β. $\frac{H_i}{d}$ corresponds to the incident wave at $x = 6$m and $\frac{H_r}{d}$ to the reflected wave at the same location. $\frac{R_u}{d}$ and $\frac{R_d}{d}$ are the runup and rundown of the wave. $K_r = \frac{H_r}{H_i}$ is the reflection coefficient.

give results that coincide to a fraction of a percent for these quantities (Figure 6). A somewhat surprising result is a double maximum found in the time variation of the total force for high waves (Figure 7). A wave of $\frac{H}{d} = \varepsilon = 0.4$ already shows sign of an extended maximum and for $\varepsilon = 0.5$ and 0.6, there is distinctively two extremes on the force curve 0.2-0.3s apart. In the case of $\varepsilon = 0.6$, the computations broke down at the time of the second maximum in a situation with essentially zero pressure and a free falling water volume near the crest. Such a double maximum of the wave force in a fairly short time constitutes a very strong loading for maritime structures. It may be noticed that experiments of the runup of steep periodic waves on a vertical wall, made by Nagai 1969 [26], showed somewhat similar features.

4.4 Reflection from a steep slope

Numerical experiments have been made for the cases of slope angles $\beta = 45°$, 70° and 90° to compare with the laboratory experiments made by Losada, et al. 1986 [24] (LVN). Our numerical wave tank was 10m long and 0.3m deep. Key figures for the computations presented in the following are given in Table 1 (discretization and computation times on an IBM 3090/200) and in Table 2 (wave data, reflected wave, runup and rundown). Due to the vectorization of the computations, the CPU-time per time step was found to increase less than proportional to the square of the number of nodes as indicated by the figures in

β	$\frac{H_i}{d}$	$\frac{H_f}{d}$	K_r	$\frac{R_u}{d}$	$\frac{R_d}{d}$
90°	0.255	0.224	0.882	0.521	0.003
90°	0.456	0.419	0.918	1.107	0.059
70°	0.259	0.232	0.893	0.558	0.001
70°	0.437	0.423	0.968	1.071	0.113
45°	0.269	0.214	0.796	0.672	0.129
45°	0.457	0.347	0.759	1.294	0.267

Table 3: Experimental results from Losada *et al.* 1986 : Solitary wave on a slope β (see Table 2).

Table 1 (378 nodes) for the computations with the long wave tank (30m) mentioned above. The experimental results are given in Table 3.

As mentioned in Section 3, in the computations, we generate the wave at $x = 0$ (i.e., 33 water depths from the slope) by Goring's procedure. The amplitude of the wavemaker is adjusted here, so that when the wave passes $x = 6$m (20 d), it has a height of $\frac{H}{d}=0.255$, 0.259, 0.269 or 0.456, 0.437, 0.457 which corresponds to those used by LVN in their experiments. The comparison between the numerical results of Table 2 and the experimental results of Table 3 shows a close agreement between the values of the maximum runups $\frac{R_u}{d}$ of the waves and a reasonable agreement between the values of the maximum rundowns $\frac{R_d}{d}$ and reflection coefficients $K_r = \frac{H_f}{H_i}$.

Figures 8 and 9 show some free surface elevations computed for the case of $\beta = 45°$ and $\frac{H}{d} = 0.457$. Figure 8 shows the incoming solitary wave propagating rightward, up to maximum runup on the slope. Figure 9 shows phases of the rush-down on the slope (the wave propagates leftward), with the development of a backward breaking, at a small distance away from the slope ($x \simeq 9.8$m). The computations broke down after the last curve shown in Figure 9. Notice that very few nodes in the computations are involved in the description of this breaking phenomenon. Clearly we reach the limits of our 2D-model in its present form, and this needs to be investigated further. Notice, however, that a somewhat similar breaking pattern was observed by LVN in the experiment corresponding to that particular wave.

Figures 10 and 11a-b present some detailed comparisons between our computed free surface elevations and those measured by LVN. In all cases, the computed and measured results have been synchronized at the time of maximum runup which, in the experiments reported, is the only well-defined time available. That means that actual comparison of the temporal development is generally limited to the time after the maximum runup. The form of the wave, however, at that time follows entirely from the generation and propagation which preceed the runup as described in the previous Section. And since the agreement with the experimental results is so good after that time, it is anticipated that it would be so also for the preceeding part of the process.

Figure 10 shows how a wave of $\frac{H}{d}=0.269$ reflects from a 45° slope. The computations are compared with the measured surface profiles at three different times, of which the first is the instant of maximum runup. Since the computed time of the lowest position of the water surface at the slope does not quite coincide with the time of the measured minimum, we have shown both. The symbols mark data points that have been developed from the original experimental records with the kind assistance of Dr. Losada, and the dotted curve is a spline fit to these points. In general, the agreement is considered good. Figure 11a shows the results for a wave of $\frac{H}{d}=0.437$, on a 70° slope. Here an extra profile (a) is available before the wave reaches the slope. As in the first case, the agreement between physical and numerical experiments is surprisingly good. Notice that even the irregularities left after the main crest has cleared the slope are quite well represented. Figure 11b further confirms this, by showing detail of the "wave tail" at an even later time. The agreement is still good, even for such small scale oscillations. This attests to the care exercised in the experiments and to the accuracy of the computational technique. The very small oscillations of the order of 1mm left behind the wave for $x \geq 8.5$m are mainly *numerical noise*. One obviously reachs here the limits of the resolution one can expect from the discretization used in this problem.

4.5 The internal velocity field during runup and rundown

The examples in the previous Section show that it is possible to reproduce surface variations recorded in experiments down to very small details and over quite extended periods of time relative to the time scale of the important events. This can only be possible if the computational solution actually represents the whole flow pattern to a high degree of accuracy. Thus it is possible to use the computational solution to analyze other properties of the flow (like the internal velocity) than those actually measured in the experiments and to have a reasonable confidence in the correctness of such predictions.

The numerical procedure that computes the development of the wave motion in space and time in our model, leaves us with both the velocity potential ϕ and $\frac{\partial \phi}{\partial n}$, known at all time steps (see Grilli 1988, 1989 *et al.* [14,15]). Therefore, the computation of the velocity at an internal point is a theoretically straight-forward evaluation of an integral. At an internal point $\mathbf{x}_o = (x_o, z_o)$, the velocity potential is given by (1) with $\alpha(\mathbf{x}_o)=1$. Differenciation of this expression with respect to x_o and z_o yields the velocity components (u, v),

$$u(\mathbf{x}_o) = \int_\Gamma [\frac{\partial \phi}{\partial n}(\mathbf{x})\frac{\partial G(\mathbf{x}, \mathbf{x}_o)}{\partial x_o} - \phi(\mathbf{x})\frac{\partial^2 G(\mathbf{x}, \mathbf{x}_o)}{\partial n \partial x_o}] d\Gamma \tag{3}$$

$$w(\mathbf{x}_o) = \int_\Gamma [\frac{\partial \phi}{\partial n}(\mathbf{x})\frac{\partial G(\mathbf{x}, \mathbf{x}_o)}{\partial z_o} - \phi(\mathbf{x})\frac{\partial^2 G(\mathbf{x}, \mathbf{x}_o)}{\partial n \partial z_o}] d\Gamma \tag{4}$$

These integrations can be performed by Gaussian quadrature, since they are regular as long as the observation point $\mathbf{x}_o$ stays inside the domain.

We have computed the internal velocity field at some points in the neighborhood and above a 45° slope, during runup and rundown of a solitary wave of $\frac{H}{d} = 0.457$. To measure velocities in that detail in the physical experiments would be a tremendous task, whereas it only requires a limited extra effort computationally. This illustrates the usefulness of combining experiments and numerical computations.

Figure 12a-d shows the velocity field at four different stages . When the first part of the wave reaches the slope, the flow field looks almost like a rotation around a point well above the incoming wave crest. Then, as the wave crest approaches the slope, the upward movement almost looks like a jet rushing up along the slope (Fig. 12a). In Fig. 12b, there is still a jet-like up-rush of the tip of the wave, but a stagnation point has developed near the toe of the slope as an indication of the reflection taking effect. Later on, this stagnation point moves up along the slope. Fig. 12c shows a remarkably interesting feature in the combination of high velocities and large accelerations around the time of the lowest position. The steep shoreward slope of the free surface in this non-distorted figure suggests horizontal accelerations close to 1 g. In a porous slope, there would be a strong outward directed pressure gradient which would result in a high risk of units being pulled out of the slope. The damage to rubble mound structures observed during experiments actually often occurs at this stage of the process. Finally, Figure 12d shows the beginning of the bakward breaking already mentioned before.

4.6 Pressure on the slope. Comparison with the SWE

One of the questions that can be asked when looking at these results for highly nonlinear wave runup is: How well will the nonlinear shallow water (NSW) equations be able to describe this flow. Those equations are the basis for Carrier & Greenspan 1958 [6] solution for periodic waves on a steep slope and were used for a solitary wave by Synolakis 1987 [35]. It is of interest to know to which extent this simpler formulation describes the actual flow situation as computed here.

It is well known that in the NSW equations representation, the pressure is assumed hydrostatic and, on a horizontal bottom, it follows implicitly that horizontal velocities are uniform over the entire depth. Even with this hypothesis, the NSW equations can be shown to apply on a steep slope. Therefore, the extent to which the solution to the exact equations computed here deviates from those ideal assumptions may indicate the degree of accuracy one can expect from the much simpler NSW approximation.

It is rather evident from looking at the flow in Figure 12 that on a steep slope the large horizontal velocities are closely associated with vertical velocities of the same order of magnitude. Therefore, vertical accelerations cannot be neglected. That implies nonhydrostatic pressure which will feed back and create horizontal velocities that are not uniform over depth. The pressure variation along the slope has been computed and divided by the hydrostatic pressure corresponding to the instantaneous water level vertically above the considered point at the slope. Results show this pressure ratio deviates quite appreciably from $+1$ (corresponding to an hydrostatic pressure as assumed in the NSW equations). In the runup phase the ratio stays within 0.7 and 1.1. During the rush-down, however, it varies between 0.6 and 1.9. The horizontal velocity profiles along verticals can be inferred from Figure 12. This leads to diagrams such as in Figure 13 where it appears that the variation is almost invariably close to being linear with a nearly constant absolute shear.

The comparison with the NSW on a steep slope is further studied in Svendsen & Grilli 1989 [33] and in the next Section for the case of a gentle slope.

5 Interaction of a solitary wave with a gentle slope

5.1 Computation of the non-breaking runup of solitary waves

The problem of the interaction of solitary waves with a gentle slope has been studied in detail by Synolakis 1987 [35] (SY). By solving the Shallow Water equations, he developped asymptotic results for the maximum runup. Both linear and nonlinear theories led to the same *runup law* $\frac{\mathcal{R}}{d} = 2.831(\cot\beta)^{\frac{1}{2}}(\frac{H}{d})^{\frac{5}{4}}$, with $\mathcal{R}$ the maximum runup, and β the slope (this expression is only valid for $\frac{H}{d} \gg \frac{H_l}{d} = (0.288\tan\beta)^2$ where H_l refers to a lower bound for the wave). SY compared the *runup law* with his experimental results and with those of Hall & Watts' 1953 [18] (H&W) and pointed out that, in many studies, runup data have been analyzed without making a distinction between breaking and non-breaking solitary waves. That distinction is essential mainly for gentle slopes where most of the data refer to breaking waves. Hence SY used a criterion of breaking during backwash, developped by Pedersen & Gjevik 1983 [28] (from simplified analytic solutions) : $\frac{H}{d} \geq \frac{H_b}{d} = 0.479(\cot\beta)^{-\frac{10}{9}}$, to sort out the non-breaking runups in the above refered experiments, which were then compared with his theoretical predictions. The agreement was good in case of gentle slopes ($\beta = 2.88°$) but deteriorated when the steepness of the slope increased ($\beta \geq 45°$). By determining the conditions where the surface slope becomes infinite (his method breaks down soon after such situations), SY finally derived a criterion of breaking during the runup $\frac{H}{d} \geq \frac{H_u}{d} = 0.818(\cot\beta)^{-\frac{10}{9}} = 1.708\frac{H_b}{d}$.

A comparison between these analytical results, our BEM results and the experimental results of LVN, H&W and SY, is presented in Table 4. Our computations are performed in a 10m long and 0.3m deep numerical wavetank. The total discretization is $N_\Gamma = 158$ (all but the 2.88° slope) to 190 (2.88° slope) nodes, with 101 nodes or 50 quasi-spline elements located on the free surface. Quadratic elements are used on the rest of the boundary. The time steps are $\Delta t = .015$ to .030s. The CPU time per time step is 1.17s to 1.71s on an IBM 3090/300.

Table 4 shows that on a 2.88° slope and for the smaller wave ($\frac{H}{d}=0.019$), SY's *runup law* predicts a runup 16% higher than his experimental result (which is here taken equal to the average of his 2 experiments for this wave : $\frac{\mathcal{R}}{d}=0.078$, 0.076 for $d=0.3097$, 0.3106m) whereas our solution is only 5% higher. Although it did not break in the experiments, the second wave ($\frac{H}{d}=0.040$) is larger than $\frac{H_u}{d}$ (equal to 0.029) and is indeed found to break during runup in both SY' and our results.

On a 15° slope and for the smaller wave ($\frac{H}{d}=0.100$), SY' *runup law* gives a runup in very good agreement (less than 1%) with our fully nonlinear solution. For the larger wave ($\frac{H}{d}= 0.200$), the discrepancy with our results is of 12%. Our results, agree with the experiments of H&W within 10% for both waves and those of SY within 10% and 22% for the two tested waves respectively. Notice that the second wave is greater than the limit of breaking during backwash estimated for this slope ($\frac{H_b}{d}= 0.111$) and is indeed found to break during the rundown in our computations. Notice, also, that Kim, *et al.* 1983 computed runups on a 15° slope (by a fairly low order BEM) for the same two waves as in Table 4 and found $\frac{\mathcal{R}_u}{d}=0.308$ and 0.766 for each of the waves respectively.

On a 45° slope, the discrepancy between SY' *runup law* and our results is of the order of 19%, whereas our results agree with the experiments of LVN within less than 3%. On a

70° slope, our runup results agree with the experiments within less than 4%. A comparison with the *runup law* is not relevant in this case, since the validity criterion of SY' solution is far too large ($\frac{H}{d}$ =0.617). If the full integral equation derived by SY (equation (3.3) in SY) is used for the 45° and 70° slopes, one gets the results named $\frac{R'}{d}$ in Table 4 (personal communication). This reduces the maximum discrepancy with the experimental results to 8% and 7% for the two slopes respectively.

Thus, the best predictions of solitary wave runups by the SW theory agree with the experimental results within 7 to 22% on 2.88° to 70° slopes (for the data in Table 4). In most of the cases our results are significantly closer to the experimental results (3 to 10%). Although the number of results available in Table 4 is limited, the discrepancies with the experimental results can tentatively be interpreted in terms of vertical accelerations, dispersive effects and friction. The two first effects are present in a fully nonlinear solution, but are neglected in the SW equation whereas friction is neglected in both solutions. On a *steep slope*, dispersive effects are negligible (small propagation distance), but vertical velocities and accelerations become important for the higher waves (as seen in Section 4.5 and 4.6). On a smooth steep slope, the influence of friction must be expected to be of small importance. Only at the uppermost tip of the runup, would some frictional effects be expected (this problem was studied by Packwood & Peregrine 1981 [27]). Results indeed show that our solution agrees with experiments within 4% (for 45° and 70° slopes). Thus, the higher discrepancy of the SW solution (8%) in this case should mostly be related to the effects of vertical accelerations. On a *gentle slope* the vertical velocities and accelerations are small. Dispersive effects, however, become important mainly for the higher waves. Friction also becomes important since longer distances of propagation are considered. Both of these effects should account for the 16% discrepancy of the SW solution on a 2.88° slope, whereas friction should mostly be responsible for the 5% discrepancy of our solution in this case. On an *intermediate slope*, all three effects should influence concurrently the SW solution, and only friction our solution. Thus one would expect discrepancies similar to those obtained on a gentle slope whereas the discrepancies obtained in both solutions on a 15° slope are some 5% larger than that. This could be due to a more important friction in the rather old H&W experiments. All this should anyway be investigated further with more waves and more slopes.

5.2   Shoaling and breaking of a solitary wave

As seen above, even small solitary waves break on a gentle slope. This kind of problem can also be addressed with our model. Shoaling and breaking of a high solitary wave is simulated in the same numerical wavetank as above, for the case of a gentle slope of angle $\beta = 2.88°$ (slope 1:20), starting at x=4m. The computational data are as above with varying time steps $\Delta t = 0.0075$s to 0.00125s.

Goring's method of solitary wave generation is again used and creates a wave of $\frac{H}{d} = 0.794$, greater than the wave of maximum energy ($\frac{H}{d} \simeq 0.78$, see Tanaka, *et al.* 1987 [36]). This wave almost stabilizes in the numerical wavetank after propagating over $10d$. Such a wave, however, was found by Tanaka *et al.* to be unstable in a long term propagation.

The shape of this highly nonlinear wave is, as expected, rather far from a perfect solitary wave (Figure 14). It sheds a strong tail of disturbances as it propagates in the

β	$\frac{H_i}{d}$	$\frac{R_u}{d}$	$\frac{R_e}{d}$	$\frac{H_b}{d}$	$\frac{R}{d}$	$\frac{H_l}{d}$	$\frac{R'}{d}$
70°	0.259	0.580	0.558	1.472	0.316	0.617	0.524
70°	0.437	1.062	1.071	1.472	0.607	0.617	1.142
45°	0.269	0.674	0.672	0.479	0.546	0.083	0.644
45°	0.457	1.257	1.294	0.479	1.064	0.083	1.194
15°	0.100	0.310	0.281	0.111	0.308	0.006	(n.a.)
15°	0.200	0.654	0.599	0.111	0.732	0.006	(n.a.)
2.88°	0.019	0.081	0.077	0.017	0.089	0.000	(n.a.)
2.88°	0.040	(—)	0.156	0.017	0.226	0.000	(n.a.)

Table 4: Runup on a slope β : The $\frac{R_u}{d}$ are the BEM results. The $\frac{R_e}{d}$ are the experimental results from Losada, *et al.* 1986 (45° and 70°), from Hall & Watts 1953 (15°) and from Synolakis 1987 (2.88°). $\frac{R}{d}$ is the asymptotic *runup law* from Synolakis 1987 and $\frac{R'}{d}$ is the corresponding full expression. $\frac{H_b}{d}$ is the criterion of breaking during backwash from Pedersen & Gjevik 1983, and $\frac{H_l}{d}$ is the validity limit of Synolakis' *runup law*. (—) means there was breaking during the runup.

tank, and exhibits low scale modulations close to its tip. After reaching the slope (x=4m) the wave (surprisingly) slightly decreases in maximum height over the next 2m. Then, it starts shoaling and eventually breaks around $x = 6.95$m.

Figures 15a and 15b show the breaking process itself. Unlike with periodic waves, the breaking is very local and limited to a small jet at the tip of the wave. Tanaka *et al.* 1987 [36] found a somewhat similar breaking pattern in their computation of the instability of solitary waves over a constant depth.

6 Conclusions

The two-dimensional computations described here have been made with a version of the Boundary Integral Equation Method which does not apply complex variables and does not assume the waves are periodic in space. This method can also be used, as it is, for any kind of bottom geometry. It turns out that sawtooth instability does not occur (even after more than 1000 time steps), and that both energy and volume inside the computational region are controled sufficiently well to not making it necessary to carry out artificial adjustments during the computations as has been necessary in e.g. some of the computations reported by Dold & Peregrine 1984 [7].

Although the method has not yet been developed to its full capacity the results show that phenomena such as generation, long term propagation, breaking and runup on steep and gentle slopes, of large amplitude waves can accurately be predicted by this method. And because of the accuracy of the computations the method can also be used to analyze flow properties such as velocity and pressure fields that could not easily be measured during experiments in a physical wavetank.

The maximum runup of solitary waves on 2.88° to 90° slopes is shown to agree well with experimental or other fully nonlinear results. Comparison of predicted free surface elevations on a steep slope with detailed experimental results shows a good agreement, even for the small scale oscillations behind the wave. The analysis of the velocity and pressure during reflection of a high solitary wave from a 45° slope shows some important discrepancies with the hypothesis and predictions of the SWE (important vertical velocity, non-hydrostatic pressure). The runups computed with the SWE also exhibit more important discrepancies with the experimental results than with our method, on both gentle and steep slopes. This is tentatively interpreted in terms of vertical acceleration, dispersive effects (both neglected in the SWE) and friction. Computation of the breaking of a solitary wave on a 1:20 slope shows that the overturning is limited to a small region at the tip of the wave.

A   Generation of a solitary wave by a piston wavemaker

In a wave of permanent form, we have at any instant (Svendsen & Justesen [34]),

$$\int_{-d}^{\eta} u \ dz = c_a \eta + Q_s + U_c d \tag{5}$$

where c_a is the propagation speed of the wave in a fixed frame of reference, $\eta(x,t)$ is the wave elevation above the still water level, Q_s is the nonlinear mass flux averaged over a wave period, and U_c, the speed of the current defined as the averaged particle velocity below wave trough level.

In a solitary wave with an infinitely long wave period, the right hand side of (5) simply reduces to $c\eta$ where c is the speed of the wave relative to the water, so that (5) becomes the simpler expression used by Goring [11] for determining the motion required by a piston wavemaker to generate a specified water surface elevation immediately in front of the wavemaker. Since the piston motion creates a depth uniform horizontal velocity $u_p(t)$ (5) reduces to,

$$u_p(d + \eta) = c\eta \tag{6}$$

which means that a required η can be generated by a piston velocity,

$$u_p(t) = \frac{c\eta}{d + \eta} \tag{7}$$

This corresponds to a piston motion $x_p(t)$ given by,

$$x_p(t) = \int_0^t \frac{c\eta}{d + \eta} \ d\tau \tag{8}$$

To generate a solitary wave profile of amplitude H in water of constant depth d with,

$$\eta(x,t) = H \operatorname{sech}^2[\frac{\kappa}{d}(x - ct)] \tag{9}$$

$$c = \sqrt{g(d + H)} \tag{10}$$

$$\kappa = \sqrt{\frac{3H}{4d}} \tag{11}$$

(9)-(11) are substituted into (8) with $x = x_p(t)$ required throughout the integration in order to account for the finite amplitude of the piston. Notice that most wave maker theories describe the solution to the problem : What is the wave motion generated by a specified paddle motion ? It may be interesting to notice that the procedure described above is the only one (known to the authors) for solving the "inverse" wave generation problem : How do we move the wavemaker to generate a wave of a specified (arbitrary) form.

Since a solitary wave profile like (9) extends to infinity in both directions, Goring introduced a way of estimating the importance of the significant horizontal extension of the wave by defining an apparent wavelength 2λ which represents the part of the wave where $\eta > 0.001H$, with $\frac{\lambda}{d} = \frac{3.8}{\kappa}$ and $3.8 = \text{arcosh}[(\frac{1}{.001})^{\frac{1}{2}}]$.

The wave generation by the wavemaker starts at $t = 0$ with the begining of the solitary wave profile considered as significant, i.e. for $x = \lambda$. Therefore, in the wave formula (9), we set $x = x_p + \lambda$ and, by integrating (8), we get,

$$x_p(t) = \frac{H}{\kappa}[\tanh \frac{\kappa}{d}(ct - x_p - \lambda) + \tanh \frac{\kappa}{d}\lambda] \tag{12}$$

This formula, implicit in x_p, is solved for any given t by Newton iterations. Then $u_p(t)$ is computed by (7) for $\eta(x_p,t)$ and, eventually, $\frac{\partial u_p(t)}{\partial t}$ is found by derivation of (7). Hence, the boundary conditions can be defined on the wavemaker.

References

[1] Brorsen, M. & Larsen, J. Source Generation of Nonlinear Gravity Waves with the Boundary Integral Method. *Coastal Engineering* **11**, 93-113, 1987.

[2] Byatt-Smith, J.G.B. The Reflection of a Solitary Wave by a Vertical Wall. *J. Fluid Mech.* **197**, 503-521, 1988.

[3] Camfield, F.E. & Street, R.L. An Investigation of the Deformation and Breaking of Solitary Waves. *Dept. of Civil Engng., Stanford University, Technical Report No.* **81**, 1967.

[4] Carrier, G.F. Gravity Waves on Water of Variable Depth. *J. Fluid Mech.* **24** (4), 641-659, 1966.

[5] Carrier, G.F. & Greenspan, H.P. Water Waves of Finite Amplitude on a Sloping Beach. *J. Fluid Mech.* **4** (1), 97-110, 1958.

[6] Cointe, R., Molin, B. & Nays, P. Nonlinear and Second-Order Transient Waves in a Rectangular Tank. In *Proc. 6th Intl. Conf. on Behavior of Off-shore Structures, Delft (B.O.S.S.),* 1988.

[7] Dold, J.W. & Peregrine, D.H. Steep Unsteady Water Waves : An Efficient Computational Scheme. In *Proc. 19th Intl. Conf. on Coastal Engineering, Houston, USA,* pp. 955-967, 1984.

[8] Dommermuth, D.G. & Yue, D.K.P. Numerical Simulation of Nonlinear Axisymmetric Flows with a Free Surface. *J. Fluid Mech.* **178**, 195-219, 1987.

[9] Fenton, J.D. & Rienecker, M.M. A Fourier Method for Solving Nonlinear Water-Wave Problems : Application to Solitary-Wave Interactions. *J. Fluid Mech.* **118**, 411-443, 1982.

[10] Gardner, C.S., Greene, J.M., Kruskal, M.D. & Miura, R.M. Method for Solving the Korteweg-de Vries Equation. *Physical Revue Letters* **19**, (19), 1095-1097, 1967.

[11] Goring D.G. Tsunamis - The Propagation of Long Waves onto a Shelf. *W.M. Keck Laboratory of Hydraulics and Water Ressources, California Institute of Technology, Report No.* **KH-R-38**, 1978.

[12] Greenhow, M. & Lin, W.M. Numerical Simulation of Nonlinear Free Surface Flows Generated by Wedge Entry and Wave-maker Motions. In *Proc. 4th Intl. Conf. on Numerical Ship Hydrody., Washington, USA*, 1985.

[13] Greenhow, M. & Lin, W.M. The Interaction of Nonlinear Free Surfaces and Bodies in Two Dimensions. *MARINTEK, Report No.* **OR/53/530030.12/01/86**, 36 pps, 1986.

[14] Grilli, S., Skourup, J. & Svendsen, I.A. The Modelling of Highly Nonlinear Waves : A Step Toward the Numerical Wave Tank. Invited paper in *Proc. 10th Intl. Conf. on Boundary Elements, Southampton, England*, Vol. 1 (ed. C.A. Brebbia), pp. 549-564. Computational Mechanics Publication. Springer Verlag, Berlin, 1988.

[15] Grilli, S., Skourup, J. & Svendsen, I.A. An Efficient Boundary Element Method for Nonlinear Water Waves. *Engineering Analysis with Boundary Elements* **6** (2), 1989.

[16] Grilli, S. & Svendsen, I.A. The Modelling of Highly Nonlinear Waves. Part 2 : Some Improvements to the Numerical Wave Tank. To appear in *Proc. 11th Intl. Conf. on Boundary Elements, Cambridge, Massachusetts, USA*. Computational Mechanics Publication. Springer Verlag, Berlin, 1989a.

[17] Grilli, S. & Svendsen, I.A. The Modelling of Nonlinear Water Wave Interaction with Maritime Structures. To appear in *Proc. 11th Intl. Conf. on Boundary Elements, Cambridge, Massachusetts, USA*. Computational Mechanics Publication. Springer Verlag, Berlin, 1989b.

[18] Hall, J.V. & Watts, J.W. Laboratory Investigation of the Vertical Rise of Solitary Waves on Impermeable Slopes. *Beach Erosion Board, US Army Corps of Engineer, Tech. Memo. No.* **33**, 14 pp, 1953.

[19] Hammack, J.L. & Segur, H. The Korteweg-de Vries Equation and Water Waves. Part 2. Comparison with Experiments. *J. Fluid Mech.* **65** (2), 289-314, 1974.

[20] Kim, S.K., Liu, P.L.-F. & Liggett, J.A. Boundary integral Equation Solutions for Solitary Wave Generation Propagation and Run-up. *Coastal Engineering* **7**, 299-317, 1983.

[21] Kravtchenko, J. Remarques sur le calcul des amplitudes de la houle lineaire engendrée par un batteur. In *Proc. 5th Intl. Conf. on Coastal Engineering*, pp. 50-61, 1954.

[22] Lin, W.M., Newman, J.N. & Yue, D.K. Nonlinear Forced Motion of Floating Bodies. In *Proc. 15th Intl. Symp. on Naval Hydrody., Hamburg, Germany*, 1984.

[23] Longuet-Higgins, M.S. & Cokelet, E.D. The Deformation of Steep Surface Waves on Water - I. A Numerical Method of Computation. *Proc. R. Soc. Lond.* **A350**, 1-26, 1976.

[24] Losada, M.A., Vidal, C. & Nunez, J. Sobre El Comportamiento de Ondas Propagádose por Perfiles de Playa en Barra y Diques Sumergidos. *Dirección General de Puertos y Costas Programa de Clima Maritimo. Universidad de Cantábria. Publicación No.* **16**, 1986.

[25] Mirie R.M. & Su C.H. Collisions Between Two Solitary Waves. Part 2. A Numerical Study. *J. Fluid Mech.* **115**, 475-492, 1982.

[26] Nagai, S. Pressures of Standing Waves on Vertical walls *J. Waterways, Harbors Div. ASCE* **95**, 53-76, 1969.

[27] Packwood, A.R. & Peregrine, D.H. Surf and Runup on Beaches : Models of Viscous Effects. *School of Mathematics, University of Bristol, Report No.* AM-81-07, 1981.

[28] Pedersen, G. & Gjevik, B. Run-up of Solitary Waves *J. Fluid Mech.* **135**, 283-299, 1983.

[29] Peregrine, D.H. Flow Due to a Vertical Plate Moving in a Channel. *Unpublished Notes,* 1972.

[30] Roberts, A.J. Transient Free-Surface Flows Generated by a Moving Vertical Plate. *Q.J. Mech. Appl. Math.* **40** (1), 129-158, 1987.

[31] Su, C.H. & Gardner, C.S. Korteweg-de Vries Equation and Generalizations. III. Derivation of the Korteweg-de Vries Equation and Burgers Equation. *J. Math. Phys.* **10** 536-539, 1969.

[32] Su, C.H. & Mirie, R.M. On Head-on Collisions between two Solitary Waves. *J. Fluid Mech.* **98** 509-525, 1980.

[33] Svendsen, I.A. & Grilli, S. Nonlinear Waves on Slopes. To appear in *J. Coastal Res.,* 1989.

[34] Svendsen, I.A. & Justesen, P. Forces on Slender Cylinders from Very High and Spilling Breakers. In *Proc. Symp. on Description and Modelling of Directional Seas,* paper No. D-7, 16 pps. Technical University of Denmark, 1984.

[35] Synolakis, C.E. The Runup of Solitary Waves. *J. Fluid Mech.* **185**, 523-545, 1987.

[36] Tanaka, M., Dold, J.W., Lewy, M. & Peregrine, D.H. Instability and Breaking of a Solitary Wave. *J. Fluid Mech.* **185**, 235-248, 1987.

[37] Vinje, T. & Brevig, P. Numerical Simulation of Breaking Waves. *Adv. Water Ressources* **4**, 77-82, 1981.

[38] Yim, B. Numerical Solution for Two-Dimensional Wedge Slamming with a Nonlinear Free surface Condition. In *Proc. 4th Intl. Conf. on Numerical Ship Hydrody., Washington, USA,* 1985.

Figures

Fig. 1a: Sketch of a cross-section in a coastal area protected by a maritime structure. g is the gentle slope in front of the structure, s the steep structure slope, a a high wave breaking in shallow water and b a smaller wave runing-up on the structure.

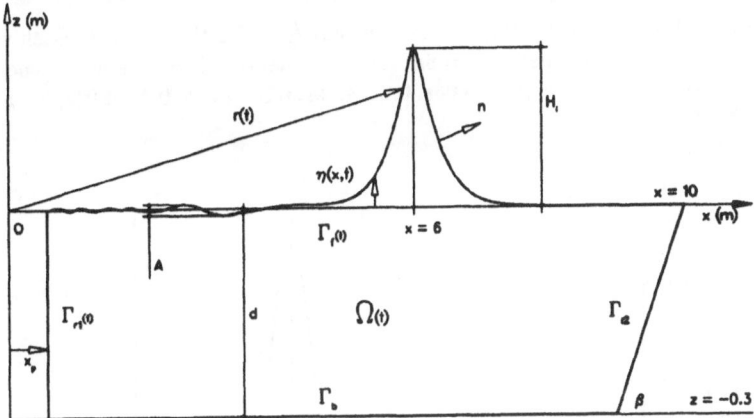

Fig. 1b: Sketch of the region used in the numerical computations. Definition of geometrical parameters.

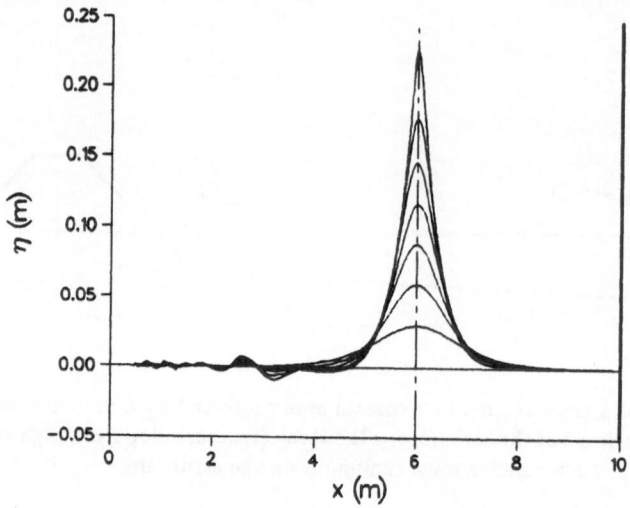

Fig. 2: Surface profile $\eta(x)$ for solitary waves of height $\frac{H}{d} = 0.1$ through 0.7. Each wave is shown at the moment the crest passes $x = 6$m (20 water depth d from the wave generator). The propagation times are : $\frac{H}{d} = 0.1 : 5.66$s, $0.2 : 4.79$s, $0.3 : 4.34$s, $0.4 : 4.04$s, $0.5 : 3.80$s, $0.6 : 3.62$s, $0.7 : 3.44$s.

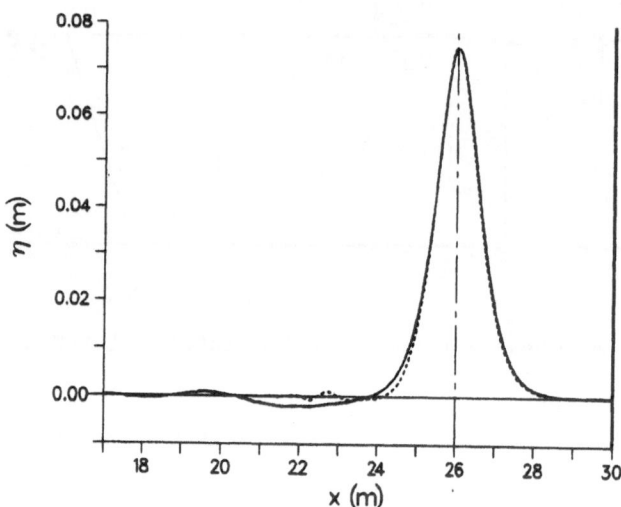

Fig. 3: Comparison between two solitary wave surface profiles generated at $x = 20$m (- - -) and $x = 0$m (——) respectively (i.e. after propagating 20 d and 86.6 d, d=0.30m). To be of the same height at $x = 26$m the two waves were generated with slightly different amplitude.

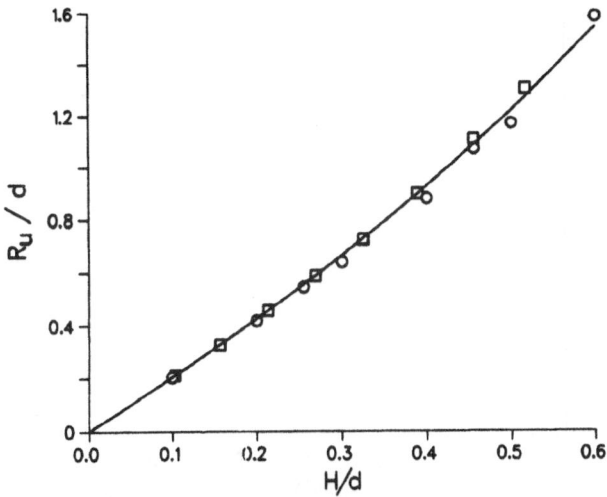

Fig. 4: Runup R_u for reflection of a solitary wave from a vertical wall versus the value of $\frac{H}{d}$ for the wave 4m (13.3 d) in front of the wall (as in Fig. 1). (——) Su & Mirie (1980) 3rd order results, ($\square$) Fenton & Rienecker (1982) Fourier Method results and (o) present BEM results.

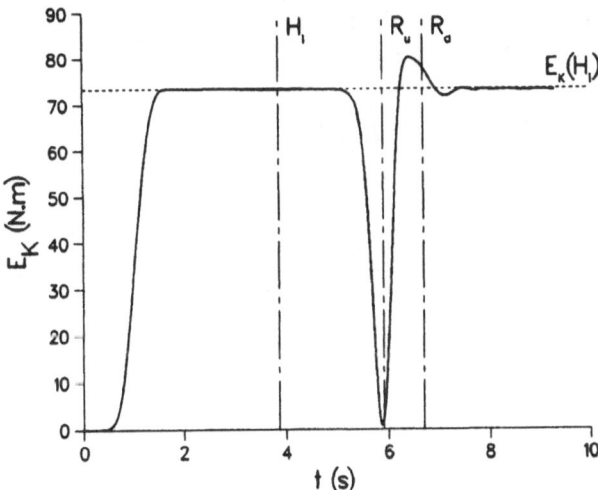

Fig. 5: Change in kinetic energy E_K during reflection of a solitary wave from a vertical wall, for a wave of $\frac{H}{d}$=0.456. Notice the kinetic energy never quite reaches zero, and the minimum occurs slightly before the instant of maximum runup.

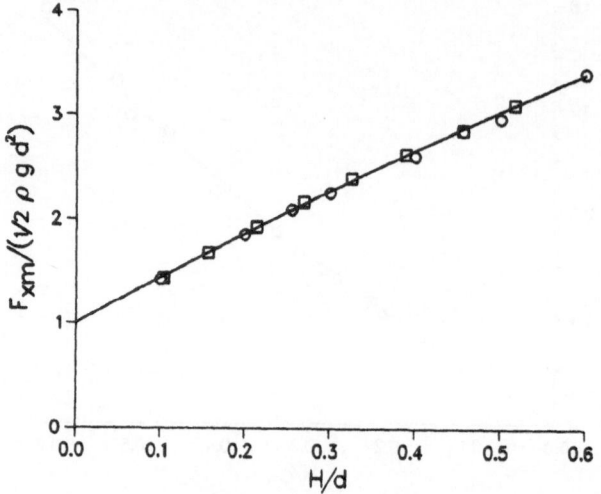

Fig. 6: Maximum horizontal pressure force F_{xm}, in dimensionless form, on a vertical wall during reflection of a solitary wave, as a function of $\frac{H}{d}$. Symbols are as in Fig. 5.

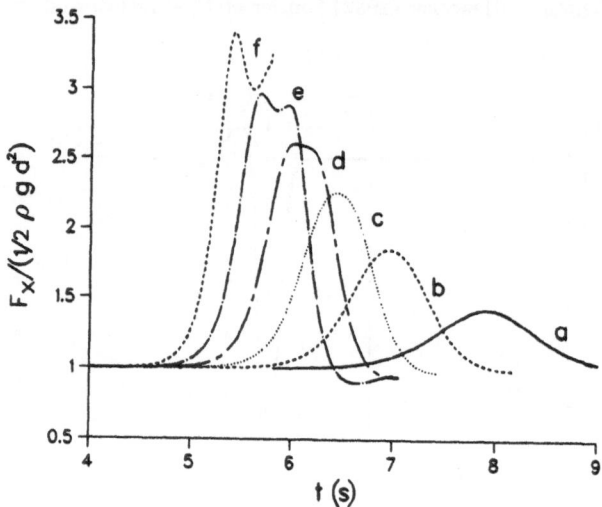

Fig. 7: Horizontal pressure force F_x, in dimensionless form, on a vertical wall during reflection of a solitary wave. $\frac{H}{d}$ = (a) : 0.1, (b) : 0.2, (c) : 0.3, (d) : 0.4, (e) : 0.5, (f) : 0.6.

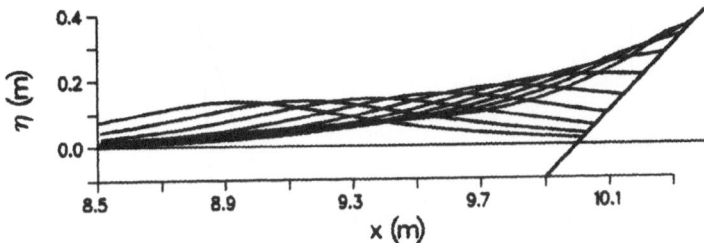

Fig. 8: Reflection of a solitary wave from a 45° slope, $\frac{H}{d}= 0.457$. Numerical results up to maximum runup. Curves are plotted every 0.1s for $t=5.3$ to 6.2s.

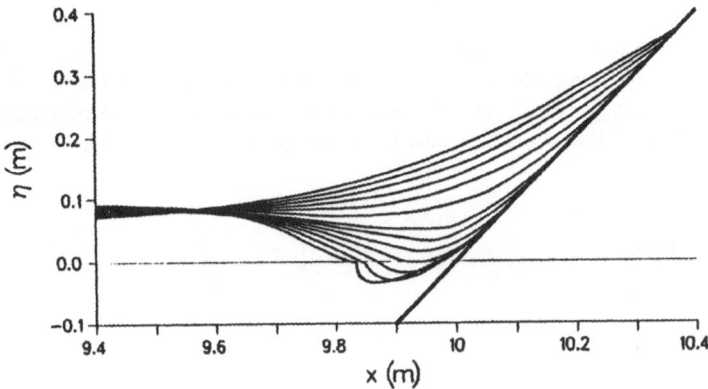

Fig. 9: Reflection of a solitary wave from a 45° slope, $\frac{H}{d}= 0.457$. Numerical results from maximum runup, down to backward breaking. Curves are plotted every 0.05s for $t=6.2$ to 6.5s and every 0.02s for $t=6.50$ to 6.64s.

410

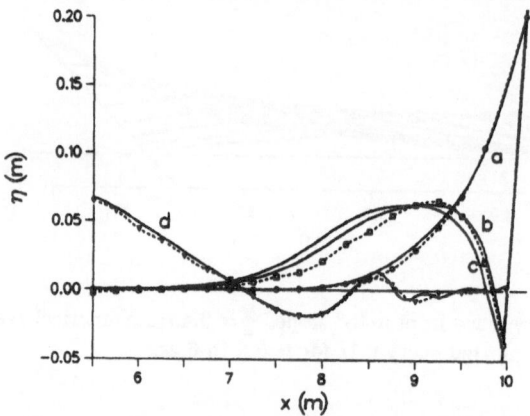

Fig. 10: Comparison between computations (——) and measurements (- - -) by Losada, et al. (1986) for reflection from a 45° slope, $\frac{H}{d} = 0.269$ with : (a) Instant of maximum runup ($t = 6.77$s in computations), (b) Instant of lowest position of water surface in experiments ($t = 7.31$s), (c) Instant of lowest surface position in computations ($t = 7.38$s in computations), (d) $t = 9.08$s.

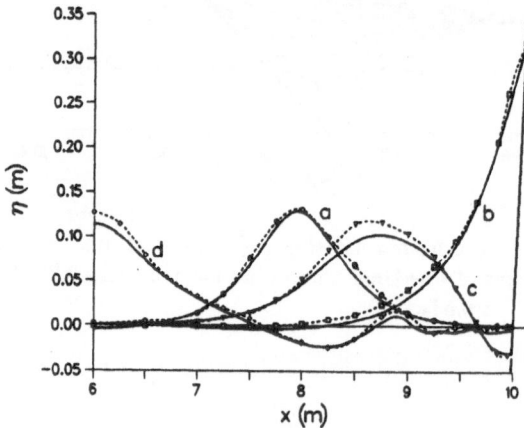

Fig. 11a: Same as Fig. 10 for a 70° slope, $\frac{H}{d} = 0.437$ with : (a) Wave profile somewhat before maximum runup ($t = 4.88$s), (b) Instant of maximum runup ($t = 6.04$s in computations), (c) Instant of lowest position of water surface in computations ($t = 6.67$s), (d) $t = 7.97$s.

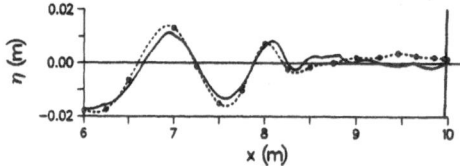

Fig. 11b: Same as Fig. 10 for a 70° slope, $\frac{H}{d} = 0.437$. Detail on the "Wave tail" profile at $t = 9.48$s

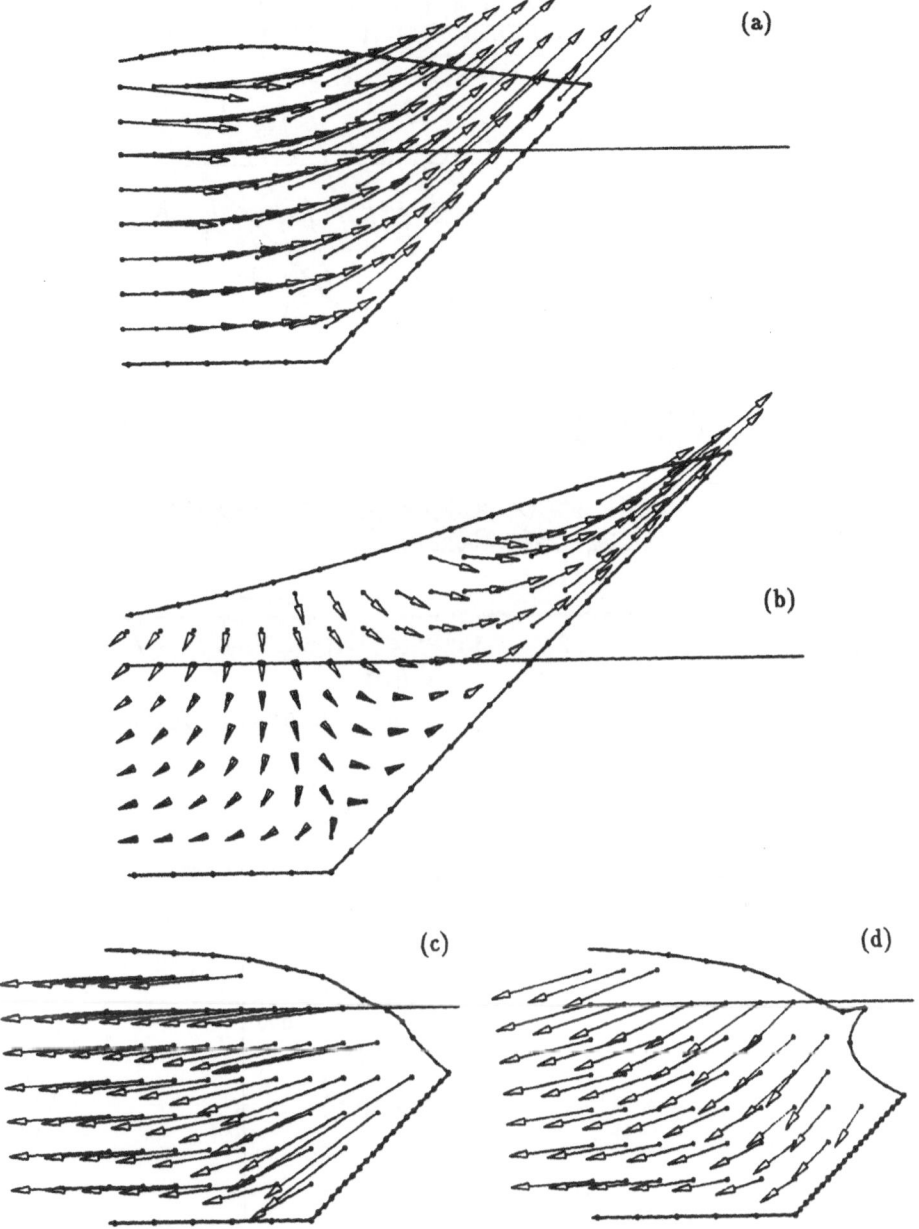

Fig. 12a-d: The internal velocity field during runup and rundown of a solitary wave with $\frac{H}{d} = 0.457$ on a 45° slope, at time (a) : 5.60s, (b) : 5.93s, (c) : 6.65s, (d) : 6.73s.

412

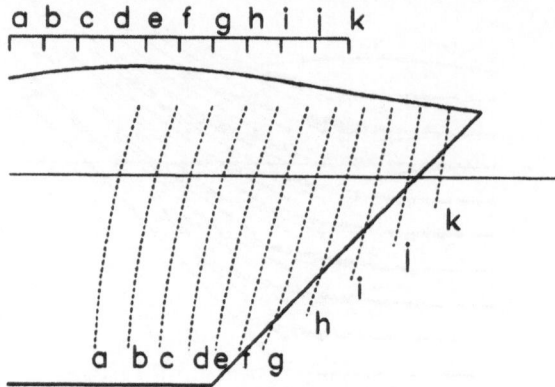

Fig 13 : (- - -) Envelopes of the horizontal velocity vectors of Fig. 12a. Vectors start at the positions of the upper letters and end on the corresponding curves.

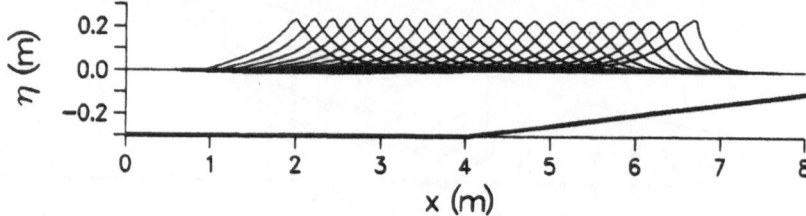

Fig. 14: Shoaling and breaking of a solitary wave of $\frac{H}{d}= 0.794$ on a 1:20 slope. Curves are plotted every 0.1s for $t= 1.6$ to 3.7s.

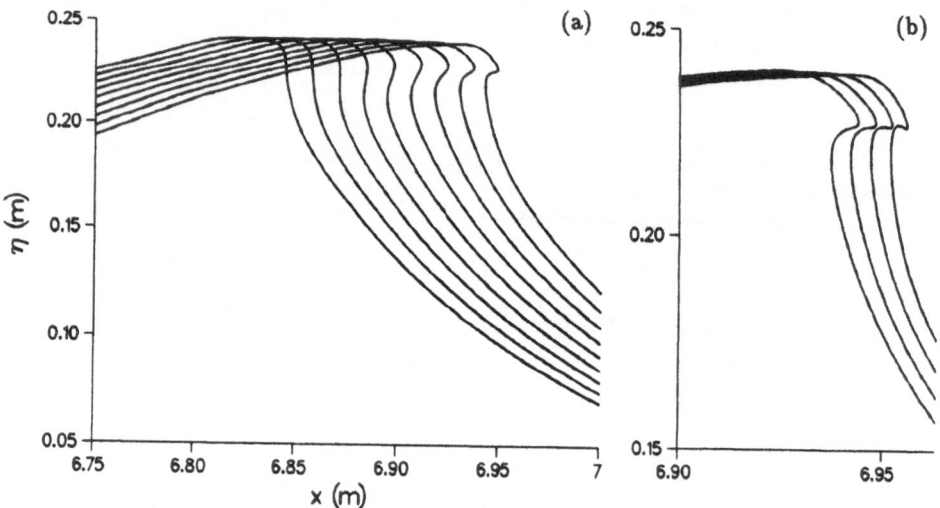

Fig. 15a-b: Details of the breaking of the same solitary wave as in Fig.14.

FIELD OBSERVATIONS OF WAVE-CURRENT INTERACTION AT THE SEA BED

R. L. SOULSBY
Hydraulics Research Ltd
Wallingford
Oxon   OX10 8BA
UK

J. D. HUMPHERY
Proudman Oceanographic Lab
Bidston Observatory
Merseyside   L43 7RA
UK

ABSTRACT.   Field measurements are presented of the mean, turbulent, and wave-induced velocities near the sea bed in combined wave-plus-current flow, with which to test theories of wave-current interaction. A technique of splitting the variance within the velocity spectrum was developed to separate wave orbital velocities from turbulent fluctuations, which were often as large as the orbital velocities. The wave boundary layer was much thinner than the lowest measuring height (10cm) and acted as a source of turbulence to the current boundary layer, analogous to an increased apparent bed roughness. This caused both the turbulent kinetic energy and the mean bed shear stress to increase strongly with the wave:current ratio. The results have implications for sediment transport, tidal dynamics, wave dissipation, and forces on pipelines.

1.   INTRODUCTION

Over most of the continental shelf, water motions due to both waves and currents occur at the sea bed. The wave and current boundary layers interact nonlinearly near the bed, and their combined effect must be taken into account in practical applications involving sediment transport or forces on submarine pipelines. A number of theoretical approaches have been proposed to predict the velocity and shear-stress structure of the combined wave-plus-current (W+C) boundary layer (e.g. Bijker, 1967; Grant and Madsen, 1979; Fredsøe, 1984; Davies et al, 1988; Myrhaug and Slaattelid, 1989). Although the theories agree qualitatively on a number of features, such as the enhancement of the bed shear stress and the apparent bottom roughness due to waves, they differ markedly in their quantitative predictions (Dyer and Soulsby, 1988). Field data to test the predictions are scarce, and, with the notable exception of the measurements by Lambrakos et al (1988), have generally been made over mobile beds, which greatly complicates their interpretation. To help remedy this shortage, field measurements are presented here of the velocities within the W+C boundary layer over an immobile rough bed.

413

A. Tørum and O. T. Gudmestad (eds.), Water Wave Kinematics, 413–428.
© 1990 *Kluwer Academic Publishers.*

2. MEASUREMENTS

The measurements were made over the period 11 to 19 October 1986 at a site (50° 35.82'N, 1° 31.51'W) in the English Channel 7.5km south west of the coast of the Isle of Wight (Fig 1). The mean water depth was 25m with a maximum tidal range of 2.0m. The sea bottom was a flat featureless mixture of gravel, sand and shell having a median grain diameter of d_{50} = 12.5mm, and relatively widely graded with d_{10} = 1.75mm and d_{90} = 27mm. Bottom photographs confirmed that the bed was generally immobile during the experiment. Temperature and salinity profiles showed negligible density stratification of the water column. Tidal currents were almost rectilinear, exhibiting a very narrow tidal ellipse which rotates clockwise. Waves approached from an approximately 40° sector to the south west. Their significant heights measured with a Waverider 80km west of the site varied between 0.29 and 1.87m.

The instruments were mounted on a self-recording frame (Fig 2) with the acronym STABLE (Sediment Transport And Boundary Layer Equipment), described by Humphery (1987). They comprised open-head

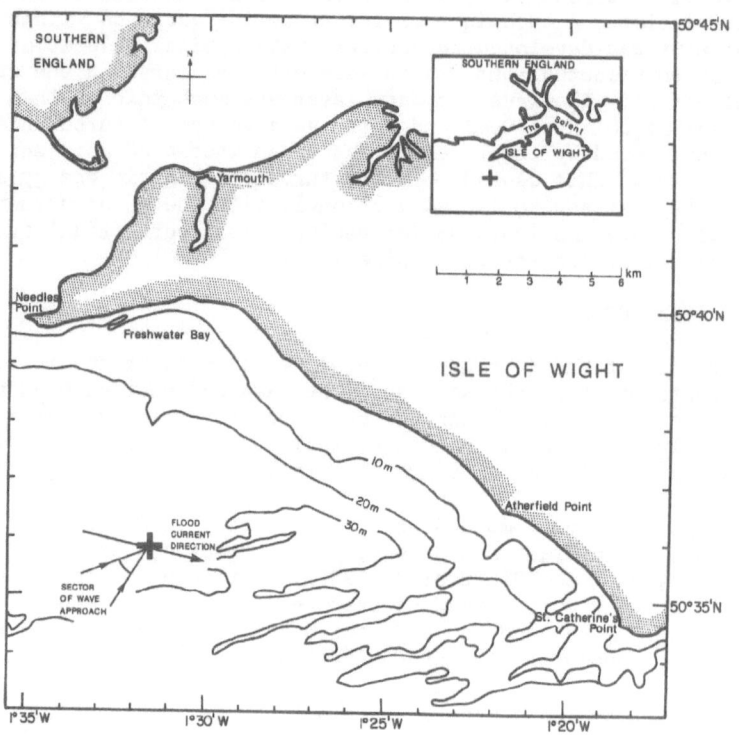

Figure 1. Plan of measurement area

Colnbrook electromagnetic current meters (EMCMs) which measured the two
horizontal components of velocity at heights z = 10 and 80cm, and all
three components by using an orthogonal pair mounted at z = 40cm. A
pressure transducer measured the tidal variation in water depth, but
failed to record the wave-induced pressures as intended. Data was
recorded on a Seadata logger set to record at 4 samples/sec during an
8.5 minute record (in logger parlance a "Burst", referred to later),
with one record logged every 3 hours. The 3 hour interval between
records gave a slow progression through the position in the tidal
cycle, so that a wide range of current speeds were obtained during the
experiment.

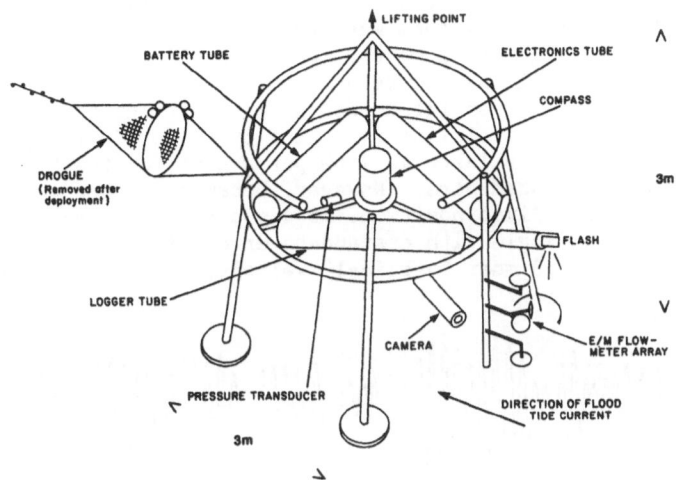

Figure 2. Layout of instrumentation system (STABLE)

STABLE was deployed using a drogue to align the axis of the
instruments with the direction of the flood tidal current. The drogue
was removed by a diver immediately after deployment, who also set the
heights of the instruments accurately above the bed, and took bottom
samples.
    The data presented here correspond to records taken when the
current was in the flood direction, for which the flow was unobstructed
by the structure of the frame. Short records and otherwise imperfect
records were also omitted, leaving 28 records for further analysis.
Data from the EMCM at z = 80cm were noisy, so only the mean velocities
from this level were used.

3. RESULTS

3.1. Preliminary Analysis

EMCMs are subject to electronic zero-drift, which can introduce errors

into the measured mean velocities. To minimise these errors the
time-varying zero-drift residuals were calculated by applying a
purpose-designed 9-point tidal filter to the time-series of record-mean
velocities over all the records (flood and ebb) in the 8 day period.
It was then assumed that the <u>true</u> residual current was zero, and the
filtered residual was subtracted from each mean velocity to leave only
the tidal component. The estimated error in the mean velocities due to
this process is ±2cms⁻¹.

A linear trend, corresponding to the tidal variation in current
speed, was removed from each velocity record. Velocities from the
orthogonal EMCMs were resolved into components along and transverse to
the axis of STABLE, and the two vertical velocity signals were
averaged. The data at z = 40cm were then rotated vertically into a
coordinate system aligned along the mean streamline. Such a rotation
is essential for obtaining accurate Reynolds stresses, particularly for
W+C flows, since the error introduced by misalignment of the EMCM
relative to the mean streamline is typically 8% per degree for a pure
current flow, and is as large as 156% per degree for the "waviest" of
the W+C records measured here. However, because the bed was immobile
and STABLE was in a fixed position, it was possible to calculate the
necessary rotation angle with a standard error of only 0.07° by
plotting the measured mean "vertical" velocity against the horizontal

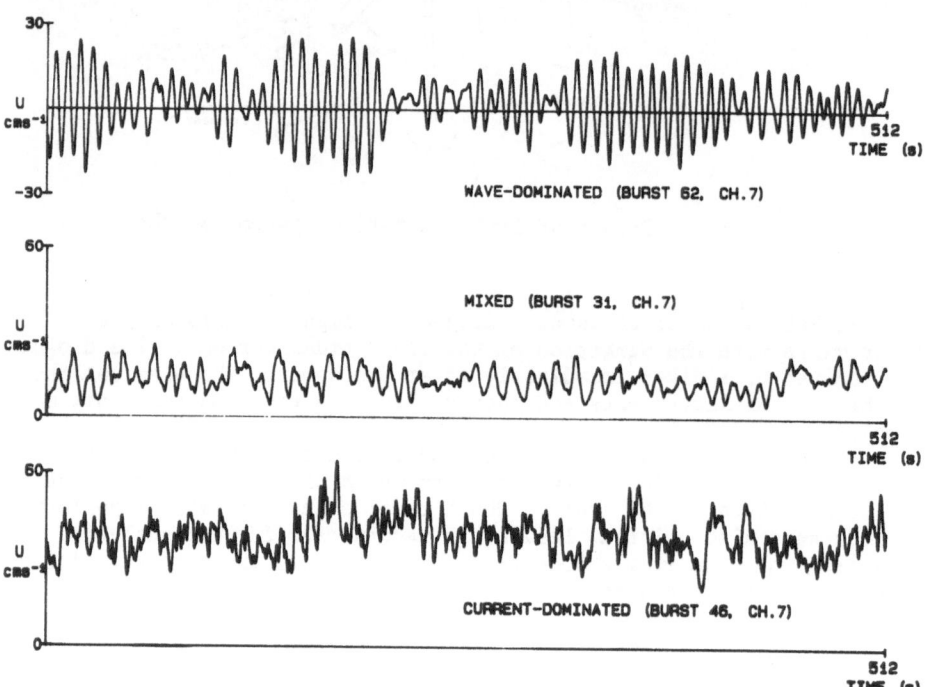

Figure 3. Examples of time-series of instantaneous streamwise
velocity component at 10cm above bed

velocity for all the records in the 8-day deployment together, and
fitting a regression line through them. Thus the standard errors in
the Reynolds stresses due to misalignment were no larger than ±11% in
the worst case, and ±0.6% in the best.

A wide range of combinations of waves and currents were obtained,
as illustrated by three examples in Fig 3. Burst 62 (top) is
wave-dominated with only a very weak mean current and a clear regular
wave signal with the orbital velocity reaching about 25cms⁻¹, and
evidence of groupiness. The waves are generated by local wind and the
peak period of the bottom velocity (since the shortest waves are
attenuated by the depth) is 7.6s; however, for most of the other
bursts the waves were dominated by Atlantic swell with periods between
10 and 15s. Burst 46 (bottom) is current-dominated, comprising a mean
velocity of 33cms⁻¹ with large turbulent fluctuations superimposed on
it, but no wave motions can be discerned by eye. Burst 31 (middle) has
a mixture of waves and currents, with a discernible wave signal which
is offset by the mean current velocity and overlain by irregular
turbulent fluctuations. Many of the bursts were of the mixed type, and
present a problem in distinguishing the wave-induced velocity from the
turbulent fluctuations.

Making an extension of the principle of Reynolds decomposition,
the instantaneous horizontal velocity U(t) in the x-direction can be
written as

$$U(t) = \bar{U} + u_w(t) + u_t(t) \tag{1}$$

where $\bar{U}$ is the time-mean velocity, t is time, $u_w(t)$ is the wave orbital
velocity, $u_t(t)$ is the turbulent fluctuation in velocity, and
$\overline{u_w} = \overline{u_t} = 0$. A similar decomposition can be made for the transverse
horizontal component V(t). Since the wave-induced vertical motions are
negligible close to the bed, the vertical velocity $w_t(t)$ can be
regarded as entirely turbulent.

3.2. Mean Velocities

For each record the mean velocities $\bar{U}$ and $\bar{V}$, respectively along and
transverse to the fore-aft axis of STABLE, were calculated. The mean
current speed $S = (\bar{U}^2 + \bar{V}^2)^{1/2}$ and direction $\phi = \tan^{-1}(\bar{V}/\bar{U})$ relative to
STABLE were calculated at each height. Since the wave boundary layer
appeared to be always thinner than 10cm (see later), the speeds could
be fitted to a logarithmic profile of the form

$$S(z) = \frac{u_*}{\kappa} \ln\left(\frac{z}{z_0}\right) \tag{2}$$

where the friction velocity u_* is related to the bed shear stress τ_0
through $\tau_0 = \rho u_*^2$, the bed roughness length z_0 is related to the
physical roughness of the bed under pure current conditions but
includes wave-enhancement under W+C conditions, von Karman's constant
$\kappa = 0.40$, and ρ is water density. The current speed $U_{100} = S(100)$ at a

height of lm was also calculated by extrapolation of the data for each record using equation (2).

The values of u_* and z_0 showed evidence of enhancement in those records with a large wave contribution, but exhibited considerable scatter due to both sampling variability and sources of error in the measurements. Consequently the results are not presented immediately, but are included in section 3.6 in an averaged form with other estimators of τ_0.

3.3. Splitting the Spectrum

There is no well-established method of separating the wave orbital velocities $u_w(t)$ from the turbulent fluctuations $u_t(t)$ in equation (1). The problem is exacerbated in the present data by the failure to log the wave-induced pressure signal. A method was therefore developed for separating the variances $\overline{u_w^2}$ and $\overline{u_t^2}$ without separating out the instantaneous time-series of $u_w(t)$ and $u_t(t)$. Since $u_t(t)$ is, by definition, not correlated with $u_w(t)$, it is seen from equation (1) that the total variance $\overline{u^2}$ of the time-series U(t) is the sum of the wave-variance and the turbulent-variance,

$$\overline{u^2} = \overline{u_w^2} + \overline{u_t^2} \tag{3}$$

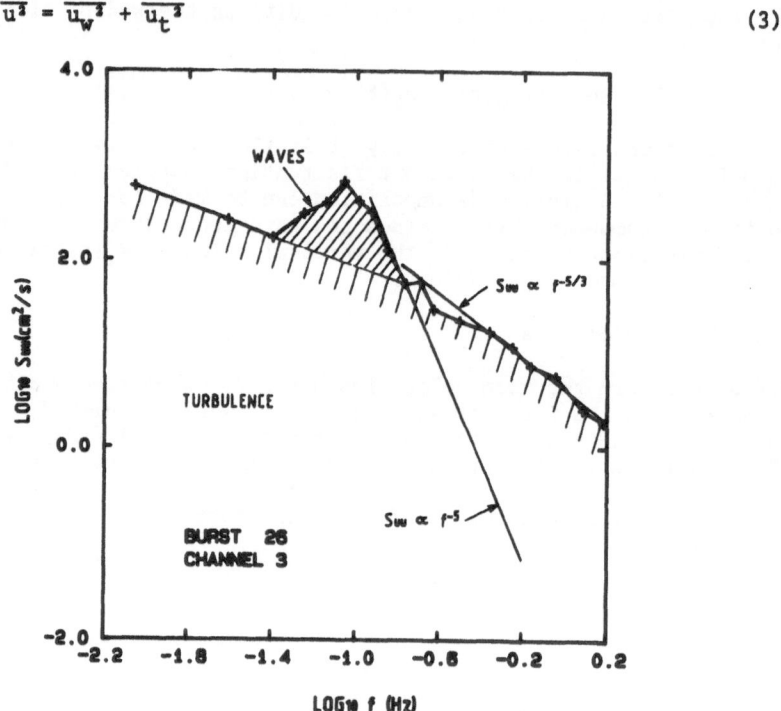

Figure 4. Log-log plot of spectrum of velocity, showing technique of splitting the variance

The area under the energy spectrum $S_{uu}(f)$, where f is frequency, is equal to the total variance $\overline{u^2}$. When the spectrum is plotted on log-log axes (Fig 4) it is seen to comprise a conventional turbulence spectrum, with a characteristic $f^{-5/3}$ power law behaviour in the inertial subrange, onto which is superimposed a wave velocity spectrum, with a peak in the region of 0.1Hz and a characteristic f^{-5} power law decay at higher wave frequencies. The wave peak typically occurs in the frequency range 1/25Hz to 1/6Hz, but spectral estimates in this range will also contain turbulent-variance. This suggests that the wave contribution should be split from the turbulent contribution by interpolating the spectrum across the base of the wave peak, as shown by the line in Figure 2. The wave-variance $\overline{u_w^2}$ is then given by the area (on linear axes) between this line and the spectrum, while the turbulent-variance $\overline{u_t^2}$ is given by the area under the line and the rest of the spectrum. The relative sizes of the contributions is better seen on the same plot on linear axes (Fig 5), which shows the wave spectrum in a form more familiar to waves specialists. The straight

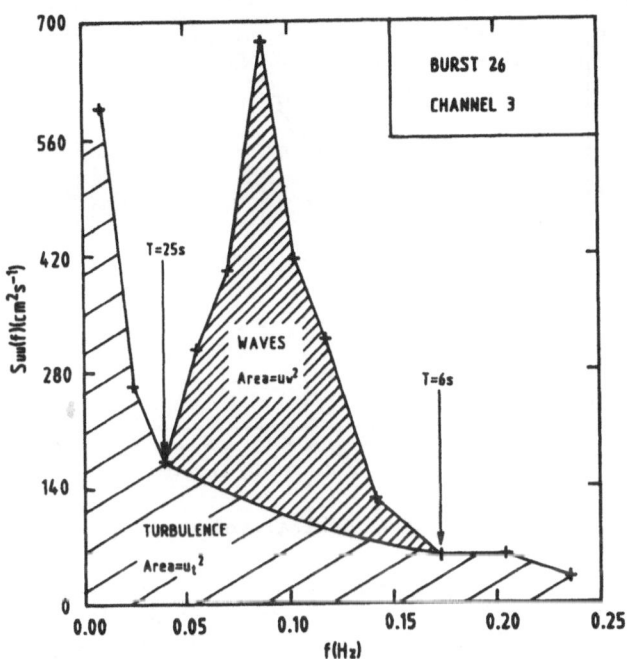

Figure 5. Plot of same spectrum as Figure 4 on linear axes and truncated at 0.25 Hz

dividing line on log axes becomes a curve on linear axes. In this example 40% of the variance was due to waves and 60% due to turbulence. In other examples the split varied from 96%:4% to 1%:99%. The splitting technique was applied to the u and v components, and at heights of 10cm and 40cm.

3.4. Wave Orbital Velocities

The u and v components of wave variance were combined at each height to give a single non-directional measure of the root-mean-square orbital velocity σ,

$$\sigma_{10} = (\overline{u_w^2} + \overline{v_w^2})_{z = 10cm} \tag{4}$$

with a similar expression for σ_{40} at z = 40cm. Plotting σ_{10} versus σ_{40} (Fig 6) shows good agreement with relatively little scatter, giving confidence in the consistency of the spectrum splitting technique. If the lower measurement lay within the wave boundary layer we would

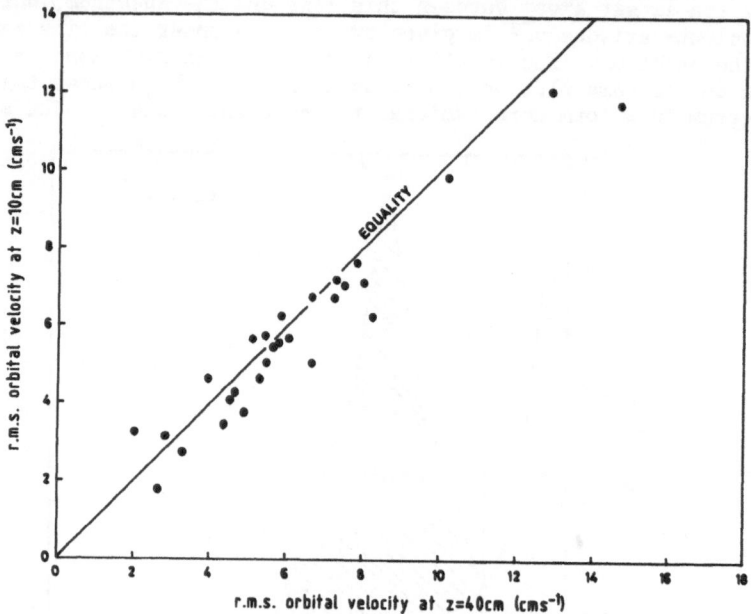

Figure 6. R.m.s. wave orbital velocity σ_{10} versus σ_{40}

expect to see a difference between σ_{10} and σ_{40}. However, calculation of the theoretical thickness of the logarithmic portion of the wave boundary layer using the semi-empirical methods of Jonsson and Carlsen (1976), and Lambrakos (1982), gives a maximum value of 1.5cm, and the height of the maximum velocity in the "overshoot" region (Lambrakos, 1982) is predicted to have a maximum value of 2.0cm. Furthermore, the inviscid attenuation between 40cm and 10cm is negligible. It is therefore reasonable to expect σ_{10} to be equal to σ_{40}, and indeed the mean ratio σ_{10}/σ_{40} is not significantly different to one. To reduce experimental scatter we therefore combine the orbital velocities at both levels as follows:

$$\overline{u_{w}^{2}}_{,\text{mean}} = \tfrac{1}{2}(\overline{u_{w}^{2}}_{,}10 + \overline{u_{w}^{2}}_{,}40) \tag{5}$$

$$\overline{v_{w}^{2}}_{,\text{mean}} = \tfrac{1}{2}(\overline{v_{w}^{2}}_{,}10 + \overline{v_{w}^{2}}_{,}40) \tag{6}$$

$$\sigma_{\text{wave}} = (\overline{u_{w}^{2}}_{,\text{mean}} + \overline{v_{w}^{2}}_{,\text{mean}})^{\tfrac{1}{2}} \tag{7}$$

$$\Psi_{\text{wave}} = \tan^{-1} (\overline{v_{w}^{2}}_{,\text{mean}} / \overline{u_{w}^{2}}_{,\text{mean}})^{\tfrac{1}{2}} \tag{8}$$

A sinusoidal wave having the same orbital velocity variance as the measured waves would have a velocity amplitude of $\sqrt{2}\,\sigma_{\text{wave}}$, and this value can be used for comparison with laboratory or theoretical monochromatic waves. Alternatively, the "significant velocity" = $2\sigma_{\text{wave}}$ may be used (eg Myrhaug and Slaattelid, 1989). The mean direction of wave propagation, Ψ_{wave}, is measured relative to the axis of STABLE. The validity of the estimate of Ψ_{wave} given by equation (8) was confirmed by comparing values with a scatter plot of the instantaneous (U,V) vectors for a few representative records. The direction θ of the waves relative to the current is given by

$$\theta = \Psi_{\text{wave}} - \phi_{40} \tag{9}$$

Values of θ lay in the range 8° to 70°.

The wave period was characterised by the peak period T_p of the wave-peak in the spectrum, taking the mean of the four values of the u and v component at the heights z = 10 and 40cm, but excluding cases where no peak was evident. Values lay in the range $6 < T_p < 16s$.

3.5. Turbulence

Values of the horizontal turbulent variances $\overline{u_t^2}$ and $\overline{v_t^2}$ were obtained by splitting the spectrum at the two heights z = 10 and 40cm, and the vertical turbulent variance $\overline{w_t^2}$ was obtained directly at z = 40cm. Low and high frequency losses occur in the variances due to spectral cut-off associated with the limited record length and digitization rate respectively. The combined loss, calculated by the method of Soulsby (1980), amounted to between 10% and 38% for $\overline{u_t^2}$ and $\overline{w_t^2}$, with similar values expected for $\overline{v_t^2}$. Nonetheless, the ratio of the components generally lay close to the standard steady boundary layer ratios (Soulsby, 1983) $\overline{u_t^2} : \overline{v_t^2} : \overline{w_t^2} = 2.4^2 : 1.9^2 : 1.2^2$, indicating that the detailed structure of the turbulence is unaffected by the wave action. The horizontal components of turbulence were significantly smaller at z = 10cm than at z = 40cm, with the mean of the ratio $(\overline{u_t^2}_{,}10 + \overline{v_t^2}_{,}10)^{\tfrac{1}{2}} : (\overline{u_t^2}_{,}40 + \overline{v_t^2}_{,}40)^{\tfrac{1}{2}} = 0.88$, though this could be partly due to greater spectral losses at the lower height.

The turbulent kinetic energy density E was calculated at z = 40cm from the definition

$$E = \tfrac{1}{2} (\overline{u_t^2} + \overline{v_t^2} + \overline{w_t^2}) \tag{10}$$

The ratio E/S^2 should take a constant value, depending on z_0, for a steady current over a fixed rough bed. This ratio is plotted in

Figure 7 against the wave:current ratio σ_{wave}/S_{40}. It is seen that E/S^2 is indeed constant for values of $\sigma_{wave}/S_{40} \leq 0.2$, but for larger values of σ_{wave}/S_{40} the energy increases rapidly as wave-generated turbulence diffuses up from the wave boundary layer, reaching a value

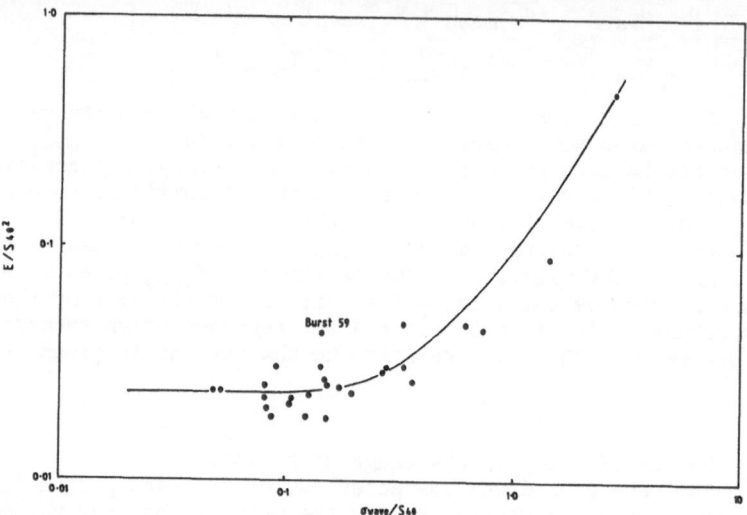

Figure 7. Normalised turbulent kinetic energy as function of wave:current ratio

10 times as large as the current-alone value for $\sigma_{wave}/S_{40} = 2$.

The turbulent kinetic energy is important for entraining and diffusing sediment, and can be predicted by turbulence-energy closure models of the W+C boundary layer (eg Davies et al, 1988).

3.6. Bed Shear Stresses

One method of estimating the bed shear stress τ_0, using the mean velocity profile, has already been described in section 3.2, and these estimates will be denoted $\tau_p = \rho u_*^2$. However, it is notoriously difficult to measure τ_0 accurately, so it is good practice to use more than one technique. One of the most direct measures of τ_0 is through the Reynolds stress $-\rho \overline{uw}$, which can be used when high-frequency turbulence measurements are available as in this case. It is not necessary to separate out the wave orbital velocities for this purpose, since the vertical wave-induced velocity is both small and in quadrature with the horizontal component, giving zero contribution to $\overline{uw}$. The streamwise ($-\rho\overline{uw}$) and transverse ($-\rho\overline{vw}$) components were calculated, and combined to give the modulus, denoted τ_r, and direction Ψ_{stress} of the stress:

$$\tau_r = \rho(\overline{uw}^2 + \overline{vw}^2)^{\frac{1}{2}} \tag{11}$$

$$\Psi_{stress} = \tan^{-1} \left(\overline{vw} / \overline{uw} \right) \tag{12}$$

Apart from cases with weak currents, Ψ_{stress} was aligned to within ±10° of the current direction ϕ_{40}, showing no evidence of the veering which is predicted by some W+C boundary layer theories. Near to slack water larger veering angles ($\Psi_{stress} - \phi_{40}$) were observed, but not in a consistent direction, and these might alternatively be attributed to the tidal dynamics, or simply to large random errors in these small stresses.

A third method of calculating τ_0 was also used, based on the similarity argument that τ_0 is proportional to E in the region near the bed where energy production equals dissipation. The same relationship should be observed in W+C as in current-alone flows, since the wave boundary layer is thin compared with z = 40cm so that it is seen at this level as an enhanced roughness. Using the constant of proportionality observed in a wide variety of flows (Soulsby, 1983), the bed shear stress τ_e deduced from the turbulent kinetic energy is

$$\tau_e = 0.19 \, \rho \, E \tag{13}$$

This method is closely related to the "dissipation" method based on the inertial subrange of the spectrum which has been used by some authors, but equation (13) is simpler to apply and relies on fewer assumptions.

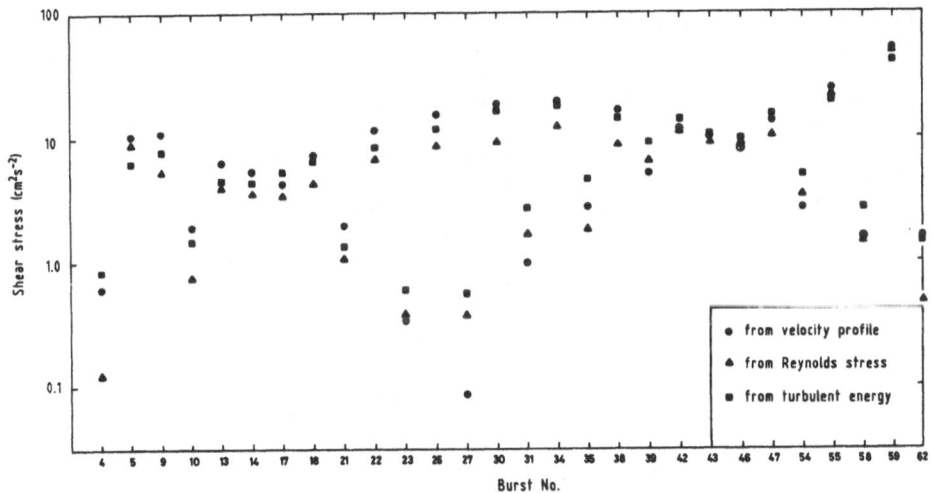

Figure 8. Comparison of three methods of calculating bed shear stress

The three estimates, τ_p, τ_r and τ_e, are independent since they are based on different aspects of the data, and all the methods have individual strengths and weaknesses. They are compared in Figure 8, and found to be in reasonable agreement for most of the bursts. The greatest disagreements occur for the weakest stresses, as might be expected. In general τ_p and τ_e are in close agreement, with τ_r taking a slightly smaller value, possibly due to low and high frequency losses (typically 5-22%) and spatial averaging due to the separation of the two EMCM heads. However, the agreement is sufficiently close that it is reasonable to average the three methods to provide a best estimate of the stress:

$$\tau_0 = \tfrac{1}{3}(\tau_p + \tau_r + \tau_e) \qquad (14)$$

4. DISCUSSION

One of the main goals of this study was the determination of the wave enhancement of τ_0, which increases the frictional resistance felt by tidal currents and hence the frictional force acting on the sea-bed sediments. Since sediment transport responds as a high power of τ_0, prediction of τ_0 in combined W+C flows has been an important aim of theories of wave-current interaction. To eliminate the tidal variation in current speed the results are presented in non-dimensional form as the drag coefficient $C_{100} = \tau_0/\rho\, U^2_{100}$, which is plotted against the wave:current ratio σ_{wave}/U_{100} in Figure 9.

The behaviour of C_{100} is seen to be similar to that of E, taking a constant value for current-dominated flows with $\sigma_{wave}/U_{100} \lesssim 0.1$. As the wave:current ratio increases beyond this value, C_{100} rises steeply in response to the increased eddy viscosity produced within the wave boundary layer, reaching values many times larger than the current-alone value of C_{100}. The data points lie scattered about a smooth curve (drawn by eye). Some of the scatter may be due to differences in the wave parameter a/z_0, where $a = \sqrt{2}\,\sigma_{wave}\,T_p/2\pi$ is the semi-orbital excursion, and in the relative wave-current direction θ. These lay in the ranges $51 < a/z_0 < 400$ and $8° < \theta < 70°$, but no discernible trends could be identified within the scatter.

A word of caution should be given, since the records with largest wave:current ratio (Bursts 62 and 27) occurred close to slack water, when tidal acceleration effects might introduce an extraneous physical process which would enhance the turbulence level and the stress. Considerable differences in direction of the mean velocities at the three heights, and also of the shear stress, were observed for these two records.

Also of note is Burst 59, which is identified as anomalous in Figures 7 and 9. This record had much the largest bed shear stress observed ($\tau_0 = 46.4$ dyn.cm^{-2}), due to a combination of peak spring tides and strong wave action. This exceeds the threshold of motion of grains finer than 7mm. It seems likely that this resulted in the finer grains in the bed being winnowed out, leaving a much rougher bed, and resulting in increased values of E and C_{100}.

A necessary input to W+C theories is the physical bed roughness

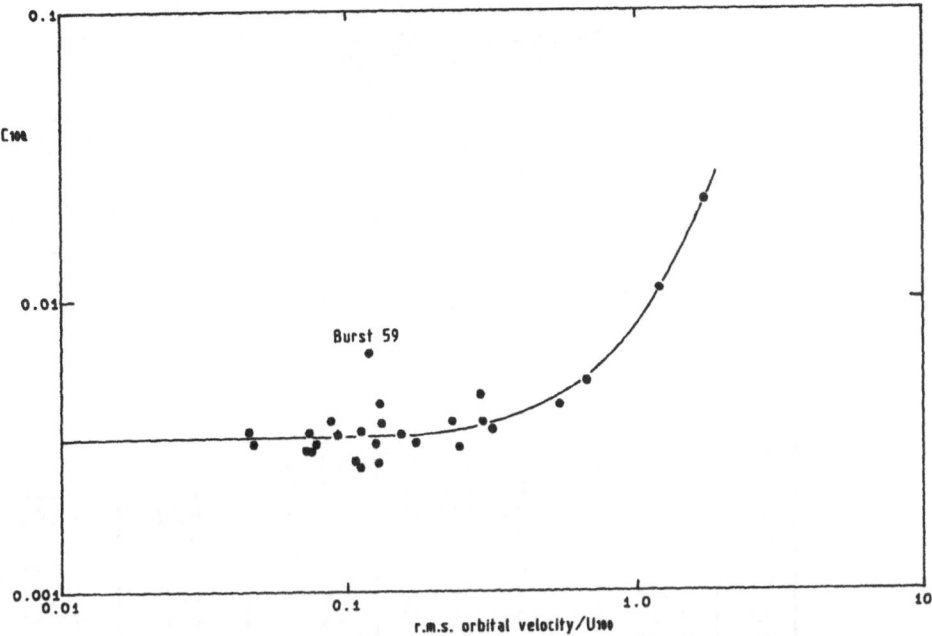

Figure 9. Drag coefficient $C_{100} = \tau_0/\rho\, U^2_{100}$ as function of wave-current ratio σ_{wave}/U_{100}

z_0. The most accurate method of evaluating z_0 is by back calculation from the mean value of $C_{100} = 0.0034$ for those records with negligible wave action, through

$$z_0 = 100 \exp(-\kappa/C^{\frac{1}{2}}_{100}) \qquad (15)$$

with $\kappa = 0.40$ and z_0 in cm, yielding a physical bed roughness of $z_0 = 0.10$cm. This value lies between the values 0.3cm for a gravel bed and 0.03cm for a sand/gravel mixture tabulated by Soulsby (1983), which is consistent with the observed bed composition. In terms of the Nikuradse roughness the observed z_0 is equal to $d_{50}/12.5$ or $d_{90}/27$.

A summary of the main quantities required for testing W+C boundary layer theories is given in Table 1. However, it is outside the scope of the present paper to make such tests. A more detailed tabulation of the data is available on request.

5. CONCLUSIONS

These field measurements of velocities near the sea-bed in combined wave and current flows have shown that:

(a) a technique of splitting the spectrum appeared to satisfactorily

TABLE 1. Summary of main measured flow variables. For all
bursts (except Burst 59) physical bed roughness
z_o = 0.10cm, water depth h = 25m.

Burst No.	Current Speed U_{100} (cm s^{-1})	R.m.s. Orb. Vel. σ_{wave} (cm s^{-1})	Period T_p (sec.)	Direction θ (deg.)	Bed Shear Stress τ_o/ρ (cm^{2}s^{-2})	T.K.E. E (cm^{2}s^{-2})
4	10.79	5.86	13.62	37.11	0.51	4.29
5	45.50	5.90	12.24	49.32	9.18	42.63
9	48.19	4.47	12.46	50.97	8.02	41.00
10	18.76	4.33	11.61	49.41	1.36	7.62
13	39.02	3.04	11.07	52.77	4.90	23.38
14	35.09	3.92	13.42	52.59	4.38	22.24
17	34.33	5.27	13.78	52.61	4.10	24.00
18	43.13	5.42	14.19	63.68	6.01	33.32
21	19.29	5.68	13.28	37.06	1.43	6.89
22	56.36	7.24	12.48	52.94	8.78	42.94
25	8.99	6.06	11.70	8.53	0.43	3.10
26	65.55	7.00	11.00	51.77	12.09	61.53
27	3.89	6.72	12.43	136.04	0.33	2.84
30	69.37	4.98	11.30	49.31	14.73	87.17
31	19.29	5.56	9.88	61.21	1.77	14.27
34	72.17	12.47	14.21	69.65	16.94	98.23
35	31.16	7.60	15.36	67.26	3.04	23.83
38	69.89	7.78	14.29	57.16	13.05	75.75
39	42.55	5.60	14.11	62.44	6.87	47.59
42	59.31	4.37	13.24	52.37	12.38	74.22
43	57.33	4.30	13.00	49.64	9.99	55.84
46	50.10	2.25	10.11	39.97	8.93	51.30
47	64.03	2.99	12.21	70.07	13.22	81.65
54	31.03	2.73	11.81	35.60	3.74	26.44
55	79.02	7.31	10.99	44.76	21.88	105.96
58	22.98	7.29	5.96	21.40	1.90	14.42
59	84.04	10.01	9.03	48.43	46.44	216.43
62	11.05	13.31	7.57	15.60	1.33	7.91

separate the variance due to wave orbital motions from that due to turbulence

(b)   the frequently-made assumption that the standard deviation of velocity in a W+C flow can be identified entirely with the waves should be viewed with caution, since the turbulent contribution can be quite large

(c)   orbital velocities at 10cm and 40cm above the bed were not significantly different, confirming that the wave boundary layer in these relatively mild conditions was thinner than 10cm

(d)   the detailed structure of the turbulence, as evidenced by the ratios of the velocity components, was not altered by the addition of wave-generated turbulence

(e)   three methods of deriving the bed shear-stress (profile, Reynolds stress, and energy) gave essentially similar values under all W+C conditions

(f)   both the bed shear-stress and the turbulent kinetic energy increased strongly with the wave:current ratio.

The results show qualitative agreement with theoretical predictions. A quantitative comparison is underway.

ACKNOWLEDGEMENTS

The valuable contribution made by Dr A D Heathershaw in the development of STABLE is gratefully acknowledged. The work was performed as part of a research programme into sediment transport by combined waves and currents funded by the Ministry of Agriculture Fisheries and Food.

REFERENCES

Bijker, E.W. (1967) 'Some considerations about scales for coastal models with movable bed', Publ. No. 50, Delft Hydraulic Lab., Delft, Netherlands, 142pp.

Davies, A.G., Soulsby, R.L. and King, H.L. (1988) 'A numerical model of the combined wave and current bottom boundary layer', J. Geophys. Res. 93 (C1), 491-508.

Dyer, K.R. and Soulsby, R.L. (1988) 'Sand transport on the continental shelf', Ann. Rev. Fluid Mech. 20, 295-324.

Fredsøe, J. (1984) 'Sediment transport in current and waves', Series Paper 35, Inst. Hydrodynamic and Hydraulic Eng., Tech. Univ. Denmark, Lyngby, 37pp.

Grant, W.D. and Madsen, O.S. (1979) 'Combined wave and current interaction with a rough bottom', J. Geophys. Res. 84 (C4), 1797-1808.

Humphery, J.D. (1987) 'STABLE - an instrument for studying current structure and sediment transport in the benthic boundary layer', Proc. Electronics for Ocean Technology Conference, Heriot-Watt Univ. Edinburgh, pub. by IERE, London.

Jonsson, I.G. and Carlsen, N.A. (1976) 'Experimental and theoretical investigations in an oscillatory turbulent boundary layer', J. Hydraulic Res. 14, 45-60.

428

Lambrakos, K.F. (1982) 'Seabed wave boundary layer measurements and analysis', J. Geophys. Res. 87 (C6), 1471-4189.

Lambrakos, K.F., Myrhaug, D. and Slaattelid, O.H. (1988) 'Seabed current boundary layers in wave-plus-current flow conditions', J. Waterway Port Coastal and Ocean Eng 114 (2), 161-174.

Myrhaug, D. and Slaattelid, O.H. (1989) 'Combined wave and current boundary layer model for fixed rough seabeds', Ocean Eng. 16 (2), 119-142.

Soulsby, R.L. (1980) 'Selecting record length and digitization rate for near-bed turbulence measurements', J. Physical Oceanog. 10 (2), 208-219.

Soulsby, R.L. (1983) 'The bottom boundary layer of shelf seas', in B. Johns (ed.), Physical Oceanography of Coastal and Shelf Seas, Elsevier Science Pub., Amsterdam, pp 189-266.

COMPUTATION OF STEEP WAVES ON A CURRENT WITH STRONG SHEAR NEAR TO THE SURFACE

JOHN R CHAPLIN
Department of Civil Engineering
City University
London EC1V 0HB
UK

ABSTRACT. This paper describes the results of computations of steep regular waves on a current which is concentrated near to the surface. The non-linear addition of wave and current leads to a significant increase in the elevation of the crest, and in the particle velocity at the crest, compared to a linear sum of the wave and current velocity components.

1. INTRODUCTION

For purposes of predicting loading on offshore structures it is desirable to have a reliable model for calculating flow kinematics in waves propagating on a current of given profile. The interaction between waves and currents is a complex process which has many features that are not well understood. This paper is concerned with the idealised state of affairs of non-linear regular waves which are assumed to exist on a stable collinear current of some arbitrary profile over the water column. It does not address the problem of the non-uniform or time dependent flow associated with the process through which a stable condition is achieved, but discusses a final steady state in which the waves are assumed to be propagating in fluid which has rotation but no viscous or turbulent shear stress. While it may apply to only idealised conditions, the value of a solution to this problem is in defining a standard which can in future form a basis for evaluating experimental results, and a benchmark for other models.

Possibly owing to experimental difficulties, there are few measurements which help the designer in determining particle velocities and accelerations in combined wave and current flows. With a given design wave and an assumed current profile, there are several ways in which the two flows might be added to provide information for loading calculations on offshore structures. A comparison of several different options is described by Eastwood et al (1987). They conclude that where there is need to resort to linear superposition, stretching the undisturbed current profile up to the instantaneous surface, with no

A. Tørum and O. T. Gudmestad (eds.), Water Wave Kinematics, 429–436.

attempt to impose conservation of mass flux, appears to be the best approximation at least in the region of the crest. This is likely to the the most critical region for wave loading calculations.

Alternatives to linear superposition which are available depend on the conditions. For small amplitude waves Thomas (1981) derived a method for linear waves on a current of an arbitrary profile. He found that for this case the predictions were in surprisingly good agreement with results obtained from an irrotational wave superimposed on a uniform current of magnitude equal to the depth-averaged mean of the desired current. For finite amplitude waves on a current of uniform shear (uniform vorticity) a suitable modification can be made to stream function wave theory as described by Dalrymple (1974a). Analytical methods for this case are also available (Simmen and Saffman 1985, and da Silva and Peregrine 1987). The stream function theory can also be modified to accommodate a current in the form of two (or more) layers of different but uniform vorticities (Dalrymple, 1974b).

For the more general problem of finite amplitude waves on a current of arbitrary profile, there seems to be only one method described in the literature and this is given by Dalrymple (1977). It is based on earlier work by Dubreil-Jacotin and uses a rather inelegant finite difference approach to approximate the eigenfunctions which form the basis of the more common stream function solution for irrotational waves. In principle Dalrymple's method can be applied to any wave/current combinations, but it is awkward to program and suffers from slow convergence rates.

The results described below were obtained from a program, based on Dalyrmple's (1977) method, that is described in detail by Chaplin (1989). The main differences between the method used here and Dalrymple's are that the stream function at the surface no longer has to be known a priori, and that a transformation is introduced to concentrate the finite difference mesh near to the surface, where the strongest vorticity and velocity gradients can be expected.

2. The computational method

The method is described fully by Chaplin (1989). The solution of the combined wave and current flow is computed iteratively, starting from initial conditions which in the cases presented below were provided either from linear theory or, for the free surface profile, from a 9th order stream function solution of the wave in the absence of any current. The computation proceeds to minimise the errors in the Dynamic Free Surface Boundary Condition by adjusting the surface profile, the value of the stream function at the surface, and the wavelength, while the flow everywhere satisfies the requirement

$$\nabla^2 \psi = \omega. \tag{1}$$

$\psi(x,y)$ is the stream function and ω the vorticity; x,y is a Cartesian reference frame moving with the wave speed. Since the flow is inviscid but rotational there is also a relationship between ψ and ω which is defined by the distribution of the undisturbed current $U(y)$:

$$\omega = f(\psi). \tag{2}$$

For any chosen undisturbed current profile the form of equation (2) may be found. The method used here is based on the assumption that the disturbance created by the waves is irrotational, so that equation (2) applies equally in the combined wave/current flow. An adjustment has to be made to allow for the effect of the mass transport introduced by the waves themselves. Equation (2) is derived for the undisturbed current when viewed from a frame of reference moving at such a speed that the total mass transport between the surface and sea bed is the same as in the combined wave/current flow when observed in the frame x,y. The derived form of equation (2) is then applied to the combined wave and current flow. Since the wave speed is a dependent parameter, the vorticity distribution has to be updated at each iteration of the solution.

The solution is obtained on a finite difference mesh (of $I \times J$ cells) which covers the domain of interest in the x,ψ plane in which the field variable y is the local elevation of the streamline whose stream function is ψ. As in stream function methods, the convergence of the solution may be measured in terms of the root mean square value of the departure, from the mean value, of the total heads calculated at the I nodes of the finite difference mesh that are on the free surface. This rms departure is below referred to as E.

Despite careful use of successive over-relaxation, the rate of convergence is slow, and the solutions given below required up to 2000 iterations.

3. Computed results for quadratic surface currents.

Results are presented below for waves of height H 16.8m, period T 14.4s, in water of depth h 34m. The current distribution is chosen so as to have zero overall mass transport, but consists of a uniform flow u_b in one direction, from the sea bed to an elevation of 32m, balanced by a current predominantly in the opposite direction over the top 2m of the water column. The current in the upper zone follows the form

$$U = u_b + \left(\frac{u_s - u_b}{4} \right) y^2, \tag{3}$$

where u_s is the current speed at the surface, and y (in metres) is the elevation above 32m from the bed. The five cases presented are defined in Table 1. In each case the mesh dimensions I and J were respectively 13 and 50.

Table 1

Computed results for waves of H = 16.8m, T = 14.4s, h = 34m, with surface currents between -2 to +2m/s

Case	$u_s - u_b$ (m/s)	E (mm)	ψ_s (m²/s)	L (m)	crest elevation (m)	surface velocities	
						crest (m/s)	trough (m/s)
1	-2	23	580.3	253.4	11.55	5.03	-4.50
2	-1	28	581.9	253.9	11.62	6.66	-3.62
3	0	27	583.4	254.5	11.70	8.41	-2.74
4	1	18	585.7	255.3	11.80	10.43	-1.86
5	2	21	588.3	255.9	12.04	13.44	-0.92

Notable results are the disproportionate increases in crest elevation and particle velocity at the crest with increasing surface current speed. These are plotted in figure 1. It is not surprising that there are substantial changes since the ratio of particle velocity at the crest to the wave speed increases from 0.35 (case 1) to 0.76 (case 5). Non-linear features can be expected to increase accordingly in importance. A corresponding change in the free surface profile near to the crest is shown in figure 2. The relative increase in elevation at the crest in case 5, and the lowering of the surface at a small distance away, are similar to features found in irrotational waves as the steepness (or degree of non-linearity) is increased. In the present case there is virtually no change in steepness, and the differences are due to the addition of the current at the surface. Figure 1 also shows the substantial non-linear increase in particle velocity at the crest, as the strength of the current increases.

The distribution of particle velocities over the water depth beneath crest and trough for cases 1 and 5 are plotted in figure 3. This shows the stretching of the surface current, which occupies slightly more than the top 9% (4.2m) of the instantaneous water column under the crest in case 5. In the undisturbed case the current (over the top 2m) extends downwards by less than 6% of the water depth.

4. Conclusions

Computations of non-linear waves propagating on water with a shallow surface current, by a method based on Dalrymple (1977), reveal significant non-linear features. For a given wave height and period,

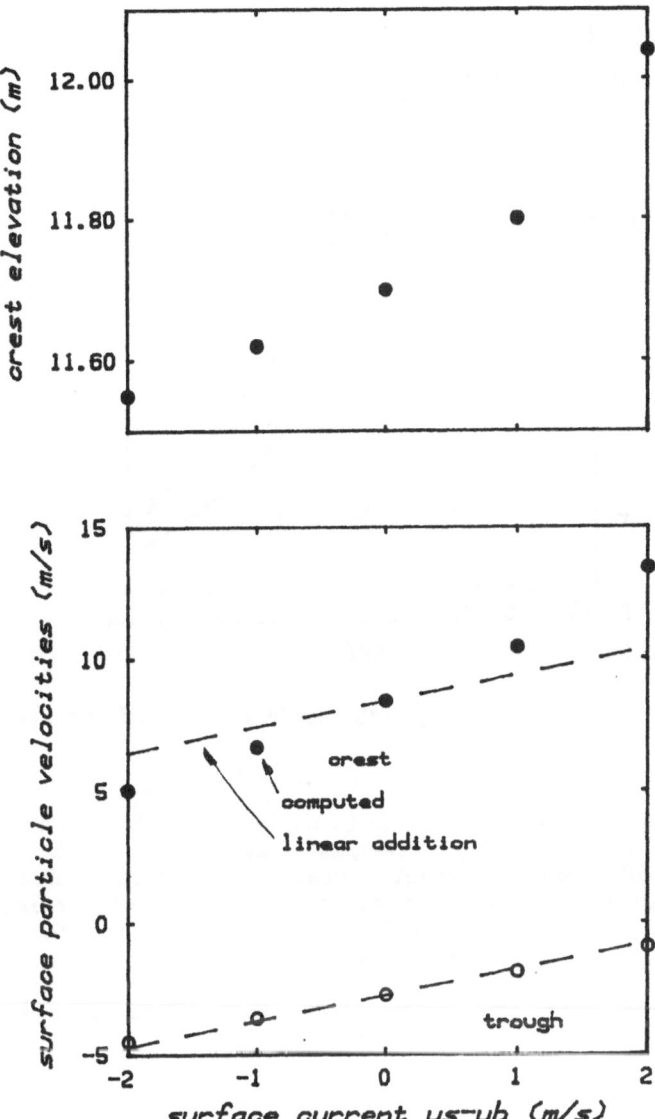

FIGURE 1. Surface particle velocities at the crest and trough, and crest elevations, as functions of the net surface current. The overall mass transport due to the current is zero.

434

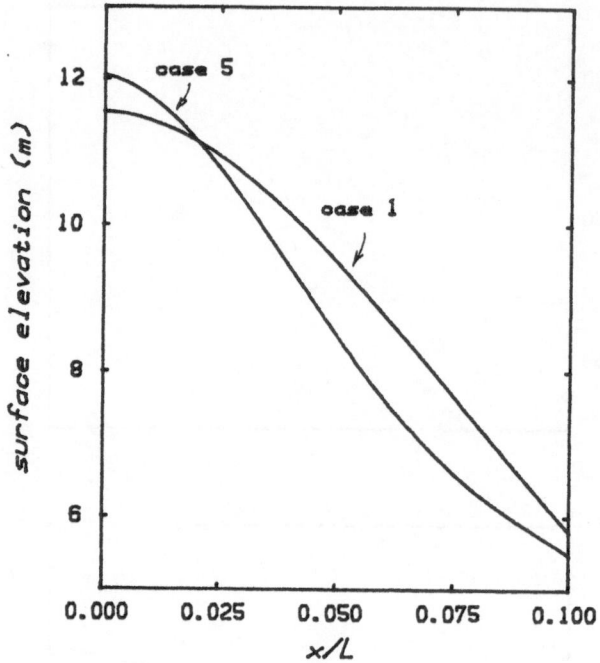

FIGURE 2. Free surface profiles in the region of the crest.

and with no net mass transport due to the current, the elevation of the crest and the particle velocity at the crest increase disproportionately with current speed, and more rapidly than predictions based on linear superposition. In view of the importance the kinematics of wave crests in loading calculations for offshore structures, these features warrant further study.

There is also a need for good quality experimental data on waves propagating on non-uniform currents.

5. Acknowledgement

The code used in the computations carried out for this paper was developed in a project funded by the UK Department of Energy.

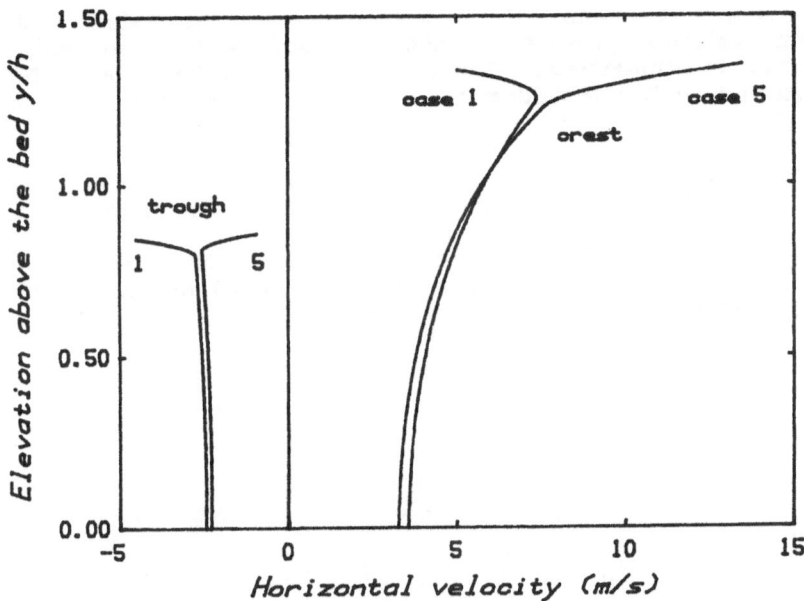

FIGURE 3. Particle velocities beneath the crest and trough

6. References

Chaplin, J R, 1989, 'The computation of non-linear waves on a current of arbitrary non-uniform profile', Report to the UK Department of Energy under contract TA 93/22/323:1.

Dalrymple, R A, 1974a, 'A finite amplitude wave on a linear shear current', Journal of Geophysical Research, vol79, pp4498-4504.

Dalrymple, R A, 1974b, 'Water waves on a bi-linear shear current', Proceedings of the 14th International Conference on Coastal Engineering, ASCE.

Dalrymple, R A, 1977, 'A numerical model for periodic finite amplitude waves on a rotational fluid', Journal of Computational Physics, vol24, pp29-42.

da Silva A F and Peregrine D H, 1987, 'Steep, steady surface waves on water of finite depth with constant vorticity', School of Mathematics, Bristol University, Report AM-87-14.

Eastwood J W, Townend I H and Watson C J H, 'The modelling of wave current velocity profiles in the offshore design process', 1987, Advances in Underwater Technology, Ocean Science and Offshore Engineering, vol12: Modelling the Offshore Environment. Society for Underwater Technology, London, pp327-342.

Simmen J A and Saffman P G, 1985, 'Steady deep water waves on a linear shear current', Studies in Applied Mathematics, vol73, pp35-57.

Thomas G P, 1981, 'Wave-current interactions: an experimental and numerical study. Part 1. Linear waves', Journal of Fluid Mechanics, vol110, pp457-474.

KINEMATICS OF FLOW ON STEEP SLOPES

A. R. GÜNBAK
Assoc.Prof.Dr.
Coastal and Harbor Eng. Lab.
Middle East Tech. Univ.
Ankara 06531 Turkey

ABSTRACT. Small scale two dimensional hydraulic model tests with
regular waves on steep impervious smooth structure slopes of 2/3, 1/2
and 1/3 showed that, maximum impact, drag and lift forces on such slopes
occurs under a specific flow kinematics when a deep run-down following
a high run-up is interacted with an incoming plunging-collapsing breaker.
This critical flow occurs at a surf similarity parameter of $2.5 < \xi < 3.5$.
The kinematics of flow at this critical flow condition together with
its probability of occurence is most interesting during the design of
dike or rubble mound slopes.

1. INTRODUCTION

An incoming short gravity wave propagating towards shallower water
depths on a sloping bottom breaks on the slope and runs up to an
elevation named wave run-up, Ru. Under the action of gravity the run-up
tongue recedes down the slope to an elevation where it meets with the
new incoming breaker. This lowest elevation of the water front on the
slope is named wave run-down, Rd. Roos and Battjes (1976), Battjes (1974)
and Günbak (1979) gave an extesive summary of the change of these
properties by changing wave and slope characteristics. They describe the
flow kinematics on the slopes using the surf similarity parameter, ξ,
described on Fig.1.
Below, a description of the flow conditions which is most critical
to consider during the design of a sloping structure against wave action
shall be given. Although the results are given from regular wave tests
with steep impervious smooth slopes, the conclusions are qualitatively
correct for a beach slope or for a rubble-mound breakwater slope.

2. FLOW CONDITIONS ON A SLOPE

Wave flume tests conducted with regular waves on smooth impervious
continuous slopes of 2/3, 1/2 and 1/3 showed the non-dimensional wave

A. Tørum and O. T. Gudmestad (eds.), Water Wave Kinematics, 437–440.

438

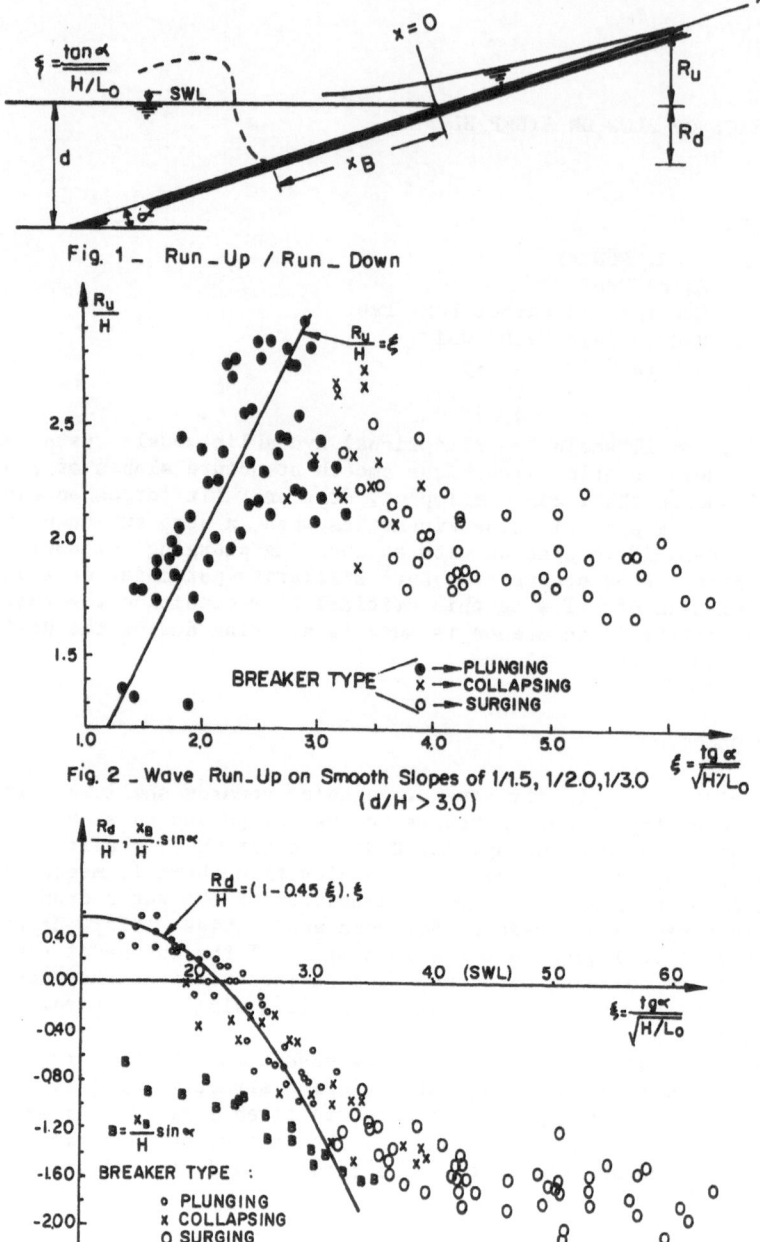

Fig. 1 _ Run_Up / Run _ Down

Fig. 2 _ Wave Run_Up on Smooth Slopes of 1/1.5, 1/2.0,1/3.0
(d/H > 3.0)

Fig. 3 _ Wave Run_Down and Breaking Point Data on Smooth Slopes

run-up and run-down variation shown on Fig.2 and Fig.3. On Fig.3 the non-dimensional breaker distance, XB/H, is also shown. The relative run-up and run-down may be expressed by the equations; Ru/H=ξ for ξ<2.7 Ru/H=1.9 for ξ>4.0 Rd/H=1-0.45 ξ for ξ<2.7 Rd/H=-1.7 for ξ> 4.0. From these Figures, it may be seen that wave run-up and run-down both increases by increasing ξ values in the range of breaking waves up to surging breakers and spreads around a constant value with surging breakers. Wave run-up shows a maximum value at the passage from collapsing to surging breakers.

Defining swash height, Hsh=Ru-Rd, as the vertical distance on the slope which becomes wet and dry during a wave cycle,

$$Hsh=0.45 \ \xi^2 H=0.072 \ gT^2 (\tan \ \alpha)^2 .... \text{ for } \ \xi< 2.7$$

$$Hsh=3.6 \ H .... \text{ for } \ \xi> 4.0$$

where g is the acceleration of gravity. It may be seen that swash height is independent of wave height with breaking waves and only dependent to wave height with surging breakers.

The above conclusions are true for milder impervious smooth slopes tested above, except the applicability of the ξ ranges may change slightly toward higher ξ values. Sandy beaches may also be included in this group.

For rubble mound breakwaters, wave run-up do not show any maximum value at the passage from breaking to surging waves but show a continuous increase. All the other general conclusions are qualitatively applicable also for such slopes. The quantities do change due to percolation and transmission.

3. CRITICAL FLOW CONDITIONS FOR STRUCTURAL DESIGN

Wave run-up usually determines the crest elevation of a structure. Impact pressures from a plunger breaker is used during the design of impervious dike slopes and it is this force dislocating an armor stone upslope on a rubble mound breakwater slope of 1/3 or milder. From Fig.3 it may be seen that there remains a layer of water from the previous run-down at low ξ values with a plunging breaker. This causes the plunging breaker to hit on a layer of water remained from the previous run-down and relatively gentle impact pressures shall be experienced by the slope. The highest pressures shall exist with plunging breakers hitting on bare slopes which are expected to occur with ξ values between 2.5 to 3.5.

The slope perpendiculer lift forces are very important for the design of dike slopes and armor stability of rubble-mound slopes. Fig.4 shows the time lapse drawings of an incoming breaker interacting with the run-down water. Schematic vectorial velocity representations are also shown. Considering results presented at Fig.2 and Fig.3, it may be seen that highest downslope velocities are expected at the lowest run-down values following a high run-up. Therefore, these highest run-down velocities which may dislocate an armor stone down slope is expected to occur with surging breakers. But with surging breakers run-down completes it-self before the next incoming wave and the interaction of a run-down and

440

run-up flow does not exist. This interaction does exist with breaking waves where the down slope oriented run-down velocities meeting the upslope oriented breaking toe velocities cause an eddy-like turn of the velocity field which has an acceleration and velocity component perpendicular to the slope as shown in Fig.4. Higher run-down velocities meeting with high incoming breakers cause larger slope perpendicular accelerations and velocities which is decisive in dislocating an armor rock or tearing the asphalt or concrete surface of a dike or pulling a lot of sediment to suspension at a beach.

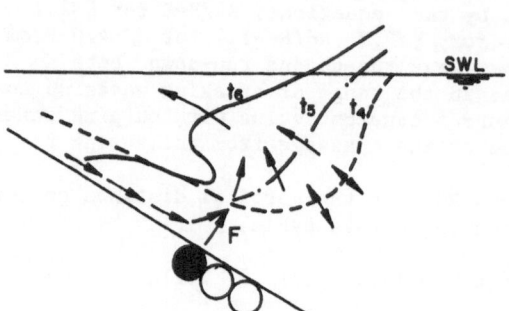

Fig. 4 _ Interaction of Run_Up _ Run_Down

Therefore, it is the kinematics of these high breaking waves which generated deep run-downs following high run-ups which is interesting from the structural point of view. It is highly necessary to be able to define these conditions for a real sea stochastically, which will be generated from a combination of at least two succeeding individual waves.

REFERENCES.

Roos a. and Battjes, J.A. (1976), 'Characteristics of flow in run-up of periodic waves, Proceedings of the 15th Conference on Coastal Engineering sponsored by ASCE, 781-795.

Battjes,J.A., (1974), 'Computation of set-up, longshore currents, run-up and overtopping due to wind generated waves, communications on Hydraulics, Dept. of Civil Engineering, Delft Univ. of Tech. Rept. no. 74-2, 1-31.

Günbak, A.R. (1974), 'Rubble Mound Breakwaters', Division of Port and Ocean Engineering Report no. 1, The University of Trondheim, Norway.

MEASUREMENT OF THE VELOCITY AND TURBULENT FIELDS GENERATED BY THE SWELL
IN THE VICINITY OF WALLS OR OBSTACLES

M. BELORGEY, A. JARNO, J. LE BAS & E. VASSELIN
Laboratory of Fluid Mechanics
Université du Havre
25 Rue Philippe Lebon   BP 540
76058 Le Havre Cédex
France

ABSTRACT. Swell develops an orbital motion of the fluid particles, excluding any technique of measurement of the velocities with a material
support, because the disturbance induced by the probe comes back to the
point of measurement.
Laser Doppler velocimetry allows a sharp measurement of instantaneous
velocities and turbulence in fluid flows, without disturbing the movement of the fluid particles.
Accordingly we have adapted this technique so as to analyse :
     - the oscillatory boundary layer generated by swell on the sea-bed
     - the swell effect on structures.
Our results qualify some modelisations.

1. INTRODUCTION

For more than a century, wave study has remained one of the few branches
of learning in which we know more from theorical calculation than from
observations of direct measurements in situ. Up to now, swell simulation in the laboratory has principally allowed to know better about the
water particles motion by a direct view of their trajectories, and to
measure in the aggragate, wave stresses upon structures inside a reduced pattern, and to appreciate their effects.

The measurement of instantaneous velocities of water particles inside swell generated in the laboratory is one aspect which has been yet
little approached. Now the knowledge of the instantaneous velocities of
fluid particles allows us to know instantaneous hydrodynamical stresses
and then have a keener approach of the swell-structure interaction problems and of the sediment carriage.

Doppler Laser velocimetry is presently the only technique allowing
the measurement of instantaneous velocities without disturbing the flow.
It has already been used by Beech (1978) to measure mean velocities in
the boundary layer on the bottom. By interpolation, this method allows
him to deduce the mean velocity profile in a cross-section of an oscillatory boundary layer. This technique has also been used by Anastasiou,
Tickel and Chaplin (1982) in the case of non-stop measures in a wave.

441

As for us, we have adapted Laser Doppler velocimetry to the particular conditions of the orbital motion of fluid particles (Bélorgey and Le Bas (1984)). Accordingly, we have created an interface between the swell detector and the Laser Doppler velocimeter associated with a computer IBM AT3 (figure 1). This system allows us to measure the instantaneous velocity components of the particles for a given wave phase, giving us access to the component mean values, along with the R.M.S. velocity fluctuations (figure 2) (Vasselin (1987)).

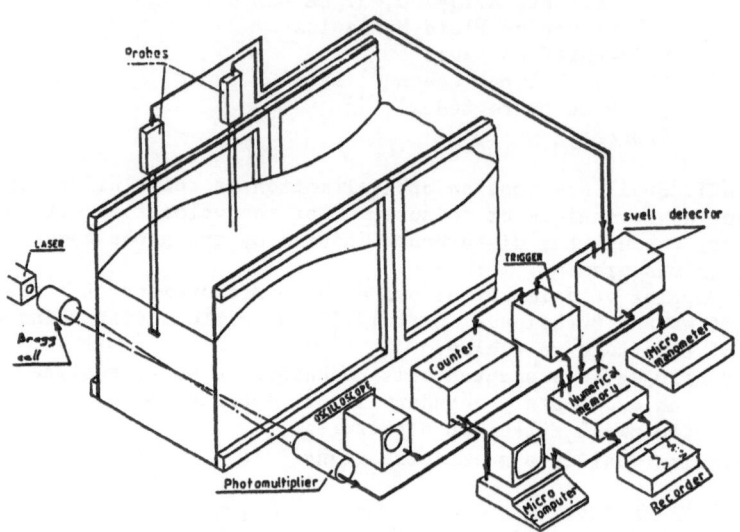

Figure 1. Diagram of the successive elements of measurement

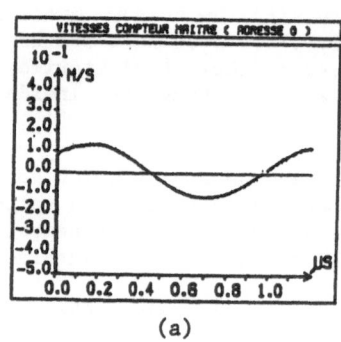

(a)

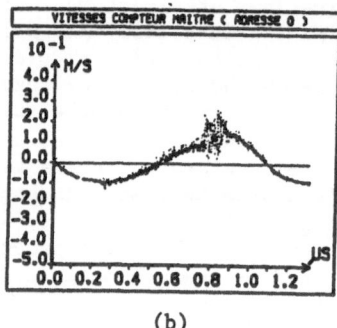

(b)

Figure 2. Instantaneous velocities
       a) : without obstacle
       b) : in a point behind a cylinder

2. STUDY OF THE OSCILLATORY BOUNDARY LAYER GENERATED BY SWELL ON SEA-BEDS

The analysis of our results from the boundary layer point of view, allows us to bring a new light on the oscillatory boundary layer phenomenon on the bottom of the sea.

For this study, we had the following swell conditions :
- Amplitude H $= 0.047$ m
- Wave length L $= 1.27$ m
- Period T $= 1.08$ s
- Water depth h $= 0.27$ m

The swell curve H/L of 0.037 was smaller than 0.048 (necessary conditions for a good stability of the waves during their propagation).

2.1. Study of the mean field and of the turbulent field

Our investigations especially relate to the shear stress and the turbulence production terms, according to the swell phase.

The study of the mean field and of the turbulent field in a point of the flow is performed by writting scalar magnitudes in the following form :

$$U_i = \overline{U}_i + u_i'$$

In this case, $\overline{U}_i$ is a periodic function of the time. This leads to the mean motion equation :

$$\rho \left(\frac{\partial \overline{U}_i}{\partial t} + \overline{U}_j \frac{\partial \overline{U}_i}{\partial x_j} \right) = \rho \, \overline{f}_i - \frac{\partial \overline{p}_i}{\partial x_i} + \frac{\partial \tau_{ij}}{\partial x_j}$$

With : (Reynolds'tension)

$$\tau_{ij} = \mu \left(\frac{\partial \overline{U}_i}{\partial x_j} + \frac{\partial \overline{U}_j}{\partial x_i} \right) - \rho \, \overline{u_i' u_j'}$$

In the case of monochromatic swell, we can consider the bidimensional motion. By choosing an axis system $(\vec{x}, \vec{y})$ such that $(0, x)$ is in the swell propagation direction and $(0, y)$ vertical ascendant, we obtain the tangential stress under the following form :

$$T_x = \mu \left(\frac{\partial \overline{u}}{\partial y} + \frac{\partial \overline{v}}{\partial x} \right) - \rho \, \overline{u'v'} \qquad \text{with } u = \vec{U} \cdot \vec{x}$$
$$v = \vec{U} \cdot \vec{y}$$

Near the wall, our measurements show that :

$$\frac{\partial \overline{v}}{\partial x} \cong 0 \qquad \text{and} \qquad - \rho \, \overline{u'v'} \gg \mu \frac{\partial \overline{u}}{\partial y}$$

In other respects, the equation of turbulent kinetic energy can be written under the following general form :

$$\underset{\text{①}}{\frac{D\overline{q^2}}{Dt}} = - 2 \ \overline{u_k' u_i'} \ \frac{\partial \overline{u_i}}{\partial x_k} \qquad \text{②}$$

$$+ 2 \ \nu \ (\overline{\frac{\partial u_i}{\partial x_k} \cdot \frac{\partial u_i}{\partial x_k}} + \overline{\frac{\partial u_i}{\partial x_k} \cdot \frac{\partial u_k}{\partial x_i}}) \qquad \text{③}$$

$$+ \frac{\partial}{\partial x_k} \{ u_k \ q^2 + \frac{2p'}{\rho} - \nu \ \frac{\partial q^2}{\partial x_k} - 2 \ \nu \ \frac{\partial \overline{u_i' u_k'}}{\partial x_i} \} \qquad \text{④}$$

The term ② called turbulence production term informs on structure nature in the motion.

In our case, with our notations, this term includes the following elements :

$$- \overline{u_k' u_i'} \ \frac{\partial \overline{u_i}}{\partial x_k} = - \ \overline{u'^2} \ \frac{\partial \overline{u}}{\partial x} - \overline{u'v'} \ \frac{\partial \overline{u}}{\partial y} - \overline{u'v'} \ \frac{\partial \overline{v}}{\partial x} - \overline{v'^2} \ \frac{\partial \overline{v}}{\partial y}$$

An analysis of the different terms magnitude has shown that the only term

$$- \overline{u'v'} \cdot \frac{\partial \overline{u}}{\partial y} \qquad \text{was preponderant.}$$

2.2. Experimental results

The first results concerning the boundary layer has been given by Bélorgey and Le Bas (1986), Le Bas (1986).

2.2.1. Mean horizontal components in the boundary layer

Figure 3 shows the evolution of the mean horizontal components of the velocities for different swell phases according to the height above the bottom.

The analysis of these results shows up not only the overflow for the phases corresponding to the crest (0°) and to the trough (180°) but also an inversion of the velocity near the bottom for phases next 90° and 270° which correspond to zero horizontal components in the free flow out of the boundary layer.

2.2.2. Analysis of the turbulent field

Analysis of the horizontal component fluctuations of the velocities in the form :

$$\frac{\sqrt{\overline{u'^2}}}{c} \qquad \text{(c : wave propagation velocity)}$$

Two types of curves are obtained by drawing the evolution of $\frac{\sqrt{\overline{u'^2}}}{c}$ in terms of $\frac{y}{h}$, for each phase of the swell.

Figure 4a shows this evolution for a section of measurements corresponding to the wave crest (phase 0°). For this phase, it can be seen that the maximum of $\sqrt{\overline{u'^2}}/c$ occurs well for a height above the bottom corresponding to the thickness of the renewal sublayer, as it occurs for a turbulent boundary layer generated by an unvarying flow. The same type of curve is found in the wave trough (180°).

Figure 4b shows the evolution of $\sqrt{\overline{u'^2}}/c$ for the swell phase where the horizontal components are near 0. In this case the curve shows two maxima of turbulence, corresponding to zones having a high velocity gradient.

2.2.3. Study of the stresses

The results given by the figure 5 allow the comparison between the turbulent stresses and the viscous stresses. It is seen that $|\rho \, \overline{u'v'}|$ is widely more important than $|\mu(\partial \overline{u}/\partial y)|$. So, the only evolution of the turbulent shear stress, according to the swell phase must be considered.

This curve shows a peculiarity ; as a matter of fact the stress sign is the same for the phase 0° (crest) and for the phase 180° (trough) whereas the mean velocities have opposed signs for the same phases. This peculiarity is confirmed by positive results relating to the terms of turbulence production under mentionned.

2.2.4. Turbulence production

The figure 6 shows the evolution of the turbulence production term $(\overline{u'v'} \, \partial \overline{u}/\partial y)$, according to the swell phase for a given height above the bottom (y = 0.25 mm). It can be noticed that for the phase corresponding to the trough (180°), the production term is negative, whereas it is positive for the crest (0°). This result confirms the peculiarity of the signs of the turbulent shear stress and we find again the notion of "negative production" of turbulence, pointed out by Béguier (1976) in unlike cases of boundary layer.

2.3. Conclusion on the turbulent boundary layer

We have presented only a part of our results in this paper, but yet they allow us to draw some conclusions :
- phenomena in the boundary layer are very different according to the successive swell phases especially for crests and troughs.
- crest passages are priviliged times for eddy production, whereas trough passages correspond more to a "relaminarisation" of the flow.
From that fact an explanation can be given concerning the transport of sediments : the mean transport is due to an important displacement during the wave crest flow (positive mean velocity and positive turbulent shear stress), then a lighter displacement in the opposite direction occurs during the wave trough flow (negative mean velocity but tur-

bulent shear stress still positive). (Figure 7).

3. DYNAMIC EFFECT OF THE SWELL ON IMMERGED STRUCTURES

Our present study related to a horizontal cylinder at low Keulegan–Carpenter number (K.C. = Um . T/D = 2.3 and Re = Um . D /ν = 4400).
The experimental conditions are shown on figure 8.

3.1. The kinematic approach

Application of the dynamic equation to a control volume around the cylinder leads to an analysis of the velocity terms influence on the total loading.

Fluid volume $\mathcal{D}$ is limited by $S \cup S_0$ closed surface. Notations are defined in figure 9.

The dynamic equation applied to $\mathcal{D}$ is :

$$\iiint_{\mathcal{D}} \rho \frac{d\vec{V}}{dt} d\omega = \iiint_{\mathcal{D}} \rho \vec{F} d\omega + \iint_S \vec{T} dS + \iint_{S_0} \vec{T} dS_0$$

A determination of the total drag T exerted by the fluid on the cylinder is obtained, projecting the dynamic equation on the x – axis.

$$T = \frac{+}{-} \iint_{S_{1/3}} \mu(\frac{\partial \overline{U}}{\partial y} + \frac{\partial \overline{V}}{\partial x}) dS_{1/3} \quad \frac{+}{-} \iint_{S_{2/4}} - p \, dS_{2/4} \quad - \iiint_{\mathcal{D}} \rho \frac{\partial \overline{U}}{\partial t} d\omega$$

$$\frac{-}{+} \iint_{S_{1/3}} \rho \, \overline{U}.\overline{V} \, dS_{1/3} \quad \frac{-}{+} \iint_{S_{1/3}} \rho \, \overline{u'v'} \, dS_{1/3} \quad \frac{-}{+} \iint_{S_{2/4}} \rho \, \overline{u'^2} \, dS_{2/4}$$

$$\frac{-}{+} \iint_{S_{2/4}} \rho \, \overline{U}^2 \, dS_{2/4}$$

where $(\overline{U},\overline{V})$ and (u',v') are the mean instantaneous velocity components and the associated fluctuations ; p is the pressure and μ the dynamic viscosity.

3.2. Velocity measurements at the wave crest

Mean instantaneous velocities at the wave crest passage above the cylinder and 0.04 s later, are presented in figure 10. Wave propagation sense is from right to left. Knowledge of the kinematic field around the cylinder at two very near instants will permit to approximate the instationnary term $\partial \overline{U}/\partial t$ appearing in force calculation when using the control volume method.

Experimental data where smoothed with an interactive B - Spline method. Streamlines were plotted for these 2 phases (figure 11) and reveal the existence of a vortex attached to the cylinder, which moves around it with a 0.1 m s^{-1} mean circonferential velocity. While rotating the vortex grows but seems to loose intensity.

The upper flow separation point displaces from 120° position at the wave crest to 132°, 0.04 s later.

3.3. In-line force analysis

Measured in-line force and calculated one using Morison's vectorial equation have been compared. The comparison leads to the conclusion that the measured shift between wave and force is lower than the predicted one. This tends to prove that inertial effects due to fluid acceleration are over estimated by the model.

Another remark concerning the total wave force can be done. Application of Morison equation gives a zero average force on a period when measurement leads to a non-zero force.

3.4. Conclusion on the effect of the swell on immerged structures

An important vortex activity, all talking place downstream the cylinder has been pointed out. Through K.C. number is small, the lower vortex development may be sufficient to permit a shedding. This phenomenon due to viscous effects may induce a significant loading on the cylinder.

Another contribution to the in-line force is due to a circulation of high intensity around the cylinder revealed by visualization.

Analysis of the real flow around the cylinder tends to demonstrate that potential theories used to calculate forces on big horizontal structures are not completely satisfying.

4. REFERENCES

Anastasiou, Tickel and Chaplin, J.R. (1982) 'Measurements of particle velocities in laboratory sole random waves', Coastal Engineering, 6, pp. 233-254.

Bélorgey M. and Le Bas J. (1984) 'Mesure par vélocimétrie Doppler laser de la vitesse instantanée des particules fluides dans une houle produite en laboratoire', La Houille Blanche, n° 5, pp. 363-368.

Bélorgey M. and Le Bas J. (1986) 'Application of the Laser Doppler Velocimetry to the study of the turbulence generated by the swell in the vicinity of walls or obstacles', Proc. of 3rd Int. Symp. on Applic. of the laser anemometry to fluid mechanics, LADOAN publishers (Lisbonne), pp. 1-6 in chap. 2.3.

Vasselin E. (1987) 'Acquisition et traitement de données pour vélocimétrie doppler à laser', Rapport de stage DEA Instrumentation et commande, Université du Havre.

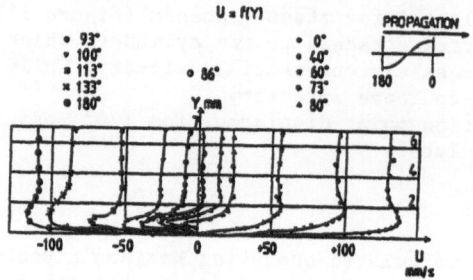

Figure 3. Horizontal velocity field for different wave phases.
0°≤∅ ≤180°

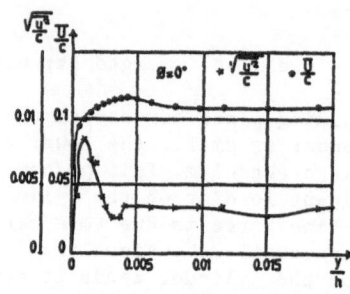

Figure 4a. Evolution of the velocity
fluctuations and of the velocity horizontal component
∅ = 0°. (C is wave velocity)

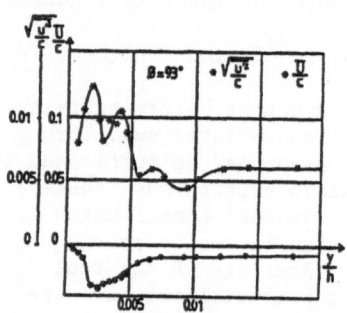

Figure 4b. Evolution of the
velocity fluctua-
tions and of the
velocity horizon-
tal component
∅ = 93°. (C is
wave velocity)

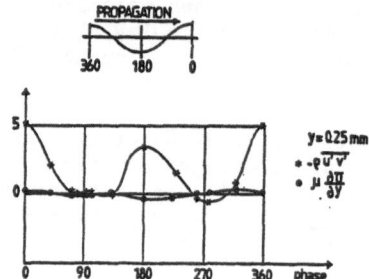

Figure 5. Comparison between viscous and
turbulent shear stress

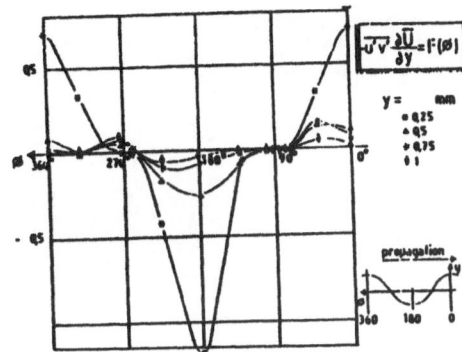

Figure 6. Evolution of the turbulent production terms according to the wave phase.

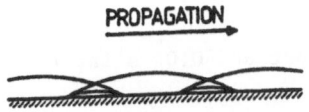

Figure 7. Sketch of sediment transport

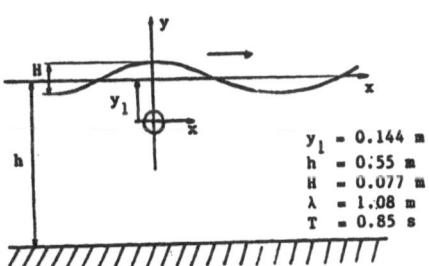

$y_1 = 0.144$ m
$h = 0.55$ m
$H = 0.077$ m
$\lambda = 1.08$ m
$T = 0.85$ s

Figure 8. Experimental conditions

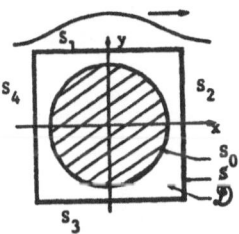

Figure 9. Definition of the control volume

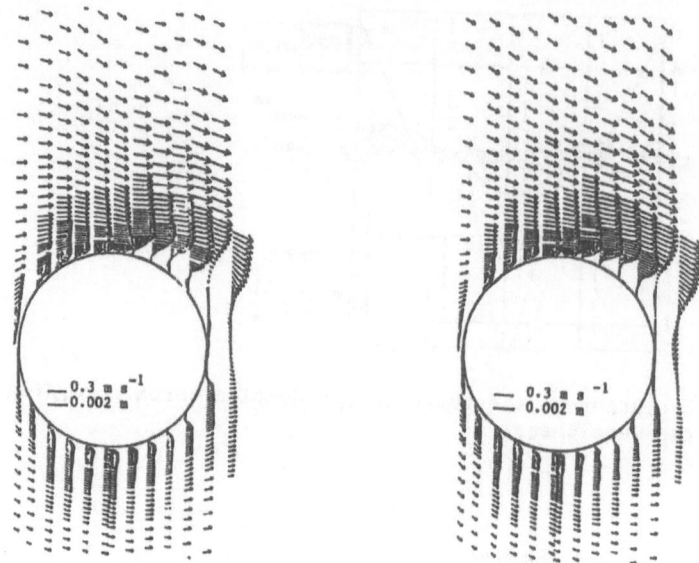

Figure 10. Velocity field at the wave crest and 0.04 s later

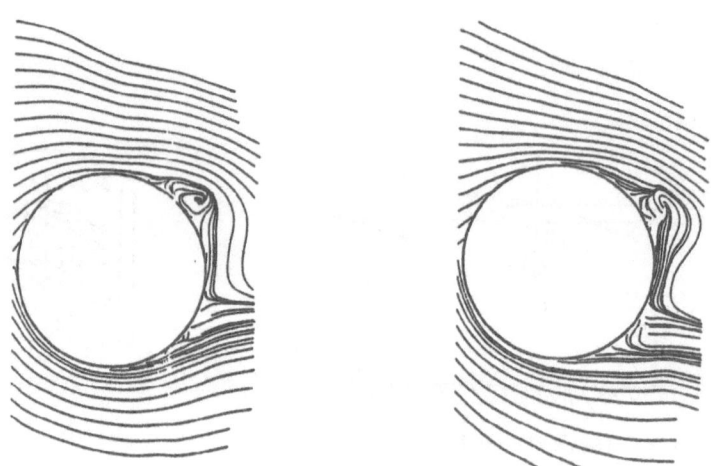

Figure 11. Streamlines at the wave crest and 0.04 s later

Breaking and Freak Waves

BREAKING WAVES

SØREN PETER KJELDSEN
MARINTEK, Norwegian Marine Technology Research Institute A/S
P.O. Box 4125
N-7002 TRONDHEIM
NORWAY

ABSTRACT. A new non-linear 3-dimensional Lagrangian theory is presented for prediction of kinematics in the crests of gravity sea
waves. The theory is compared with measurements made in laboratory
experiments, and may be used to calculate wave loads and slamming
caused by freak waves. 10 freak waves have been measured on the Norwegian Continental Shelf. In 2 cases, two freak waves occurred in
rapid succession within the same wave group. 9 of the 10 freak waves
occurred without any influence from the sea bottom.

1. INTRODUCTION

The study of breaking waves was initiated in Norway after some severe
ship accidents had occurred. In the period 1970-78 not less than 26
Norwegian vessels were lost due to capsizings. Altogether 72 lives
were lost. In 13 of the accidents surviving members of the crews
reported to the Courts of Inquiry that their vessel was unexpectedly
hit by an extreme steep and elevated wave that suddenly capsized the
vessel. The largest of these vessels had a length of 70 m and a
displacement of 500 tons (KJELDSEN 1983). In the 13 remaining accidents there were no survivors. Courts of Inquiry concluded that these
vessels disappeared in very bad weather, the exact series of events
being unknown, but that a sudden capsizing in steep waves was the most
probable cause since no emergency calls were given. On January 3rd
1984 a wave crest peaking 20 m above mean water level damaged a wall
of the control room at the 2/4-A platform at the EKOFISK field. There
a thin metal sheet was destroyed and water flooded the room behind.
Due to water in the electrical circuit, power was lost and production
was stopped for a period of 24 hours, (KJELDSEN 1984 a).

A. Tørum and O. T. Gudmestad (eds.), Water Wave Kinematics, 453–473.
© 1990 Kluwer Academic Publishers.

2. NON-LINEAR 3-DIMENSIONAL LAGRANGIAN THEORY

A theory is developed to predict kinematics in the crest of gravity waves. Input to the mathematical model are 3 time series measured at the same position x_0, y_0, namely the surface elevation $\eta(x_0, y_0, t)$ and the slopes of the wave surface $S_x(x_0, y_0, t) = \partial\eta/\partial x$ and $S_y(x_0, y_0, t) = \partial\eta/\partial y$ as they can be measured with a pitch-and-roll directional wave buoy, or alternatively with measurements of wave slope within a wave basin. Output from the mathematical model are 3 time series giving the (3-D) position of a particle at the free surface, 3 time series giving the local particle velocities for this particle in a Lagrangian description, and 3 time series giving the local particle accelerations.

When the slopes of the free surface are measured in the orthogonal (x-y) directions, then a mean wave propagation direction θ can be computed:

$$\tan 2\theta = \frac{2\, \overline{\frac{\partial\eta}{\partial x} \cdot \frac{\partial\eta}{\partial y}}}{(\overline{\frac{\partial\eta}{\partial y}})^2 - (\overline{\frac{\partial\eta}{\partial x}})^2} \qquad (2.1)$$

in which θ is positive anticlockwise from the x-axis and the overbar represents time averages. (LONGUET-HIGGINS, 1957).

The surface elevation is assumed to consist of a non-linear Stokes wave propagating in the direction θ and a large number of linear waves propagating in all directions:

$$\eta(t) = a_p \cdot \cos \omega_p t + \tfrac{1}{2} K_p a_p^2 \cdot \cos 2\omega_p t + 3/8 K_p^2 \cdot a_p^3 \cdot \cos 3\omega_p t$$

$$+ \sum_{i=1}^{n} a_i \cos(\omega_i t + \phi_i) \qquad (2.2)$$

Her ω_p is given as:

$$\omega_p = 2\pi f_p \qquad (2.3)$$

where f_p is the peak frequency in the wave spectrum, a_p is the wave amplitude corresponding to the peak frequency, and K_p is the wave number calculated from the non-linear dispersion relation:

$$gK_p (1 + K_p^2 a_p^2 + \ldots) = \omega_p^2 \qquad (2.4)$$

We now use a coordinate system with the xy-plane at the mean water level, and the vertical z-axis positive upwards. The position of a particle at the free surface of the wave $\bar{r}(t) = (x(t), y(t), z(t))$ can now be computed at any instant, see Fig. 1.

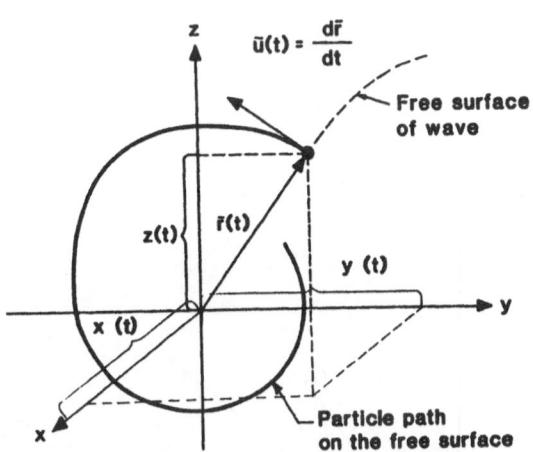

Fig. 1. Lagrangian description of wave kinematics.

The ratio between the vertical and the horizontal excursions of a particle at the free surface on a Stokes wave depends on wave steepness:

$$\alpha = \frac{e_z}{\sqrt{e_x^2 + e_y^2}} = f(a_p K_p) \qquad (2.5)$$

This ratio is 1.0 for the linear wave components. The horizontal and vertical particle motions in the linear wave components are in quadrature. They thus form a Hilbert-transformable pair of signals. The resulting vertical motion summed up for all frequencies is already given as:

$$z(t) = \eta(t) \qquad (2.6)$$

We can now compute the projection $\bar{r}_{xy}$ of the vector $\bar{r}$ on the xy-plane using the Hilbert transform:

$$\mid \bar{r}_{xy}(t) \mid = H\left[\eta(t)\right] = \frac{1}{\pi} \int\limits_{-\infty}^{\infty} \frac{\eta(\tau)}{t-\tau} d\tau \tag{2.7}$$

The vector $\bar{r}_{xy}$ in the xy-plane is shown in Fig. 2. The angle β can be found from:

$$\beta(t) = \frac{\pi}{2} - Arctg\, \frac{S_x(t)}{S_y(t)} \tag{2.8}$$

We then obtain:

$$x(t) = \mid \bar{r}_{xy} \mid \cos \beta(t) \tag{2.9}$$

$$y(t) = \mid \bar{r}_{xy} \mid \sin \beta(t) \tag{2.10}$$

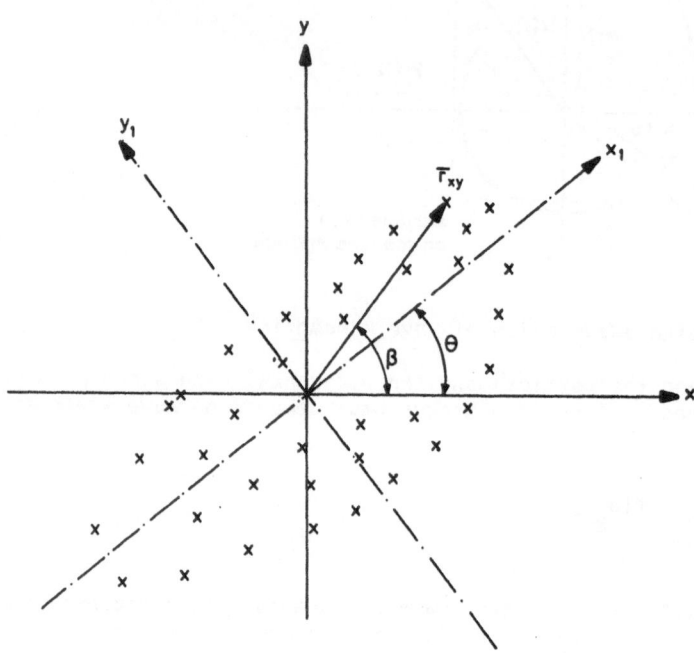

Fig. 2. Scatter ellipse.

When the vector $\bar{r}_{xy}$ is plotted at constant time intervals a scatter ellipse will occur as shown in Fig. 2. The major axis of the ellipse is then in the direction θ. A new coordinate grid x_1, y_1 can then be introduced where the axis x_1 is in the direction θ. A measure of the scatter along the major axis x_1 of the ellipse is then:

$$S_{x_1} = \sqrt{\overline{(\frac{\partial \eta}{\partial x_1})^2} - \left[\overline{\frac{\partial \eta}{\partial x_1}}\right]^2} \qquad (2.11)$$

while the scatter along the minor axis is:

$$S_{y_1} = \sqrt{\overline{(\frac{\partial \eta}{\partial y_1})^2} - \left[\overline{\frac{\partial \eta}{\partial y_1}}\right]^2} \qquad (2.12)$$

The ratio :

$$M = \frac{S_{x_1}}{S_{y_1}} \qquad (2.13)$$

provides a measure of spreading in the wave direction. If $M = 1$ then there is no preferred direction and the scatter is spherically symmetric. If $M = \infty$, then the motion is a uniform wave in the direction θ.

The position of a particle at the free surface is now given by the vector $\bar{r}(t) = (x(t), y(t), z(t))$ and the local particle velocity vector can then be calculated as:

$$\frac{d\bar{r}}{dt} = (\dot{x}(t), \dot{y}(t), \dot{z}(t)) \qquad (2.14)$$

Further the local particle acceleration vector can be calculated as:

$$\frac{d^2\bar{r}}{dt^2} = (\ddot{x}(t), \ddot{y}(t), \ddot{z}(t)) \qquad (2.15)$$

Thus 6 time series are obtained for particle velocities and accelerations. Examples of results are given in section 5.

3. EVIDENCE OF FREAK WAVES

It is particularly important for engineers to obtain information on rare waves with a damaging potential (freak waves), and such information is valuable both for the naval architects dealing with seakeeping, manoeuvering and stability of vessels, and for the engineers dealing with design of offshore structures, determination of

458

deck levels and computations of wave loads and slamming. First a definition of a freak wave is needed.

We define a freak wave as a wave with a wave height that is larger
than 2 times significant wave height. Both zero-downcross and zero-
upcross wave heights should be controlled in order to determine
whether a wave is a freak wave or not. A freak wave might appear as a
non-breaking or a breaking wave. In our studies we developed a
counting technique for successive trough/crest events as they could be
observed in time series from oceanographic measurements and we
established joint probability density distributions of crest front
steepness and wave heights, see Fig. 3.

Applying this method it became evident, that most data grouped in a
quite regular pattern, and further that a few cycles representing
maximum conditions suddenly dropped outside of this regular pattern
during gales. Thus this new counting technique is most useful for
identifying outstanding episodic events that do not follow the pattern
of the main part of the gathered data. For example, Fig. 4 shows the
time series containing 2 steep elevated waves as identified in Fig. 3.
We define freak waves as such outstanding events.

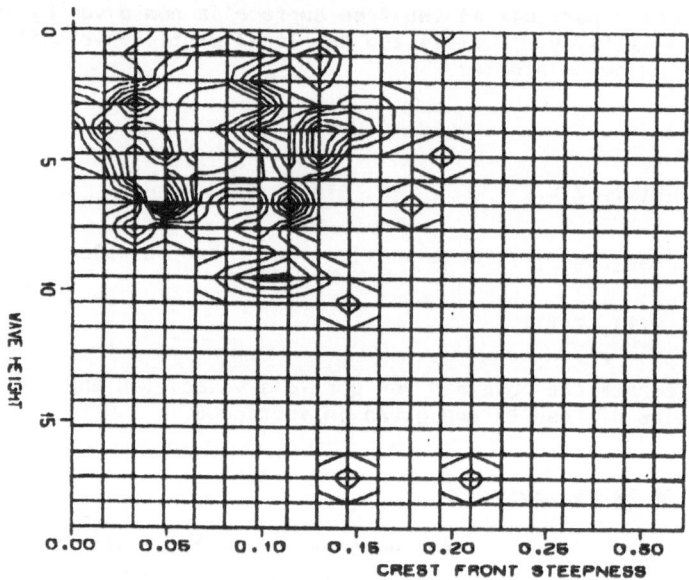

Fig. 3. Joint probability density distribution for wave height
(metres) and crest front steepness, at the FRIGG Field on 85.11.06 at
06 GMT. The significant wave height was 8.49 m.

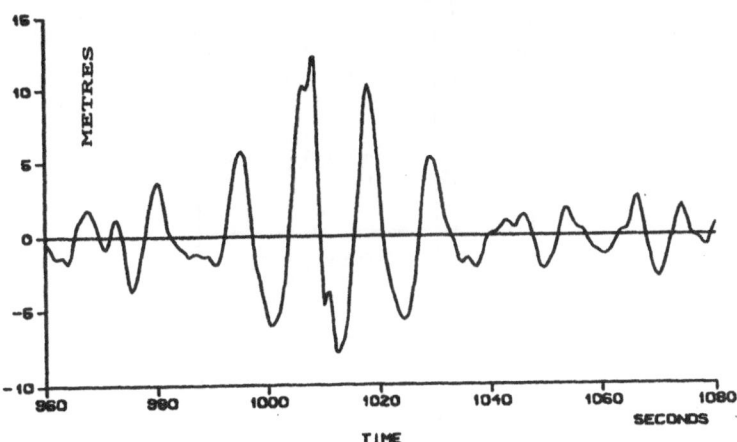

Fig. 4. Time series containing 2 freak waves as identified in the JPD-distribution shown in Fig. 3.

At the FRIGG Field on the Norwegian Continental Shelf we have observed 8 events of this kind measured with a vertical down-scanning wave radar installed on the jacket QP. At this site the water depth is 99.4 m measured below the lowest astronomical tide level (L.A.T.). The most interesting case is the one shown in Fig. 3 and Fig. 4. Here two freak waves with wave heights 18.29 m and 18.15 m suddenly occur together, following each other within the same wave group. The sea state had a significant wave height of only 8.49 m at this time. The two freak waves occur with zero-downcross periods 12.07 sec. and 11.29 sec., corresponding to wave lengths 227 m and 199 m. Model experiments with waves in wave tanks have shown that the influence of the sea bottom on the waves is governed by the ratio h/L, where h is the water depth and L is the wave length. When this ratio is larger than 0.5, there is no influence from the sea bottom. For the first wave we obtain h/L = 0.438 and for the second wave h/L = 0.5. Thus we can conclude that the influence from the bottom is very marginal. However, the small modulation seen on the crest of the first freak wave is similar to modulations seen on breaking or broken waves in laboratory experiments. For a discussion of the possibility that the first wave is a broken wave based on wave asymmetry parameters: ϵ, λ and μ see section 6. It is most remarkable that the first freak wave occurs with a crest height $\eta' = 12.24$ m, and a ratio $\eta'/H_{1/3} = 1.44$.

Based on the Rayleigh distribution of wave heights and linear wave theory we would expect: $H_{max}/H_{1/3} = 2.00$ and $\eta'_{max}/H_{1/3} = 1.00$. Table 1 gives a summary of the events where freak waves have been measured. Also event 7 is an example where two freak waves are following each

Table 1.

SUMMARY TABLE OF FREAK WAVE EVENTS.

EVENT No.	TIME Date	TIME Hour, min, s	$\frac{H_{zd}}{H_{1/3}}$	$\frac{H_{zu}}{H_{1/3}}$	$\frac{\eta'}{H_{1/3}}$	WAVE PARAMETERS H_{zd} (m)	H_{zu} (m)	T_{zd} (s)	η' (m)	ϵ	λ	μ	SEA STATE $H_{1/3}$ (m)	COMMENT
1	83.01.08	03, 11, 09	2.22	1.95	1.52	10.34	9.07	8.78	7.07	0.248	0.855	0.687	4.66	
2	83.01.09	00, 11, 19	2.31	2.41	1.86	11.40	11.86	10.31	9.16	0.247	0.512	0.804	4.93	
3	83.01.11	03, 10, 57	2.04	1.74	1.18	6.86	5.85	10.05	3.96	0.112	1.13	0.580	3.36	
4	83.01.12	03, 0, 17	2.02	2.15	1.32	6.68	7.09	11.22	4.35	0.0839	0.557	0.648	3.30	
5	83.01.15	03, 15, 13	2.29	2.26	1.68	9.77	9.64	7.39	7.19	0.308	0.512	0.738	4.27	
6	85.11.06	06, 16, 48	2.15	2.35	1.44	18.29	19.98	12.07	12.24	0.141	0.232	0.669	8.49	2 freak waves !
			2.14	1.86	1.21	18.15	15.83	11.29	10.24	0.215	0.922	0.564	"	
7	85.11.07	06, 10, 29	1.95	2.18	1.19	5.52	6.17	9.31	3.38	0.0949	0.731	0.612	2.83	
8	85.11.09	06, 12, 45	2.14	1.59	1.16	6.07	4.50	9.43	3.29	0.106	1.14	0.542	"	2 freak waves !
			1.98	2.23	1.29	11.35	12.78	10.95	7.44	0.163	0.778	0.655	5.74	

other within the same wave group. Event 4 is a case where the observed freak wave occur within a wave group, but in this case, there is only one freak wave within the group. In the remaining cases, the freak waves occur as single elevated waves. Thus the study of wave groups is very important for understanding the freak wave phenomena.

The analysis of the freak wave events that has been performed show that it is necessary to include both the zero-downcross wave height H_{zd}, and the zero-upcross wave height H_{zu} in the statistical treatment of the freak wave phenomena. For instance, in event 6 the first freak wave occurs with H_{zd} = 18.29 m while H_{zu} = 19.98 m for the same wave! The choice between H_{zd} and H_{zu} introduces a significant difference in the wave statistics. The discrepancy of 1.69 m is very significant.

The conclusion that can be drawn by examining Table 1 is, that there are 8 events where freak waves occurred, and in two of these events two freak waves followed each other and occurred successively within the same wave group. Thus the total number of freak waves that was found was 10. 9 of these waves occurred without any influence of the sea bottom. The sea bottom had a very marginal influence on the 10th wave. It might be a breaking or a broken wave.

4. LABORATORY REPRODUCTION OF FREAK WAVE EVENTS

Most experiments in wave tanks or basins are performed with regular waves and/or irregular sea spectra. Both methods are very time consuming. With regular waves, even with a great number of individual runs, sharp peaks of transfer functions may be lost. With irregular sea tests the statistical analysis requires at least 20 minutes testing time to limit excessive scattering of the results.

A more severe drawback for statistical test techniques is, that in some cases it is not possible to reconstruct a marine accident in the laboratory. In some cases even after long periods of testing no waves with a damaging or capsizing potential was seen. In other cases such waves suddenly occurred in an uncontrolled way, at a position far away from the model that was tested. This was the background for the development of a transient test technique, (KJELDSEN, 1982).

Laboratory techniques for the generation of non-linear transient wave groups have now been used for a considerable time by MARINTEK for testing of both vessels and offshore structures.

A freak wave event is modelled with a δ-function, and wave dispersion is controlled by a non-linear term depending on wave amplitude. With such a technique it is possible to focus a number of waves in such a way that a single elevated wave suddenly occurs at one prescribed time

and position, see Fig. 5. Both 2- and 3-dimensional wave focusing have
been achieved. As shown in Fig. 5 a non-linear transient wave group is
generated and travels towards the focusing position (target). In this
process the number of individual waves decreases within the group.
Tuning of a large number of experiments have given the following
results:

- The leading wave in the wave group might break before the target
  is reached.
- The single elevated wave at the focal point can occur as a non-
  breaking wave.
- The single elevated wave at the focal point can break very
  slightly as a "spilling breaker". In this case a very small jet
  develops at the wave crest.

- The single elevated wave at the focal point can break as a very
  violent "plunging breaker". In this case a very large jet develops at the wave crest with a splash down in the wave trough
  ahead of the crest.

Various command signals are developed, and thus the phenomena can be
well controlled. The wave group observed in Fig. 4 seems to be very
similar to the wave group observed on wave gauge No. 4 in Fig. 5. Thus
the focusing of wave groups in space seems to be very important for an
understanding of the freak wave phenomena.

It should be reiterated that the wave generation technique which is
developed has been used to generate all kinds of conditions, from non-
breaking wave groups, non-breaking single waves, to very slightly
breaking "spilling breakers" and violent breaking "plunging breakers".
Table 2 gives a summary of various experiments that have been performed with the transient test technique.

KJELDSEN (1983) describes how experiments with transient waves can be
performed, combined with an irregular wave experiment, in such a way
that a prescribed wave spectrum can be reproduced. Also GÜNTHER &
BERGMANN (1986) give examples on how transient wave groups can be used
for testing of offshore structures.

One of the interesting experimental programmes that has been performed
at MARINTEK is programme I shown in Table 2. Local wave forces were
measured at 26 positions with shear force transducers on the leg of a
steel jacket structure. 15 of these force transducers were above mean
water level (MWL). Tests in scale 1:25 were performed with regular
waves, irregular waves and focused 2 and 3-dimensional breaking waves.
The performance of the entire experimental programme showed that
extreme wave load intensities are associated with transient 2- or
3-dimensional breaking waves, of relatively short wave periods, and

**DISTANCE FROM
WAVE GENERATOR:**

x = 10 m.

x = 25 m.

x = 26 m.

x = 35 m.

x = 60 m.

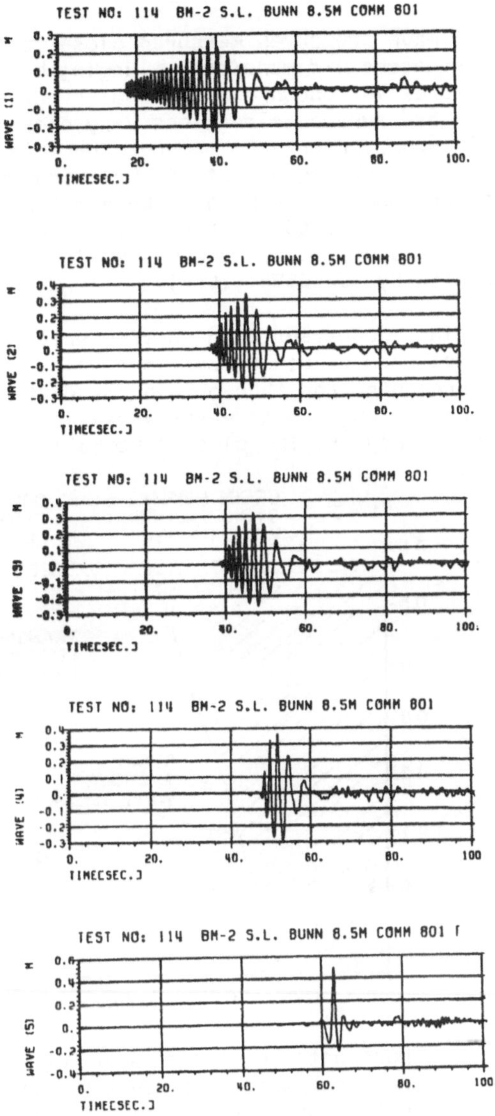

Fig. 5. Example of non-linear focusing in the ocean basin of many transient waves leading to one single freak wave. (Note the change of vertical scale from top to bottom).

not with the highest waves in the simulated sea states.

Local wave pressures measured close to the free surface in the crests
of transient 2-dimensional plunging breakers exceeded local wave
pressures in monochromatic regular waves by a factor of 5. The regular
waves had surface elevations very close to predictions made from Stokes' second-order wave theory. The total integrated forces and the
total overturning moments in breaking waves exceeded those measured in
regular monochromatic waves by a factor of 3, as reported by KJELDSEN,
TØRUM. DEAN, (1986). Further, the particle velocities were measured in
the crests of the focused waves, in a "WAVE-FOLLOWER" experiment
reported by KJELDSEN (1984).

Assuming a constant drag coefficient, it is then possible to combine
the measured values of particle velocities at the crest, and all force
measurements, and thus obtain the distribution of horizontal particle
velocities above mean water level in a transient plunging breaker as
shown in Fig. 6. The plot is normalized with respect to phase velocity.

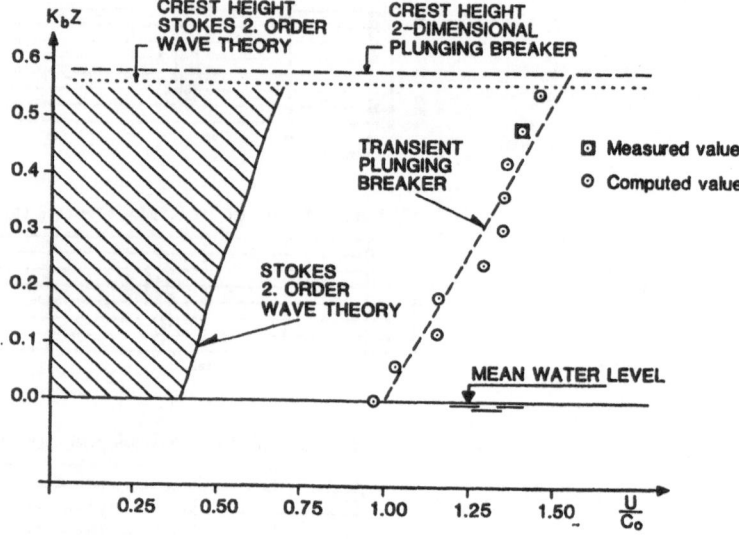

Fig. 6.   Horizontal particle velocities in a 2-dimensional transient
          wave compared with theoretical predictions of particle velo-
          cities obtained by application of Stokes' 2. order wave
          theory. RUN No. 305. KJELDSEN & AKRE (1985).

We see that the local horizontal particle velocities in the transient
wave are nearly twice as large in magnitude as the local particle
velocities obtained from Stokes' 2. order wave theory for a wave with

Table 2.

DEVELOPMENT OF WAVE GENERATION TECHNIQUE

EXPERIMENTAL PROGRAMME	PURPOSE	LABORATORY FACILITY
A	Perform deep water breaking waves in an experiment using convergence walls (1:23). Detect shock pressures on a small vertical plate, and on a tilted plate hit from below by deep water breaking crests.	Wave Flume - NHL
B	Develop command signal from non-linear theory that creates a plunging breaker in deep waters.	Wave Flume - NHL
C	Give quantitative results of particle velocities in the crests of breaking waves using high-speed film technique.	Wave Flume - NHL
D	Investigate the amplification of a freak wave in an opposing current.	Wave Flume - NHL
E	Investigate slamming on concrete platform. Deterministic and stochastic command signals with wind superposed.	Wind - Wave Flume NHL
F	Develop new command signals for various types of breaking waves. ("Spilling Breakers and Plunging Breakers"). Investigate instability and capsizing of fishing trawlers and freighters in beam seas and in broaching modes.	Large Wave Flume MARINTEK
G	Develop new command signals. Tests with semisubmersible "OCEAN RANGER". Damaged stability condition with large angle of list. Development of command signals containing wave groups matching critical period for pitch followed by a freak wave with various phases, relative to platform movements.	Large Ocean Simulating Basin MARINTEK.
H	Subsidence of EKOFISK field. Test drag force on deck module of jacket structure due to flooding by breaking wave.	Large Ocean Simulating Basin MARINTEK
I	Develop new command signals for generation of 3-dimensional Breaking Waves. Mapping of Local Wave Forces above mean water level on leg of steel jacket structure.	Large Ocean Simulating Basin MARINTEK

equal steepness. Particular attention should be given to the fact that this is the case for the entire zone between wave crest level and mean water level. It should also be emphasized that transient waves attain horizontal particle velocities that exceed the phase velocity in nearly the entire zone above the mean water level. This vertical distribution must be taken into account properly when wave forces are derived for design applications. If a high horizontal particle velocity is considered only to occur in a very local zone in the upper part of a crest of a breaking wave, then wave forces in transient waves of the kind encountered here might be seriously underestimated. The results given in Fig. 6 should be regarded as extreme values, representative for one of the worst cases that can be expected.

5. RESULTS FROM THE MATHEMATICAL MODEL

Fig. 7 gives an example of results obtained with the mathematical model described in section 2. It shows the experimental observed surface elevation measured in one of the tests with the leg of the steel jacket. Below is the computed energy envelope η^2, the energy flux and the computed time series of the horizontal particle velocity $\dot{x}(t)$. Wave B is a breaking wave. It gave much higher wave forces on the structure, than the following wave, even in this case, where the following wave has a much larger wave period, and a much larger wave height. Application of the mathematical model using the Hilbert transform technique can explain this result. The computations show, that the maximum particle velocity at the wave crest in the breaking wave was 3.20 m/sec., but in the wave behind, it was only 1.74 m/sec. The measured drag force was therefore larger in the breaking wave. Further, it can be seen that the computed particle velocity of the breaking wave exceeds the phase velocity of the breaking wave. The phase velocity of the breaking wave is calculated from the zero-downcross wave period $T_{zd} = 1.54$ sec. and becomes $C_0 = 2.40$ m/sec. We then obtain $\dot{x}_{max}/C_0 = 1.33$. This predicted result is in reasonable agreement with the actual measurements shown in Fig. 6. This shows the feasibility of using the mathematical model with the Hilbert transform technique to analyse time series from oceanographical measurements at sea in order to investigate whether waves at sea are breaking or not.

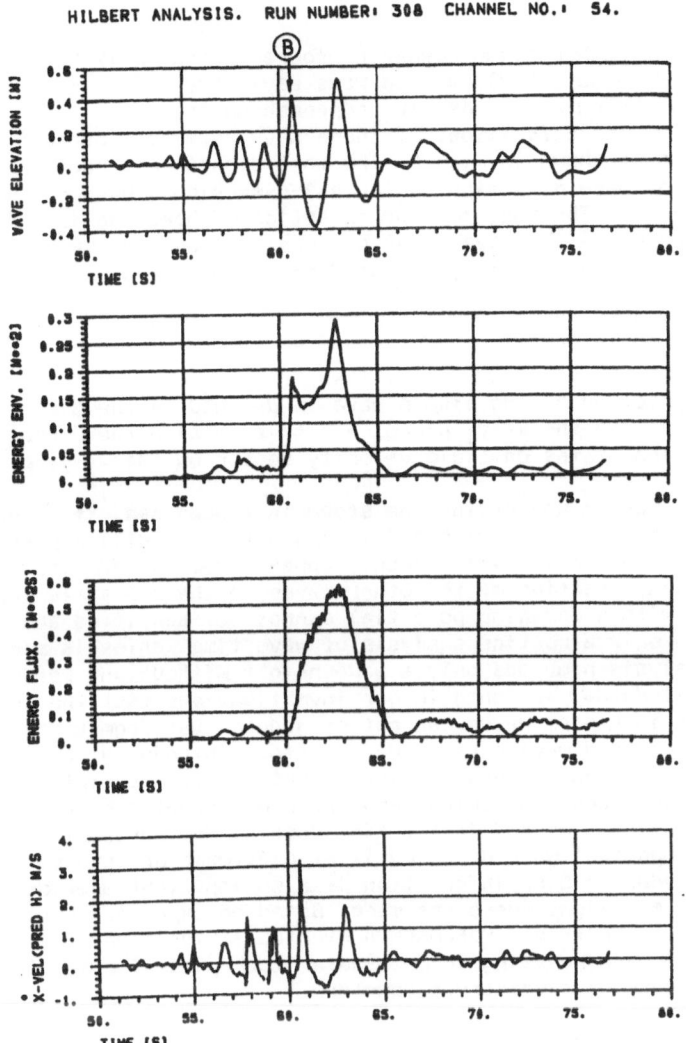

Fig. 7. Measured surface elevation and computed energy envelope, energy flux and horizontal particle velocity at the wave crest. The computed particle velocity exceeds phase velocity for the breaking wave B.

6. WAVE ASYMMETRY

As a part of the Norwegian research programme "SHIPS IN ROUGH SEAS", KJELDSEN & MYRHAUG (1978) defined wave asymmetry parameters in order to describe random waves that are approaching breaking in a quantitative way. The definitions of these wave asymmetry parameters are now extended as shown in Fig. 8. The most significant addition is the definition of wave direction θ_B for a 3-dimensional breaking wave in a directional sea. The position for inception of breaking is defined as the position where the wave front becomes vertical. At this position the wave direction in the x_1-y_1 coordinate system can be defined as:

$$\theta_B = \theta - \beta \tag{6.1}$$

where θ is calculated from eq (2.1) and β is calculated from eq (2.8). Thus the direction of the single wave is defined as the direction of the r_{xy} vector in the x_1-y_1 plane, and this is also the direction of the projection of the particle velocity vector on the x_1-y_1 plane.

For a small vessel weathering the storm in a head sea, it is very important whether a wave with a damaging potential will strike it on the bow, or whether it will suddenly appear from an odd direction far from the mean direction of the other waves in the sea state. However, rare waves with a damaging potential cannot be identified and documented when only a routine analysis of wave time series is made. The routine analysis provides only a wave height without any reference to a horizontal datum. We found in our investigations that the inception of breaking in random waves can not be judged just from a measurement of wave height and wave period. We therefore introduced the crest front steepness, that gives a unique quantitative description of the wave as it approaches breaking, and the crest front steepness is defined both in space and in the time domain. The horizontal asymmetry factor is important when deck levels of platforms are evaluated in design procedures. Such information is also important when designs of breakwaters and lighthouses are made. Based on experiments and analysis of high-speed films, KJELDSEN & MYRHAUG (1979) found that the inception of wave breaking in deep water took place for the following values of the wave asymmetry parameters:

$$
\begin{aligned}
0.32 &< \epsilon_x &< 0.78 \\
0.90 &< \lambda &< 2.18 \\
0.84 &< \mu &< 0.95
\end{aligned}
\tag{6.2}
$$

DEFINITION OF WAVE DIRECTION θ_B FOR A SINGLE BREAKING WAVE :

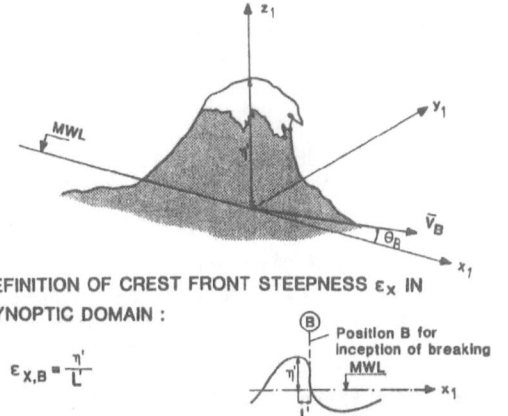

DEFINITION OF CREST FRONT STEEPNESS ε_x IN SYNOPTIC DOMAIN :

$$\varepsilon_{x,B} = \frac{\eta'}{L'}$$

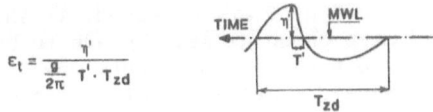

DEFINITION OF CREST FRONT STEEPNESS ε_t IN TIME DOMAIN :

$$\varepsilon_t = \frac{\eta'}{\frac{g}{2\pi} T' \cdot T_{zd}}$$

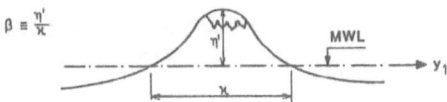

DEFINITION OF CREST LENGTH $\varkappa$ AND 3-D CREST SHAPE FACTOR β IN SYNOPTIC DOMAIN :

$$\beta = \frac{\eta'}{\varkappa}$$

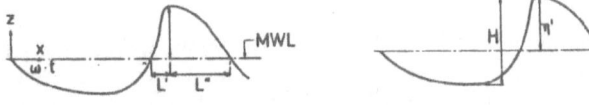

VERTICAL ASYMMETRY FACTOR

$$\lambda = \frac{\varepsilon}{\delta} = \frac{L''}{L'}$$

HORIZONTAL ASYMMETRY FACTOR

$$\mu = \frac{\eta'}{H}$$

Fig. 8. Definition of direction, wave crest length, crest front steepness and wave asymmetry factors for a 3-dimensional breaking wave in a directional sea.

The asymmetry parameters can be applied to judge various types of breaking modes. The highest values of the parameters given above are thus valid for plunging breakers in deep waters.

Experiments with transient wave groups performed at MARINTEK showed that the inception of breaking occurred for:

$$\mu > 0.77 \tag{6.3}$$

measured in the time domain, (KJELDSEN 1982). This is in good agreement with an investigation by DUNCAN, WALLENDORF AND JOHNSON (1987).

The ISSC (International Ship and Offshore Structures Congress) (1988) has now recommended the use of these parameters in research work. Also IAHR, the International Assosiation for Hydraulic Research has given such a recommendation.

We will now discuss whether the waves shown in Fig. 4, could be breaking waves or not. The modulation on the first wave crest is similar to modulations seen on broken waves in laboratory experiments. However, none of the asymmetry parameters are large enough to prove that it is at the position in space, where the inception of breaking takes place. When Fig. 4 and Fig. 5 are compared, it should be mentioned that the observed wave group at sea appears to be somewhat similar to the wave group observed in the laboratory 35 m from the wave generator. We will therefore mention that the possibility exists, that the wave group shown in Fig. 4 is a result of a non-linear wave-wave interaction, and that the group is focused at a position in space different from the one where the measurement is taken.

In a careful controlled experiment KJELDSEN et al. (1980) showed, that the wave-wave interaction between only 2 waves is sufficient to create a plunging breaker in deep water.

This experimental result is in good agreement with theoretical computations of wave instability. (LONGUET-HIGGINS 1978).

7. FORECASTS

Areas known to be particularly prone to freak waves are the Nova Scotia shelf, the Bermuda rise, the waters of Greenland, the coast of Northwest India and the waters of Southeast Africa, where the Agulhas current opposes the main direction of the wave dispersion.

Based on 10 years research on ship accidents and freak wave encounters, a mathematical model has been established in Norway that predicts encounter probabilities for freak waves both in deep and shallow waters, (KJELDSEN, 1989). The mathematical model has been used both for hindcasting and for forecasting of dangerous waves in 24 particular exposed areas along the Norwegian coast. There is a good agreement between the predictions made from this model and damages and accidents on vessels in the actual areas.

8. CONCLUSIONS

1) 10 freak waves have been measured at the FRIGG Field. The ratios between the wave heights of the freak waves and the significant wave heights were in the range $2.0 < H_{zu}/H_{1/3} < 2.4$.

2) Event No. 6 is the most interesting. In this case two freak waves occurred successively, following each other within the same wave group. One of these waves occurred with a wave height 20 m in a sea, where the significant wave height was only 8.49 m. The ratio between the crest height and the significant wave height was in this case 1.44. This is far above all theoretical predictions.

3) 9 of the 10 observed freak waves occurred completely without any influence from the bottom.

4) The scientific study of wave groups is important for the understanding of the freak wave phenomena.

5) The mathematical model based on the Hilbert transform can be applied to determine whether a wave is breaking or not, based on a kinematic breaking criteria.

6) The application of data from wave radars combined with the use of higher order Stokes theory appear to give sufficiently good results for running laboratory tests and for engineering and seakeeping analyses.

9. REFERENCES

DUNCAN, J.H., WALLENDORF, L.A., JOHNSON, B., 1987: "An Experimental Investigation of the Kinematics of Breaking Waves". Report No. EW-7-87. United States Naval Academy. Annapolis, Maryland, USA.

GÜNTHER, F.C., BERGMANN, J., 1986: "Gaussian Wave Packets- A new Approach to Seakeeping Tests of Ocean Structures". Applied Ocean Research. Vol. 8, No. 4.

ISSC, 1988: Proceedings Int. Ships and Offshore Structures Congress, Vol. 1 p 19, Copenhagen, Denmark, Aug. 1988.

KJELDSEN, S.P., 1982: "2- and 3-dimensional Deterministic Freak Waves". Proc. 18th Int. Conference on Coastal Engineering, Cape Town, South Africa.

KJELDSEN, S.P., 1983: "Determination of Severe Wave Conditions for Ocean Systems in a 3-dimensional Irregular Seaway". Proc. 8th Congress of the Pan-American Institute of Naval Engineering, Sept. 83, Washington D.C., USA.

KJELDSEN, S.P., 1984 a: "Dangerous Wave Groups". Norwegian Maritime Research Vol. 12. No. 2. pp 4-16.

KJELDSEN, S.P., 1984 b: "The Experimental Verification of Numerical Models of Plunging Breakers". Proc. 19th Int. Conference on Coastal Engineering, Houston, Texas, USA.

KJELDSEN, S.P., 1989: "The Practical Value of Directional Ocean Wave Spectra". Proc. Symposium on Measuring, Modeling, Predicting and Applying Directional Ocean Wave Spectra. Applied Physics Laboratory. The Johns Hopkins University, Maryland, USA, April 1989.

KJELDSEN, S.P. & MYRHAUG, D., 1978: "Kinematics and Dynamics of Breaking Waves". MARINTEK Report No. STF60 A 78100, Trondheim, Norway.

KJELDSEN, S.P., MYRHAUG, D., 1979: "Breaking waves in deep waters and resulting wave forces". Proc. 11th Offshore Technology Conference. Houston, Texas, USA.

KJELDSEN, S.P., TØRUM, A., DEAN, R.G., 1986: "Wave Forces on Vertical Piles caused by 2- and 3-dimensional Breaking Waves". Proc. 20th Int. Conference on Coastal Engineering, Taipei, Taiwan.

KJELDSEN, S.P., VINJE, T., MYRHAUG D., BREVIG, P., 1980: "Kinematics of deep water breaking waves". Proc. 12th Offshore Technology Conference. Houston, Texas, USA.

KJELDSEN, S.P., AKRE, A., 1985: "Wave Forces on Vertical Piles near the free surface caused by 2-dimensional and 3-dimensional Breaking Waves". MARINTEK Report No. PR53 530012.04 01 85, Trondheim, Norway.

LONGUET-HIGGINS, M.S., 1957: "The statistical analysis of a random, moving surface". Phil. Trans. Roy. Soc. Vol. A-249, pp 321-387.

LONGUET-HIGGINS, M.S. 1978: "The instabilities of gravity waves of finite amplitude in deep water. I. Superharmonics, II. Subharmonics". Proc. Roy. Soc., Ser. A 360, pp. 471-505.

SCHWARTZ, M., BENNET, W.R., STEIN, S., 1966: "Communication Systems and Techniques", McGraw-Hill, New York.

10 ACKNOWLEDGEMENTS

I would like to thank C.T. STANSBERG for development of software
related to the application of the Hilbert transform and J. SKJELBREIA
for assistance with analysis of wave time series measured at the FRIGG
Field.

Computations of Breaking Waves

D.H. PEREGRINE
Dept.of Mathematics
Bristol University
Bristol BS8 1TW
England

ABSTRACT This is a brief survey of the use of programs which can model steep unsteady waves including their overturning as they break. Examples and applications are discussed, together with their limitations. Suggestions are made for their engineering use and for checking their accuracy.

1. Introduction

Since the first successful computations describing the initial overturning of a breaking wave by Longuet–Higgins and Cokelet (1976) numerous investigators have followed in their wake. With a few exceptions boundary–integral schemes have been used for solving Laplace's equation and time integration is usually implemented following fluid particles in the free surface. Other approaches do not appear to have been analysed in such detail, nor do they appear able to achieve the same surface resolution, so they will not be considered here.

There are a wide range of different numerical implementations of boundary–integral schemes for steep unsteady water waves. A detailed discussion of mathematical aspects is not attempted here, especially as almost every investigator has some further variation. The approach of Longuet–Higgins and Cokelet (1976) has been more or less abandoned due to its liability to "zig–zag" instabilities. The method of Vinje and Brevik (1981) is much less susceptible to instabilities and is readily extended to a variety of applications. The examples of wedge slamming and body emergence given by Greenhow (1987,1988) illustrate the robustness of this method even when its accuracy is becoming suspect. Baker, Meiron and Orszag (1982) gave a good discussion of boundary–integral methods which stimulated the development of a remarkably accurate and efficient program by Dold and Peregrine (1986a). this latter program has now been extended in various ways, the study of solitary–wave instability by Tanaka, Dold, Lewy and Peregrine (1987) providing examples of its accuracy and the way in which an efficient program can easily be used for integrations with long space and time scales.

Here we shall briefly survey the various problems to which these programs are applied and then discuss their value for applications and research. Some of the difficulties which need to be overcome to extend their application and versatility are discussed, together with some aspects that need consideration if this type of calculation is to become an estabilished part of engineering design procedures.

A. Tørum and O. T. Gudmestad (eds.), Water Wave Kinematics, 475–490.

2. Demonstrated examples

In the first demonstration of any program, it has been usual to cause breaking in a simple manner. This can easily be achieved by choosing initial conditions with sufficient energy. For example large–amplitude sine waves with velocity potential given by linear theory for steady propagation break, though sometimes with somewhat unusual shapes, see figure 1. More realistic examples are obtained by forcing by a pressure distribution (Longuet–Higgins and Cokelet, 1976) or by using the shape and surface velocity potential of an exact steadily propagating wave but using a smaller depth of water (New, Peregrine and McIver, 1985).

By introducing computational discretization points on rigid boundaries Vinje and Brevik (1981) first modelled waves with rigid obstacles in the computational domain. These obstacles can be protrusions from the bed, in mid water or be partly immersed, as floating or fixed bodies. This category includes the important cases of waves meeting a beach and wave generation by a moving paddle. Although the local behaviour at a moving contact line on a moving surface has given problems, successful implementations are reported (Lin, Newman and Yue, 1984, and New, 1983b) and Greenhow's (1987,1988) examples where there are fluid particles separating from the rigid surface to form incipient spray are of special note.. Greater efficiency in computation is obtained for a wide range of examples when points on the rigid boundaries are eliminated by using conformal transformations to change boundaries into a plane or circle. This allows the use of an image of the free surface. The example shown in figure 2 was computed in this way.

Several numerical studies have followed the nonlinear evolution of unstable wave–trains. Steep nonlinear waves are liable to several instabilities. Many two–dimensional instabilities have now been investigated. Whether or not instabilities lead to breaking is of particular interest. The rapidly growing alternate–crest instability of Longuet–Higgins (1987) is shown by Longuet–Higgins and Cokelet (1978) to lead to breaking in all the deep–water cases considered. This instability is restricted to wave steepnesses greater than $ak = 0.41$. Another instability of very steep waves is the Tanaka (1983, 1986) instability of waves steeper than those with maximum energy density. This has been demonstrated for both deep–water and solitary waves, so may be expected to also occur for periodic waves on finite water depth. Tanaka et al (1987) found the expected evolution to breaking, but were surprised to find that changing the sign of the most unstable eigenfunction led to an evolution with decreasing wave amplitude and no breaking. Jillians' (1989) study of deep–water waves shows a similar dichotomy of behaviour but the periodic nature of the waves permits an eventual evolution to breaking even for the mode which initially grew "away" from breaking.

Deep–water waves of moderate steepness are unstable to long modulations in the Benjamin–Feir (1967) instability. Dold and Peregrine (1986b) give a preliminary report of their investigations of the boundary between waves for which the modulation grows and breaking occurs and those which evolve through a maximum modulation to a recurrence of a near–uniform state. See figures 3 and 4. The continuing evolution of these latter waves is one example of a further type of study. The initial development of a bore from a long shallow–water wave is another study of wave evolution. Teles da Silva and Peregrine (1989) follow this evolution looking at the characters of undulations to guide early identification of those waves which are going to break. See figures 5 and 6. Yet another evolution, often used in wave flumes, is to focus waves at a point in space time. Dommermuth et al (1987) compare experiment and computation for such a case, and see Skyner's contribution to this workshop.

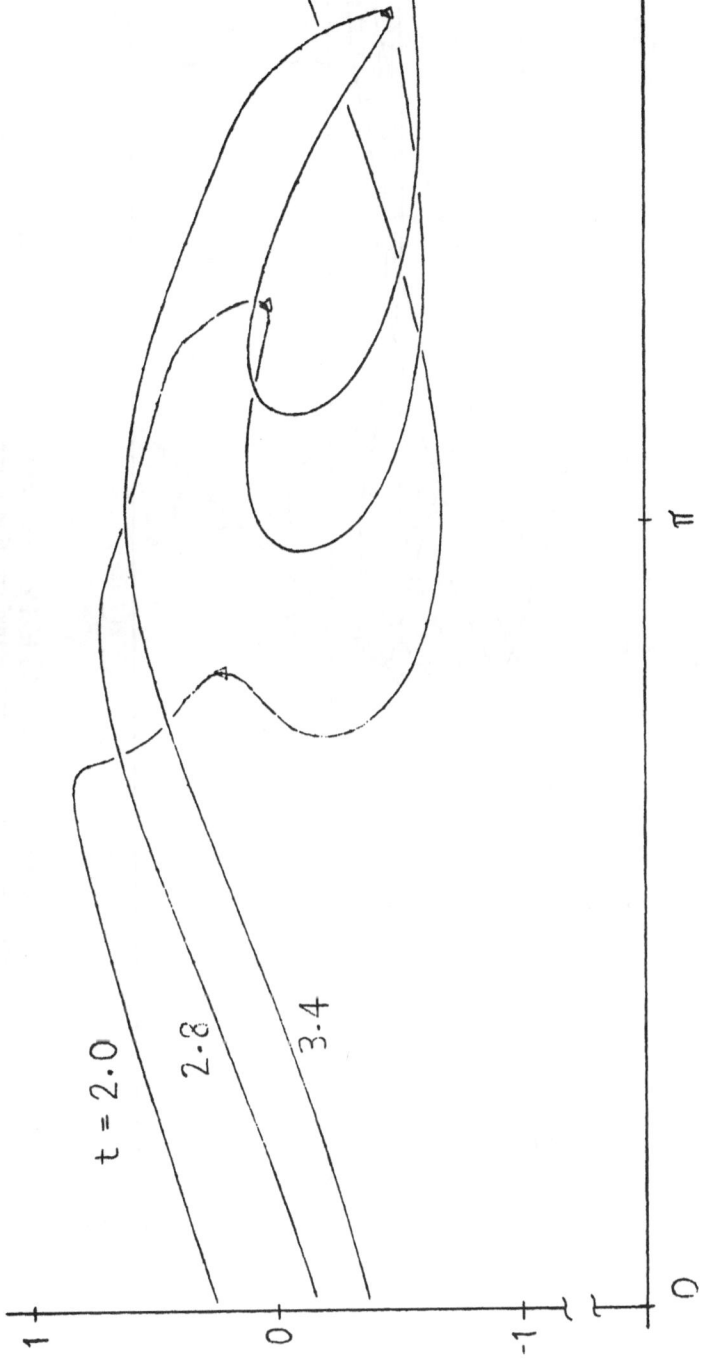

Figure 1. Evolution of the initial conditions $\eta = \cos x$, and $\phi = \sin x$ on the free surface. (Courtesy of J.W.Dold).

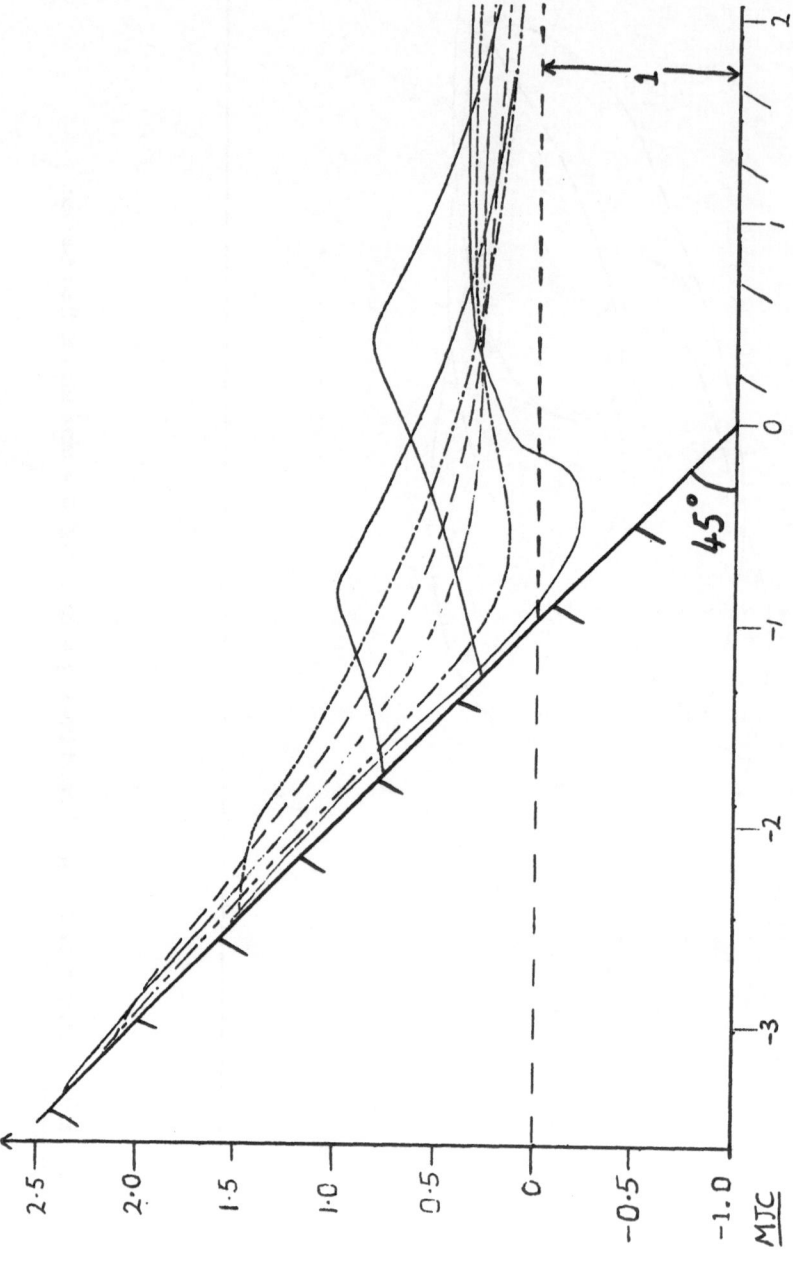

Figure 2. Solitary wave of amplitude 0.8h incident on a 45° slope from water of depth h. Computation performed with a conformal transformation mapping all the bed into a plane. (Courtesy of M.J.Cooker).

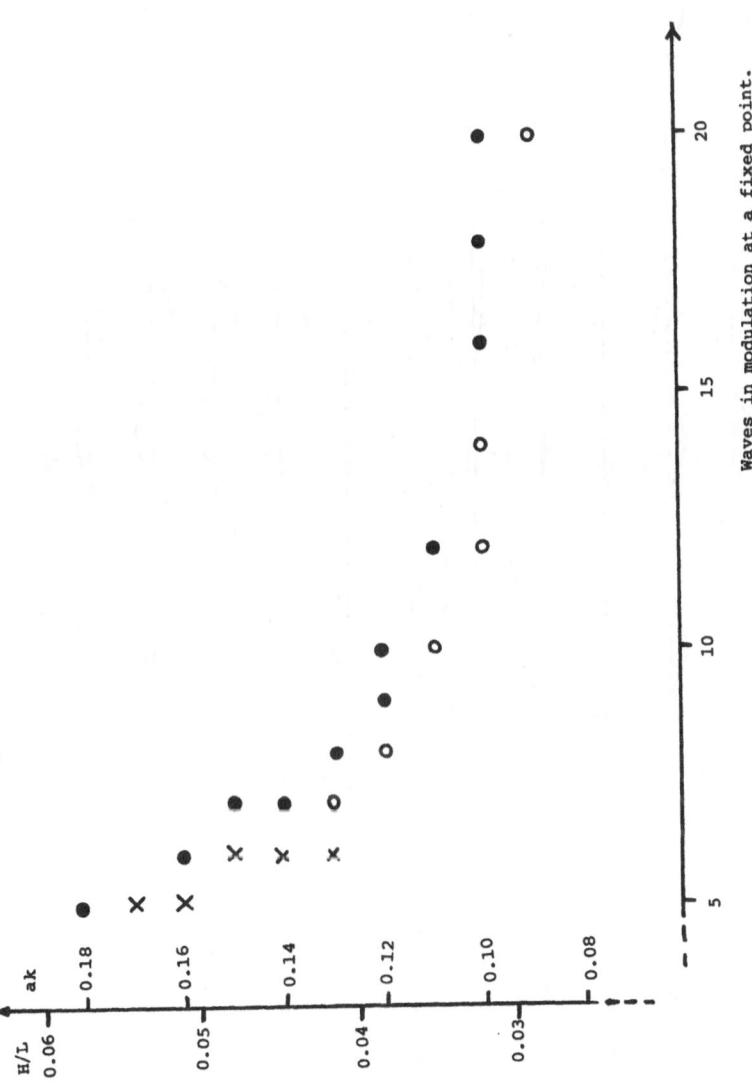

Figure 3. Summary of numerical computations of the evolution of a small modulation on an otherwise uniform wave train. Filled circles denote computations in which a wave broke. Open circles denote cases where wave evolution continued with strong modulations and no breaking. Crosses denote examples where the modulation did grow.

480

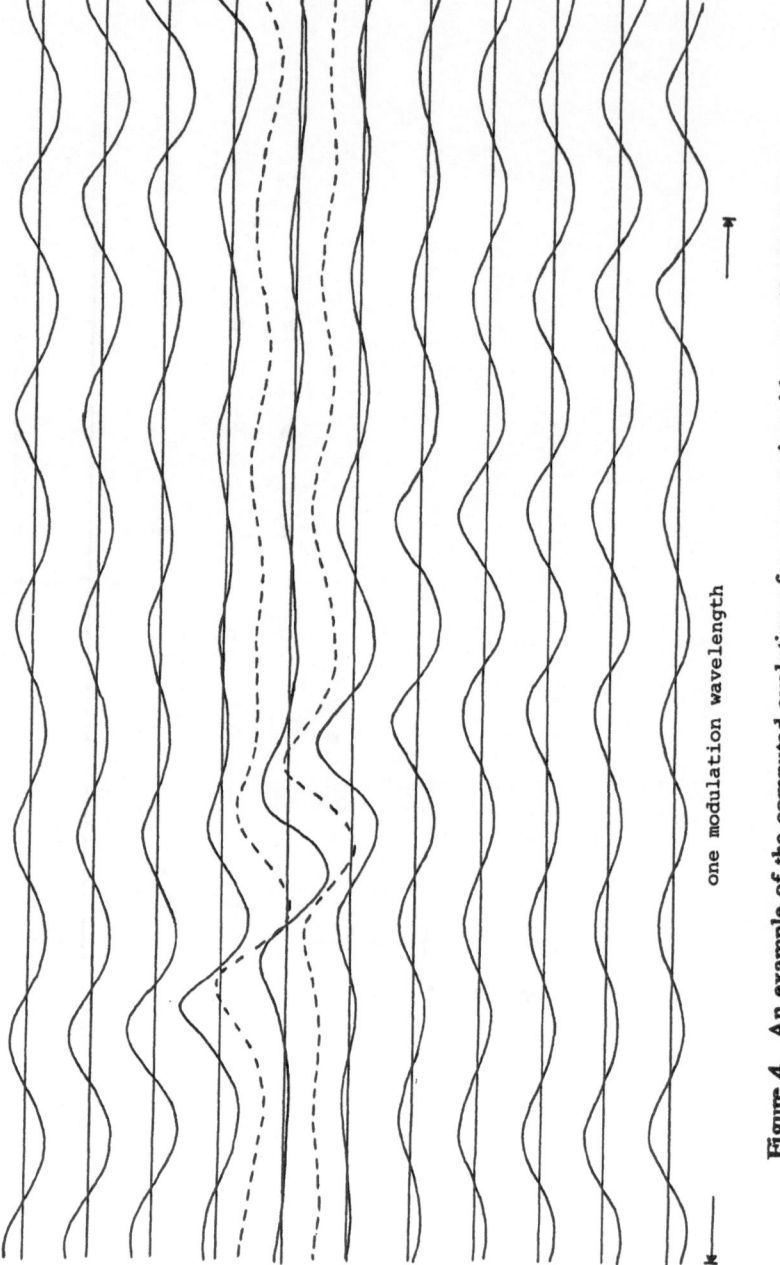

one modulation wavelength

Figure 4. An example of the computed evolution of a wave train with a small modulation. The waves become strongly modulated and then an almost plane wave recurs. The interval between most wave profiles is 20 periods of the waves. The initial wave steepness ak = 0.11. A time series would show 10 waves in a modulation since group velocity is only half the phase velocity. (Courtesy of J.W.Dold).

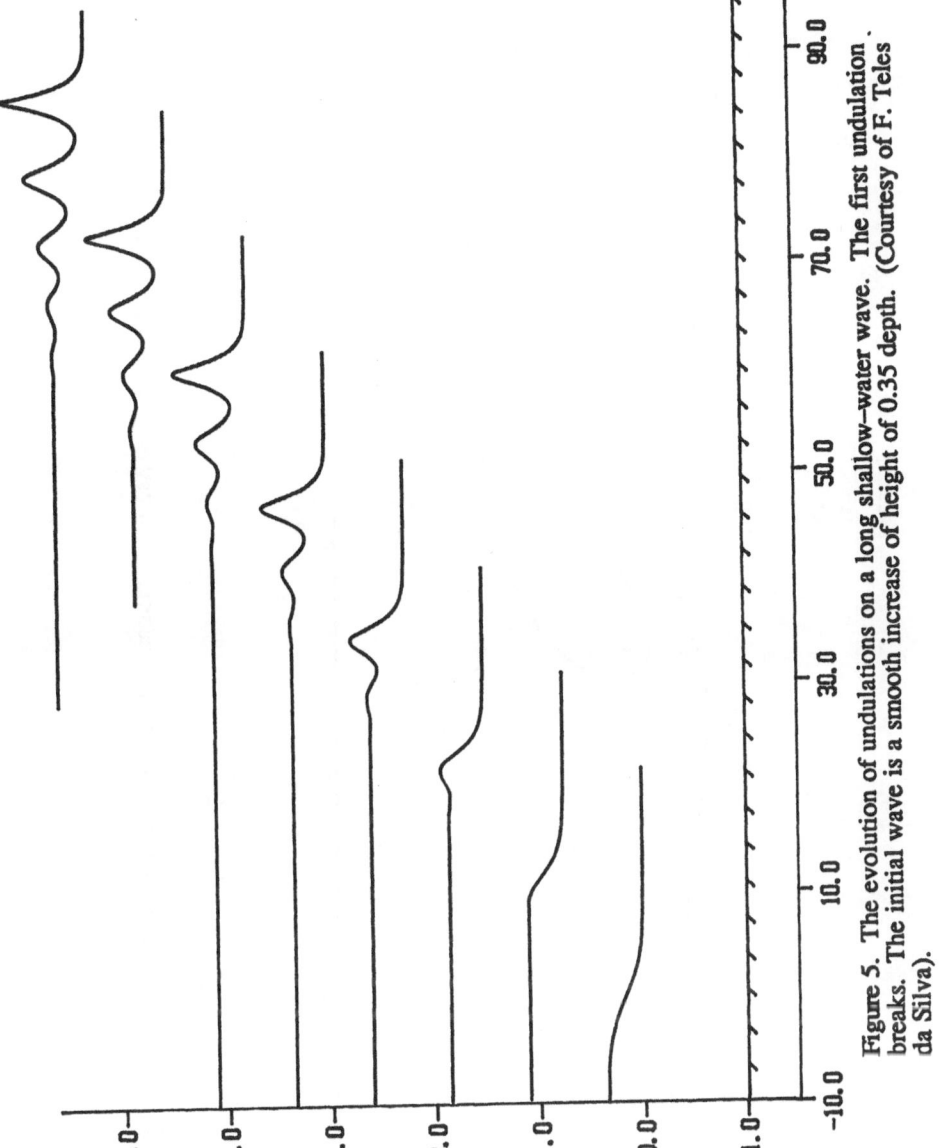

Figure 5. The evolution of undulations on a long shallow–water wave. The first undulation breaks. The initial wave is a smooth increase of height of 0.35 depth. (Courtesy of F. Teles da Silva).

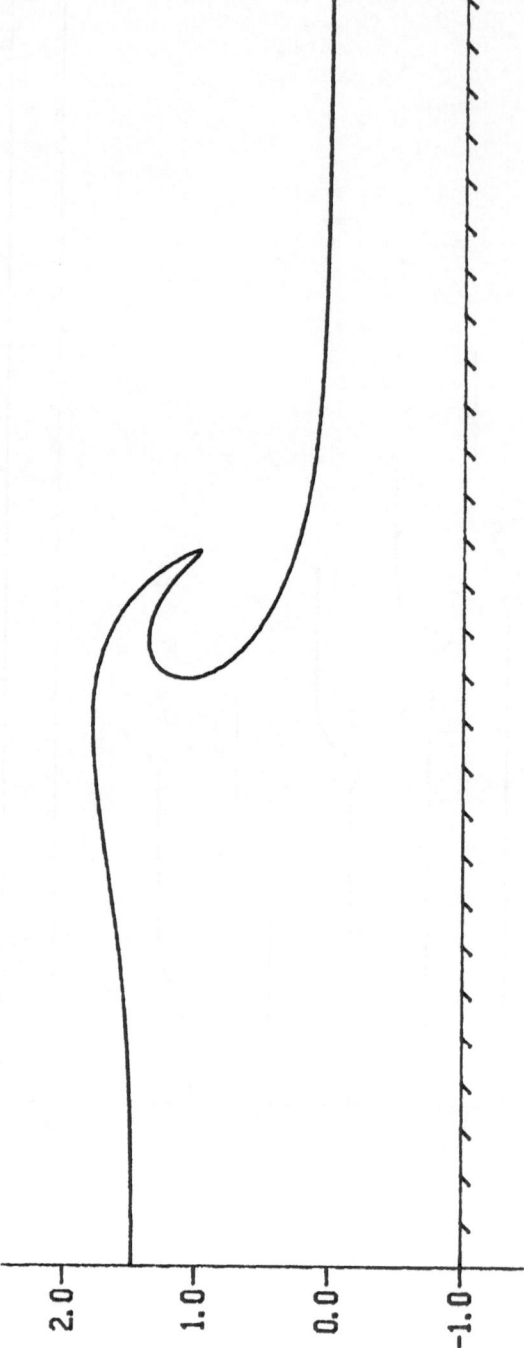

Figure 6. A large amplitude shallow water wave breaking. (Courtesy of F. Teles da Silva).

3. Applications

The most immediate application of unsteady steep wave computations is provision of hydrodynamic data. This can be velocity and pressure fields, beneath a single wave, or in the more complex situations where multiple waves are modelled. Particle paths can also be found, e.g. see Skourup's contribution to this volume. New, Peregrine and McIver (1985) give detailed diagrams of selected breaking waves from which surface data can be found. Pressure on bounding surfaces can be integrated to give forces and moments, however, this is an area where care is needed since flow separation is likely off curved surfaces and corners leading to substantial variations between a real flow and the calculated potential flow.

Publications tend to be about particular applications, except for those simply demonstrating that a particular scheme works. I have been disappointed that relatively few steps that have been made towards understanding the breaking process itself. Peregrine, Cokelet and McIver (1980) drew attention to the region of high particle acceleration that occurs beneath the emerging jet. This stimulated New (1983a) to examine the region carefully which lead to recognition that it resembled flow around an unsteady ellipse. An exact solution was found for this flow which Greenhow (1983) combined with further exact solutions of Longuet–Higgins (1983) to give an approximate analytical model. The Tanaka instability is the means by which a travelling wave of near limiting steepness may be triggered to break. However, the clear change in the evolution as breaking is approached in this and other cases suggests breaking is a phenomenon which is not closely related to its cause.

On the other hand there are plenty of examples where the boundary–integral programs are used to find where and when waves break. Instability and evolution studies have already been mentioned. Waves on beaches are another study area, though insofar as these reproduce experimental results there is little to be learned except that which comes from the ready availability of full hydrodynamic behaviour. Such studies show that several parameters may be needed to describe the breaking process. For example, table 1 gives the direction of the initial jet for the waves illustrated by Longuet–Higgins and Cokelet (1976). Although these were forced by pressure this ceases well before breaking occurs so they can be taken as representative of the breaking process. The wide range of direction and speed of the initial jets is matched by variations in size. Other computations confirm this variety.

Table 1 Initial velocity of overturning jet

These values are for velocity at the point at which the surface first becomes vertical for waves which have been forced for half a period by pressure distributions of the given maximum amplitude, following Longuet–Higgins and Cokelet (1976).

Pressure amplitude	Velocity components
0.05	1.27, 0.21
0.10	1.46, 0.36
0.146	1.56, 0.30
0.15	1.67, 0.19
0.2	1.67, 0.00

Results are more interesting when computation gives completely unexpected behaviour. Cooker and Peregrine's (1989) study of solitary waves meeting a submerged semi–circular cylinder is a case of this. For no combination of wave amplitude and cylinder radius do waves break before the top of the cylinder as was expected. Many waves break well beyond the cylinder and some break backwards towards the cylinder after passing over it.

One feature of unsteady waves which is clarified by examining very steep waves is the phenomenon of a double maximum of force and pressures found experimentally on vertical walls. (eg Nagai, 1969). For a very steep solitary wave against a vertical wall it is clear that the first maximum of pressure occurs as the fluid at the crest is projected into near free fall. The pressure then reduces until it must increase again when the crest fluid falls back and decelerates. See figures 7 and 8.

4.   Present limitations of unsteady wave computations

This discussion is confined to two–dimensional waves since practical three–dimensional wave programs are only in an early stage of developoment.

(a) *Splashing and turbulence.*

The most severe problem with present programs is that they either stop, or give unrealistic solutions once breaking commences. Given enough computation it is possible to follow an overturning jet until it touches on the surface in front of the wave, when splashing starts. Such spashing has yet to be successfully modelled, Peregrine (1981) gives a very simple splash model. In fact we still need experimental information on the nature of the water motion in the splashing and vortices that result. Peregrine's (1983) review drew attention to this problem and recent papers by Bonmarin (1989) and Tallent, Yamashita and Tsuchiya (1989, and in this volume) give detailed analysis of plunging breakers.

The study of spilling breakers has advanced further. Measurements have been presented by Duncan (1981, 1983), Battjes and Sakai (1981), Stive (1980) and Nadaoka et al (1988). The commencement of theoretical modelling can be seen in the work of Peregrine and Svendsen (1978), Madsen and Svendsen (1982), Svendsen and Madsen (1984), Tulin and Cointe (1987), Cointe (1987) and Banner (1987). However, as one might expect for such a strongly turbulent flow, there is as yet no agreed definitive approach. On the other hand some of the models can be applied to unsteady computations.

(b)   *Radiation condition*

A nonlinear radiation condition has yet to be devised which is suitable for waves on any water depth. This is a handicap in computations where the computational domain must be extended beyond the region of interest to avoid reflection from its boundary. Two–dimensional shallow–water waves can easily be dealt with since the characteristics of the shallow–water equations give the appropriate right– and left–propagating waves. For shorter waves, there is, as yet, no way of determining which parts of the wave motion at a vertical section correspond to waves propagating in specified directions. At present the techniques used in these cases include ad hoc truncations of the computational domain once radiating waves have travelled far enough, and the use of sponge layers, e.g. Baker, Meirion and Orszag (1989), and in the axisymmetric case, where waves decrease in amplitude as they propagate outward, linear theory is used (Dommermuth and Yue, 1987).

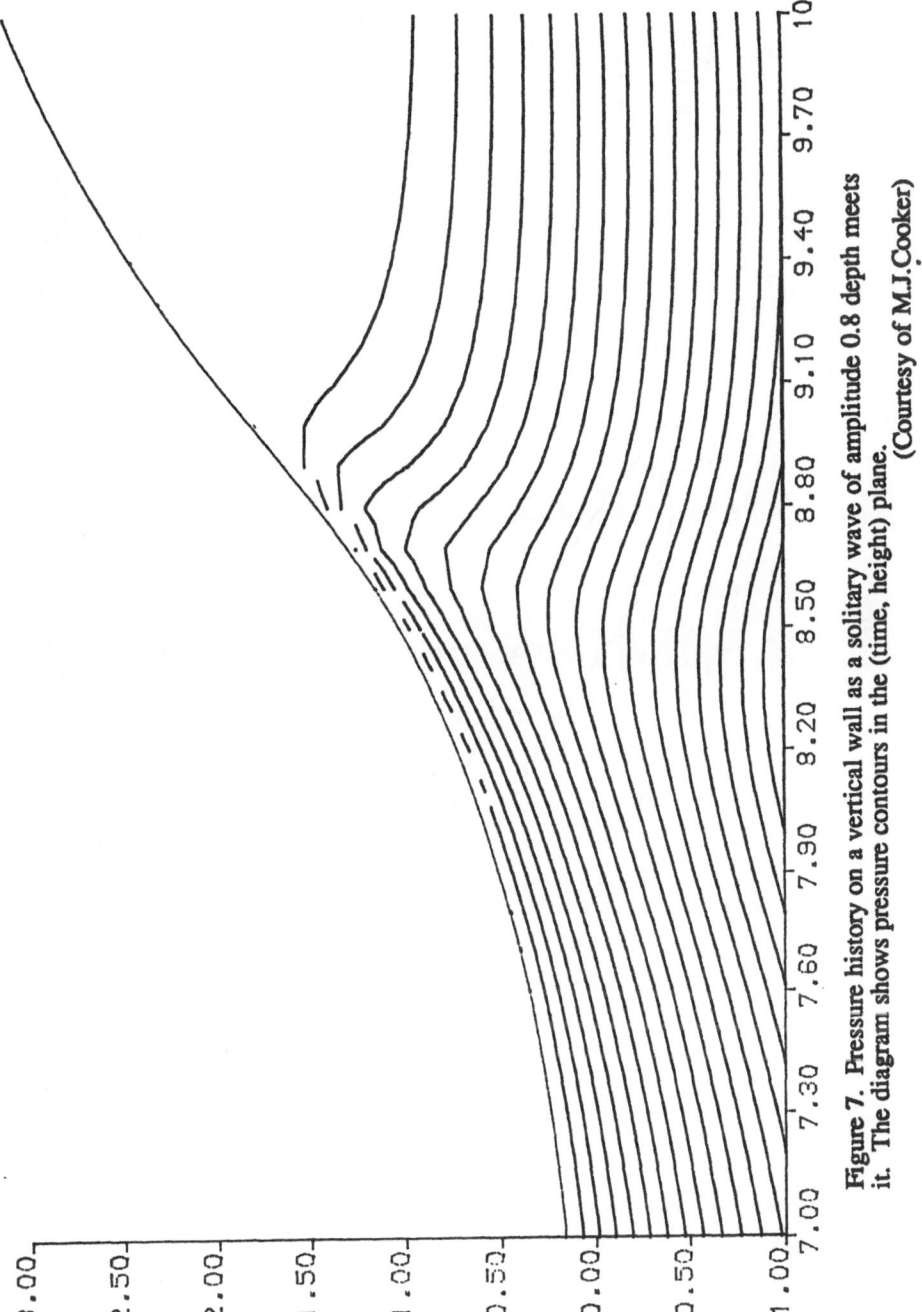

Figure 7. Pressure history on a vertical wall as a solitary wave of amplitude 0.8 depth meets it. The diagram shows pressure contours in the (time, height) plane.

(Courtesy of M.J.Cooker)

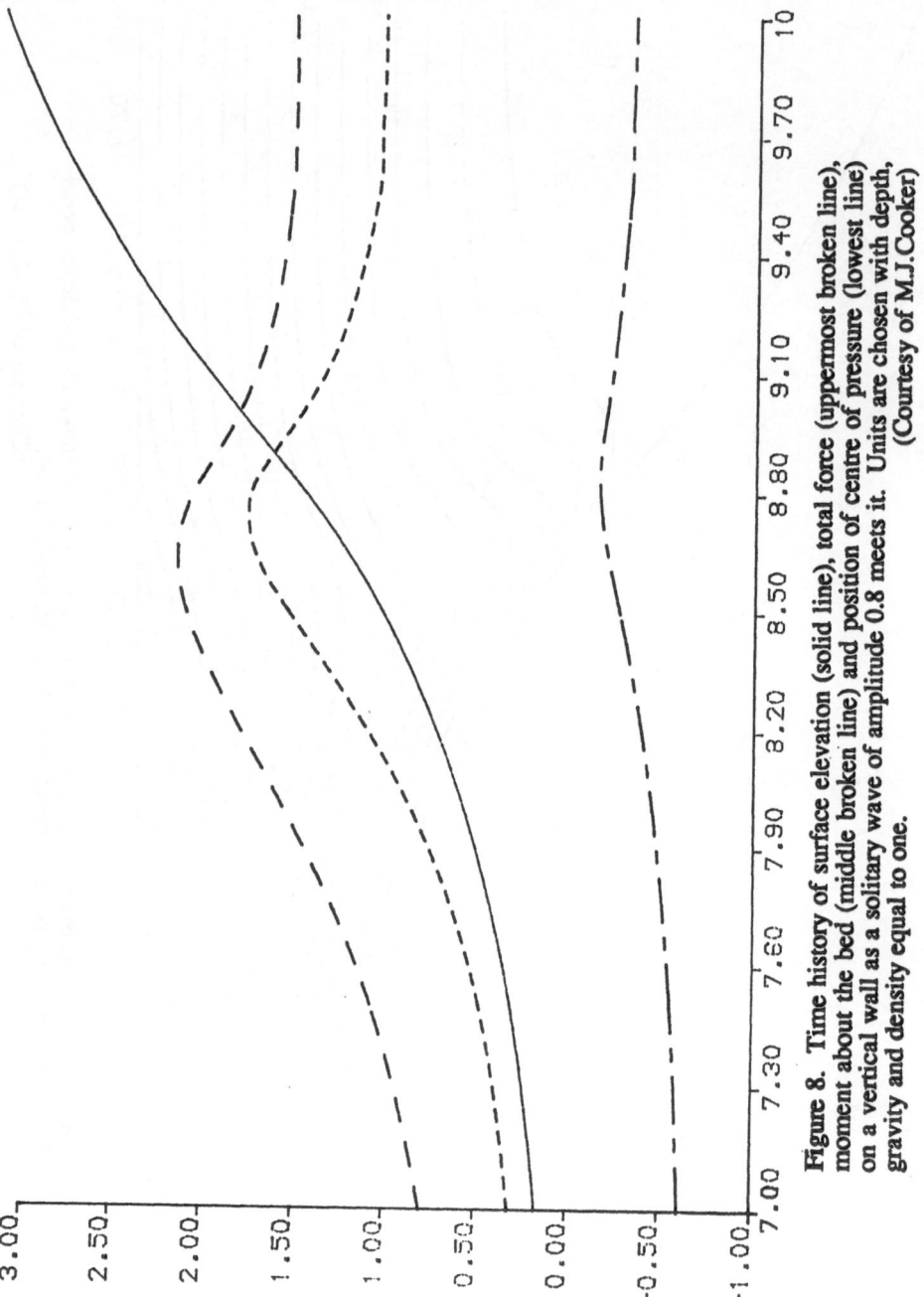

Figure 8. Time history of surface elevation (solid line), total force (uppermost broken line), moment about the bed (middle broken line) and position of centre of pressure (lowest line) on a vertical wall as a solitary wave of amplitude 0.8 meets it. Units are chosen with depth, gravity and density equal to one.
(Courtesy of M.J.Cooker)

5.   Programs for a practical engineer

The dominant feature from a user's point of view is that compared with periodic waves, unsteady waves such as breaking waves cannot be simply codified.   For example, periodic waves are fully described once water depth and current, wave period, height and direction are specified.   An unsteady wave can be started with any shape for the water surface and any distribution of tangential surface velocity component.   Guidance for users in the form of standard initial conditions is highly desirable.   Standard initial conditions also have a role in allowing designers to readily specify the wave conditions used to customers, to sub-contractors and to insurers.   These initial conditions then lead to "standard breaking waves".

Once a standard breaking wave is chosen there is a further extension to the design process.   In regular periodic waves, every point at a given depth experiences the same hydrodynamic environment as the waves pass by.   In unsteady waves every point experiences different hydrodynamics.   For example, there is only one point in space–time at which a particular wave breaks.   This means that designers may need data presented in various different ways; perhaps emphasising regions of maximum velocity, pressure gradient and fluid acceleration.   Graphical presentation is valuable here but is very often still specific to particular computers.

Current programs stop once the overturning jet becomes too fine to resolve accurately.   This raises the question: "How far should the breaking be followed ?", since this limitation is an important restriction on the use of the program.   In most cases a suitable stopping point may be when the jet is sufficiently well–formed that the surface beneath it has a horizontal tangent.   By then the horizontal velocity of the jet is likely to be near its maximum, with most further acceleration of the jet–tip being simply due to gravity.   In addition by that time the geometrical size of the jet is becoming clear.   It is probably best to develop a program which is self–adapting in terms of point numbers and accuracy until it reaches this, or some other, predetermined point.   Further computation is possible but only with rapidly increasing demands on computer resources.   Research is required to model spilling breakers or the splashes and vortices of plunging breakers.

As a specific example, consider a program used in the manner of New, McIver and Peregrine (1985).   An initial periodic steadily travelling wave of amplitude a_0 is chosen; but instead of computing with the appropriate depth h_0 the same initial conditions for the surface are used on a smaller water depth h_1.   This introduces only one extra parameter and is easy to understand.   If results are required for depth h_1 then various values of a_0 and h_0 may be chosen.   Its simplicity gives much to recommend it, even so once wavelength and water depth are selected there is still a two–parameter, (a_0, h_0) space of solutions to explore.

6.   Accuracy

Programs can be checked to high accuracy by using a steep, steady wave for initial conditions.   We find an integral–equation method to be most versatile and effective for finding the steep–steady wave solutions.   An extension of the methods used by Tanaka (1983, 1986) described by Teles da Silva and Peregrine (1988) is the method we use at present.

On the other hand, accuracy is probably more important and harder to verify in the overturning region of the crest, where numerical resolution usually becomes rapidly worse.   Checks between completely different programs, such as that described by McIver and Peregrine (1981) are very valuable, but perhaps the best check is a local analysis of the motion. It is clear that in a slender jet there can be practically no significant water pressure acting on the tip of the jet which must therefore be in free fall. That is the tip should have acceleration equal to that due to gravity.

Research support from the U.K. Science and Engineering Research Council is gratefully acknowledged.

References

Baker, G.R., Meiron, D.I., & Orszag, S.A. (1982) Generalized vortex methods for free–surface flow problems. J.Fluid Mech. 123, 477–501.

Baker, G.R., Meiron, D.I., & Orszag, S.A. (1989) Generalized vortex methods for free–surface flow problems. II Radiating waves. To appear J.Sci.Computing.

Banner, M.L. (1987) Surging characteristics of spilling zones of quasi–steady breaking waves. Proc.IUTAM Sympos.Nonlinear Water Waves. Eds., K. Horidawa & H. Maruo. Springer–Verlag 151–158.

Battjes, J.A. & Sakai, T. (1981) Velocity field in a steady breaker. J.Fluid Mech. 111, 421–437.

Benjamin, T.B. & Feir, J.E. (1967) The disintegration of wavetrains in deep water Part I. Theory. J.Fluid Mech. 27, 417–430.

Bonmarin, P. (1989) Geometric properties of deep–water breaking waves. J.Fluid Mech. to appear.

Cointe, R. (1987) A theory of breakers and breaking waves. Ph.D. dissertation. Univ.Calif., Santa Barbara, USA.

Cooker, M. & Peregrine, D.H. (1988) Solitary waves passing over submerged breakwaters. Proc.21st Internat.Conf.Coastal Engng. ASCE.

Dold, J.W. & Peregrine, D.H. (1986a) An efficient boundary–integral method for steep unsteady water waves. In "Numerical Methods for Fluid Dynamics II" (Eds. K.W. Morton & M.J. Baines). 671–679 Clarendon, Oxford.

Dold, J.W. & Peregrine, D.H. (1986b) Water wave modulation. Proc.20th Internat.Conf.Coastal Engng. I, 163–175.

Dommermuth, D.G., Yue, D.K.P., Rapp, R.J., Chan, F.S. & Melville, W.K. (1987) Deep water breaking waves; a comparison between potential theory and experiments. J.Fluid Mech. 89, 432–442.

Duncan, J.H. (1981) An experimental investigation of breaking waves produced by a towed hydrofoil. Proc.Roy.Soc.Lond. A 377, 331–348.

Duncan, J.H. (1983) The breaking and non–breaking wave resistance of a two–dimensional hydrofoil. J.Fluid Mech. 126, 507–520.

Greenhow, M. (1983) Free–surface flows related to breaking waves. J.Fluid Mech. 134, 259–275.

Greenhow, M. (1987) Wedge entry into initially calm water. Appl.Ocean Res. 4, 214–223.

Greenhow, M. (1988) Water–entry and –exit of a horizontal circular cylinder. Appl.Ocean Res. 10, 199–206.

Jillians, W.J. (1989) The superharmonic instability of Stokes waves in deep water. J.Fluid Mech. To appear.

Lin, W.–M., Newman, J.N. & Yue, D.K. (1984) Nonlinear forced motions of floating bodies. 15th Sympos.Naval Hydrodynamics, Hamburg.

Longuet–Higgins, M.S. (1978) The instabilities of gravity waves of finite amplitude in deep water. II Sub–harmonics. Proc.Roy.Soc.Lond. A 360, 489–505.

Longuet–Higgins, M.S. (1983) Rotating hyperbolic flow: particle trajectories and parametric representation. Quart.J.Mech.Appl.Math. 36, 247–270.

Longuet–Higgins, M.S. & Cokelet, E.D. (1976) The deformation of steep surface waves on water. I A numerical method of computation. Proc.Roy.Soc.Lond. A 350, 1–26.

Longuet–Higgins, M.S. & Cokelet, E.D. (1978) The deformation of steep surface waves on water. II. Gowthof normal–mode instabilities. Proc.Roy.Soc.Lond. A 364, 1–28.

Madsen, P.A. & Svendsen, I.A. (1982) Turbulent bores and hydraulic jumps. J.Fluid Mech. 129, 1–25.

McIver, P. & Peregrine, D.H. (1981) Comparison of numerical and analytical results for waves that are starting to break. Int.Sympos.Hydrodynamics in Ocean Engng. Trondheim, Norway. 203–215.

Nagai, S. (1969) Pressures of standing waves on vertical wall. Proc.ASCE, J.Waterways & Harbors Div. 95, WW1 53–76.

Nadaoka, K., Hino, M. & Koyano, Y. (1989) Structure of turbulent flow field under breaking waves in the surf zone. J. Fluid Mech. To appear.

New, A. (1983a) A class of elliptical free surface flows. J.Fluid Mech. 130, 219–239.

New, A. (1983b) On the breaking of water waves. Ph.D. dissertation, Bristol Univ.

New, A., McIver, P. & Peregrine, D.H. (1985) Computations of breaking waves. J.Fluid Mech. 150, 233–251.

Peregrine, D.H. (1981) The fascination of fluid mechanics. J.Fluid Mech. 106, 59–80.

Peregrine, D.H. (1983) Breaking waves on beaches. Ann.Rev.Fluid Mech. 15, 149–178.

Peregrine, D.H., Cokelet, E.D. & McIver, P. (1980) The fluid mechanics of waves approaching breaking. Proc.17th Coastal Engng.Conf.ASCE. Sydney. Vol 1, 512–528.

Peregrine, D.H. & Svendsen, I.A. (1978) Spilling breakers, bores and hydraulic jumps. Proc.16th Coastal Engng.Conf. 1, 540–550.

Svendsen, I.A. & Madsen, P.A. (1984) A turbulent bore on a beach. J.Fluid Mech. 148, 73–96.

Stive, M.J.F. (1980) Velocity and pressure field of spilling breakers. Proc.17th Coastal Engng.Conf. ASCE 1, 547–566.

Tanaka, M. (1983) The stability of steep gravity waves. J.Phys.Soc Japan 52, 3047–3055.

Tanaka, M. (1986) The stability of solitary waves. Phys.Fluids 29, 650–655.

Tanaka, M., Dold, J.W., Lewy, M. & Peregrine, D.H. (1987) Instability and breaking of a solitary wave. J.Fluid Mech. 185, 235–248.

Tallent, J.R., Yamashita, Y. & Tsuchiya, Y. (1989) Formation and evolution of large eddies by breaking water waves. Submitted for publication.

Teles da Silva, A.F. & Peregrine, D.H. (1988) Steep, steady, surface waves on water of finite depth with constant vorticity. J.Fluid Mech. 195, 281–302.

Tulin, M.P. & Cointe, R. (1986) A theory of spilling breeakers. Proc.16th Sympos.Naval Hydrodynamics, Nat.Acad.Press, Washington DC. 98–105.

Vinje, T. & Brevik, P. (1981) Numerical simulation of breaking waves. Adv.Water Resources 4, 77–82.

A COMPARISON OF TIME-STEPPING NUMERICAL PREDICTIONS WITH WHOLE-FIELD FLOW MEASUREMENT IN BREAKING WAVES

D.J. SKYNER, C. GRAY AND C.A. GREATED
Physics Department
Edinburgh University
Edinburgh
EH9 3JZ
SCOTLAND

ABSTRACT. Instantaneous, full-field velocity measurements under the crest of a laboratory generated breaking wave are presented and compared with a fully non-linear time-stepping numerical model. A plunging breaker is generated, on constant water depth, using the numerical model, then reproduced in a wave flume by matching wave amplitude timeseries at a position just before the breaking. The generation of the waves is achieved by means of a computer controlled wave paddle and measurements of the flow made with a full-field photographic technique (PIV). The photographic records of the flow are analysed on an automatic rig and the measurements are shown to compare well with the numerical calculations.

1 Introduction

Breaking waves represent one of the most extreme events in a marine environment and a knowledge of their internal kinematics is necessary for the estimation of drag and inertial forces on offshore structures.

A number of robust, fully non-linear time-stepping numerical models are now available for two dimensional waves [1] [2], but they can be difficult to apply to practical situations and are limited in their application to the early stages of breaking. Experimental studies have been undertaken, for comparison with results from numerical schemes [3] and for waves that cannot be realistically modelled numerically [4] [5]. A potentially fruitful area of research is the comparison of numerical predictions with experimental measurements, along with an assessment of the relative merits of the two approaches.

1.1 COMPARISON OF NUMERICAL AND EXPERIMENTAL WAVES

If the fluid flow is inviscid, irrotational and incompressible, and the numerical model accurately time-steps the boundaries, then similar waves would propagate identically in an experiment as numerically. By trying to achieve a comparison our first intention is to show that the full-field measurement technique (PIV) can give sensible measurements in the inaccessible parts of extreme waves. If the match is good then the numerical method is also validated. Furthermore, experimental measurements can continue past breaking, where the current numerical schemes stop, so that a good match up to this point would provide a useful starting point for future numerical modellers.

A. Tørum and O. T. Gudmestad (eds.), Water Wave Kinematics, 491–508.

The main difficulty encountered in trying to compare waves propagated numerically and experimentally was that the available numerical model required the wave height and velocity potential to be defined in space at a particular time, whereas experimentally a driving timeseries is supplied to the wavemaker at a particular position.

Our first attempt at a comparison was to generate, by trial and error, a plunging breaker in the laboratory, then to try to replicate it numerically by calculating the starting profile and velocity potential from a wavegauge record. However, linear theory proved to be inadequate for the task as the wavefield is too non-linear in its propagation towards breaking.

A more successful scheme was found to be the iterative selection of a starting profile which would propagate numerically into a plunging breaker and then to attempt its reproduction in the wave tank. This method has the advantage that a timeseries can be obtained from the computations for a particular place, which then should not be too difficult to reproduce experimentally. The main difficulty is that the wave height is related to the paddle motion via a transfer function, which is only easily determined if the motion is sufficiently linear.

1.2 FULL-FIELD VELOCITY MEASUREMENT

Laser Doppler Anemometry (LDA) has been successfully used to make precise velocity measurements under water waves [6], but some problems are encountered in its application. Most of these complications have been solved by the appropriate implementation of established computer algorithms for analysing the Doppler signal and detailed studies of velocity distributions have been made and utilised for structural design purposes [7]. Nevertheless a fundamental limit exists in the use of LDA and other point measurement devices used in water wave studies.

Generally, a whole vector plot of particle velocities under a specified wave is required for design purposes and as a sensitivity test of theoretical models. So, for point measurement techniques, complete vector plots can only be constructed in cases where the wavefield being studied can be faithfully repeated many times while moving the probe to a series of positions within the flow field. Accurate repetition of a wave profile relies on the quality of the wave generator and for truly random waves it is impossible to obtain full-field information from point measurement procedures.

Recently Particle Image Velocimetry (PIV) [8] [9] has been successfully applied to the study of water wave motions. This is a non-intrusive velocity measurement technique which allows a complete two-dimensional flow fields to be captured at a single instant. The water is seeded with small neutrally buoyant particles and illuminated with a plane sheet of pulsed laser light. Using a conventional camera operating with a shutter time exceeding the pulse period, photographic film is exposed to two or more images of the moving particles. The displacement of the particle pairs determines the flow velocity at any particular point, and is measured by means of optical and digital processing of the developed film.

The application of the PIV technique to breaking waves is described. Waves are generated in a flume that has been purpose built for PIV analysis with an illumination system based on a 20 W CW Argon Ion laser. Interpretation of the photographic flow records is by means of an automatic analysis rig controlled by a microcomputer.

2 The Generation of Breaking Waves

Because water waves are dispersive on finite depth, extreme waves can be generated by choosing a wave-packet which will self-focus. Most methods for choosing the starting wave-group assume linear theory, possibly augmented by the inclusion of a non-linear phase

speed [11]. However, since the wave-propagation becomes increasingly non-linear as breaking approaches the starting conditions are somewhat arbitrary and normally require their
parameters to be adjusted until the required breaker is formed.

We decided to attempt the generation of breaking waves by the superposition of sinusoidal
components, with enough parameters in the specification of the spectrum to allow some fine
adjustment of the breaker's form. Equation 1 describes a wavefield composed of discrete
sinusoidal components at equally spaced frequencies.

$$a(x,t) = \Re\left\{ \sum_{n=N_0}^{N_1} A_n e^{ik_n x + i\omega_n t + i\phi_n} \right\} \tag{1}$$

where

$$\omega_n = 2\pi n\, \delta\omega \tag{2}$$

and

$$\omega^2 = gk \tanh kh \tag{3}$$

In order to limit the number of parameters the spectral shape was selected such that the
components had amplitudes given by

$$A_n = (a_0 + a_1\omega) \exp\left(-\left| \frac{\omega - \omega_1}{\omega_0} \right|^s \right) \tag{4}$$

And phases ϕ_n chosen so that, if linear theory applied, the components would have some
common phase Φ at $x = 0, t = 0$.

The spectral form (4) was selected as covering the 'flat top' spectral shape often used in
the generation of freak waves [3][12], while allowing the spectrum to tail off smoothly at
the extremes.

3 Numerical Model

The numerical model available for comparison with experimental measurements was the
fully non-linear time-stepping scheme developed by Dold and Peregrine [2] at Bristol University. The model assumes the flow to be inviscid, incompressible and irrotational. With
these assumptions the calculation of the fluid motion is reduced to the evaluation of discrete
points along the surface, resulting in a computational efficient scheme. The implementation
available to us assumes periodic boundary conditions and requires the coordinates of points
along an initial surface along with their velocity potentials, and will time-step the surface
motion until breaking occurs.

Having decided on a spectral form and chosen initial parameters making use of experience
in generating demonstration 'freak' waves in the laboratory, starting profiles and velocity
potentials were calculated with linear theory for the numerical model.

In the model the computational points move with the surface. The resulting flexibility in
the distribution of the computational points can be used to advantage as resources can be
concentrated on regions of the wavefield where the most rapid changes are expected. The
point distribution was made proportional to the local wavenumber, with enhancement in
the region where breaking was expected.

The model requires periodic boundary conditions and for the horizontal extent to be
rescaled to 2π, and other quantities by the appropriate Froude scaling. Care was taken that
all the the wavepackets considered tailed off naturally before the boundaries. The point

distribution function was also forced to be continuous across the boundaries by employing a smoothing algorithm.

Each numerical wavegroup tested was allowed to propagate until it 'broke' or passed the place where breaking was expected. The spectral parameters, mainly overall size, were then altered until a plunging breaker was produced which just avoided breaking at the penultimate crest. With 200 computational points it was possible to try another iteration every few hours, and after couple of dozen iterations a suitable breaker was arrived upon. The chosen wave was then rerun with 400 computational points. The profile of the wave as it nears breaking, used to judge its suitability, is plotted in figure 2.

The spectral shape finally selected is shown in figure 1 and the resulting starting profile and velocity potential plotted in figure 3 along with the distribution function used to obtain the computational co-ordinates.

From the program output a wave amplitude timeseries was extracted by interpolation at $x = -1.0\,\text{m}$. This is about 1.5 m before the breaking position and was chosen to be a suitable place to attempt the experimental match, given the limited length of the wave flume. In order to obtain a record which contained most of the passage of the wavepacket the timeseries was extented both before and after the time period covered by the numerical model, using linear theory. This extension was found to be necessary if Fourier analysis was to be used in the experimental reproduction, although the linear theory only applies correctly at the start of the record. The composite timeseries is shown in figure 4.

4 Experimental Measurements

4.1 WAVE TANK

The experimental measurements of breaking waves were made in a 6 m long wave flume, sketched in figure 5. Waves are generated by a hinged wavemaker, and propagate through the PIV measuring region before being absorbed by an expanded metal beach. The experimental arrangement includes a wavegauge so that the composition of the wavefield can be established and tank repeatability checked. The wavemaker is of the absorbing type [10], with its force on the water made equal to the sum of a drive signal and filtered velocity signal, leading to the simultaneous generation of waves and absorption of reflections respectively. The wave drive signal is produced by a microcomputer which can sample the wavegauge and trigger the camera while generating the wavefield.

Photographic recording of the flow beneath the waves requires optical access into the flume, so the central 2 metre section of the tank has 20mm thick glass walls and a glass base to allow passage of the illuminating laser sheet.

4.2 REPLICATION OF NUMERICAL WAVE

The wave height timeseries obtain from the numerical model provided the point at which the experimental wavefield was required to match the numerical. With a knowledge of the expected breaking position and taking into account the restriction of the PIV measuring region a wire wavegauge was placed 1.300m in front of the wavemaker. This position is about $2\frac{1}{2}$ times the hinge depth away from the wavemaker which was judged to be sufficiently far to avoid transient wave effects [13].

The replication of the timeseries is most easily undertaken by calculating wavemaker drive signals in the frequency domain, so the composite numerical timeseries was Fourier transformed, and the resulting spectrum is shown in figure 6. The transfer function of the wavemaker, the wave amplitude and phase of the travelling wave generated by a unit sinusoidal drive signal, is experimentally measured and is only valid if the waves are linear. As the waves are markedly non-linear at the position where the timeseries was to be replicated, an iterative scheme was attempted for the generation of the required record.

The initial wavemaker drive spectrum was calculated from the linear transfer function and tried in the tank. The frequency limits for this spectrum are somewhat arbitrary as it is always likely that some spurious components will be generated outside the chosen bounds, and it was found that the form of the breaker was quite sensitive to the top frequency limit. The wavegauge was sampled as the wavegroup passed and the effect of the anti-aliasing filter in the sampling system removed computationally. The later portion of the wave record, after the breaking, contains reflections and spurious high frequencies from the wavemaker's final large motion which are difficult to disentangle in a non-linear wavefield. Therefore, the experimental wave record was patched with the linearly expected timeseries used to extend the numerical record. One advantage of this scheme is that the same operations are done to both numerical and experimental wave timeseries before Fourier Transformation. In the frequency domain the required and obtained spectra were compared. If the match was judged inadequate, then each component in the drive spectrum was modified in both amplitude and phase by multiplication by the complex quantity $R(\omega)$ given by

$$R(\omega) = \sqrt{\frac{A(\omega)_{required}}{A(\omega)_{obtained}}} \qquad (5)$$

This modifier has the effect of stepping halfway to that required in both amplitude and phase and was used because the simple ratio was found to overshoot.

Iterations were continued until the obtained spectrum was judged to match that required sufficiently closely. In the iterations the lower frequencies were found to converge first, which is expected because the higher frequencies are partially composed of their harmonics. The final spectrum in shown in figure 6 and associated timeseries in figure 8, while the spectra obtain during the six steps of the iterative process are plotted in figure 7.

4.3 BREAKING WAVE EXPERIMENTS

Having established a wavemaker drive signal which would replicate the numerical wave at one position the wave was generated in the flume and photographs of the internal kinematics taken at number of phases of the wave through breaking. Because the camera took a couple of seconds to wind on, the different phases were recorded by allowing the tank to settle, then repeating the wavefield and triggering the camera at the appropriate time.

The photographs were taken at a series of times from the start of the wavefield incremented by $1/40$ s on each occasion. The time that the camera takes to open its shutter having been triggered was measured to be 74 ms. Because the real time control of the tank is based around a 160 Hz interrupt rate this was rounded to 75 ms and subtracted from the required measurement time to give the trigger time.

Repeatability of the wavefield is illustrated in figure 9 and the experimentally measured profile of the wave as it approaches breaking in figure 10.

4.4 PIV PHOTOGRAPHY

The region of interest beneath the waves is illuminated by a sheet of light and photographed from the side. In order that the PIV photograph represents a sufficiently instantaneous record of the flow within the wave the shutter time of the recording camera and the time duration between illuminating pulses must be as short as possible. This also imposes the requirement that the pulse time be very short. Lourenco [14] quotes a minimum pulse separation to pulse time ratio of 20, and although in practice smaller ratios than this can be used, generally the larger the ratio the better the quality of the flow records. The measurements described in this paper were made using a CW laser as greater control is afforded over pulse separation than with conventional pulsed lasers, permitting this same light source to be used for PIV measurements on a variety of other flows [15] [16]. However, for a sheet illumination system that is pulsed by mechanical or electrooptical means there will be a severe constraint upon the light power in each pulse if the pulse time is very short. In the Edinburgh system this difficulty is avoided by employing a *Scanning Beam* system of flow illumination whereby the beam from the laser is narrowed to approximately 1.5 mm using a telescope arrangement, and then reflected off a multi-faced rotating mirror and a parabolic recollimating dish to scan through the area of interest [17]. This manner of illumination ensures that the light power from the laser entering the flow is maximised and that a very short but intense pulse of light is imparted to the tracer particles within the flow. In our current system each illuminating pulse is approximately 200 times shorter than the time between pulses, which for a pulse separation of 1 ms gives a pulse time of $5\,\mu s$ ensuring that streaking of the particle images is negligible; this avoids the associated problems of streaked photographs [15].

The water within the wave tank was seeded with Conifer Pollen which has an average diameter of $70\,\mu m$. These particles are used because, when wet, their density is very close to that of water ensuring that within the timescale of an experiment there is no noticeable settling of the tracers as may be experienced with more traditional seeding materials such as Aluminium powder. This is important as any settling time will result in an inadequate density of tracers in the upper measurement region of a passing wave, eventually resulting in signal dropout where a good record of the flow is particularly important. The closeness of the seeding density to that of the water also ensures that the tracer particles faithfully follow the flow even under high flow accelerations.

Seeding concentrations are maintained at a level that will ensure a high density of non-overlapping particles on the film record. This is important so that within the small region in which a point measurement of the particle separation is calculated there are three or more particle pairs allowing unambiguous determination of the flow direction and magnitude. For a light sheet of thickness 1.5 mm, an image / object magnification of 0.102, and a circular probe diameter of 0.75 mm, a minimum seeding density of 2×10^7 is required.

In order to record the whole crest region of the breaking wave in a single exposure it is necessary to have a large illuminated measurement region. Consequently, a large film format is required so that the image / object magnification is maximised to allow good resolution of the velocity measurements from the final flow record. A Hasselblad ELM camera body with a 120 film back was used giving $28.25\,cm^2$ of exposable film for each flow photograph. The illuminated tracer particles are imaged onto the film plane by a planar T 80 mm focal length f 2.8 lens system. The high quality of the lens ensures that the imaged particles are not excessively enlarged over the diffraction limit of the lens by optical abberrations.

Sinha [18], points out that the refraction of a water/glass/air boundary will lead to distortion of the image plane when photographing a planar region within a glass tank of fluid with a different refactive index to air. This effect is more pronounced the greater the angle of incidence of light from the optical axis of the imaging optics. In order to account for such distortions, which would lead to significant positional errors on analysis of the flow record, a grid scribed on perspex was photographed in the same position as the illuminated region of the flow. Precise measurement of the positions of the intersection points of the grid on the developed film were made using a travelling microscope with the film positioned on a light table. Relating these measurements back to the known dimensions of the perspex grid gave a map of the geometric distortion due to refactive index differences, the limited optical quality of the tank walls, and any small distortions from the lens. The photographic emulsion used was T-Max 100 ASA which has good spectral sensitivity between 4700 nm and 5500 nm which covers all the lasing lines of the Argon Ion laser used. Each film is developed for 7 minutes at 20° C with agitation by multiple inversion every 30 seconds to reduce particle migration due to developer starvation. The film is washed by dilution then fixed for $3\frac{1}{2}$ minutes, again at 20° C to avoid reticulation. The resolution achieved with this film is good (200 lines/mm) even though it has considerable sensitivity over the wavelengths used to illuminate the flow allowing good image definition and little effect upon particle image shape or size.

4.5 PIV ANALYSIS

Typical PIV photographs contain an enormous amount of precise velocity information relating to the recorded flow; this is visible by inspection of the film under magnification, as the physical separation of particle images recorded from successive illuminations. The separation of 'related images' must be measured without confusion from other nearby images and combined with the parameters M (the image / object magnification and Δt (the time between illuminating pulses) to determine the particle separation and thus the flow velocity at that point. Manual analysis of flow records has been achieved with apparently good results [19] but for the large quantity of data contained in the wave records presented here precise automatic analysis is required.

The system used at Edinburgh is based on a 32 bit microcomputer and interprets the film point by point over a dense grid using methods of optical and digital analysis. The film is mounted on a two-dimensional microtranslation stage allowing motion vertically and horizontally in the plane of the photograph. At points typically separated by 0.5-2.0 mm the film is coherently illuminated over a small circular region (0.75 mm diameter) and the diffracted light passing through the photograph is collected and observed at the front focal plane of a biconvex lens as displacement speckle fringes or Young's fringes [19]. The orientation and separation of such fringes is related to the separation and orientation of the particle images within the probed region of film. If three or more particle image pairs exist within this region then the most common particle separation is that induced by the flow and so the fringes generated will unambiguously represent the flow velocity. The automatic analysis of displacement fringes has received much attention in order to determine a fast, reliable, and accurate method of resolving their orientation and frequency.

Semi-automatic techniques have been used based on one-dimensional Fourier transformation of the intensity distribution after they have been averaged in the direction of the fringes. This, of course, requires an operator to align the diffraction pattern. Fully automatic methods have been developed [20] [21], utilising one-dimensional autocorrelations in several directions to determine both fringe frequency and orientation, but for fringes that are particularly noisy the failure rate may be high. The fringes generated from most PIV photographs are typically of very low visibility due to a number of factors including,

  - Refractive index and thickness variations across the film backing
  - Film grain noise
  - Noticeable cross interference due to small numbers of particle pairs
  - Small variations in particle velocity across the measurement area due to turbulence and/or large velocity gradient
  - Reduction in particle image contrast due to air entrainment, etc.

This necessitates the use of a fringe analysis algorithm that uses all the information present in the fringe intensity distribution. Huntley [22], describes a system of fringe evaluation based on the two-dimensional Fourier transformation of the digitised fringes which achieves 100% success for fringes of visibility greater than 7%. Even though the calculation of the two-dimensional transform is time-consuming the scheme is adopted because of its ability to successfully process extremely noisy fringes.

Digitisation of the fringes is achieved using a Panasonic CCD camera. The digitised intensity pattern is transferred to the computer's hard disc until all the fringe data from the film is saved. This allows the CCD camera and the probe laser (2 mW HeNe) to be switched off while the more time-consuming fringe analysis is being performed, prolonging the life of the camera and preserving the fringe data for visual inspection should a particular point measurement fail. The two-dimensional Fourier transform is calculated by performing a Fast Hartley transform [23] on the rows and then on the columns of the digitised data. This gives a significant reduction in the computation time over a Fast Fourier Transform as the data has no imaginary part on starting the calculation. Subtraction of the fringe pedestal prior to transformation zeroes the d.c. component and significantly aids the resolution of low fringe frequencies [24]. Location of the discrete signal peak is then a simple search for maximum data point in the transform plane. The peak is the sum of a number of peaks associated with the individual particle pairs. The average position of these contributory peaks is found by calculating the centre of mass of the data in the vicinity of the maximum. The position k_x, k_y calculated in this way gives the spatial frequency components in the horizontal and vertical directions which when multiplied by a calibration value and divided by the illumination interval gives the horizontal and vertical components of velocity. The visibility of the fringes is also calculated from the height of the signal over the area under the fringe intensity data, allowing an estimate of the measured velocity's reliability.

Figure 12 illustrates the fringe aquisition apparatus, which analyses photographic negatives under computer control.

Within each set of data generated from a flow photograph there is a subset of measurement points which are not consistent with those velocity values about them. Their orientation and/or magnitude may differ significantly from neighbouring values because their fringe frequency values are anomalous and generated by detection of a peak other than the peak due to the separated particle images. The reasons for such failures are noted above and occur more frequently the poorer the quality of the fringes. The success rate quoted above from Huntley [22] is consistent with measurements made with the system described but a significant fraction of fringes measured from wave records have visibilities well below 7%. By considering the measured fringe visibility and the fringe frequency and orientation it is possible to identify most of the spurious vectors. These values are removed and in places where sufficient numbers of valid neighbours exist an interpolated value substituted.

5   Results and Conclusions

Figure 11 contains an overlay of experimentally measured and numerically generated surface profiles, at equivalent phases, as the wave approaches breaking. While the overall match of position and size is good, the detail around the plunging tip is not. It was found, during the replication of the numerical wave, that the exact form of the breaking wave was very sensitive to changes in the generated wave spectrum, so that the discrepancies in figure 11 could be explained by a 1% error in the wavegauge calibration. In future experiments the sensitivity to small changes will be explored and the drive signal to the wavemaker modified until a good match of surface profiles is acheived at the plunging phase.

Despite the inexact match of the surface p·ofiles the photographic records of the internal velocities were analysed and a typical example is plotted in figure 13. For comparison numerically generated internal velocities for an earlier phase of the wave, with a similar surface profile, are plotted in figure 14. ·

6 Acknowledgement

The authors gratefully acknowledge the funding provided by the Marine Technology Directorate U.K. for this project. They also wish to thank Prof. D.H. Peregrine et al of the Mathematics Dept., Bristol University who developed the time-stepping computer program and allowed the authors to run this at Edinburgh.

References

[1] Longuet-Higgins, M.S. and Cokelet, E.D. (1976) *The Deformation of Steep Surface Waves in Water. A Numerical Method of Computation.* Proc. Royal Soc. London, A. 350, 1-26.

[2] Dold, J.W. and Peregrine, D. P. (1986) *Numerical Methods for Fluid Dynamics II.* Authors Morton, K.W. and Baines, M.J. Oxford University Press, 671-679.

[3] Dommermuth, D.G., Yue, D.K.P., Lin, W.M., Rapp, R.J., Chan, E.S. and Melville, W.K. (1988) *Deep-water Plunging breakers: a comparison between potential theory and experiments.* J. Fluid Mech., 189, 423-442.

[4] Van Dorn, W.G. and Pazan, S.E. (1975) *Laboratory Investigation of Wave Breaking, Part 2, Deep Water Waves.* Research Report, Advanced Ocean Engineering Laboratory, Scripps Institute of Oceanography, University of California, San Diego, S10 Ref. No. 75-21, AOFL Report £71.

[5] Kjeldsen, S.P., Vinge R. and Brevig, P. (1980) *Kinematics of Deep Water Breaking Waves.* Proc. 12th Annual Offshore Technology Conference, Texas, 317-325.

[6] Easson, W.J. and Greated, C.A. (1984) *Breaking Wave Forces and Velocity Fields.* Coastal Engineering, 8, 233-241.

[7] Easson, W.J. (1987) *Velocity and Force Measurement in the Splash Zone.* J. of Strain, Feb, 15-18.

[8] Gray, C. Greated, C.A. and Fancy, N.A. (1987) *The Application of PIV to Measurements Under Waves.* 2nd Int. Conf. on Laser Anemometry, Strathclyde.

[9] Gray, C. and Greated, C.A. (1988) *The Application of PIV to the Study of Water Waves.* Optic and Lasers in Engineering, 9.

[10] Salter S.H. (1982) *Absorbing Wave Makers and Wide Tanks.* Proc. Conf. Directional Wave Spectra Applications, ASCE, 185-200.

[11] Kjeldsen, S.P. (1982) *2 and 3 Dimensional Deterministic Freak Waves.* Norwegian Hydrodynamic Laboratories Report, 1983.

[12] Greenhow, M., Vinje, T., Brevig, P. and Tanylor, J. *A theoretical and experimental study of the capsize of Salter's Duck in extreme waves.* J. Fluid Mech., 118, 221-239.

[13] Hyun, J.S. (1976) *Theory for Hinged Wavemakers of Finite Draft in Water of Constant Depth.*

[14] Lourenco, L. (1986) *Theory and Application of Particle Image Displacement Velocimetry.* in Lecture Series 1986-09, Von Karman Inst. of Fluid Dynamics, Brussels.

[15] Sharpe, J.P., Gray, C., Greated, C.A. and Campbell, M. (1989) *Measurements of Acoustic Streaming using PIV.* Acoustica.

[16] McCluskey, D.R., Easson, W.J., Greated, C.A. and Glass D. (1989) *Use of Particle Image Velocimetry to Study Roping in Pneumatic Conveyence.* 4th European Symposium in Particle Characterisation, Nuremberg, W.Germany.

[17] Gray, C. and Greated, C.A. (1988) *A Scanning Beam System for the Two Dimensional Illumination of Flow Fields.* Von Karman Inst. of Fluid Dynamics, Belgium.

[18] Sinha, S.K. (1988) *Improving the Accuracy and Resolution of Particle Image or Laser Speckle Velocimetry.* Experiments in Fluids, 6, 67-68.

[19] Burch, J.M. and Tokarski, J.M.J. (1968) *Production of Multiple Beam Fringes from Photographic Scatters.* Optica Acta, Vol 15, No. 2, 101-11.

[20] Robinson, D.W. (1983) *Automatic Fringe Analysis with a Computer Image Processing System.* Applied Optics, 22, 2169-2176.

[21] Moraitis, C.S., Buchlin, J.M. and Reithmuller, M.L. (1987) *Improved Autocorrelation Analysis of Fringe Images for Laser Speckle Velocimetry.* ICIASF June 22-25, 1987.

[22] Huntley, J.M. (1986) *An Image Processing System for the Analysis of Speckle Photographs.* J. Physics. E. Sci. Intrum., 19.

[23] Bracewell, R. N. (1986) *The Fourier Transform and its Applications.* McGraw-Hill.

[24] Pickering, C.J.D. and Halliwell, N.A. (1985) *Particle Image Velocimetry: Fringe Visibility and Pedestal Removal.* Applied Optics, Vol 24, No. 6, 2474-2476.

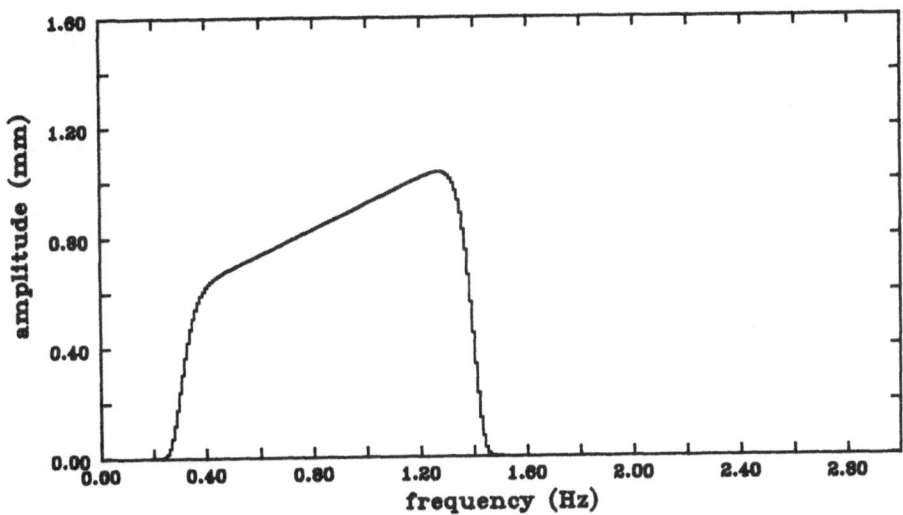

Figure 1: Wave amplitude spectrum found to produce a suitable numerical breaker.

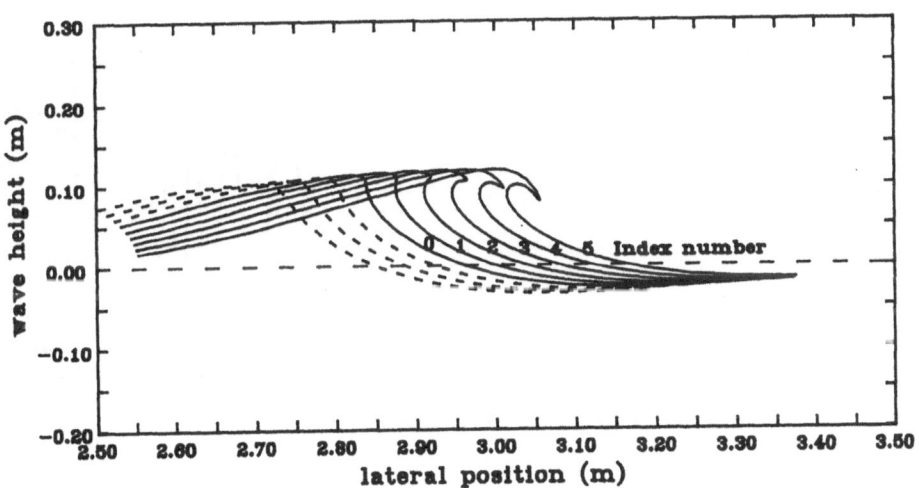

Figure 2: Numerical profile as wave approaches breaking.

502

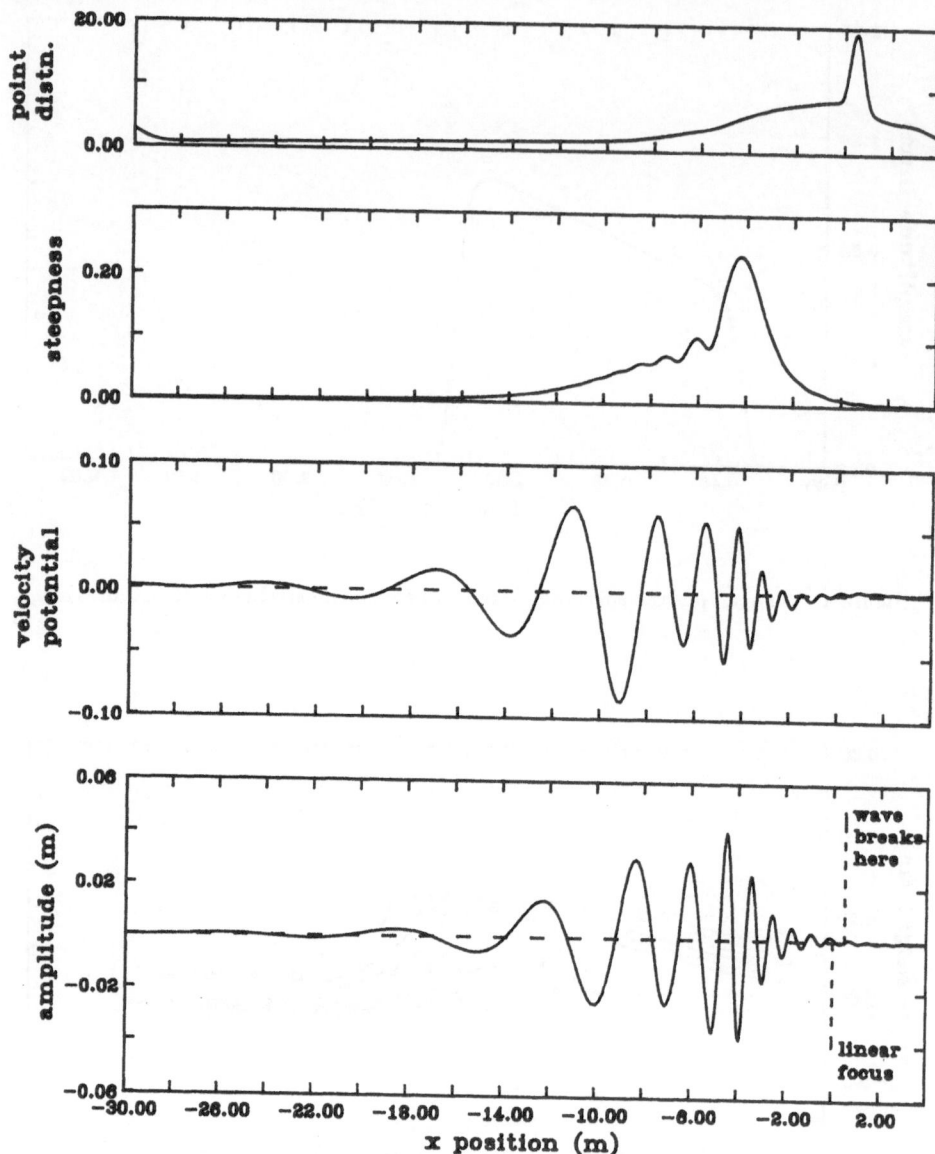

Figure 3: Starting profile and velocity potential calculated by linear theory 6 seconds before expected linear focus.

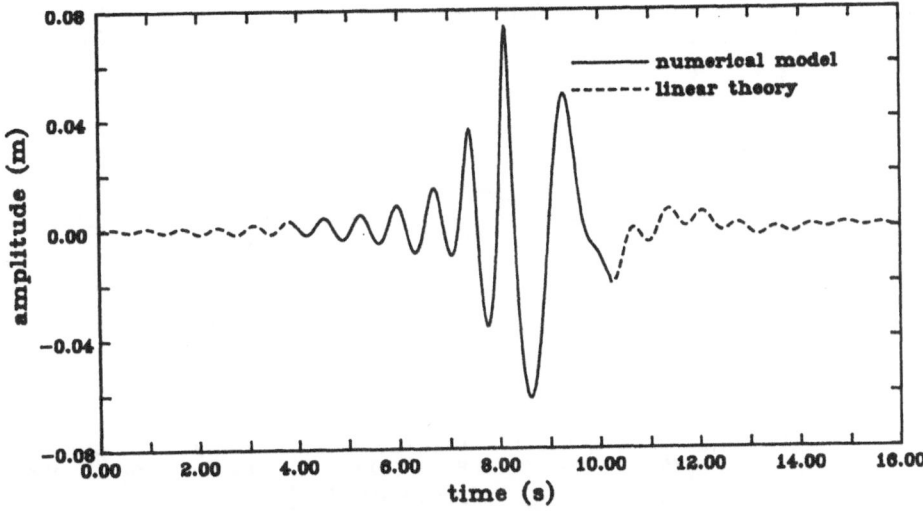

Figure 4: Composite wave amplitude timeseries from the numerical model and linear theory.

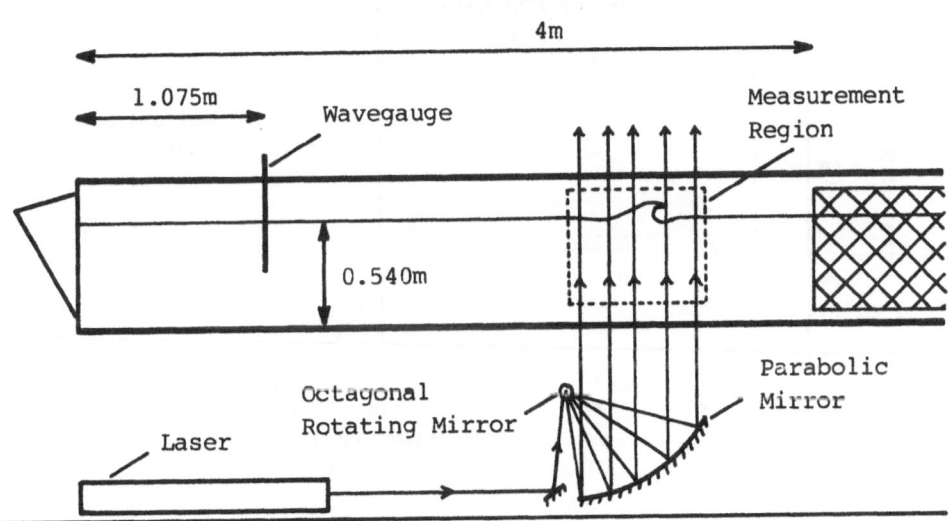

Figure 5: Wave flume used for the PIV analysis of breaking waves.

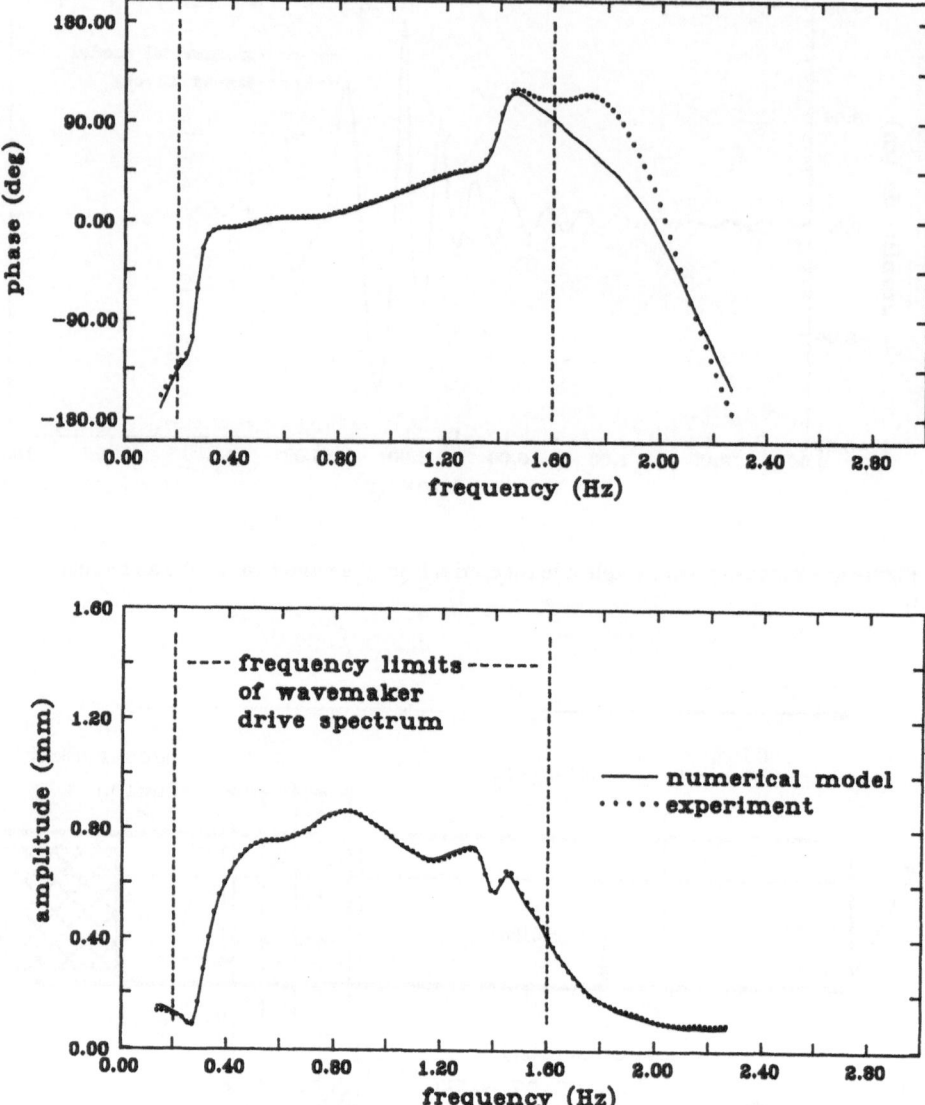

Figure 6: Experimental wave spectrum found by iteration compared to required spectrum.

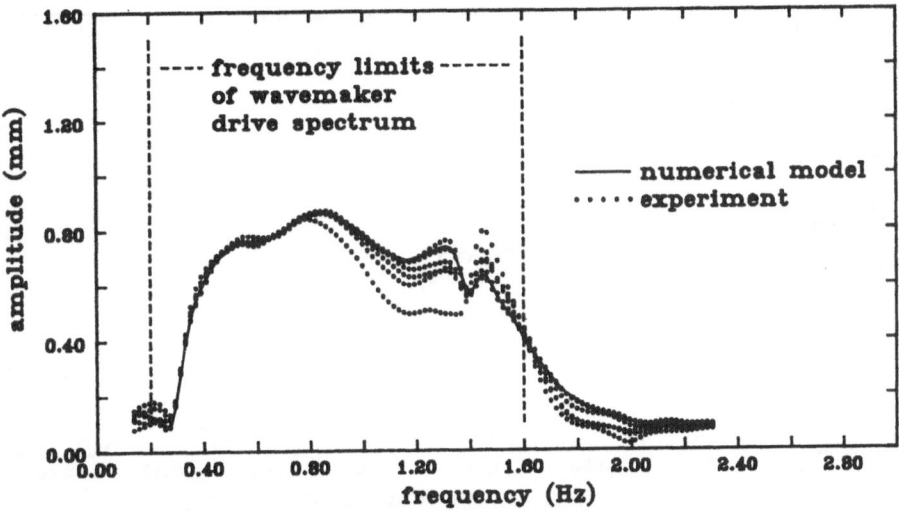

Figure 7: Experimental iterations towards the required wave amplitude spectrum.

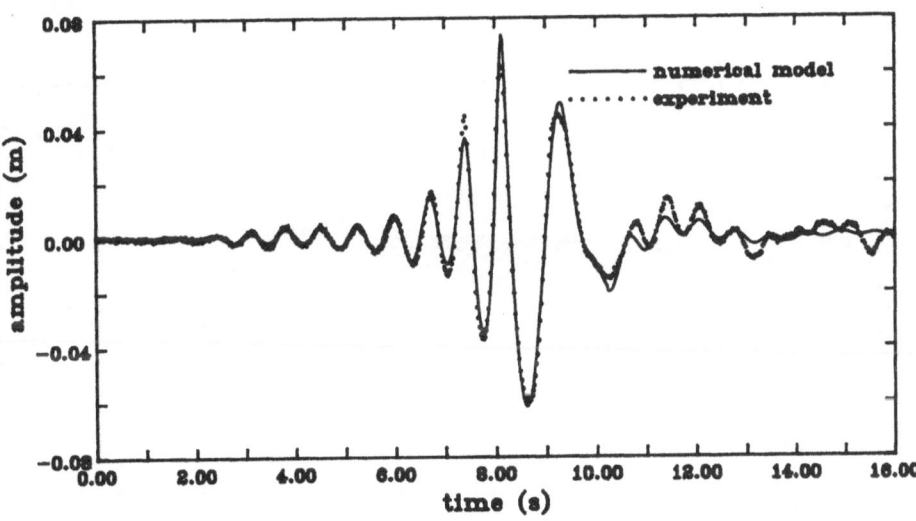

Figure 8: Experimentally achieved wave amplitude timeseries compared to the timeseries generated by the numerical model.

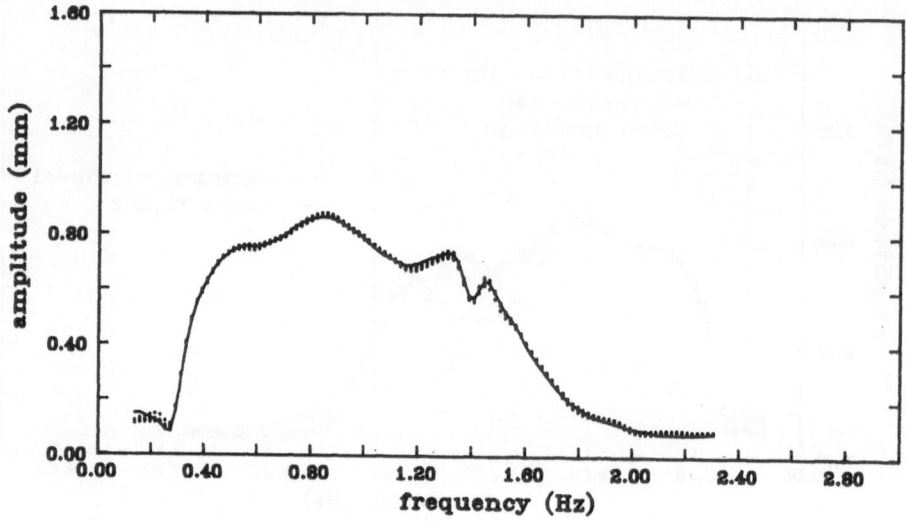

Figure 9: Repeatability of the experimentally measured wave spectrum.

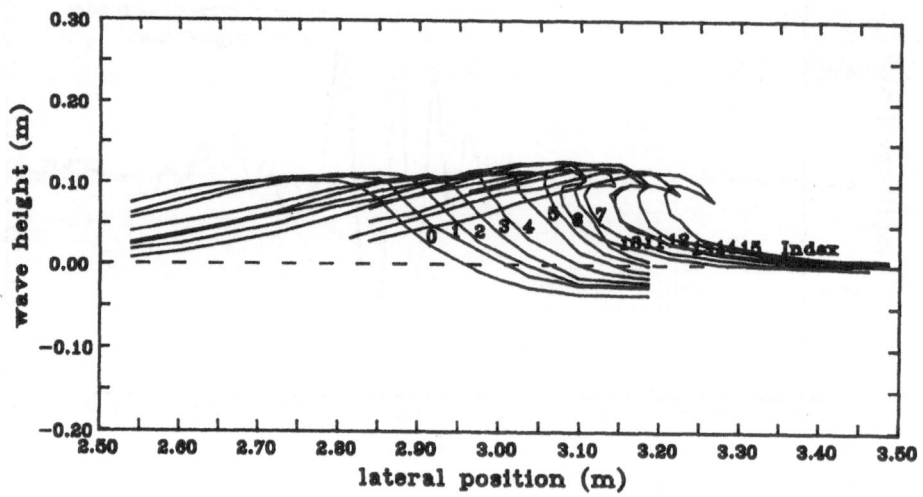

Figure 10: Experimentally measured profile of the wave approaching breaking.

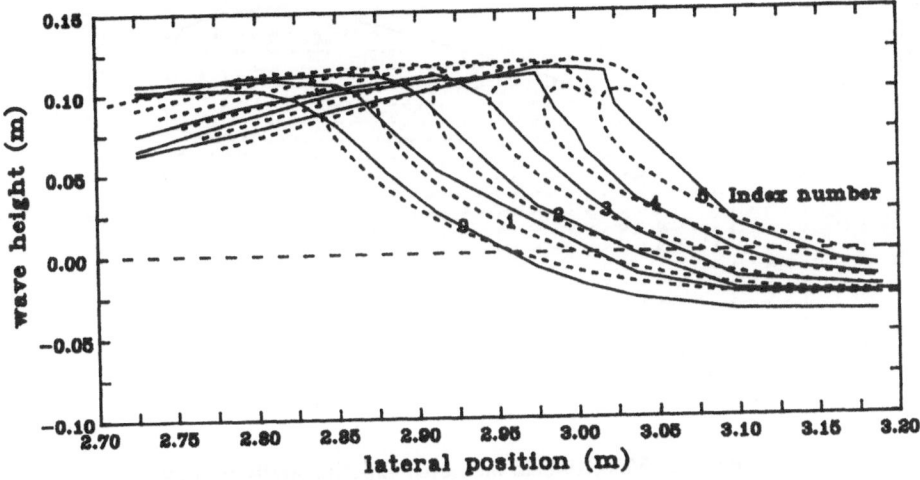

Figure 11: Overlay of numerical and experimental wave profiles.

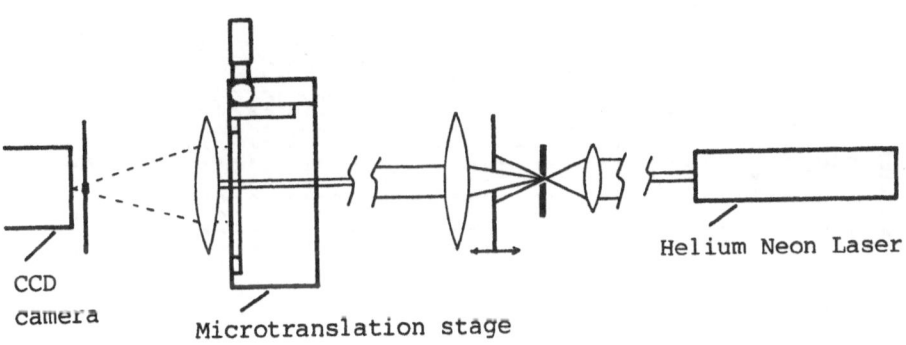

Figure 12: Generation of the speckle displacement fringes and digisation in the Fourier plane of the lens.

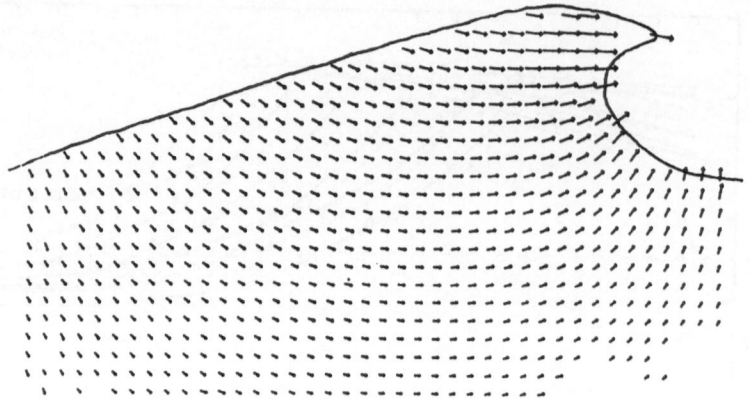

Figure 13: Velocity field measured experimentally by PIV.

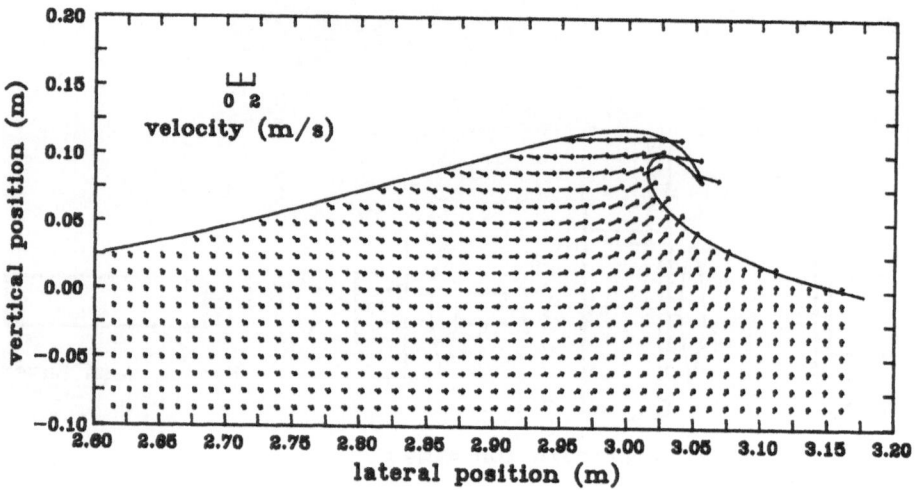

Figure 14: Numerically generated velocity field for an *earlier phase* of the wave.

TRANSFORMATION CHARACTERISTICS OF BREAKING WATER WAVES

James R. Tallent, Takao Yamashita, and Yoshito Tsuchiya
Disaster Prevention Research Institute, Kyoto University,
Ujishi, Kyoto, Japan

ABSTRACT. The purpose of this study is to examine the physical processes of wave energy decay in breaking, and to determine how this process varies for changes in initial wave and beach conditions. Visualization methods are the primary source of data acquisition. Wave energy decay is discussed in terms of the intermittent formation of large scale vortex like motions (eddies). These surface generated eddies are observed to possess an organized structure in their formation and decay. Detail observations of the eddy generation and decay mechanisms are presented and discussed. The length and time scales of the large scale eddies are measured for a range of initial wave conditions and for two bed slopes. It is shown that the large scale vortex like motions generated in the surface region of a breaking wave play a very important role in wave energy dissipation. Furthermore, characteristics of the eddy are shown to depend on the initial wave properties, beach slope, and surf zone location.

1. INTRODUCTION

The fluid motion in a wave undergoes a rapid transformation from organized to turbulent motion following the onset of wave breaking. The turbulence generated is generally assumed to be of a chaotic nature and the length scale of turbulence is usually approximated on the basis of boundary conditions (depth, surf zone length, etc.) or on non-breaking wave properties (wave height, water particle excursion, etc.). Furthermore, surf zone turbulence is known to vary according to the breaker type, such as is reflected in the horizontal mixing term used in the theory of longshore currents. Breaking waves can be classified into three categories based on elevation decay characteristics, they are: spilling, plunging, and surging. The variation in turbulence scale and intensity with breaker type, is especially noted in the region immediately forward of the breaking point. In the case of wave breaking induced by near shore shoaling this region is known as the 'Transition Region' or 'Outer Zone'. The Transition region is typically identified by the wave elevation decay, wave profile and mean water level characteristics (Figure 1).

Miller (1976) is frequently cited as the first to document the intermittent nature of breaking waves as they propagate across the surf zone. His findings are significance in that a coherent breaking wave structure, and hence turbulent structure, was for the first time recognized. Peregrine (1981 & 1983) focused his attention on developing a numerical model for the wave crest evolution, and presented a qualitative description of processes and mechanisms occurring in the region immediately following the impact of the overturn with the trough. Basco (1985 &

A. Tørum and O. T. Gudmestad (eds.), Water Wave Kinematics, 509–523.
© 1990 Kluwer Academic Publishers.

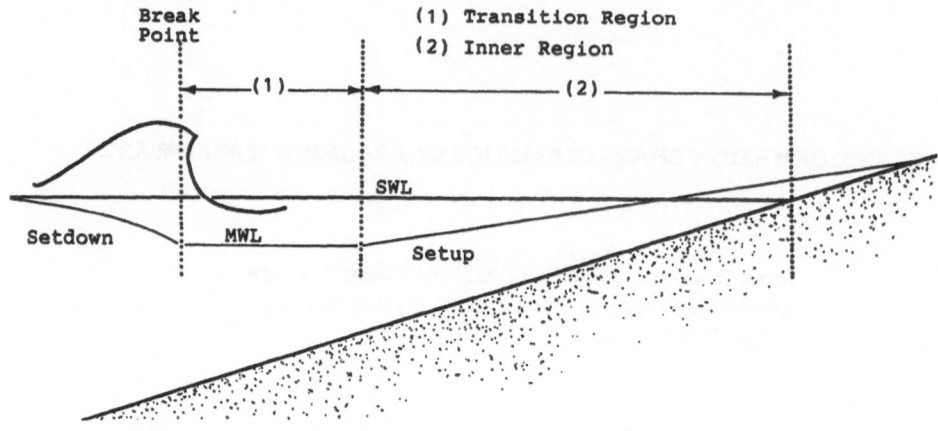

Figure. 1 Transition and Inner region classification based on fluctuations in the Mean Crest and Mean Water Level.

1988) presented a qualitative description of a plunging breaker and recently developed a mathematical model for the Transition region based on the conservation of total momentum, however, experimental verification is yet to be accomplished. Should the transformation of a breaking water wave be deterministic for given initial wave and beach conditions, significant improvements could be made in developing an understanding of the cross shore energy, mass, and momentum fluxes.

Experimental investigations of the surface region from the breaking point shoreward have proven to be most difficult due to the lack of equipment suitable to this highly aerated region. However, recent advances in the development of flow visualization techniques have enabled researchers to detect and describe the nature of phenomenon occurring in very difficult environments. The purpose of this study is to employ such visualization techniques to observe and describe the evolution of the wave breaking phenomena.

2. PROCEDURE AND EXPERIMENTAL SETUP

The characteristics of breaking water waves are investigated using a 36.8 m long wave flume equipped with a hydraulically driven piston type wave generator. Wave breaking is induced by shoaling on a sloping bottom which is varied from 1/20 to 1/50. The water depth is held constant throughout the investigation at 40 cm. The surface fluctuations and internal flow structure of the breaking waves are observed using a carriage mounted high speed video (HSV) camera operating at 200 frames per second. The camera carriage speed is adjusted to the approximate wave celerity. Breaking wave cross-shore position and elevation measurements are extracted using a 5x5 cm grid placed on the inside of the glass wall. Incident wave elevation data was collected using capacitance type wave gauges.

In Section 3, the evolution of the breaking wave is investigated in the region immediately following the initiation of breaking where significant losses of wave elevation and energy occur.

Solitary plunging waves are used to minimize observational difficulty and to establish an ideal evolution model. To enhance the visibility of the air/water interface and the near surface internal flow structure a black lamp lighting system, neutrally buoyant fluorescent particle tracers (Dia. = 3 mm), and fluorescent dye injections are used in combination.

In Section 4, the transformation characteristics of laboratory generated regular waves are investigated for a series of initial wave conditions and for two bed slopes. The quantities measured and presented in this section are based on a ten wave average. To perform this investigation with minimal error it is extremely important to establish a consistent breaking point. However, when using a piston type wave generator second order wave harmonics (Fontanet wave) are produced at the wave-paddle/bed interface which prevent the breaking point from stabilizing. Therefore, a submerged sill of height approximately 10 percent the water depth is placed close to the wave paddle resulting in a significant reduction in the Fontanet component, this technique is discussed by Hulsbergen (1974).

3. SOLITARY PLUNGING BREAKER CHARACTERISTICS

Consider the plunging breaker frame series shown in Figures 2 and 3 These two figures were constructed by a frame by frame analysis of the high speed video tape. The total time over which the frame series extends for Figure 2 and 3 is 1.11 and 0.34 second, respectively. The breaking point wave elevation, H_B, is approximately 25% larger in Figure 2, than in Figure 3. Salient features considered crucial in the initial transformation are:

1) the wave crest overturn,
2) the initial penetration of the fluid jet into the trough,
3) the formation of a fluid splash-up,
4) the formation of a large scale vortex like motion (eddy),
5) the interaction between the eddy and the wave,
6) the bifurcation of overturning fluids forming a pair of counter rotating eddies,
7) the interaction between the eddy pair and the trough fluids, which leads to the creation of a second pair of counter rotating eddies.

Wave crest overturning (1) is presently the most successfully investigated aspect of the breaking wave evolution. Several mathematical models exist based on potential-flow theory which roughly satisfy the observed phenomena, Longuet-Higgins and Cokelet (1976,1978), Dold and Peregrine (1984). Furthermore, New (1983) experimentally investigated the characteristic shape of the region trapped between the overturning jet and the wave surface. He found that the boundaries of this region could be approximated by an ellipse with an axis ratio of $\sqrt{3}$. The region immediately shoreward of the initial wave crest overturn, Transition region, has until recently remained a 'black box' topic, due to the rapidity of the phenomenon and the departure from the domain of potential theory. This investigation concentrates on the wave evolution following the impact of the overturning Jet with the trough.

Consider the wave profile corresponding to Fig. 2-Frame 6, and Fig. 3-Frame 3. At the plunge time, the tip of the jet is far forward the wave crest, and trough penetration initiates. The initial volume of tip fluid entering the trough is relatively small and since the trough fluids are highly accelerated towards the wave crest, the jet fluids rapidly recirculate towards the wave crest. This event has one very pronounced effect on the wave profile, the angle θ_E (defined in Fig. 3-Frame

512

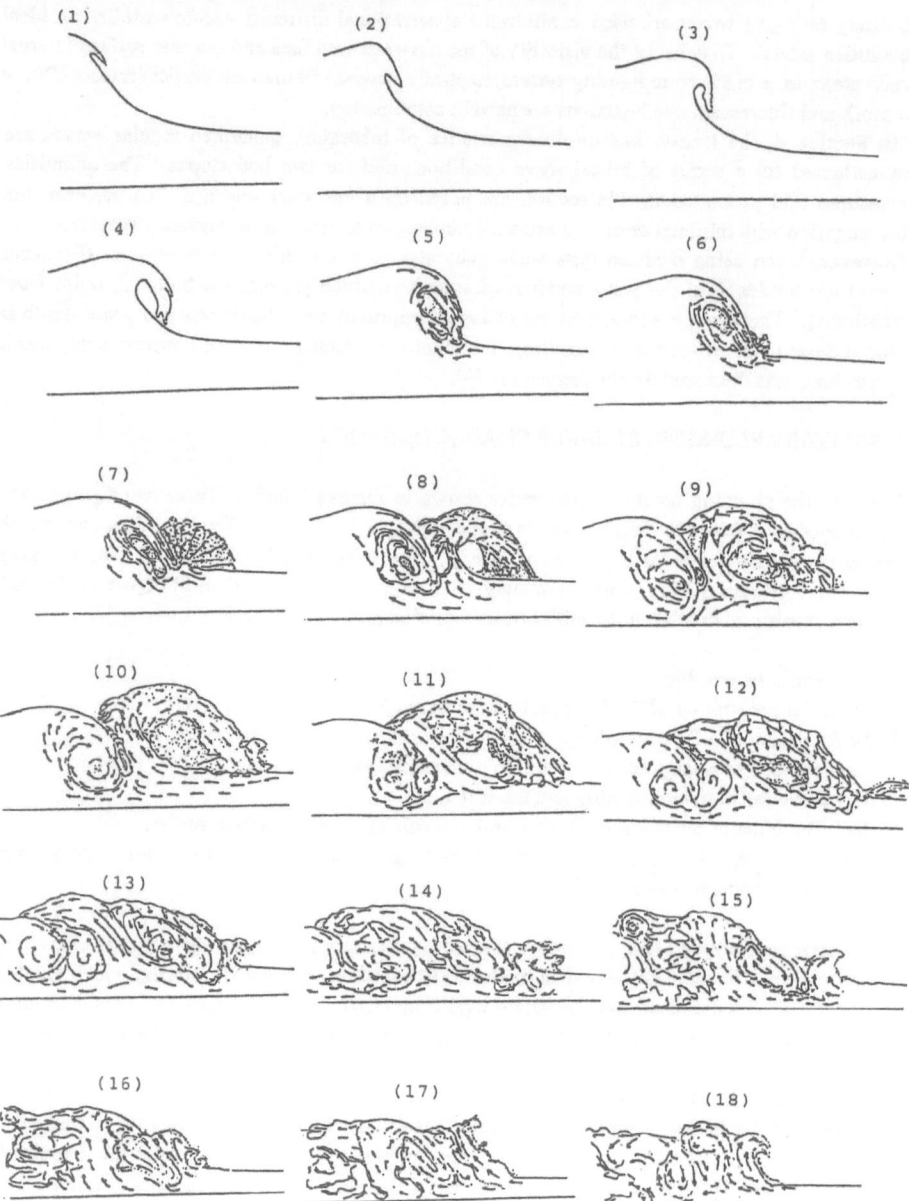

Figure. 2 Frame series od breaking solitary wave (Case A).

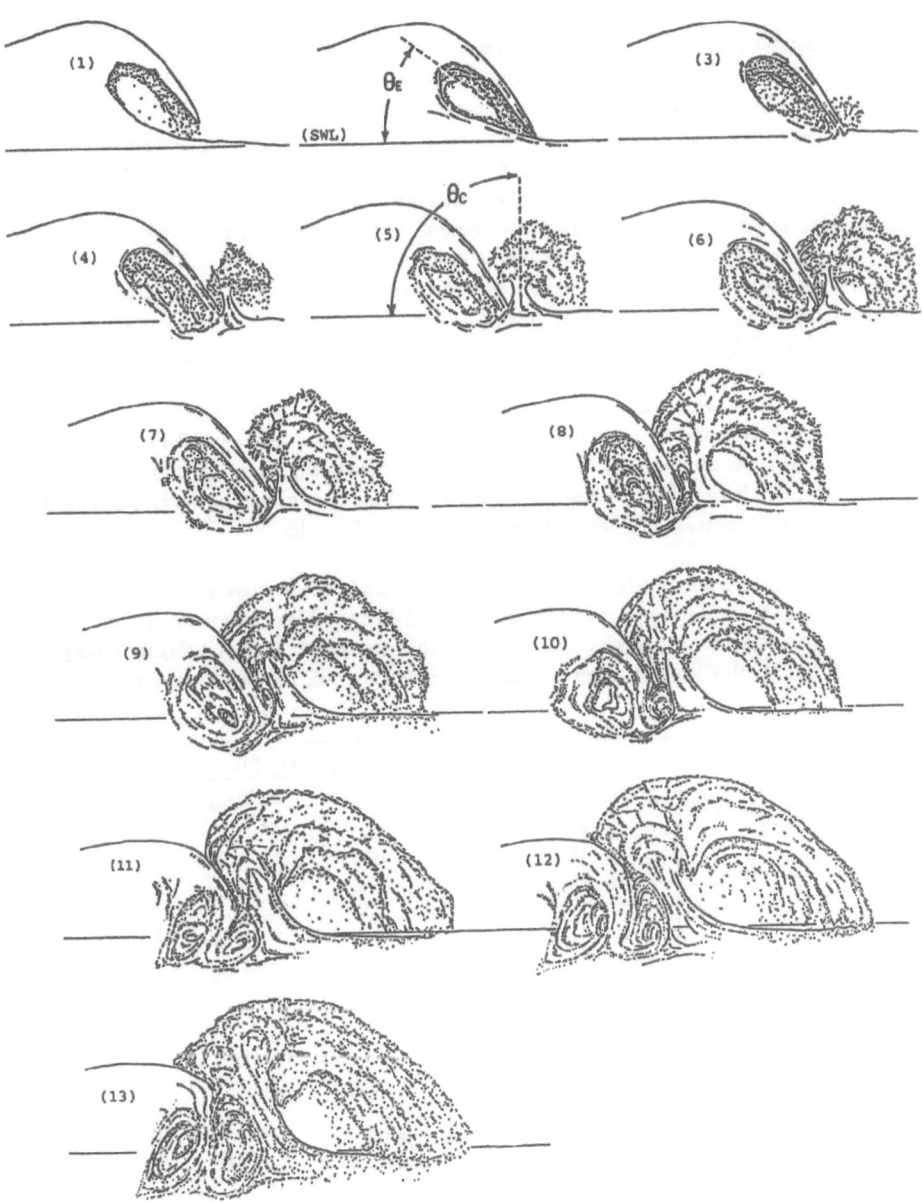

Figure. 3 Frame series of breaking solitary wave (Case B).

2) begins to increase. This phenomnon possibly occurs due to the wave energy imbalance created by the fluid mass circulation at the toe of the wave.

As the wave propagates forward from the plunge time and θ_E continues to increase, a fluid splash-up initiates at the location of the jet impact. The splash-up phenomenon typically emerges as a fine fan shaped 'spray', followed by the eruption of a near vertical and sometimes slightly shoreward inclined splash-up 'core'. The two terms used here to describe the splash-up, 'spray' and 'core', are selected based on observations of high and low air content, respectively. The spray, which is a small volume of fluid as compared to the core, appears to originate from both the overturn and trough fluids, Fig. 2-Frame 7. The core, however, is observed to originate from the undisturbed trough fluids alone, Fig. 2-Frame 8. The origin of the splash-up fluid and its development appear to be closely related to θ_E. As θ_E increases the splash-up grows in elevation (Fig. 3, Frames 1-10). The splash-up spray initiates at small angles of θ_E, and the splash-up core at larger angles. A simple mathematical model of splash-up formation due to the impact of a fluid jet can be seen in Peregrine (1981).

In Figure 4 the wave and splash-up crest elevation trends for the Figure 3 Frame series are shown as time proceeds from the breaking point. At t=0.44 sec, the wave elevation begins to decays at a rate similar to that occurring during the initial crest overturn. Several observations are made just prior to this rapid wave elevation decay, they include:

1) The eddy shape is nearly elliptical and θ_E is approximately 45-50 degrees.
2) The angle of the splash-up core θ_C (defined in Fig. 3-Frame 5) is nearly vertical.
3) A very small counter rotating vortex initiates at the intersection of the impinging fluid jet and the splash-up. At this location of strong shear and opposing fluid particle trajectories vorticity is created.

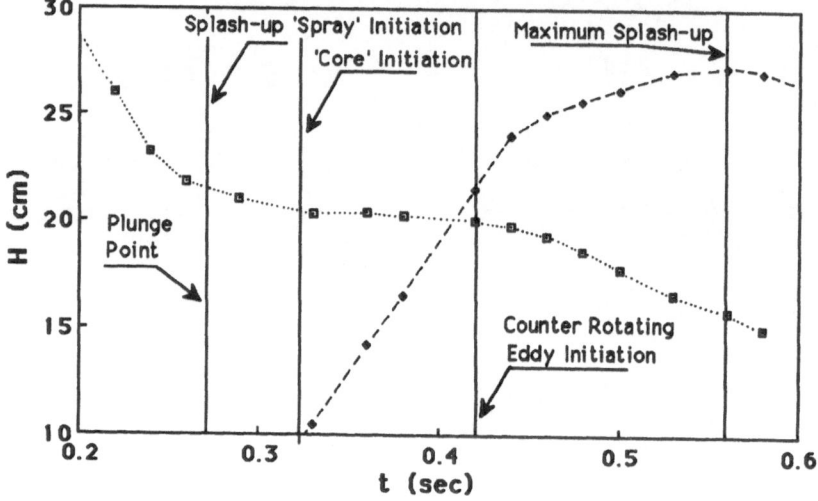

Figure. 4 Wave and splash-up crest level measurements for Case B solitary wave frame series.

To avoid more confusion, the clockwise rotating vortex (eddy) will be referred to as the 'Primary Eddy' and the counter rotating vortex as the 'Secondary Eddy'. Detail observations indicate that the small counter rotating vortex which formed at the tip of the overturn, triggers a bifurcation in the overturning fluids into both the primary and secondary eddies. Consequently, the direction of shear on the splash-up core instantly reverses causing the splash-up core to fall back on to the pair of eddies. The secondary eddy rapidly grows in size and hence wave elevation rapidly decreases, Fig. 2, Frames 9-12 and in Fig. 3, Frames 8-12.

The chain of events which follow the rapid Secondary eddy growth are summarized as follows:

1) a rapid decline in the splash-up core angle, θ_c, Fig. 3, Frames 8-12,

2) a rapid decrease in the splash-up elevation growth, Figure 4,

3) rapid wave elevation decay, Figure 4, and

4) finally, the Secondary eddy is observed to grow towards the surface under the influence of the backwards surging splash-up core, as can be observed in Fig. 2, Frames 12-14. It is postulated that a significant amount of energy first held by the primary eddy is transferred to the splash-up via the secondary eddy.

Once the fluid splash-up reaches it's maximum elevation and begins to collapse back down into the trough a new fluid circulation initiates, Fig. 2, Frames 10-13. This new circulation closely resembles that of the previously described primary eddy, yet at a slightly smaller scale. Thus, a subsequent pair of counter rotating eddies are formed which eventually initiates a second fluid splash-up, and the process continues. Reference should be made to the schematic diagram of Figure 5 idealizing the transformation process of the first two full vortex creation cycles and part of a third. Salient features of this schematic diagram include: Frame 1- plunge time; Frame 3- a) initiation of splash-up core, b) initiation of near constant wave elevation and c) Primary eddy angle, θ_E, initiates forward rotation; Frame 5- a) initiation of a small counter rotating eddy, and b) the splash-up core initiates a backwards fall onto the Primary and Secondary eddies, i.e., θ_c decreases; Frame 7- a) surge of wave crest fluids over the Primary eddy, b) initiation of rapid wave elevation decay, c) initiation of rapid area increase of Secondary eddy, and d) the Primary eddy takes the form of a slanted or oblique eddy; Frame 10- a) maximum elevation of splash-up; Frame 11- a) initiation of fluid circulation below collapsing fluid splash-up; Frame 12- a) initiation of a second splash-up core, Frame 14- a) θ_{CI} increases to the extent that a forward crest surge initiates, and b) initiation of a rapid crest elevation decline; Frame 15- a) second splash-up collapses initiating the circulation that will form the third eddy, b) initiation of splash-up spray.

4. LARGE SCALE EDDY CHARACTERISTICS OF PERIODIC WATER WAVES

In order to develop a more realistic model of large scale eddy formation by breaking water waves it is necessary to employ regular waves under various initial conditions. To initiate this investigation several test case are performed in the laboratory over a range of initial wave steepness, Ho/Lo, and for two beach slopes, 1/20 and 1/50. Based on the parameter definition sketch shown in Figure 6, the series of large scale vortices generated by breaking periodic water waves are investigated. Three length scales are used to describe the characteristics of a fully developed eddy. The term 'fully developed' is used to indicate maximum measurable length scales. The length scales are described as: the maximum amplitude of the splash-up, H_s, measured from the instantaneous trough level; the maximum vertical diameter of the vortex, D_e, including the penetration below the instantaneous trough level; and the horizontal distance over

516

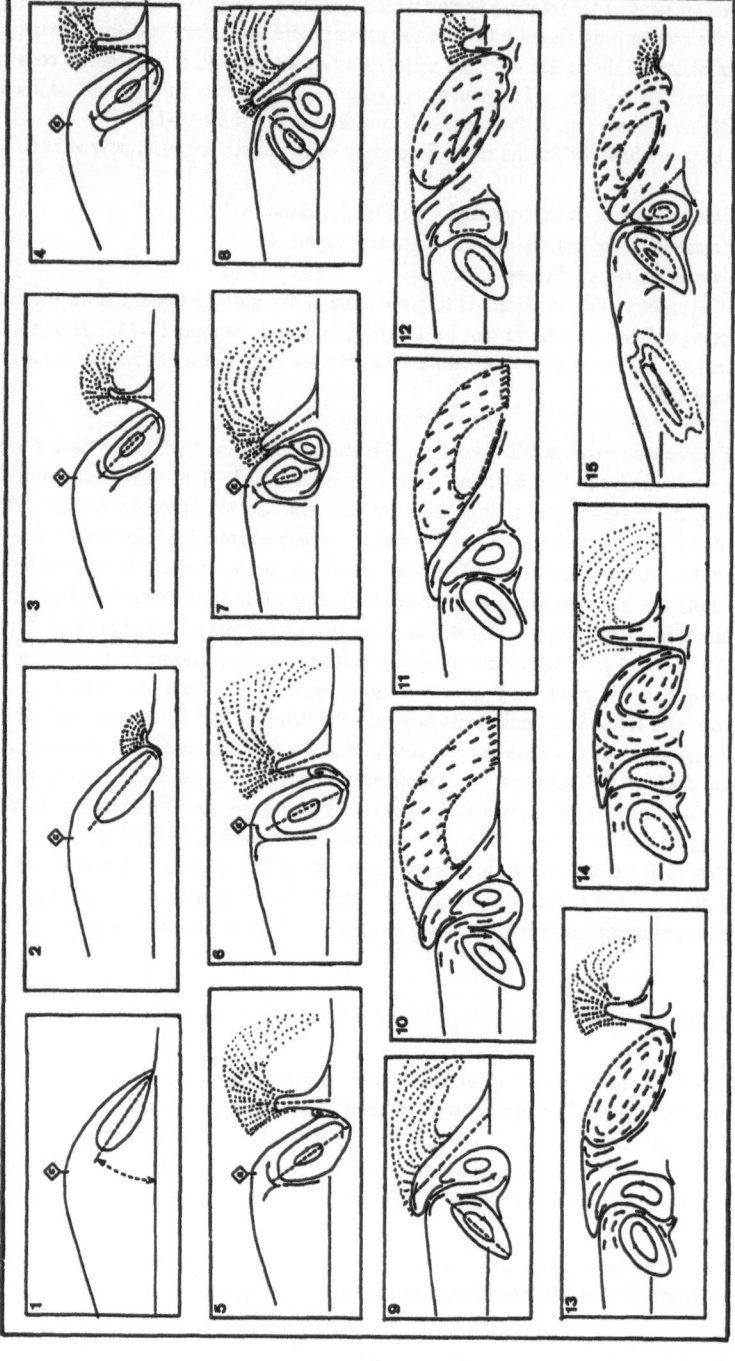

Figure. 5　Schematic diagram of the wave transformation in the outer region.　Wave crest position is denoted by c.

which the eddy travels while it is forming, Le. The horizontal length scale measurement for the Nth eddy, Le$_N$, initiates at the location of the Nth splash-up and terminates at the initiation of the N+1 splash-up. A time scale, Te, is also measured over this travel distance.

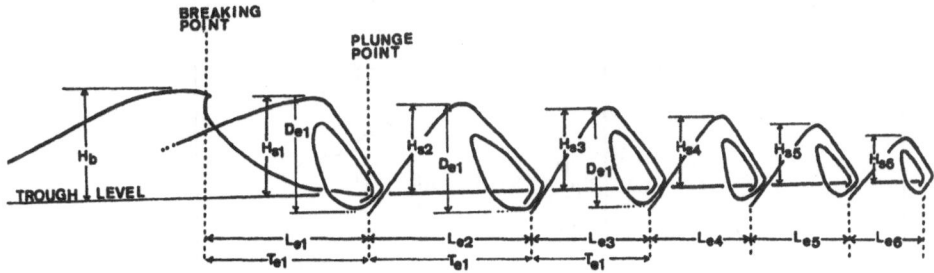

Figure. 6 Definition sketch of length and time scales measured for the eddy series.

In Figures 7 and 8, wave, splash-up, and eddy crest elevations are plotted versus time from the initiation of observable wave crest turbulence, where time is normalized by wave the period, Tw, and elevation by breaking wave height, Hb. Turbulence and air entrainment at the wave crest are observed to initiate long before the wave reaches maximum elevation. In Figures 7 and 8, the surface fluctuations associated with the formation of ten and seven eddies are shown, respectively. Recall that the first eddy is formed by the wave and is not associated with a splash-up, therefore in Figure 7 it is possible to count nine splash-ups, totaling ten eddies. In both cases the wave is visually verified as spilling or spilling/plunging transition, i.e., breaking initiated at the crest of the wave in the form of turbulence; the classical crest overturn is not observed. Initial wave steepness, Ho/Lo, for Figures 7 and 8 are measured as 0.061, and 0.042, respectively, and the bottom slope is 1/20.

In each of these figures there is an inset figure which shows the trends of the measured length and time scale parameters: Hs, De, and Te. First examine Figure 7, the ordinate, $\sum$Te/Tw, is a running sum of the eddy formation period normalized by the wave period, and the abscissa is the eddy diameter, De, and maximum splash-up elevation, Hs, normalized by the breaking wave elevation, Hb. For the first eddy, the maximum amplitude, Hs, is interpreted as the wave crest to trough height measured at the time of the fully developed eddy. For example, in the Figure 6 inset, the first eddy is fully formed at approximately $\sum$Te/Tw=0.43, and the length scales associated with this eddy are Hs/Hb=0.73, and De/Hb=0.45. Several conclusions are made from Figure 6 and video observations, they include:

1) Maximum eddy diameter, De/Hb, and splash-up elevation, Hs/Hb, occur at the second eddy, N=2, following this time the eddy and splash-up magnitude tend to follow a near exponential decay.
2) The value of Hs/Hb remains larger than De/Hb for a time slightly less than one wave period from the initiation of breaking, $\sum$Te/Tw<1.0. Once the value of Hs/Hb and De/Hb become equal, eddy penetration below the trough level initiates.
3) Observations indicate that the splash-up and eddy formation characteristics are gradually

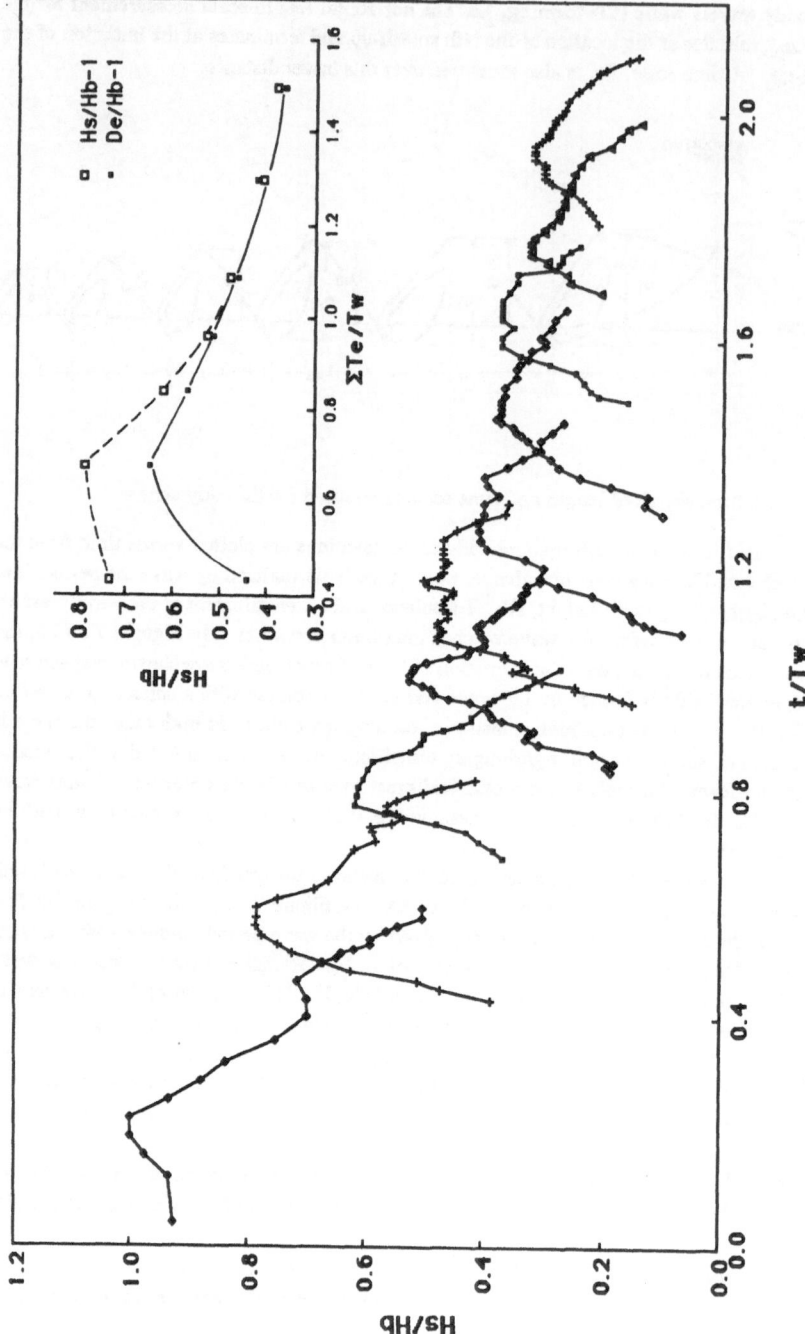

Figure. 7 Wave and splash-up crest level measurements for spilling type periodic water wave (Ho/Lo=0.045). The insert figure describes the length and time scale characteristics of the splash-up and eddy.

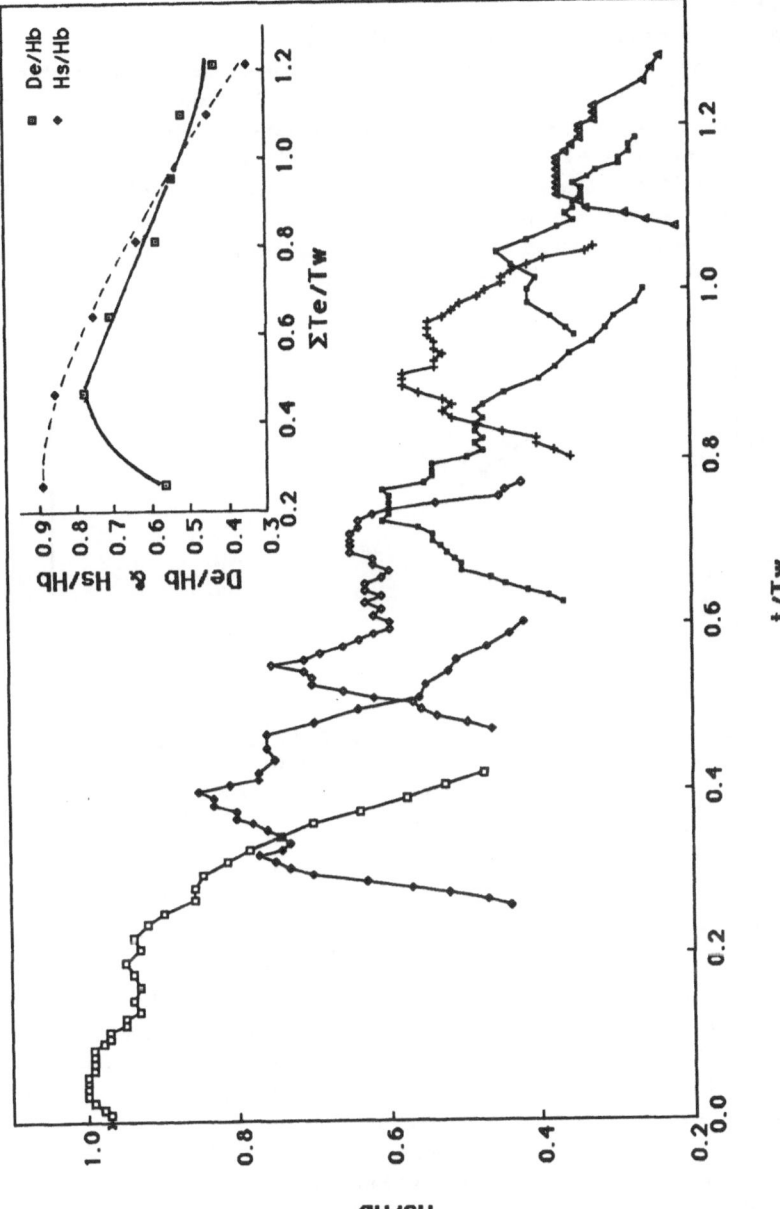

Figure. 8 Wave and splash-up crest level measurements for spilling/plunging transition type periodic water wave (Ho/Lo=0.026). The insert figure describes the length and time scale characteristics of the splash-up and eddy.

changing as the number of eddies increase. As N increases the eddy penetration angle θ_E decreases and splash-up angle θ_c increases (Figure 5), consequently, mass flux from eddy to eddy increases with N.

In Figure 8, similar eddy formation and decay characteristics are observed, however, the relative magnitudes of the first few eddy diameters and splash-ups are much larger than the corresponding ones of Figure 6. Also, the relative eddy formation time, Te/Tw, for the first eddy is much less in the case of Figure 8. The relationship between wave steepness and eddy size is more clearly shown in Figure 9. In this figure, five different values of Hb/Lo are compared, the bottom slope is held constant at 1/20. The eddy diameters of the first four eddies (N=1,2,3,4) are averaged and plotted versus the breaking point wave steepness, which varies from 0.01 to 0.08. The linear decay in De/Hb from 0.8 to 0.5 is clearly observed as wave steepness increases, furthermore, it appears that an upper limit exist at De/Hb=0.8.

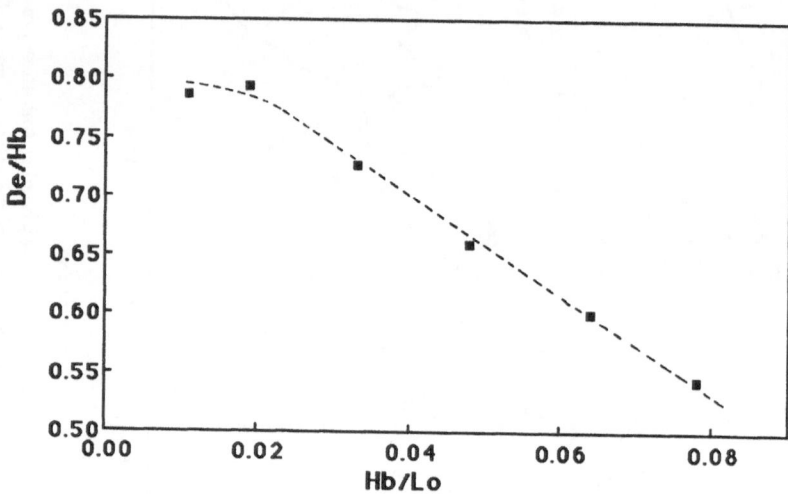

Figure. 9 Relationship between average eddy length scale and wave steepness.

The bottom slope effect on large scale eddy formation can be see in Figure 10. In this figure eddy diameter is shown versus time from the breaking point for two waves with the same initial conditions yet breaking on two different slopes, 1/20 and 1/50. The most significant difference is observed in the formation of the first eddy. On the 1/50 slope, the wave crest overturn occurs at a much smaller scale than on the 1/20 slope, consequently, a much smaller eddy is formed. The eddy size tends to decay at a similar rate for the second, third, and forth eddies, however beyond the fourth eddy, the rate of decay appears to decrease for the 1/50 slope.

Finally, in Figure 11, the normalized eddy diameter, De/Hb, versus normalized depth at which the eddy forms, he/h , is presented for all data. The surf similarity parameter, $\xi = i/(\sqrt{Ho/Lo})$, is used to define the initial condition. This figure illustrates the combined bottom-slope/wave-steepness effect on the cross shore eddy amplitude decay. From Figure 11, it is observed that as ξ decreases the fraction of the surfzone over which large scale eddies are generated decreases. In several cases, on the 1/50 bottom slope, the breaking process ceased

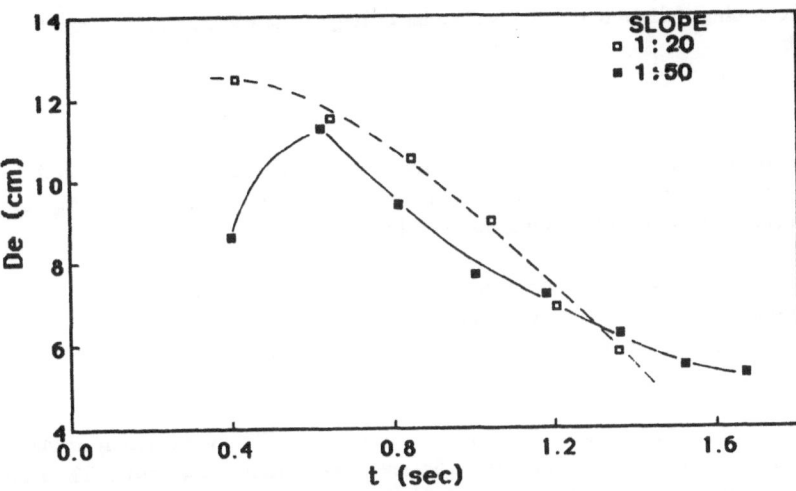

Figure. 10 Comparison of eddy length scale for two waves of equil steepness (Ho/Lo=0.014) breaking on a bottom slope of 1/20 and 1/50.

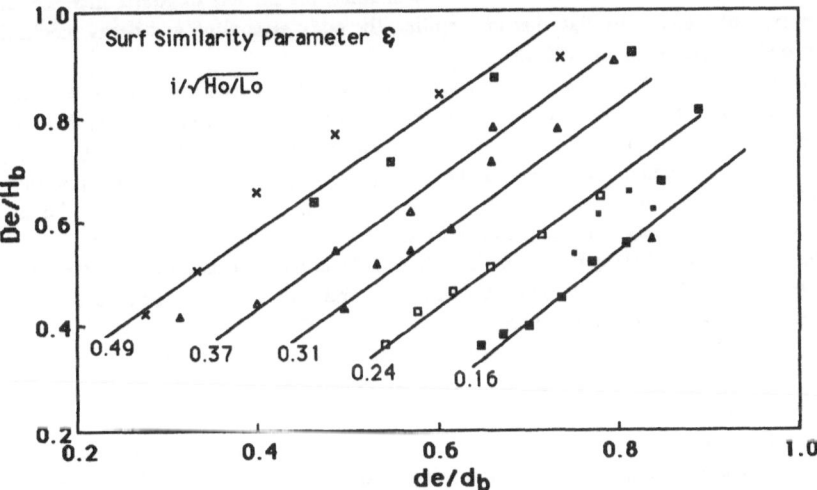

Figure. 11 Cross shore characteristics of the eddy length scale decay for a 1/20 and 1/50 bottom slope. The surf similarity parameter is used to classify the initial conditions.

(wave reforming) following the observed series of large scale eddies, this is a topic for future research.

5. Conclusions

In Section 3 the transformation characteristics of solitary waves were examined in the region immediately forward the breaking point. Several important mechanisms governing the wave energy decay were identified and discussed, they include:

1) The formation of the fluid splash-up at plunge point of the crest overturn.
2 The formation of a large scale vortex like motion (eddy).
3) The bifurcation of overturning fluids forming a pair of counter rotating eddies.
4) The generation of a series of eddies formed by events very similar to the first.

The fluid in the splash-up was observed to originate from either the overturn or the trough depending on the angle of overturn impact, θ_E, and stage of eddy development. The splash-up reached maximum elevation immediately following the formation of a counter rotating eddy, furthermore, the rate of wave elevation decay was observed to increase following this time. A second eddy was observed to initiate as a result of the first-eddy/trough interaction, and, at a later time by the interaction with the remaining wave form. Therefore, the eddy energy is supplied in a two phase process. As the eddy size becomes smaller, the number of eddies increase, or the breaker type becomes more like that of a spiller, the eddy generation and decay mechanisms described above are gradually relaxed.

In Section 4, periodic water waves were employed to investigate the formation characteristics of large scale eddies. The series of eddies were defined in terms of four parameters, they are: the maximum amplitude of the splash-up, Hs, measured from the instantaneous trough level; the maximum vertical diameter of the vortex, De, including the penetration below the instantaneous trough level; the horizontal distance over which the eddy travels during its formation, Le; and the time required for the eddy to travel the distance Le.

It was shown that eddy length and time scales are a function wave steepness and bottom slope. As wave steepness increases the splash-up amplitude and eddy diameter decrease, however, the rate at which the eddy series decay becomes more gradual. Changes in the bottom slope affect the eddy formation in a way similar to that of wave steepness, i.e., increasing the bottom slope results in larger eddy sizes and faster decay rates. However, the bottom slope does not appear to play as important a role as the wave steepness.

Within this study observations and phenomena interpretations were presented describing the mechanisms leading to and resulting from the evolution of a breaking wave. Results indicate that a coherent breaking wave structure exists which can be described in terms of a series of large scale vortex-like motions. Significantly more research is needed in this direction to develop the foundation on which a model can be constructed, however, it appears to be a promising avenue.

REFERENCES

Basco, D.R. (1985) 'A Qualitative Description of Wave Breaking', J. Waterway, Port, Coastal and Ocean Engr. ASCE. 111, WW2: 171-188

Basco, D.R. (1988) 'On the Partition of Horizontal Momentum Between Between Velocity and

Pressure Components Through the Transition Region', Proc. 21st. Conf. on Coastal Engr., Vol 1, 682-697

Dold, J.W., Peregrine, D.H. (1984) 'Steep Unsteady Water Waves: An Efficient Computational Scheme', Proc. 19th. Int. Conf. on Coastal Engr., Vol 1, 955-967

Longuet-Higgins, M.S., Cokelet, E.D. (1976) 'The Deformation of Steep Surface Waves on Water. I: A Numerical Method of Computation', Proc. R. Soc. London. A350: 1

Hulsbergen, C.H. (1974) 'Origin, Effect and Suppression of Secondary Waves', Proc. 14th Conf. on Coastal Engr., Vol 1, 392-411

Miller, R.L. (1976) 'The Role of Vortices in surf zone Prediction: Sedimentation and Wave Forces', Beach and Nearshore sedimentation, Soc. Econ. Paleontol. Mineral. Spec. Publ. 24, 81-98

New, A.L. (1983) 'A Class of Elliptical Free-Surface Flows', J. Fluid Mech. 130, 219

Peregrine, D.H. (1981) 'The Fascination of Fluid Mechanics', J. Fluid Mech. 106, 59

Peregrine, D.H. (1983) 'Breaking Waves on Beaches', Ann. Rev. Fluid Mech. 15, 149-78

TURBULENCE GENERATION IN A BORE

Harry H.Yeh
Department of Civil Engineering
University of Washington
Seattle, WA 98195
USA

ABSTRACT. It is shown, from the vortex dynamics view point, that the baroclinic torque is the primary mechanism to create turbulence on the front face of a bore. This mechanism is known to be important in a stratified fluid, but seldom discussed for a homogeneous fluid involved with a free surface because the free surface is often thought to be a barotropic surface. Nonetheless, finite magnitude of vorticity can be created at the interface of inviscid homogeneous fluids (air and water) if they are viewed as a gas-liquid two-layer system. With this vorticity generation mechanism, turbulence generation should take place along a finite region of the bore front, but not at a point of the front toe.

1. Introduction

When waves approach the shore, they often break offshore and form quasi-steady bores near the shore; the whole front faces of the waves are covered with turbulence. Based on the fact that turbulence is vortical motions and the vorticity in a homogeneous fluid is a conservative quantity, the mechanisms of turbulence generation in a bore are examined in this paper.

Let us first consider a 'pure' bore as shown in Fig. 1 that is a single uniform bore propagating into a quiescent water body on a horizontal and impermeable bottom boundary. The phenomenon of a bore is often considered to be equivalent to that of a hydraulic jump (which is a stationary transition from subcritical to supercritical flow). It is because under the one-dimensional flow assumption, the mathematical description of a bore motion can be reduced identically to that of a stationary hydraulic jump by taking the Galilean transformation with the bore propagation velocity. The one-dimensional flow assumption is evidently inappropriate when turbulence is examined in detail. Yeh and Mok (1989) pointed out that the fundamental difference between a bore and hydraulic jump arises from their boundary conditions. The boundary conditions are the ones that cause the shear-flow field and turbulence of a bore to be different from that of a hydraulic jump. In fact, according to Peregrine & Svendsen (1978), turbulence in a bore resembles that of a wake, whereas it behaves like a jet in a hydraulic jump, as pointed out by Rajaratnam (1965).

The turbulent front face of a bore is considered to form a recirculating mean flow or a 'surface roller' as shown in Fig. 1. Turbulence associated with the surface roller was examined by Peregrine & Svendsen (1978). Based on their laboratory experiments, Peregrine and Svendsen proposed that turbulence is initiated at the toe of the bore front, and the initial stage of turbulent flow resembles that of a mixing layer. A question is that, if

A. Tørum and O. T. Gudmestad (eds.), Water Wave Kinematics, 525–534.
© 1990 *Kluwer Academic Publishers.*

this is the case, then, where and how is the vorticity generated. In the case of a mixing layer, however, vorticity is introduced to the fluid domain from the upstream boundary layers such as that created along a split plate illustrated in Fig. 2. There is no vorticity source upstream in a 'pure' bore propagating into a quiescent water body; i.e. the fluid must initially be irrotational. Hence, vorticity must be created at the front in itself but is not advected from upstream; from this point of view, turbulence formation in a bore cannot be analogous to that of a mixing layer. In the case of a hydraulic jump as illustrated in Fig. 3, a weak vortex sheet can exist along the water surface due to the upstream boundary condition, such as vorticity generated at a sluice gate or a nozzle.

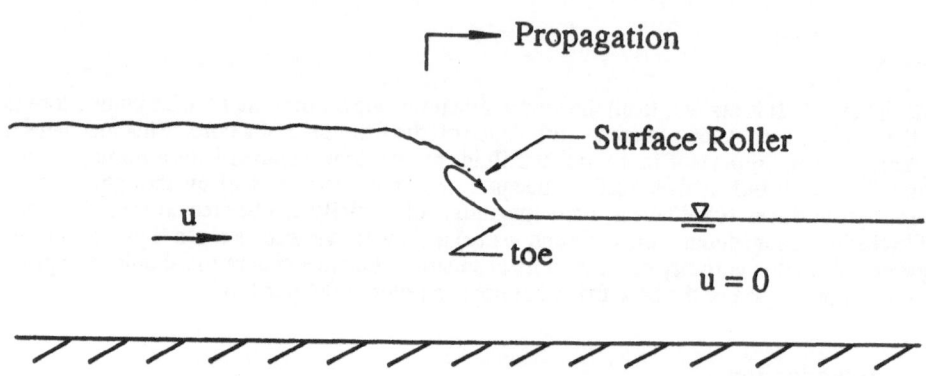

Figure 1. A sketch of a bore propagating into a quiescent water body.

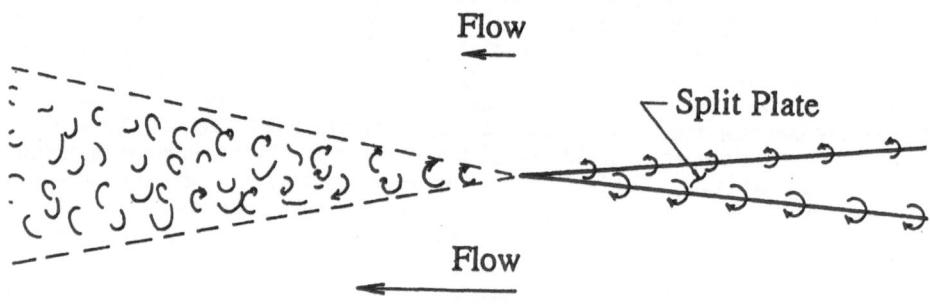

Figure 2. Vorticity advection into a mixing layer.

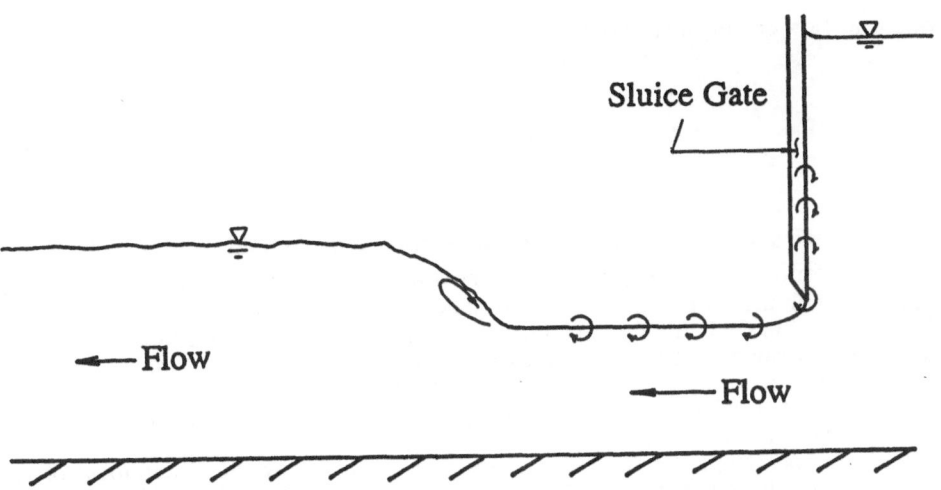

Figure 3. Vorticity advection into a hydraulic jump

2. Background

By taking the curl of the Navier-Stokes equation, the vorticity equation can be expressed as

$$\frac{D\omega}{Dt} = \omega \cdot \nabla u + \nu \nabla^2 \omega - \nabla\left(\frac{1}{\rho}\right) \times \nabla P, \tag{1}$$

where ω is the vorticity (i.e. **curl u**), **u** is the fluid particle velocity, ν is the kinematic viscosity of the fluid, ρ is the fluid density, P is the pressure, the operator ∇ is the gradient, and the bold letter represents a vector. The term on the left-hand side of (1) represents the rate of change of vorticity as following a fluid parcel, and the terms on the right-hand side represents, respectively, the rate of change of vorticity due to bending and stretching of a vortex line, diffusion of vorticity by viscous force, and production of vorticity due to the baroclinic effect. If the fluid is inviscid and the density is either constant or a function of pressure only (i.e. homentropic), then the last two terms in (1) are dropped out. Then Kelvin's theorem is valid, i.e.

$$\frac{D\Gamma}{Dt} = 0, \tag{2}$$

where the flow circulation Γ is defined as

$$\Gamma = \int_c \mathbf{u} \cdot d\mathbf{x} = \int_s \omega \cdot d\mathbf{A}. \tag{3}$$

In (3), s is the surface enclosed by an arbitrary curve c in the fluid domain, and A is the surface vector of s. Equation (2) states that the flow circulation, Γ, around the closed curve c moving with the fluid remains constant. Specifically, vorticity cannot be created nor dissipated within the inviscid fluid domain. Note that even in a viscous fluid, (2) is still a good approximation if the contour curve c does not pass a region where viscous effects are important (e.g. boundary layers). In fact, it is at the boundaries and not in the fluid domain that the vorticity can be created. The vorticity is then introduced into the fluid domain via the viscous diffusion.

Vorticity generation at a 'free' surface can be explained by the requirement of the condition of zero tangential stress at the boundary (Batchelor, 1968). Unlike a solid boundary where a no-slip condition is imposed, the velocity gradients vanish at a free surface. Even with this weak requirement, a boundary layer must be formed near the free surface and the generated vorticity diffuses into the fluid by viscous forces. It is emphasized that in the case of fluid with constant density or of a homentropic flow field, the boundaries are the ones which can create vorticity and viscosity is the one which introduces the vorticity into the fluid domain. The viscous diffusion is a slow process, although vorticity can be advected, stretched, and bended quickly once it is introduced into the fluid domain. Hence, this mechanism does not provide a plausible explanation for immediate appearance of strong vorticity in the surface roller of a bore, especially considering the fact that there is no vorticity at all in the region ahead of the bore front.

In the case of a non-homogeneous fluid, vorticity can be produced within the fluid domain whenever the fluid is displaced from a state in which ∇P and $\nabla \rho$ are parallel; i.e. the last term in (1) is non-zero. Kelvin's theorem, (2), can be generalized as

$$\frac{D\Gamma}{Dt} = - \int_s \left(\nabla\left(\frac{1}{\rho}\right) \times \nabla P \right) \cdot d\mathbf{A}. \tag{4}$$

This is often called Bjerknes' theorem (see, for example, Lamb, 1932, Art 166a). Basically, this vorticity is generated by the torque caused by the pressure force acting on a fluid parcel in which the center of mass is deviated from its centroid owing to the nonhomogeneity of the fluid. Hence this mechanism is often termed the 'baroclinic torque'. Due to the baroclinic torque, it is well known that internal waves in a stratified fluid are rotational flow phenomena.

3. Vorticity Generation at Air-Water Interface

A majority of water-wave problems are formulated based on the assumptions of irrotational motions of a Newtonian, incompressible, and homogeneous fluid with a constant gas pressure on the free surface. Suppose that the free surface boundary is viewed as an interface of two fluids, viz. the air and water, which is of course physically better represented than the water-vacuum interface. The kinematic boundary condition, i.e. conservation of mass across the interface at $B(x, t) = 0$, is

$$[\rho(\mathbf{V} \cdot \mathbf{n} - \mathbf{u} \cdot \mathbf{n})] = 0, \tag{5}$$

where, as sketched in Fig. 4, $\mathbf{V}$ is the velocity of the boundary, $\mathbf{n}$ is the unit normal vector on the interface, and the bracket [ ] represents

$$[f(\xi)] = \lim_{\substack{\xi \to x \in B \\ \xi \in D_a}} f(\xi) - \lim_{\substack{\xi \to x \in B \\ \xi \in D_w}} f(\xi), \tag{6}$$

where D_a and D_w represent the fluid domains of air and water, respectively, and, hereinafter, the letter subscripts 'a' and 'w' denote properties in the air and water, respectively. On a material boundary like the air-water interface, (5) is reduced to

$$\mathbf{V} \cdot \mathbf{n} = \mathbf{u} \cdot \mathbf{n}, \tag{7}$$

which must satisfy for both water and air domains, separately.

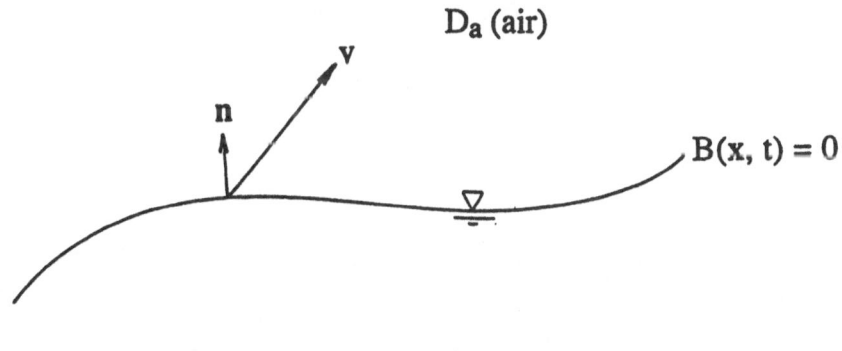

Figure 4. A sketch of the free water-surface boundary.

530

The dynamic boundary condition for an incompressible fluid is

$$[\mathcal{T}_{ij} \cdot \mathbf{n}] = 0, \tag{8}$$

where $\mathcal{T}_{ij}$ is the stress tensor, i.e. both normal and tangential stresses must be continuous across the interface. If we consider only the normal stress and neglecting the surface tension effect, then (8) is reduced to be

$$P_a = P_w. \tag{9}$$

It says that the pressure must be continuous, but does not need to be constant along the interface.

Based on the inviscid model, the equation of motion of the air in the tangential direction, λ, at the interface can be written as

$$\frac{Du_a}{Dt} = -\frac{1}{\rho_a}\frac{\partial P_a}{\partial \lambda} - g\frac{\partial \eta}{\partial \lambda}, \tag{10}$$

where η is the vertical displacement of the interface from an arbitrary datum, u_a is the air-flow velocity in the λ direction, and the curvature of the interface was assumed to be negligibly small. Because of the requirement of the kinematic boundary condition (7), the inertial effect (the left-hand side of (10)) does not vanish in the case of the interface being in motion. Even without the inertial effect, the pressure gradient in the tangential direction, λ, does not vanish as long as the interface is not horizontal, but having the order of magnitude of

$$\frac{\partial P_a}{\partial \lambda} \approx O\left(\rho_a g \frac{\partial \eta}{\partial \lambda}\right) \ll 1. \tag{11}$$

Note that the pressure gradient in (11) is caused basically by the weight of the air and is very small in comparison with any terms in the equation of motion in the water. This and (9) appear to provide the justification for the assumption of constant pressure along the air-water interface, and consequently, the interface is considered to be barotropic, i.e. ∇P and $\nabla \rho$ are parallel. (Note that $\nabla \rho$ always points in the direction normal to the interface.)

However, it is important to remember that $\frac{\partial P_a}{\partial \lambda} = 0$ is not exact but an approximation as demonstrated in (11) and, as discussed later, this does indeed contribute to the vorticity generation at the interface.

Now let us estimate the rate of change in circulation at the air-water interface using the Bjerknes theorem, (4). Taking an integration contour for the circulation as shown in Fig. 5, and taking a limit of the thickness, δ, of the narrow strip area enclosed by the contour, (4) can be reduced to be

$$\frac{D\Gamma}{Dt} \approx - \int_S \lim_{\delta \to 0} \frac{1}{\delta} \left(\frac{1}{\rho_w} - \frac{1}{\rho_a} \right) \left(\rho_a g \frac{\partial \eta}{\partial \lambda} \right) \delta \, d\lambda,$$

$$\approx O \left(\frac{\rho_w - \rho_a}{\rho_w} g \, \Delta\eta \right),$$

$$\approx O(g \, \Delta\eta) \quad \text{for } \rho_w \gg \rho_a, \tag{12}$$

where (11) was used for the pressure gradient in the direction perpendicular to the gradient of density, and $\Delta\eta$ is the elevation difference in the integration contour. Hence, even though the pressure gradient along the interface is infinitesimal as $\rho_a \to 0$, the production of vorticity at the interface is finite because of the singularity in $\nabla\left(\frac{1}{\rho}\right)$.

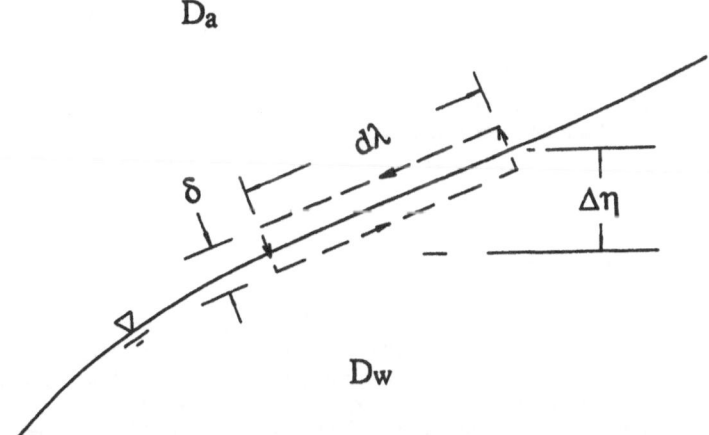

Figure 5. Integration contour, c, for flow circulation at a free surface boundary.

A physical explanation for the creation of vorticity at the interface can be provided as follows. Consider a spherical fluid parcel shown in Fig. 6a. If the fluid is inviscid and homogeneous, then there is no mechanism to rotate this fluid parcel since the lines of action of pressure force must pass through the center of mass. On the other hand, a fluid parcel at the interface cannot have a spherical shape; as shown in Fig. 6b, the shape is no longer arbitrary. The center of mass is shifted from the center of the sphere. Because the pressure force is always normal to the surface, pressure gradient can cause rotation of the fluid parcel. This physical argument together with the Bjerknes theorem, (12), demonstrates that the air-water interface is not barotropic but is a baroclinic surface which produces finite magnitude of vorticity, at least, $O\,(g\,\Delta\eta)$. Note that the order of magnitude in (12) is based only on the static effect of the air, i.e. the weight of the air. When the air-water interface is in motion so that the inertial effects become important, then, the generation of vorticity by this mechanism can be more significant than the estimation of (12).

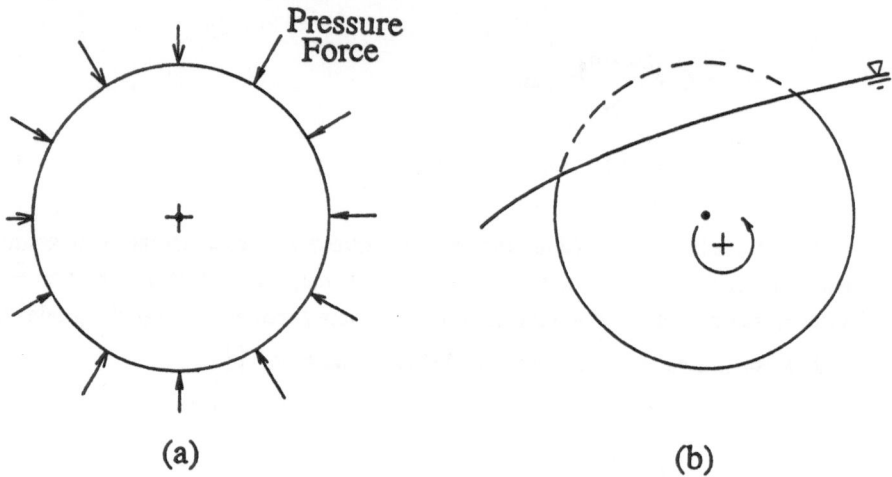

Figure 6. A homogeneous fluid parcel with center of mass indicated by $+$. The lines of action of pressure force pass through centroid of spherical volume, ●. a) no mechanism to cause rotation without an interface; b) torque to cause rotation of a fluid parcel with an interface.

4. Turbulence in a Bore

As discussed earlier, turbulence generation at the front face of a bore cannot be similar to that of a mixing layer since there is no vorticity initially at the free surface ahead of the bore front, i.e. the water there is quiescent. Considering the length of the bore front and the air-water interface there, vorticity generation by the viscous forces cannot be significant. Hence, turbulence generation on the front surface must be caused by the vorticity created by the baroclinic torque described in Sec 3. It is emphasized that, from a fluid mechanics point of view, there is no other mechanism to create vorticity in a fluid. If we neglect the

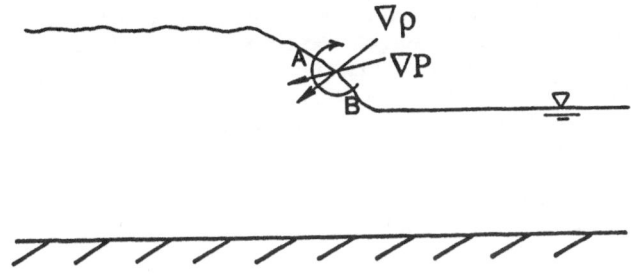

Figure 7. Vorticity creation on the front face of a bore due to the baroclinic torque.

effects of entrained air bubbles in a bore, then, the gradient of fluid density is normal to the free water surface. As shown in Fig. 7, in order to create the surface-roller eddies, the gradient of pressure must point towards the horizontal direction; there must exist a slight pressure difference between points A and B, $P_A > P_B$, as indicated in the figure. As demonstrated in (12), the pressure difference does not need to be large in order to create significant vorticity by the baroclinic torque. Evidently, this turbulence generation mechanism is different from that of a mixing layer, and the turbulence is generated not at the point of the front toe but along the entire front face.

In the case of successive bores approaching the shore just as in the natural coastal condition, the water in front of a bore is no longer quiescent, but instead, there exists a strong backwash flow created by the predecessor. The backwash flow ahead of an incoming bore contains vorticity in the bottom boundary layer. When the backwash flow meets the incident bore, the vorticity is detached from the boundary layer due to the flow separation caused by the adverse pressure field (the flow depth increases by the bore). This type of turbulent generation was reported and termed as a "backwash vortex" by Matsunaga & Honji (1980). Hence, in the natural coastal condition, turbulence in a bore is generated both at the front surface by the baroclinic torque and at the bottom boundary by flow separation of the backwash boundary layer that is initially created by the viscous effect.

5. Conclusion

Even though the component of ∇P in the direction normal to $\nabla \rho$ is infinitesimal at the air-water interface, the magnitude of $\nabla\left(\frac{1}{\rho}\right) \times \nabla P$, i.e. the baroclinic torque, is found to be finite owing to the singularity in $\nabla \rho$. (While the direction of $\nabla \rho$ is always normal to the interface, it is an approximation that the direction of ∇P is also normal to the interface; the actual direction of ∇P deviates from the normal to the interface.) Hence, generally accepted notion of the air-water interface being barotropic is found to be false, and the free water surface is a baroclinic surface, instead.

It is shown that turbulence appeared on the front face of a bore is originated from the vorticity created by the baroclinic torque. As a consequence of this vorticity generation mechanism, the turbulence generation must take place in a finite region of the front face, but not at a point of the front toe as proposed by Peregrine and Svendsen (1978). Furthermore, the vortex dynamics discussed in this paper does not support the conjecture that the turbulence generation is analogous to that of a mixing layer.

Lastly, the vorticity generation mechanism by the baroclinic torque is not only applicable to the case of bores but also provides enlightening explanations for various phenomena involving a liquid-gas interface. Such topics are discussed in Yeh and Breidenthal (1989).

The work for this paper was supported by the U.S. National Science Foundation (CES-8715450) and the Office of Naval Research (N00014-87-K-0815).

References

Matsunaga, N. & Honji, H. (1980) 'The backwash vortex', J. Fluid Mech. 99, 813-815.

Peregrine, D.H. & Svendsen, I.A. (1978) 'Spilling breakers, bores and hydraulic jumps', Proc. Coastal Eng. Conf., 16th, 540-550.

Rajaratnam, N. (1965) 'The hydraulic jump as a wall jet', J. Hydraul. Div., ASCE, 91, 107-132

Yeh, H.H. & Mok, K.M. (1989) 'On turbulence in bores', Phys. Fluids (submitted).

Yeh, H.H. & Breidenthal, R.E. (1989) 'Vorticity generation at a liquid-gas interface', (in preparation).

FREAK WAVE KINEMATICS

STIG E. SAND [1]
Danop
Slotsmarken 10
DK-2970 Hørsholm
Denmark

N.E. OTTESEN HANSEN
LICengineering
Ehlersvej 24
DK-2900 Hellerup
Denmark

PER KLINTING
Danish Hydraulic Institute
Agern Alle 5
DK-2970 Hørsholm
Denmark

OVE T. GUDMESTAD
Statoil
Forus, P.O. Box 300
N-4002 Stavanger
Norway

MARTIN J. STERNDORFF [2]
Mærsk Olie & Gas A/S
Rønnegade 2
DK-2100 København K
Denmark

ABSTRACT. Prototype records of extreme single waves recorded on the Danish Continental Shelf have been analysed in detail. Statistical calculations show that these waves do not belong to the traditional short term statistical distributions used for ocean waves. The waves are too high, too asymmetric and too steep.

[1] As of May 1st, 1989 Danish Maritime Institute, Hjortekærsvej 99, DK-2800 Lyngby, Denmark.

[2] As of October 1, 1988 Danish Hydraulic Institute.

A. Tørum and O. T. Gudmestad (eds.), Water Wave Kinematics, 535–549.

Emphasis has been placed on deriving a kinematic model for the extreme wave phenomenon in order to predict forces on offshore structures. A series of model tests aiming at reproducing the recorded waves from nature has been carried out to support the kinematic investigations. It has been found that the horizontal velocities above MWL are higher than predicted by generally accepted kinematic theoreties, whereas the velocities in the lower part of the water column is somewhat lower than expected. A directional (crossing seas) model is therefore proposed. It includes a spilling breaking velocity profile above the MWL and a modified Wheeler profile below the MWL in order to describe the extreme wave kinematics.

1. INTRODUCTION

Evidences of extreme single waves - freak waves - have become increasingly substantial during the last decade. It may be the development of more accurate field measuring equipment that has led to more frequent observation of this phenomenon in combination with a more detailed examination of prototype records once the awareness of such extremes is made. Section 2 illustrates several of these extreme waves and documents that this is a deep water as well as a shallow water phenomenon. Klinting and Sand (1987) analysed some of the freak waves from a 13 hours storm in September 1983 recorded at the Gorm Field at the Danish Sector of the North Sea. They concluded that asymmetry, steepness, crest height and directional composition were unique. Studies by LICengineering (LIC) and Danish Hydraulic Institute (DHI) have further clarified some of the basic characteristics of the phenomenon, and the present paper is mainly concerned with these results. To-day, however, it is still not quite clear how the freak waves are generated and what kind of mathematical tool will be adequate for an accurate description of the surface, kinematics, etc.

In the search for an explanation of the phenomenon similar extreme cases have been looked for in the literature. Yue and Mei (1980) consider the diffraction of a steady Stokes wave train by a thin wedge and shows that significant non-linear effects create a wave jump, i.e. a change in amplitude of the incident uniform waves. This actually corresponds to the Mach stem phenomenon, which was further studied by Peregrine (1983).

Dependent on the wedge and jump angles typical amplifications of more or less than two appear, but values as high as four can be found theoretically. Such extreme values have, however, never been confirmed by experiments.

Furthermore, the upper limit for steepness reasonably dealt with by the Non-Linear Schrödinger (NLS) equation is considered to be about $ak \approx 0.2$.

A focussing effect might be of particular interest to study in the present context. Peregrine et al. (1988) described the linear and the non-linear effects of wave rays crossing at a central point. Numerical integration of the NLS by means of a high-order explicit scheme was used to describe the situation. In the case of linear waves it is easy to show that the focussing tendency will be counterbalanced by a diffraction effect.

The non-linear description showed a further defocussing effect thus leaving resulting total amplitudes at a rather moderate level. A very convincing match with wave tank tests was obtained and reported in the above-mentioned paper. The focussing of N waves each from an angular interval of a can be compared to the Mach stem effect and the wedge with angle a.

The most classical instability theory is probably the one due to Benjamin & Feir (1967). They showed that disturbances (modulations) may grow, even for wave steepnesses well below the maximum. However, in the present case we are not looking for the growth of a modulation - we are looking for a phenomenon that makes a large wave grow even bigger on the expense (perhaps) of neighbouring waves. Dold and Peregrine (1986) found an instability from modulating a regular wave train of typically 10 waves. When following the waves for a hundred periods or more a concentration of wave energy took place, thus leading to wave breaking. This could occur even with an initial steepness as low as 3%. Figure 1 shows a regular wave train with ak = 0.12 after 61 periods.

Figure 1. Surface profile of a 5 wave modulation after 61 periods. Steepness ak = 0.12, vertical exaggeration is 5 times (Dold and Peregrine, 1986).

In addition to the numerical attempts to deal with steep and breaking waves several wave tanks have experimented with the generation of extreme waves. Thus, Isaacson (1984) and Kjeldsen (1984) describe methods and wave generator control strategy for the generation of steep breaking waves at a predetermined position in a tank. With respect to the freak wave phenomenon in nature it is, however, doubtful that the basic mechanism is the same, since the wave tank philosophy will be under the influence of given boundaries, paddle set-up, etc.

2. FREAK WAVE EVIDENCES

The recordings taken at the Gorm Field in the Danish Sector of the North Sea have been particularly useful in identifying the freak wave phenomenon. Figures 2, 3 and 4 show records from the 24th November, 1981, 17th November, 1984 and the 27th November, 1984 of surface elevations at the Gorm Field. However, as identicated earlier similar extreme situations have been experienced at several other locations. Table 1 outlines the records dealt with in the present study. It should be noted that in several of the records more than one freak wave appears per storm.

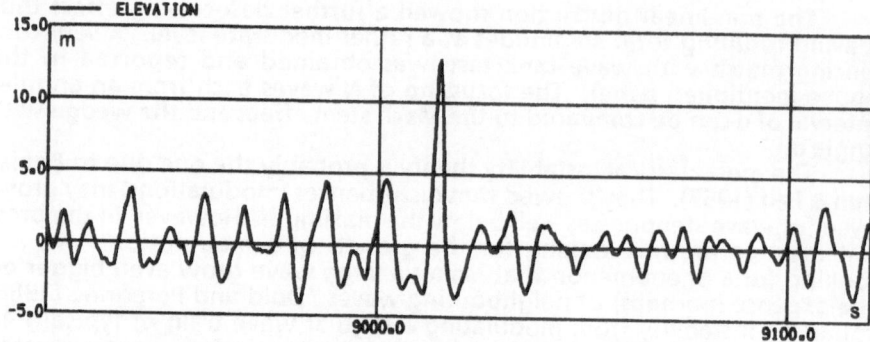

Figure 2. Freak wave A recorded on the 24th November, 1981 at the Gorm Field in the Danish Sector of the North Sea.

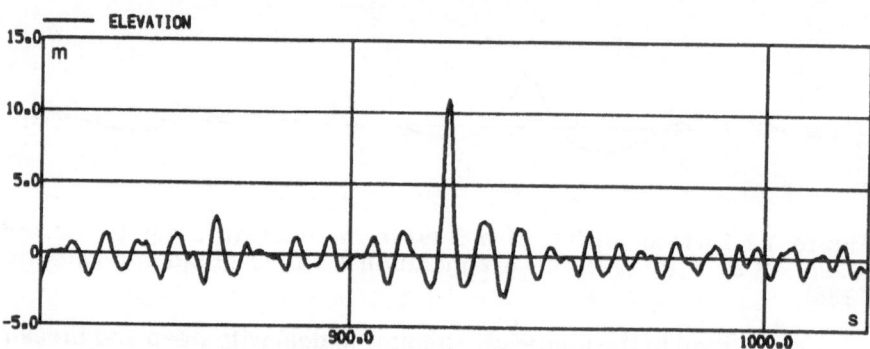

Figure 3. Freak wave B recorded on the 17th November, 1984 at the Gorm Field in the Danish Sector of the North Sea.

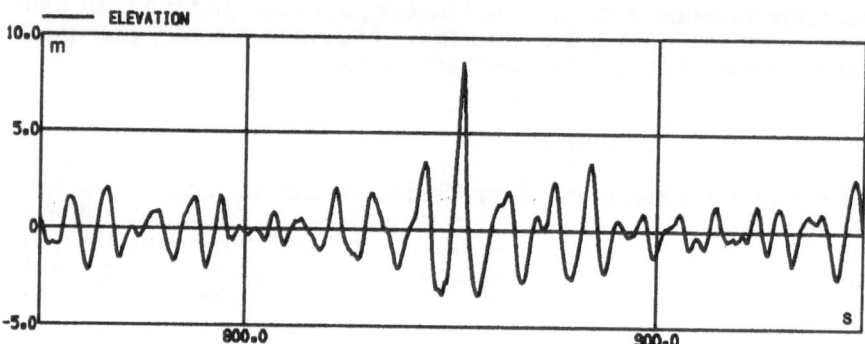

Figure 4. Freak wave C recorded on the 27th November, 1984 at the Gorm Field in the Danish Sector of the North Sea.

TABLE 1. List of some of the available freak wave records.

Location	h m	H_s m	H_{max} m	H_{max}/H_s	Registration	Year	H_{max}/H_s Theory
Cork, Eire	20	5,0	12,8	2,6	Waverider	1969	-
Hanstholm, DK	20	2	6-7	3	Visual	1985	1.59
Hanstholm, DK	40	3,5	7,6	2,2	Waverider	1985	1.59
Gorm Field, DK	40	6,8	17,8	2,6	Radar	1981	1.97
Gorm Field, DK	40	7,8	16,5	2,1	Radar	1981	1.97
Ekofish, N	70		20-22	> 2,5	Damage	1984	-
Gulf of Mexico	100	10,4	19,4	1,9	Wave staff	1969	-
Gulf of Mexico	350	10,0	23,0	2,3	Wave staff	1969	-
Gorm Field	40	5,0	12,0	2,4	Radar	1984	1.97
Gorm Field	40	5,0	11,3	2,3	Radar	1984	1.97
Gorm Field	40	5,0	11,0	2,2	Radar	1984	1.97
Gorm Field	40	4,8	13,1	2,7	Radar	1984	1.97

It is particularly important to be aware of deficiencies and necessary compensations/corrections related to the choice of recording instruments. Thus, for the widely used waverider it should be noted that apart from a tendency to avoid crests (and prefer riding on flanks) the electrical integration introduces a frequency-dependent amplification and phase shift. This is less severe for the wave radar and the electromagnetic current meter. However, proper data conditioning and restoring have to take place prior to detailed investigation of individual waves.

3. GORM FIELD RECORDS

At both the Dan and the Gorm Fields in the Danish Sector of the North Sea DHI has been recording wind, waves and temperature for Mærsk Oil and Gas A/S, who is operating the platforms. The instrument set-up at the location with the longest continuous recording periods is shown in Figure 5. It is essential to note that waves are measured by means of a wave radar shooting downwards from a bridge to the flare tower (D), and that horizontal velocities are available at three levels (two component Marsh McBirney type current meters). During September-December, 1983 several continuous records of durations up to 12-15 hours were taken, and in September-December, 1984 several of up to 5-7 hours got taped.

All of these records have been scanned for freak waves, and all the identified extreme events have been subject to a detailed study. This paper will only concentrate on a few typical examples, e.g. wave A, B and C in Figures 2, 3 and 4 and otherwise describe the general findings.

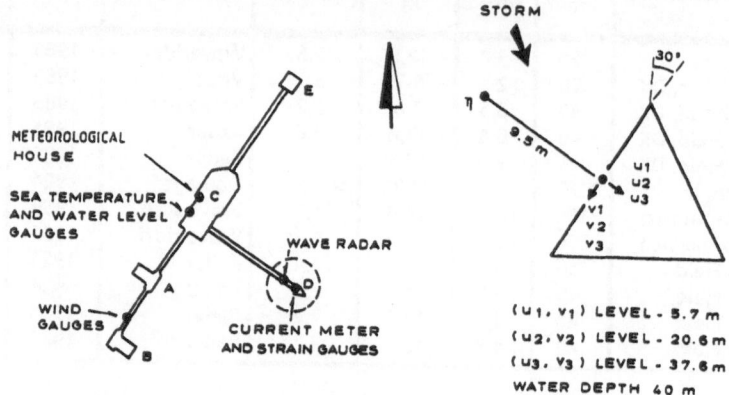

Figure 5. Instrument set-up at the Gorm Field in the Danish Sector of the North Sea.

4. STATISTICS

In investigating the freak wave phenomenon a simple check against traditional statistical distributions is mandatory. This should reveal whether these extreme waves can be considered logic representations of the very tail of the Weibull distribution. Figure 6 shows the wave height distribution in one of the North Sea storms against the Rayleigh distribution. Obviously, at least two to three of the highest waves deviate significantly from the traditional distribution. Also, Table 1 comprises a column showing the most probable maximum wave height, which is clearly exceeded by the recorded one.

Some of the 5 hours storms - corresponding to 2400 waves - have been analysed in further detail. In some of the series more than one freak wave appears. Based on the Rayleigh distribution the probability of j freak waves in n waves is:

$$P\{\,j\,freaks < H_n\} = \sum_{i=o}^{j-1} \binom{n}{i} (1 - exp(-\frac{H_n^2}{8\,m_o}))^{n-i} \, exp\,i\,(-\frac{H_n^2}{8\,m_o})$$

The probability distribution function for j = 1 and 2 freak waves in one series is plotted in Figure 7. The H/Hs ratios for some of the measured extremes are shown for three different storms. Again the Rayleigh distribution cannot be said to be suitable for description of the freak wave phenomenon.

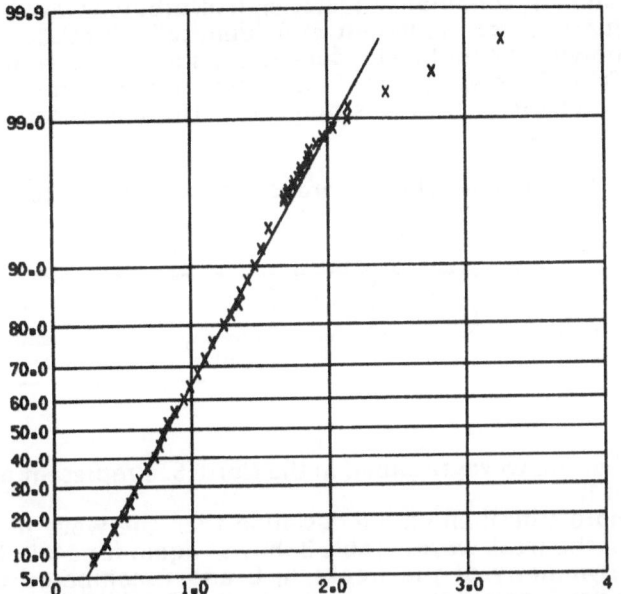

Figure 6. Wave height distribution normalized by $\sqrt{8m_0}$ compared to Rayleigh formula (power 2).

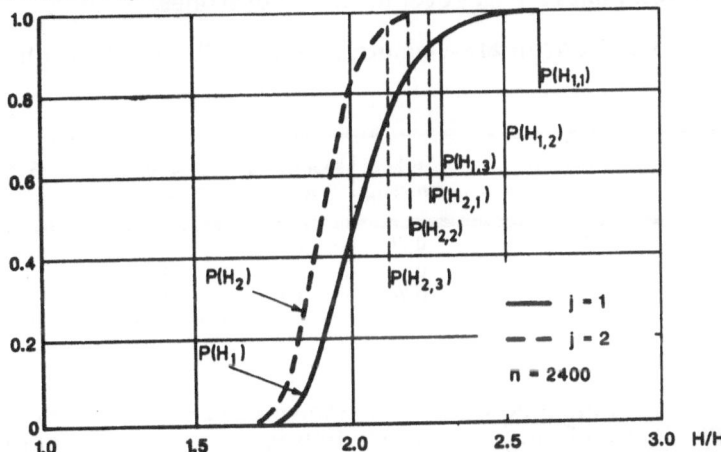

Figure 7. Probability distribution function for 1 and 2 freaks (highest and second highest) in the same storm (2400 waves) compared to measured values (vertical lines).

Based on stationary conditions and Rayleigh distributed wave heights, the probability function for the largest of N identically distributed wave heights is well known. The probability function can be expressed in terms of H_{max}/H_s, and Table 2 shows a series of such ratios and the corresponding probability of exceedence, Q, of this quantity. The computations are carried out for 2400 waves.

TABLE 2. Wave height ratio and probability of exceedence.

H_{max}/H_s	2.0	2.2	2.4	2.5	2.6	2.7	2.8
Q	0.45	0.14	$2 \cdot 10^{-2}$	$9 \cdot 10^{-3}$	$3 \cdot 10^{-3}$	$1 \cdot 10^{-3}$	$4 \cdot 10^{-4}$

It appears that the waves recorded in the North Sea represent a probability of less than 10^{-3}.

However, more critical and more deviating than the wave height is the crest height of the freak waves. Table 3 shows - again for 2400 waves - the horizontal asymmetry of the freaks A, B and C compared to the maximum Stokes wave asymmetry according to Cokelet (1977), and on a statistical basis also the crest height to the significant height for the Rayleigh distribution (most probable value) and the Stokes 5th order value. The measured values are way over the theoretical ones.

TABLE 3. Horizontal asymmetry and normalized crest heights.

Freak Wave Ref.	Measured μ_H	Stokes max μ_H	Measured a_c/H_s	Rayleigh a_c/H_s	Stokes 5th a_c/H_s
A	0.66	0.76	1.51	0.99	1.22
B	0.79	0.76	2.08	0.99	1.22
C	0.73	0.76	1.74	0.99	1.22

To improve the theoretical models the probability distributions should be derived for non-Gaussian processes. Hence, proper account for the skewness of natural waves will increase the predicted crest heights.

Defining the skewness as $\gamma = m_3/(m_2)^{1.5}$ a new probability distribution function can be derived (Haver 1989). Table 4 shows the a_c/H_s values for different steepnesses of which 0.3 is considered rather extreme for e.g. the Nowegian Sector. However, again the measured values of 1.5 to 2.1 clearly exceed the theoretical ones even at the 99% level.

TABLE 4. Normalized crest height for various values of skewness.

Level	$\gamma = 0$	$\gamma = 0.1$	$\gamma = 0.2$	$\gamma = 0.3$
Most probable	0.99	1.05	1.12	1.18
90%	1.13	1.21	1.30	1.36
99%	1.27	1.37	1.47	1.57

Another element to study is the crest front steepness, s'_c. From a large number of storms recorded on the Norwegian continental shelf Kjeldsen (1981) formed the joint probability of wave height and crest front steepness. It was found that the maximum rms value of s'_c was 0.122 and that the combination:

$$1.7 \leqq H/H_s \leqq 1.8 \text{ and } s'_c = 0.3$$

has the probability $1.6 \cdot 10^{-6}$. These values can again be compared to our values. First of all the H/H_s ratio often lies around 2.5 for the freak waves, and secondly the three crest front steepnesses of waves A, B and C are found to be 0.36, 0.28 and 0.24, respectively. That is, again exceeding even extreme values recorded and processed from a large Norwegian data base.

5. LABORATORY REPRODUCTION

As a supplement to the analyses of the prototype records from the Gorm Field including wave elevations and three levels (below MWL) of horizontal velocities a model test series was carried out by DHI. The aim was to re-produce in the offshore test basin some of the freak wave events recorded in the North Sea and to study the horizontal velocities above MWL in par-ticular.

In the generation of the actual time series from the North Sea it was confirmed that the freak waves are indeed very special. Great difficulties arose in reproducing these extremely high and very steep waves with both traditional linear theory as well as irregular second order theory. Typically, the crest lost height and the smallest disturbances (odd reflection patterns) would cause the wave to break before the instrument set-up. Further, reasons for the problems were the extreme directions present in the freak wave (i.e. larger than ±45 degrees), the steepness, instability, reflections, etc.

However, in the most successful cases the records were generated as a directional as well as a uni-directional wave field and the velocity profile above the MWL was measured by means of a movable device developed at DHI.

A typical result of the velocity measurements is shown in Figure 8, which represents the freak wave B (from Figure 3). However, as mentioned above the prototype and the model wave elevations were not identical, but it was nonetheless valuable to examine the steepest and highest wave possibly generated in the model. The measured data points are compared to Stokes 5th order profile. As a general result it was found that the Stokes profile underestimates the crest velocities. The same result was obtained with the theory of kinematic boundary conditions fit (KBCF) by Forristall (1985).

Thus, the basic conclusion of the model test series was that horizontal velocities below MWL are moderate and reasonably predictable from conventional theories, whereas the velocities in the crest are higher than expected. The broken line in Figure 8 represents a spilling breaking theory as developed by Longuet-Higgins (1973). Under the assumption that the freak wave is spilling breaking, which with respect to the earlier steepness calculations is probably a reasonable assumption, the velocity profile in the crest can be written as:

$$u(r) = c - \sqrt{gr}$$

in which c is the wave celerity and r the vertical distance downwards from the top. It appears from Figure 8 that the match is reasonably good.

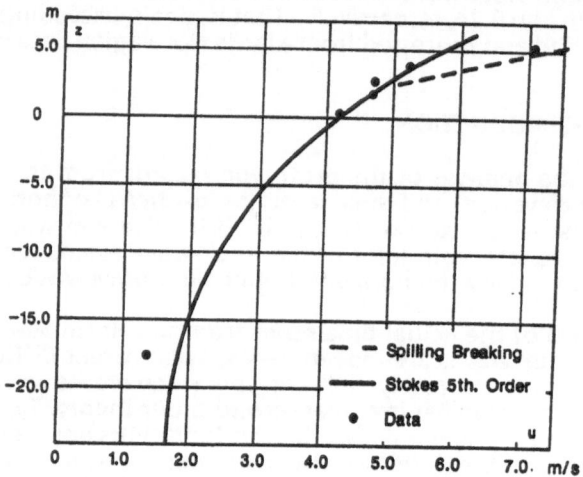

Figure 8. Freak wave B generated in the laboratory. The measured velocities are compared to Stokes 5th order theory and to a spilling breaking theory.

The results of the corresponding directional generation of the same prototype freak waves did not deviate drastically from the results described above. Typically, the crest velocities were reduced by about 10%, whereas changes in the profile below MWL could hardly be found.

6. 2-D KINEMATIC MODEL

The prototype records of elevations and horizontal velocities of freak waves as well as the supplementary study in the laboratory of the velocity profile above the MWL should give some basic information as to the physics and composition of the phenomenon. A 2-D and a 3-D model have been considered and linear, non-linear and instability theories have been applied. Some of the thoughts related to 2-D waves are given below.

From Figure 9 it is clear that although the freak wave elevation is large the horizontal velocities are nearly unaffected. This is in line with the model test results. One way of obtaining large velocities above MWL and moderate below is by matching a spilling breaking wave theory with that of a linear or non-linear traditional progressive wave theory below the MWL; e.g. as indicted in Figure 8. As a representative of the tratidional theories e.g. Stokes profile, Forristall KBCF, Wheeler and Gudmestad (1988) approaches have been considered.

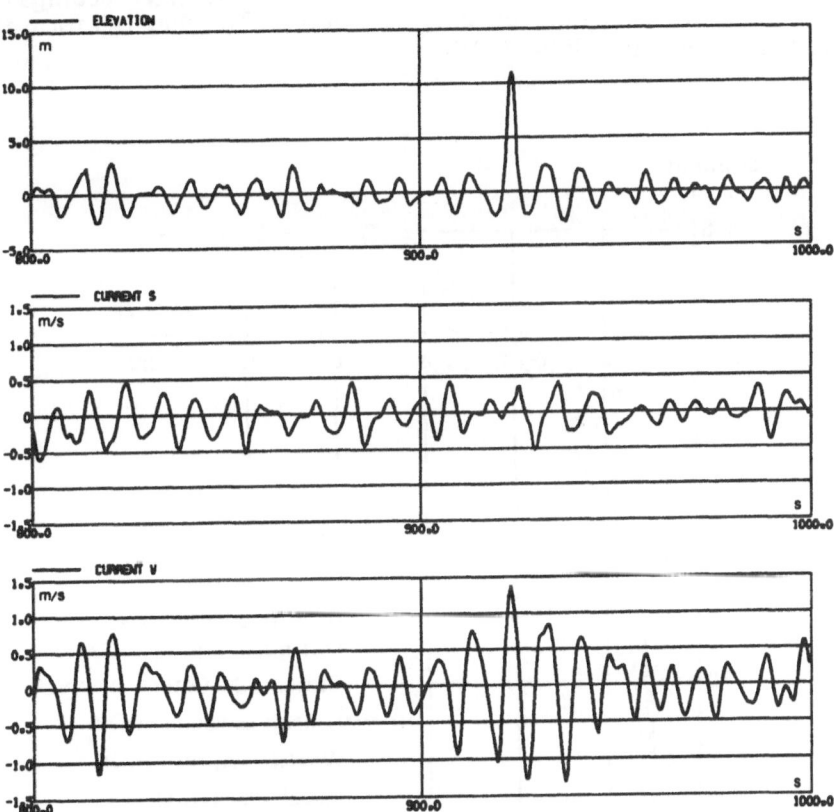

Figure 9. Time series of elevations and two components of horizontal velocities (level -5.7 m) of wave B.

546

As an illustration of the general match between theory and Gorm measurements Figure 10 has been drawn on the basis of freak wave C. Measured velocities at two levels are indicated and in addition the strain gauge measurements (at +5 to +10) have been converted to velocities under the assumption of the drag coefficient being between 0.7 and 0.5 (air entrainment), cf. Hoerner (1965) and Dean et al. (1985). The theoretical predictions are made from Stokes 5th order theory, a Wheeler approximation and a spilling breaking theory as described earlier.

For comparison the model test results have also been scaled and plotted. The range of velocities covered by four freak waves is shown as the shaded area in Figure 10.

It is seen that the Stokes and Wheeler profiles underpredict the velocities in the crest and that the slope of the spilling breaking profile seems more correct. The model test results is in between because they represent close-to-breaking and non-breaking waves in the laboratory. However, although a reasonable match could be obtained with the measured crest values (strain-gauge measurements) the two current meter readings at -5.7 and -37.6 are clearly exceeded by all theories. This may call for a strong directional model of freak waves.

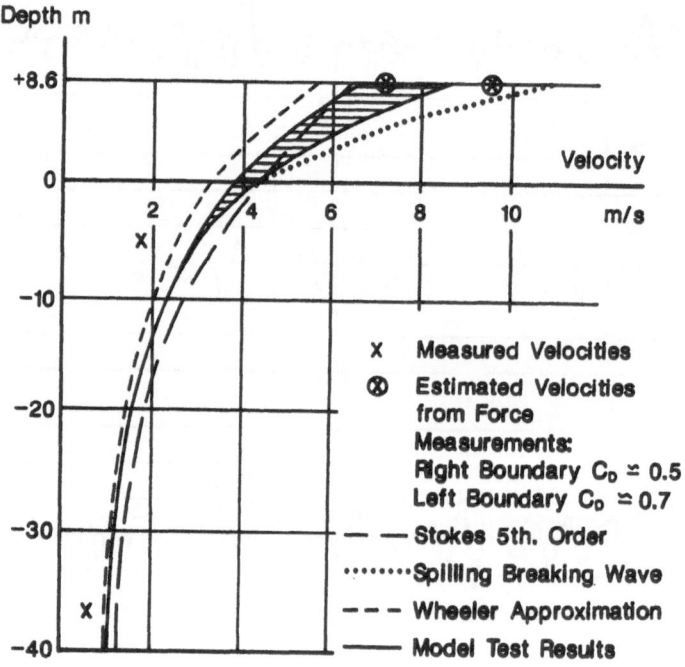

Figure 10. Predicted and measured velocity profile of freak wave C. Range of model test results is shown for comparison.

A check has been made as to the influence of bounded long waves. Since this is a quadratic quantity and the freak waves are extremely high a reduction of the horizontal velocities below MWL can be expected. A second-order bounded long-wave computation was performed on the series with freak wave C. The gap between measured and predicted velocities at level -5.7 m showed to be reduced from about 35% to now 20%. Hence, significant improvements cannot be obtained by accounting for the bounded long waves.

In the explanation of freak waves the theories discussed in the Introduction have been thoroughly studied. It is seen, however, that e.g. a higher order Stokes description will not suffice no matter the order of the perturbations, because the gap between actual crest elevation and a first or second-order description is extremely large. Furthermore, the classical instability theory seems to be of a different nature than the one prevailing here. Perhaps the wave jump and the long time evolution of wave groups could point in the right direction, but derivations for irregular (and perhaps directional) waves have to be completed. In addition steepnesses of ak = 0.2 is generally considered the maximum for successful application of the NLS equation. This should be compared with values of ak for freak wave A, B and C of 0.28 to 0.30.

7. 3-D KINEMATIC MODEL

The present investigations point towards a freak wave model with large velocities in the crest and rather low velocities, i.e. below predictions of traditional wave theories, in the water column below MWL. Since there has been no obvious model in the literature that may be applied for the phenomenon it has initially been attempted to compose the freak wave of linear elements.

A directional wave field with high frequency components building up the crest would indeed not have a strong influence on velocities below MWL. A model consisting of an ordinary directional sea accounting for about half the crest height of the freak wave superimposed by high frequency components focusing at the position of one of the high waves in the ordinary sea was actually described by Klinting and Sand (1987) as a result of directional analyses of prototype records. The focussing effects may be formed by a moving low pressure system. Indeed analyses of the development of the sea and the track of the wind systems indicate that in the North Sea the storm direction often changes from North towards West thus generating waves (swell) travelling South superposed by (high frequencies) local waves travelling East.

Anyway, it does not seem impossible to develop a crossing seas model which will match the tendency of measured quantities in Figure 10, and further to predict reasonably well the horizontal velocities by means of a spilling breaking theory above MWL and e.g. a Wheeler profile below MWL. Although this may be a likely freak wave model further research is strongly required.

8. FUTURE RESEARCH

The aim of the present paper has been to identify some of the typical features of the freak waves available in the DHI data base. A simple model has been proposed as an initial attempt to describe the kinematics. The future goal can be formulated as three main areas of concern, viz.:

- Well documented kinematic theory - emphasis above MWL.
- Reliable statistical model of freak wave occurrence.
- Ultimate and accidental limit state design guidelines.

However, as a next step towards this goal the data base must be extended and further statistical as well as individual freak wave analyses must be performed. Furthermore, an understanding of the origin of the phenomenon may arise from detailed study of wind and air pressure fields - their extension and progression - relative to the occurrence and character of freak waves. Following the solution of the generation problem further determination of parameters influencing the phenomenon can take place. This may apart from wind systems be spectral width, peak period, average steepness, etc.. Subsequently, sufficient information should be available for guidelines to be laid down for the design of offshore structures with respect to freak waves.

9. ACKNOWLEDGEMENT

The financial support and the permission to publish part of the jointly sponsored study leave the authors grateful to Mærsk Oil & Gas A/S and to the Statoil-Group consisting of Statoil Efterforskning og Produktion A/S, BHP Petroleum (Denmark) Inc., Total Marine Denmark, DENERCO K/S, LD Energi A/S, EAC Energy A/S, and Dansk Olie- og Gasproduktion A/S.

10. REFERENCES

Benjamin, T.B. and Feir, J.E. (1967) 'The Disintegration of Wave Trains on Deep Water', Journ. Fluid Mechanics, Vol. 17, 417-430.

Cokelet, E.D. (1977) 'Steep Gravity Waves in Water at Arbitrary Uniform Depth', Phil Trans. Royal Soc. of London, Ser. A, Vol. 286, No. 1335, 183-220.

Danish Hydraulic Institute (1989) 'Documentation of Extreme Wave Kinematics - Model Tests', Report for the Danish Statoil-Group (2nd licensing round, proprietory report).

Dean, R.G., A. Tørum and Kjeldsen, S.P. (1985) 'Symp. Separated Flow around Marine Structures', Norwegian Inst. of Technology, Trondheim.

Dold, J.W. and Peregrine, D.H. (1986) 'Water-Wave Modulation', 20th International Conf. Coastal Engineering, ASCE, Taipei.

Forristall, G.Z. (1985) 'Irregular Wave Kinematics from a Kinematic Boundary Condition Fit (KBCF)', Journ. Applied Ocean Research, Vol. 7, No. 4, 202-212.

Gudmestad, O.T. (1988) 'A New Approach for Estimating Irregular Deep Water Wave Kinematics', Submitted to Journ. Applied Ocean Research.

Haver, S. (1989) 'Probability Distribution of Highest Crest in a Non-Gaussian Process', Private Communication.

Hoerner, S.H. (1965) 'Fluid Dynamic Drag, Practical Information on Hydrodynamic Drag and Hydrodynamic Resistance', Washington.

Isaacson, M. (1984), 'Synthetic Evolution of an Extreme Wave', Report for National Research Council, Ottawa, Canada, Part 1 & 2.

Kjeldsen, S.P. (1984) 'Dangerous Wave Groups', Norwegian Maritime Research, Vol. 12, No. 2, 4-16.

Kjeldsen, S.P. (1981) 'Design Waves', Norwegian Hydrodynamic Laboratories, Rep. No. NHL 1 82008, 62.

Klinting, P. and Sand S.E. (1987) 'Analysis of Prototype Freak Waves', Proc. Spec. Conf. Nearshore Hydrodynamics, Univ. of Delaware, Newark, 15.

LICengineering and Danish Hydraulic Institute (1989) 'Wave Kinematics - Extreme Waves'. Report for the Danish Statoil-Group and Mærsk Olie & Gas A/S (Proprietory Report).

Longuet-Higgins, M.S. (1973) 'A Model of Flow Separation at a Free Surface', Journ. Fluid Mechanics, Vol. 57, Part 1, 139-148.

Peregrine, D.H., Skyner D., Stiassnie, M. and Dodd, N. (1988) 'Non-linear Effects on Focussed Water Waves', Proc. 21st Conf. Coastal Engineering, ASCE, Spain.

Peregrine, D.H. (1983) 'Wave Jumps and Caustics in the Propagation of Finite-Amplitude Water Waves', Journ. Fluid Mechanics, Vol. 136, 435-452.

Yue, D.K.P. and Mei, C.C. (1980) 'Forward Diffraction of Stokes Waves by a Thin Wedge', Journ. Fluid Mechanics, Vol. 99, Part 1, 33-52.

MICROCOMPUTER CAPABILITIES - NUMERICAL SIMULATION OF A BREAKING WAVE

B. SCHAEFFER
8, Allée Bonaparte
91080 - Courcouronnes
France

ABSTRACT. A new microcomputer program using a simple finite differences algorithm has been devised and applied to the simulation of the movement of a water wave generated by a piston-type wavemaker moving at a constant speed in a channel.

A grid is applied on the fluid domain, dividing it into quadrilateral cells where the pressure is constant and in quadrilateral elements obtained by joining the four immediate neighbours of the current node of the grid. During a time increment, an element is considered as a translating solid on which the surface forces are constant along each side. The body forces are constant in the whole element. After having computed the resultant force, Newton's law is applied and the displacement obtained by integrating twice the acceleration, using first order finite differences.

The computing method has been validated by comparison with experimental and numerical results from the literature involving large free surface motion. The crashing of a wave has been successfully simulated. By varying depth and wavemaker speed, it was found that, while the base of the wave has a speed depending only on the depth, the crest has a speed twice that of the wavemaker. The screen of the microcomputer was filmed by a camera triggered by the microcomputer.

1. INTRODUCTION

The movement of water waves has been studied for one century by many authors (Longuet -Higgins (1980), Larras (1979)). The first reported numerical calculations of a crashing wave seems to be those of W.E. Pracht, in Fernbach and Taub(1970), by numerical techniques (Eulerian description and finite differences). Recent papers try to compute the shape of a breaking wave by analytical methods (Greenhow (1983), Peregrine (1983), New et al. (1985), Baker et al. (1982)). With the advent of microcomputers, it may become easier to use simple iterative methods resulting in small programs than analytical methods needing heavy intellectual efforts or sophisticated numerical methods on large computers (Schaeffer (1988). The computing times on microcomputers may still be long, but the gap with large computers is becoming smaller every year. The method described in this paper has been validated on numerical examples from the literature (Welford and Ganaba (1981), Pedersen and Gjevik (1983)) and will be applied here to very large amplitude gravity waves.

A. Tørum and O. T. Gudmestad (eds.), Water Wave Kinematics, 551–560.

2. BASIC EQUATIONS

List of symbols:

c : speed of sound
C : boundary between free surface and inside parts of Σ'
d : diagonal vector of a cell
D : domain of the physical space
dl : line element
ds : surface element
dv : volume element
F : force
g : acceleration of gravity vector
i : index
K : bulk modulus
l : length of a side of a "solid element"
m : mass
n : outer normal unit vector
p : pressure
R : resultant force
t : time
t : tangent unit vector
v : velocity vector
x : position vector
° : initial value

Γ : acceleration vector
Δt : time increment

ρ : specific mass
Ω : domain occupyied by fluid particles
Ω' : domain occupyied by fluid particles having a free surface portion
Σ : boundary of domain Ω
Σ' : boundary of domain Ω'

Let us consider (fig. 1) a domain Ω, bounded by a surface Σ, occupied by a set of fluid particles that we follow in their movement in a fluid domain D of the physical space . The fluid contained in Ω moves under the action of volume forces (gravity), surface forces (pressure) and line forces (surface tension). If the domain is at the boundary of two different media, like Ω', there is a surface tension along the part of the boundary Σ' which separates the element from the other medium. For small elements we may suppose that the pressure is constant in the element and may neglect rotation. The surface tension, if any, is an external force only at the limit C between the interior and free surface parts of the frontier Σ' .

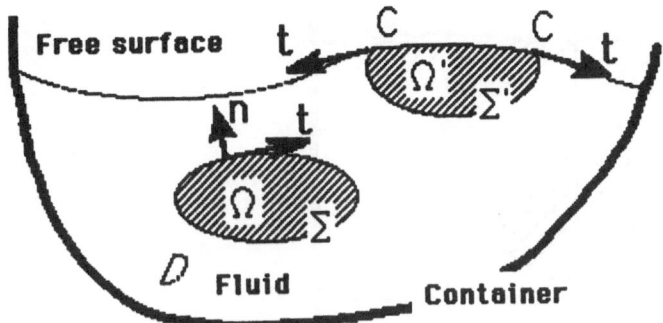

Figure 1. Physical domain of the fluid dynamic problem.

Applying Newton's second law of motion in the integral form, we may write:

$$(1) \qquad \iiint_{\Omega} \rho \Gamma \ dV = \iiint_{\Omega} \rho g \ dV - \iint_{\Sigma} p \ n \ ds + \int_{C} T \ t \ dl$$

If the acceleration is constant in the fluid domain Ω, equation (1) may be simplifyied:

$$(2) \qquad \Gamma = g \ - \ \frac{1}{m} \iint_{\Sigma} p \ n \ ds \ + \frac{1}{m} \int_{C} T \ t \ dl$$

where

$$(3) \quad m = \iiint_{\Omega} \rho \ dV = \text{constant}$$

From this equation, provided the variables on the right-hand side are known, the new position of the center of gravity is obtained by integrating twice equation (2). For a liquid, the surface tension T is constant and the pressure p is given by the equation of state.

3. NUMERICAL MODEL

The equation (1) of the movement is solved by discretisation of time and fluid space. Time is divided into constant time intervals Δt. A two-dimensional grid (not necessarily rectangular) divides the fluid, assumed to be of unit thickness, in quadrilateral cells in which the pressure is constant. With each node of the grid is associated a "solid element" obtained by joining the four immediate neighbours of the node. This quadrilateral element is considered as a solid moving without rotation during a time increment under the action of the body and surface forces. Hence, the node has the same movement as the center of gravity of the element. A cell is shared by two adjacent elements (fig. 2) and therefore Newton's third law of action and reaction is satisfyied. The difference with classical finite difference or finite elements methods is that the "solid element" is a contour and not a set of nodes.

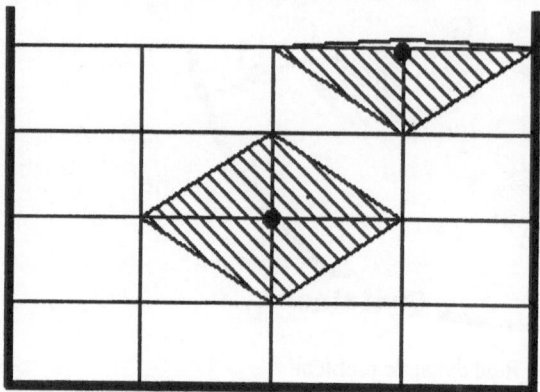

Figure 2. Cells of uniform pressure (rectangular) and "solid" elements (hatched losanges) inside the fluid and at the free surface.

The elements on the boundary are "contracted", that is, the part of the element outside the fluid domain has zero mass, zero volume but prescribed pressure. With this method, it is not necessary to define special elements on the boundary. On the free surface, for each element, the action of the surface tension is described by two forces acting along the two sides of the element coinciding with the free surface.

Figure 3 shows how the continuous medium is divided. The displacements and accelerations are computed on the nodes. A mesh is a quadrilateral of material where the pressure is constant. An element is the contour on which is applied Newton's law, the resultant force being applied to the central node.

The pressure in a cell is proportional to the ratio of the change in area to the initial area of the cell. For a quarilateral, the pressure is simply related to the cross products of the diagonals of the cell:

$$(4) \qquad p = -K \left(\frac{d_1 \times d_2}{d_1^\circ \times d_2^\circ} - 1 \right)$$

Constant pressure is assumed on the free surface. On rigid boundaries, there is free gliding (except at the lower corners) with normal speed equal to the normal speed of the boundary, if the boundary is moving, zero otherwise. Surface tension is constant in absolute magnitude and acts tangentially to the free surface.

The resultant effort applied on an element is computed by adding vectorially the weight and the forces applied on each of the four sides of the element. These forces F_i are normal to the sides of the element:

$$(5) \qquad F_i = - p_i \, n_i l_i$$

The resultant force on a solid element is

(6) $$R = \sum_{i=1}^{4} F_i$$

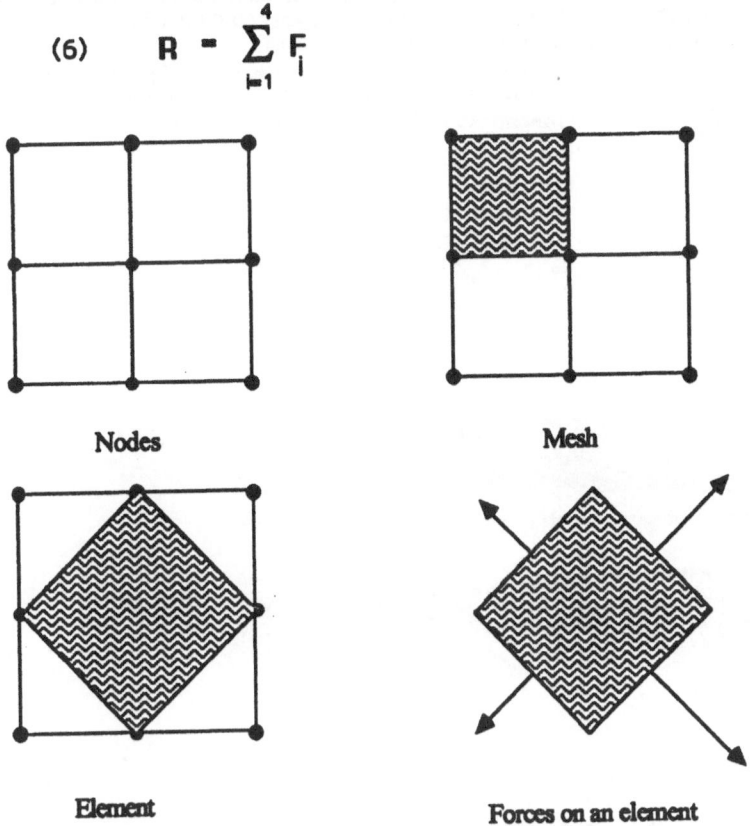

Figure 3. A "solid element" is made from four half meshes.

At each time step, the acceleration, computed by application of Newton's law,

(7) $$\Gamma = \frac{R}{m} + g$$

where m is the mass of an element, is integrated numerically twice according to the formulæ:

(8) $V(t) = V(t-\Delta t) + \Gamma(t)\,\Delta t$

(9) $X(t) = X(t-\Delta t) + V(t)\,\Delta t$

The fluid is initially at rest. At time 0, the external pressure and the gravity are instantaneously applied. It was necessary to use surface tensions much larger than the value for water. The general stability condition was that the Courant number, computed with the speed of sound (not the wave speed) had to be less than one for the smallest cell dimension. In order to minimize the computing time, small values of sound speed were used.

The computation was programmed with the help of UCSD Pascal on an Apple][and later on a Macintosh microcomputer. With the latter, the calculation takes 40 ms per computing cycle and per node instead of 200 ms. On a Macintosh II, if compiled to the arithmetic coprocessor, this time is reduced to a few milliseconds.

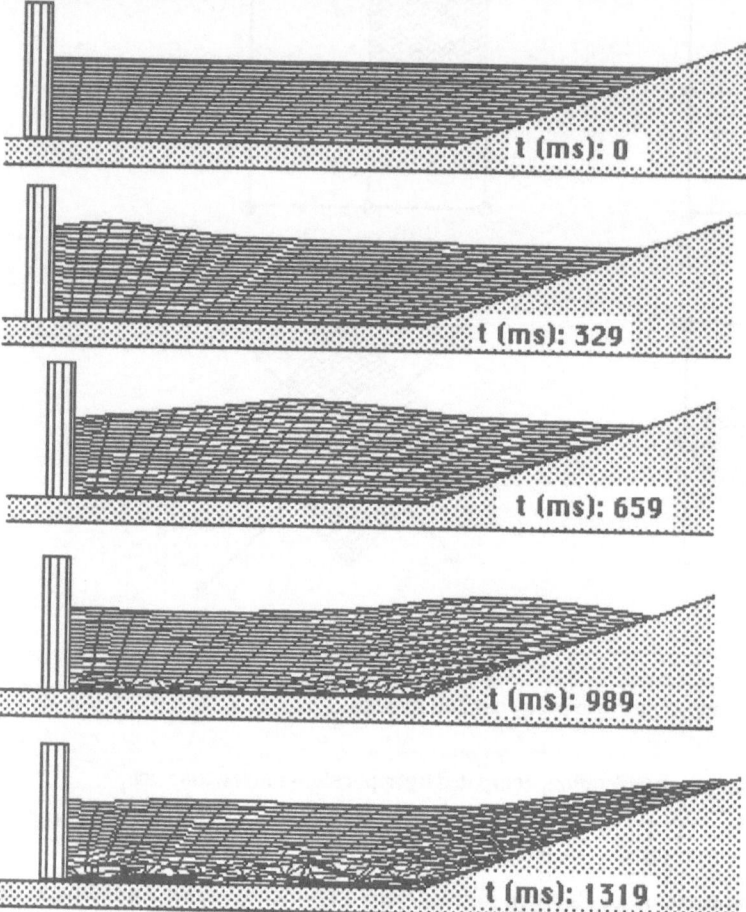

Figure 4. Run-up of a solitary wave generated by the movement of a wavemaker at a speed of 0.2 m/s during 0.5 second with an initial water depth of 0.15 m. and an external pressure of 1 kPa. The speed of sound is 10 m/s, the surface tension is 10 N/m and the specific mass is 1000 kg/m^3 (Schaeffer (1985)).

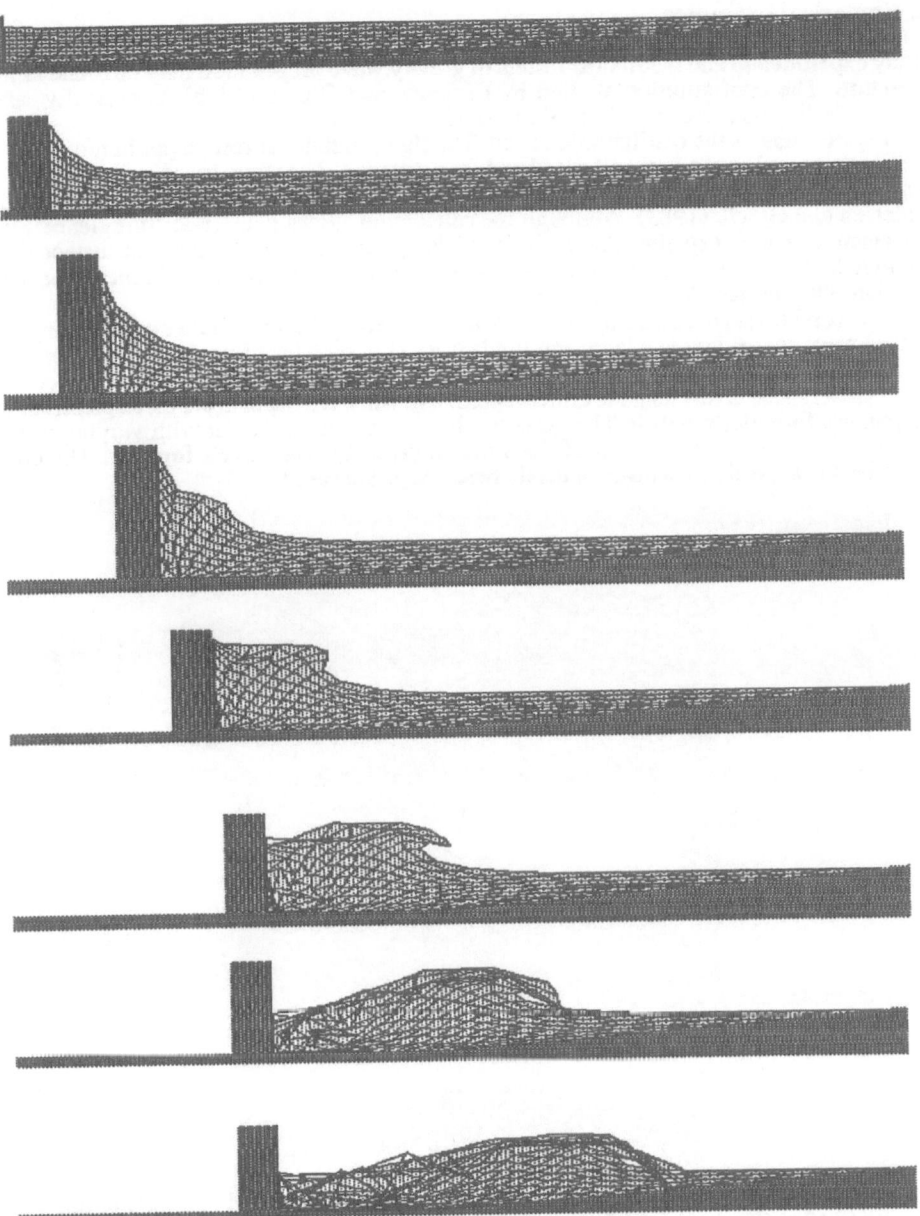

Figure 5. Simulation of a breaking wave. Time sequences of wave overturning with the same conditions as on figure 4 except that the speed of the wavemaker is 1 m/s instead of 0.2 m/s.

558

4. Numerical examples

Many experimental and theoretical results of gravity wave propagation may be found in the literature. The configuration studied by Pedersen and Gjevik (1983) numerically and experimentally was chosen to validate the method.

Figure 4 shows the configuration used. The fluid, initially at rest, is put in motion by a wavemaker moving at a constant speed and then stopping. The wave propagates from left to right towards the beach. The figure may be compared directly with the photographs taken by Pedersen and Gjevik (1983). Although the calculation shows numerical "turbulence", the agreement with their experimental results is satisfying. In this particular case, the method does not give better results than theirs except that these calculations are truly bidimensional and therefore allow the simulation of wave overturning.

A second calculation was performed with a speed of the wavemaker approaching the critical wave speed, giving a large amplitude wave (fig. 5). The influence of depth has been visualised on fig. 6, showing calculations for three different depths. Wave breaking occurs when piston speed is less than the critical speed, e.g. when the depth is 0.1 m, as predicted by Lagrange's formula, $v = \sqrt{gh}$. The same has been done on fig. 7, but with varying piston speed. The speed of the bottom of the wave is given by Lagrange's formula. The crest propagates at a speed independant of depth, twice the piston speed.

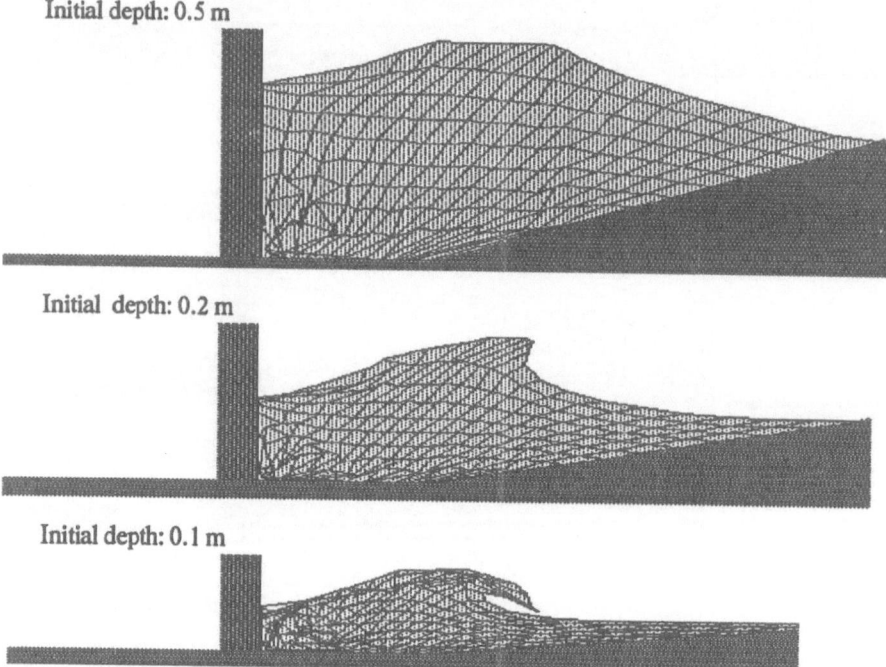

Figure 6. Influence of depth for a piston moving at a constant speed of 1 m/s, at the same moment (0.58 s). The bottom of the wave moves at a speed increasing with depth, but the crest propagates at the same speed, independant of depth.

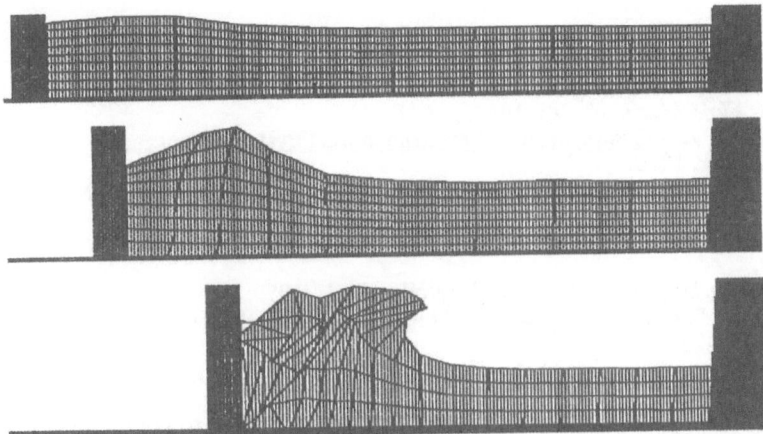

Figure 7. Influence of piston speed (0.1 - 0.5 - 1 m/s) on wave propagation at the same time (t= 0.58 s). Depth: 0.1 m. Wave speed is twice the piston speed.

5. DISCUSSION

The incompressibility hypothesis is usually considered as a simplifying assumption, but it is also possible to use the reverse approximation, that is to replace the slightly compressible fluid by a more compressible fluid, provided that the volume change may be neglected. It will be true if the Mach number is smaller than, for example 0.1.

The main stability criterion is the Courant-Friedrichs-Lewy criterion restricting the distance a wave travels in one time increment to less than one space interval (Hirt (1968)). It applies to the sound speed, not to the gravity wave speed, for a Mach number smaller than one. Numerical instabilities, associated to other criterions, appear after a few thousand iterations.

6. CONCLUSIONS

An explicit Lagrangian prediction algorithm for two-dimensional compressible flow with free surface has been formulated and applied to solitary wave generation, propagation, run-up and overturning. In spite of its simplicity, the method takes into account non-linearity, compressibility, free-surface movement and surface tension. For practical reasons (stability and computing time), the numerical values for the physical properties may be different from those of water, but this has only a slight incidence on the numerical results.

Another advantage of the simplicity of the method its ability to work on microcomputers. It should be useful for the simulation of the two-dimensional problems of fluid mechanics that can be solved in the Lagrangian description.

7. REFERENCES

Baker, G.R., Meiron D.I. and Orszag S.A., "Generalized vortex methods for free-surface flow problems", *J. Fluid Mech.*, n° 123, 1982, p. 477.

Fernbach, S. and Taub, A. - Computers and their Role in the Physical Sciences, Gordon and Breach, New York, 1970.

Greenhow, M. - 'Free-surface flows related to breaking waves', J. Fluid Mech., 1983, 134, 259.

Hirt C.W. - 'Heuristic Stability Theory for Finite-Difference Equations', J. Comp. Phys., 1968, 2, 339.

Larras, J., *"Physique de la Houle et des lames"*, Eyrolles, Paris, 1979.

Longuet-Higgins M.S., "The unsolved problem of breaking waves", *Proc. Conf. Coastal Eng.*, *17th*, 1980, pp. 1-28.

New A.L., McIver P. and Peregrine D.H. - 'Computations of overturning waves', J. Fluid Mech., 1985, 150, 233.

Newell, A.C., "The History of the Soliton", *J. Appl. Mech.*, n° 50, 1983, p. 1127.

Pedersen G., Gjevik B. - 'Run-up of solitary waves', J. Fluid Mech., 1983, 135, 283.

Peregrine, D.H. - 'Breaking Waves on Beaches', Ann. Rev. Fluid Mech., 1983, 15, 149.

Schaeffer B., "A lagrangian 'solid' element method for large amplitude movement of a compressible fluid with free surface", *4th Int. Conf. Num. Meth. Laminar Turbulent Flow*, Swansea 1985.

Schaeffer B., "Possibilités des microordinateurs - simulation numérique d'une vague déferlante, dont le mouvement en profondeur et le profil sont calculés par microordinateur" ATMA, 88ème session, Paris, 1988.

Welford, L.C., Ganaba, T.H. - 'A finite element method with a hybrid Lagrange line for fluid mechanics problems involving large free surface motion', Int. J. Num. Methods Eng., 1981, 17, 1201.

FREQUENCY DOWN-SHIFT THROUGH SELF MODU-LATION AND BREAKING

KARSTEN TRULSEN AND KRISTIAN B. DYSTHE
University of Tromsø
P.O.Box 953
N-9001 TROMSØ
NORWAY

ABSTRACT. We simulate the development of a moderately steep wavetrain, using evolution equations correct to fourth order in the wave steepness (see Dysthe (1979) and Lo and Mei (1985)), with an added term simulating effects of wave breaking. It is found that the breaking damps the developing sidebands selectively, such that the most unstable lower sideband comes out of the modulation–breaking process being the dominant one. The reason for this, in our opinion, is the tendency towards spatial localization of the part of the wavetrain contributing to the upper sidebands. This "high-frequency" part of the signal seems to concentrate around the steepest portions of the wavetrain, where breaking occurs. This is found both in our simulations and in experimental records (Melville 1983).

1   Introduction.

Evolution of trains of surface gravity waves has been extensively studied for the last 25 years. It started with the prediction (Lighthill (1965)) and subsequent measurements (Feir (1967)) of the so called modulational instability of Stokes waves. Today this instability, and the subsequent development of the unstable wavetrain are well documented experimentally, and fairly well understood theoretically.

There is one rather characteristic phenomenon, however, that has remained unexplained. This is the frequency down-shift that seems to have been first reported by Lake *et. al.* (1977). In their experiment a continuous wavetrain of frequency ω is generated at one end of a wavetank. At some distance from the source, the modulational instability appears as the growth of essentially two sidebands $\omega \pm \Delta\omega$. The resulting groups, (or bunching of the wavetrain) develop a characteristic asymmetric form. Then at still larger distance from the source, an approximate recurrence of the continuous wavetrain seems to occur. A more careful inspection, however, reveals that the frequency has been lowered from ω to $\omega - \Delta\omega$. Here $\Delta\omega$ is the most unstable modulation frequency occuring at $\Delta\omega/\omega \simeq ka_0$, where k and a_0 are the wavenumber and initial amplitude of the wavetrain respectively.

This down-shift does not always happen. The initial steepness, ka_0, of the wavetrain

A. Tørum and O. T. Gudmestad (eds.), Water Wave Kinematics, 561–572.

must be sufficiently large. (Melville (1982) gives $ka_0 > 0.16$.) It seems that whenever down-shift occurs, there is also wavebreaking.

On the theoretical side it was demonstrated by Lo and Mei (1985), using evolution equations derived by a perturbation technique (correct to fourth order in the steepness ka_0), that the two sidebands develop asymmetrically with the lower one dominating. Finally. however, they saw an approximate recurrence, without a down-shift. Adding a term representing viscous damping to this equation only resulted in a lowering in all the amplitudes involved. It did not selectively favour the lower sideband to any noticeable extent.

In the present paper we are trying out the assumption that breaking is an essential ingredient in the process of down-shifting the frequency. We believe that the role of breaking is to selectively damp the upper side band. This seems possible because: 1) Spatial separation of the different frequency components occurs during the evolution of the modulational instability (see e.g. Melville (1983)), and 2) It so happens that the steepest parts of the wavetrain correspond to the upper sideband.

Starting with the equations used by Lo and Mei (1985), we add a term simulating the effect of breaking. This term disappears whenever the local wave steepness is below some critical value. When, however, the steepness becomes supercritical, the "breaking term" makes it relax towards this critical value.

The numerical simulations show a definite trend towards frequency down-shift, in accordance with the ideas sketched above.

The plan of the paper is the following: A brief review of the fourth order nonlinear Schrödinger equation and its properties with respect to stability, is presented in section 2. Numerical simulations of this equation, corresponding to two different initial conditions. are presented. Then in section 3, the "breaking term" is introduced. The same simulations are then repeated, but now also including wave breaking.

2   The model without breaking.

2.1   Derivation of the governing equations.

As our starting point, we take the equations

$$\nabla^2 \phi = 0 \quad \text{for} \quad -h \leq z \leq \zeta(\mathbf{x}, t) \tag{1}$$

$$\frac{\partial^2 \phi}{\partial t^2} + g\frac{\partial \phi}{\partial z} + \frac{\partial}{\partial t}(\nabla\phi)^2 + \frac{1}{2}\nabla\phi \cdot \nabla(\nabla\phi)^2 = 0 \quad \text{for} \quad z = \zeta(\mathbf{x}, t) \tag{2}$$

$$\frac{\partial \zeta}{\partial t} + \nabla\phi \cdot \nabla\zeta = \frac{\partial \phi}{\partial z} \quad \text{for} \quad z = \zeta(\mathbf{x}, t) \tag{3}$$

which govern the irrotational flow of an incompressible. inviscid fluid with a free surface. The Laplace equation (1) is the condition of incompressibility. Equation (2) is the total time derivative of the Bernoulli equation and (3) is the kinematic surface condition. Here $\phi(\mathbf{x}, z, t)$ is the velocity potential and $\zeta(\mathbf{x}, t)$ is the surface elevation associated with the wave motion. We shall consider deep water waves ($h \to \infty$) and g is the gravitational acceleration. The horizontal coordinates are $\mathbf{x} = (x, y)$. z is the vertical coordinate and t is time.

For the slowly evolving wavetrain, we assume an expansion for ϕ and ζ:

$$\phi = \bar{\phi} + Ae^{kz+i\theta} + A_2e^{2(kz+i\theta)} + \cdots + c.c. \tag{4}$$

$$\zeta = \bar{\zeta} + Be^{i\theta} + B_2e^{2i\theta} + \cdots + c.c. \tag{5}$$

where $\theta = \mathbf{k}\cdot\mathbf{x} - \omega t$ and $k = |\mathbf{k}|$. $\bar{\phi}$ and $\bar{\zeta}$ are real functions, representing the mean flow and surface elevation brought about by the radiation stress of the wave. The coefficients A, A_n, B and B_n are complex.

The wave steepness, $\epsilon = ka$, where k is the wave-number and a is the amplitude $a = |\zeta|$, is used as an ordering parameter. We make the assumption that A is of order $O(\epsilon)$, and that all coefficients are slowly varying on scales ϵt and ϵx. When the coefficients are constant, the solution takes the form of a Stokes expansion.

By expanding (2) and (3) to fourth order in ϵ, and retaining only one horizontal dimension, we get the modified nonlinear Schrödinger equation of Dysthe (1979):

$$\frac{\partial A}{\partial t} + \frac{\omega}{2k}\frac{\partial A}{\partial x} + \frac{1}{8}i\frac{\omega}{k^2}\frac{\partial^2 A}{\partial x^2} + 2i\frac{k^4}{\omega}A|A|^2$$

$$- \frac{1}{16}\frac{\omega}{k^3}\frac{\partial^3 A}{\partial x^3} - \frac{k^3}{\omega}A^2\frac{\partial A^*}{\partial x} + 6\frac{k^3}{\omega}|A|^2\frac{\partial A}{\partial x} + ikA\frac{\partial\bar{\phi}}{\partial x}\bigg|_{z=0} = 0. \tag{6}$$

We now introduce dimensionless variables and transform to a coordinate system moving with the linear group velocity. Following Lo and Mei (1985), we introduce the following transformations

$$\left.\begin{array}{rclrcl} A & \to & -\frac{ia_0\omega}{2k}A, & \bar{\phi} & \to & a_0^2\omega\bar{\phi}, \\ \epsilon\gamma(2kx - \omega t) & \to & \xi, & \epsilon^2 kx & \to & \eta, \\ \epsilon k\gamma z & \to & z, & \epsilon k\gamma h & \to & h, \\ B & \to & a_0 B, & B_n & \to & a_0 B_n, \\ \bar{\zeta} & \to & a_0\bar{\zeta}, & \zeta & \to & a_0\zeta \end{array}\right\} \tag{7}$$

where now $\epsilon = ka_0$. All simulations will be done with periodic boundary conditions on ξ, and γ is a scale factor to normalize the domain of ξ to $(0, 2\pi)$.

The transformed equations read

$$\frac{\partial A}{\partial\eta} + i\gamma^2\frac{\partial^2 A}{\partial\xi^2} + iA|A|^2 + 8\epsilon\gamma|A|^2\frac{\partial A}{\partial\xi} + 4i\epsilon\gamma A\frac{\partial\bar{\phi}}{\partial\xi}\bigg|_{z=0} = 0. \tag{8}$$

$$4\frac{\partial^2\bar{\phi}}{\partial\xi^2} + \frac{\partial^2\bar{\phi}}{\partial z^2} = 0 \quad \text{for} \quad -\infty < z \leq 0. \tag{9}$$

$$\frac{\partial\bar{\phi}}{\partial z} = \frac{\partial}{\partial\xi}|A|^2 \quad \text{for} \quad z = 0 \tag{10}$$

$$\frac{\partial\bar{\phi}}{\partial z} = 0 \quad \text{for} \quad z \to -\infty. \tag{11}$$

These equations are the same as those used by Lo and Mei (1985).

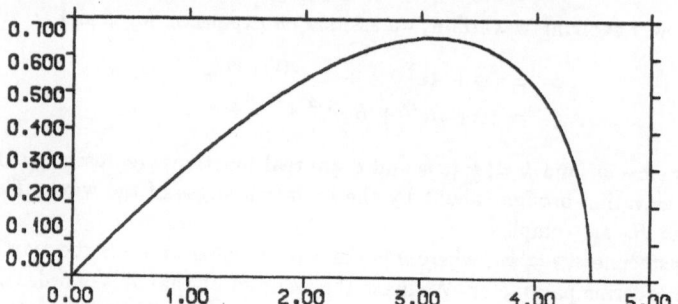

Figure 1: Growth rate versus wavenumber of perturbation.

While η corresponds to physical space, we shall in the following refer to η as a time variable, due to the parabolic form of (8).

The surface elevation in the new coordinates, is

$$
\zeta(\xi,\eta) = \epsilon^2\gamma\frac{\partial\bar{\phi}}{\partial\xi}\bigg|_{z=0} + \left(\frac{1}{2}A - \frac{i\epsilon\gamma}{2}\frac{\partial A}{\partial\xi} - \frac{3\epsilon^2}{16}A|A|^2\right)e^{i\theta} +
$$
$$
\left(\frac{\epsilon}{4}A^2 - i\epsilon^2\gamma A\frac{\partial A}{\partial\xi}\right)e^{2i\theta} + \left(\frac{3\epsilon^2}{16}A^3\right)e^{3i\theta} + c.c.,
\tag{12}
$$

where the phase function becomes $\theta = \xi/\epsilon\gamma - \eta/\epsilon^2$.

2.2 Linear stability analysis.

Equations (8)–(11) have the ξ-independent solution $A = A_0 e^{-iA_0^2\eta}$, where A_0 is real. This corresponds physically to an unmodulated carrier wave. This solution can be shown to be unstable to small perturbations of amplitude and phase.

Let the ξ-independent solution be perturbed as $A = A_0(1+a)e^{i(\theta-A_0^2\eta)}$, where a and θ are small real perturbations of amplitude and phase respectively. We solve equations (8)–(11) by linearizing in the three variables a, θ and $\bar{\phi}$, and by guessing a solution of the form

$$
a,\theta \propto \exp[i(\lambda\xi - \Omega\eta)], \quad \bar{\phi} \propto \exp[Kz + i(\lambda\xi - \Omega\eta)].
\tag{13}
$$

This gives the dispersion relation

$$
\Omega = 8\lambda\epsilon\gamma A_0^2 \pm i|\lambda|\gamma\sqrt{2A_0^2 - \lambda^2\gamma^2 - 4|\lambda|\epsilon\gamma A_0^2}.
\tag{14}
$$

The ξ-independent solution is unstable to sideband perturbations which make the radicand positive. In figure 1 we show the growth rate $\mathrm{Im}\,\Omega$ of the unstable mode of perturbations with wavenumber λ, corresponding to $\epsilon = 0.23, \gamma = 0.229$ and $A_0 = 1.0$.

Due to the periodic boundary condition that we impose on ξ, only integral values of the wavenumber λ are applicable. In the situation of figure 1, there are four unstable sidebands, and the third sideband is the most unstable one.

2.3 Numerical results without breaking.

Simulations corresponding to the time development of two different initial wave profiles will be presented in this paper. All numerical results shown use the parameter values $\epsilon = 0.23$ and $\gamma = 0.229$. This corresponds to a carrier of 19 wavelengths across the 2π periodic domain in ξ. (For details on the numerical scheme, see Trulsen (1989).)

The first initial wave profile is taken after the experiments of Keller as referenced by Lo and Mei (1985), and consists of two "sidebands" of equal magnitude with no carrier wave present. This initial profile was also chosen by these authors in their simulations. As the initial wave profile, we require the first harmonic to be $B(\xi) = 0.5\exp(i\xi) + 0.5\exp(-i\xi)$, which can be shown to correspond to an initial value for A of $A(\xi) = 0.483\exp(i\xi) + 0.537\exp(-i\xi)$. Unfortunately, the experimental results only go as far as $\eta = 3.35$. This is about the point in the development where the features we are looking for begin to appear.

In figure 2 we show the time development of the surface from $\eta = 0$ to $\eta = 15$. One can clearly see the development through one recurrence cycle of length about 10. In figure 3 we present selected Fourier components of the surface displacement, numbered relative to the carrier with wavenumber 19. Since all nonlinear terms in (8) are cubic in A, only odd Fourier components will arise in the first harmonic. Thus we present the first and third upper and lower "sidebands". Figure 3 was also presented by Lo and Mei (1985).

The second initial wave profile is taken as a carrier wave, with all unstable sidebands present. We shall use

$$\zeta(\xi) = 2\text{Re}\left\{\left[0.5 + 0.005\sum_{k=1}^{6}(e^{ik\xi} + e^{-ik\xi})\right]e^{i\xi/\epsilon\gamma}\right\}. \tag{15}$$

Figure 4 shows the time development of this initial wave profile. The development of some Fourier components is shown in figure 5. No sign of recurrence can be seen in this case. Indeed, there was no sign of recurrence, even when the simulation in figure 4 was taken through a much longer time than is presented here.

The most important thing to be observed from these two time developments, is that the wave steepness becomes much larger than the theoretical and experimental limitation, at which waves begin to break. As reported by Lo and Mei, Keller did observe breaking at his last probe at $\eta = 3.35$. This is just at the beginning of a time regime of heavy modulation in figure 2.

3 The model for damping by wave breaking.

3.1 Localization of sidebands and wave breaking.

An initially uniform wavetrain evolves according to the equations (8)–(11) by developing modulations in amplitude and phase. We denote the first harmonic contribution to the surface elevation ζ (see equation (5)) by $Be^{i\theta} + c.c.$, where B is the complex amplitude, and $\theta = \xi/\epsilon\gamma - \eta/\epsilon^2$. Writing $B = |B|\exp i\psi$ we can exhibit the amplitude envelope $|B|$ and the wavenumber modulation $\psi_\xi = \partial\psi/\partial\xi$. Typical snapshots are shown in figure 6. A similar snapshot from the initial part of the modulation process would have shown the

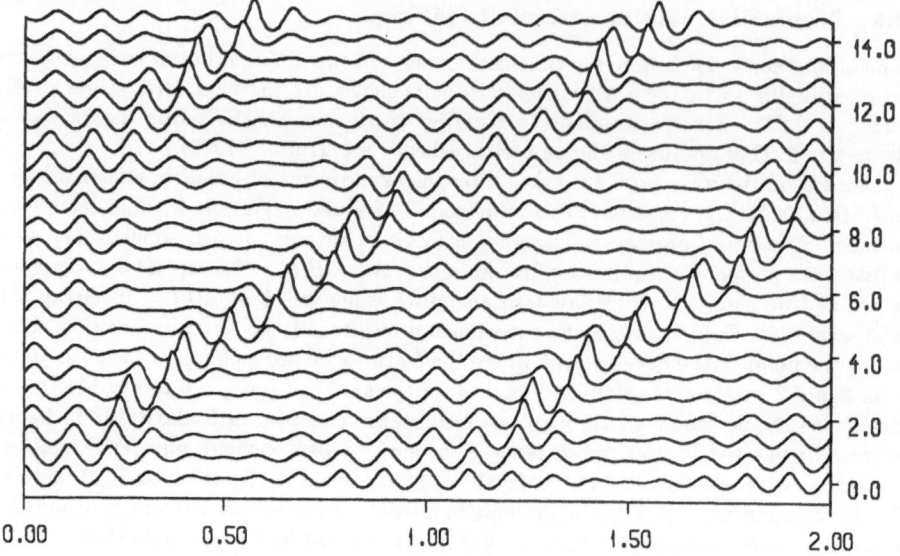

Figure 2: Time development of wave profiles, corresponding to the initial condition of Lo and Mei. $\xi/2\pi$ is taken along the horizontal axis and η is taken along the vertical axis.

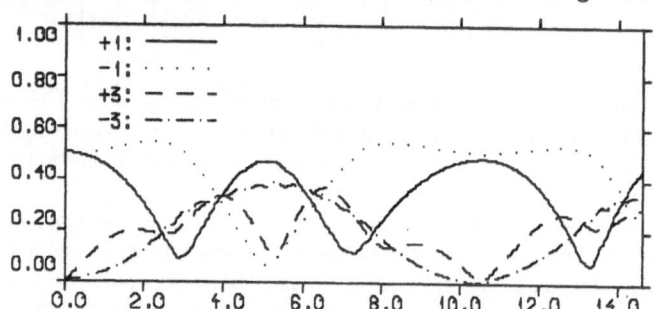

Figure 3: Time development of spectral components, corresponding to the initial condition of Lo and Mei. Component n corresponds to the nth sideband relative to the carrier $1/\epsilon\gamma$.

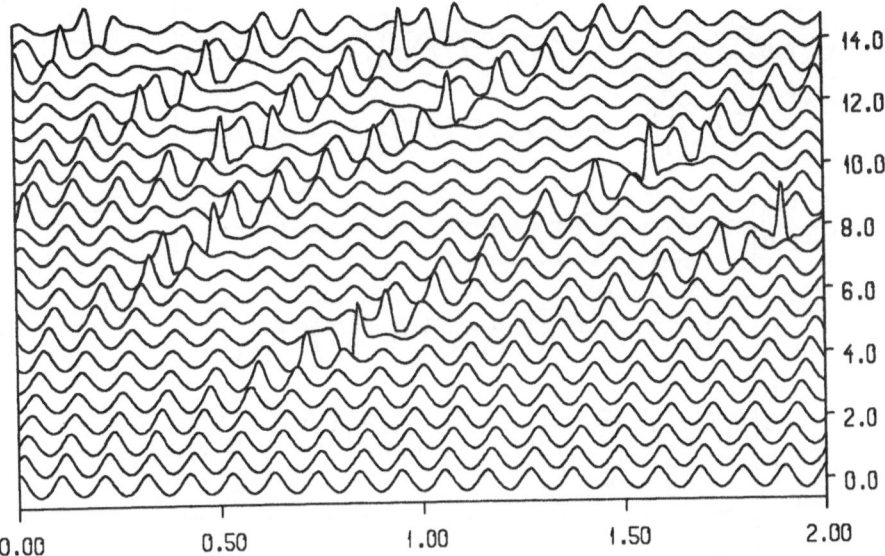

Figure 4: Wave profiles corresponding to the perturbed carrier wave.

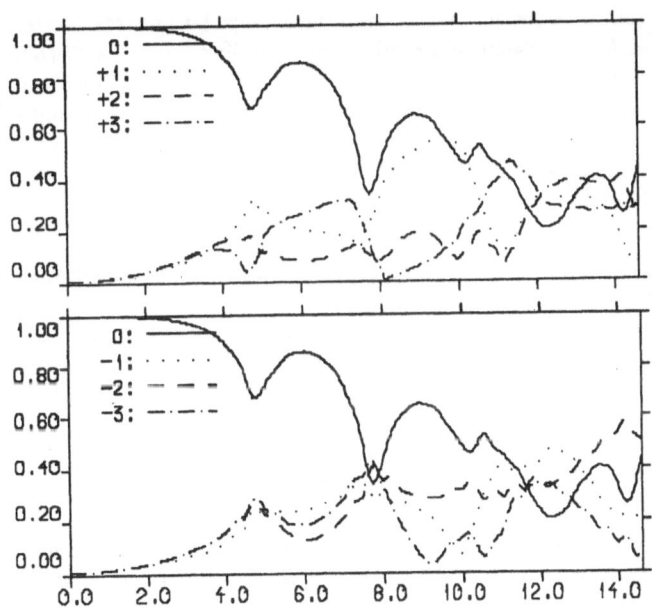

Figure 5: Spectral development for the three upper and lower sidebands, corresponding to the perturbed carrier wave.

568

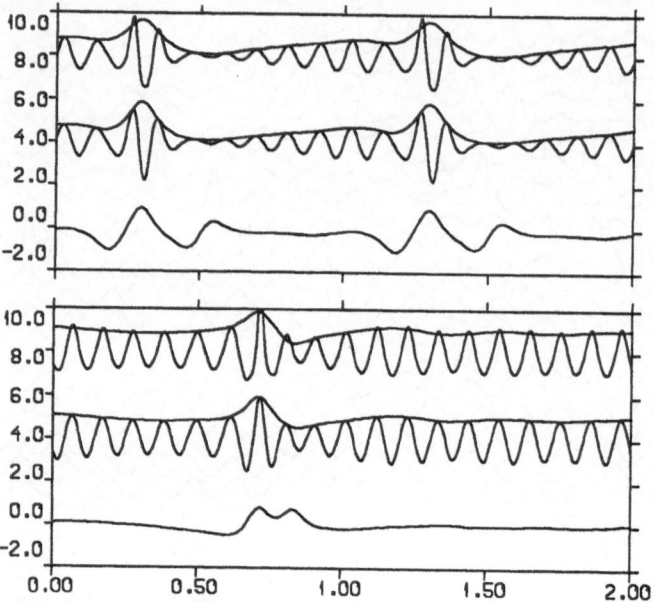

Figure 6: Modulation of wave for initial conditions after Lo and Mei corresponding to $\eta = 2.66$ (top), and for the perturbed carrier corresponding to $\eta = 3.99$ (bottom). The upper profile is the surface displacement, ζ, superimposed with $|A|$, both with zero level shifted to 8.0. The middel profile is the first harmonic, $Be^{i\theta} + c.c.$, superimposed with its amplitude, $2|B|$, both with zero level shifted to 4.0. The lowest graph is the wavenumber of the modulation, ψ_ξ, where $B = |B| \exp i\psi$. The wavenumber is scaled by .2, all the other graphs are drawn to scale.

amplitude and phase modulations to be exactly 90° out of phase through the group, as is found explicitly from the stability analysis.

The development of amplitude and phase modulations can also be extracted from experimental records of $\zeta(t)$ (by bandpass filtering around the first harmonic, and taking a Hilbert transform of the filtered signal). This has been done by Melville (1983) for wavetrains evolving through modulational instability. His records are rather similar to those in figure 6.

A remarkable feature is the fact that the steepest parts of the group correspond to positive values of the excess wavenumber ψ_ξ. At the same time, the dominant contribution to the upper sideband comes from those parts of the wavetrain where $\psi_\xi > 0$. Since breaking is localized to the steepest parts of the wavetrain, this seems to be a mechanism for selective damping of the upper sideband.

Investigating these ideas, we shall proceed to add a term in equation (8) intended to simulate the damping effect of wavebreaking.

3.2 The damping model.

We now introduce a source term, S, on the right hand side of (8), which we write on the following form

$$\frac{\partial A}{\partial \eta} + C = S \equiv -\frac{1}{\tau}A\left[\left(\frac{|A|}{A_0}\right)^r - 1\right]H(|A| - A_0) \qquad (16)$$

where H is the Heaviside unit step function

$$H(x) = \begin{cases} 0 & x < 0 \\ 1 & x \geq 0 \end{cases}. \qquad (17)$$

As seen, the added source term only enters the evolution equation for the amplitude of A.

The parameter A_0 gives the critical value for $|A|$, corresponding to the wave steepness when breaking starts. The corresponding critical "steepness" will be ϵA_0. The two parameters r and τ determine the rate by which the supercritical value of $|A|$ relaxes toward A_0.

Leaving out the terms corresponding to C in (16), we have the following solution for the amplitude of A:

$$|A| = A_0(1 - ce^{-\frac{r\eta}{\tau}})^{-\frac{1}{r}} \qquad (18)$$

where c is a constant. This shows that $|A|$ relaxes toward A_0 with a typical relaxation time τ/r.

This relaxation time should be of the same order of magnitude as the duration of a typical breaking event, which is much smaller than a typical time for evolution of the wave envelope (in the absence of the source term S). As long as this condition is satisfied, it is not expected that the effects of the source term will be sensitive to the values of τ and r chosen. This is confirmed by our numerical simulations. The important parameter in S is therefore A_0. The simulated wave steepness is larger than $\epsilon|A|$, as can be appreciated in figure 6. The critical "steepness" ϵA_0 should therefore be smaller than the limiting steepness of $\simeq 0.39$ (see Melville (1982)). In our simulations we have used $\epsilon A_0 = 0.35$.

3.3 Numerical results with wavebreaking.

The same initial waveprofiles as described in section 2.3 are now simulated with the additional term in equation (16). We present simulations corresponding to $\tau = 1/8$, $\epsilon A_0 = 0.35$ and $r = 2$. As already mentioned, τ and r are not critical, thus results corresponding to one typical set of parameter values are enough to get a good impression of the qualitative behaviour of the model. Figures 7 and 8 show the development of the initial profile of Lo and Mei. The lower sideband becomes dominant after a time corresponding to one "recurrence" cycle of the undamped development.

The simulation of the perturbed carrier with damping is more interesting since it involves more degrees of freedom. The results are presented in figures 9 and 10. The most unstable lower sideband becomes dominant, with the original carrier damped down to a level comparable to all the other sidebands. The second upper sideband, which is next to the most unstable one, is less damped than the other sidebands. It turnes out that this is a common feature for different choices of the initial perturbation.

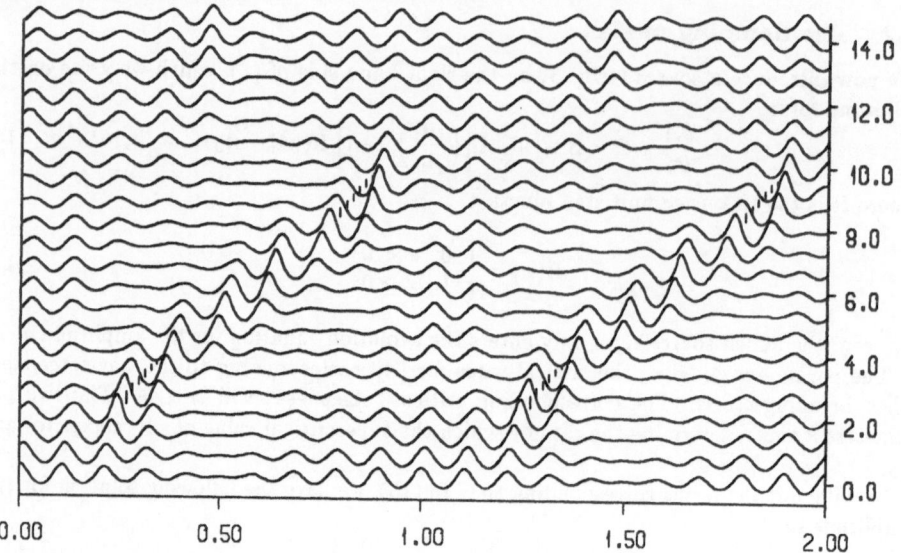

Figure 7: Wave profiles corresponding to the initial condition of Lo and Mei, with breaking. The vertically hatched areas show the localization of the breaking.

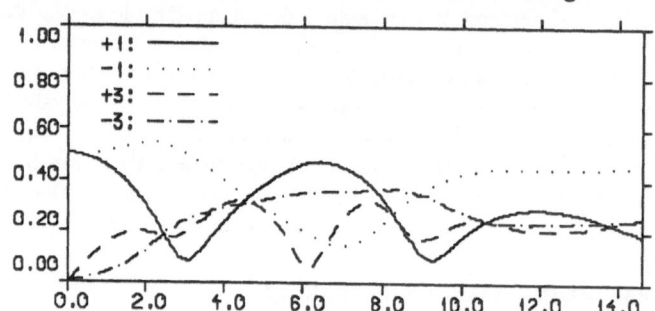

Figure 8: Spectral development corresponding to the initial condition of Lo and Mei, with breaking.

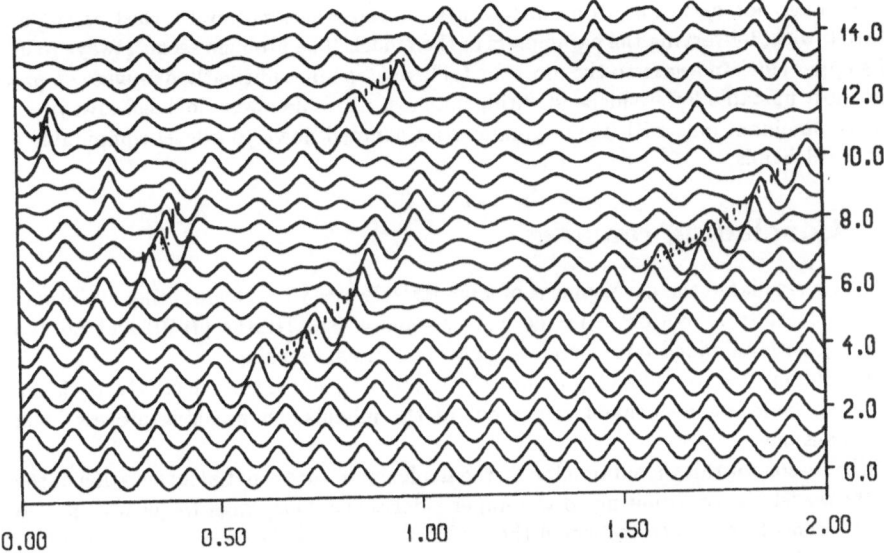

Figure 9: Wave profiles corresponding to the perturbed carrier wave, with breaking.

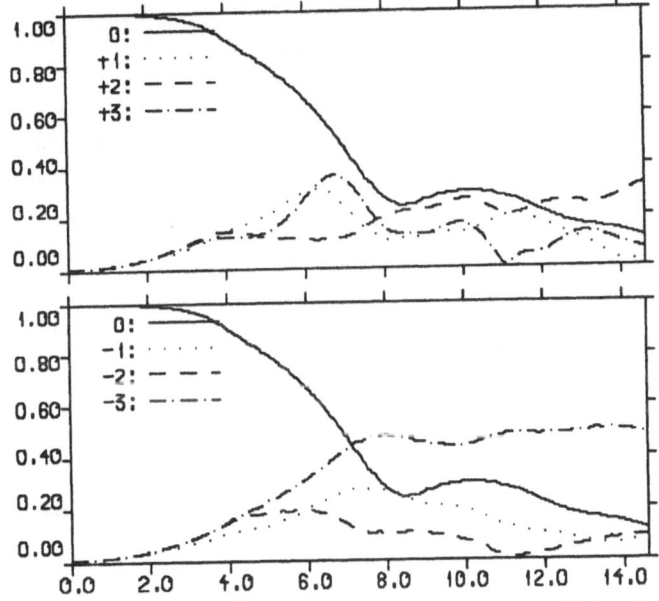

Figure 10: Spectral development for the three upper and lower sidebands, corresponding to the perturbed carrier wave, with breaking.

We have not been able to produce a new carrier wave, as unperturbed as the one we started with, by varying the parameters in the model. The problem seems always to be that the upper sideband next to the most unstable one, is not sufficiently damped compared to the most unstable lower sideband. Still, the most unstable lower sideband always ends up as the dominant one, with the original carrier damped down to a level comparable to the other sidebands.

4 Concluding remarks.

We have simulated the development of a moderately steep wavetrain, using evolution equations correct to fourth order in the wave steepness (see Dysthe (1979) and Lo and Mei (1985)), with an added term simulating effects of wave breaking.

It is found that the breaking damps the developing sidebands selectively, such that the most unstable lower sideband comes out of the modulation–breaking process being the dominant one.

The reason for this, in our opinion, is the tendency towards spatial localization of the part of the wavetrain contributing to the upper sidebands. This "high-frequency" part of the signal seems to concentrate around the steepest portions of the wavetrain, where breaking occurs. This is found both in our simulations and in experimental records (Melville 1983).

REFERENCES

DYSTHE, K. B.   1979   Note on a modification to the nonlinear Schrödinger equation for application to deep water waves. *Proc. R. Soc. Lond.* A **369**, 105–114.

FEIR, J. E.   1967   Discussion: some results from wave pulse experiments. *Proc. R. Soc. Lond.* A **299**, 54–58.

LAKE, B. M., YUEN, H. C., RUNGALDIER, H. & FERGUSON, W. E.   1977   Nonlinear deep-water waves: theory and experiment. Part 2: Evolution of a continuous wave train. *J. Fluid Mech.* **83**, 49–74.

LIGHTHILL, M. J.   1965   Contributions to the theory of waves in nonlinear dispersive systems. *J. Inst. Math. Appl.* **1**, 269–306.

LO, E. & MEI, C. C.   1985   A numerical study of water-wave modulation based on a higher-order nonlinear Schrödinger equation. *J. Fluid Mech.* **150**, 395–416.

MELVILLE, W. K.   1982   The instability and breaking of deep-water waves. *J. Fluid Mech.* **115**, 165–185.

MELVILLE, W. K.   1983   Wave modulation and breakdown. *J. Fluid Mech.* **128**, 489–506.

TRULSEN, K.   1989   *(Master thesis.)* University of Tromsø.

EXTREME WAVES IN LABORATORY GENERATED IRREGULAR WAVE TRAINS

C.T. STANSBERG
Norwegian Marine Technology Research Institute A/S
P.O. Box 4125 Valentinlyst
N-7002 Trondheim
Norway

ABSTRACT.   Large waves in experimental model scale random
wave trains are investigated. Statistical distributions of
crest heights, troughs and peak-to-peak wave heights are
compared to the Rayleigh model and to linear numerical wave
data. Wave heights are seen to agree well with the Rayleigh
distribution. The largest individual waves from the experiment show asymmetric behaviour with higher crests and more
shallow troughs than predicted by the linear theory. This is
seen both from Weibull distribution plots and from wave
profiles of individual large waves in the time domain.
Particle velocities in extreme waves, estimated by linear
transformations of wave elevation measurements, are particularly asymmetric and peaked. Proper design rules for extreme
waves should be further developed. More knowledge on full
scale waves must form a central part of this development.

1. INTRODUCTION.

The proper description of a design sea state will often
depend on the actual application. In many cases, simple
parameters like the significant wave height and the central
or peak wave period are sufficient, while in most cases the
shape of the spectrum is included as well. Based on the
superposition of linear plane wave components, estimates of
the expected largest wave during a storm may then be found by
the Rayleigh model for wave height distribution. This is a
widely accepted method because it often gives reasonable
results for deep water waves, and because it is simple. If
the extreme wave heights, or crest heights, are of particular
importance, however, one has to remember the fact that the
linear wave model assumes small wave amplitudes only. High

A. Tørum and O. T. Gudmestad (eds.), Water Wave Kinematics, 573–589.
© 1990 Kluwer Academic Publishers.

(steep) waves are non-linear, deviating from linear waves in both shape and kinematics /1/. The asymmetric shape with high crests and shallow troughs is the most visual effect, while the increased water particle velocity is perhaps of equally great importance to marine structures. Examples of significantly non-linear large waves in deep water have definitely been observed in the real ocean /2-3/. Thus a design wave or sea state should often also include additional information to the spectrum. This could either be a specification of the non-linearity itself, or simply a specification of the expected maximum wave height.

To include non-linear effects in irregular wave descriptions is, however, a complex matter. The theoretical and numerical tools are very complicated, and the data available from real, irregular ocean waves have not yet resulted in uniquely defined guidelines. The purpose of the present paper is to describe the occurrence and the physical properties of large waves in long-crested irregular deep water wave trains generated in a laboratory wave basin. Thus it is a purely empirical study, where the randomly occuring largest waves in model scale experimental records are "picked out" and considered in some detail. Although laboratory experiments alone cannot give us perfect and complete knowledge on real ocean waves, such data are certainly of some value when combined with and compared to available field and theoretical/numerical data.

2. NUMERICAL AND PHYSICAL MODELLING OF RANDOM WAVE TRAINS.

The simulation of irregular water waves is a comprehensive subject in itself, and will not be the main content of this study. But in order to describe the conditions under which the extreme waves are obtained, we shall briefly summarize the main principles and practical details of the wave simulation carried out for this study. See also ref. /4-6/.

The basic principle used in this work, is the simulation of "random realisations" of specified sea states /4/. Thus a set of finite, but long, arbitrarily selected samples (time windows) of infinitely long and stationary random wave elevation trains with specified spectra were simulated as digitized signals on a computer. These are obtained as linear superpositions of harmonic components, as expressed by eq. (1):

$$x(t_n) = \sum_{k=-k}^{k} N(f_k) \exp[j2\pi f_k \cdot t_n] \quad , \quad n = 1, \ldots, N \qquad (1)$$

$$N(f_k) \equiv A(f_k) \cdot \exp[j\emptyset(f_k)]$$

where
x_n = wave elevation
t_n = time step no. n
f_k = frequency no. k
$N(f)$ = complex Fourier coefficient (2-sided)
$A(f) \equiv |N(f)|$ = spectral amplitude (2-sided)
$\emptyset(f)$ = phase of $N(f)$

The practical calculation of x is carried out by inverse FFT (Fast Fourier Transform). The spectral amplitudes $A_k \equiv A(f_k)$ and phases $\emptyset_k \equiv \emptyset(f_k)$ are modelled as random variables, such that each value A_k is chosen from a Rayleigh distribution, and each value $\emptyset_k$ is chosen from a uniform distribution in $[0, 2\pi]$. This simulates the "random" properties of the realisations.

The signals obtained directly from eq.(1) are referred to as "the linear model" in the analysis to follow. Such time series are then used as input via linear transfer functions to a wavemaker in the 50 m wide and 80 m long MARINTEK Ocean Basin, shown in fig. 1. A large hydraulic, double-flap wavemaker was used to generate long-crested irregular waves (fig. 2). The wavemaker motion can be assumed to be linear. Thus all possible non-linearities in the experimentally recorded wave elevation signals are generated by nature itself in the water region between the wavemaker and the measuring points (located 35m away). The bottom was set at 4.9m depth.

Pierson-Moskowitz and JONSWAP spectra were modelled. See fig. 3, where the input and output shapes are shown for some of the numerical and experimental records. The actual significant wave heights were in the range $\approx$ 0.10 - 0.50 m, and the typical peak wave periods covered the range 1.5 - 2.5s. Table 1 describes all the tests included in this presentation. In the inverse FFT, 4096 frequencies were used, and the generation time step as well as the acquisition sampling period was 0.050s. To obtain long enough repetion periods of the time series, several blocks with "new" sets of random amplitudes and phases were added together in a careful manner. In this study, time series repeating after 16384 points (i.e. 2 blocks) were generated. Thus the records included $\approx$ 400 - 700 individual waves.

Fig. 4 shows 2 short samples of time series from the study,

576

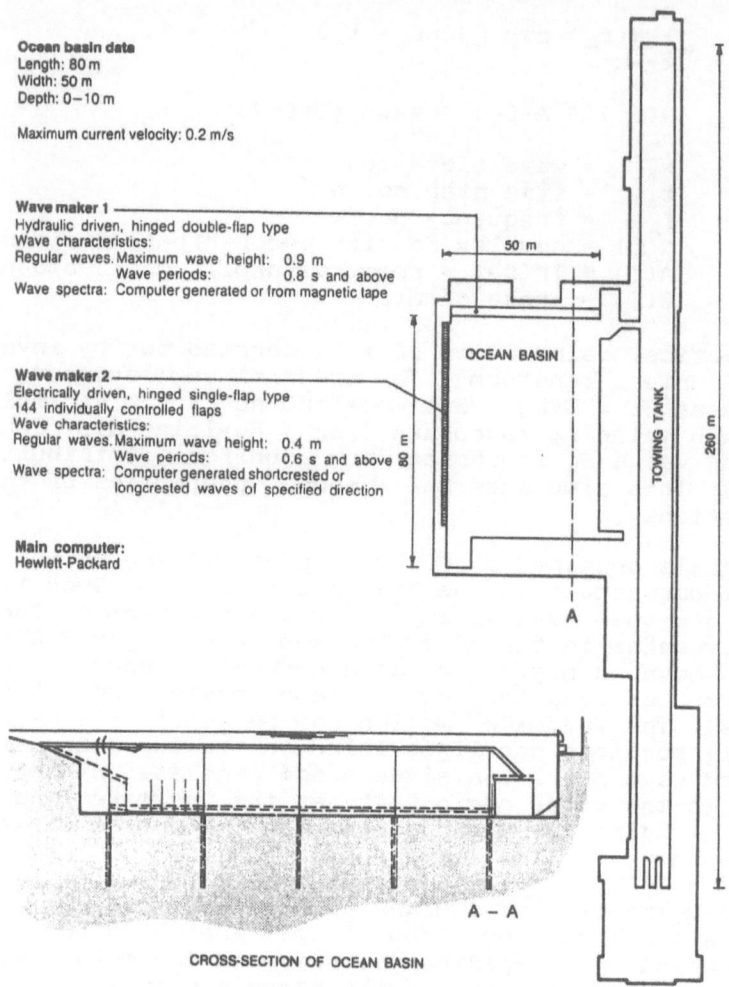

Ocean basin data
Length: 80 m
Width: 50 m
Depth: 0–10 m

Maximum current velocity: 0.2 m/s

Wave maker 1
Hydraulic driven, hinged double-flap type
Wave characteristics:
Regular waves. Maximum wave height: 0.9 m
                    Wave periods: 0.8 s and above
Wave spectra: Computer generated or from magnetic tape

Wave maker 2
Electrically driven, hinged single-flap type
144 individually controlled flaps
Wave characteristics:
Regular waves. Maximum wave height: 0.4 m
                    Wave periods: 0.6 s and above
Wave spectra: Computer generated shortcrested or
                    longcrested waves of specified direction

Main computer:
Hewlett-Packard

50 m

OCEAN BASIN

TOWING TANK

260 m

80 m

A

A – A

CROSS-SECTION OF OCEAN BASIN

Fig.1.   The MARINTEK Ocean Laboratories. In this study,
          Wavemaker 1 was used. Measurements were made close
          to the centre of the Ocean Basin.

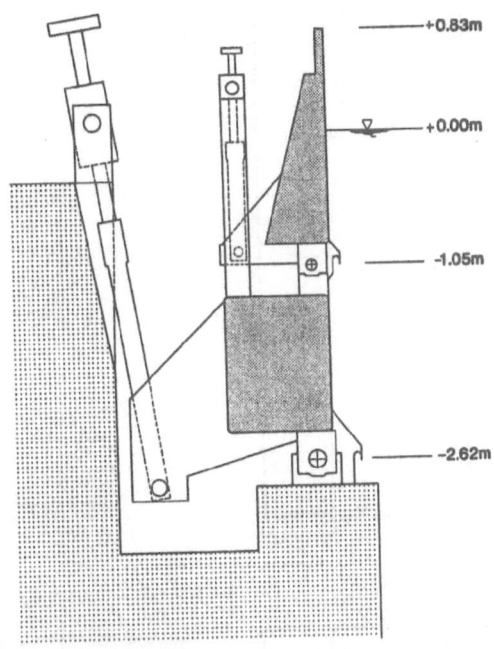

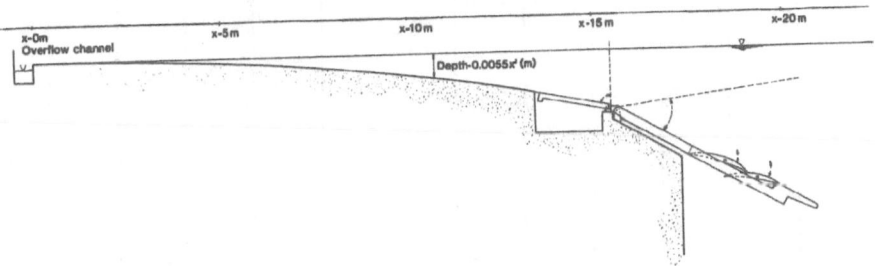

Fig.2. Cross-sections of the hydraulic double-flap wavemaker and of the parabolic beach.

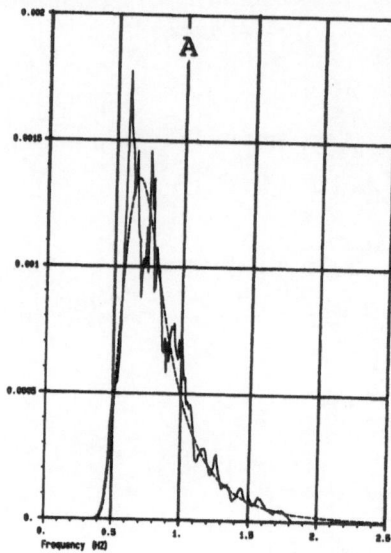

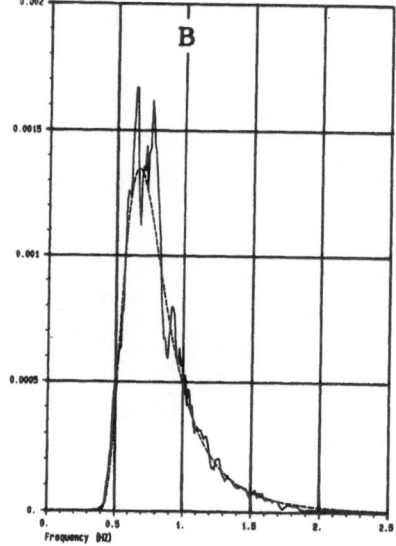

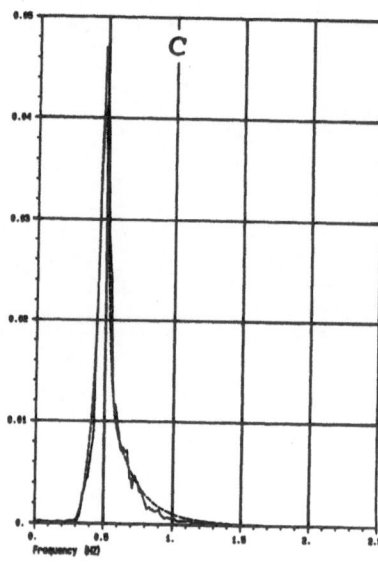

Fig.3. Power spectra of wave records, 3 examples:
A) Linear numerical wave, Pierson-Moskowitz spectrum
B) Experimental wave, moderate Pierson-Moskowitz spectrum
C) Experimental wave, JONSWAP storm spectrum
Dotted lines indicate input spectra.

TABLE 1. A list of test runs included

Test no.	Peak Period Tp [s]	Significant wave height Hmo [m]	JONSWAP peakedness Gamma	No. of waves N	Further comments
244	1.50	0.10	1.0	694	Numerical wave
338	1.50	0.10	1.0	686	Experimental wave, moderate
339	2.00	0.20	1.0	521	Experimental wave, moderate
344	1.50	0.18	3.3	569	Experimental wave, storm
340	2.00	0.34	3.3	441	Experimental wave, storm
342	2.50	0.53	3.3	354	Experimental wave, storm

TABLE 2. Statistical parameters of crests and wave heights.

Test no.	St.dev. σ [m]	Sign. crest value A-1/3 [m]		Sign.wave height H-1/3 [m]		Max. crest value Amax [m]		Max wave height hmax [m]	
		Measur.	Rayleigh	Measur.	Rayleigh	Measur.	Rayleigh	Measur.	Rayleigh
244	0.025	0.050	0.050	0.097	0.101	0.106	0.094	0.195	0.188
338	0.026	0.054	0.052	0.101	0.103	0.127	0.098	0.207	0.196
339	0.049	0.103	0.098	0.193	0.196	0.234	0.181	0.358	0.362
344	0.046	0.104	0.092	0.187	0.184	0.243	0.177	0.376	0.344
340	0.085	0.199	0.169	0.342	0.338	0.376	0.309	0.594	0.618
342	0.133	0.308	0.266	0.541	0.532	0.560	0.477	0.964	0.954

"Sign." means "Significant" (the average of the highest 1/3).
"Rayleigh" means estimate based on the Rayleigh model.

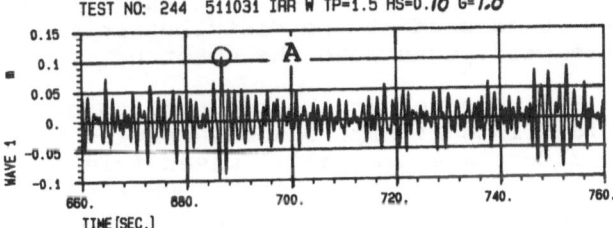

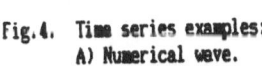

Fig.4. Time series examples:
A) Numerical wave.
B) Experimental wave.

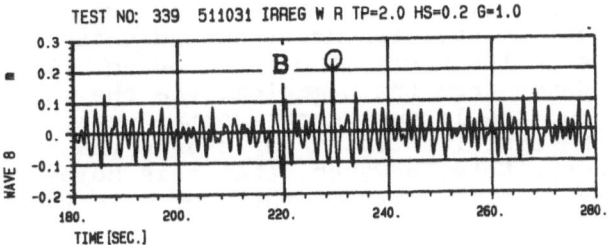

one numerical and one experimental. The following analysis is concentrated on the occurence and properties of the largest of the waves. Typical examples are indicated on the figure. These are randomly occuring in the wave trains, since the phases of the spectral amplitudes are random (i.e. in no manner arranged in order to produce particularly large waves). The comparison between the numerical and the experimental extreme wave properties is of particular interest in the study. Fig. 4 indicates slightly more assymetric behaviour in the physical than in the numerical model. This tendency will be a central point in the analysis.

3. PROBABILITY DISTRIBUTIONS OF WAVE CRESTS, TROUGHS AND HEIGHTS.

Large or extreme waves in random wave trains are statistically unstable, which means that large uncertainties are connected with single events of this type. Different finite-record realisations of the same sea state gives a significant variability in the maximum wave crest etc. This results from general statistical properties of random signals. In order to obtain statistically reliable information on the occurence of these properties, we have processed the wave elevation records by a mean-crossing analysis, reducing the records to sets of crest, through and peak-to-peak values. Probability distributions of these values are plotted with Weibull axes and compared to the Rayleigh distribution model:

$$P(A \geq a) = \exp \left[- a^2 / (2\sigma^2) \right] \qquad (2)$$

where $\sigma \equiv$ the standard deviation of the corresponding time series. For linear waves, the Rayleigh model is normally assumed to predict the distribution of mean-crossing peaks quite well, even when the spectrum is not very narrow /7-8/.

The presentation of the probability distributions is here divided into 2: Fig. 5 shows results for numerical waves and for moderate experimental waves (Pierson-Moskowitz spectra). The typical steepness here is Hmo/L(Tpeak) ≈ 0.025, where Hmo = significant wave height, L = wave length and Tpeak= peak wave period. Fig. 6 shows results for the experimental "storm" waves, with JONSWAP spectra and typical steepness ≈ 0.05. The Rayleigh model is indicated as a whole straight line in the plots, and based on eq. (2). We see a clear tendency from the numerical via the moderate to the "storm" waves: the largest experimental waves get more asymmetric, with significantly larger crests than predicted by linear theory. Note, however, that this holds for the largest waves only. The majority of the crests in the records, in

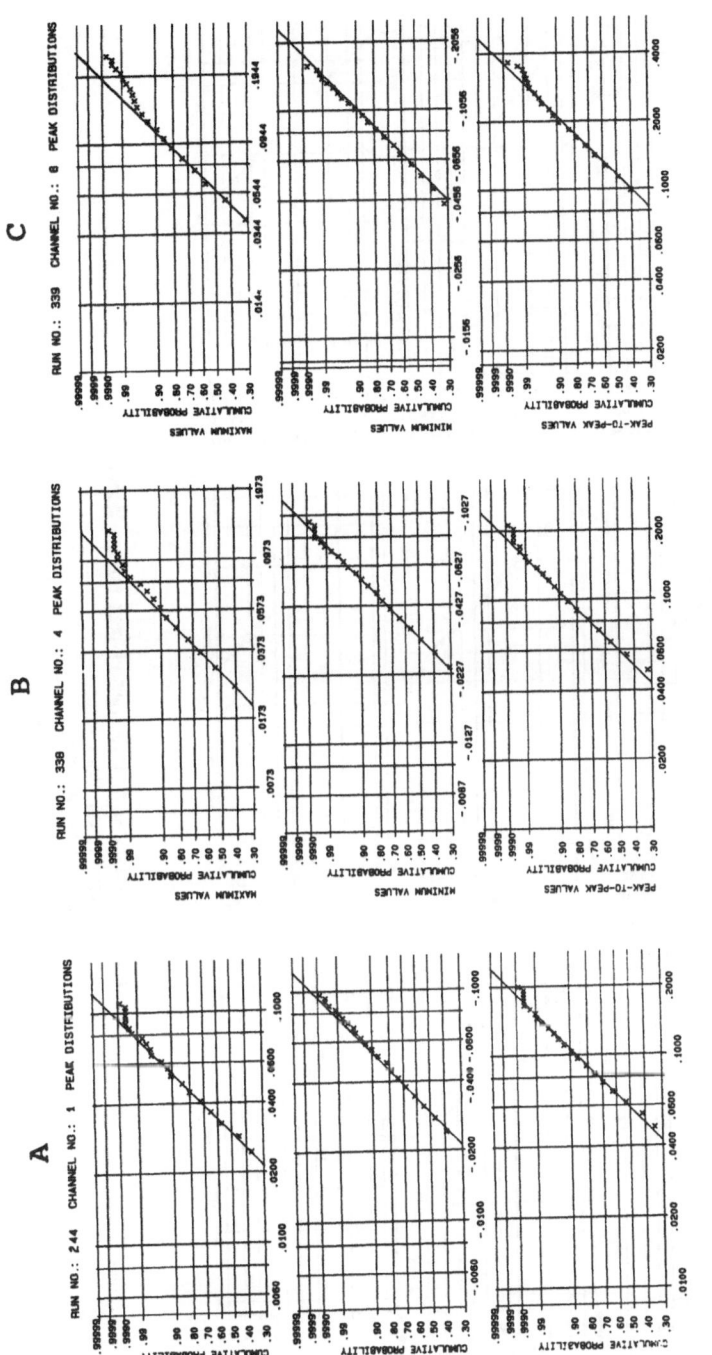

Fig.5. Statistical distributions of peak values, Weibull axes.
A) Linear numerical wave
B) Moderate experimental waves   Tp=1.5s   Hmo=0.10m
C) Moderate experimental waves   Tp=2.0s   Hmo=0.20m

582

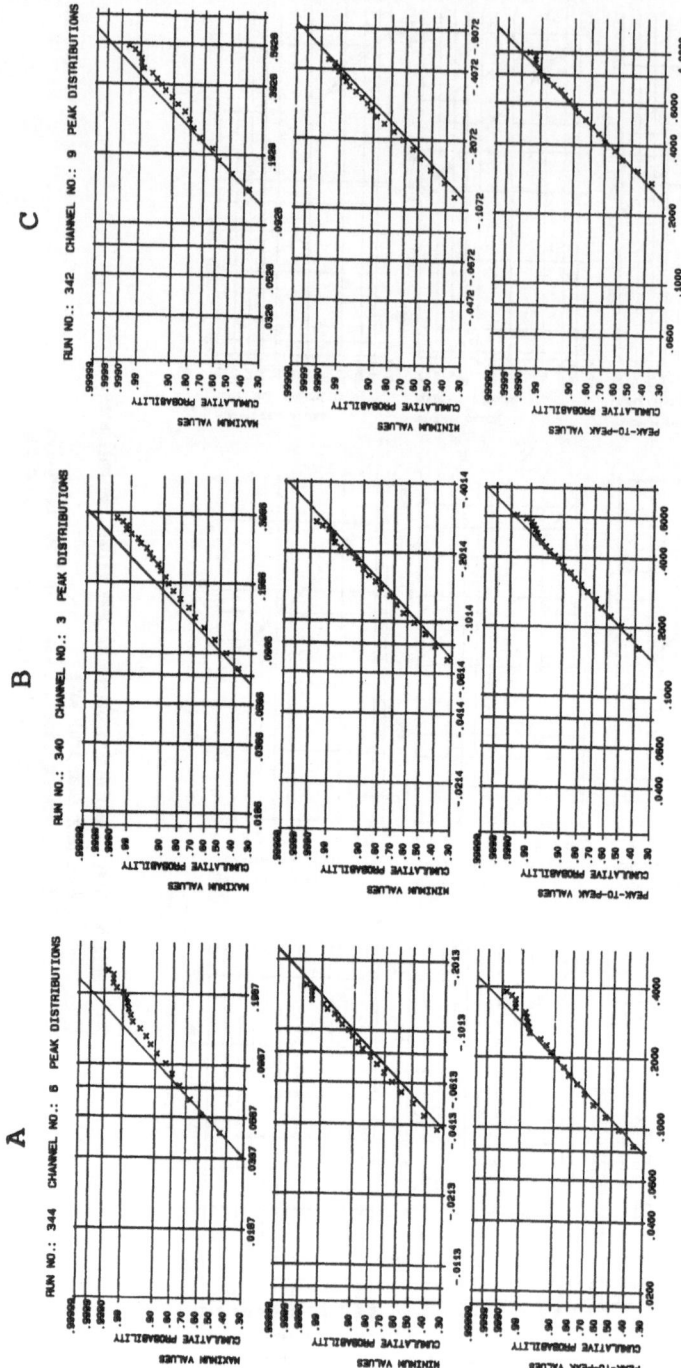

Fig.6. Statistical distributions of peak values, Weibull axes.
Experimental storm waves.
A) Tp=1.5s Hmo=0.18m
B) Tp=2.0s Hmo=0.34m
C) Tp=2.5s Hmo=0.53m

particular for the moderate sea state, do not deviate
considerably from the simple model. Table 2 also shows this.
We see in particular from this table that for the moderate
waves, the significant crest values (A-1/3) are very close to
the linear values. It is also interesting to note the fact
seen from both figures 4-5 and table 2 that the largest peak-
to-peak wave heights compare very well with the linear
prediction.

The fact that the numerical peak values compare well with the
expected Rayleigh curve means that the experimental
deviations from this curve may be interpreted as mainly
arising from physical non-linear processes between the
wavemaker and the measuring points.

4. WAVE PROFILES AND OTHER CHARACTERISTICS OF SELECTED LARGE WAVES.

From the statistical analysis in section 3 we observe that
the asymmetric behaviour of the largest waves is reasonably
representative for the group of large waves in these records.
The physical properties of selected single extreme waves are
therefore considered to be of interest, and the main part of
Following analysis is simply a presentation of recorded wave
elevation profiles in the time domain. A few results from
studies of the wave-group-bound long waves (the "set-down"
effect) and of the particle velocitiy estimated from the
measured wave elevation, are also included.

Fig. 7 shows the time-domain profiles of the largest waves in
the numerical and the 2 "moderate" experimental records. It
is clearly observed that the 2 experimental cases show more
asymmetric behaviour and more peaked crest profiles than the
numerical case. This effect is confirmed by the "storm" waves
in fig. 8, although we see that deep troughs may still occur
in large experimental waves. The "assymetry factor", defined
as the ratio between the maximum crest A and the connected
peak-to-peak value H, is definitely a variable quantity,
which also depends on whether the wave height is defined as
an up-crossing or down-crossing value. This study indicates
that a factor in the range to 0.6 - 0.65 is not unusual for
the extreme waves. The study also indicates that the maximum
crests often occurs in the first part of a wave group, with
the succeeding trough normally being deeper than the preceding one.

Fig. 9a shows the non-linear long wave component ("set-down")
of the wave shown in fig. 7c. The corresponding square wave
envelope signal (based on the slowly varying Hilbert envelope

584

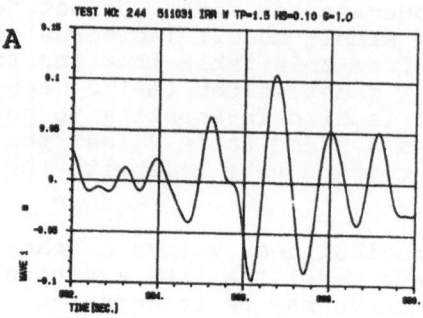

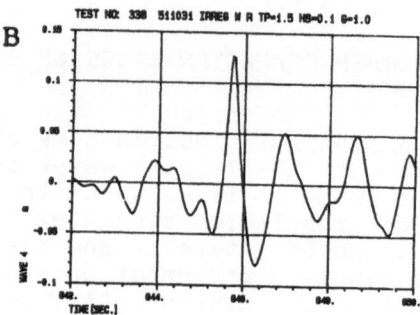

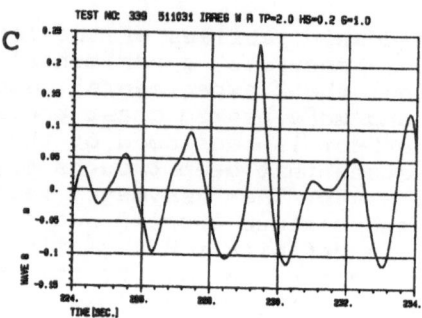

Fig.7. Time domain extreme wave profiles.
A) Linear numerical wave record.
B-C) Moderate experimental wave records

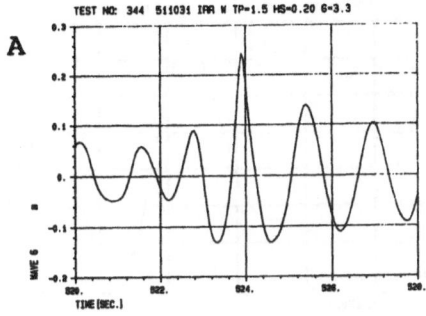

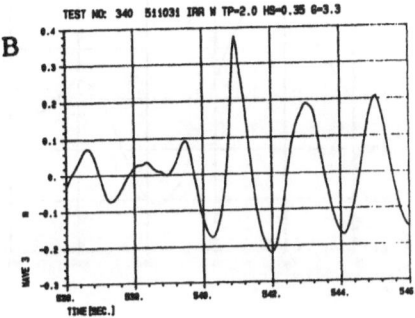

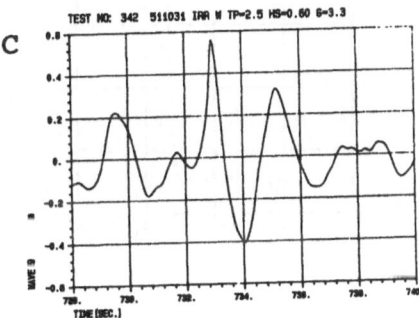

Fig.8. Time domain extreme wave profiles, storm wave records.

586

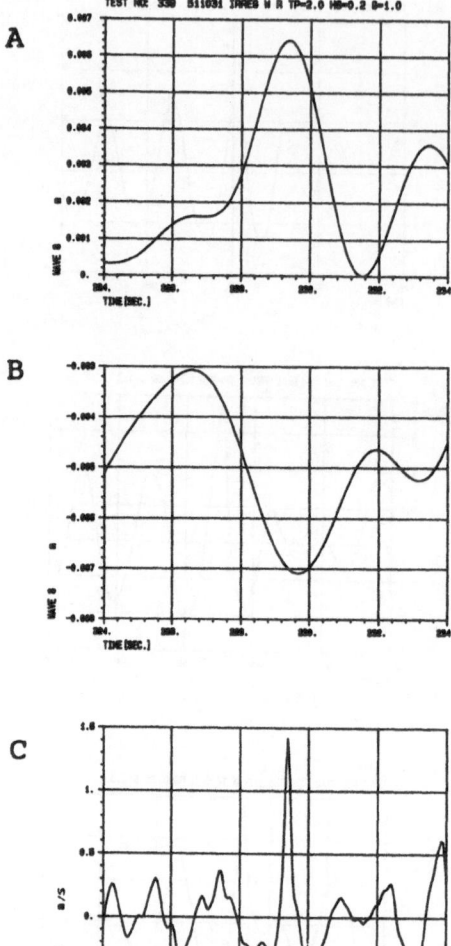

Fig.9. Additional properties of the wave in fig. 7C:
A) Wave envelope
B) Long wave
C) Estimated horizontal particle velocity.

/9/) is shown in fig. 9b. A certain correlation is observed
petween this and the long wave, as predicted by several
authors, e.g. in /10/. Studies of long waves in the other
experimental waves in figures 7 and 8 have further qualitatively confirmed this effect.

Fig. 9c shows an estimate of the horizontal water particle
velocity in the wave shown in fig. 7c. This estimate is based
on the superposition of a circular orbital motion of each
frequency component, for which the amplitude and phase is
found by a linear transformation of the measured wave
elevation to velocity around still water level. This simple
estimate certainly has its limitations when the velocity over
still water level is concerned, and should be tested against
real velocity measurements. But we believe that the strongly
asymetric behaviour expressed in the figure has some real
meaning and should be noted. For large numerical waves (ex.
Fig. 7a), such particular velocity asymmetrics are not
observed, since the numerical model in eq. (1) includes no
systematic asymmetry in the sign of the signal.

5. CONCLUSIONS.

S random wave trains, of which 5 laboratory generated and 1
numerical, have been analysed with respect to the occurence
and properties of their largest wave crests and wave heights.
It is observed that while the numerical data, based on a
linear wave model, follow the Rayleigh distributions quite
well for both crests and wave heights, the steepest of the
experimentally generated waves show an assymetric behaviour.
Thus for individual waves with steepness H/L larger than ≈
0.04, the crests are usually higher than predicted by the
Rayleigh theory. They are also significantly more peaked,
which in turn leads to extra high particle velocity predictions. The experimental wave _heights_ however, seem to
follow the linear (Rayleigh) model very well, due to shallow
troughs. Non-linear group-bound long waves, although small in
deep water, slightly reduce the experimental asymmetri.

An asymmetric behaviour of experimental steep waves is in
general an expected effect, both from a theoretical point of
view /1/, and from recent full scale measurements with fixed
sensors /2-3/. We believe that the general information in the
results presented in the previous sections is a reliable
contribution in the development of better extreme wave design
conditions. Some points, however, should be further investigated in order to make conclusions from model scale extreme
wave simulations even more reliable. First, possible scale
effects in non-linear extreme waves are not clearly enough

documented. Next, the 3-5 percent amplitude reflection from the laboratory beach may have a slight influence. Furthermore, the generation of "genuine" non-linear waves may be improved by including non-linear wavemaker transfer functions. Finally, the effects of wind and current are probably not negligible in a complete wave description. Whether or not these points are significant, should be clarified by more complete information from full scale data.

6. REFERENCES.

1.   Any serious textbook on water waves, e.g.:
     Kinsman, B., "Wind Waves", Prentice-Hall, New Jersey,
     U.S.A., 1965.

2.   Kjeldsen, S.P.K., "Breaking Waves", presented at the
     NATO Advanced Research Workshop Water Wave Kinematics,
     Molde, Norway, May 1989.

3.   Sand, S.E., et al., "Freak Wave Kinematics", presented
     the NATO Advanced Research Workshop Water Wave Kine-
     matics, Molde, Norway, May 1989.

4.   Stansberg, C.T., "On the Use, Modelling and Analysis of
     Irregular Water Surface Waves", Proceedings, Vol. 2, the
     8th OMAE Conference, The Hague, Netherlands, March 1989.

5.   Funke, E.R. and Mansard, E.P.D., "A Rationale for the
     use of the Deterministic Approach to Laboratory Wave
     Generation", Proceedings, IAHR seminar on Wave Analysis
     and Generation in Laboratory Bains, Lausanne, Switzer-
     land, August 1987.

6.   Stansberg, C.T. and Kjeldsen, S.P.K., "MARINTEK Wave
     Catalogue 1988", MARINTEK Report No. 511031.00,
     Trondheim, Norway, 1988.

7.   Cartwright, D.E. and Longuet-Higgins, M.S., "The
     Statistical Distribution of Maxima of a Random
     function", Proceedings of the Royal Society of London,
     Series A, Vol. 237, p. 212, 1956.

8.   Chakrabarti, S.K. and Cooley, R.P., "Statistical
     Distribution of Periods and Heights of Ocean Waves",
     Journal of Geophysical Research, Vol. 82, No. 9, p.
     1363, 1977.

9. Stansberg, C.T., "Statistical Analysis of slow-Drift Responses", ASME <u>Journal of Energy Resources Technology, Vol. 105</u>, pp. 310-317, 1983.

10. Barthel, V. et al., "Group Bounded Long Waves in Physical Models", <u>Ocean Engineering, Vol. 10</u>, No. 4, 1983.

Advances in Impulsively Generated Water Waves

by
Bernard Le Mehaute, Shen Wang and Tarang Khangaoankar
Rosenstiel School of Marine and Atmospheric Science
University of Miami
Miami, Florida 33149

and

Douglas Outlaw
Waterways Experiment Station
Vicksburg, Mississippi 39180

Keywords/Abstract: Impulsively generated water waves/nonlinear waves/
underwater explosion/tsunami/drops

Some recent development on the analysis of surface water waves
generated by locally impulsive disturbances is presented in this paper.
While the classical Cauchy-Poisson approach to the problem has various
practical applications, the theoretical analysis has been limited to
either leading wave or the trailing wave solution in the farfield.
Direct numerical integration except for very simple form of disturbance
has difficulties due to the highly oscillatory kernel function of the
Cauchy-Poisson integral. The new development shows that the method of
Fourier transform may remove all the difficulties which have been
faced either analytically or numerically and the solution is uniformly
valid for both leading and trailing waves, deep and shallow water and
for either near or far field, within the linear definition. An inverse
application of this method may also determine the infield initial
disturbances at the origin. Nonlinear approximate solution to the
same problem in the nearfield, where linear solutions are not exactly
valid is presented. Appropriate KdV equation in radial coordinate
system are solved linearly using Fourier transform techniques but
nonlinear corrections are added using implicit finite difference
scheme. Wave data from experiments of TNT explosion and plate dropping
in shallow water are presented. Substantiation of the theory is
demonstrated by comparing the theoretical calculation with the experimental results.

A. Tørum and O. T. Gudmestad (eds.), Water Wave Kinematics, 591–608.

INTRODUCTION

This paper deals with the classical Cauchy-Poisson problem (Stoker, 1957), i.e. with water waves generated by a local disturbance of the free surface in an unbounded body of water. The subject matter has been abundantly investigated in the past with regard to tsunami and explosion generated water waves (Kranzer and Keller, 1951; Le Mehaute, 1970). The classical theoretical approaches were essentially limited to the linear approximation and to the far field, based on the stationary phase approximation. Accordingly the solutions were only valid for the trailing waves in deep or intermediate water depth. The leading wave solution was resolved by Kajiura (1963) in the form of an Airy function; but it was not valid for the trailing waves (Whitham, 1974). The specific limitations of each method, have been detailed by Wang et al. (1987) and will not be repeated here.

The purpose of this paper is to present a review of advances which have been done lately in regard to the propagation of axysymmetrical impulsively generated water waves. This includes

1) A general approach to the linear solution which is valid regardless of the definition of the initial disturbance, for the infield, nearfield and the farfield as well, for the leading wave and the trailing waves, in deep or shallow water, and (with some limitations) on a nonuniform bathymetry.

2) The inverse problem is also resolved. Given an impulsive generated wave as function of time or a wave record, the initial disturbance which causes this wave is determined.

3) A nonlinear solution valid for the nearfield in shallow water.

1. Linear Solution of Transient Wave by FFT

The linear solutions are defined in dimensionless terms with respect to a length L, which is either the water depth $D*(L = D*)$ or a radius of the initial disturbance $(L = R*)$, in a radial coordinate system r, z, θ with axisymmetry $(\partial/\partial\theta = 0)$. The vertical axis z is upwards from the SWL. The solution is defined by a potential function $\phi(r,z,t)$ which satisfies continuity. The general solution is given by a linear superposition of harmonic components of frequency σ and wave number k, as (Stoker, 1957)

$$\phi(r,z,t) = \int_0^\infty J_0(k_r) \frac{\cosh k(1+z)}{\cosh k} [A(k) \sin \sigma t + B(k) \cos \sigma t]k \, dk$$

$$\eta = -\phi_t|_{z=0} \tag{1}$$

The dispersion relationships are:

$$\sigma^2 = k \tanh k \tag{2}$$

$$\sigma = k(1 + \tau k^2) \tag{3}$$

in the case of gravity wave, and in the case of capillary-gravity wave respectively. In the latter case, the deep water assumption is justified. Also

$$\tau = \tau^*/\rho g \, R^{*2} \tag{4}$$

and τ^* is the surface tension, ρ the density and g the gravity acceleration.

The coefficients $A(k)$ and $B(k)$ are obtained from the initial free surface conditions at $t = t_0 = 0$. These conditions are defined by a localized free surface disturbance $\lambda(r_0)$ defined as

1) a free surface velocity:

$$\lambda(r_0) = w(r_0,t_0) = - \phi_{tt}(r_0,t_0) \tag{5}$$

2) a free surface elevation:

$$\lambda(r_0) = \eta_0(r_0,t_0) = - \phi_t(r_0,t_0) \tag{6}$$

3) an impulse: $\lambda(r_0) = I(r_0,t_0) =$

$$p(r_0) \, \delta(t - t_0) = \phi(r_0,t_0) \tag{7}$$

Inserting these into (1) at $t = t_0 = 0$ and $z = 0$, gives sucessively

$$\lambda(r_0) = \int_0^\infty k \, dk \, J_0(kr) \begin{array}{c} \sigma^2 \, B(k) \\ - \sigma A(k) \\ B(k) \end{array} \tag{8}$$

which could be inverted by virtue of the Fourier-Bessel Theorem, so that, introducing the notation function $H(k)$

$$H(k) = \begin{array}{c} B(k)\sigma^2 \\ -A(k)\sigma \\ +B(k) \end{array} = \int_0^R \lambda(r_0) \, J_0(kr_0) \, r_0 \, dr_0 \tag{9}$$

Replacing dk by $d\sigma/V(k)$ where V is the group velocity, one has finally after some arithmetic

$$
\begin{matrix} \eta^W \\ \eta^\eta \\ \eta^I \end{matrix} \quad (r,t) = 2/\pi \int_{o}^{\infty} F(\sigma) \begin{matrix} \sin \sigma t \\ \cos \sigma t \\ \sin \sigma t \end{matrix} \, d\sigma \tag{10}
$$

where w, η, I refer to conditions 5,6,7, respectively, and

$$
F(\sigma) = \begin{matrix} \sigma^{-1} \\ 1 \\ -\sigma \end{matrix} \quad \pi/2 \, J_0(kr) \, \frac{k \, H(k)}{V(k)} \tag{11}
$$

Note eq. (9) is independent of the physical meaning of $\lambda(r_0)$, but depends only upon its mathematical definition. It is a Hankel Transform which can be integrated analytically or in the general case numerically depending upon the definition of $\lambda(r_0)$. Many other pertinent mathematical definitions for $\lambda(r_0)$ can be found in Le Mehaute et al. (1987a). The wide variety of analytical solutions to $H(k)$, combined with the possibility of three physical interpretations, and their linear superposition with an arbitrary multiplier, allows us to cover practically all physical problems of interest. If not, note since the integral $H(k)$ (eq. 9) is finite, its numerical integration is straightforward.

Eq.(10) is recognized as an inverse sine or cosine Fourier Transform of $F(\sigma)$, which can be obtained by (FFT^{-1}) so that

$$
\eta(r,t) = FFT^{-1} [F(\sigma] \tag{12}
$$

Defined as such, the solution is absolutely general and not subjected to the limitations of the previous approaches. An example is presented in Figure 1, in shallow water case (R large). The original disturbance is a hyperbolic dome. It is compared to the corresponding stationary phase solution, which is evidently not valid for the leading waves.

Analytical solutions (eq. 10) can also be found directly in many cases, where $H(k)$ is analytically solvable.

1) When r = 0, then $J_0(kr) = 1$, and since dispersion requires time and distance, $\sigma \cong k$. A large number of solutions of this type are found in Le Mehaute et al. (1987a), giving an insight on the wave movement at the origin.

2) When r is large, and for sufficiently large value of k, then J_0 can be replaced by the first term of its asymptotic value, and the method of the stationary phase is valid, as done in the past.

3) In the case where k is small (leading wave), then σ is replced by $(k - k^3/6)$. An Airy integral is obtained (Whitham, 1974).

Finally in the case of a 3-D bathymetry, the wave motion no longer exhibits cylindrical symmetry and the wave pattern is affected by refraction and shoaling.

If the bottom slope is gentle, and the depth varies little over the size of the original disturbance, the FFT method is still applicable, provided the source size R is small compared to the distance from the origin, and J_0 in eq. (11) is replaced by (Le Mehaute et al., 1987b)

$$S(k) = K_R(\sigma) \, K_S(\sigma) \, [\frac{2}{\pi k \, s(\sigma)}]^{1/2} \cos (\int_0^s d(ks) - \frac{\pi}{4}) \tag{13}$$

where K_R is the refraction coefficient and K_s the shoaling coefficient between the origin and a point at a distance from the origin. $s(\sigma)$ is the distance along the wave ray between the origin and that point. The movement near the origin can also be resolved along a wave ray perpendicular to the bottom contours of a plane bathymetry. This wave ray is straight and the wave on both sides exhibits perfect symmetry, $J_0(kr)$ in eq. (11), is then replaced by

$$S(k) = K_R \, K_S \, J_0 \, [\int_0^s d(ks)]. \tag{14}$$

Corresponding results obtained by FFT are presented in Figure 2.

The results of the present method have also been confronted with experimental results obtained on a 3-D scale model at CERC. An initially circular transient wave was generated by a programmable wave paddle (Le Mehaute et al., 1988c). Fig. 3 is a typical comparison of the obtained results. Note that the experimental wave is steeper that the linear theoretical wave, due to nonlinear convective effects. This will be the subject of a following section.

2. Original Disturbance From Wave Records

Finally the inverse problem can also be resolved by FFT, given the history in time $\eta(r_1,t))$ at $r = r_1$. Then

$$F(\sigma) = \int_0^\infty \begin{matrix} \eta^W(r,t) \sin \sigma t \\ \eta^\eta(r,t) \cos \sigma t \\ \eta^I(r_1,t) \sin \sigma t \end{matrix} \ dt \tag{15}$$

Therefore, inserting equation (11), one obtains

$$H(k) = 2s \ \pi s \ \ CFFT^+ \ [\eta(r,t)] \ \frac{V(k)}{k} \ \frac{1}{J_0(k_r)} \ \begin{matrix} \sigma \\ 1 \\ -\sigma^{-1} \end{matrix} \tag{16}$$

and by applying the Fourier–Bessel Theorem to equation (9)

$$\lambda(r_0) = \int_0^\infty k \ dk \ J_0(k \ r_0) \ H(k)$$

If instead of carrying out a SFFT or a CFFT, one carries a total FFT, then one is able to separate the relative effect of two types of initial disturbances, either η and w, or η and I.

Indeed, adding the three effects yields

$$\eta(r,t) = \int_0^\infty J_0(kr)[H_\eta(k)\cos \sigma t + (H_w(k)\sigma^{-1} + H_I(k)\sigma)\sin \sigma t]k \ dk \tag{17}$$

So that defining

$$F_1(\sigma) = \frac{\pi k J_0(kr)}{V(k)} H_\eta(k) \tag{18}$$

$$F_2(\sigma) = \frac{\pi k J_0(kr)}{V(k)} [\frac{H_w(k)}{\sigma} + H_I(k)\sigma] \tag{19}$$

and

$$A(\sigma = F_1(\sigma) + i \, F_2(\sigma) \tag{20}$$

Then $\eta(r,t)$ can be defined by a Fourier Transform such as

$$\eta(r,t) \, \frac{1}{\pi} \, \text{Re} \int_0^\infty A(\sigma)e^{i\sigma t} \, d\sigma \tag{21}$$

Inversely, at a fixed distance r from the origin:

$$A(\sigma) = \int_0^\infty \eta(t)e^{-i\sigma t} \, dt \tag{22}$$

Then

$$F_1(\sigma) = \text{Re} \, A(\sigma) \tag{23}$$

$$F_2(\sigma) = I_m \, A(\sigma)$$

$F_1(\sigma)$ allow the determination of $H_\eta(k)$ straightforwardly from equation (13). In the case of $F_2(\sigma)$ a choice based on other (physical) consideration has to be done concerning the relative influence of the impulse vs velocity as initial conditions. For example, if one assumes that $I(r_0,o) = 0$, then $H_w(k)$ is determined straightforwardly from equation (14). Applying the Fourier-Bessel theorem to equations (9), inserting the expressions for $H(k)$ from (13) and (14), and noticing that $J_0(kr)$ is the modulating functions for $F(\sigma)$, so that the envelopes

$$E[F(\sigma)] = \frac{F(\sigma)}{J_0(kr)} \tag{24}$$

yields finally

$$\eta_0(r_0) = \int_0^\infty \frac{1}{\pi} \, E[F_1(\sigma)] \, V(k) \, J_0(kr) \, dk \tag{25}$$

$$w_s(r_0) = \int_0^\infty \sigma \, E[F_2(\sigma] \, V(k) \, J_0(kr) \, dk. \tag{26}$$

This method has been applied to wave records obtained from a TNT explosion in a shallow pond by varying systematically the water depth, the yields and the depth of burst. The numerical results, as well as matching theoretical formulations which can be Hankel transformed are shown on Figure 4. Then, in the present case it is found that

$$\eta_0(ro) = A_\eta [\frac{4}{3} (\frac{r_0}{R_\eta})^2 - \frac{1}{3} (\frac{r_0}{R_\eta})^4 - 1] \tag{27}$$

$$w_s(r_0) = A_w \exp [- (\frac{r_0}{R_w})^2] \tag{28}$$

where $A_\eta = 0.402$, $R_\eta = 1.25$, $A_w = 0.803$, $R_w = 0.8$. The corresponding Hankel Transforms are

$$H_\eta(k) = \frac{4A_\eta}{k^2} J_4(kR_\eta) \tag{29}$$

$$H_w(k) = \frac{A_w R^2}{2} \exp [- (\frac{kR_w}{2})^2] \tag{30}$$

A comparison of the theoretical model with experiments is shown on Figure 5. The remaining discrepancies can partly be explaiend by a wave reflection of the leading wave interferring with the trailing direct wave, small water depth. The four coefficients A_η, R_η, A_w, R_w are then parameterized as function of yields, water depth, and depth of burst.

3. Nonlinear Impulsive Wave in Shallow Water by Split-Step Fourier Algorithm.

When a disturbance of large size $R(R = R^*/D^*)$ and large amplitude occurs, the nonlinear convective effect can no longer be neglected. This is the case of explosion in shallow water. The hydrodynamics problem is then divided into four domains.

1) Near the explosion, the bubble expansion and collapse are resolved by hydrocodes, with a compressible phase followed by a non-compressible phase after the schock wave separate away from the free surface. The compressible phase is treated as a nondissipative potential flow by the Boundary Integral Equation Method (BIEM).

2) Then dissipative "base surge" appears, i.e. a circular bore travelling outwardly, and decaying rapidly with distance.

3) The bore give rise to a nonbreaking dispersive transient wave pattern of KdV type.

4) As the wave amplitude keeps decaying with time and distance, the linear solution such as described in the previous section becomes valid.

The Boussinesq equations in a radial coordinate system which translates the third problem have been established by Wu (1975). These are in

$$\eta_t + \frac{1}{2} \left[((\epsilon\eta) + 1) \; \phi_r \right]_r = 0 \tag{31}$$

$$u_t + \epsilon u \, u_r + \eta_r = \frac{\mu^2}{3} \left[\frac{1}{r}(r \cdot \phi_r)_r \right]_{rt} \tag{32}$$

for which a numerical scheme can be established. Here, $\epsilon = A/D <$ 1, $\mu = k\,D < 1$, and $U = \epsilon/\mu^2 \cong 0$ (1). However, the problem can be more conveniently resolved by transforming eq. (14) into their KdV equivalent and uses the split-step FFT method.

The KdV equation for free surface water waves was first derived by Miles (1978). In dimensional form the KdV equation may be expressed in two canonical forms.

$$\eta_t + \sqrt{gh} \; (1 + \frac{3\eta}{2h}) \; \eta_r + \frac{\eta}{2r} \sqrt{gh} + \frac{h^2}{6} \sqrt{gh} \; \eta_{rrr} = 0 \tag{33}$$

and

$$\eta_r + \frac{1}{\sqrt{gh}} \; (1 - \frac{3\eta}{2h}) \; \eta_t + \frac{\eta}{2r} - \frac{h^2}{6} \; \eta_{ttt} \; \frac{1}{(\sqrt{gh})^3} = 0 \tag{34}$$

The above two equations can be solved by finite difference method. However for numerical stability of the nonlinear equations, the increments Δr, and Δt have to be made small. Since the nonlinear region for explosion generated waves in shallow water may be expected to extend over large distances, FDM to solve the KdV equations is cost prohibitive.

Tappert [1975] proposed Split Step Fourier Method (SSFM) for numerical solutions of KdV type equations. The method has since found an extensive applications in the field of acoustics. We have applied the SSFM for the water wave problem for the first time.

The essence of the solution is to propagate the wave using only the linear part of the equations, then add a nonlinear correction to the time history for every step advanced in r, by solving the non-linear part of the equation.

Step 1 - Linear Propagation:

Taking only the linear terms, the dispersive equations is

$$\eta_r + \frac{\eta_t}{\sqrt{gh}} + \frac{\eta}{2r} - \frac{h^2}{6(\sqrt{gh})^3} \eta_{ttt} = 0 .$$ (35)

using Fourier analysis,

$$\eta(r + \Delta r, t) = \sqrt{\frac{r}{r + \Delta r}} + [\frac{1}{2\pi} \int_{-\infty}^{\infty} \overline{\eta}(r, \sigma) e^{-i(\alpha)} e^{i\sigma t} d\sigma]$$ (36)

Time history at increment of Δr is given by,

$$\eta(r + \Delta r, t) = \sqrt{\frac{r}{r + \Delta r}} F^{-1} [\eta(\overline{r}, \sigma) e^{-i\alpha}]$$ (37)

where

Δr = increment in range

$$\overline{\eta}(r, \sigma) = \int \eta(r, t) e^{-i\sigma t} dt$$ (38)

$$\alpha = [\frac{h^2}{6(\sqrt{gh})^3} \sigma^3 + \frac{\sigma}{\sqrt{gh}}] \Delta R$$ (39)

Step 2 - Nonlinear Correction:

Now considering the non-linear part,

$$\eta r - \frac{3}{2h} \frac{\eta \, \eta_t}{\sqrt{gh}} = 0 \qquad\qquad (40)$$

The step r is repeated using the time series of   from
the linear solution to compute, $(r + r,t)$ by FDM solution of 28.

Thus far each step advanced in r, two operations, i) linear propagation and ii) nonlinear correction are carried out.

This method propagates the solution using FFT. So large distances
may be covered efficiently. The nonlinear correction may be done
using an implicit FDM scheme to make the solution unconditionally stable.

A large number of comparisons between theoretical results obtained
by neglecting the nonlinear terms and taking them into acount with
experimental results have been made. These experiments were done at
CERC with 10 lbs of TNT explosion in shallow water (1 to 4.0 feet).
Also a steel circular plate of 3 feet diameter, 1 inch thick was
dropped from height of 3 to 6 feet, into shallow depth (0.2 feet to 1
foot). Fig. (6) shows experimental wave records (10 lb charge),
plotted with the numerical predictions. The first wave record is
the input for the numerical model. Similar comparison of theory and
experiment is shown in Fig. (7), for the experiments with the falling
circular plate. Comparisons with larger yield explosion in shallow
water were also done. Since the numerical model is reversible, it is
possible to input farfield linear records in to KdV model to step back
towards the source of explosion. In all cases the theoretical results
have been substantiated by the experimental results, as shown in
Figures 6 and 7.

Acknowledgements

These studies were in major part sponsored by the Defense Nuclear
Agency. The writers are indebted to Lt. C. Carlin for her guidance,
support and encouragement. The experiments were carried out at the
Waterways Experiment Station, Vicksburg, Miss. upon approval of the
Office of the Chief of the U.S. Army of Engineers.

References

Dorrestein, R. (1951) General linearized theory of the effect of
    surface films per water ripple, Kan. Ned. Akad. Wet. B. 54,
    pp. 260-272.

Kajiura, K. (1963) The leading wave of a tsunami, Bull. Earthquake Res. Inst. 41, pp. 535-571.

Khangaoankar, T. and B. Le Mehaute (1989). Original disturbances from wave records. (submitted for publication).

Kranzer, H.C. and Keller, J.B. (1963) Water wave produced by explosion, J. Applied Physics, Vol. 30, No. 3.

Le Mehaute, B. (1970) Explosion generated water waves, 8th Symposium Naval Hydrodynamics, Pasadena, Calif., August.

Le Mehaute, B., S. Wang and (1987a) Cavities, domes and spikes, J. of Hydraulic Research, Vol. 25, No. 5, pp. 503-602.

Le Mehaute, B. and Soldate, M. (1987b) Transformation of transient waves on sloped bathymetry, Proc. of Conference of Coastal Hydrodynamics, Newark, Delaware, July.

Le Mehaute, B., C.C. Lu and D. Outlaw (1988c) Generation of transient wave by snake paddle. J. of Ocean and Applied Physics. (Pending)

Le Mehaute, B. (1988) Gravity capillary rings generated by water drops, J. of Fluid Mechanics, Vol. 197, pp. 415-427.

Miles, (1978) Axisymmetric Boussinesq Wave, J. Fluid Mech. 80, 149.

Stoker, J.J. (1957) <u>Water Waves</u>, Interscience Publishers, NY, pp. 149-196.

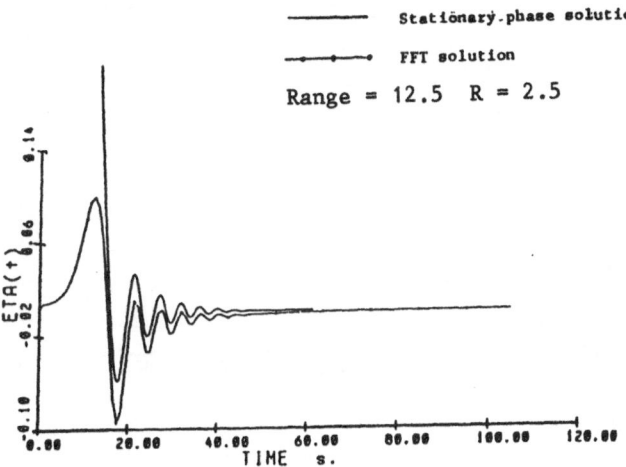

Fig. 1   Comparison of solution obtained by FFT and by the method
of stationary phase (R = 2.5, r = 12.5).

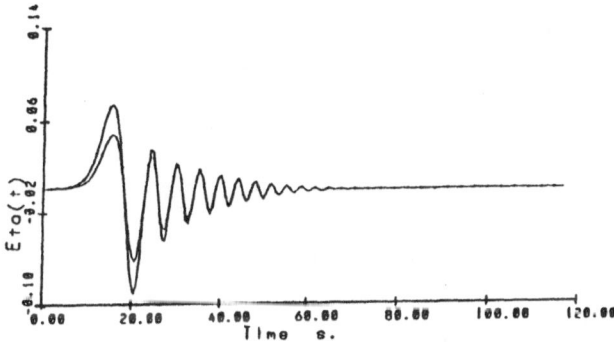

Fig. 2   Comparison of transient wave on a slope with the same
wave on a horizontal sea floor.

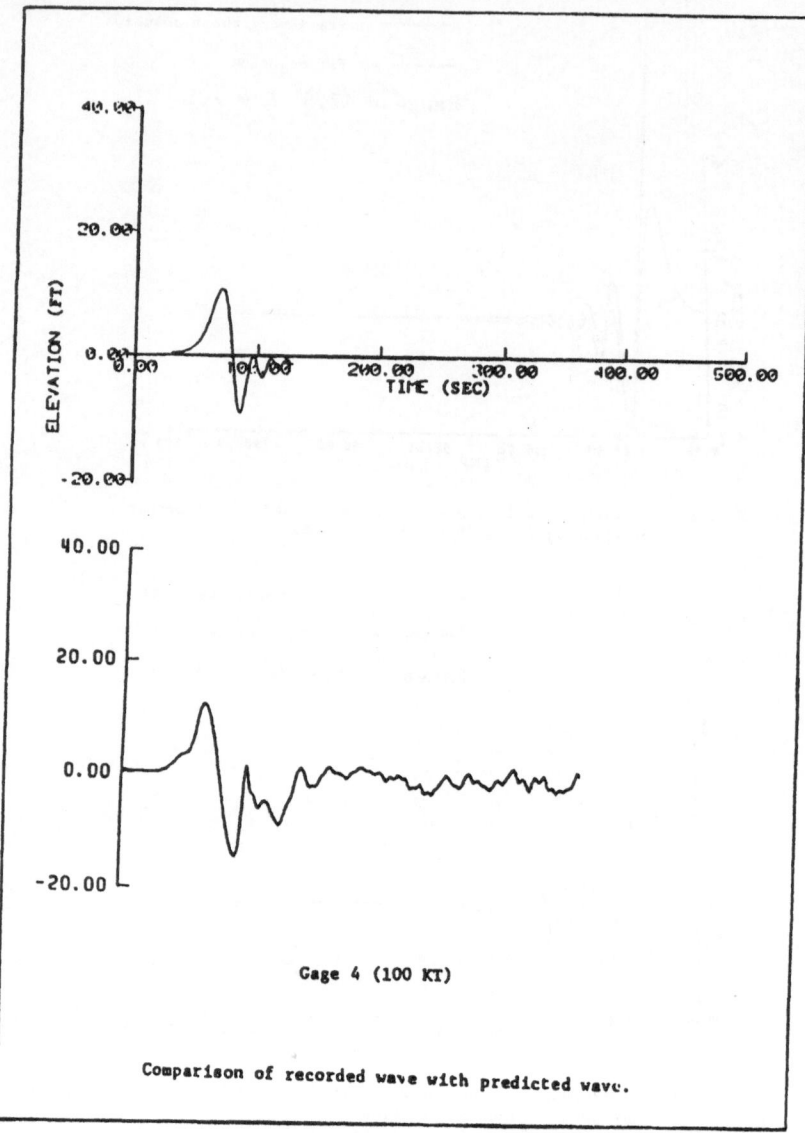

Fig. 3 Comparison of experimental wave records on a 3-D
bathymetry with theoretical prediction obtained by FFT.

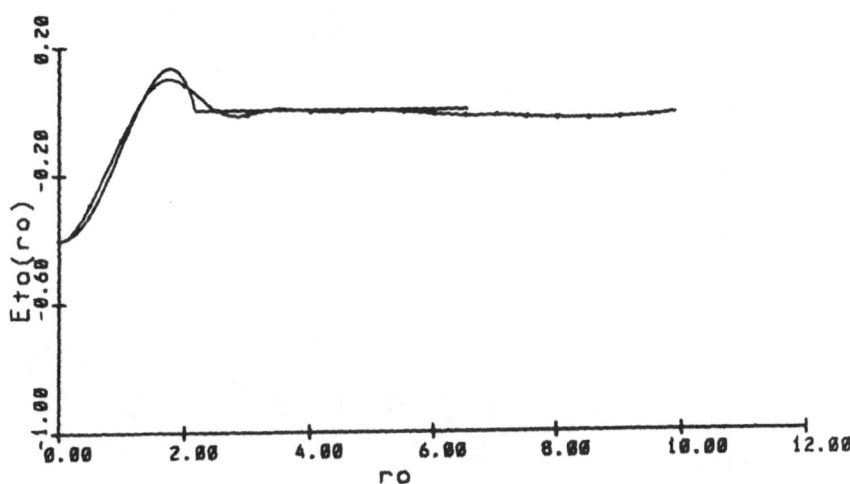

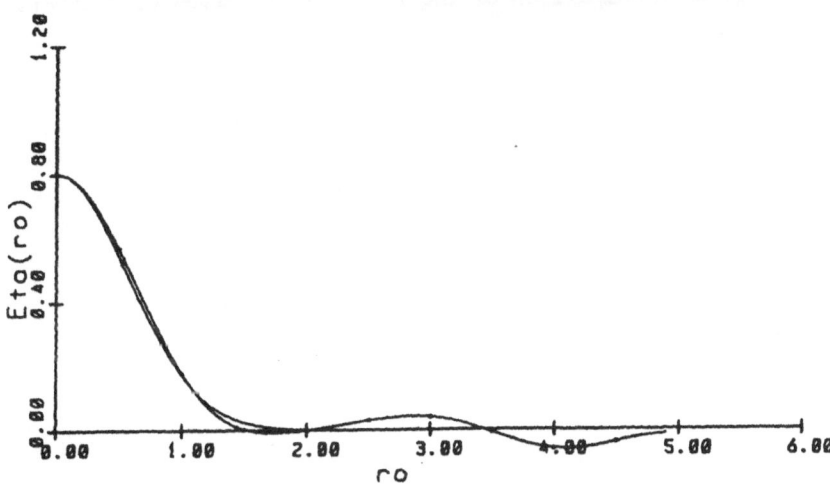

Fig. 4 Numerically obtained free surfaces disturbances from wave
records (velocity and free surface elevations) and
analytical matching curves.

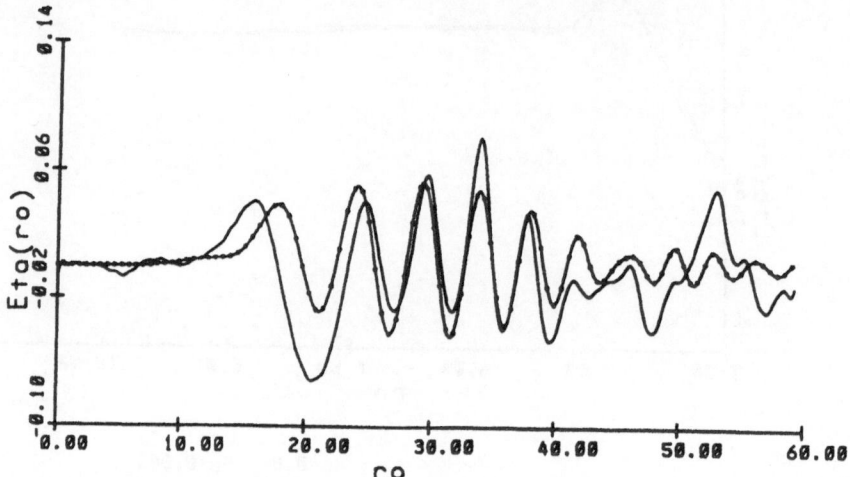

Fig. 5 A comparison of the theoretical curves with experiments.

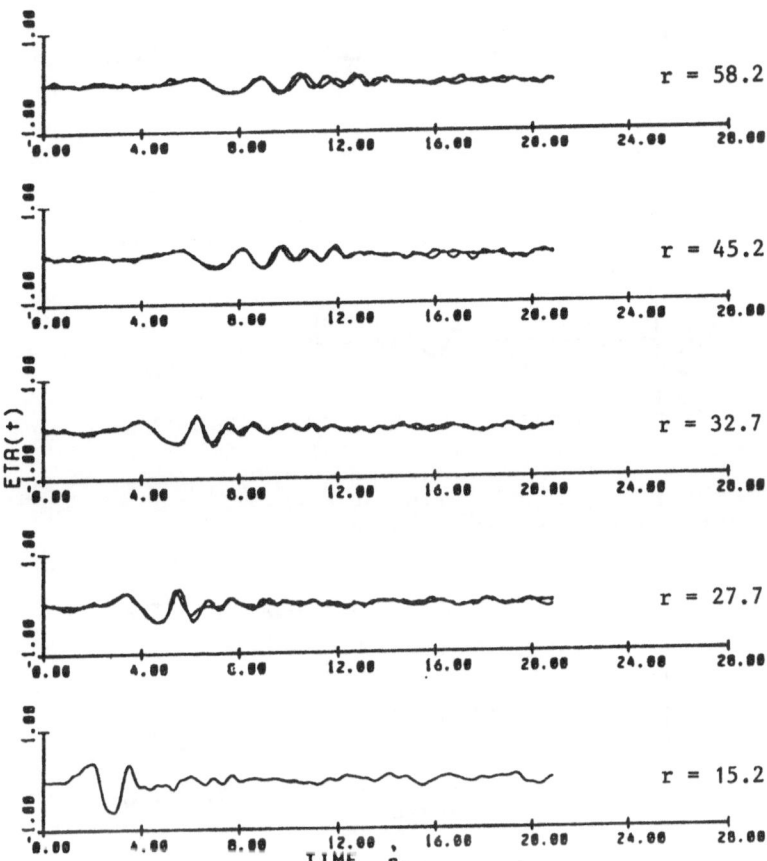

Fig. 6 Evolution of nonlinear KdV explosion generated wave in shallow water and comparison with experimental data of a TNT explosion.

608

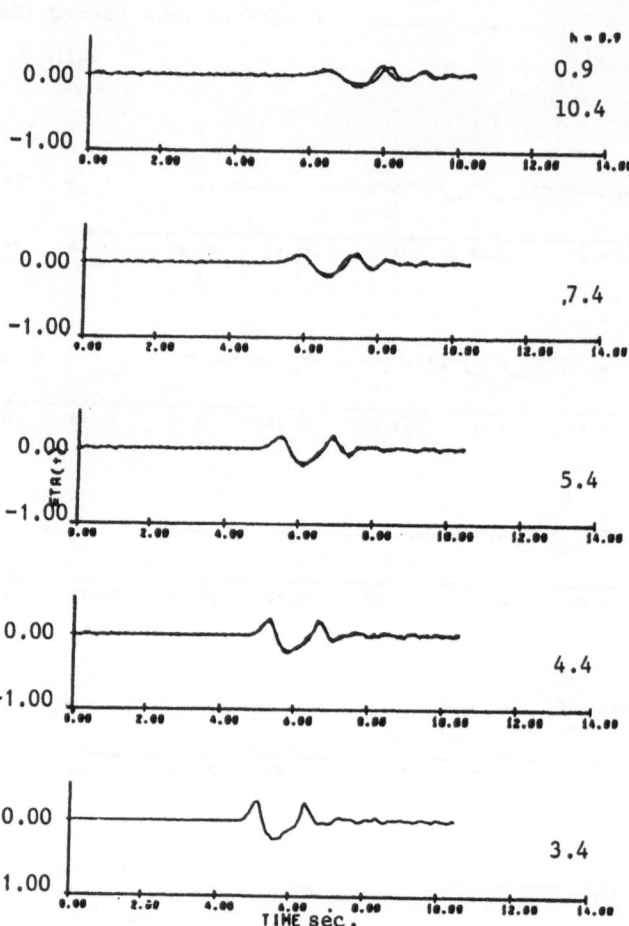

Fig. 7 Evolution of nonlinear explosion generated wave in
shallow water and comparison with wave generated by
the drop of a cylindrical plate.

FREAK WAVES: A POSSIBLE EXPLANATION

R. G. DEAN
Coastal and Oceanographic Engineering Department
University of Florida
336 Weil Hall
Gainesville, Florida 32611 U.S.A.

ABSTRACT. A possible explanation is proposed for the occurrence of freak waves, which are defined as waves with larger heights than expected based on the Rayleigh distribution. The suggested cause is due to nonlinearities of superposition of waves which are not accounted for in the Rayleigh distribution. When two waves combine, if their fundamental components add linearly, it can be shown that the combined wave height increases by more than the sum of the fundamental components. The argument does not address the correctness of linear addition of the fundamental components nor does it include energy closure. An example is presented illustrating the concept.

1. Introduction

Freak waves are those occurring within a sequence of waves which have been identified as being higher than can be expected from the Rayleigh distribution for wave heights. Because the Rayleigh distribution does not predict a limiting wave height, it is not possible by examining a particular wave to conclude on the basis of its height that it is a "freak wave". Rather, it is necessary to consider many candidates and to evaluate whether or not, taken in their aggregate, these wave heights are higher than the expected values.

2. Background

The evidence appears fairly strong that freak waves exist in nature (Sand, et al., 1989) and under laboratory conditions (Stansberg, 1989). As noted above, the existence of "freak waves" must be judged on the basis of their exceedance of probabilities as based on the Rayleigh distribution. Figure 1 presents the ratio of the highest wave in a train of M waves, H_m, to the significant wave height, H_s, of the train. As an example of the nature of the distribution, for a wave train consisting of 2,000 waves, the most probable maximum wave height is approximately 2 times the

A. Tørum and O. T. Gudmestad (eds.), Water Wave Kinematics, 609–612.
© 1990 Kluwer Academic Publishers.

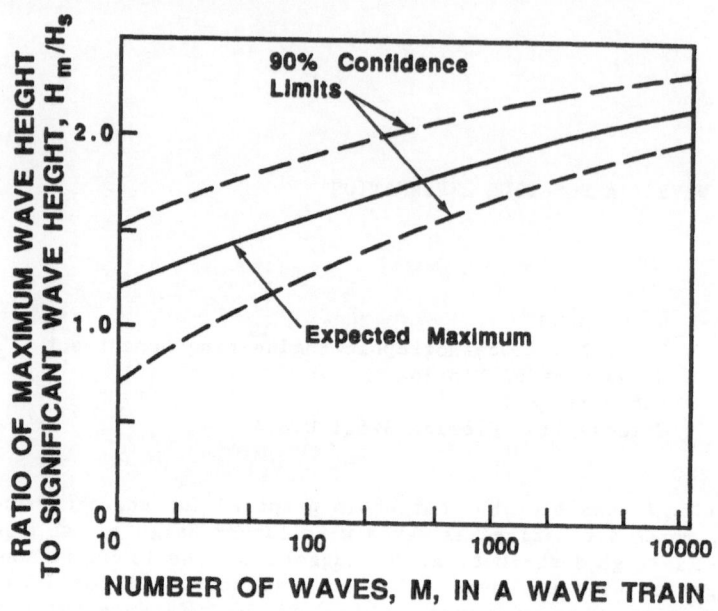

Figure 1. Rayleigh distribution results for expected maximum wave height in a train of M waves. Maximum wave height expressed as a ratio to significant wave height.

significant height; however, there is a 5% chance that this ratio will be less than 1.8 and a 5% chance that it will be greater than 2.2.

3. Possible Cause of Freak Waves

In general, it appears that both wave nonlinearities and directionality should be considered as possible causes of freak waves. However Stansberg (1989) has carried out a laboratory study and concluded that freak waves can be generated in the laboratory under essentially two dimensional conditions. The hypothesis presented below concentrates on nonlinearities as a possible cause of freak waves.

First, we recall that the Rayleigh distribution considers the waves to be superposed linearly and that when they are combined, their crest elevations, and essentially their wave heights are added linearly. Next consider a single nonlinear wave component, $\eta_1(t)$, given by

$$\eta_1(t) = \sum_{n=1}^{N} a_{1_n} \cos \sigma_{1_n} t \tag{1}$$

and recall that according to nonlinear theory

$$a_{1_n} \alpha (a_{1_1})^n \tag{2}$$

The wave height, H_1, associated with $\eta_1(t)$ is

$$H_1 = 2 \sum_{n=1}^{N/2} a_{1_{2n-1}} \tag{3}$$

i.e., only the odd components contribute to the wave height.

Extending the above model to two components and their (nonlinear) superposition, the combined water surface displacement, η_T, is

$$\eta_T = \eta_1 + \eta_2 = \sum a_{T_n} \cos \sigma_{T_n} t \tag{4}$$

and, as before,

$$a_{T_n} \alpha (a_{T_1})^n \alpha (a_{1_1} + a_{2_1})^n \tag{5}$$

The wave height of the combined wave system is given by Eq. (3), as

$$H_T = 2 \sum_{n=1}^{N/2} a_{T_{2n-1}} \tag{6}$$

The two leading terms of Eq. (6) are

$$H_T = 2(a_{1_1} + a_{2_1}) + 2F_3(a_{1_1} + a_{3_1})^3 \tag{7}$$

where F_3 is a nonlinear interaction coefficient. The height, H_T, is greater than the heights obtained by the Rayleigh distribution due to several factors, foremost of which are the cross-product terms in the second term of Eq. (7).

In summary, when waves superimpose, they do not combine linearly and the nature of the nonlinear interaction is such that there is an increase in the combined wave heights over and above that which would occur if the combination were linear.

4. Example

In order to illustrate the mechanism under consideration, a numerical example was carried out using a program which allows one or two waves to be represented including all nonlinearities. A wave having a (nonlinear) height of 16.22 ft was selected. The fundamental for this wave was doubled and the associated wave height was determined to be 34.43 ft, i.e. 1.99 ft greater than twice the height of the fundamental wave. However, the effect on the crest elevation was even greater. The crest elevation of the individual wave with a height of 16.22 ft was 10.53 ft whereas the crest elevation of the wave with twice the fundamental component was 24.27, i.e. 3.21 ft higher than twice that of the individual wave.

5. Summary and Concluding Remarks

A possible explanation is presented for freak waves as defined as those waves having heights greater than consistent with the Rayleigh distribution. The suggested cause is due to the nonlinear interaction terms associated with the sum of the fundamental components of the primary waves being combined. The proposed model is not complete as it does not include energy closure. Stated somewhat differently, the model presented here assumes that the fundamental components of two primary waves are added linearly, whereas it is likely that these fundamental components would be decreased somewhat in the interaction process.

6. References

Sand, S.E., Ottesen Hansen, N.E., Klinting, P., Gudmestad, O.T., and Sterndorf, M.J. (1989) 'Freak wave kinematics', Proceedings, NATO Advanced Research Workshop on Water Wave Kinematics, Molde, Norway.

Stansberg, C.T. (1989) 'Extreme waves in laboratory generated irregular wave trains', Proceedings, NATO Advanced Research Workshop on Water Wave Kinematics, Molde, Norway.

CURRENT-WAVE INTERACTIONS OBSERVED IN THE LABRADOR SEA EXTREME WAVES
EXPERIMENT

SØREN PETER KJELDSEN
MARINTEK, Norwegian Marine Technology Research Institute A/S
P.O. Box 4125
N-7002 TRONDHEIM
NORWAY

1. Introduction

The Labrador Sea Extreme Waves Experiment "LEWEX", is a comprehensive
international effort, with the purpose to improve seakeeping capabilities of vessels. Two topics are identified where a scientific approach
is needed in order to achieve a better understanding of phenomena and
better prediction. These are:
- Full scale measurements, experimental simulations and theoretical
  modelling of directional seas.
- Sea loads, slamming, and green seas impact on vessels in directional
  extreme seas. This also includes prediction of extreme motions and
  seakeeping in extreme conditions.
  The international research programme, LEWEX has been initiated with
participation of the following nations: United States, Canada, Norway,
Netherland, France, West Germany, Spain, United Kingdom. The LEWEX-
project has support from 17 institutions in these countries, and has a
variety of resources available for its activities. Major components
include Dutch and Canadian research vessels, Canadian and American
research aircrafts, the American Geosat spacecraft and the advanced Norwegian Ocean Simulating Basin at MARINTEK in Trondheim. Wave models from
9 North American and European institutions complement 19 independent in-
situ, ship-borne, air-borne and satellite-borne instruments applied
simultaneously with the purpose to obtain syncronized measurements of
wave spectra in the winter season in one of the most hostile and rough
areas of the North Atlantic, the Labrador Sea, north-east of, and south-
west of the grand banks of New Foundland (see Fig. 1). The scientific
planning of this particular experiment was made by BALES et al. (1987).

2. Syncronized measurements of directional sea spectra

The major scope of the LEWEX-project is to evaluate applicability and
usefulness of directional wave data for engineering design purposes. The
largest responses are not always encountered in the most extreme sea
state. It turns out that a combination of less severe directional sea

A. Tørum and O. T. Gudmestad (eds.), Water Wave Kinematics, 613–619.

states superposed upon each other, in some cases lead to larger
responses. Thus, bimodal and trimodal directional seas are studied with
respect to their generation, propagation current-wave refraction and
dissipation. A very extensive data bank has been established at Applied
Physics Laboratory at the John Hopkins University in Maryland, U.S.A.
that contains all measured wave spectra. This data bank contains
situations where at least four independent directional sea systems (a
wind sea and 3 swells) are superposed upon each other. The performance
of the deployed directional instruments under such complicated conditions is not well known. Therefore, one of the main scientific objectives of the LEWEX-project is to recommend/identify new procedures where
existing techniques of measurement or analysis are inadequate.

This is not a simple task. In principle the LEWEX-experiment contained
the following four different systems for measurements of directional sea
states:
- Moored oceanographic buoys.
- Freely floating oceanographic buoys.
- Air- and space-borne radars (such as SCR, SAR and ROWS. All 3 instru-
ments were found capable of monitoring several important aspects of the
directional spectrum. In LEWEX all 3 instruments were air-borne but SAR
and ROWS are capable of operating from orbit.)
- Ship-borne radars and wave gauge.
At least five methods for analysis of directional spectra are
available. These are:
- The Direct Integral Method.
- The Fourier Series Expansion Method.
- The Maximum Likelyhood Method.
- The Maximum Entropy Method.
- The Variational Method.

3. Current-Wave Interactions

In the first phase of the LEWEX experiment all instruments were applied
in the same sea state at a position 50°N 47,5°W from 19th-20th of March.
The measured directional spectra are all correlated with a common key
sensor. The WAVESCAN directional wave buoy is chosen as key sensor.
(Transfer functions between results from this buoy and results from a
laser wave gauge array installed on a fixed platform is available).
Measurements x_1, x_2, ...x_J of the same parameter x are then obtained
with J instruments with varying accuracy. An assessment of the accuracy
of each individual instrument can then be obtained from a study of the
standard deviations σ_1, σ_2....σ_J, obtained from correlations with
measurements made with the key sensor. A weight factor w_i is then
defined for each instrument with the following equation:

$$w_i = \frac{1}{\sigma_i^2}$$

(3-1)

The results obtained from all instruments can then be combined in a weighted mean value:

$$m_x = \frac{\sum\limits_{i=1}^{J} w_i \, x_i}{\sum\limits_{i=1}^{J} w_i} \qquad (3\text{-}2)$$

m_x is then the best estimate that can be obtained for the true value of x. The relative accuracy of m_x then becomes:

$$\frac{\sigma_{m_x}}{m_x} = \frac{\sqrt{\sum\limits_{i=1}^{J} w_i}}{\sum\limits_{i=1}^{J} w_i \, x_i} \qquad (3\text{-}3)$$

This is much smaller than the relative accuracy σ_i/x_i of results obtained from a single instrument.

In the second phase of the LEWEX experiment all instruments were deployed simultaneously across the current shear between the Labrador current and the Gulf Stream as Fig. 1 shows. At this location a strong spatial variation of sea states was observed due to current-wave refraction. Here all calibrated sensors each with an associated weight factor could be operated as a tool for mapping of spatial and temporal variations in the directional wave fields caused by current-wave interactions. Fig. 2 gives examples of some of the measured directional wave spectra in the current shear. The current shear zone is a very unstable region and large eddies are from time to time released from the mean current and travels independently. Such eddies are tracked from satellites. Examples of computations of refraction of directional wave spectra within such eddies are shown in SAEVERAAS, KJELDSEN, NAEROEY 1988. Further work on development of non-linear algorithms for current-wave interactions are in progress. The measurements shown in Fig. 2 are made close to a busy traffic route between European and North American ports. (The site for the loss of the "TITANIC" is nearby). During the LEWEX experiment a freak wave was measured here with a ship-borne wave gauge. This freak wave was obviously caused by strongly non-linear current-wave interactions. The area is known as dangerous and weather damage is quite often reported by vessels passing this site. It is therefore important that algorithms developed for current-wave interactions are incorporated in mathematical models used for operational wave forecasting. These models are dealt with in section 4.

4. Mathematical models for operational wave forecasting

Results from 9 mathematical models are compared and evaluated within the LEWEX research programme. All these are used today for operational forecasting of directional wave spectra. Several of the models are in use for global wave forecasting and hindcasting. The mathematical models can be classified in 3 categories. The first generation wave models are developed with the basic assumption, that there is no interaction between frequencies. The basic formulation of wave growth and decay follows that of PIERSON 1982. First generation models are GSOWM and ODGP.

In the revised energy balance of the second-generation wave models, the independent evolution of individual wave components is effectively prevented by the coupling through the non-linear energy transfer. For a normal, relatively slowly varying wind field, the redistribution of energy by the nonlinear transfer is sufficiently rapid relative to advection and the other source functions that a universal, quasi-equilibrium spectral distribution of the JONSWAP form is established whose shape is largely independent of these other processes and which can be approximately characterized by a single, slowly changing scale parameter, such as the total energy E or peak frequency f_p (HASSELMANN et al., 1976). Second generation models are BMO, HYPA, NOAA and VAG. The second generation wave models may be expected to encounter difficulties in the windsea-swell transition regime of the spectrum in which the nonlinear energy redistribution is neither neglible nor dominant. Such transition regimes arise whenever the wind speed decreases or the wind direction turns. The third generation wave models integrates implicitly the basic transport equation which describes the evolution of the two-dimensional wave spectrum according to:

$$\frac{\partial E}{\partial t} + \nabla \cdot (\vec{C}_g E) = S_{in} + S_{nl} + S_{ds} \qquad (4\text{-}1)$$

The left-hand side of this equation describes propagation effects. Here $\vec{C}_g$ is the group velocity. The physical processes affecting the wave state are modelled by the three source terms on the right-hand side. S_{in} represents the source function describing the transfer of energy from wind to waves, S_{nl} the nonlinear transfer due to resonant four-wave interactions, and S_{ds} the dissipation resulting from whitecapping and turbulence. It was not possible to model the energy balance in this way until an efficient and accurate algorithm to compute the non-linear transfer was developed and the dissipation was reliably parameterized. The 3G-WAM model is an example of a third generation wave model. It runs daily at a cray computer at the European Centre for Medium-Range Weather Forecasts, and computes over the entire globe two-dimensional ocean wave spectra $E(f, \theta, x, t)$, which are functions of wave frequency f, direction θ, position x and time t. The foundations of this model were laid at the Max Plack Institute for Meteorology in Hamburg (WAMDIG, 1988).

Other third generation wave models that participate in the LEWEX research programme are the NASA-model and the BIO-model. All models are run both in forecast and hindcast modes, stored in the LEWEX data bank and correlated with in-situ measurements. The standard deviations σ_i for each of the 9 models can then be obtained, and it is then possible to define a weight factor for each model from eq (3-1). With asses to results from all models it is then possible to prepare a weighted forecast. Such weighted forecasts can be used in weather-routing-expert-systems for vessels of the kind treated by KJELDSEN (1989). Such a weighted forecast can also be an efficient tool when large scale marine operations are performed at sea. (Such as search and rescue). A comprehensive collection of publications from the LEWEX research programme is available from APL (1989).

5. Conclusions

1) Algorithms for current-wave interactions should be incorporated in operational wave models, and used in daily forecasting in particular in wave exposed areas such as the one shown in Fig. 1.
2) With assess to results from several models, a weighted wave forecast can be prepared.
3) Results from satellite-borne, ship-borne and in-situ measurements can be combined and used together when a weight factor depending on accuracy are defined for each wave sensor.

6. References

APPLIED PHYSICS LABORATORY 1989: "Proceedings from the LEWEX-Symposium on Measuring, Modelling, Predicting and Applying Directional Ocean Wave Spectra" held at APL april 89. Technical Digest Vol 11. No 2.The John Hopkins University, Maryland, U.S.A. (in press).

BALES S., BEALE R.C., FREEMAN G.F. 1987: "LEWEX-Science Plan".Report from DTNSRDC, Bethesda, Maryland, U.S.A.

HASSELMANN K., ROSS D.B., MÜLLER P., SELL W., 1976: "A Parametrical Wave Prediction Model". J. Phys. Oceanogr. 6. pp 201-208.

KJELDSEN S.P., KROGSTAD H.E., OLSEN R.B., 1988: "Some Results from the Labrador Sea Extreme Waves Experiment". Proc. 21st Int. Conf. on Coastal Engineering. Malaga, Spain.

KJELDSEN S.P. 1989: "The Practical Value of Directional Ocean Wave Spectra". Technical Digest Vol 11 No 2. The John Hopkins University, Maryland, U.S.A. (in press).

PIERSON W.J., 1982: The Spectral Ocean Wave Model (SOWM), A Northern Hemisphere Computer Model for Specifying and Forecasting Ocean Wave Spectra. Technical Report DTNSRDC-82/011, David W. Taylor Naval Ship Research and Development Center, Bethesda, MD 20084, 186 pp.

SAEVERAAS N., KJELDSEN S.P., NAEROEY A. 1988: "Report from the Court of Inquiry made after the loss of M/S "SUN COAST" at Stad in Norway, 2nd December 1984". Department of Justice. (To be obtained from MARINTEK A/S, Trondheim, Norway).

WAMDIG 1988: S. Hasselmann, K. Hasselmann, E. Bauer, P.A.E.M. Janssen, G.J. Komen, L. Bertotti, P. Lionelli, A. Guillaume, V.C. Cardone, J.A. Greenwood, M. Reistad, L. Zambresky and J.A. Ewing, 1988: "The WAM model, a third generation ocean wave prediction model". JPO, Vol. 18, No. 12.

618

7. Acknowledgements

The LEWEX-project is organized within NATO as a multi-lateral scientific research programme. Two working groups are responsible for the performance of the LEWEX-project. These are RSG-1 and RSG-2. A list of participating scientists is given by BALES et al. (1987).

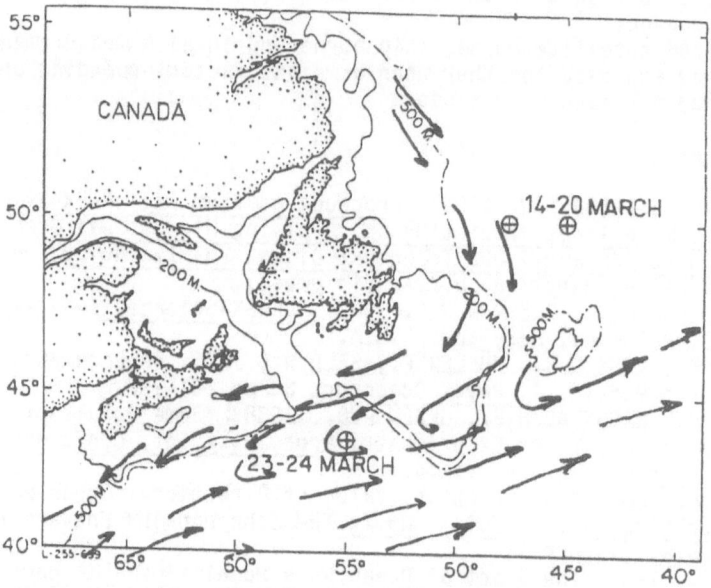

Fig. 1. Positions for measurements of directional ocean wave spectra used in the LEWEX research programme. On the 23-24th of March 1987 directional spectra were obtained in a position 55°W, 43°N in the current shear between the cold Labrador current coming from the North, and the warm Gulf Stream coming from the south.

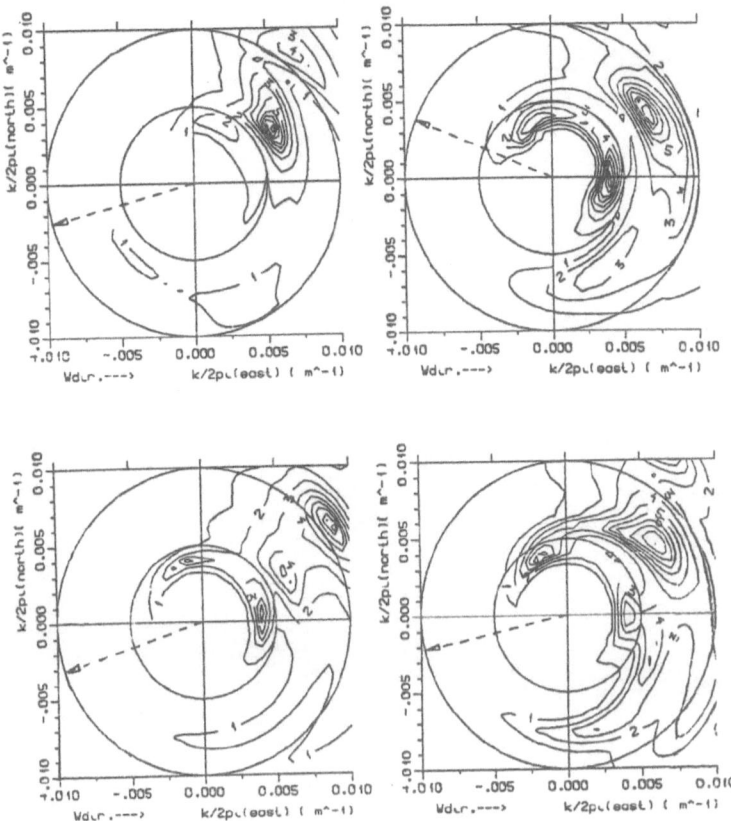

Fig.2.   Examples of directional ocean wave spectra measured in the
current shear on the 24th of March at 0600 GMT, 0730 GMT, 0900 GMT, 1030
GMT. The polar plots is a new standard for presentation of directional
information, developed within the LEWEX research programme. The plot is
oriented with North at the top and shows the direction the waves are
coming from. The outher circle corresponds to a wave length 100 m, while
the inner circle corresponds to a wave length 200 m. The spectral density is shown as contour plots. The contours are computed for 12 levels
in a linear scale. The measured wind direction is indicated with an
arrow. (From KJELDSEN et. al. 1988).

Discussion on Session 5: Breaking and Freak Waves

Mr R. M. Webb

From the presentations of Kjeldsen and Sand during this session one might be forgiven for expecting that "Freak" waves are to be found in almost any storm one cares to investigate. Were this to be so it would overthrow the accepted relationships between H_s and H_{max} and the waves could not be considered as "Freak". However there did appear to be many instances of such waves and all data owners should perhaps now have a close look at their data to see if undetected "Freaks" exist at their location.

Some time ago I initiated a study into this using the measured data available at West Sole in the southern North Sea. For this we selected the ten most severe storms between 1974 and 1986 and found that the individual wave heights on the digitised charts followed a Gluhouski distribution. This work was published as "Wave height distributions in shallow water" by C. J. Martin, C. K. Grant and R. A. Sproson at the Oceanology International Conference, Brighton in 1987. A copy of this paper is attached.

You will note that in this study we were not considering asymmetry and adopted Tann's conservative approximation that the height of a zero up crossing wave is twice the height of the crest from the mean line. Looking again at fig 6 from the paper in the light of the NATO workshop presentations there is perhaps one point, with 11 m measured wave height, where the divergence from the expected value might possibly be considered as falling into the realms of a "freak" wave. This however depends on the definition of the term, and might well be seen as lying within the error bounds of the distribution. I do not have the original data or detailed report to check on these matters now, but interested parties might wish to contact Dr Colin Grant at BP.

A. Tørum and O. T. Gudmestad (eds.), Water Wave Kinematics, 621–622.
© 1990 Kluwer Academic Publishers.

Mr R. M. Webb

Mr Tallent, in Figs 3(b), 4(a) and 4(b), 5 and 6 appeared to imply that the splash up after a breaker can reach a greater elevation than the crest elevation at breaking. If this is so then it is of concern to designers when considering deck elevations for platforms and piers subject to wave breaking. Of perhaps equal concern is the degree of aeration of the water in the splash up of a shallow water breaking wave. ie the greater the percentage of air the less concern. Similarly designers are interested to have information on the degree of aeration in the breaking waves themselves. This is probably most likely to occur in deep water breaking waves where "white horses" abound well before severe breaking might be encountered. The author's views on these matters would be of great interest.

Measurements

FLOW VISUALIZATION BY VIBRATING CAMERA
Simultaneous registration of pathlines and velocity field

G.F. CLAUSS, Prof. Dr.-Ing.
Institut für Schiffs- und Meerestechnik
Technische Universität Berlin
Salzufer 17 - 19, 1000 Berlin 10, F.R.G.

ABSTRACT. A novel technique for flow visualization is presented. As usual, the flow field is photographed by a time exposure. In front of the camera, however, a slightly tilted mirror is rotating distorting the particle paths periodically. As a result the pathlines show a trochoidal modulation with significant peaks marking subsequent revolutions of the mirror. Thus, particle paths and the accompanying velocity distribution are registered simultaneously. Calibration is integrated in the exposure. The technique also allows for the identification of flow direction. The paper illustrates the "vibrating" camera technique on the basis of some typical flow phenomena.

1. Introduction

The detailed knowledge of flow characteristics is the prime condition of analyzing structure-fluid interactions. Especially with unsteady flows the time-dependent variation of the flow field covering a large area must be documented to specify the local velocities and accelerations at selected positions. As those parameteres are characterized by magnitude, phase and direction the usual flow visualization techniques yield insufficient data.

In the special case of a hydrodynamically transparent offshore structure moving in the seaway the total fluid force dF on an element of a circular component (diameter D) is calculated by the modified Morison equation

$$dF = \varrho \frac{\pi D^2}{4} ds \frac{\partial v_N}{\partial t} \;+\; C_a \varrho \frac{\pi D^2}{4} ds \frac{\partial u_{RN}}{\partial t} \;+\; C_d \frac{\varrho}{2} D ds |u_{RN}| u_{RN},$$

| Froude-Krylov force | hydrodynamic mass force | frictional force |

625

A. Tørum and O. T. Gudmestad (eds.), Water Wave Kinematics, 625–632.

in which ρ is the fluid density, C_a and C_d are the added mass and drag coefficients, respectively.

The Froude-Krylov force depends on the wave particle acceleration normal to the cylinder axis. The hydrodynamic mass force and the frictional force result from the normal components of relative acceleration and velocity, respectively, which are calculated from wave particle and body motions /1/. All accelerations and velocities refer to the axis of the element. Evidently, the accurate knowledge of these parameters is crucial and their derivation from surface elevations by an appropriate wave theory is not sufficient in many cases.

In general, flow visualization techniques provide specific information on pathlines, sometimes supplemented by puzzling velocity field data. An excellent review of current methods is summarized in /2/.

2. Vibrating camera technique

This paper presents a new technique of flow visualization using a "vibrating" camera for simultaneous registration of particle paths and the velocity field /3/. As usual, the flow field is photographed by a time exposure. In front of the camera, however, a rotating mirror is installed, and the picture is taken via this arrangement. As the mirror is slightly tilted a reeling motion is observed. Thus, a fixed particle is photographed as a circle and a moving particle is represented as a trochoidal curve with significant peaks marking subsequent revolutions of the mirror. Tilt and frequency of rotation modulate the light beam registered by the camera. Consequently, these parameters must be adjusted according to the observed flow phenomena. If the motion is photographed with an open shutter the trochoid represents the particle path, the distances of individual peaks mark the instantaneous velocity at consecutive locations. Different particle paths are cleary recognized and separated even at unsteady flow conditions. Fig. 1 shows the flow field behind a harmonically moving flap which is towed through water with constant speed. For calibration a scale is provided and a lamp is connected to the towing carriage. The trochoidal representation of its path serves as a velocity standard. By comparison to other trochoids of surface particles the velocity field is easily evaluated. Of course, this analysis should be done at a large scale, using the photographs as slides. Note that this method also allows for the identification of the flow direction. Hence, even the sense of rotation of the vortices can be evaluated.

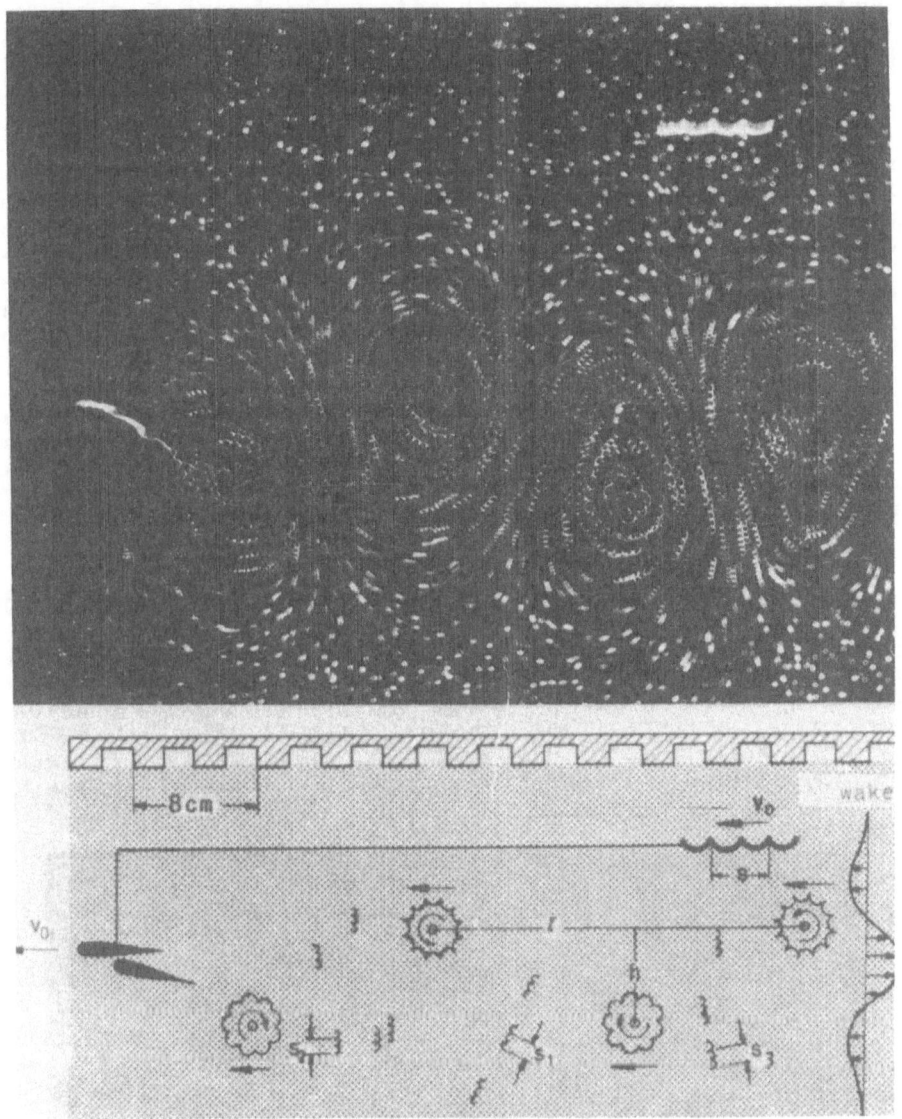

Fig. 1   Flow field behind a harmonically moving flap

3. Illuminated particles with adjustable density

For specific visualization problems tailor made particles have been developed. The "particles" are composed of plexiglass pipe sections - typically 10 - 15 mm in diameter and 20 mm long - with watertight seals on both ends. One of the seal is a metal screw which can be tightened or loosened to adjust the buoyancy. The pipe sections can be equipped with a battery and a micro-lamp to serve as a floating illuminated particle. Depending on the diameter and the threat of the screw a very sensitive adjustment of the displacement of the particle is achieved. Even during the experiments the average density is easily adapted to the density of the surrounding fluid simply by turning the screw. Thus, neutrally buoyant particles are feasible. Of course, the same technique also allows for the production of very small "passive" particles if the void is not used for specific equipment. In this case the luminosity is provided by fluorescent or phosphorescent substances.

4. Visualization of orbital motions in waves and wave groups

As a second example, Figs. 2 and 3 show particle paths and associated velocity marks of orbital motions beneath the water surface. The photos clearly demonstrate the constant magnitude of the particle velocity in a harmonic deep water wave. In these experiments high-density polyethy-

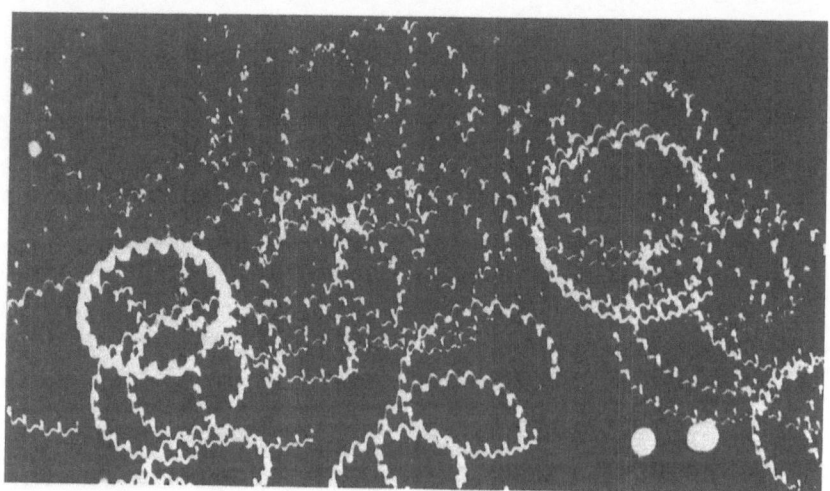

Fig. 2   Orbital motions in a wave field

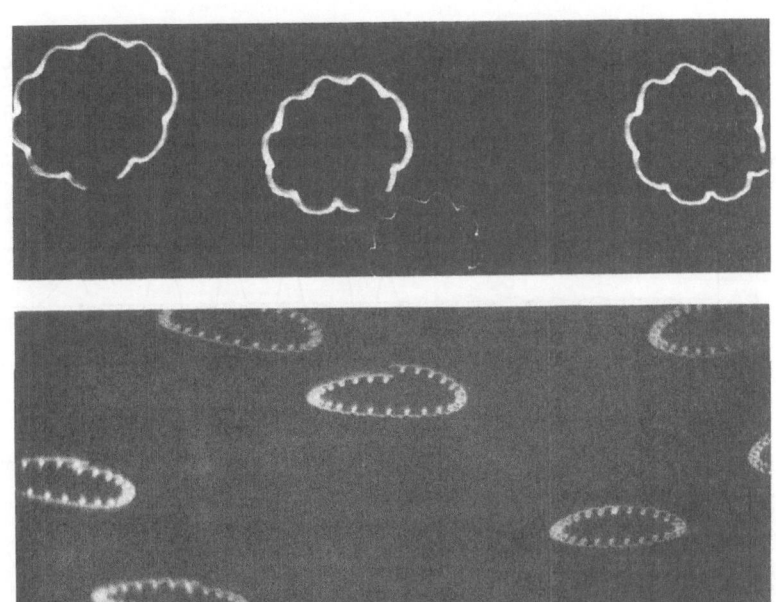

Fig. 3   **Selected particle paths in deep and shallow water waves**

lene particles (Hoechst trade mark HOSTALEN) with a density of ρ = 0.995 + 0.005 g/cm³ were used. Only a narrow vertical slice of the wave tank was illuminated by a filament quartz lamp focused by a cylindrical lens to select the relevant test section. Further experiments are planned with particles placed at predetermined positions to investigate fluid-structure interactions.

The experiments in regular waves are complemented by tests with special transient wave trains of short duration, so called Gaussian wave packets. Based on fundamental derivations of Coulson /4/ and Kinsman /5/ a very efficient experimental technique was developed composing wave groups by an infinite number of harmonic components with a Gaussian shaped amplitude spectrum /6-8/. Fig. 4 shows such a wave train and the accompanying Fourier spectrum resulting from a FFT-analysis of the surface elevation.

Seakeeping tests with Gaussian wave packets show considerable advantages:

- the wave train is well defined at any location of the tank, any test is exactly repeatable,
- the wave elevation and spectrum can be standardized or easily adapted to any specific problem,

- the duration of the test is very short; reflections from beach or wave board do not interfere the results. At the culmination point the length of the wave train is small. This facilitates short duration seakeeping and manoeuvring tests,
- the results show high resolution and are in good agreement with regular wave test data.

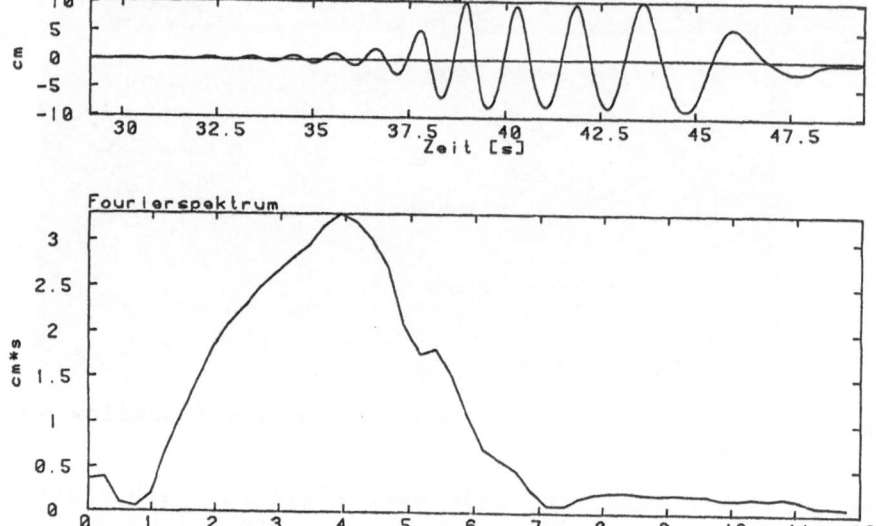

Fig. 4 Gaussian wave packet and associated Fourier spectrum

In comparison to the random wave technique, the Fourier transform of the wave train is unique, the spectra are deduced without any statistical scattering. This also results in smooth transfer functions. The controlled repeatability of arbitrary wave groups allows the definition of standard wave packets which could be adapted at any towing tank, world wide, for comparative tests.

Seakeeping tests with Gaussian wave groups have been accomplished successfully with a variety of offshore structures /6-8/. At present studies are underway to investigate the velocity and acceleration field under these transient waves. Fig. 5 shows typical particle paths at the surface. The photo on top is just taken with an open shutter, the photo below shows path lines which are modulated by the reeling mirror. Note the convective flow component. Further investigations shall reveal whether the velocity and acceleration field is the vectorial superposition of the component wave fields which define the Gaussian wave packet.

Fig. 5   Particle paths in Gaussian wave packets

5. Conclusions

For the understanding of fluid-structure interactions the knowledge of the velocity and acceleration fields is of crucial importance. Their analysis closes the gap between surface elevations and fluid forces. The paper presents a novel technique of modulating the photographed path lines by a reeling mirror in front of the camera. With this method pathlines and the velocity fields can be registered simultaneously. The photos taken by the "vibrating" camera present particle paths and the velocity distribution and give information on flow direction and calibration.

632

6. Acknowledgements

This paper is partly based on a research project funded by the German Research Association (DFG). The financial support is gratefully acknowledged. The author also is indepted to Dipl.Ing. M. Suprayan for his assistance.

7. References

/1/ Clauss, G., Lehmann, E. and Östergaard, C. (1988) Meerestechnische Konstruktionen, Springer-Verlag, Berlin, Heidelberg, New York, London, Paris, Tokyo

/2/ Gad-el-Hak, M. (1986) 'Review of flow visualization techniques for unsteady flows', in C. Véret (ed.) Flow Visualization IV, Hemisphere Publishing Corporation, Washington, New York, London (1987)

/3/ Hertel, H., Affeld, K. and Clauss, G. (1969) 'Die vibrierende Kamera - ein photographisches Verfahren zur Geschwindigkeitsmessung', VDI-Zeitschrift 111, Nr. 8

/4/ Coulson, C.A. (1949) Waves, a mathematical account of the common types of wave motion, Oliver and Boyd, Interscience Publ. Inc., New York

/5/ Kinsman, B. (1965) Wind waves, their generation and propagation on the ocean surface, Prentice-Hall Inc., Englewood Cliffs, NJ.

/6/ Clauss, G. and Bergmann, J. (1986) 'Gaussian wave packets - a new approach to seakeeping tests of ocean structures', Applied Ocean Research, Vol. 10, No. 2

/7/ Bergmann, J. (1985) 'Gaußsche Wellenpakete - Ein Verfahren zur Analyse des Seegangsverhaltens meerestechnischer Konstruktionen, PhD Thesis, published at the Institute of Naval Architecture and Ocean Engineering, Technical University Berlin

/8/ Clauss, G. and Riekert, T. (1988) 'Generation of wave groups for model testing and some applications in coastal and offshore engineering', in 2nd International Symposium on Wave Research and Coastal Engineering, University of Hannover, F.R.G., Oct. 12-14, 1988

MEASUREMENTS OF WAVE KINEMATICS IN THE WADIC PROJECT

Harald E. Krogstad
Section for Industrial Mathematics
SINTEF-Group
N-7034 Trondheim
Norway

Stephen F. Barstow
Oceanographic Company of Norway A/S
Pir-senteret
N-7005 Trondheim
Norway

ABSTRACT.   This note gives a short   summary of wave kinematics
measurements made during the WADIC project at   the Edda platform in the
Ekofisk field in the North Sea during winter 1985-86.

1. INTRODUCTION

The Wave Direction Measurement Calibration Project (WADIC) was initiated
by Chevron Oil Field Research Company and supported by most oil
companies operating in the North Sea as a part of their accompanying R&D
activity in Norway. WADIC had the primary objective to evaluate
operational directional wave measurement systems under severe open ocean
wave conditions (Allender et al., 1989).
    The wave measurements   included an   instrumented tower   with
three-axial current meters and a co-located array of laser altimeters
for accurate profiling of the ocean surface. These measurements provided
data for a separate study of wave kinematics which was carried out with
additional support from the UK Dept. of Energy.
    The objective of the wave kinematics study was to evaluate the use
of directional, Gaussian, linear wave theory (DGLWT) in predicting the
near surface wave kinematics.
    This short note summarises some of the conclusions from the study.
A more complete description may be found in Vartdal et al. (1989) and
Krogstad (1989).

A. Tørum and O. T. Gudmestad (eds.), Water Wave Kinematics, 633–640.
© 1990 *Kluwer Academic Publishers.*

2. THE WAVE KINEMATICS MEASUREMENTS

The WADIC field experiment took place over a four month period during winter 1985-86 at the Edda platform on the Ekofisk field in the central North Sea.

Figure 1 shows the Edda platform in the North Sea where an instrument tower holding eight current meters was attached to the NE corner. The original plans were for two current meters above the free surface, but the sea floor subsidence at Ekofisk caused the tower to be positioned about 1.5 m lower than expected. The highest current meter was a Marsh-McBirney (MMB) electronic current meter (1.5 m above MSL), whereas the remainder were Simrad Ultrasonic Meters (UCM-10). To the side of the tower, five Thorn EMI laser altimeters were mounted in a circle with diameter 7 m for monitoring the surface elevation. All instruments were logged on an HP-1000 central computer. The sampling interval was 2 Hz, normally for a 40 minutes period every 3 hours, but hourly during storms. Figure 2 shows an overview of the sea states experienced during the field experiment.

An example of a data series is shown in Fig. 3. It may be seen that the current measurements in the splash zone occasionally suffer from spikes in connection with the transition from water to air and vice versa.

3. SUMMARY OF RESULTS

Of a total of 920 available records, six were selected for the wave kinematics study. These correspond to moderate to high sea states with one record having a significant wave height close to 11 m.

There are numerous ways by which the validity of DGLWT may be checked. The present study employed comparisons of distribution functions, distribution of extrema and spectral intercomparisons. There exists a well developed theory for the simulation and prediction of multivariate Gaussian stochastic processes and the somewhat novel idea of utilizing this to check the validity of DGLWT was suggested to the project team by WADIC's scientific advisor, Prof. Leon E. Borgman, University of Wyoming (Borgman, 1990).

Although the ocean surface is close to being Gaussian, it is also evident that wave crests tend to be more peaked and troughs more rounded than they would be if the surface was completely Gaussian. This was confirmed by the laser measurements whereas this was not statistically significant for the subsurface current meters. In addition, the "crest" and "trough" distributions for the current meters followed Gaussian theory.

An example of conditional predictions vs. measurements is shown in Fig. 4. The predictions are based on the directional spectrum and the measured time series from one of the lasers.

For sensors in the splash zone, comparisons of spectra and distribution functions do not, of course, make any sense. However, the simulation procedures may nevertheless be applied by simulating time series for a current meter which is periodically in air as if the meter

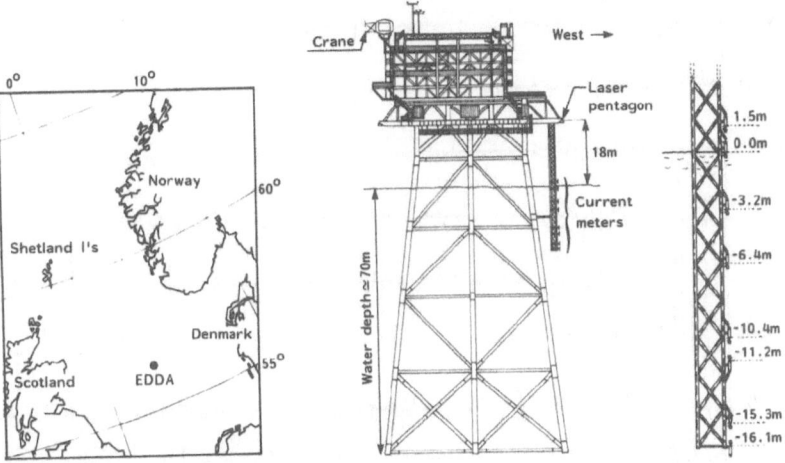

Figure 1. The Edda Platform operated by Phillips Petroleum Co. Norway A/S and the tower holding the current meters.

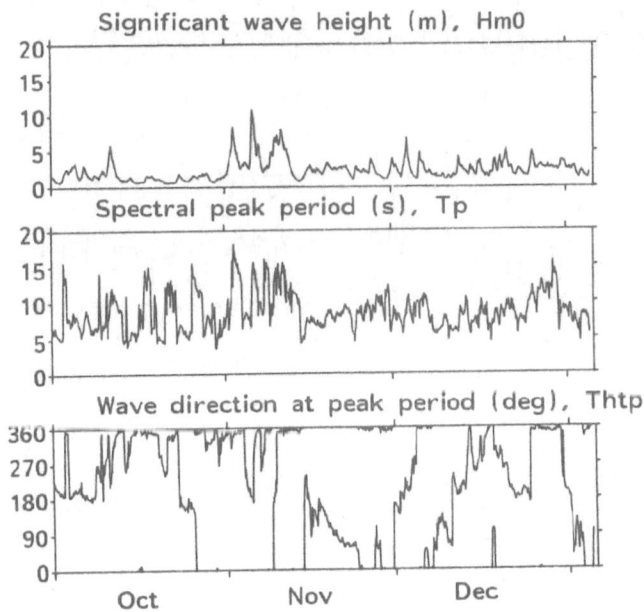

Figure 2. Overview of the sea states during the WADIC field experiment.

636

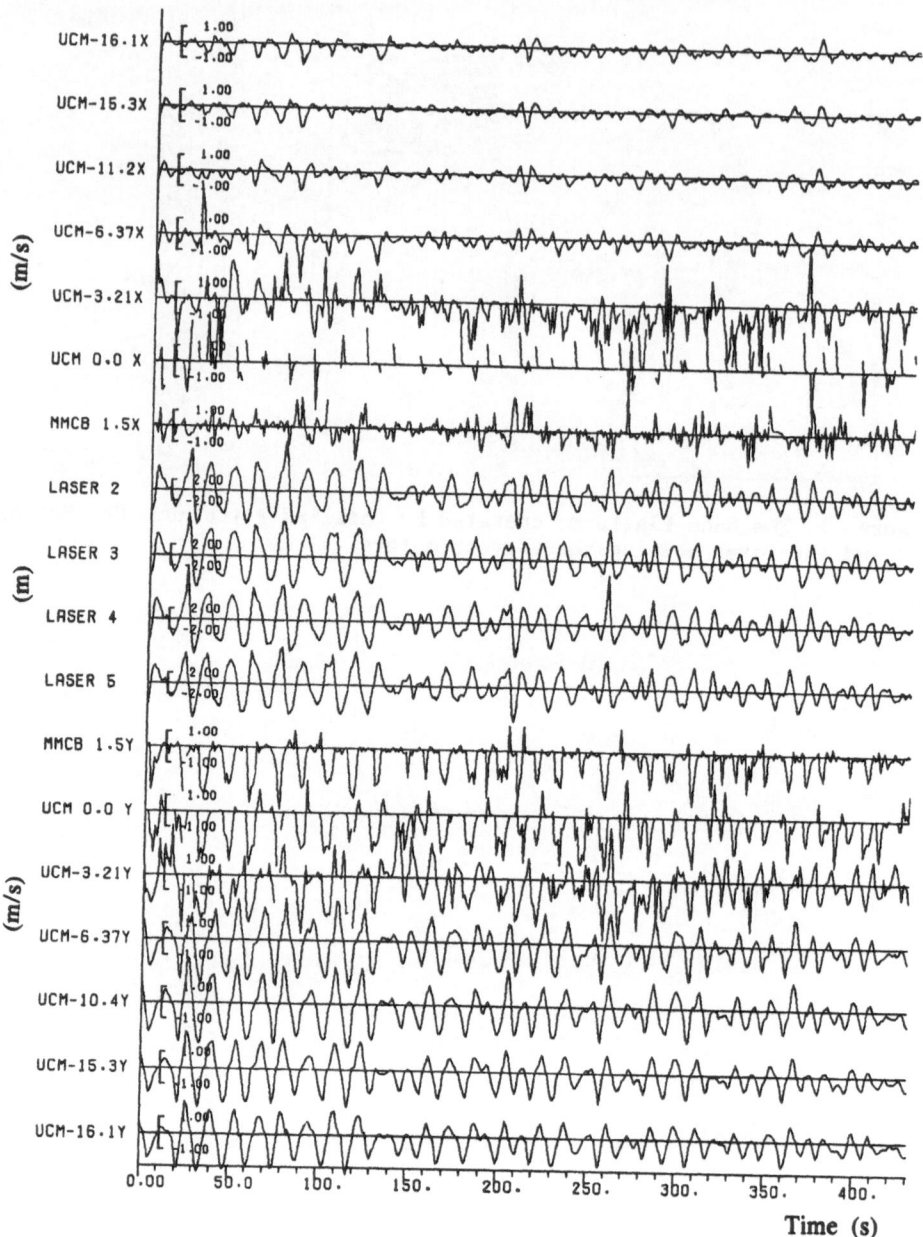

Figure 3. Examples of time series obtained by the current meters and the lasers. (6 Nov. 0020 GMT). The height of the current measurements in metres is indicated to the left.

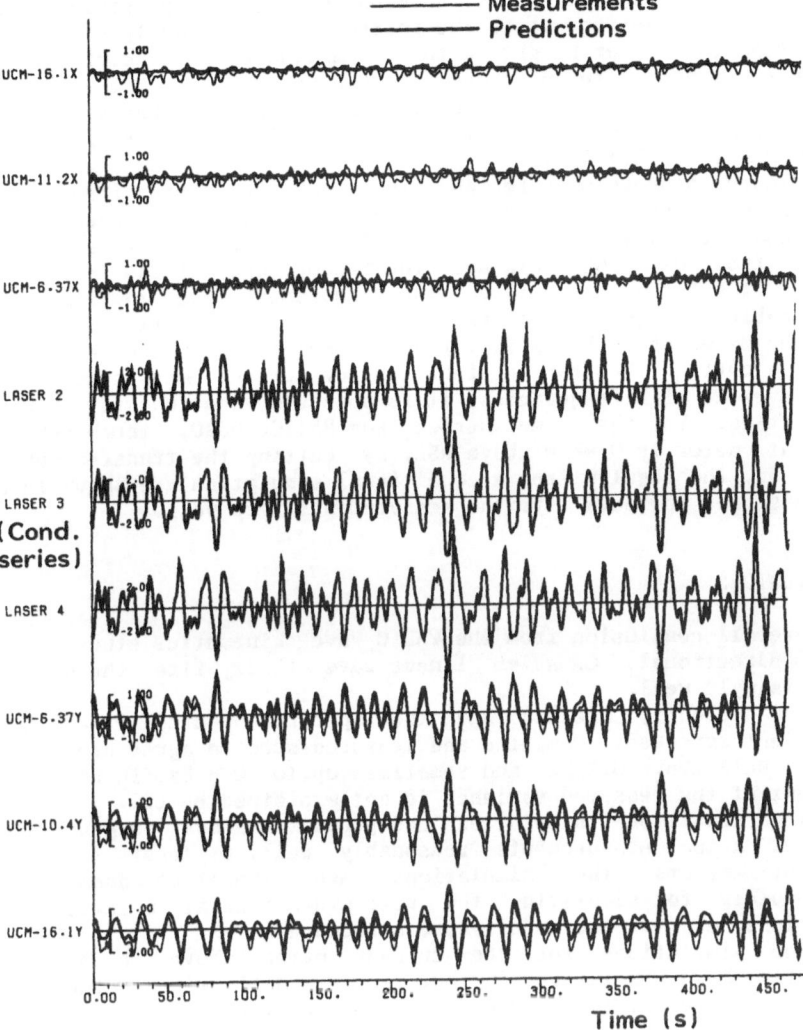

Figure 4. Measured time series vs. predictions based on the directional spectrum and the surface measurements from one laser. Data from 6 Nov. at 0420 GMT.

was completely submerged. As a second step it is possible to discard parts of the series where the instrument is actually in free air by comparing the sensor height to the instantaneous water surface.

For a sensor situated above the mean water level, the high frequency components will be amplified by the factor exp(kz). For the WADIC data, the exponential factor is not dramatically large as for z = 1.25 m and f = 0.5 Hz, exp(kz) ≃ 3.5.

Figure 5 shows measured and simulated time series from the current meters in the splash zone for the record with the most severe sea state. The conditionally simulated water level at the meter is also shown. Apart from the UCM@-3.21x and the x-component of the Marsh-McBirney meter, the agreement between the measurements and the simulations is reasonable. In general, it was found that the conditional simulations for the meter above MSL slightly over-predicted the measurements.

By cutting the transfer function for current meters above the mean sea level at high frequencies, it is possible to attenuate the simulated velocities. For the time series from 851106/0020, the Marsh-McBirney current meter is 0.98 m above MSL. By cutting the transfer function at 0.20 Hz, the results from a conditional simulation are shown in Fig. 6. In this case the data fits the measurements very well.

CONCLUSIONS

An overall conclusion from the WADIC wave kinematics study seems to be that directional, Gaussian linear wave theory fits the measurements surprisingly well.

The velocity spectra and the spectra computed from DGLWT show excellent agreement. Computed and measured spectra agree closely in most cases well above 0.2 Hz and sometimes up to 0.3 Hz. In general, 5-10 percent of the measured variance is not explained by LWT.

The conditional simulations and predictions are in general able to predict actual measurements reasonably well, at least for the deeper current meters. The simulations have therefore demonstrated a methodology for simulating the wave kinematics at a large number of locations, e.g. around an offshore structure.

The simulations for the current meter above the MWL came out slightly high. An interesting but simple modification to the simulation program is to cut the corresponding velocity transfer function at the high frequency end. In this way the data were found to give a better match in one particular case, but the lack of more meters above MSL prevented a closer study of this possibility.

The study has demonstrated the high quality of the WADIC data set and it is obvious that the data still contains much un-revealed information worthy of more research.

ACKNOWLEDGEMENTS

We would like to thank the following oil companies and institutions for their financial support to WADIC:

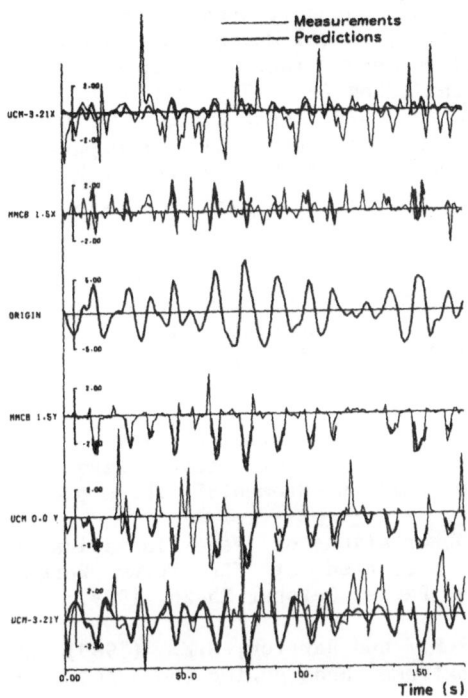

Figure 5. Example of splash zone predictions (conditional on Laser 3) and measured time series. Data from 6 Nov. at 0020 GMT.

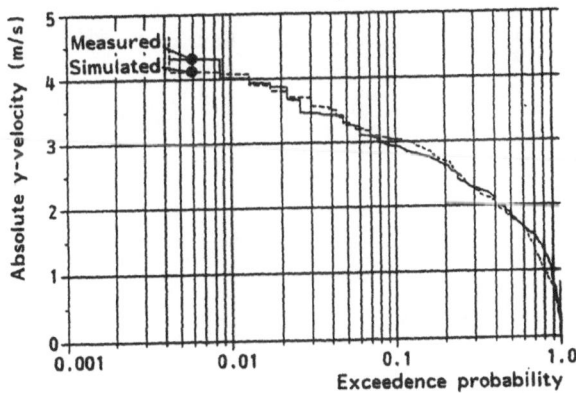

Figure 6. Probability of exceedence plots for measured vs. conditionally simulated values for the Marsh-McBirney current meter where the transfer function has been cut at 0.2 Hz. (6 Nov. 0020 GMT).

Chevron Oil Field Research Company, Conoco Norway Inc., A/S Norske Shell, BP Petroleum Co. Norway, Unocal Norge A/S, Arco Norway Inc., Esso Norge a.s, Mobil Exploration Norway Inc., Phillips Petroleum Co. Norway A/S and Co-Venturers (including Fina Exploration Norway Inc., Norsk Agip A/S, Elf Aquitaine Norge A/S, Norsk Hydro A/S, Den Norske Stats Oljeselskap A/S, and Total Marine Norsk A/S), UK Department of Energy, and Continental Shelf and Petroleum Technology Research Institute Ltd.

REFERENCES

Allender, J., Audunson, T., Barstow, S.F., Bjerken, S., Krogstad, H.E., Steinbakke, P., Vartdal, L., Borgman, L.E. and Graham, C. (1989) "The WADIC Project: A Comprehensive Field Evaluation of Directional Wave Instrumentation", To appear in Ocean Engineering.

Borgman, L.E., Allender, J., Krogstad, H.E., Barstow, S.F., Audunson, T. (1990) "Conditional Simulation of Ocean Wave Kinematics and Comparisons with Storm Field Measurements". These proceedings.

Krogstad, H.E. (1989) "Observations of Wave Kinematics in the WADIC Experiment", To be presented at E&P Forum Workshop on Wave Kinematics and Loading, Paris, October 25-26, 1989.

Vartdal, L., Krogstad, H.E. and Barstow, S.F. (1989) "Measurement of Wave Properties in Extreme Seas during the WADIC Experiment, OTC 5964, Houston 1989.

MEASURING THE LONG WAVE KINEMATICS IN THE PO-WAVES PROJECT

MANUEL A. B. MARCOS RITA
Laboratório Nacional de Engenharia Civil
Departamento de Hidráulica
Av. do Brasil 101
P-1799 Lisboa Codex, Portugal

ABSTRACT. The NATO PO-WAVES project is in progress in Portugal since the end of 1987. One of the objectives is the establishment of a method for determining the long wave climate at the entrance of a harbour, for which a vast measuring field campaign of short and long waves, as well as the observation of moored ships, is starting to be carried out at Sines harbour. Results will allow a better understanding of the role of wave grouping in the occurrence of long waves in harbours and their effects on moored ship behaviour.

1. INTRODUCTION

In Portuguese harbours numerous problems have arisen due to the presence of long waves which, in the case of Sines, have been solved only by removing the ships from their berths to the open sea. In 1985/86, for instance, total berth downtime was approximately 5,000 hours for 665 ships, with an average value per berth of 625 hours.

The construction of port facilities at ever more exposed locations such as Sines has caused the behaviour of moored ships to be carefully analised by both researchers and design engineers. The problem has two aspects, one related to the ships and the way they are moored and the other related to the waves:

- Increasing the size of the ships while keeping the relatively soft mooring systems makes the mass-spring system more sensitive to the exciting loads at lower frequencies, as the resonance frequencies of the moored ship shift to lower values; at these values, hydrodynamic damping is so small that high amplitudes of motion are reached and, as a consequence, high

A. Tørum and O. T. Gudmestad (eds.), Water Wave Kinematics, 641–644.
© 1990 Kluwer Academic Publishers.

mooring line forces and fender forces are caused. This will have a negative impact on the cargo handling efficiency and on the safety of the moored ship.
- Low frequency wave excitation is caused by slow varying wave drift forces, which is a typical effect of long wave activity.

Therefore, proper knowledge of long wave activity at the harbour entrance will allow the design of better lay-outs and better estimates of berth downtime.

Apart from the usual "free" long waves which can be found along the coast, long waves can also be "bounded" to storm waves. In fact, storm waves propagate in groups of varying lengths and heights that give rise to long periodic oscillations of the water level, i.e., to long waves.

The mathematical relationship between wave grouping and bounded long waves has been derived by several authors, and it is believed that they, or at least some of them, describe long waves rather well. However, proper verifications in nature have not been made so far.

In 1984, Delft Hydraulics and LNEC, using a combination of results from limited measurements made at Sines and from a physical model, tried to establish an approach to the ocurrence of long waves at Sines based on the work of Sand (1981) and Sand (1982). It was then concluded, Vis et al. (1985), that to study the problem in more detail and with a higher accuracy much more measurements made by more sophisticated equipments were needed.

2. THE PO-WAVES PROJECT

In 1987, a proposal for a reserch project on the wave climatology of the Portuguese coast was submitted to NATO Scientific Affairs Division for partial funding, under the Science for Stability Programme (1987-92).

This project - PO-WAVES - is conducted by the Portuguese Hydrographic Institute and by LNEC and it consists of 2 sub-projects, one of which (sub-project B) includes a reserch on long waves and their effects on moored ships, which will be based on measurements made at Sines harbour and its vicinity.

The collected data will allow the following objectives to be accomplished:

- To devise a method of establishing a first approximation long wave climate at the entrance of a harbour, for design purposes;
- To determine the correlation between the moored ship

behaviour and the combined action of short and long waves in order to estimate berth downtime;
- To validate the results of wave disturbance tests (physical model), as far as the ship motions and mooring forces are concerned, by comparison with field measurements;
- To validate the results of a mathematical model of moored ship behaviour.

3. MEASUREMENTS AND EQUIPMENTS

In this sub-project it is proposed to measure:

- The short waves, including the directional spreading, near the harbour entrance (approx. 50 meters of water depth) and outside the harbour (approx. 100 meters of water deph), with two WAVEC buoys from DATAWELL;
- The long waves, including the directional spreading, near the harbour entrance (close to the location were one of the WAVECs will be installed) and close to the shore at a water depth of about 25 meters. These long waves will be measured with two SEA DATA directional pressure sensors, mounted on the sea bed, coupled to data loggers with capacity for 30-35 days of data.
- The short waves at a water depth of approx. 25 meters, close to the location where one of the directional pressure sensors will be installed, with a DATAWELL waverider buoy. The directions of these waves will be obtained from the directional pressure sensor.
- The long waves inside the harbour, with 3 SEA DATA non-directional pressure sensors, each mounted onto the wall of a caisson (being part of berths 7/8 and 4/5), coupled to a telemetric system installed on top of the caisson.
- The motions of the ships moored to berths 4 and 5 using:
    - 2 laser rangefinders for surge motion detection, one for each berth, aligned with the main horizontal axes of the ships;
    - 4 accoustic ranging systems, for sway and yaw motions, 2 in each berth, laying across the main horizontal axes of the ships;
- The mooring line forces using the already existing force detection systems at berths 4 and 5 (which need to be calibrated).

The signals of all the above mentioned sensors, except those of the directional pressure sensors, will be collected in the Centralized Command Room of the harbour, where a HP 300 series computer will sample them and store them, in real time, on a high capacity magnetic tape cartridge system.

At LNEC a similar computer system will enable pre--processing of the recorded data after which the signals will be sent on line to the mainframe computer of LNEC (VAX 8700) where they can be further processed.

Measurements will be compared with results from physical and mathematical models in order to verify the validity of these models and to study the behaviour of the moored ships.

To this end, at the LNEC a physical model of Sines harbour will be constructed and an integral software package will be developed to simulate wave propagation and the resulting moored ship motions.

The 2 WAVEC buoys were deployed in the beginning of 1989, and have been since then, after a testing period, in full operation.

The directional pressure sensor to be placed at a water depth of approx. 50 meters (SEA DATA's WAVEPRO DR/XP) and one of the 3 non-directional pressure sensors, to be placed onto the wall of the second caisson of berth 4/5 (SEA DATA's WAVEPRO NR/XP) were delivered in February 1989 and their deployment is scheduled for the end of May 1989.

The WAVEPRO DR/XP sensor utilizes a quartz pressure sensor for wave detection and an electromagnetic current sensor for measuring instantaneous current velocity which is converted to north and east averages by a data cruncher. Resolution of this sensor is 0.15 cm/sec.

Due to restrictions in funding availability, only in the beginning of 1991 will all the equipments be in simultaneous operation.

4. REFERENCES

Sand, S. E. (1981) 'Long waves in directional seas', manus-script submitted to the Journal of Coastal Enginnering, Elsevier Scientific Publishing Company, Amsterdam.

Sand, S. E. (1982) 'Wave grouping described by bounded long waves', Ocean Engineering, Vol.9, N.6.

Vis F. C., Mol A., Rita M.A.B.M. and Deelen, C. (1985) 'Long waves and harbour design', International Conference on Numerical and Hydraulic Modelling of Ports and Harbours, Birmingham.

EXPERIMENTAL FACILITY FOR PROGRESSIVE EDGE WAVES

Harry H. Yeh and Kai-Meng Mok
Department of Civil Engineering
University of Washington
Seattle, WA 98195
U.S.A.

1. Introduction

Edge waves are one type of wave phenomena that may be an especially important factor for understanding coastal hydrodynamics. These waves are one of a class of trapped wave phenomena which can occur nearshore by wave refraction owing to the variable water depth. Edge waves propagate parallel to the shoreline with their crests pointing offshore while their amplitude is maximum at the shore and decays asymptotically to zero in the offshore direction. Thus, energy of edge waves is concentrated and confined to the near-shore region. Because edge-wave motions are boundary dominated flows, they appear to be influenced by the fluid viscosity and to deviate from the prediction of the inviscid theory (Yeh, 1985;1986;1987).

The experimental wave basin was designed specifically for the investigation of the fundamental mechanics of progressive edge waves in a real fluid environment. The primary objective of the study is to measure detailed flow fields created by 'clean' edge waves; the measurements include that in the bottom boundary layer and flow interactions within the swash zone, i.e. near the air-water-beach contact line. Because of the nature of the study, the basin was constructed with extremely high precision.

2. Edge-Wave Basin

A view of the edge-wave basin and the schematic drawings are shown in Figs. 1 and 2, respectively. The dimensions of the basin are 13.4 m long, 5.5 m wide, and 0.9 m deep. The entire tank is elevated so that the bottom floor of the tank is located 1.1 m from the laboratory floor. This setup together with the glass beach (described below) enable us to measure a velocity field from below using the laser-doppler anemometer (LDA) as well as various techniques of flow visualization.

The steel frame structure having the height adjustable column members was designed and erected with the aid of a visible laser level instrument so as to level the basin floor to within 1.6 mm of tolerance from the horizontal plane for the fully loaded condition. The tank floor and side walls were constructed with 2.86 cm thick sheets of "oriented stranded boards", bolted down to the steel frame, and then lined with fiberglass laminate. Five layers of chopped strand fiberglass mat were used to achieve the stiff and water-tight surface of the tank. The finished floor is very close to the horizontal (± 1.6 mm tolerance). Considering the size of the tank, this precision of the tank is more than satisfactory. In

645

A. Tørum and O. T. Gudmestad (eds.), Water Wave Kinematics, 645–648.
© 1990 *Kluwer Academic Publishers.*

order to achieve the precision of the beach construction, the region of the toe of the glass beach section was made absolutely horizontal by pouring resin; the liquid resin formed the smooth horizontal plane by gravity.

The central section of the beach is constructed of a 2.4 m long, 3 m wide, and 1.27 cm thick tempered glass plate. As shown in Fig. 2, the glass plate has a water tight seal to provide the dry observation section. Outside of the observation section (the glass beach section), the beach modules are constructed with 1.27 cm thick Extren sheets supported by the Extren frame. Extren is a proprietary combination of fiberglass reinforcements and thermosetting polyester system; it is relatively stiff, resistant against corrosion, and is hydrophilic (wettable). The wettability of the beach material is crucial to the edge-wave experiments in order to minimize the effects of the air-water-beach contact line. This consideration is important since the maximum energy of an edge wave is located at the shoreline.

The slope of the beach, β, is set $15°$ from the horizontal. With this slope, which can be considered to be mild, $\beta << O(1)$, edge waves of Stokes' mode ($n = 0$) and two other higher modes ($n = 1$, and 2) are possible (Ursell, 1952). For a milder slope, other higher modes are possible and the separation of the modes from the measured raw data would become difficult. In addition, it is also advantageous that the slope $\beta = 15°$ is an angle of the form $\beta = \pi/2m$, m = integer (in this case m = 6), for which the explicit analytical solution for the leaky mode is available. (For the slope $\beta \neq \pi/2m$, no explicit solution based on the full water-wave theory has been found.)

Edge waves are generated directly at one end of the beach by the wave paddle as shown in Fig. 2. The overall wave-generating system consists of three components: the wave paddle, the hydraulic power unit, and the electrical servo system. The hydraulic power unit drives the wave paddle, while the electrical servo system controls the paddle motion. A wedge-shaped wave paddle is hinged offshore and oscillated in the longshore direction about the axis perpendicular to the beach surface. Note that this linear approximation of the paddle motion for edge waves with exponential offshore decay is analogous to that of a "plunger" or "flap" type wavemaker used for the generation of deep-water waves. (As discussed in Yeh (1987), the characteristics of edge waves are analogous to that of two-dimensional deep-water waves.) Furthermore, a uniform paddle velocity with respect to a line perpendicular to the beach surface coincides with the theoretical velocity field of linear edge waves. With the exception of the wave paddle, the entire wave-generating system is installed directly on the laboratory floor outside the tank; thus the mechanical vibration associated with the system does not disturb the water in the tank. Even the wave paddle itself is supported by the structure outside the tank so that any unwanted vibration caused by the paddle motion does not transmit to the beach surface, the tank walls or floor.

The top of the side walls is made of steel tubing which can support an instrument carriage that can traverse along the tank. There are two drains in the tank as shown in Fig. 2; one is used simply to drain the water from the tank or to recirculate the water through the filtering system The other drain outlet is placed at 71 cm above the tank floor in order to skim the water surface to avoid excessive surface contamination of the water.

The flow field in the tank can be measured directly through the glass-plate beach with an optics device, e.g. the backscattering LDA. I believe that this arrangement is the only way to accurately measure the velocity field with the LDA since a focal point of the laser beams can be sufficiently short and the refraction effects can be minimized. In addition, because of the large and dry working area underneath the glass beach, this setup has an advantage for traversing readily the LDA measuring points as well as for flow visualization. Note that the LDA cannot be used from above because of the variable refraction caused by the air-water interface curvature created by the wave motion, and a conventional intrusive flow meter (e.g. an electro-magnetic flow meter) is inadequate because velocity measurements are required in water of very shallow depth.

It is emphasized that the tank is constructed with extremely high precision in spite of its relatively large size. A wave tank of this size is essential to generate clean and measurable edge waves. Scaled down waves generated in a smaller size tank would cause many adverse effects; e.g. 1) excessively strong viscous effects, 2) unwanted disturbances such as leaky-mode waves which might be reflected back from the side walls, and 3) inaccurate measurements of the wave field because edge waves decay exponentially offshore. A weakly nonlinear edge wave, say $ak = 0.2$, of the Stokes mode having the wave length 2 m on the beach $\beta = 15°$ has the wave amplitude of 3.6 mm at 50 cm offshore, i.e. the fundamental investigation of edge waves requires measurements of small scale motions in a large fluid domain. Moreover, one of the primary objectives is to explore the fluid mechanics associated with the flow near the air-water-beach contact line, including the boundary layer formation and its interaction with the free surface, and the surface tension effects due to the contact line. Such experiments require measurements in the spatial resolution of the order of millimeters.

The edge-wave basin was designed and constructed with the aid of Messrs. Arn Thoreen, Donald Romain, and Elton Daly. The financial supports of the University of Washington and the Office of Naval Research (N00014-87-K-0815) are appreciated.

References

Ursell, F. (1952) 'Edge waves on a sloping beach', Proc. R. Soc. Lond. A 214, 79-97.
Yeh, H.H. (1985) 'Nonlinear progressive edge waves: their instability and evolution', J. Fluid Mech. 152, 479-499.
Yeh, H.H. (1986) 'Experimental study of standing edge waves', J. Fluid Mech. 168, 291-304.
Yeh, H.H. (1987) 'A note on edge waves', in R.A. Dalrymple (ed), Coastal Hydrodynamics, Amer. Soc. Civil Engr., New York, pp. 256-269.

Figure 1. A view of edge-wave basin.

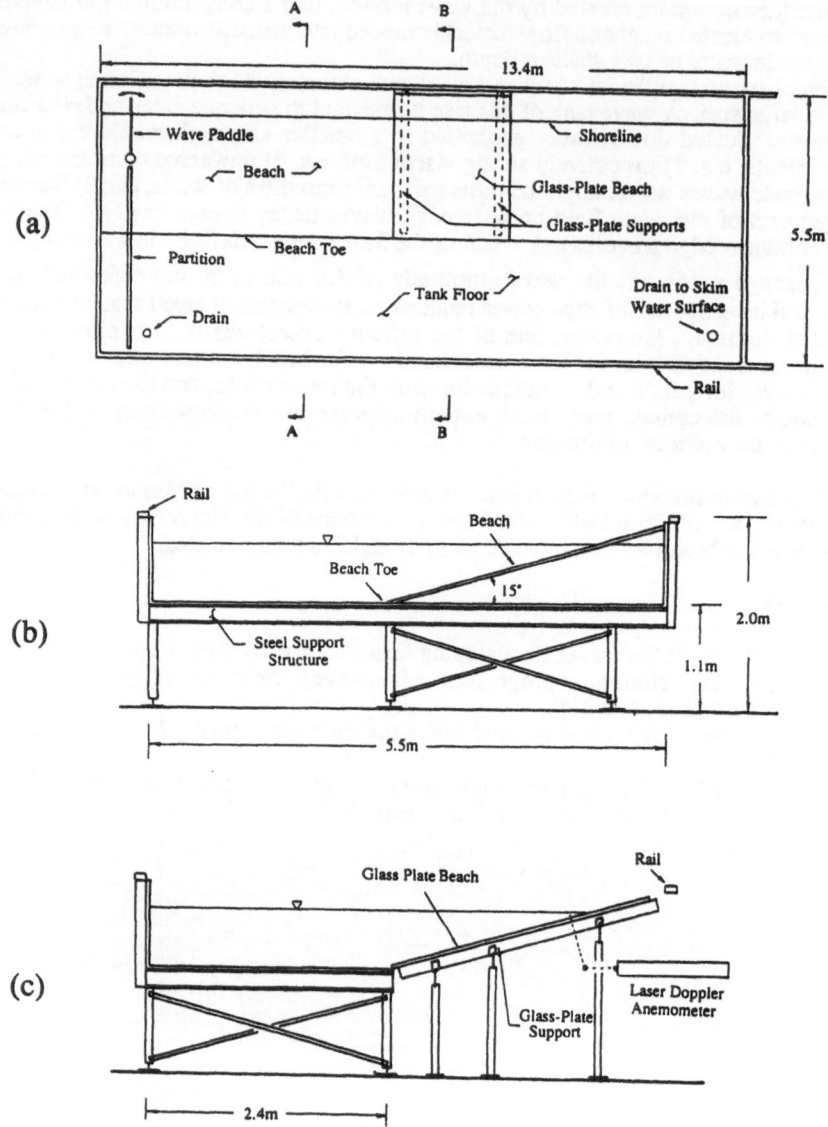

Figure 2. Schematic drawings of edge-wave basin. a) plan view; b) elevation view, Section A-A; c) Section B-B.

Forces

MORISON TYPE WAVE LOADING

Professor Geir Moe
The Norwegian Institute of Technology
Division of Structural Engineering
N-7034 Trondheim - NTH
Norway

ABSTRACT. A review has been made of some of the regular flow applications of the Morison formula, then some new material concerning the force coefficients in irregular wave situations has been presented.
   The regular flow cases considered are fixed circular cylinders subjected to
* Straight and oblique rectilinear, harmonic flows
* Orbital flows with the cylinder in the plane of the particle orbit
* Flows rotating around the cylinder axis
* Coexisting waves and currents
* Surface proximity effects
   The irregular flow cases have been treated as random processes, obtaining spectra and distributions of maxima. Also seastates have been characterized by certain dimensionless numbers, and it has been attempted to correlate the force coefficients to these numbers.

1. INTRODUCTION

Determination of wave forces on structures has for some time been a central problem in Offshore Engineering.

   Most of the effects occurring in <u>unseparated flows</u> have been understood for several decades and practical solutions are now available for most geometries and flows, even though problems still exist in some cases, e.g. for members in the splash zone and with nonlinear effects such as the so-called drift forces. The separated flow cases, by constrast, are only partly understood and requires a great deal of empirism to be dealt with.

   The present paper is on forces in separated flows about cylindrical objects. Forces in this regime are usually predicted by the so-called Morison formula, which in its most basic form can be written

$$f = C_m \varrho \pi R^2 \dot{u} + C_d \varrho R u|u| \qquad (1.1)$$

in which ϱ is water density, $R = D/2$ is cylinder radius, and $u = u(t)$ and $\dot{u} = \partial u/\partial t$ are water particle velocity and acceleration respectively. The latter is to be determined in undisturbed flow, at the position of the cylinder, i.e. the influence of the cylinder on the flow is to be neglected. C_m and C_d are dimensionless constants, estimation of their values is at the heart of the present problem.

   Formula (1.1) is named after the lead author of a 1950 paper, Morison et al. (1950). It is commonly referred to as "The Morison equation", but the present author prefers the term "the Morison formula". This is indeed a minor distinction, but "formula" may be more indicative of the degree of empirism involved.

   The Morison formula occupies a central role in many conference proceedings (e.g. "BOSS", "OTC" and "ASME/OMAE"), in periodicals, and also

A. Tørum and O. T. Gudmestad (eds.), Water Wave Kinematics, 651–677.

in at least two recent textbooks, namely Sarpkaya and Isaacson (1981) and Chakrabarti (1987). It is probably a practical impossibility to prepare a reasonably complete reference list in this area. Instead an attempt will be made to identify different important cases, according to the physics involved. Then some selected references will be given as the need arises.

In spite of its widespred popularity, the Morison formula is often misused. One particularly common mistake is to use a C_m coefficient applicable only for a potential flow situation for separated flow situations. Also C_d coefficients applicable for a steady flow is often used in distinctly unsteady flow situations. Both approaches can be seen even in rules and regulations, yet it should be intuitively obvious that the inertia force in a separated flow will be different from a case where the flow do not separate. Likewise the steady current drag is one limiting situation, and zero form drag in the rapidly changing case another. Between these limiting cases there is a range in which neither is valid. Another common misunderstanding concerns the range of applicability of this formula. Many users will claim that the $C_d \simeq 0$ limit is reached at a certain fixed value of diameter to wavelength, instead of relating it to the Keulegan-Carpenter number. (See later.)

2. CLASSIFICATION OF CASES ANALYSED BY THE MORISON FORMULA

2.1. GENERAL

The cases that are analysed by use of the Morison formula may be classified in several ways, depending on tradition, field of application and personal preferences.

Very broadly speaking one should consider two aspects:
1. Structural geometry and stiffness, etc.
2. Particle velocity kinematics.

At the outset a dimensional analysis involving a rigid, circular cylinder in a "wavy" flow may be very useful in identifying the dimensionless parameters involved. However, even this apparently simple basic case may be defined in several ways depending mainly on the flow situation involved. E.g. Sarpkaya & Isaacson (1981) consider a rectilinear harmonically oscillating flow, obtaining the following four dimensionless groups.

$$K = u_0 \, T/D \quad \text{(Keulegan-Carpenters No.)} \tag{2.1}$$

$$Re = u_0 \, D/\nu \quad \text{(Reynolds No.)} \tag{2.2}$$

$$k/D \qquad \text{(relative roughness)} \tag{2.3}$$

$$t/T$$

in which u_0 is the particle velocity amplitude, T is the oscillatory period, t is time, ν is the kinematic viscosity, and k is the (average) diameter of the surface roughness elements. Sarpkaya then proceeds to neglect the t/T parameter, recommending that (1.1) is used with suitable time-invariant averages for C_d and C_m.

Newman (1977) considers long crested, regular, harmonic waves and lists the following additional dimensionless groups:

$$\eta_0 /L \qquad \text{(wave "steepness")} \tag{2.4}$$

h/L	(waterdepth parameter)	(2.5)
β	(angle of incidence)	(2.6)

in which η_a is surface amplitude, L is wavelength, h is waterdepth and β is the angle of incidence of the incoming wave relative to a characteristic orientation of the body. For circular cylinders in deep water waves, the latter should be the angle between the plane of wave propagation and the cylinder axis. However, even this six parameter list is not at all complete. For one thing the submergence of the body, z, has not been covered. And further, in finite or shallow water depths, the cylinder orientation relative to the waves requires two angles, since the particle orbit will now be elliptic. Also, the forces on a cylinder will be influenced by proximity to a boundary, e.g. the free surface, the bottom or another structure. And finally one must bear in mind that the structure can be different from a fixed, "very long" circular cylinder and the flow different from a regular unidirectional, harmonic wave.

On the other hand, in many cases some of the parameters will be of minor influence. Thus the situation treated by Sarpkaya, namely a purely unidirectional sinusoidal flow far away from any boundaries and acting on a circular cylinder perpendicular to the flow may serve as a reference case. It covers some very essential parameters that may be of a dominating influence in a great number of applications. Other situations are then treated as additional considerations and discussed in addition to the basic case.

The cylinder on which the forces are to be determined, may for example be a member of a jacket, a marine riser, a mooring line, or part of e.g. a fish farming installation. In some of these situations the member may be considered rigid, in other cases "flexible".

2.2. EFFECTS OF CYLINDER RIGIDITY

Rigid members such as members of a jacket will not respond appreciably to individual vortices created by separation of the flow, but only to the integrated effects of the wave forces on all members. (Global response.) Then, especially if the motion of the structure is quite small relative to the particle motion, experiments performed with rigid cylinders will be relevant, (except possibly in determination of hydrodynamic drag damping).

Flexible members such as mooring lines, cables, flexible risers, etc. may respond directly to the local flow pattern around it, interacting with the flow, i.e. having so-called hydroelastic motions. Reliable predictions of these situations cannot at present be made. Consider briefly the related, but simpler situation of a flexible cylinder in a current. The findings from a series of tests with circular cylindric segments that were spring supported both in the in-line and the transverse direction has been summarized by Moe & Wu (1989). The in-line natural frequency was twice the transverse natural frequency. When the cylinder was subjected to a steady, uniform current, then vortex induced vibrations occurred throughout the lock-in range, and these were nearly sinusoidal, but with a slight amplitude modulation. The force, however, was quite irregular, see Fig. 1. An attempt was made to repeat the experiments with forced sinusoidal motions with amplitude and frequency equal to the average motion amplitude and frequency. The resulting force history was now quite different from the previous case, even though some key macro features came out very similar, see Moe & Wu (1989). A striking example of the difference between the two cases can be seen in Fig. 2, showing the correlation coefficient between the forces measured on two force sleeves, 6 diameters apart, Wu (1989). It can be seen that the correlation coefficient in the self-excited tests are sometimes much higher, sometimes much lower than for the forced vibration tests.

Considering how similar the two motion histories are, this leads one to conclude that even very small pertubations in the motion patterns may have a profound effect on the forces, and then also on the flow patterns. The variation in axial correlation thus implies different three-dimensional behaviour, possibly associated with flow instabilities. Such effects must by the way be expected to be extremely difficult to model numerically.

2.3. OTHER MEMBER PROPERTIES

The effects of cylinder roughness has already been mentioned. Due to space limitations very little data will be given for this case herein.

Member proximity efects. Such structures as riser bundles or riser "ribbons" will be in very close proximity to each other. In other situations one or more small members may be attached to the surface of a large tank, a ship or e.g. the leg of a drilling rig. The forces in such situations may exhibit large deviations from the sum of the forces acting on the members if no coupling effects had been present.

2.4. PARTICLE KINEMATICS, RELATIVE TO CYLINDER

The following flow effects may be very important in determining the magnitude of the forces on a cylinder in a "wavy flow":

A.   Orientation of cylinder axis relative to the plane of wave propagation. (I.e. relative to a plane through a particle orbit). There are two basic cases: a horizontal cylinder parallel to the wave front, and a cylinder perpendicular to that direction (e.g. vertical).

B.   The shape of the particle orbit. This may be given as the ratio of the two semiaxes.

C.   The waves may be unidirectional and regular, consisting of one or several frequency components - or unaxial and irregular, described by random process theory.

D.   The wave may be as under point C, but directional spread may be included.

E.   A current (steady, uniform?) may be added.

F.   Very close to the surface, special effects are present. (See Section 4.2.).

G.   Flow-field turbulence may be important.

H.   In a stratified fluid the forces on horizontal members may be profoundly changed. (Will not be pursued herein.)

3. MORISON FORCES - REGULAR WAVES

3.1. GENERAL

Very much work has been done to determine the force coefficients, C_a and C_d, for the "basic case" introduced in Section 2.1. (Harmonic, rectilinear flow in an infinite fluid.) Here the force coefficients are functions of K, Re and roughness, (usually assumed to be of a hard fouling type and then characterized by k/D). Thus

$$[C_d, C_m] = f(K, Re, k/D) \qquad (3.1)$$

Values of these parameters have been given for a reasonable wide parameter range by Sarpkaya and Isaacson (1981), see Figs. 3 and 4. Often an alternative choice of parameters is used, in which

$$\beta = Re/K \qquad (3.2)$$

replaces the Reynolds number.

It is particularly striking to see from Fig. 3 that the C_m value plunges from the potential flow value 2.0 to well below 1.0 in an intermediate Keulegan Carpenter range (say 10 to 20) and at subcritical Reynolds numbers. The C_d variation goes in the opposite direction. This implies a phase shift of the total force whose magnitude turns out to change much less than the drag and inertia components do. A so-called total force coefficient, C_F, defined by

$$C_F = f_{max}/\rho R u_o^2 \qquad (3.3)$$

is therefore often used. Alternatively the force maximum and the velocity amplitude are replaced by their respective rms values, e.g. Stansby et al. (1983) or Bearman (1988). Force coefficient curves will usually be more smooth and the C_F datapoints will usually show less scatter, than the corresponding C_d and C_m values. On the other hand, C_F does not give all the necessary information about the force, since the phase and indeed the full force trace, are needed in a structural analysis. The C_F approach should therefore never be used as more than a supplement to the Morison formula. For comparison a normalized form of (1.1) for harmonic flows may be written

$$u = u_o \cos\omega t$$

$$f/\rho R u_o^2 = C_d \cos\omega t \cdot |\cos\omega t| - C_m (\pi^2/K)\sin\omega t \qquad (3.4a)$$

in which K is given by (2.1). Taking the derivative of the above function one finds that a maximum occurs at a time t given by

$$\sin\omega t = -\frac{\pi^2}{2K} \cdot \frac{C_m}{C_d} \qquad (if > -1) \qquad (3.4b)$$

$$\cos\omega t = 0 \qquad (else) \qquad (3.4c)$$

In the latter case one has

$$C_F = C_m(\pi^2/K) \qquad (\omega t = 1.5 \cdot \pi) \qquad (3.4d)$$

In the former case

$$C_F = C_d[1 + (\frac{\pi^2}{2K} \frac{C_m}{C_d})^2] \qquad (3.4e)$$

3.2. FLOWS ON INCLINED CYLINDERS

Several investigations have been made with rectilinear, oscillatory flows acting at an oblique angle against a circular cylinder. The best approach is to use the velocity component perpendicular to the cylinder in the Morison formula, the so-called "cross-flow principle", which is known to work well for steady flows.

Garrison (1985) and Sarpkaya (1985) have investigated some subcritical cases, showing that the inertia term can be predicted accurately based on the normal component of the flow, except in the range 10 < K < 20 where the inclined cylinder inertia coefficient remains almost constant, while it in perpendicular flows drops off dramatically. The drag coefficient is well predicted by the cross flow principle at high Keulegan-Carpenter numbers, but shows some deviations at low K numbers. The effects of superimposing an axial flow may be to trigger a more three-dimensional flow structure, also influencing the separation process. Loosely speaking this appears to smoothen out the more peculiar features seen in the basic flow case.

3.3. ORBITAL FLOW EFFECTS - VERTICAL CYLINDERS

For subcritical Reynolds number flows several investigations have been published on the effects of orbital flow when the pile is in the plane of the particle velocity ellipse. This ellipse is characterized by v_0/u_0, in which u_0 and v_0 are the horizontal and vertical particle velocity amplitudes respectively. Stansby & Al. (1983) and Sarpkaya (1984) both found that the v_0/u_0 ratio had a strong influence on the force coefficients in the Keulegan-Carpenter range 10 to 30. When v_0/u_0 increased, then C_d decreased and C_m increased. Sarpkaya's tests were done in a U-tube with forced transverse cylinder motions while Stansby's tests were made in a wave flume, but the results appeared to be in very close agreement. In constrast, Chackrabarti has presented test results from a wave flume in which the results showed a considerable scatter, but no clear correlation between force coefficients and the v_0/u_0 ratio were found.

In Fig. 5 is shown a collection by Bearman (1988) of high Reynolds number results. (Re in the range 10^5 to $5 \cdot 10^5$.) Two of the experimental cases involve vertical cylinders, namely the triangles, Nath (1985) considering a regular wave situation, and the squares, Bearman & Al. (1985) for a case of irregular very narrow banded and steep waves. The other three series of tests are with vertical cylinders oscillated horizontally in water otherwise at rest. It is not easy to see any indications of the effects of orbital flow here. That is not quite conclusive, however, in view of the range of Reynolds numbers that were present in the tests, and the complexity of such experiments.

Tentatively one may conclude that the cross flow principle seems to be approximately correct for an orbital flow when the pile axis is in the plane of the velocity ellipse.

3.4. ROTATING FLOWS AROUND CYLINDERS

For a submerged horizontal circular cylinder parallel to a wave front the incoming wave velocity vector will rotate around the cylinder axis and the wake will also rotate following essentially the water particle paths, so that, in pure harmonic waves, the cylinder will be passing through its own wake. In a rectilinear oscillatory flow, the cylinder also passes through its own wake, but then it reenters the wake going in the opposite direction, and encounters vortices of the opposite sense. For the rotating wake case the pile reenters the wake going in the same direction as before. The situation is thus similar to a row of cylinders in a so-called tandem arrangement, i.e. the drag should be expected to decrease. This is indeed confirmed experimentally.

This problem has been investigated either in waves, or by rotating a cylinder in water otherwise at rest. These situations will be nearly equal. However, a pressure gradient is present in the wave case, and over the cylinder surface it integrates to the so-called Froude-Kylov force. For the rotating cylinder case this term is absent, but it can be easily calculated, and thus also corrected for. The rotating cylinder experiment results in very accurate motion kinematics, while laboratory generated water waves will have appreciable deviation from its idealized shape, and will often also, have elliptic paths. Some results from Teng and Nath (1985) are reproduced in Fig. 6. The drag coefficient is considerably lower than for the harmonic rectilinear case. Holmes and Chaplin (1978) have published rotating cylinder results, and here the results are even lower than the averages in Fig. 6.

The drag force is per definition aligned with the instantaneous value of the particle velocity. For a pure rotating flow, $v_0 = u_0$, the particle acceleration vector will be perpendicular to the velocity vector. Thus the force component perpendicular to the velocity vector will be identified as the inertia term. The part of the wake nearest to the cylinder will be roughly parallel to the particle velocity vector, thus oscillatory lift will be included in the inertia force. Chaplin (1985) has shown that also an asymmetric lift force will occur due to the different magnitudes of the vorticity created on the two sides of the cylinder. This will give rise to an outwardly directed force opposing the force due to particle acceleration. The net result will therefore be a much smaller inertia coefficient than for the corresponding rectilinear flow. Fig. 7 shows a collection of inertia coefficients in rotating flow compared with Sarpkaya's rectilinear flow results. The values are very low except at Reynolds numbers well above critical, where one might speculate that turbulence makes the wake irregular. The other rotating flow cases in Fig. 7 are from low Reynolds numbers wave tests. The Teng & Nath (1985) results in Fig. 6 are from a flow with much less perfect particle kinematics, and show higher and more scattered C_d values. One might argue that these conditions are more indicative of real flows in the ocean. In Bearman & Al. (1985) measurements on a horizontal cylinder in narrow banded steep waves are given, and an attempt is made to fit a rotating flow Morison formulation to the force trace by a least squares fit. The results were discouraging. The same problem for a low Keulegan-Carpenter regime has been addressed in Chaplin (1988). He introduced two sets of drag and inertia coefficients. For the x direction he wrote

$$F_x = C_{dx} \varrho R u(u^2 + w^2)^{1/2} + C_{my} \varrho \pi R^2 \dot{v} \qquad (3.5)$$

Now the force traces for individual waves can be reasonably well fitted. However, the coefficients of the inertia components C_{mx} and C_{my} still exhibits excessive scatter when plotted against the corresponding Keulegan-Carpenter numbers K_x and K_y. By considering the circulation Chaplin introduced the following expressions:

$$C_{mx} = 2 - r K_y^2 \qquad (3.6a)$$

$$C_{my} = 2 - r K_x^2 \qquad (3.6b)$$

in which r is an empirical function of the orbit parameter v_0/u_0. Now

658

the fit is reasonable, see Fig. 8.

3.5. COEXISTING WAVES AND CURRENTS

The case of a rectilinear oscillatory flow in-line with a steady, uniform current has been investigated extensively. For subcritical Reynolds numbers a study was made by Verley and Moe (1978), and some of these results can be found in Sarpkaya and Isaacson (1981) and Chakrabarti (1987). The dimensionless parameters used were

$$[C_{dc}, C_{mc}] = f(K, \beta, U_r, k/D) \tag{3.7a}$$

$$U_r = U T/D \tag{3.7b}$$

in which U_r is the only new parameter relative to the "basic case" (3.1). $T = 2\pi/\omega$ is the period of the oscillatory flow and U is the steady current velocity. The Morison formula were based on the relative velocity between current U and the oscillatory pile velocity $\dot{x}$.

$$\dot{x} = -u = -u_0 \cos\omega t \tag{3.8a}$$

$$f = C_{dc}\varrho R(U+u)|U+u| + C_{mc} \varrho \pi R^2 \dot{u} \tag{3.8b}$$

This approach was denoted "the relative velocity formulation". An alternative formulation based on independent flow fields were also employed, but was found to offer no clear advantages.

Iwagaki & Al. (1983) made similar experiments, obtaining similar results. They also tried to find a new definition for K, so that K and U_r could be combined into one parameter, and they found that the best expression was

$$K_2 = (2\pi/D) \int_{t^*}^{T/2} (U + u_0 \cos\omega t)dt , \quad u_0 > U \tag{3.9a}$$

$$t^* = Arc \cos(U/u_0)/\omega \tag{3.9b}$$

$$K_2 = \pi U T/D \qquad u_0 < U \tag{3.9c}$$

Using this Keulegan-Carpenter number they found that the results collapsed reasonably well on the zero-current C_d and C_m curves. They also explored the following alternative formulation

$$K_1 = (U+u)T/D \tag{3.10}$$

By introducing suitable characteristic values of time, velocity etc. in the Navier Stokes equation a forth version of the Keulegan-Carpenter number may be formulated

$$K_3 = [(U+u_0)^2/u_0]T/D \tag{3.11}$$

Sarpkaya & Al. (1984) probably were the first to introduce K_3, but they did not give any details on how this number was derived. They presented wave-current test results that covered a wider parameter range, and reports to have tried to plot these versus K_1, K_2 and K_3. But they concluded that the force coefficients basically depended on the original four parameter list (3.7a), and that the attempts to combine K and U_r had been unsuccessful.

Some high Reynolds number results can be found in Rodenbusch & Gutierrez (1983).

3.6. SURFACE PROXIMITY EFFECTS

Vertical surface piercing piles will experience maximum velocity at the water surface, and the Morison formula will then predict the largest forces at this location. However, the pressure at the water surface must everywhere be equal to the atmospheric pressure, so the force at this location must be zero. This apparent contradiction is a consequence of the use of what in effect is a strip theory and thus implicitly assumes very gradual axial changes, at a point where the axial force function is in fact discontinuous. Near the surface a three-dimensional flow will arise, and that will cause a smoothing out of the dynamic pressure and a force decrease. Furthermore there will be a run-up at the front stagnation point and a drawdown at the rear side.

This problem has been discussed by Dean & Al. (1981). The run-up was estimated to $u^2/2g$ and an empirical reduction function expressed in terms of the distance from the undisturbed free surface s, divided by $u^2/2g$, was presented. Fig. 9 shows one figure from the Dean & Al. paper. The predicted run-up appears to be extremely large, say about 10 m on a pile of only 1.2 m diameter. In Dean & Al. (1985) some preliminary results from laboratory tests with regular waves are presented.

4. MORISON TYPE FORCE FOR IRREGULAR WAVES

4.1. INTRODUCTION

In principle there are two problems to be dealt with in this area:

1. If the Morison formula in the forms (1.1), (3.5), (3.8) or some extension hereof, is indeed valid with well defined constants C_m and C_d, then the problem is to use random process theory to determine the variance and cross spectra of the Morison force, the statistics of extreme forces and so on.

And for vibration analysis of structures also hydrodynamic drag damping and added hydrodynamic mass must be derived from the Morison formula using e.g. equivalent linearization techniques.

2. The determination of the force coefficients C_d and C_m for an irregular wave case cannot be expected to follow from any regular flow situation, no more than e.g. the coexisting current and wave case can be predicted from the regular wave case.

The first problem attracts a great deal of knowledgeable analysts, and has therefore been covered quite well. The second problem, by contrast, has received much less attention, and will therefore be the main topic of this review.

4.2. FITTING OF MODELS TO EXPERIMENTS

A very common method of analysis has been to approximate the waves, or perhaps more commonly, the particle velocity traces, as piecewise sinusoidal, usually on a halfwave by halfwave basis. In an interesting early simulation study Borgman (1969) showed that the calculation of the force coefficients by this method gives very wide scatter even for typical ocean waves, starting from a "perfect" force trace generated by the Morison formula with given C_d and C_m values. A method of this type has been used by Heideman et al. (1979), for the OTS data.

Several other methods exists. Bostrøm (1987) used the following statistical methods:

1. Fitting the probability density of forces by the method of moments.
2. Fitting the force peak density distribution by the method of maximum likelihood.
3. Least square fitting of the force spectrum.
4. Fitting the cross-spectra between surface elevation and force.
5. Fitting the time series of forces by an extended Kalman filter.

There are undoubtedly a large number of other approaches available. Thus Bearman et al. (1985) used a time averaged version of the following expression

$$C_D = 2E[F(t) \cdot u(t)]/\{\varrho DE[|u^3(t)|]\} \qquad (4.1)$$

$$C_M = 4E[F(t) \cdot \dot{u}(t)]/\{\varrho \pi D^2 E[\dot{u}^2(t)]\} \qquad (4.2)$$

It appears that the two expressions are valid provided u(t) is a stationary process. This may be seen by multiplying (1.1) with u and $\dot{u}$ respectively and taking the expectation to both sides. Bearman et al. use (4.1) and (4.2) for very short time series, namely on a wave by wave basis. Then stationarity cannot be invoked, but E(u $\dot{u}$) and E(u|u|u) will probably still be almost zero and (4.1) and (4.2) will therefore constitute a simple and numerically stable method.

4.3. MODELLING WAVE DIRECTIONALITY

A plot of measured horizontal particle velocity and horizontal force are given in Fig. 10 from Sparboom (1986). The particle acceleration is the rate of change of the particle velocity vector, and its horizontal component (in directional seas) will therefore almost always differ from the direction of the horizontal velocity. The horizontal wave force vector will not be alligned with either of the two, nor with the vector sum of the horizontal drag and inertia terms according to a vector form of the Morison formula. One reason for this is the lift forces. More importantly it should be remembered that separation takes time. In regular waves at a Keulegan-Carpenter number 10-20 one pair of vortices are shed per half cycle. If the incoming particle velocity were to change its direction, the position of already shed vortices would change only gradually. Thus for moderate values of K (say 40 or less?) the wake direction would probably be better charac+erized by the average particle velocity direction than by its instantanous value. Bostrøm (1987) accordingly determines the principal particle velocity direction which in the cases investigated by him, differs very little from the principal wave force direction. He then substitutes the components of particle velocity, acceleration and force, all in the principal wave direction, into the Morison formula, determining C_d and C_m.

Rodenbusch & al (1980), in contrast, used a so-called "velocity tracking method", based on the instantaneous velocity direction and take the acceleration and force components in the same direction. The Morison formula is then used to determine C_d and C_m. Thus the magnitude of u|u| will always be taken fully into account, while only part of the total force will be considered. Hence the C_d value will be

smaller than with the Bostrøm approach. For small K-numbers in which
the acceleration forces tend to dominate, the method becomes quite
uncertain since neither acceleration nor force act along the direction
used.

Bishop (1984) starts with the Morison's formula on vector form with
particle velocity and acceleration $u_t = [u \ v]$ and $\dot{u}_t$ respectively. Thus

$$|u_t| = (u^2 + v^2)^{1/2}$$

$$F = C_d \varrho R u_t |u_t| + C_m \pi \varrho R^2 \dot{u}_t \tag{4.3}$$

Squaring and averaging both sides he got

$$E[F^2] = (C_d \varrho R)^2 E[u_t^4] + (C_m \varrho \pi \varrho R^2)^2 E[\dot{u}_t^2] \tag{4.4}$$

because $E[u_t \cdot |u_t| \cdot \dot{u}_t]$ will be negligible. He sets $C_d = 1.0$ in the
inertia dominated regime, and then C_m can be calculated from (4.4). In
the drag domiated regime, he choses C_m and calculates C_d from (4.4).

Sparboom (1986) uses essentially the same procedure, but he estab-
lishes two equations for C_d and C_m by taking the time averages of (4.4)
for two neighbouring time intervals. He then proceeds to average the
estimates of C_d and C_m for several intervals. The problem with this
approach is that a "perfect result" would yield exactly the same values
in both intervals for $E[F^2]$, $E[u_t^4]$ and $E[u_t^2]$, making the 2 equations
singular. The "Method of moments" uses (4.4) and an expression for
$E[F^4]$, obtaining two independent equations for C_d and C_m, Pierson and
Holmes (1965). (Here one may assume that u_t is zero mean and
unidirectional.)

4.4. NON-DIMENSIONAL NUMBERS

To extend of the nondimensional numbers to irregular waves one may go
back to the Navier Stokes equation, introducing characteristic time,
length and velocity, t_c, D and u_c respectively. This is well known to
result in the occurence of certain non-dimensional numbers as
multipliers to dimensionless, order-one differential expressions.
Among these multipiers are the Reynolds number $Re = u_c D/\nu$ and the
Keulegan-Carpenter number $K = u_c t_c /D$. In stationary flow $t_c \to \infty$ and u_c
= U so that one gets the customary expression for Re. In harmonic
flows $t_c = T$ and $u_c = u_0$ and again the well known expression for
oscillating Reynolds number and K are recovered. For regular waves and
a current, K_3 of (3.11) follows. In irregular, zero mean flows the
following characteristic velocity and time have been suggested, Moe &
Bostrøm (1989), Overvik (1988).

$$u_c = \sqrt{2} \ \sigma_u \tag{4.5}$$

$$t_c = 2\pi \ \sigma_u / \sigma_{\dot{u}} \tag{4.6}$$

The latter is the average zero upcrossing period of the particle
velocity, T_z. Substituting, one obtains

$$Re = \sqrt{2} \ \sigma_u D/\nu \tag{4.7}$$

$$K = 2\pi \ \sqrt{2} \ \sigma_u^2 /(\sigma_{\dot{u}} \ D) \tag{4.8}$$

Bishop (1984) has introduced the following Keulegan-Carpenter number.

$$K_\ast = \frac{4\pi}{\sqrt{3}\,D}[E(u_t^4)/E(\dot{u}_t^2)]^{1/2}$$
(4.9)

For a plane, purely harmonic case $E[u_t^4] = 3\,u_0^4/8$, and $E[\dot{u}_t^2] = u_0^2/2$, thus from (4.9)

$$K_{\ast,reg} = 2\pi\,u_0^2/(D\,\dot{u}_0) = u_0\,T/D$$

reduces to the standard definision. In irregular waves $\dot{u}$ and u will both approach the normal probability distributions. Then $E[u^4] = 3\,\sigma_u^4$ and $E[u^2] = \sigma_u^2$, so that

$$K_{\ast,norm} = 4\pi\,\sigma_u^2/(D\,\sigma_{\dot{u}})$$
(4.10)

Comparing (4.10) and (4.8) it is seen that these two versions of the Keulegan-Carpenter number while equal for regular sine-waves differ by a factor of $\sqrt{2}$ for normally distributed particle velocities and accelerations. The first expression (4.8) relates back to the ratio between the temporal change of particle velocity to the spatial, i.e. it is a comparison of the partial derivative with time, $\partial u/\partial t$, to the transport derivatives, $u \cdot \partial u/\partial x$ and $v \cdot \partial u/\partial y$. This number should then be expected to relate to whether or not separation will start for a representative half cycle, or how developed separation will become. The Bishop definition (4.9) compares the contributions from the drag and inertia terms to the mean square force. The drag term is quadratic, and amplifies a large velocity more than the linear inertia term does. The ratio between the mean square drag and mean square inertia terms is therefore larger (by a factor $\sqrt{2}$), than the parameter predicting degree of separation for the "average" wave. Thus both approaches make sense, and it is not easy to see which is to be preferred.

4.5. DESIGN WAVE LOADING SITUATIONS - TYPICAL VALUES OF K AND RE

Surface waves in the ocean are irregular, have directional spread and are often idealized as stationary stochastic processes within a short time span, typically in the order of a few hours, so-called short term situations. These are typically characterized by significant wave height H_s and average zero crossing period, T_z. For a given H_s and T_z a wave spectrum may be specified e.g. using the ISSC form of the Pierson-Moskowitz spectrum. Describing also some additional variables, such as the main direction and spread of the waves, simultaneous direction and magnitude of the currents, etc. one has in principle a description of a design seastate.

The design seastate in Norwegian waters may typically be $H_s = 16$ m and $T_z = 15$ s. Then $\sigma_\eta = 4$ m, and fairly close to the surface one may use $\sigma_u \approx \omega_0\,\sigma_\eta = (2\pi/15)4 = 1.7$ m/s and $\sigma_{\dot{u}}/\sigma_u = \omega_0' \approx 2\pi/12.5 = 0.5$ rad/s. Then from (4.8) and (4.7) with D in metres:

$$K \approx 30/D \qquad (60 - 15)$$
(4.11a)

$$Re \approx 1.8 \cdot 10^6 \cdot D \qquad (9 \cdot 10^5 - 3.6 \cdot 10^6)$$
(4.11b)

The ranges in the parentheses are for D = 0.5 to 2 m.

In fatigue calculations in principle all seastates will contribute. The most important contributions will come from situations where H_s is less than its extreme value, possibly about 5-7 m in the North Sea and with T_z = 8-9 sec, say.

Thus the design seastates of most interest in practical design has postcritical Reynolds numbers and Keulegan-Carpenter numbers in the range 4 to 60.

4.6. CONSEQUENCES OF WAVE THEORIES FOR THE CHOISE OF WAVE FORCE COEFFICIENTS

Real waves are nonlinear, and the steeper the waves the more pronounced will the nonlinearities become. A multitude of nonlinear theories exists for regular waves, while for irregular waves the present day knowledge is much more sketchy. However, some very promising research is currently going on in this field, Gudmestad & Al (1988).

A statistical description of nonlinear, deep water waves has been performed by Longuet-Higgins (1963). The probability density of the surface elevation was found to be of a Gram-Charlier series type, its first term being Gaussian, the next of third order, representing the occurence of higher peaks and lower throughs than in the linear wave case.

In short, several wave theories may be used to estimate the water particle velocities, especially for deterministic and periodic waves, but also for nonperiodic, deterministic waves, and even with probabilistic wave descriptions. The point to be made, however, is that the wave theories must be applied with wave coefficients determined in the same manner. If Stokes fifth order waves are to be used, then wave coefficients should be obtained from experiments in a similar flow, in which the particle kinematics is also determined from Stokes fifth order waves. Using these values for u and $\dot{u}$ in the Morison formula, C_d and C_m can be determined e.g. by the least squares method.

This remark also applies for the force coefficients to be used in an random vibration analysis of a wave excited platform. Then experiments in irregular waves are needed, and they should preferably be analysed using the same random process description as that to be used in the design analysis.

5. SOME LARGE REYNOLDS NUMBER, IRREGULAR WAVE CASES

5.1. GENERAL

Considerable efforts has gone into this field, much of it on proprietary projects. The results have not always been given full disclosure, nor exposure to in-depth scrutiny through discussions. The data analyses and the interpretation of results have been done by a number of methods, often precluding direct comparisons between the finding of different projects.

Some large Reynolds number, irregular wave tests are the following:

1. Wave project I and II. Chevron & others.

Literature: Aagaard & Dean (1969), Evans (1969), Wheeler (1969), Dean & Al. (1981).

2. The Offshore Test Structure (OTS). Exxon.
Literature: Haring & Al. (1978), Geminder & Pomonik (1979), Heideman & Al. (1979), and recently Bostrøm (1987), Bostrøm & Overvik (1986) and Moe & Bostrøm (1989).

3. Christchurch Bay Tower (CBT), Department of Energy, UK.
Literature: Bishop (1984), Moe & Bostrøm (1989).

4. Forschungsstelle Nordeney, German Research Council.
Literature: Sparboom (1986).

5. Delft/NMI wave flume experiments.
Literature: Bearman & Al. (1985), Bearman (1988).

6. SSPA Ocean Basin Tests, Shell.
Literature: Rodenbusch & Gutierrez (1983), Rodenbusch & Källstrøm (1986).

Space limitation precludes an extensive discussion of all these projects.

Wave project I and II. Wave project I and II have been analysed in somewhat different ways. Aagaard & Dean (1969), Dean & Al. (1981) used essentially a deterministic approach in which the particle kinematics were predicted from the irregular water surface trace by means of a (Dean) stream function theory. No directional information was available, so the waves were assumed to be unidirectional, and aligned with the average direction of the force during a given wave cycle(?) The highest waves were analysed, using the time history of the particle velocity and the measured forces, with the Morison formula, estimating the drag and inertia coefficients by a mean square technique. The results were sorted in classes, according to the Reynolds numbers of the individual waves. The results show a considerable amount of scatter, see Fig. 11. The data probably covered a large range of Keulegan-Carpenter numbers, and the dependency on this parameter could explain part of the scatter. Also directional spread, and possibly a current component might be involved.

Research station Nordeney. "Forschungsstelle Nordeney" covers work by Sparboom (1986) on a pile of 0.70 m in diameter in a waterdepth of 3-5 m, i.e. shallow water. The data might be influenced by free surface effects, and showed no clear dependency on the Reynolds or Keulegan-Carpenter numbers. Conservative values were given as C_m = 1.7 and C_d = 1.0 in the surface zone. The analysis procedure has been explained earlier in conjunction with equation (4.4).

The Delft/NMI wave flume experiments. These experiments were carried out in the so-called delta flume of the Delft Hydraulics Laboratory, see Bearman & Al. (1985). Experiments on horizontal and vertical cylinders in regular and irregular waves were carried out. The present summary will concentrate on the case of irregular waves acting on a vertical cylinder.

The test cylinder had a diameter of 0.5 m and was placed in a 5 m deep wave flume and equipped with two instrumented force sleeves, the one analysed by Bearman & Al. (1985) was at 2.5 m depth.

The irregular waves had a significant height of 0.8 m and a peak frequency f_p = 0.167 Hz.

From the plot of the velocity time series given in Fig. 6a of Bearman & Al. (1985) it appears that the velocity process was quite narrow-banded. The number of velocity peaks in the trace is namely

very nearly equal to the number of zero upcrossings, say abour 1-2%
more peaks than upcrossings. Then the waves, the particle velocity and
the particle acceleration will all have nearly the same period. This
is markedly different from the situations in the upper part of a real
ocean wave. The velocity time history exhibits pronounced nonlinearities. The force coefficients were calculated by formulae (4.1)
and (4.2). The analyses were performed in a deterministic spirit,
covering the 11 largest waves plus 5 moderate waves, and Keulegan-
Carpenter and Reynolds numbers were determined for each individual
wave.
     The results have been presented together with the regular wave
results in Fig. 5. The correlation to regular wave results may or may
not be coincidental, but it should be remembered that this narrow-
banded situation is quite different from the particle kinematics in
real ocean waves at small or moderate depths.

The Shell SSPA Ocean Basin tests. These experiments were carried out
in the 2.8 m deep Ocean Basin at SSPA in Gothenburg, Sweden, using a
pile of 1.0 m diameter. The forces from random directional waves
acting on fixed piles were simulated by moving the cylinder in water
initially at rest. The centre of the instrumented section was in the
SSPA tests only 1.5 diameter below the still water level, and a reduced
pressure build-up should be expected, as discussed in Section 3.6. In
order to eliminate this effect the cylinder was fitted with a plate
just below the still water level. In Fig. 12 some results for regular
oscillatory motions as quoted by Sarpkaya is given together with some
of Sarpkaya's own results. They seem to agree well.
     The irregular wave data has been analysed on a wave by wave basis
using the so-called velocity tracking method mentioned in Section 4.3.
Some results are given in Fig. 13. The C_d values are comparatively
low, but this must be seen in conjunction with the use of velocity
tracking, as commented in Section 4.3.

5.2. OCEAN TEST STRUCTURE (OTS)

In the OTS tests the wave forces on short segments of the 0.41 m
diameter vertical legs were measured at 4.6 m depth for 3072 s
duration. Instrumentation problems for the wave force measurements
were considerable, and mechanical constraints from marine growth were
also a problem in some cases. Ten records have been re-analysed in
Trondheim, 1 with "slight fouling", 5 on smooth cylinders and 4 with
marine growth. The sea states were similar for all these records,
resulting in "global" Reynolds number and Keulegan-Carpenter numbers in
the narrow range

$$1.7 \cdot 10^5 < Re < 2.6 \cdot 10^5$$

$$10 < K < 15$$

Bostrøm (1987) used the following Keulegan-Carpenter number $K = 2\pi \sqrt{2}\sigma_u /D$ which for narrow band deep water waves would be equal to the
value in (4.8). The current was very small in all cases, but the waves
had considerable directional spread.
     Using the halfwave method mentioned before, Heideman & Al. (1979)
found the force coefficients in Fig. 14.
     Some of the results of the Bostrøm (1987) analyses are given in
Figs. 15 and 16. It is seen that the consistency within each sea state

is remarkable. That means that for each sea state the peak distribution of force, and the force spectra can be excellently approximated with constant C_d and C_m coefficients. There is no need for frequency or amplitude dependent coefficients. On the other hand the variation of C_d and C_m between seastates were somewhat more difficult to predict, and also it turned out that different prediction methods yielded somewhat different force coefficients for the same wave record. In Table 1 the C_d and C_m values for the 5 smooth cylinder cases are summarized.

TABLE 1   Force coefficients for the 5 smooth cylinder sea states. Drag coefficients, C_d, upper 4 rows, inertia coefficient, C_m bottom 3 rows. "Moment" denotes method of moments, MLM denotes maximum likelihood method and LSM denotes least squares method. F is force and $\hat{F}$ is force peaks, $p(\cdot)$ is probability density and $S(\cdot)$ is spectral density.

		2	3	4	5	6	
	$Re \cdot 10^{-5}$	2.16	2.21	2.11	2.21	2.42	
	K	10.1	10.2	11.8	12.0	14.5	
1	$p(F)$, moment	1.42	1.13	1.07	1.05	1.01	
2a	$p(\hat{F})$, MLM(a)	1.35	1.01	1.03	1.13	1.01	
2b	$p(\hat{F})$, MLM(b)	1.45	1.16	1.10	1.12	1.03	C_d
3	$S_F(\omega)$, LSM	1.40	1.07	1.22	1.12	1.06	
1	$p(F)$, moment	1.69	1.51	1.66	1.61	1.53	
2	$p(\hat{F})$, MLM	1.85	1.65	1.83	1.69	1.63	C_m
3	$S_F(\omega)$, LSM	1.77	1.64	1.67	1.68	1.63	

It can be seen that the largest differences in the estimates for any force coefficient at any given seastate is 0.19 and that in most cases the difference is about 0.10. Unfortunately as already mentioned, the range in K is very limited, only 10 to 15, while for practical design at least the range 5 to 50 is required. The Reynolds number is critical or somewhat higher, but there are probably no points at the post-critical plateau. All the coefficients are calculated on the basis of velocity measurements, since the surface elevation measurements are reported to be "too high", Bostrøm (1987). These results are plotted in Fig. 17 together with some regular wave results in a similar range.

5.3   CHRISTCHURCH BAY DATA

Results from the "small pile" of the second Christchurch Bay tower have been analysed by Bishop (1984) using (4.4) and the procedure outlined before. His results are given in Fig. 18. Some of these data sets were reanalysed using Bostrøm's procedure. It turned out that all the data provided to us had very large constant current values for a current running roughly perpendicular to the average direction of wave propagation, see Table 2.

TABLE 2   Seastates and results - uppermost sleeve (CBT)

Run number	σ_u (m/s)	$\sigma_{\dot{u}}$ (m/s²)	Principal wave Direction (deg)	Current U (m/s)	Dir α (deg)	K	C_d	C_m
301	0.637	0.620	11.4	.49	69	12.1	1.08	1.20
302	0.681	0.614	10.1	.56	70	14.0	0.99	1.11
309	0.576	0.603	6.9	.43	64	10.1	1.01	1.04
310	0.590	0.585	10.4	.58	232	11.0	1.03	0.79
303	0.404	0.424	16.2	.41	63	7.1	1.22	1.19
311	0.429	0.448	15.0	.61	235	7.6	1.09	1.37
306	0.260	0.305	14.9	.59	243	4.1	*	1.56
316	0.227	0.244	11.6	.46	59	3.9	0.65	1.57
313				.62	250			
315				.68	71			
308				.59	70			
314				.38	242			

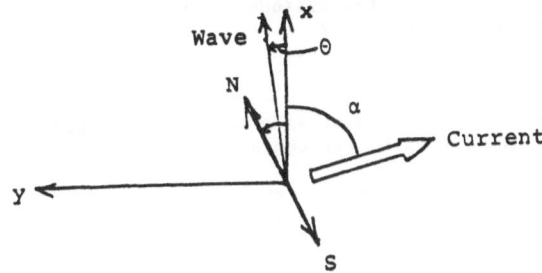

This case is therefore not representative for the condition most
offshore structure are designed for. The results from Method 3, (least
squares spectral fit) is also given in Table 2. This is for the
uppermost force sleeve, i.e. at "level 3", deeper down the current will
dominate more. The vector based analysis procedure used by Bishop (and
most other researchers) has a drag term based on the absolute value of
the velocity times the vector wave velocity, i.e. roughly $(U^2+ u^2)^{1/2}u$.
However, it is not at all obvious that a cross flow current will really
increase the drag in the wave direction. If it does not, then this
procedure will give a C_d value that decreases with depth, as is indeed
the case in Fig. 18.

In conclusion it must therefore be said that Fig. 19 may not be
applicable for cases in which there are no strong cross-currents.

The Moe & Bostrøm (1989) results get worse as the current dominates
more. Thus only the 5 first cases in Table 2 in which $\sigma_u \geq 0.98$ U
appears to be trust-worthy, and are plotted in Fig. 17.

It also turns out that the particle velocity probe may have been
partically in the wake of the test cylinder, therefore the wave staff
input has beeen preferred. The variance spectra of velocity,
acceleration and forces have also been generated and are presented in
Moe & Bostrøm (1989). As an example is given the spectra for the runs
309 and 310, see Fig. 19. The comparison between measured and estimated
spectra is very good, though not quite as spectacular as for the OTS

data in Fig. 15. The resulting force coefficients are plotted together with the results from the OTS measurements and related regular wave data in Fig. 17.

CONCLUSIONS

The Morison fomula is an empirical relationship that needs to be validated for each combination of structure and flow. Due to space limitations or insufficient knowledge only some cases have been discussed herein, namely

STRUCTURE

1. A circular cylinder, usually smooth. Rigidly supported. Groups of cylinders have not been discussed.

2. The following situations of flow relative to cylinder have been considered.
A. A basic case: Infinite fluid, rectilinear, oscillatory flow perpendicular to the cylinder axis.
B. Inclined cylinders. Rectilinear, oscillatory flow.
C. Orbital flow effects - vertical cylinders. Regular waves.
D. Rotating flow around the cylinder.
E. Coexisting regular waves and a current.
F. Surface proximity effects.

Curves of force coefficients are available in the literature for most of the above situations, and some are copied herein.

3. Special efforts have been made to examine irregular, high Reynolds number flows that resembles those for which offshore structures must be designed.

4. Several irregular wave cases have been reviewed. Traditional deterministic methods of analysis seem to give results with large uncertainties. Use of standard random process theory, on the other hand, gives spectacular fit between predicted and measured force spectra and also for the distribution of the force peaks during a given seastate.

5. The goal is to find parameters characteristing seastates, so that C_d and C_m can be specified with reasonable accuracy for design situations. The data investigated so far are consistent within themselves see Fig. 17, but a very much larger parameter range is needed. These results have so far not been verified by other researchers, and should therefore be considered as tentative.

REFERENCES

Aagaard, P.M. and Dean, R.G., "Wave Forces; Data Analysis and Engineering Calculation Method", Proc. First Offshore Technology Conference, Dallas, 1969, pp. 95-106.

Bearman, P.W., Chaplin, J.R., Graham, J.M.R., Kostense, J.K., Hall, P.F. and Klopman, G., "The Loading on a Cylinder in Post-Critical Flow beneath Periodic and Random Waves", Proc. 4th Int.Conf. on the Behaviour of Off-shore Structures. Elsevier Science Publishers, B.V. 1985, pp. 213-225.

Bearman, P.W., "Wave Loading Experiments on Circular Cylinders at Large Scale", Proc. of the Int.Conf. on the Behaviour of Offshore Structures (BOSS), Trondheim, June 1988, pp. 471-487.

Bishop, J.R., "Wave Force Investigations at the Second Christchurch Bay Tower", National Maritime Institute Report No. P.177, OT-0-82100, 1984.

Borgman, L.E., Ocean Wave Simulation for Engineering Design, J. of Waterways and Harbors Div., ASCE, Vol. 95, No. WW4, pp. 557-583, 1969.

Bostrøm, T., Hydrodynamic Force Coefficients in Random Wave Conditions, Dr.ing. Thesis, Section of Structural Engineering, Norw.Inst. of Techn., March 1987.

Bostrøm, T. and Overvik, T., Hydrodynamic Force Coefficients in Random Wave Conditions, Proc. Offshore Mechanics and Arctic Engineering (OMAE) Symposium, Vol. 1, pp. 136-143, 1986.

Chakrabarti, S.K., Hydrodynamics of Offshore Structures, Computational Mechanics Publications, Boston, 1987.

Chaplin, J.R., "Loading on a Cylinder in Uniform Elliptical Orbital Flow", Dept. of Civil Eng., University of Liverpool Report MCE/JUL85, 1985.

Chaplin, J.R., Morison Inertia Coefficients in Orbital flows, Waterway, Port, Coastal & Ocean Eng., Vol. 111, No. 2, Mars 1985, pp. 201-215.

Chaplin, J.R., "Non-Linear Forces on Horizontal Cylinders in the Inertia Regime in Waves at High Reynolds Numbers, Proceedings of the Int.Conf. on Behaviour of Offshore Structures, Ed: Moan, T., Janbu, N. & Faltinsen, O., Trondheim, Norway, June 1988.

Dean, R.G, Dalrymple, R.A., Hudspeth, R.T., "Force Coefficients From Wave Project I and II Data Including Free Surface Effects, Soc. of Petroleum Eng. J., Dec. 1981, pp 779-786.

Dean, R.G., Tørum, A., Kjeldsen, S.P., Wave Forces on a Pile in the Surface Zone from the Wave Crest to the Wave Through, NTNF Research Report, Programme for Marine Structures, Report 1.7, April 1985.

Evans, D.J., "Analysis of Wave Force Data", Proc. First Offshore Technology Conference, Dallas, 1969, pp. 51-70.

Garrison, C.J., A Review of Drag and Inertia Forces on Circular Cylinders, Offshore Technology Conference (OTC), Paper No. 3760, May 1980, pp. 205-218.

Garrison, C.J., Comments on Cross-flow Principle and Morison's Equation, J. of Waterway, Port, Coastal and Ocean Engineering, Vol. 111, No. 6, Nov. 1985.

Garrison, C.J., Field, J.B. and May, M.D., "Drag and Inertia Forces on a Cylinder in Periodic Flow", J. of the Waterway, Port, Coastal and Ocean Division, Vol. 103, No. WW2, May, pp. 193-204, 1977.

Geminder, R. and Pomonik, G.M., The Ocean Test Structure Measurement

System, Civil Eng. in the Oceans IV, ASCE, pp. 1010-1029, 1979.

Gudmestad, O.T., Johnsen, J.M., Skjelbreia, J., Tørum, A., "Regular Water Wave Kinematics", Proc. of Int.Conf. on Behaviour of Offshore Structures, Vol. 2, pp. 789-804, Trondheim, June 1988.

Haring, R.E., Shumway, D.H., Spencer, L.P., Pearce, B.K., Operation of an Ocean Test Structure, EUR 22, European Offshore Petroleum Conference and Exhibition, London, 1978.

Heideman, J.C., Olsen, O.A. and Johansson, P.I., Local Wave Force Coefficients, Civil Eng. in the Oceans IV, ASCE, pp. 684-699, 1979.

Holmes, P. and Chaplin, J.R., Wave Loads on Horizontal Cylinders, Proceedings of the 16th International Conference on Coastal Engineering, Vol. 3, Hamburg, Germany, Aug. 1978, p. 2449.

Iwagaki, Y., Asano, T. and Nagai, F., Hydrodynamic Forces on a Circular Cylinder placed in Wave-current Co-existing Fields, Memoirs of the Faculty of Engineering, Kyoto University, Japan 1983, XLV (1), 11-23.

Longuet-Higgins, M.S., "The effects of non-linearities on statistical distributions in the theory of sea waves", J. Fluid Mech., 17(3), 1963, pp. 459-480.

Moe, G. and Bostrøm, T., "Wave Force Coefficients in Irregular Waves: The Christchurchbay Data", Division of Structural Engineering, The Norwegian Institute of Technology, Trondheim, Report No. R-6-88, 1989.

Moe, G., Wu, Z.-J., "The lift force on a vibrating cylinder in a current", 8th International Conference on Offshore Mechanics and Arctic Engineering, The Hague, Netherlands, OMAE 1989, Mar 1989.

Morison, J.R., O'Brien, M.P., Johnson, J.W. and Schaaf, S.A., "The Force Exerted by Surface Waves on Piles", Petrol. Trans., AIME, Vol. 189, pp. 149-154, 1950.

Nath, J.H., "High Reynolds Number Wave Force Investigation in a Wave Flume", Naval Civil Engineering Laboratory, Port Hueneme, California, Report CR 85.004, 1985.

Newman, J.N., Marine Hydromechanics, MIT Press, 1977.

Overvik, T., Private communications, 1988.
Pierson, W.J. and Holmes, P., "Irregular Wave Forces on a Pile", J. of the Waterways and Harbors Div., ASCE, Vol. 91, No. WW4, pp. 1-10, 1965.

Rodenbusch, G. and Källström, C., "Forces on a Large Cylinder in Random 2-Dimensional Flow", Proceedings of the 18th Annual Offshore Technology Conference, Houston, May 1986.

Rodenbusch, G. and Gutierrez, C.A., "Forces on Cylinders in Two-Dimensional Flows", Technical Progress Report, Vol. 1, BRC 13-83, May 1983, Bellaire Research Center (Shell Development Company), Houston, Tx.

Sarpkaya, T., "A Critical Assessment of the Methods of Analysis of Offshore Structures After Ten Years of Basic and Applied Research".

Proc.Int.Symp. on Water Wave Research, University of Hannover, SFB 205, Germany, 1985.

Sarpkaya, T., disc. (of Comments on Cross-Flow Principle and Morison's Equation, by C.J. Garrison, WW Nov. 85, pp 1075-1079), WW Nov. 85 p 1087.

Sarpkaya, T., "Vortex Shedding and Resistance in Harmonic Flow about Smooth and Rough Circular Cylinders at High Reynolds Numbers", Technical Report No. NPS-59SL76021, Feb. 1976, Naval Postgraduate School, Monterey, Ca.

Sarpkaya, T., "Quasi-2-D Forces on a Vertical Cylinder in Waves", Discussion to paper by Stansby & Al (1983), J. of Waterway, Port, Coastal and Ocean Enginering, ASCE, Vol. 110, No. 1, Feb. 1984, pp. 120-123.

Sarpkaya, T., Bakmis, C. and Storm, M.A., "Hydrodynamic forces from combined wave and current flow on smooth and rough circular cylinders at high Reynolds number", Proceedings of the Sixteenth Offshore Technology Conference, Houston, Texas, OTC 4830, May 1984, pp. 455-462.

Sarpkaya, T. and Isacson, M., "Mechanics of Wave on Offshore Structures", Van Nostrand Reinhold, New York, 1981.

Sparboom, U.C., "Über die Seegangsbelastung Lotrechter Zylindrischer Pfähle im Flachwasserbereich", Dr.ing. Thesis, Tech.Univ. Carolo-Wilhelmina of Braunschweig, Jan. 1986.

Stansby, P.K., Bullock, G.N., Short, I., "Quasi-2-D Forces on a Vertical Cylinder in Waves", J. of Waterway, Port, Coastal and Ocean Engineering, ASCE, Vol. 109, No. 1, pp. 128-132, Feb. 1983.

Teng, C.-C., Nath, J.H., Force on Horizontal Cylinders Towed in Waves, J. of Waterways, Port, Coastal and Ocean Eng., ASCE, Vol. 111, No. 6, Nov. 1985.

Verley, R.L.P. and Moe, G., "The Forces on a Cylinder Oscillating in a Current". River and Harbour Laboratory, The Norwegian Institute of Technology, Report No. STF60 A 79061, 1979.

Wheeler, J.D., "Method for Calculating Forces Produced by Irregular Waves", Proc. First Offshore Technology Conference, Dallas, 1969, pp. 71-82.

Wu, Z.-J., "Current Induced Vibrations of Flexible Cylinder", Dr.ing. Thesis, Civil Engineering, The Norwegian Institute of Technology, to appear, 1989.

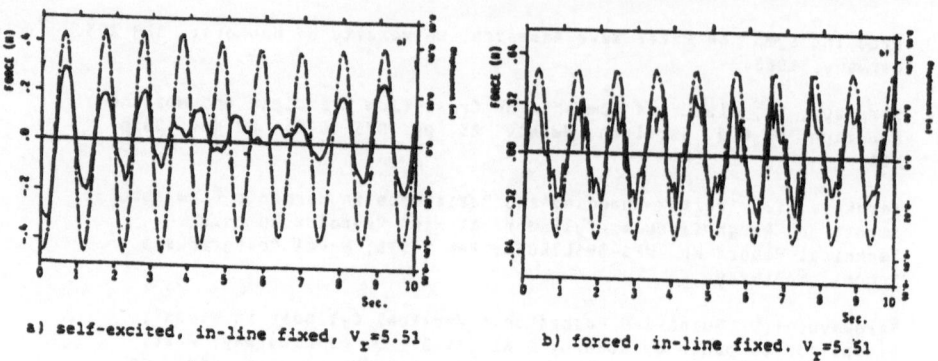

a) self-excited, in-line fixed, V_r=5.51   b) forced, in-line fixed, V_r=5.51

Fig. 1.   Force time series (full line) and displacement time series (broken line.) Moe & Wu (1989).

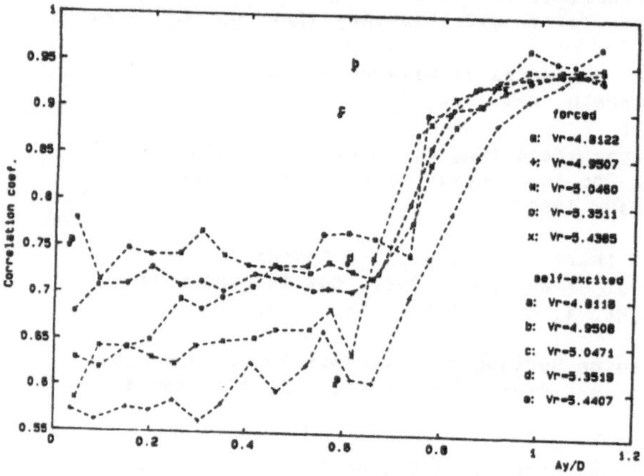

Fig. 2. Spanwise correlation coeff of lift forces on 2 sleeves 6 diamenters apart. In-line spring-supported, low Reynolds numbers. Wu (1989).

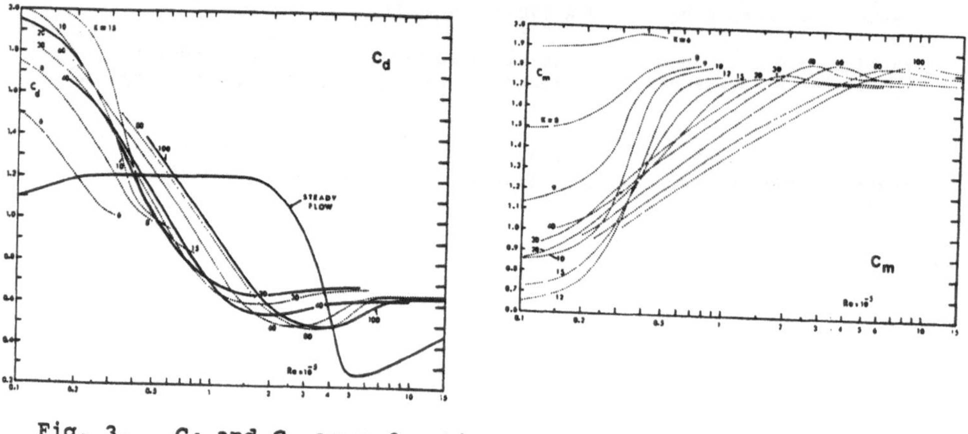

Fig. 3. C_d and C_m as a function of Reynolds number at various Keulegan-Carpenter numbers Sarpkaya (1976).

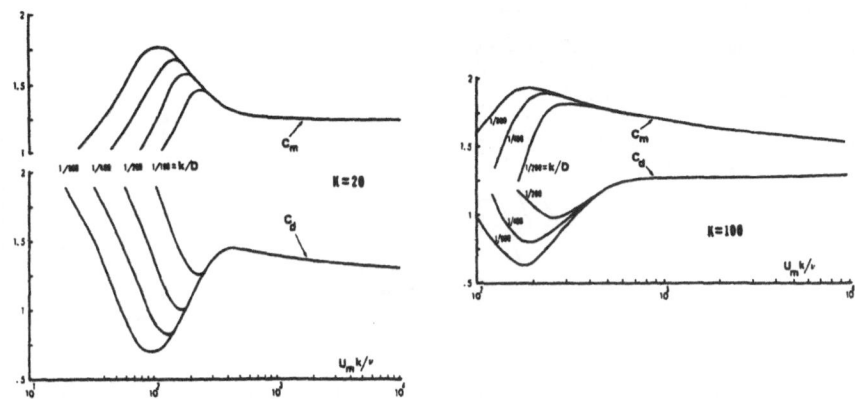

Fig. 4.   Recommended C_d and C_m for K = 20 and K = 100. Sarpkaya & Isaacson (1981).

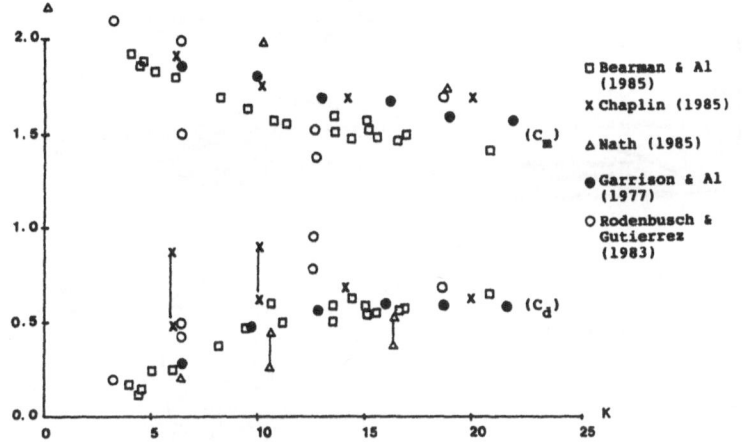

Fig. 5. C_d & C_m versus Keulegan Carpenter number, supercritical Reynolds numbers. From Bearman & Al (1985).

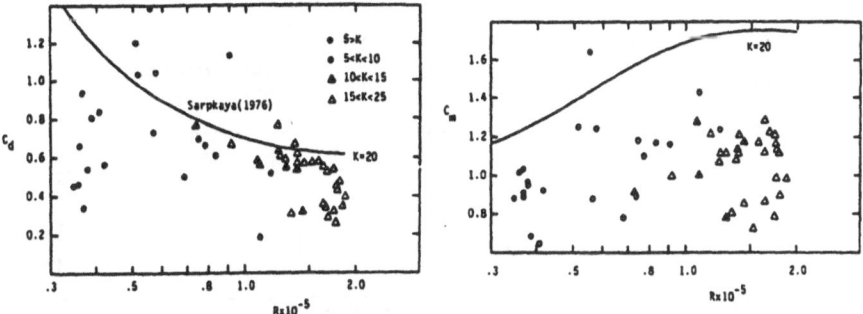

Fig. 6. C_d and C_m in rotating flow. (Horizontal cylinders in waves.) Teng & Nath (1985).

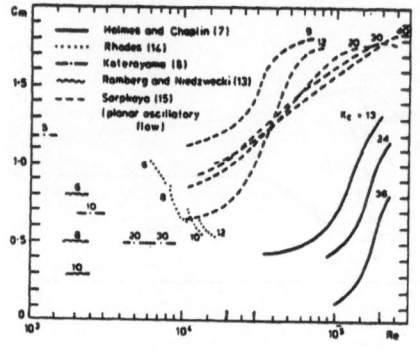

Fig. 7.   Inertia coefficient
          in rotating and plane
          flow. From Chaplin
          (1985).

Fig. 8.   Inertia coefficient
          as a function of K_x,
          K_y (see text). From
          Chaplin (1988).

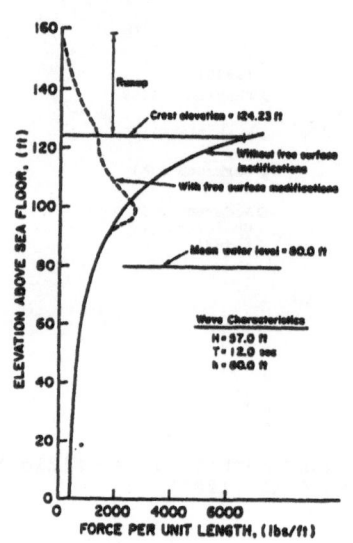

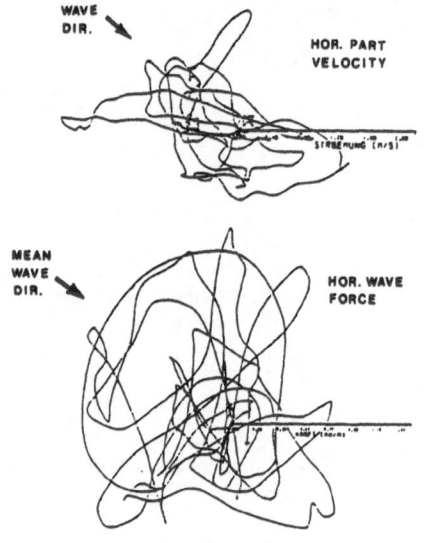

Fig. 9.   Force per unit length
          on a vertical pile.
          Dean & Al (1981).

Fig. 10. Current and Force
          vector plots.
          Sparboom (1986).

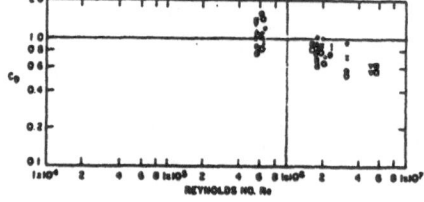

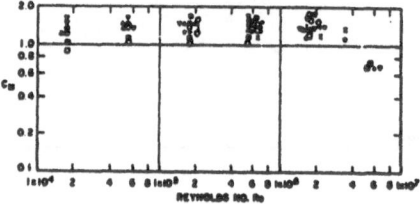

Fig. 11. C_d & C_m for Wave Force project I & II Dean & Al (1981).

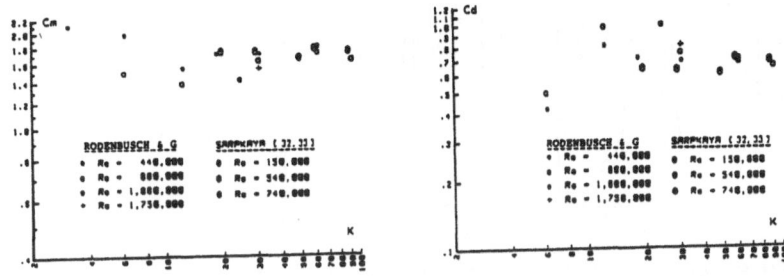

Fig. 12.   Force coeff on smooth cylinders. Rectilinear sinusoidal
            motions. From Sarpkaya (1985).

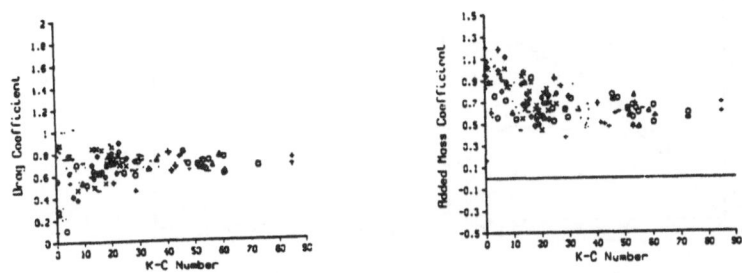

Fig. 13. C_d & C_m in irregular seas (simulated). From Rodenbusch
            & Källstrøm (1986).

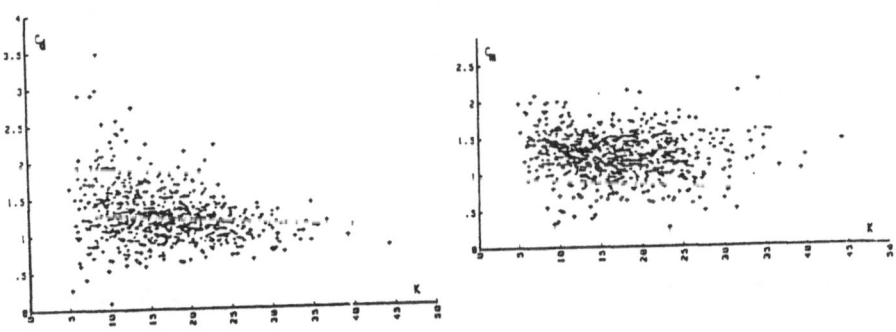

Fig. 14. C_d & C_m, halfwave approach used on the OTS data.
            Heideman & Al (1979).

676

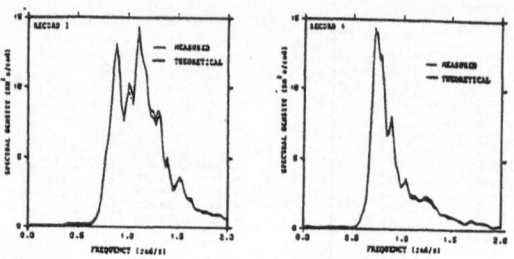

Fig. 15. Force spectra from two records in the OTS data.
Thin line: From measured force.
Thick line: From measured particle velocity.
Moe & Bostrøm (1989).

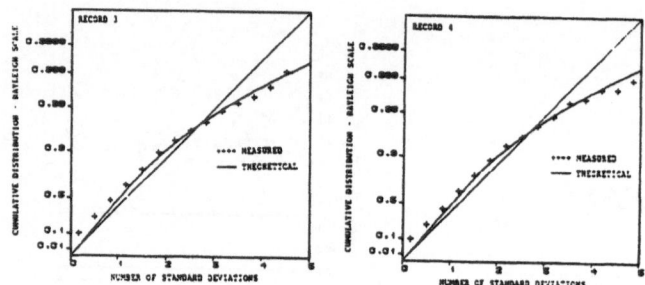

Fig. 16. Peak force distribution - OTS data. Fitted by MLM.
Bostrøm (1987).

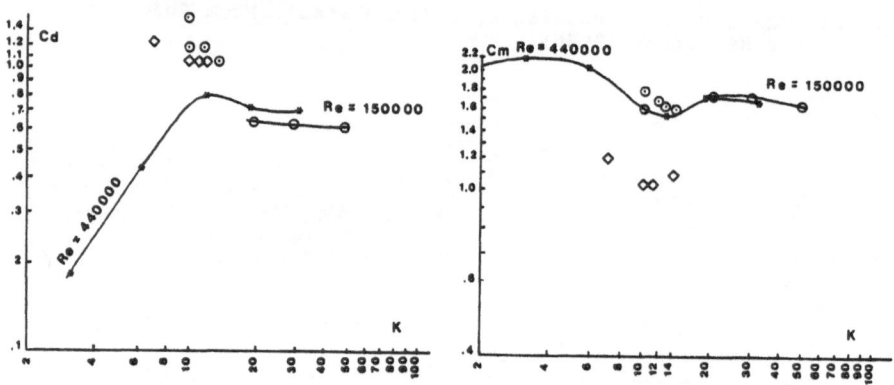

Fig. 17. Drag & inertia coeff. for smooth cylinders. Re 2.11-
2.42.10^5. Full curves, results given in Fig. 13.
Diamonds: CBT results.
Note: A strong cross current is present and data may
not be relevant for offshore structurs.
Circles: OTS results.
Moe & Bostrøm (1989).

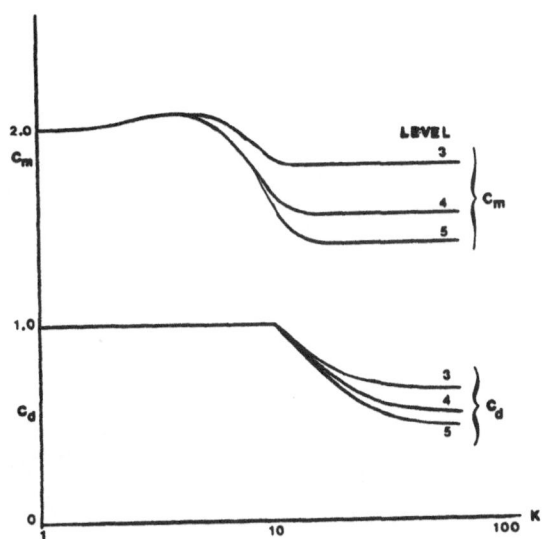

Fig. 18.   Bishop's recommended values for C_d & C_m based on the Christchurch bay measuring programme. From Bishop (1984).

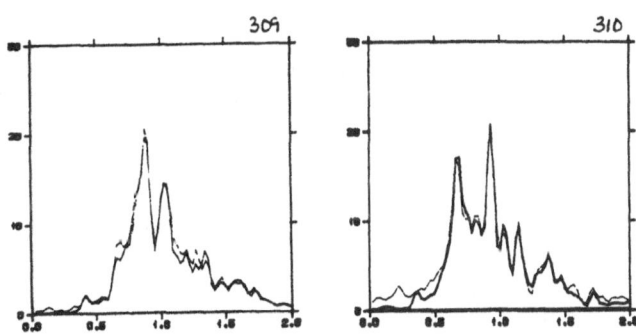

Fig. 19.   In-line wave force spectrum.
Fat line: From wave elevation measurements.
Thin line: Directly from force sleeve measurements.
Moe & Bostrøm (1989).

THE LOADING ON A VERTICAL CYLINDER IN RANDOM WAVES AT HIGH REYNOLDS NUMBERS

GERT KLOPMAN and JAN K. KOSTENSE
Delft Hydraulics
P.O. Box 152
8300 AD Emmeloord
The Netherlands

ABSTRACT. The inertia, drag, inline and transverse force coefficients have been determined for a vertical cylinder of 0.5 m diameter with a relative surface roughness $k/D \cong 1.2 \times 10^{-4}$ in random waves. Waves with a significant wave height of 1.5 m and a peak period of 5.9 s were generated in a large wave channel, at a mean water depth of 5.0 m. Reynolds numbers up to 5.3×10^5 have been achieved for Keulegan-Carpenter numbers up to 15.5. Comparisons are made with results obtained in periodic waves.

1. Introduction

The wave loading on circular cylinders is a research topic since quite some time. For the extrapolation of laboratory data to full scale offshore conditions it is essential to obtain these data in post-critical flow. The required Reynolds numbers to reach this flow regime initiated research under more schematized conditions. Especially research in oscillating water tunnels has contributed to the knowledge about wave loading on tubular members of offshore structures. Since some time large wave channels are available, in which also post-critical conditions can be achieved.

Comparisons between the periodic flow results of several high Reynolds number experiments in oscillating water tunnels, planar oscillating cylinders and vertical cylinders in wave channels are given by Bearman (1988) and Sarpkaya (1987), showing quite a good agreement for the force coefficients. Justesen (1988) describes the results of another experiment with an oscillating cylinder at high Reynolds numbers. These studies show that the inline wave forces on vertical cylinders can be described reasonably well by the classical Morison equation. Wave forces on horizontal cylinders are however poorly described by Morison's equation, as found by Bearman et al (1985a) and Chaplin (1988). An improved description for wave forces on horizontal cylinders at very low Keulegan-Carpenter numbers is given by Chaplin (1988).

This paper describes the results of a random wave loading experiment on a vertical cylinder of $D = 0.5$ m diameter, with small surface roughness (roughness height k to diameter ratio $k/D = 1.2 \times 10^{-4}$),

A. Tørum and O. T. Gudmestad (eds.), Water Wave Kinematics, 679–699.

undertaken in Delft Hydraulics' Delta Flume. The experiment is one out of a series in the Delta Flume, both in regular and random waves, funded by the UK SERC through the Marine Technology Directorate:
- smooth vertical cylinder (Bearman et al ,1985a),
- smooth horizontal cylinder (Bearman et al ,1985a; Chaplin, 1988),
- slightly rough vertical cylinder, k/D = 1.2 x 10^{-4} (Bearman ,1988),
- rough vertical cylinder, k/D = 0.038 (Wolfram & Theophanatos, 1989a; Wolfram et al 1989b), and
- flexible vertical cylinder (Bearman, 1988).

Apart from the last one, all above test series refer to fixed circular cylinders.

Bliek & Klopman (1988) report on a non-linear spectral analysis of several of the random wave tests for fixed cylinders, showing the limited validity of Morison's equation in describing forces on smooth and small roughness vertical cylinders, as well as forces on horizontal cylinders. In this paper a more classical time-domain analysis is presented, of both inline and transverse forces on a vertical cylinder, giving values of various force coefficients as a function of the Keulegan-Carpenter number.

2. Experimental Arrangement

The Delta flume has a length of 230 m, a width of 5 m and a depth of 7 m. The water depth during the experiments was 5.0 m. Waves are generated by a piston type wave board, and are absorbed by a 1:6 concrete slope at the end of the channel (maximum reflection coefficients of about 10%). Re-reflection of waves against the wave board is prevented by an active wave absorption system.

The fixed vertical cylinder of 0.5 m diameter was mounted 55 m from the wave board. The cylinder surface was covered with 240 grade waterproof emery cloth, resulting in a roughness height k of approximately 60 μm (equivalent sand roughness $k_N \cong$ 120 μm), giving a relative roughness ratio k/D of 1.2 x 10^{-4}. Inline and transverse forces were measured simultaneously at two strain-gauged force sleeves of 0.25 m height, positioned 2.5 m and 3.5 m above the bottom.

Velocities were measured by electromagnetic flow meters, two of which with disc-shaped heads were mounted near the wall at the same height as the centres of the force sleeves and in the same cross section as the cylinder axis. Two capacitance-type wave gauges were mounted 43.6 m and 48.0 m from the wave board.

3. Results and Discussion

Random waves with a JONSWAP water surface spectral density were generated, using linear wave theory for wave generation. The significant wave height of the generated waves was 1.5 m, and the period corresponding with the peak of the spectrum was T_p = 5.9 s. The peak enhancement factor γ was 3.3. The duration of the test was 189 minutes, and the signals were sampled at 10 Hz. Before further analysis, trends due to instrument drift were removed from the various signals.

Only the velocities and forces at the lower force sleeve, 2.5 m above the bottom, will be presented. When the horizontal velocity signal is divided into waves, with a wave defined by the period T between two successive downward mean level crossings of the velocity signal, the Keulegan-Carpenter numbers KC in this test varied between about 0.1 and 15.5. The Keulegan-Carpenter number is defined as KC = UT/D, with U the amplitude of the horizontal velocity Fourier component with period T. The maximum Reynolds number Re ≡ UD/ν is about 5.3 x 10^5, with ν the kinematic viscosity of the water. According to Bearman (1988) the conditions in the Delta flume for a 0.5 m cylinder in these wave conditions correspond to post-critical flow.

Spectral densities of the water surface elevation, the horizontal velocity, the inline force and the transverse force at the lower force sleeve 2.5 m above the bottom are presented in figure 1. Due to the non-linearity of the processes, the spectra do not give a complete representation of the signals. In the surface elevation spectrum, a secondary peak can be seen at twice the peak frequency, due to non-linear effects resulting in sharper water surface crests and flatter troughs. The horizontal velocity is more nearly a linear process, which can also be seen in its spectrum. The inline force has a secondary peak at 0.35 Hz, corresponding to the peak in the transverse force spectrum. The transverse force spectrum does not have a peak at the peak frequency of the horizontal velocity, but at roughly twice this frequency. The transverse force is a non-stationary process, which means a spectrum is of very limited value. The non-stationary behaviour probably leads to the broad bandedness of the lift spectrum.

Figure 2 shows time histories of the horizontal velocity, the inline force and the transverse at the lower force sleeve for a representative part of 5 minutes. Obviously the maximum transverse forces can exceed the maximum inline forces. Also the fact that the transverse force is a non-stationary process, only triggered at high enough KC, i.e. high enough velocity, can clearly be seen. The minimum KC to trigger the transverse force oscillations seems to be about KC = 7 or 8, in agreement with earlier findings for a smooth cylinder (Bearman et al, 1985a).

3.1. INLINE FORCE: INERTIA AND DRAG COEFFICIENTS

The inline force F(t) was fitted on a wave by wave basis to Morison's equation:

$$F(t) = C_M \frac{\pi}{4} \rho D^2 \frac{du(t)}{dt} + C_D \frac{1}{2} \rho D u(t)|u(t)| , \qquad (1)$$

where all the symbols have their usual meaning, and a wave is defined as the period between two successive downward mean level crossings of the velocity signal. The mean velocity for this test was -0.034 m/s at the lower force sleeve position and is due to a return current below wave trough elevation, compensating for the net mass flux in wave propagation direction in the zone between wave trough and crest elevation. The inertia coefficients C_M and drag coefficients C_D were determined with the least squares method, resulting in the following system of two equations from which C_M and C_D can easily be solved:

$$C_M \frac{\pi}{4} \rho D^2 < \frac{du}{dt} \frac{du}{dt} > \quad + C_D \frac{1}{2} \rho D < \frac{du}{dt} u |u| > = < F \frac{du}{dt} > , \qquad (2a)$$

$$C_M \frac{\pi}{4} \rho D^2 < \frac{du}{dt} u |u| > + C_D \frac{1}{2} \rho D < u^4 > \qquad = < F u |u| > , \qquad (2b)$$

where $<\cdot>$ denotes time averaging over one wave period. The off-diagonal term $< du/dt \; u \; |u| >$ can be shown to be equal to:

$$< \frac{du}{dt} u |u| > = < \frac{1}{3} \frac{d|u|^3}{dt} > = \frac{1}{3} \frac{|u(t_0+T)|^3 - |u(t_0)|^3}{T} , \qquad (3)$$

for a continuous signal averaged over a period from $t = t_0$ to $t = t_0+T$, and is theoretically equal to zero for our definition of an individual wave. In our analysis the off-diagonal terms were taken into account and treated the same way as the other time averages, since they do not have to be exact equal to zero numerically. The time derivative du/dt of the velocity was determined by applying a Fast Fourier Transform (FFT) to the complete velocity signal, multiplying the Fourier coefficients with the frequency response function of the differentiation operator and transforming back to the time domain by inverse FFT. In order to prevent high frequency noise components in the velocity signal to blow up during differentiation, the velocity signal was low-pass filtered at 0.8 Hz, also using FFT's. The forces were low-pass filtered at 1.5 Hz. The coefficients for the 2052 analyzed waves are presented in figure 3. The scatter in C_M and for KC > 5 in C_D is quite small, considering the fact these coefficients are for individual waves and the results for periodic tests are usually the average of many waves. The scatter in C_D at KC < 5 is due to the fact that the drag is contributing hardly to the total force in this range.

In order to be able to compare the random wave results with periodic flow results, KC classes of width $\Delta KC = 1$ were defined, and the median value (i.e. 50% of the values is lower than the median) of the coefficients falling in each class were determined. The median value was chosen because it was believed to be less sensitive for outliers than the arithmetic mean value. In figure 4 a comparison is made with the regular wave results for a fixed vertical cylinder, both for a smooth surface (Bearman et al, 1985a) and for the same relative roughness ratio $k/D = 1.2 \times 10^{-4}$ (Bearman, 1988). For visual aid, a smooth line is drawn through the present random wave test results. The drag coefficients C_D are rather low, when compared with the lower force sleeve results in periodic waves for the same surface roughness, and agree best with the smooth cylinder C_D values. The inertia coefficient C_M is lower than the regular wave tests values for KC < 9, and does not seem to approach the theoretical value of $C_M = 2$ at KC = 0. This is probably caused by the randomness of the waves: in periodic waves no flow separation will occur at very low KC numbers, while in random waves the flow around the cylinder is influenced by vortices shed from the cylinder during preceding higher waves with higher KC numbers.

At low KC numbers, in case of unseparated laminar flow, the first terms of a perturbation series for the values of C_M and C_D have been derived theoretically by Wang (1968). Comparisons with experimental data for smooth cylinders from sinusoidal oscillating water tunnels have been made by Bearman et al (1985b) and Sarpkaya (1986). For the present high Reynolds number test, C_M should theoretically be equal to 2.0, which value was previously seen not to be reached for the present analyzed test. A wave by wave comparison with the theoretical value for C_D:

$$C_D = \frac{3}{2} \frac{\pi^3}{KC} \left[(\pi\beta)^{-1/2} + (\pi\beta)^{-1} - \frac{1}{4} (\pi\beta)^{-3/2} + \ldots \right] , \qquad (4)$$

with $\beta \equiv D^2/(\nu T)$, is shown in figure 5. The β parameter of the waves with very low KC values was about 10^5. The theoretical value is seen to be the lower limit for KC < 0.4. The median value of C_D, using classes of $\Delta KC = 0.25$, is shown in figure 6. The behaviour of C_D is similar to those found by Bearman et al (1985b) and Sarpkaya (1986) for periodic flows at lower β values. Sarpkaya's results showed that below the Hall critical KC number, for the present case equal to about 0.3, the C_D values approach the theoretical line of Wang, equation (4). That this is not the case for the present random wave test might by due to the randomness of the signal and the roughness of the cylinder surface. The increase of C_D above KC $\cong$ 2.5 due to flow separation is also in close agreement with the high Reynolds number periodic flow results in U-tubes.

3.2. INLINE FORCE: TOTAL FORCE COEFFICIENTS

A coefficient which shows very little scatter as a function of KC for periodic flows, but does not give information about the phase dependencies between force and velocity, is the rms inline force coefficient C_{Frms}, defined for random waves on a wave by wave basis by:

$$C_{Frms} = \frac{2 F_{rms}}{\rho D (u_{rms})^2} , \qquad (5)$$

with $F_{rms} = < f \cdot f >^{1/2}$ and $u_{rms} = < u \cdot u >^{1/2}$. For random waves however, quite some scatter is present at KC < 6, as can be seen from figure 7. A first harmonic force coefficient C_{F1}, defined on a wave by wave basis as:

$$C_{F1} = \frac{2 F_1}{\rho D U^2} , \qquad (6)$$

with F_1 the amplitude of the inline force Fourier coefficient with period T, and U the amplitude of the horizontal velocity Fourier coefficient with period T, shows remarkable little scatter, see figure 8. This suggests a close coupling between the first harmonic of the force at the local wave frequency, and the kinetic energy of the flow. However, this coefficient C_{F1} has only limited value for engineering purposes.

Bearman et al (1985b) give a theoretical limit for C_{Frms} at low KC numbers, $C_{Frms} = (2\sqrt{2}) \pi^2/KC$, which is seen to be approximately an upper limit for C_{Frms} in random waves from figure 9. The coefficient C_{F1} is in close agreement with its low KC number limit, $C_{F1} = 2\pi^2/KC$, as can be seen from figure 10. This is remarkable considering the observation from figure 4 that the C_M limit is approximately 1.87, much below its theoretical value of 2.0.

3.3. TRANSVERSE FORCE: TOTAL FORCE COEFFICIENTS

As can be seen in figure 2, the maxima in the local transverse force can exceed the maxima in the local inline force. However, while the inline forces are well correlated along the vertical cylinder axis, more has to be known about the correlation of the transverse forces along the cylinder axis, in order to be able to say something about total forces on a cylinder of arbitrary length.

As is well known from transverse forces in steady and periodic flow, the transverse force is not periodic, but varies highly from cycle to cycle. Therefore, it is not surprising to see large scatter in the wave by wave transverse force coefficient C_{Lrms} in figure 11, defined as:

$$C_{Lrms} = \frac{2\ L_{rms}}{\rho\ D\ (u_{rms})^2}\ , \tag{7}$$

with L_{rms} the rms transverse force during the considered wave period. The behaviour of C_{Lrms} becomes more clear by taking its median, as presented in figure 12. Also shown is the upper quartile value (75% of the values is smaller than the upper quartile). For KC approaching zero, C_{Lrms} becomes very large, as opposed to what one would expect for unseparated flow. This has to be attributed to influences of preceding waves with larger KC, e.g. vortices of previous waves sweeping past the cylinder. Also the maximum of the median C_{Lrms} around KC $\cong$ 11 is far below the values found in regular waves.

The influence of the horizontal flow velocity and vortex shedding history can be taken into account in a crude way by plotting the C_{Lrms} value in a certain wave versus the KC number of the preceding wave. This has been done in figure 13, showing a much better behaviour of C_{Lrms} at very low KC, and closer agreement with the regular wave test data. If the transverse force indeed depends upon the flow history, it might be useful to modify existing equations for the transverse force, such as the one applied to the smooth cylinder random wave data by Graham (1987).

4. Conclusions

The forces on a vertical circular cylinder with a relative roughness ratio $k/D \cong 1.2 \times 10^{-4}$ due to random waves at high Reynolds numbers (up to 5.3×10^{5}) have been investigated experimentally. The following conclusions can be drawn:

1. Inertia and drag coefficients obtained with a wave by wave analysis show relative little scatter as a function of KC, taking into account the difficulties in obtaining C_D values at very low KC numbers (KC < 5) due to the very little contribution of drag to the total force, and the fact that regular wave results are always presented as averages over many waves.

2. The median C_D is in close agreement with C_D obtained with a smooth cylinder in regular waves (Bearman et al, 1985a). C_D shows the same behaviour at low KC as found in oscillating water tunnels (Bearman et al, 1985b; Sarpkaya, 1986), suggesting separation occurs at KC $\cong$ 2.5.

3. Median C_M is below the values obtained in regular waves for KC less than about 9, and does not seem to approach the theoretical limit of 2.0. This is possibly due to the influence of vortices shed during previous waves (with higher KC) on the forces in waves with low KC numbers.

4. The rms total inline force coefficient $C_{Frms} \equiv 2F_{rms}/(\rho D U_{rms}^2)$ shows quite some scatter for KC smaller than 6. The theoretical value for low KC, as given by Bearman et al (1985b), is at the upper bound of the wave by wave C_{Frms} values.

5. The first harmonic force coefficient $C_{F1} \equiv 2F_1/(\rho D U^2)$, as a function of KC, shows very little scatter. This suggests a direct and almost instantaneous response of the amplitude of the fundamental harmonic of the force F_1 to changes in the kinetic energy $\rho U^2/2$ of the flow.

6. The median rms transverse force coefficient C_{Lrms} as a function of KC is lower than the rms transverse force coefficients found in regular waves near KC = 11. When C_{Lrms} of each individual wave is plotted versus KC of the preceding wave, a much better agreement is found with regular wave results, suggesting the transverse force is influenced by the velocity and vortex shedding histories.

7. The present time-domain analysis is complementary to the non-linear frequency-domain analysis of Bliek & Klopman (1988). They concluded a limited validity of Morison's equation in describing wave forces on vertical cylinders. Nevertheless, the present analysis shows that if the Morison equation is used as a starting basis, valuable comparisons can be made with regular wave data.

Acknowledgements

All test series mentioned in the introduction were funded by the UK Science and Engineering Research Council through the Marine Technology Directorate's Fluid Loading Programme. The random wave tests were financed by Delft Hydraulics.
The authors would like to thank all British researchers involved in the experiments, especially Prof. P.W. Bearman and Prof. J.R. Chaplin, for their continuous initiatives and cooperation.

686

References

Bearman, P.W., Chaplin, J.R., Graham, J.M.R., Kostense, J.K., Hall, P.F. and Klopman, G. (1985a) "The loading on a cylinder in post-critical flow beneath periodic and random waves", *Proc. 4th Int. Conf. on the Behaviour of Offshore Structures*, Delft, The Netherlands, pp. 213-225.

Bearman, P.W., Downie, M.J., Graham, J.M.R., and Obasaju, E.D. (1985b) "Forces on cylinders in viscous oscillatory flow at low Keulegan-Carpenter numbers", *J. Fluid Mech.* **154**, pp. 337-356.

Bearman, P.W. (1988) "Wave loading experiments on circular cylinders at large scale", *Proc. 5th Int. Conf. on the Behaviour of Offshore Structures* **2**, Trondheim, Norway, pp. 471-487.

Bliek, A. and Klopman, G. (1988) "Non-linear frequency modelling of wave forces on large vertical and horizontal cylinders in random waves", *Proc. 5th Int. Conf. on the Behaviour of Offshore Structures* **2**, Trondheim, Norway, pp. 821-840.

Chaplin, J.R. (1988) "Non-linear forces on horizontal cylinders in the inertia regime in waves at high Reynolds numbers", *Proc. 5th Int. Conf. on the Behaviour of Offshore Structures* **2**, Trondheim, Norway, pp. 505-518.

Graham, J.M.R. (1987) "Transverse forces on cylinders in random waves", *Proc. Int. Conf. on Flow Induced Vibrations*, Bowness-on-Windermere, England, pp. 191-201.

Justesen, P. (1988) "Hydrodynamic forces on large cylinders in high Reynolds number oscillatory flow", *Proc. 5th Int. Conf. on the Behaviour of Offshore Structures* **2**, Trondheim, Norway, pp. 805-819.

Sarpkaya, T. (1986) "Force on a circular cylinder in viscous oscillatory flow at low Keulegan-Carpenter numbers", *J. Fluid Mech.* **165**, pp. 61-71.

Sarpkaya, T. (1987) "Oscillating flow about smooth and rough cylinders", *Proc. 6th Int. Conf. on Offshore Mech. and Arctic Engng.* **2**, Houston, pp. 113-121.

Wang, C.-Y. (1968) "On high-frequency oscillating viscous flows", *J. Fluid Mech.* **32**, pp. 55-68.

Wolfram, J. and Theophanatos, A. (1989a) " The loading of heavily roughened cylinders in waves and linear oscillatory flow", *Proc. 8th Int. Conf. on Offshore Mech. and Arctic Engng.* **2**, The Hague, The Netherlands, pp. 183-190.

Wolfram, J., Javidan, P. and Theophanatos, A. (1989b) " Vortex shedding and lift forces on heavily roughened cylinders of various aspect ratios in planar oscillatory flow", *Proc. 8th Int. Conf. on Offshore Mech. and Arctic Engng.* **2**, The Hague, The Netherlands, pp. 269-278.

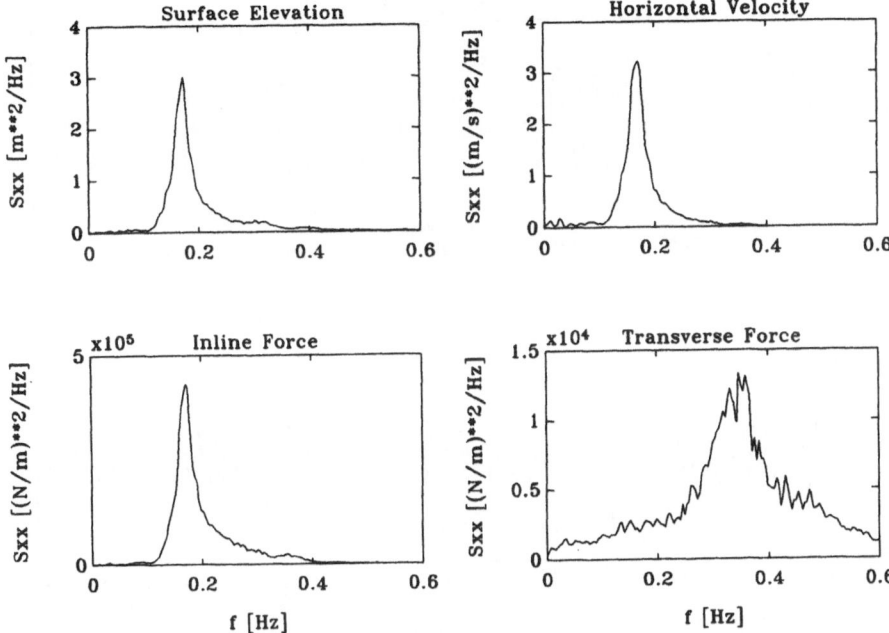

Figure 1   Spectral energy densities of the free surface elevation, horizontal velocity, inline force and transverse force.

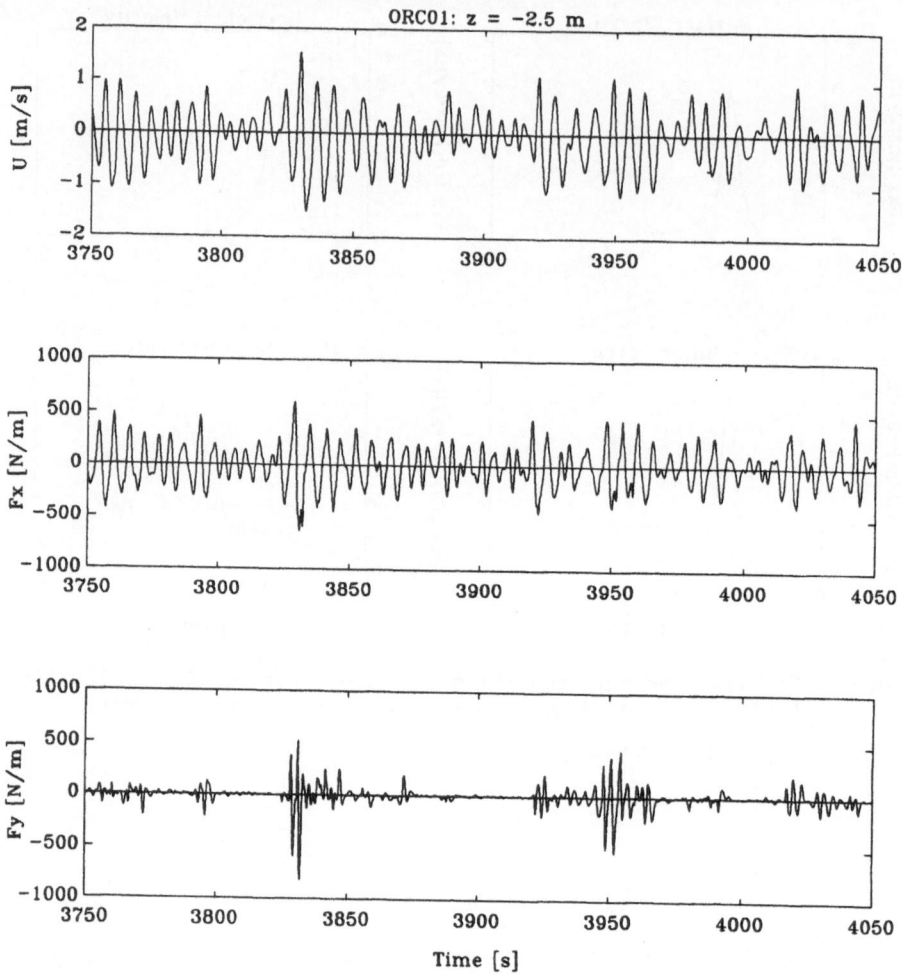

Figure 2   Time histories of the horizontal velocity, inline force and transverse force.

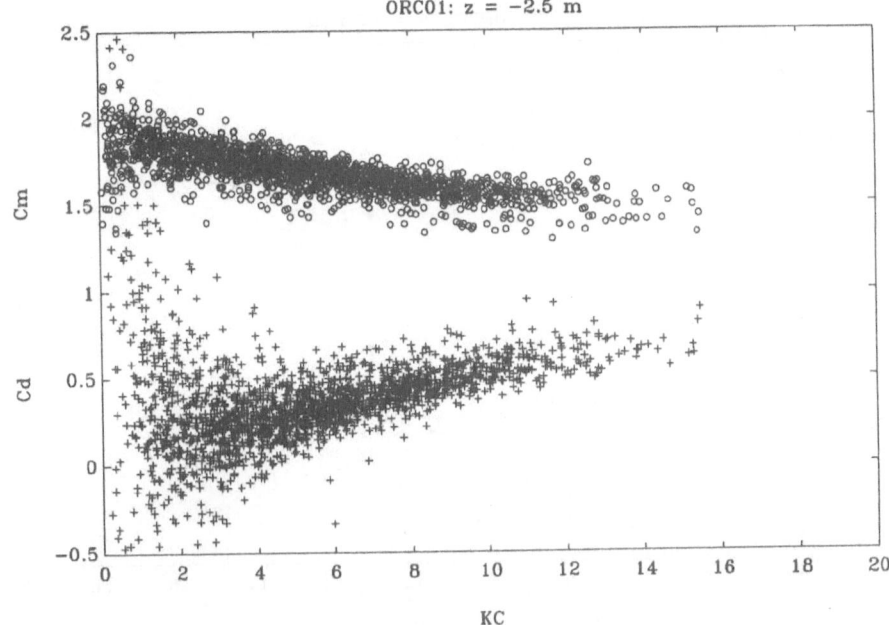

ORC01: z = −2.5 m

Figure 3 Cᴅ and Cᴍ versus Keulegan–Carpenter number; wave by wave analysis of 2052 waves.

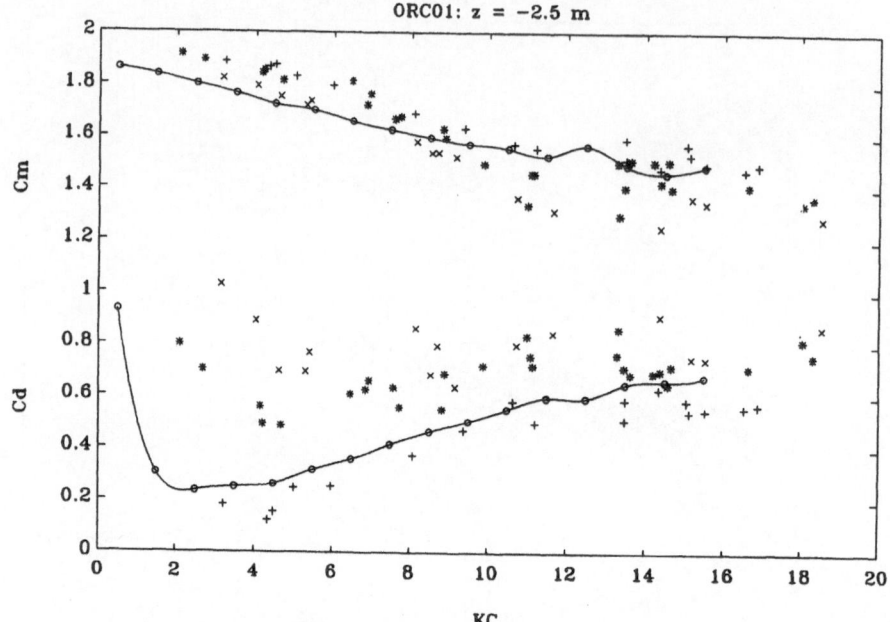

Figure 4   C_D and C_M versus Keulegan-Carpenter number; ——⊖——: median
values of random wave test, lower sleeve; +: smooth cylinder
in regular waves, lower sleeve, Bearman et al (1985a);
x: rough cylinder in regular waves, upper sleeve, Bearman
(1988); *: idem, lower sleeve.

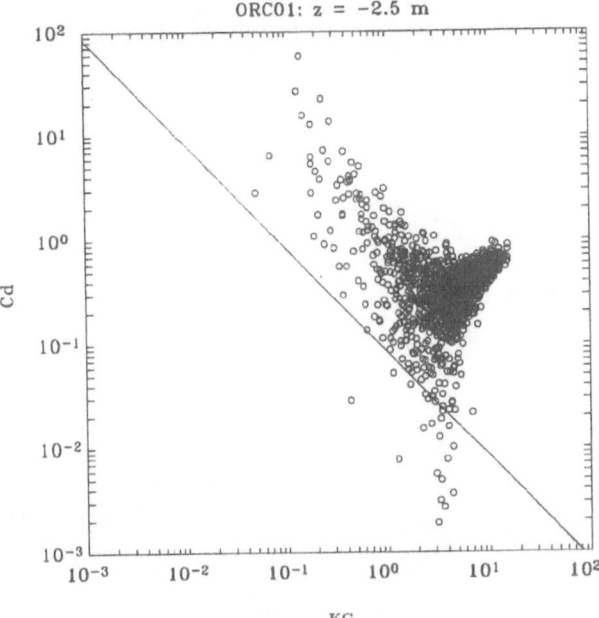

Figure 5 C_D versus Keulegan–Carpenter number; o: experiment, wave by wave analysis; ———: theory, Wang (1968).

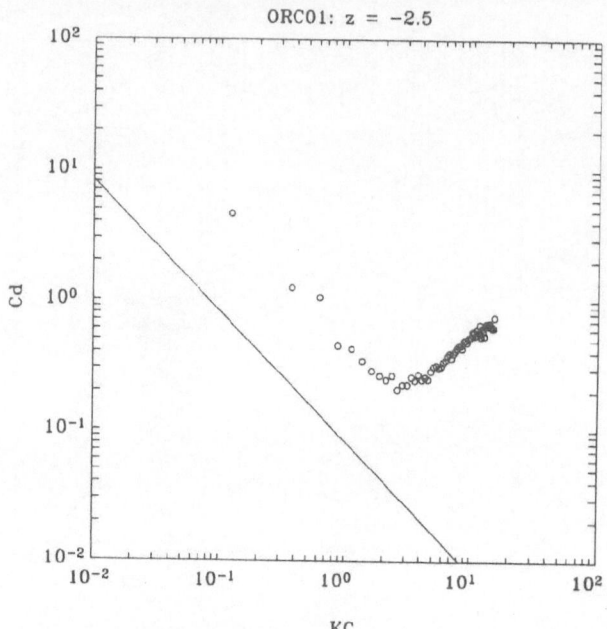

Figure 6 C_D versus Keulegan-Carpenter number; o: experiment, median value; ———: theory, Wang (1968).

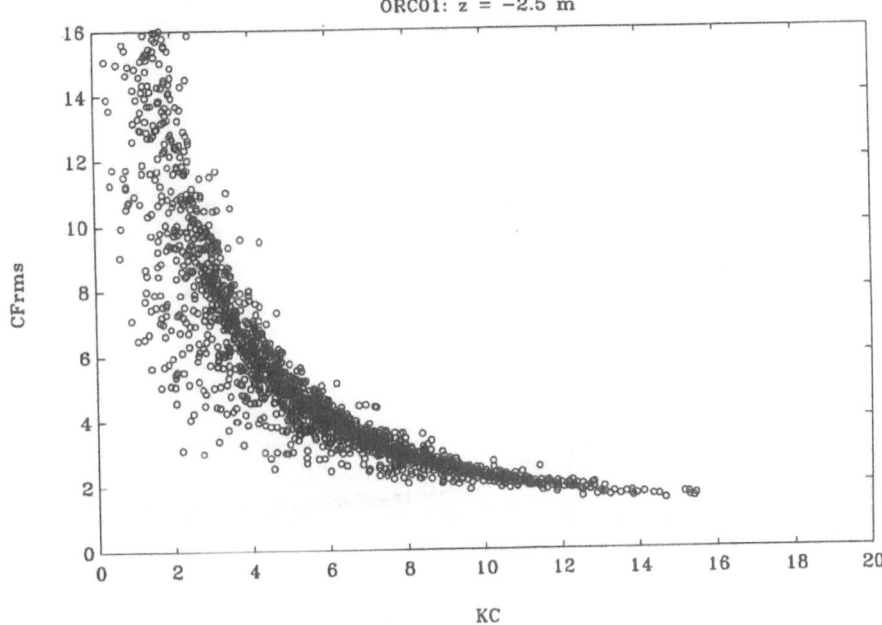

ORC01: z = −2.5 m

Figure 7 CFrms versus Keulegan–Carpenter number; wave by wave analysis.

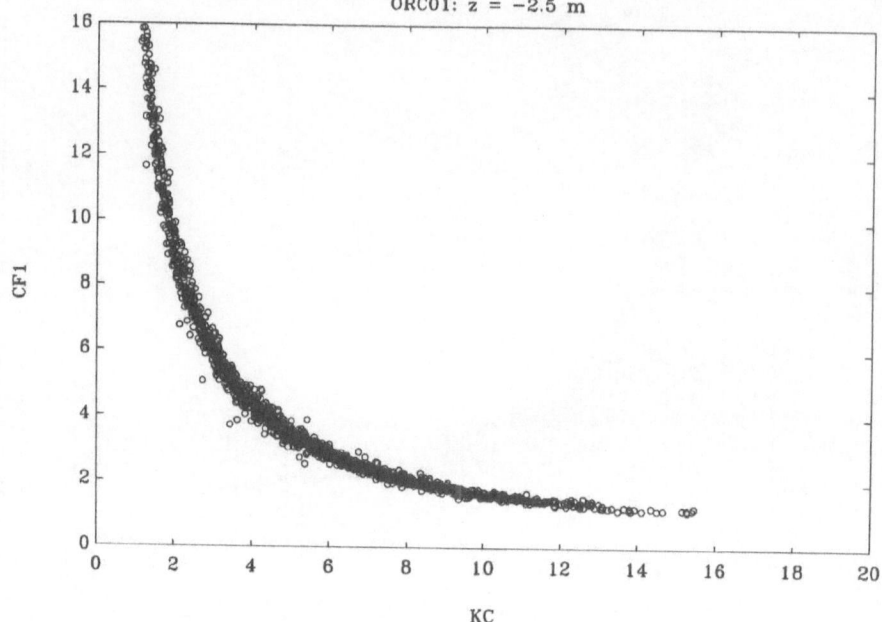

Figure 8 CF1 versus Keulegan-Carpenter number; wave by wave analysis.

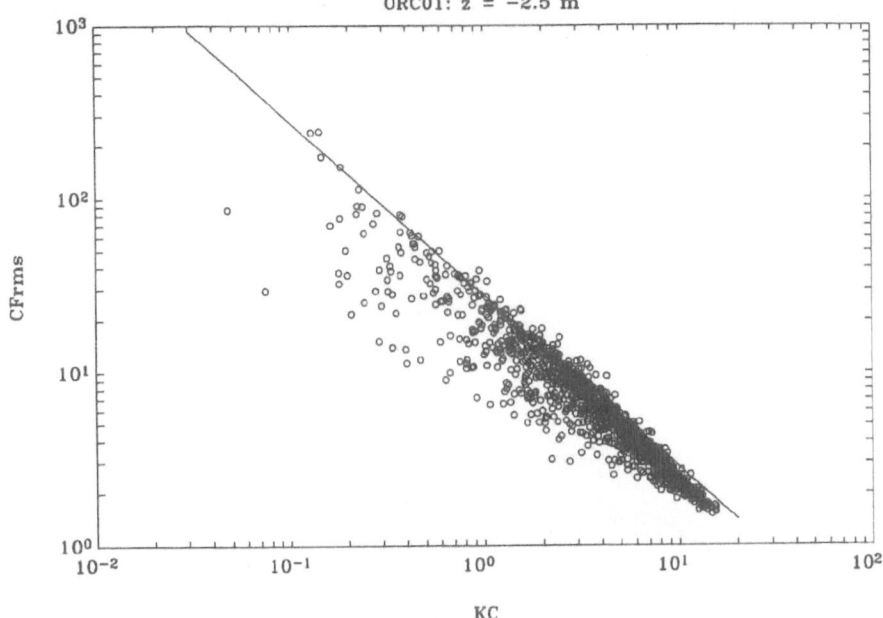

Figure 9 CFrms versus Keulegan-Carpenter number; wave by wave analysis.

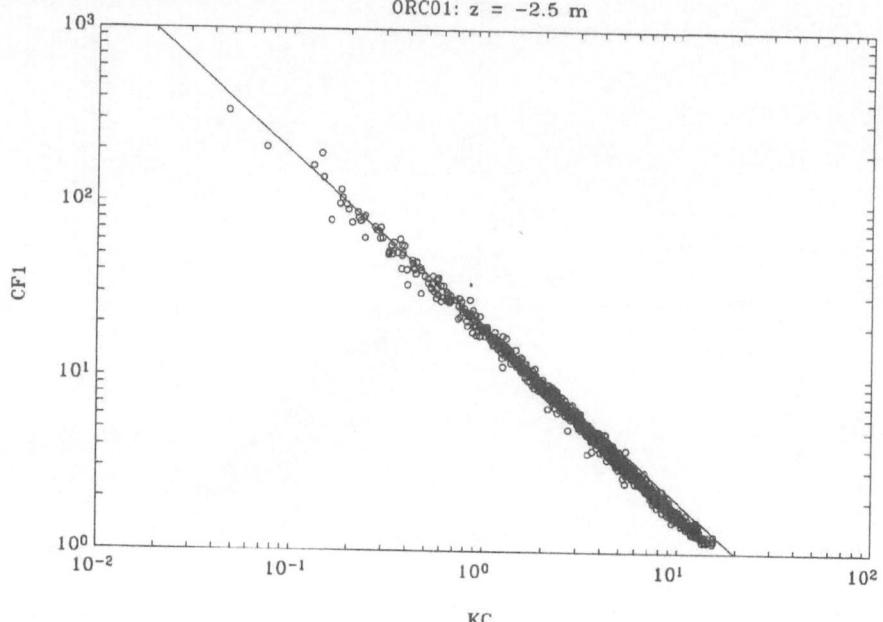

Figure 10 Cғı versus Keulegan-Carpenter number; wave by wave analysis.

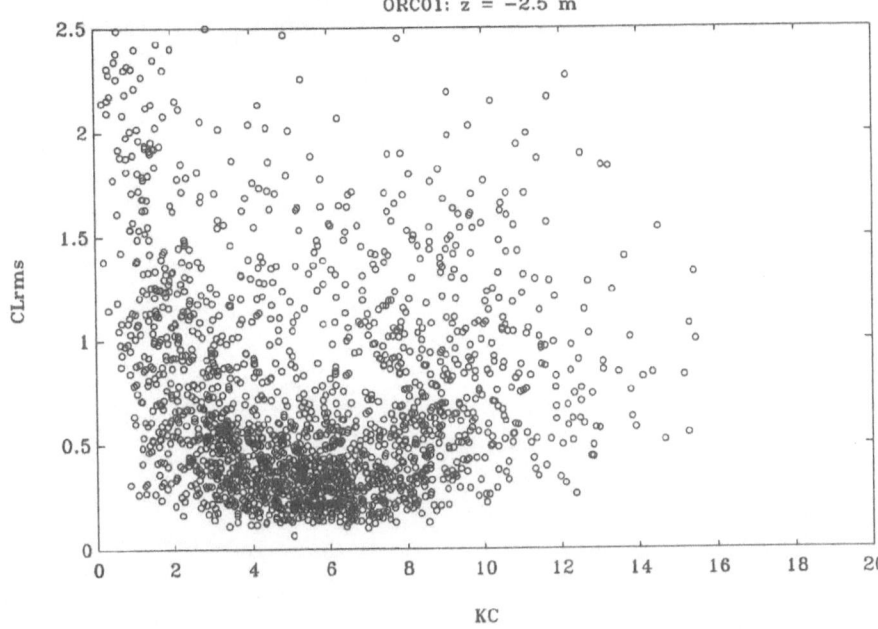

Figure 11 CLrms versus Keulegan-Carpenter number; wave by wave analysis.

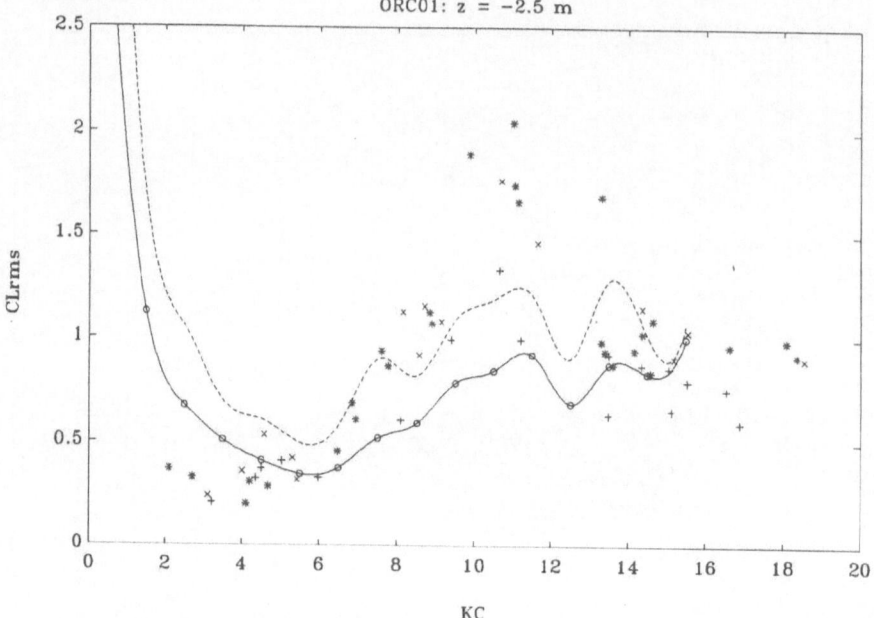

Figure 12 CLrms versus Keulegan-Carpenter number; —⊖—: median values
of random wave test, lower sleeve; -----: idem, upper
quartile values; +: smooth cylinder in regular waves, lower
sleeve, Bearman et al (1985a); x: rough cylinder in regular
waves, upper sleeve, Bearman (1988); *: idem, lower sleeve.

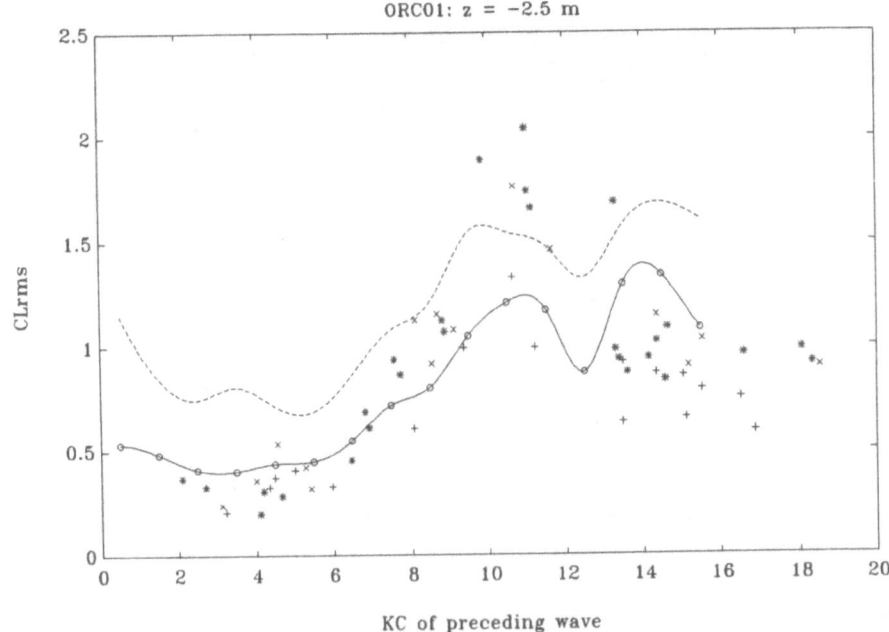

Figure 13 CLrms versus Keulegan-Carpenter number of the preceding wave; symbols as in figure 12.

LABORATORY WAVES AND ASSOCIATED FORCES

K. F. Lambrakos and L. D. Finn
Exxon Production Research Co.
P. O. Box 2189
Houston, Texas   77001
U. S. A.

ABSTRACT. Regular waves generated in the laboratory with a hinged wave maker show significant changes in wave form during propagation. The generated waves modulate as they propagate ranging from waves very nearly sinusoidal to highly nonlinear waves with higher crests than Stokes waves of same height and period. The forces and velocities measured for these waves do not exhibit modulational features comparable to those in the waves. The modulation of the waves appears to result from the presence of free harmonic wave components generated by the wave maker, nonlinear interactions of the harmonic components in the waves and possibly wave reflections.

1.   Introduction

The properties of waves generated in the laboratory are of great interest to offshore engineers and scientists since the majority of experiments with waves are performed in the laboratory. It is also of importance to understand differences between laboratory and field waves since laboratory test results are often used in the field (e.g., results on hydrodynamic forces).

The properties of laboratory waves have been studied experimentally (Iwaki and Sakai, 1970; Hansen and Svendsen, 1974) and theoretically (Madsen, 1971; Flick and Guza, 1980) by many researchers over the years. Hansen and Svendsen report results from laboratory experiments with a piston type wave generator performing sinusoidal motion. The waves produced modulated over the length of the wave flume and were described by a second order Stokes wave plus a free second harmonic wave. They also indicate other higher order effects in these waves.

The results presented in this paper were obtained from relatively larger scale laboratory tests with a flap-type wave maker. They include measurements of the wave profile and associated flow velocities and forces on a single cylinder and a model offshore platform. The interpretation of the results reported is limited.

A. Tørum and O. T. Gudmestad (eds.), Water Wave Kinematics, 701–709.
© 1990 Kluwer Academic Publishers.

2. Experimental Set-Up

The tests were performed in a wave basin that measures 90 m in length, 14.6 m in width, and 4.6 m in depth. The water depth in these tests was 3.7 m and the waves were generated by a straight flap hinged at the bottom (Figure 1). The scale of the waves and the platform structure was 1:48; the model platform cross-section was 1.73 m x 1.28 m. A cylinder extended vertically from -1.72 m to -0.5 m from mean water level (MWL); the cylinder diameter was 4.83 cm. The horizontal flow velocity was measured with an electromagnetic current meter at about 0.30 m below MWL and interior to the platform. A wave staff in-line with the current meter and cylinder measured the wave profile. The instruments and models were attached to a towing carriage.

3. Results

The waves show strong modulational features over the length of the wave basin. The strong modulation in the waves is not reflected in the velocity and force measurements. All measurements are given in full scale. Also, for all the data the wave generator performed sinusoidal motions.

A wave profile measured at a fixed location in the wave basin is given in Figure 2. The average wave height is about 21.0 m and the period is 13.5 sec. The profile shows minor modulation, and appears relatively constant overall. However, a wave profile measured during tow at 0.96 m/s through the basin shows large modulations (Figure 3); the average wave height here is about the same as in Figure 2. A modulating wave profile was also measured during tow at 0.96 m/s for a wave height of 28.0 m and period of 14.2 sec. (Figure 4). The modulation occurs at about the same time interval as for the wave in Figure 3, but higher nonlinearities in the wave shape are evident here. The wave heights given on the Figures are average wave heights over the record length.

As seen from Figures 3 and 4, the waves change in shape from nonlinear Stokes waves during the highest peaks to inverted less nonlinear Stokes waves during lowest peaks. This is shown more clearly in Figures 5 and 6 where measured waves from Figure 3 are compared to waves of the same height and period calculated with second order EXVPD theory (Lambrakos, 1981). It should be noted that the measured wave in Figure 5 is of higher order than second, and this suggests possible nonlinear inter-actions among harmonic wave components.

The measured total hydrodynamic force on the model platform from the wave in Figure 2 (fixed location in the basin) is given in Figure 7. Consistent with the wave profile, the force is fairly regular with minor modulation. The total force from the wave in Figure 3 during towing is given in Figure 8. In contrast with the wave profile, the force modulation is fairly small. It should also be observed that the highest force peak does not occur during the highest wave peaks. The findings for the wave in Figure 3 are true for the wave in Figure 4, also. The velocity measured interior to the platform (Figure 9) shows relatively small variation with time as compared to the wave profile. Also, the force measured on the single cylinder (Figure 10) shows small variation and does not exhibit the modulational features of the wave profile.

The crest and trough durations in the wave profile are plotted versus their order of occurrence in the wave record in Figure 11. The durations at the highest peaks confirm the highly nonlinear Stokes wave feature since the sharp crests last much less than the flat troughs. At the lowest peaks the crests last somewhat longer than the troughs, and this suggests inverted Stokes waves. The variation of the crest and trough areas is given in Figure 12. The area of the crest and the area of the trough for the same wave are about equal, but both crest and trough areas modulate.

4. Conclusions

The measured laboratory waves modulate over the length of the wave basin. The modulation in the waves is not reflected in the flow velocities nor in the resulting forces on a single cylinder and a model offshore platform.

On the basis of limited analysis of these data and other published results, it appears that the measured wave features are due to the presence of free harmonic wave components, nonlinear interaction of harmonic components in the waves, and also reflections.

5. References

Flick, R. E., and Guza, R. T. (1980) "Paddle generated waves in laboratory channels", Journal of the Waterway, Port, Coastal and Ocean Division, ASCE, 106 (WW1), 79-97.

Hansen, J. B. and Svendsen, I. A. (1974) "Laboratory generation of waves of constant form", Proceedings of the Fourteenth Coastal Engineering Conference, Vol. 1, 321-339.

Iwagaki, Y., and Sakai, T. (1970) "Horizontal water particle velocity of finite amplitude waves", Proceedings of the Twelfth Conference on Coastal Engineering, ASCE, Washington, D. C., 309-325.

Lambrakos, K. F. (1981) "The extended velocity potential versus Stokes wave representation", Journal of Geophysical Research, Vol. 86, No. C7, 6473-6480.

Madsen, O. S. (1971) "On the generation of long waves", Journal of Geophysical Research, 76 (36), 8672-8683.

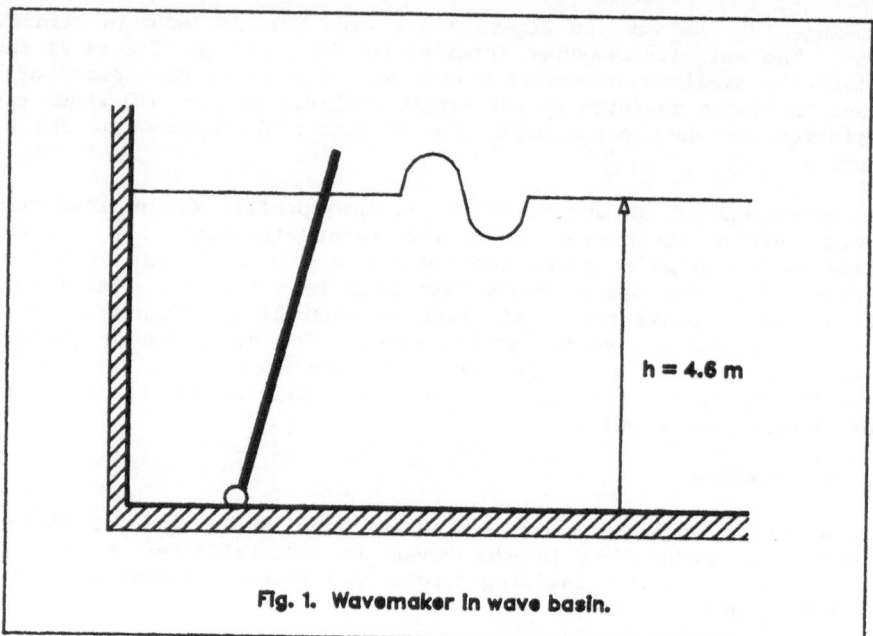

Fig. 1. Wavemaker in wave basin.

h = 4.6 m

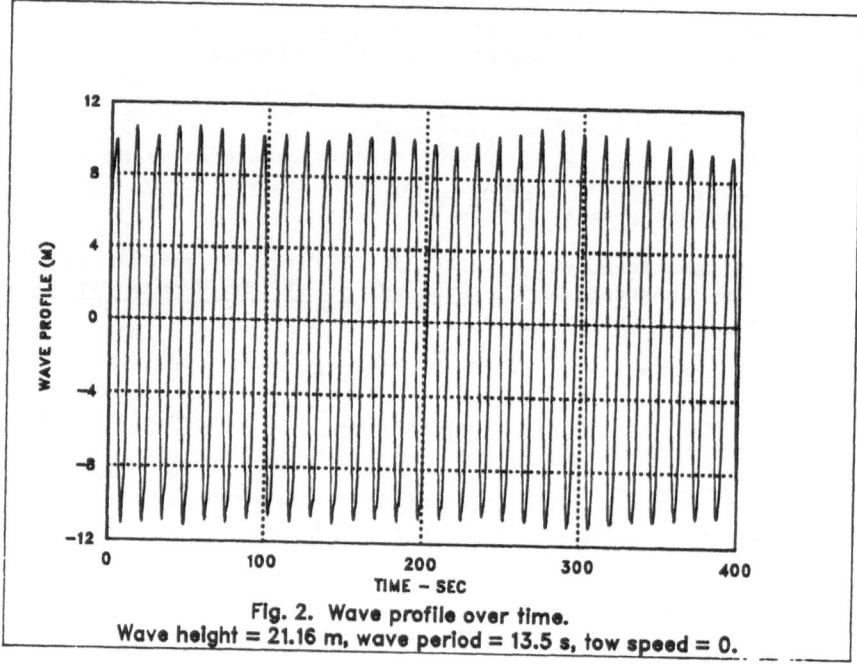

Fig. 2. Wave profile over time.
Wave height = 21.16 m, wave period = 13.5 s, tow speed = 0.

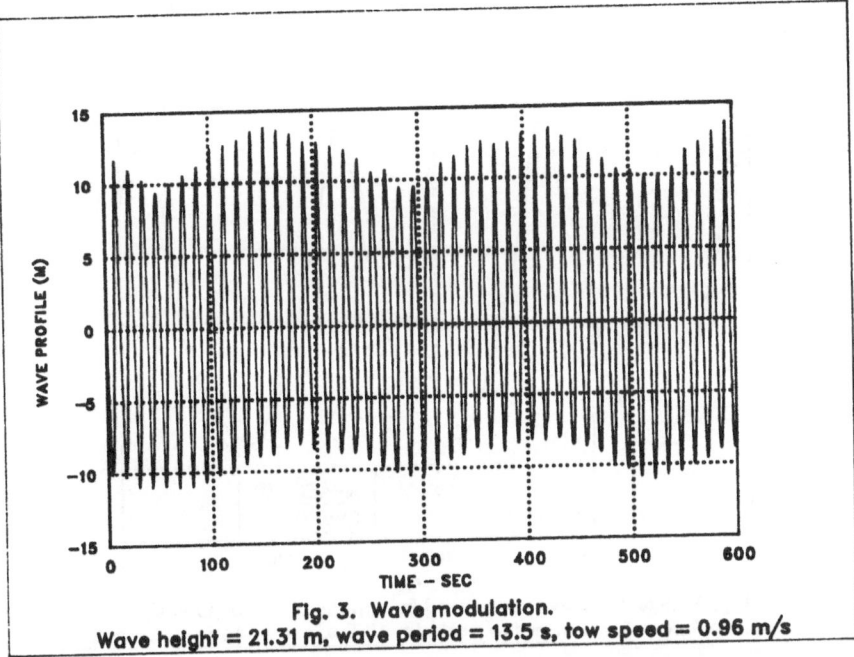

Fig. 3. Wave modulation.
Wave height = 21.31 m, wave period = 13.5 s, tow speed = 0.96 m/s

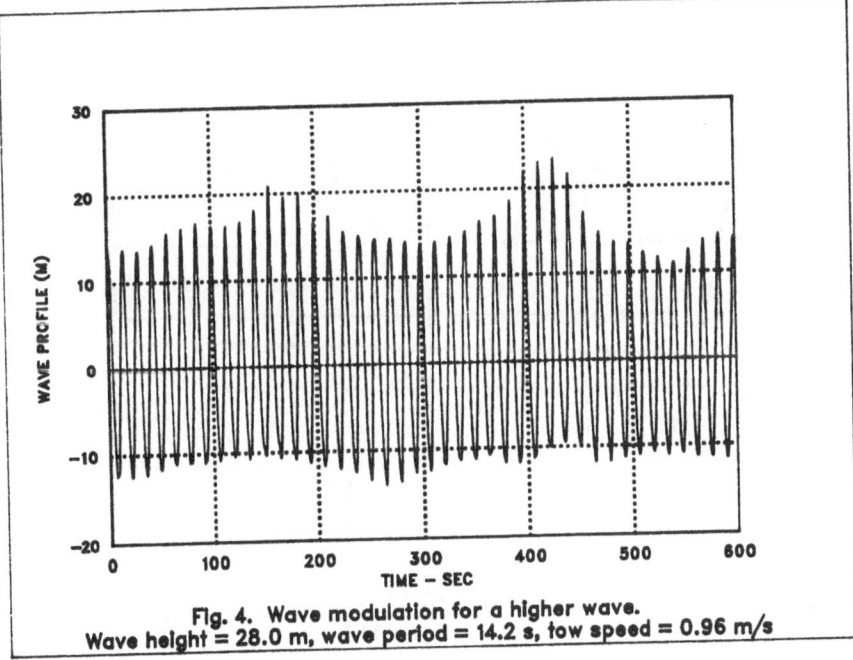

Fig. 4. Wave modulation for a higher wave.
Wave height = 28.0 m, wave period = 14.2 s, tow speed = 0.96 m/s

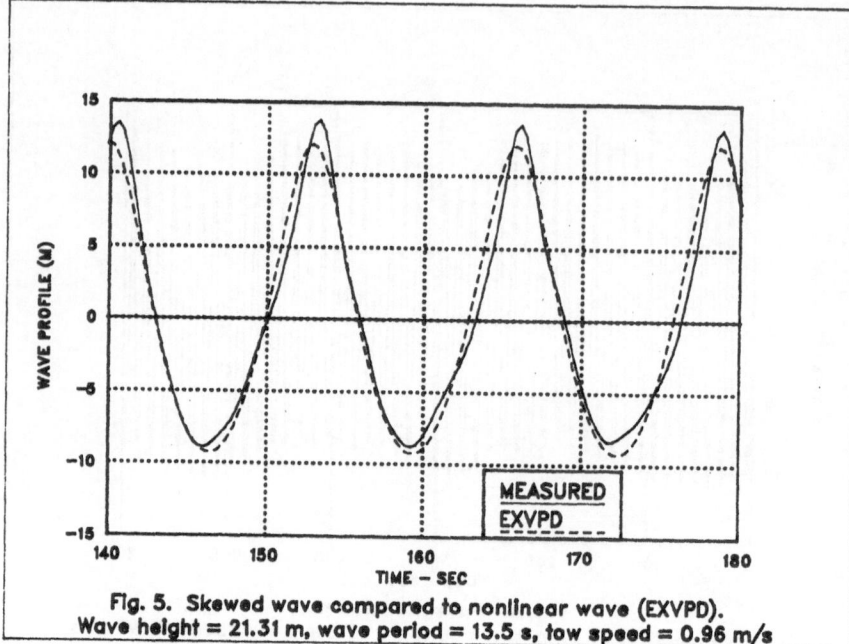

Fig. 5. Skewed wave compared to nonlinear wave (EXVPD).
Wave height = 21.31 m, wave period = 13.5 s, tow speed = 0.96 m/s

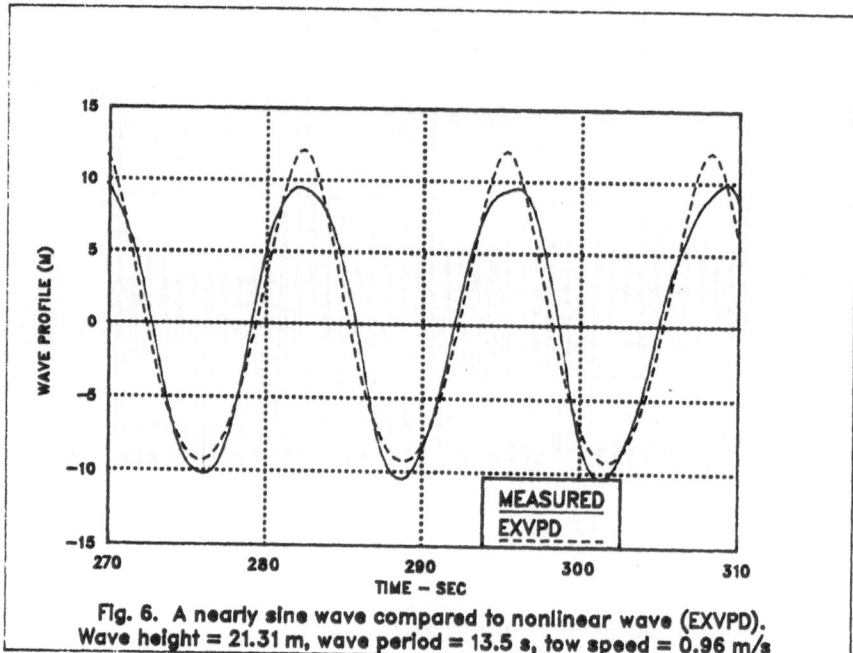

Fig. 6. A nearly sine wave compared to nonlinear wave (EXVPD).
Wave height = 21.31 m, wave period = 13.5 s, tow speed = 0.96 m/s

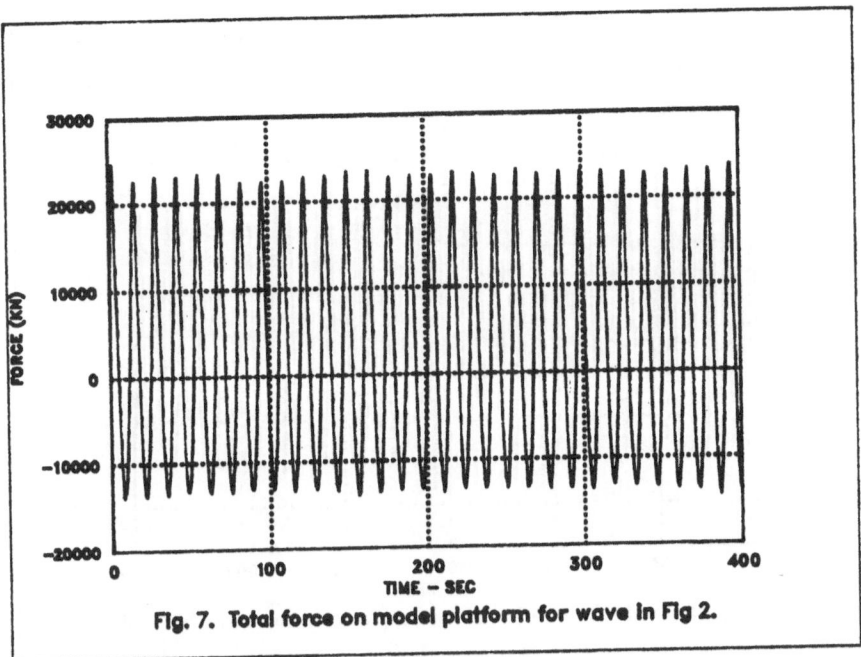

Fig. 7. Total force on model platform for wave in Fig 2.

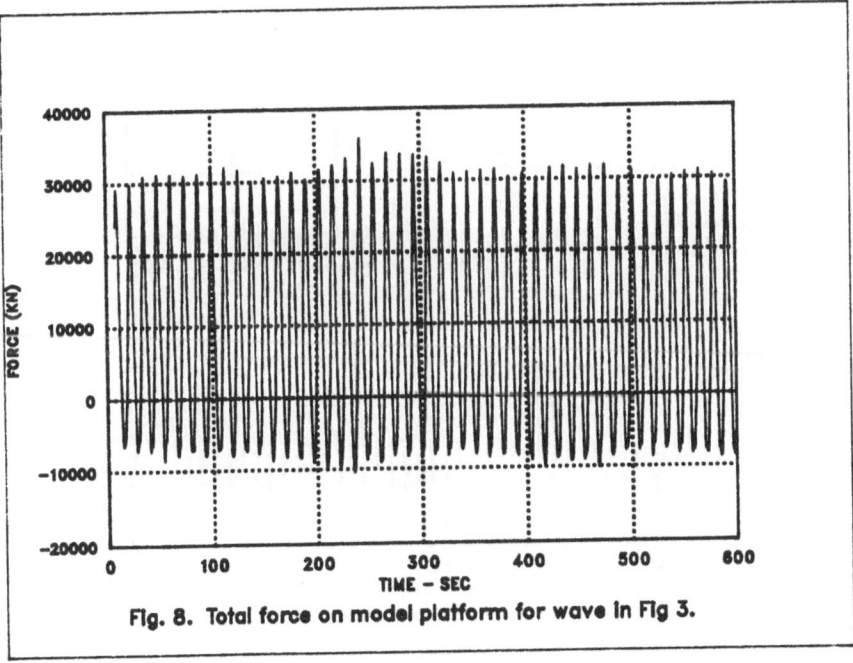

Fig. 8. Total force on model platform for wave in Fig 3.

708

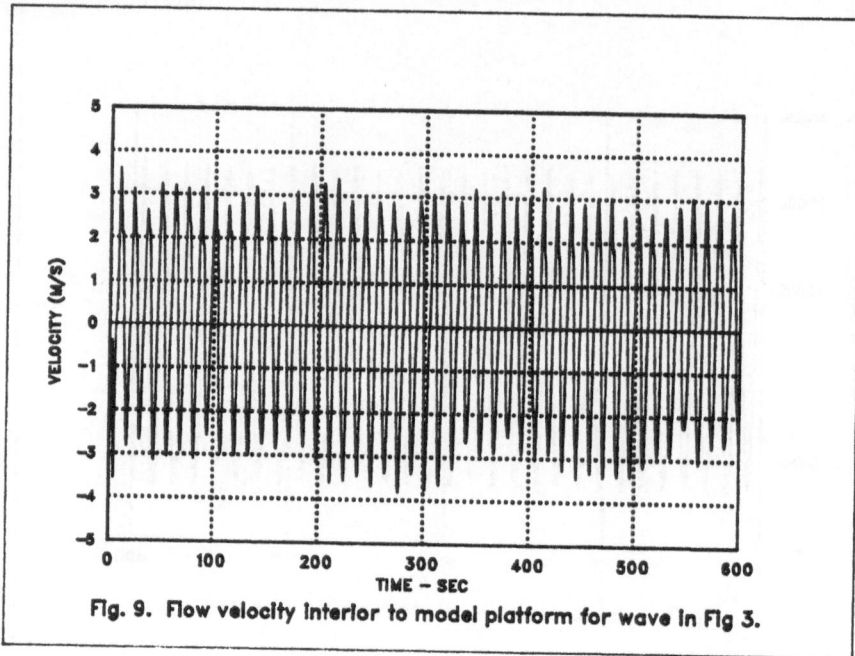

Fig. 9. Flow velocity interior to model platform for wave in Fig 3.

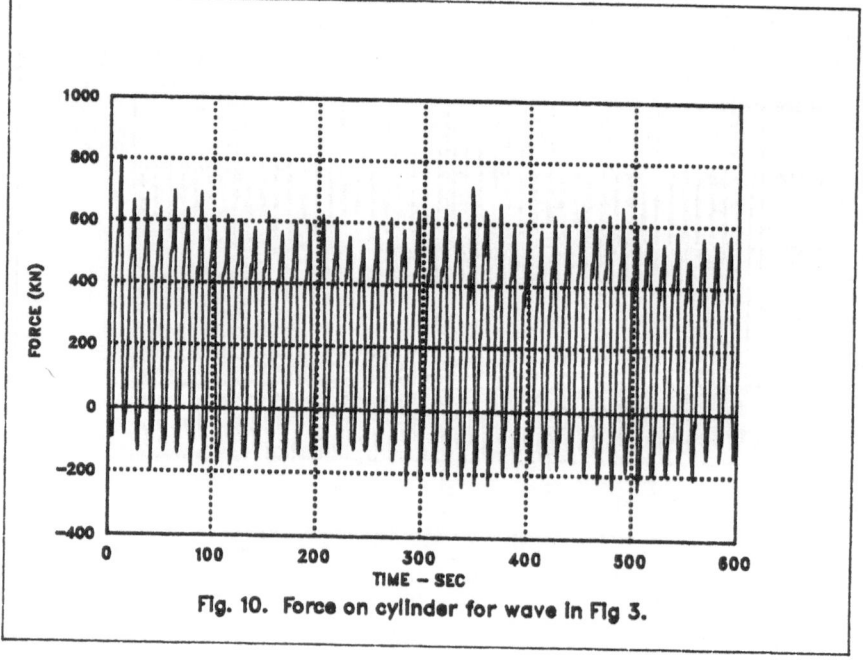

Fig. 10. Force on cylinder for wave in Fig 3.

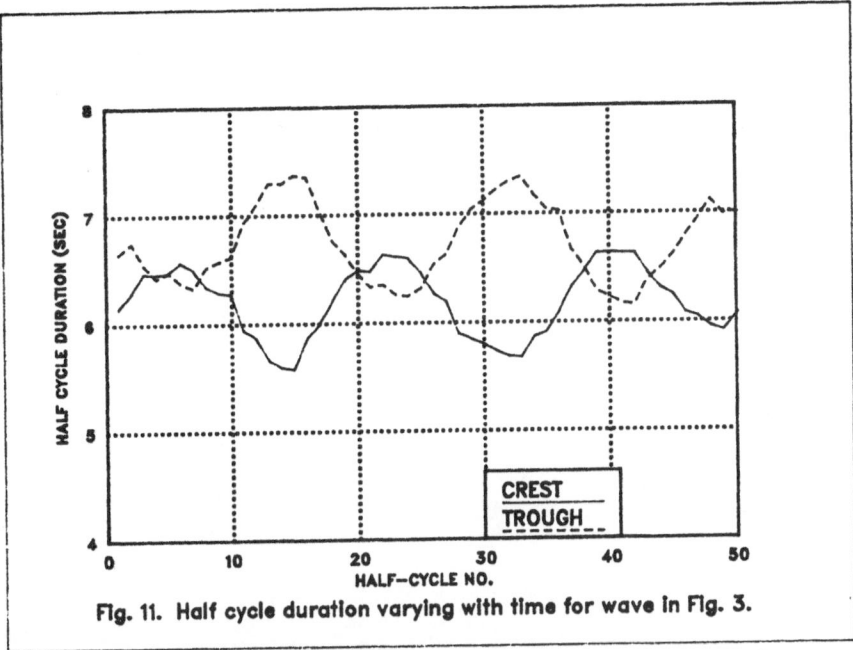

Fig. 11. Half cycle duration varying with time for wave in Fig. 3.

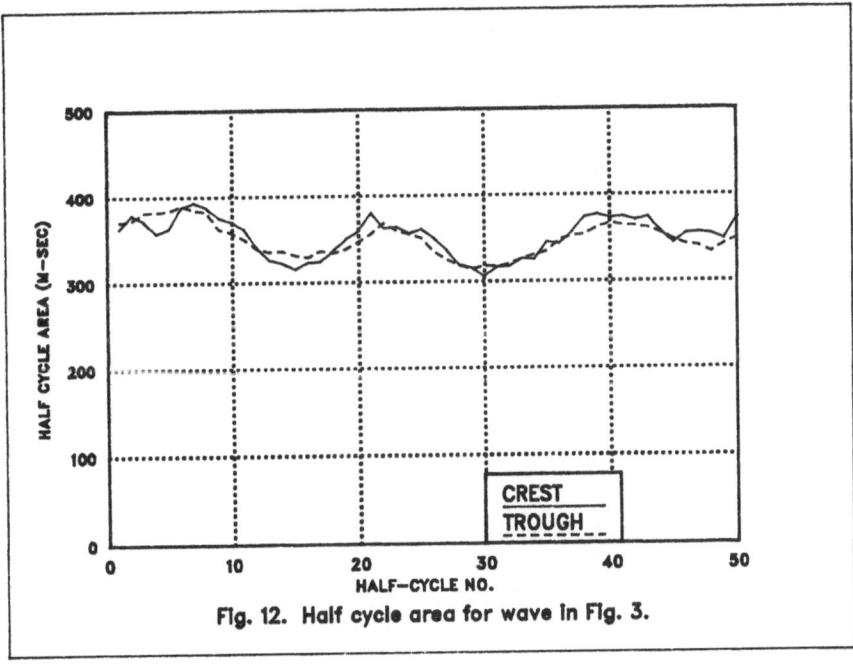

Fig. 12. Half cycle area for wave in Fig. 3.

A LABORATORY STUDY OF THE KINEMATICS AND FORCING DUE TO FIVE SIMILAR LARGE WAVES

ANDREW CORNETT
Research Officer, Hydraulics Laboratory
National Research Council of Canada
Ottawa Ontario K1A-0R6
CANADA

ABSTRACT: During a model study of a large oil production structure, five large long-crested irregular wave transients with similar wave heights and periods were found to exert very different uplift forces on the deck and wave deflectors. This paper reports on a subsequent investigation into the characteristics of these five non-breaking wave transients, including definition of the wave profile in the spatial domain, calculation of wave parameters as a function of location of the wave crest, and measurement and prediction of horizontal fluid velocities below the still water level. The forces exerted by these five waves on a segmented column located at different locations in the test area are also discussed. Considerable variability in the magnitude of forcing was observed between the five wave transients, and also at different locations due to the same wave.

1. INTRODUCTION

Large long-crested irregular waves of similar height and period can cause very different forcing on ocean structures. This point was made clear during a series of wave loading tests on the production platform proposed for the development of the Hibernia oil field in the North Atlantic off Canada's East coast. In these run-up and deck clearance tests, uplift forces on wave deflectors and the underside of the deck structure were measured in response to wave loading by five different versions of the design wave for the site. Some of the results of these tests have been summarized by Mogridge et al. (1989). It was found that these five similar waves caused repeatable uplift forces on the wave deflectors of the gravity based structure (GBS) which consistently differed by as much as 300 %. Such differences are not readily explained by existing theories of wave kinematics and structural forcing.

Characteristics of the five large waves used in this study and the resulting force maxima on a curved wave deflector are listed in Table 1. These wave characteristics were measured in the test basin without the model in place, and represent incident wave conditions at the front face of the GBS structure. Note that wave #5 produced the largest force but did not have the largest wave height or crest elevation. None of the five waves became unstable as they propagated through the test site, so the

711

A. Tørum and O. T. Gudmestad (eds.), Water Wave Kinematics, 711–729.
© 1990 Kluwer Academic Publishers.

Table 1. Wave and Force Data from GBS Run-up Tests (Mogridge et al. (1989))

WAVE NO.	Hmax (m)	CREST ELEV. (m)	Tmax (s)	Fmax[*] (x10^3 kN)
1	28.3	18.2	14.8	10.2
2	29.1	16.8	16.1	7.2
3	28.4	15.5	15.8	6.7
4	29.3	18.2	15.3	11.8
5	28.2	17.7	15.6	21.8

[*] - Force maxima on the curved deflector of column #2 averaged over a number of tests.

---

differences in forcing cannot be attributed to breaking of the incident wave.

The uplift force acting on a wave deflector is undeniably the result of a complicated interaction between the incident wave and the structure. It would not be surprising if a change in the geometry of the structure caused a dramatic change in the force caused by an incident wave. What is surprising is the fact that five waves with such similar characteristics can also lead to such dramatic differences in forcing. Although the five waves have similar characteristics, they are each unique wave transients comprised of different components which are continually passing into and out of phase as they propagate. They are also preceded by waves of varying height and period, and it is possible that the difference in forcing is due to some interaction between the wave transient and its predecessor. A good question might be, "What aspect of the evolution or composition of wave transient #5 causes greater uplift forces?"

Faced with these surprising and significant results, a research project was initiated to explore the behaviour of these five irregular wave transients in more detail. The question of wave interaction with a complex structure such as the GBS was put aside, and the focus of research was applied to the evolution and kinematics of these five wave transients. Specific objectives of this research were:

- Define the water surface profile of each wave transient as a function of space at discrete points in time.

- Trace the evolution of each wave transient as it propagates through the test site. Quantify this evolution using different wave steepness parameters.

- Measure the force exerted by each wave transient on a vertical cylinder at different locations in the test site. If possible, correlate the force maxima to spatial or temporal wave parameters.

- Measure the kinematics under each wave transient and relate these to the free surface.

2. EXPERIMENTS

Experiments were conducted in the Multidirectional Wave Basin of the NRC Hydraulics Laboratory in Ottawa, Canada, to discover more about the behaviour of these five long-crested wave transients. The facility is 30 m wide, 20 m long and 3 m deep. A water depth of 2.0 m was used in the present study. A sixty segment wave machine along the North wall was used to generate the wave transients which were absorbed by progressive wave absorbers located along the three other walls. Jamieson and Mansard (1987) indicate that reflections from these passive absorbers are less than 5 % over a wide range of wave height and period.

The same wave machine control signals used in the GBS run-up study were used to reproduce the five wave transients. The irregular wave transients were originally selected from a large number of long wave records synthesized using the random phase method as implemented by Funke and Mansard (1984), in which the amplitude for each frequency component is chosen deterministically based on a target energy spectral density, and the phase of each component is selected at random between 0 and 2π. To minimize the influence of reflected wave energy on the incident waves, each transient was created in the wave basin using the 'snapshot' technique of wave generation, in which the shortest possible burst of wave generator control signal is used. The duration of the control signal is computed so that the highest frequency component present in the wave synthesis is able to propagate to the test site and contribute in the formation of the wave transient. In addition to the reduction of reflected wave energy, the 'snapshot' technique has the added benefit of reducing the settling time necessary between successive tests. Forty-five second long wave machine control signals were used to generate each wave transient.

The waves were measured using a linear array of 12 capacitance wire wave gauges and 2 bi-directional electro-magnetic current meters. The twelve wave probes were located at 30 cm spacings in line with the direction of wave propagation. The two current meters were located in line with the tenth wave probe, at elevations of 1.46 m and 1.66 m above the floor of the basin, and were oriented to measure orbital velocities in the vertical plane. This array of wave probes and current meters was mounted on a mobile platform that was moved to three different stations throughout the test program so that the water surface could be mapped over an 8.1 m distance at 0.3 m spacing. The wave probes spanned the interval between 6.0 m and 14.1 m from the wave generator.

A 17 cm diameter column was used to measure the forcing exerted by each of the wave transients at 19 different positions within the test site. Separate tests were carried out to measure wave loading on the column at each position. The column was positioned at 30 cm intervals between 8.7 m and 14.1 m from the wave generator. The column was comprised of 9 independent segments of various heights, each capable of sensing hydrodynamic forcing in two orthogonal directions. Figure 1 shows an elevation view of the column and the arrangement of the nine segments.

714

The instrumented portion of the column extended from 91 cm above the still
water level to 59 cm below. The midpoints of segments 7 and 9 are at the
same elevations as the bi-directional current meters. The function of the
column has been reported by Cornett (1987).

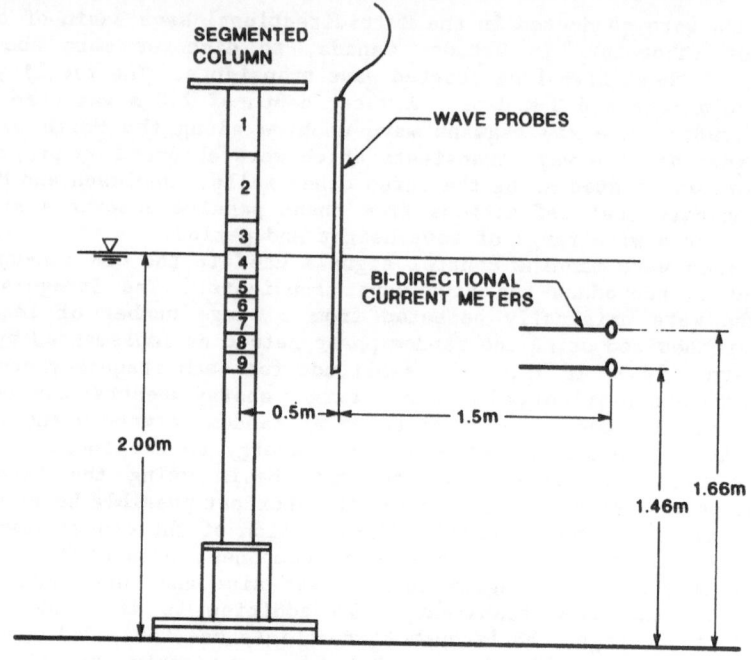

Figure 1. Sketch of the Segmented Column

Figure 2 shows a photograph of the column, wave probes, and current
meters installed in the basin before testing. The wave generator can be
seen in the rear and the progressive wave absorbers are in the foreground.
Figure 3 shows the basin during a test in which the waves are propagating
from upper-left to lower-right. During each test, 32 channels of data
were sampled at a rate of 20 Hz for a duration of 30 s.

3. SPATIAL DEFINITION OF WAVE PROFILE

Spatial definitions of wave transients #1 through #5 at four instants of
time, each separated by 0.5 s, are presented in Figures 4 through 8,
respectively. It is clear that the shape of each wave transient changes
significantly as the wave propagates across the test area. The crest
elevations of waves #1, #3, and #4 are increasing, the crest elevation of
wave #2 is decreasing, and the crest elevation of wave #5 remains
constant. The average celerity of each wave transient can be estimated
by measuring the distance covered by the wave crest in the 1.5 s duration
shown in Figures 4 through 8. This is more difficult for waves #3 and #4
because the wave crest is not always well defined. The average celerity

Figure 2. Photograph of the Multidirectional Wave Basin

Figure 3. Photograph of a large wave transient

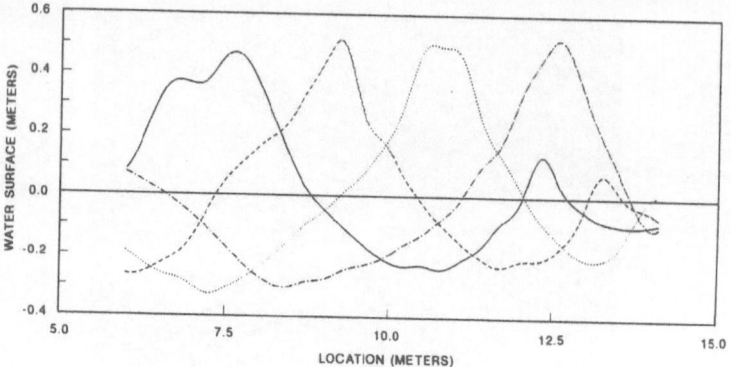

Figure 4. Spatial definition of wave #1

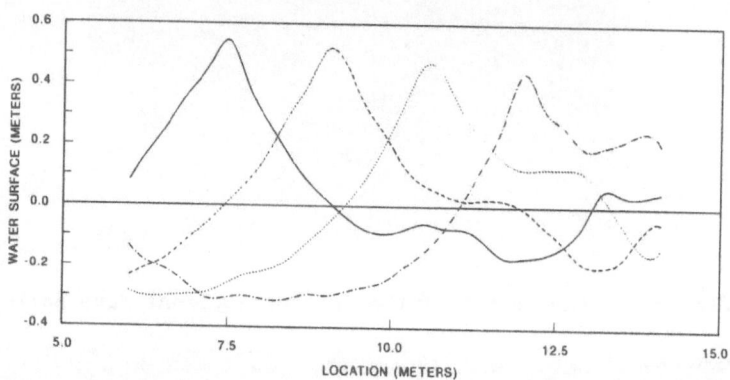

Figure 5. Spatial definition of wave #2

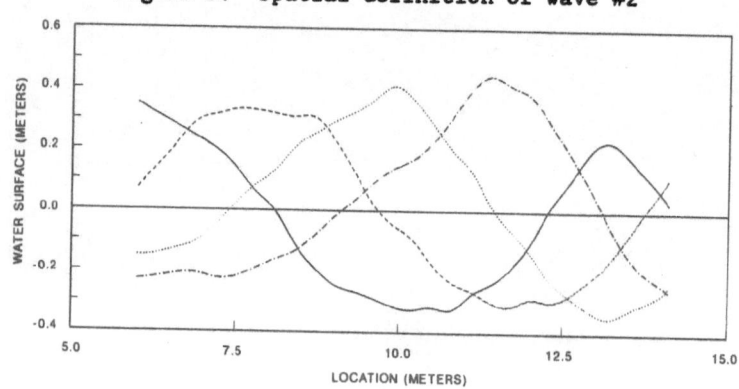

Figure 6. Spatial definition of wave #3

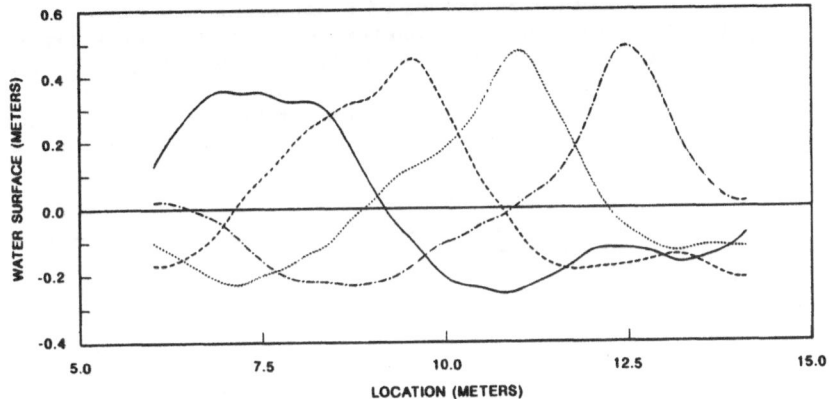

Figure 7. Spatial definition of wave #4

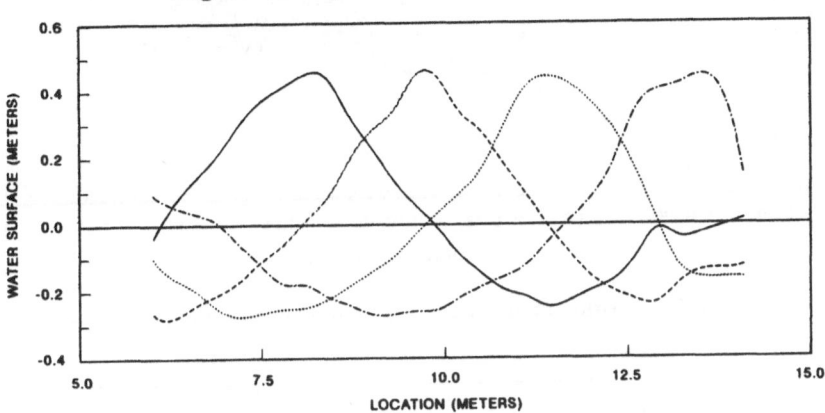

Figure 8. Spatial definition of wave #5

of wave #5 is estimated by this technique to be 3.53 m/s. This is somewhat greater than the estimate of group velocity given by linear wave theory equal to 1.96 m/s (based on assuming an average wave period of 2.2 s in a water depth of 2.0 m). The linear theory estimate of wave celerity for this wave is 3.25 m/s. Estimates of the wave length of wave #5 based on the spatial definition of the wave profile vary from 6.6 m to 7.8 m depending on whether the measure is made between troughs or zero-downcrossings. The linear theory estimate of wave length is 7.1 m.

A parameter representing the average slope of the crest front can be formed by dividing the crest elevation above mean water by the distance between the wave crest and the zero-upcrossing ahead of the crest. It is interesting to compare crest front slope obtained from the spatial domain in this way to estimates of crest front steepness obtained from time domain measurements of the same waves. Figures 9 and 10 show this comparison as a function of location for waves #1 and #2. In these figures, the lines represent estimates of crest front steepness calculated in the time domain using the downcrossing, upcrossing, and average wave periods, while the dots represent crest front slope computed from the spatial definition of each wave. The data from wave #2 indicates that measurements of crest front steepness from the time domain can be in error by as much as 100%. The data from wave #1 are included to indicate a more typical degree of agreement.

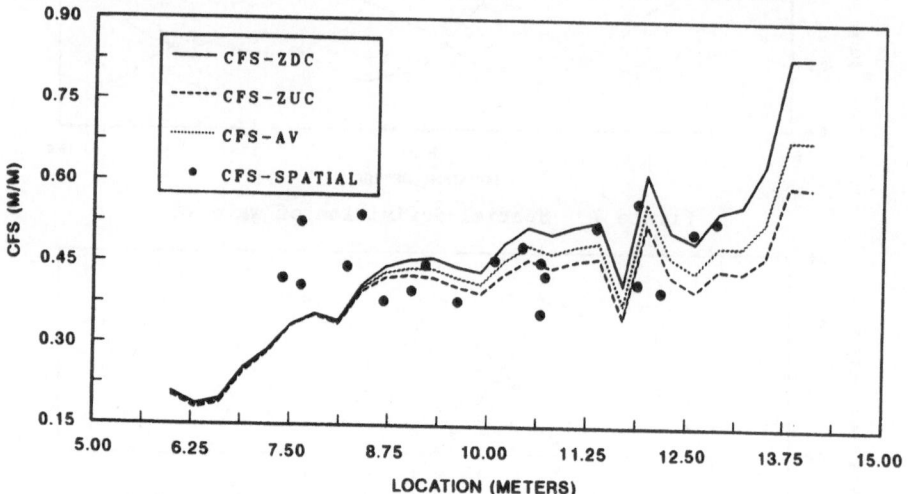

Figure 9.   Crest front slope and steepness for wave #1

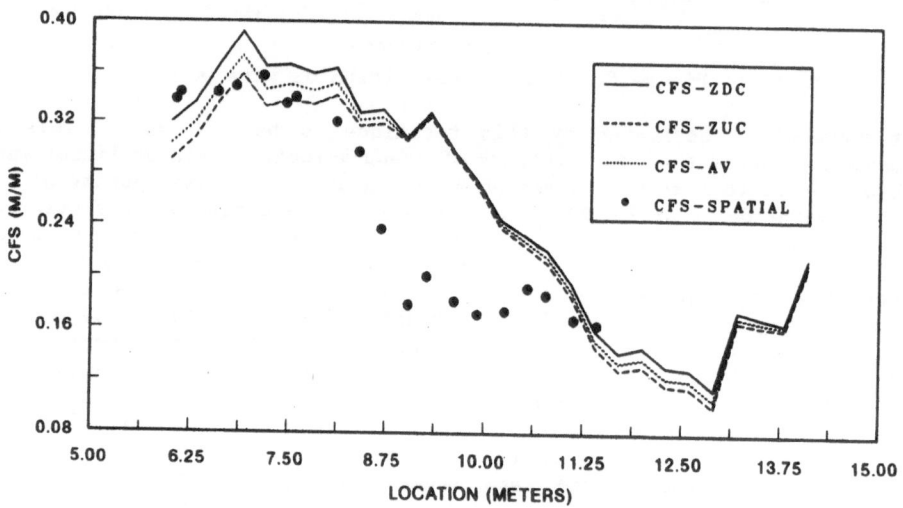

Figure 10.   Crest front slope and steepness for wave #2

4. SPATIAL VARIATION OF TEMPORAL WAVE PARAMETERS

Many wave characteristics have been computed from the temporal definition
of each wave transient obtained at 28 positions between 6 and 14.1 meters.
Some of these are shown in Figures 11 through 13. Vertical asymmetry, as
defined by Kjeldsen and Myrhaug (1978), and the IAHR (1986), is shown in
Figure 11. The forward tilting of each wave transient, as defined by Goda
(1986) is shown in Figure 12. Forward tilting of wave crests has been
computed over a downcrossing wave period according to

$$\beta_3 = \frac{1}{N-1}\Sigma(\eta'-\overline{\eta}')^3 \; / \; [\frac{1}{N-1}\Sigma(\eta'-\overline{\eta}')^2]^{3/2} \tag{1}$$

where η' denotes the time derivative of the water surface. A positive
value of forward tilting indicates a steep crest front and a gradual crest
rear, while forward tilting equal to zero indicates vertical symmetry.

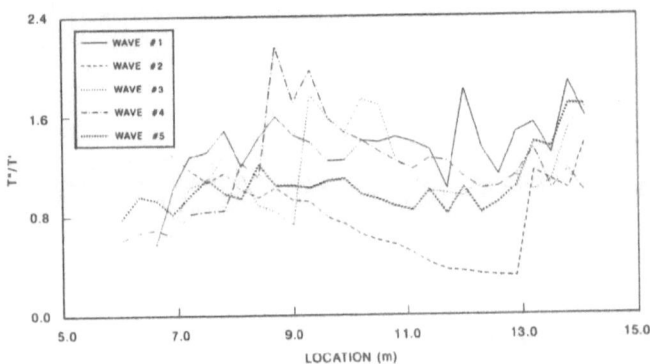

Figure 11. Vertical asymmetry versus location for five waves

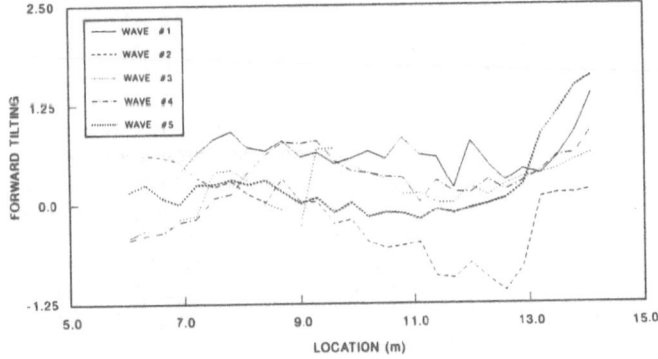

Figure 12. Forward tilting versus location for five waves

Figure 13 shows the maximum value of the time derivative of the water surface for each wave as a function of location. The derivative is computed at each point of the measured water surface time series using coefficients of interpolating cubic splines. This derivative is a measure of the instantaneous water surface steepness whereas the crest front steepness and crest front slope presented in Figures 9 and 10 represent values averaged over the entire crest front. The maximum of the instantaneous steepness can be more than twice the average crest front steepness. Note that wave #1 has the greatest maximum surface derivative and crest front steepness over most of the test site. This observation agrees with the measure of crest front slope computed in the spatial domain and reported in Figure 9.

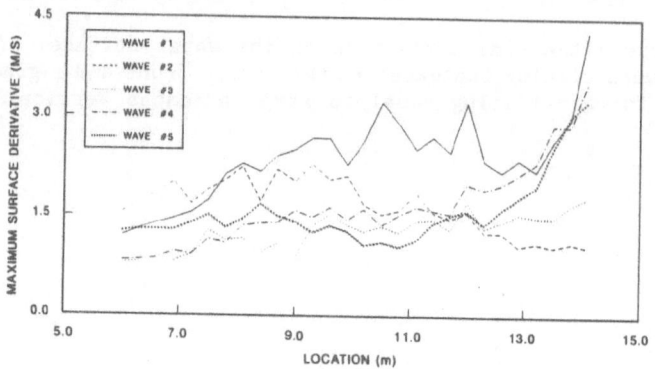

Figure 13. Maximum surface derivative versus location for five waves

5. SPATIAL VARIATION OF FORCE

Forces caused by the five wave transients acting on a 17 cm diameter column were measured at 9 elevations and 19 different locations between 8.7 m and 14.1 m. Assuming a wave period of 2.25 s to be representative of the wave transients, the Kuelegan-Carpenter number at the mean water level for this column is 13.5. This corresponds to the domain in which inertia and drag forcing are both important. Typically, those segments below the mean water level (segments 4-9) are dominated by inertial forcing, and the positive force peak on these segments occurs when horizontal fluid acceleration is greatest. This occurs near the zero upcrossing preceding the wave crest. Those segments above the mean water level (segments 1-3), which are exposed to loading only during the passage of a wave crest, are more likely to be dominated by drag forcing. The peak force acting on these upper segments often occurs beneath the crest of the passing wave. In some cases, when the crest front of the wave transient is sufficiently steep, water slapping against the face of the column can cause sudden forces of much greater magnitude than either the drag or inertial forcing mechanisms.

The positive force maxima from segments 2, 3, and 7, as well as from the signal representing the sum of the forces from all 9 segments, have been plotted in Figures 14 through 17, respectively. Note that the sum of the maxima from each segment will be greater than the maximum of the signal representing loading on the whole instrumented portion of the column. This is because the individual maxima from each segment do not generally occur at the same time. A general overview of the data in these figures indicates that surprisingly large variations exist in the peak force caused by a single wave transient as it propagates. Even greater variations exist in the forcing caused by different, although similar, waves.

Consider the force maxima exerted on segment 2, shown in Figure 14. Segment 2 is 35 cm in height and its midpoint is 38.5 cm above the still water level. The force caused by wave #5 has increased from 45 N to 181 N over a distance of only 60 cm! This tremendous increase is likely caused by the unpredictable action of wave slapping. The magnitude of this increase is similar to the difference in forcing between non-breaking and breaking waves reported by: Kjeldsen, Tørum and Dean (1986); Reddish and Basco (1987); and Kjeldsen (1987). Note that the maximum force caused by wave #1 occurs near 11.5 m. This is roughly the same location at which the crest front slope from the spatial domain definition is greatest (see Figure 9).

The force maxima acting on segment 3 are shown in Figure 15. This segment is 15 cm high and its midpoint is 13.5 cm above the still water level. The maximum variability of force due to a single wave transient (wave #5) is between 25 N and 55 N. The range of variation, including all five transients, is between 61 N (wave #1) and 19 N (wave #2). Drag forcing and occasional wave slapping dominate the force maxima recorded by segments 2 and 3.

Force maxima on segment #7, which is fully submerged and dominated by inertial loading, are shown in Figure 16. Between 10 m and 13 m, wave #2 generates significantly lower forces. Beyond 12 m, the force from wave #4 is decreasing while the forces from waves #1, #3 and #5 are increasing. Again, the variation in force maximum with location due to a single wave transient can be significant. The force maxima caused by the five waves range between 8 N and 15 N.

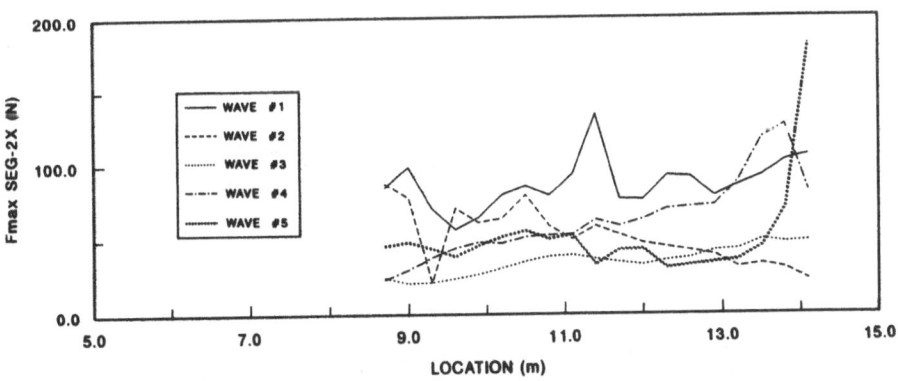

Figure 14. Force maxima on segment #2 versus location for five waves

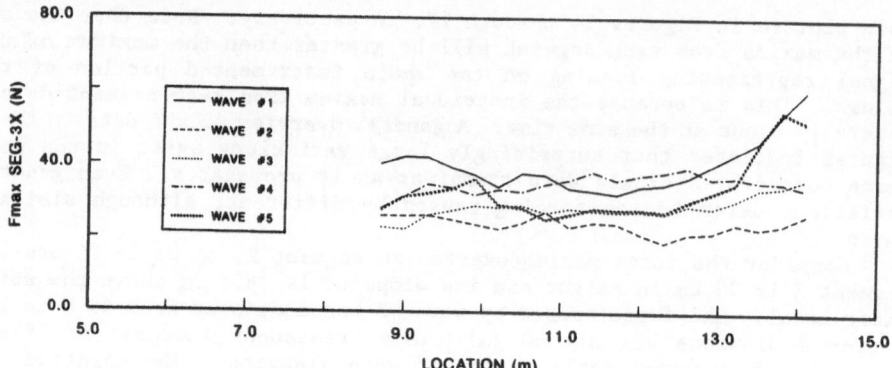

Figure 15. Force maxima on segment #3 versus location for five waves

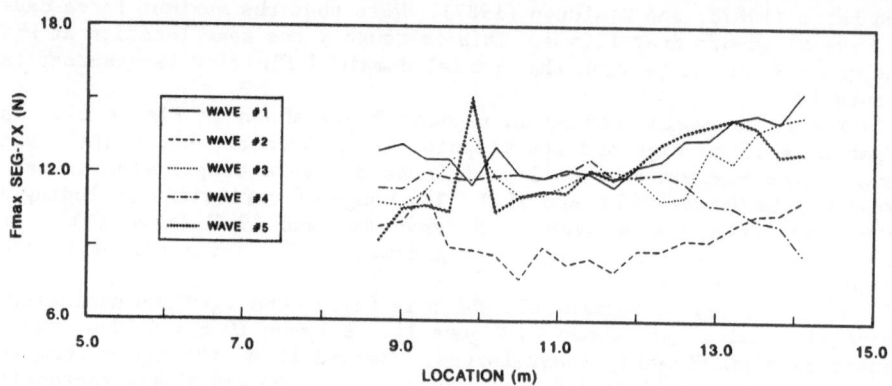

Figure 16. Force maxima on segment #7 versus location for five waves

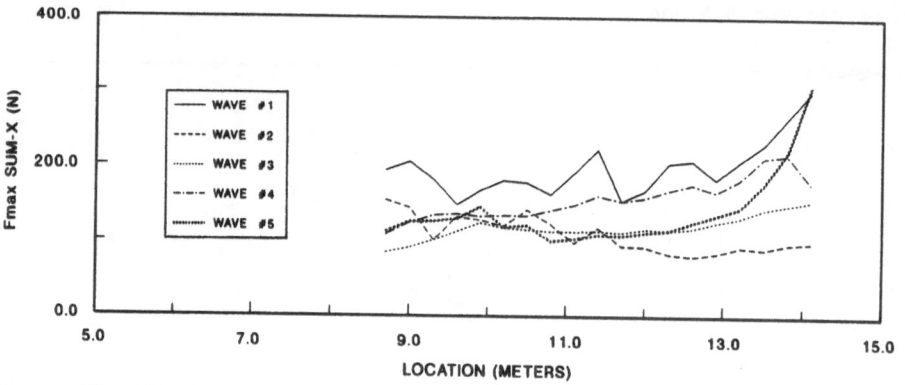

Figure 17. force maxima on all segments versus location for five waves

Figure 17 depicts the maxima of the force acting on the instrumented portion of the column as a function of column location. The largest forces are caused by wave #1, except at the 14.1 m location, where the force due to wave #5 is slightly larger. The greatest difference between the five waves also exists at the 14.1 m location where wave #2 causes 98 N of force and wave #5 causes 308 N of force. Much of this additional force can be attributed to free surface effects acting on the upper segments.

6. REPEATABILITY OF FORCE DATA

The test in which wave transient #5 acts on the column located at 11.4 m was repeated five times to assess the repeatability of the force data. Table 3 contains the force maxima recorded by segments #2, #3, #7 and #9, as well as the sum of the forcing on the instrumented portion of the column, in each of these trials. Except for the load acting on segment #2, the forcing and measurement mechanism is very repeatable. The differences in the force maxima experienced by segment #2 are likely due to small variations in the free surface of the wave crest which cause different degrees of wave slapping.

Table 2. Repeatability of Force Data

TRIAL NO.	FORCE-2 (N)	FORCE-3 (N)	FORCE-7 (N)	FORCE-9 (N)	FORCE-SUM-1:9 (N)
1	33.20	28.31	9.57	7.15	108.95
2	38.13	27.16	9.41	7.06	105.03
3	56.54	27.47	8.92	6.82	103.80
4	47.39	27.99	9.19	7.04	108.12
5	46.56	28.82	9.36	7.06	110.16

7. RELATIONS BETWEEN FORCE AND WAVE PARAMETERS

To investigate whether variations in the force experienced by the column could be related to changes in the shape of the wave transient, many plots were produced comparing wave characteristics to measured force maxima. Several of these are reproduced in Figure 18 which shows force maxima on segment #3 caused by wave #5 plotted against five wave crest parameters. These results indicate that for wave transient #5, the force peaks are positively correlated to: the maximum time derivative of the water surface, the forward tilting, the crest front steepness, the vertical asymmetry, and the crest elevation. Forces caused by the other waves are generally less well correlated with the same wave crest parameters. These parameters indicate whether a particular wave profile has the potential to exert unusually large forces on the upper portion of the column, but are insufficient to predict the magnitude of the force. One conclusion

724

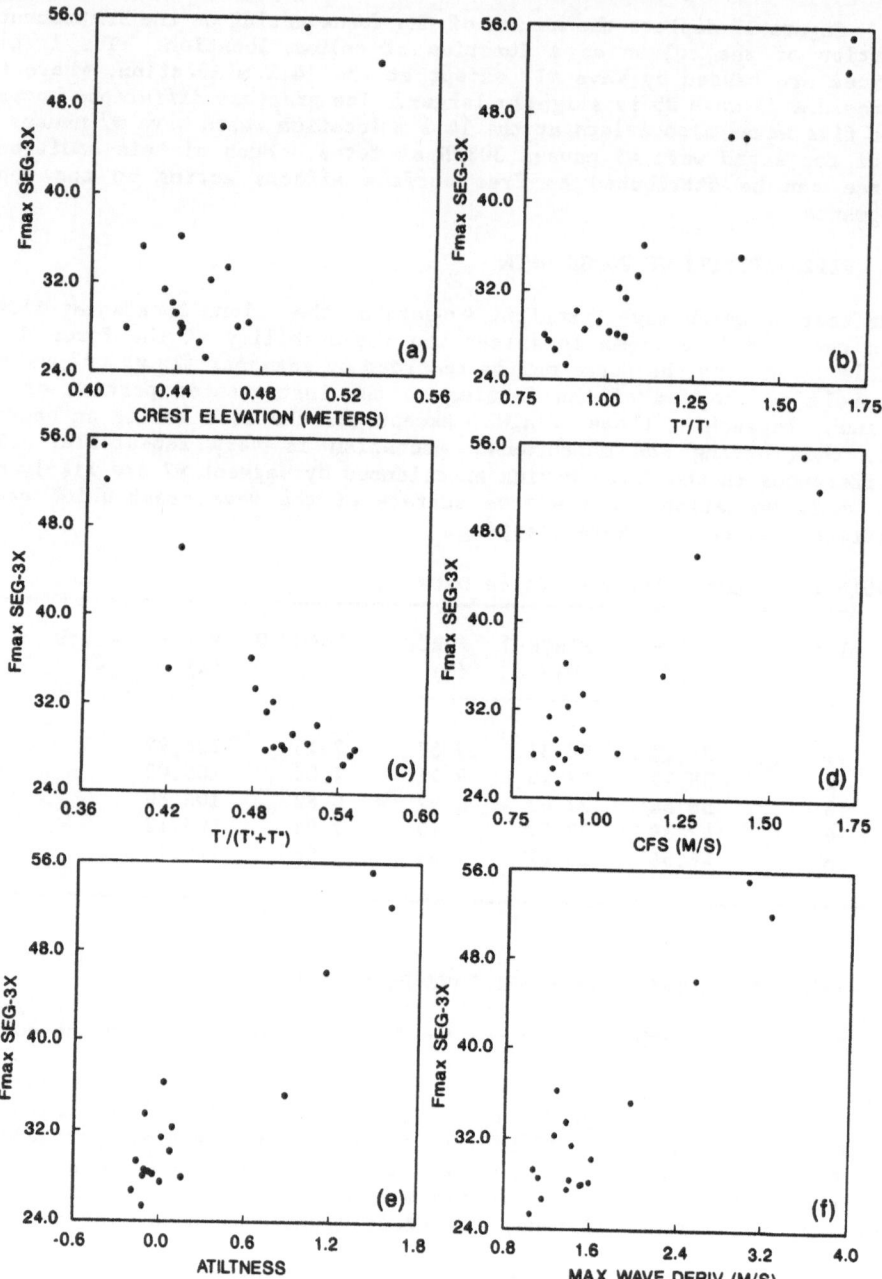

Figure 18.   Forces on segment #3 versus selected wave parameters - wave #5

that may be drawn from inspection of the complete data set is that downcrossing wave height and period are insufficient to accurately estimate the force caused by these wave transients.

8. MEASURED AND PREDICTED VELOCITIES

Another approach to learn more about these five wave transients is to examine their kinematics. Measurements of the orbital velocities under these waves were obtained at two elevations and three different locations using bi-directional electro-magnetic current meters. Initial attempts to predict the horizontal velocity from the measured water surface time series have been encouraging.

Three different estimates of horizontal fluid velocity have been achieved. The first of these uses FFT decomposition of the entire measured time series to obtain a large number of frequency components, each with small amplitude, that represent the irregular free surface. Linear wave theory is applied at each frequency to obtain an equal number of small amplitude components which represent fluid velocity at a desired depth. An inverse FFT is used to return to the time domain with a linear estimate of the fluid velocity at the target depth. This technique will be called the 'linear' estimate because linear wave theory is used.

The second technique employs the stream function wave theory of Reinecker and Fenton (1981). The zero downcrossing wave height and period of the transient are used to obtain the horizontal fluid velocity that would result from a perfectly symmetric periodic wave with the same height and period. This estimate will be called the 'Fenton' estimate.

The third method employs the stream function wave theory of Reinecker and Fenton combined with a technique to optimally fit a small number of sinusoids to the measured free surface. The downcrossing wave cycle including the transient wave crest is extracted from the wave record. An iterative scheme is employed to determine the frequency, amplitude and the phase of the sinusoid which best fits the measured wave. Once this component has been determined, it is removed from the measured signal and the search for the component which best fits the remainder is pursued. Less than six components are sufficient to provide a reasonable fit to any of the five wave transients. Figure 19 shows the match between the measured and synthesized water surface time series of wave #4 at the 12.9 m location. The frequency, amplitude and phase of the four leading components of the synthesized time series are printed below the figure. Note that the three higher frequency components are approximate second, third and fourth harmonics of the fundamental. The amplitude of the second harmonic presented here is significantly greater than what one would expect using the bounded second order harmonic theory discussed by Mansard, Sand and Klinting (1988).

An estimate of the horizontal fluid velocity under the wave transient can been obtained by using stream function theory to predict the non-linear kinematics under each sinusoidal component, and then summing these with the correct phase relationship. This method is dubbed the 'n-component' estimate where n represents the number of components used in the synthesis. The 'n-component' technique shares a disadvantage with the 'linear' method; in both cases, the components used in the synthesis may

726

not represent the actual components that comprise the natural wave. However, the 'n-component' method should offer good estimates of kinematics under individual non-linear waves, so long as the water surface is well matched by the chosen set of sinusoids.

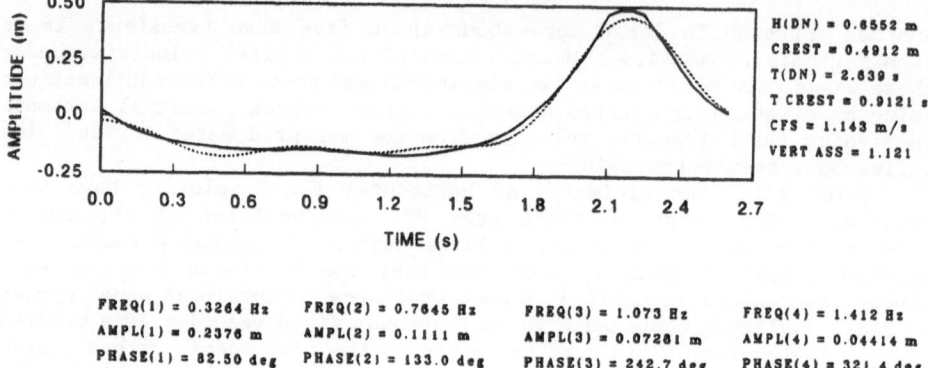

FREQ(1) = 0.3264 Hz   FREQ(2) = 0.7645 Hz   FREQ(3) = 1.073 Hz   FREQ(4) = 1.412 Hz
AMPL(1) = 0.2376 m   AMPL(2) = 0.1111 m   AMPL(3) = 0.07281 m   AMPL(4) = 0.04414 m
PHASE(1) = 82.50 deg   PHASE(2) = 133.0 deg   PHASE(3) = 242.7 deg   PHASE(4) = 321.4 deg

Figure 19.   Approximation of a wave transient by four sinusoids

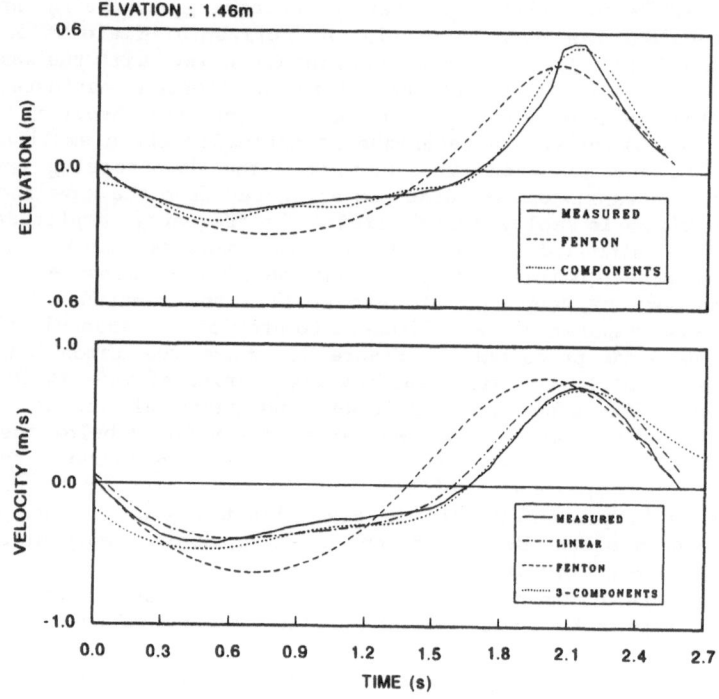

Figure 20.   Estimates of horizontal velocity from the measured free surface

Figure 20 shows the result when each of the three methods is applied to estimate the horizontal velocity at a depth of 1.46 m under wave #4. The upper window of this figure depicts the water surface measured at 13.5 m, as predicted using stream function theory, and as predicted using four optimally fit sinusoidal components. The lower window depicts the measured horizontal velocity along with the estimates obtained using the three methods discussed above. Both the 'linear' and '3-component' estimates agree well with the measured signal. The 'fenton' method is not well suited to estimating kinematics under such a nonlinear wave transient.

9. SUMMARY

1. The spatial profile of large wave transients have been observed to vary significantly as they propagate over a single wave length.

2. Wave parameters computed in the time domain, such as vertical asymmetry, forward tilting, and the maximum value of the time derivative of the water surface can be used to characterize these changes in wave profile. However, estimates of crest front steepness computed from the time domain assuming a characteristic wave length can be in error by up to 100%.

3. Large differences in the force exerted on a vertical column by a single wave transient were measured at different locations in the test area. It is generally not possible to correlate the peak positive force acting on the whole column to any one wave parameter; however, the force maxima on the portion of the column within the crest of the wave transient can be strongly influenced by the steepness of the wave crest front.

4. The five similar wave transients discussed here have caused very different forces on both a complicated oil production structure and a simple vertical cylinder. Their unusual behaviour cannot be attributed exclusively to the details of wave/structure interaction.

5. Large differences in the force exerted on a vertical column were observed between the five similar wave transients. Wave parameters traditionally used to define the design wave, such as wave height and wave period, are not sufficient to specify the most forceful or damaging wave event. The selection of design wave events based on determining the wave height and wave period of a given return period may not necessarily give the most severe conditions. The large degree of variability in the force exerted by five different realizations of the same design wave event indicate that model testing with a single design wave transient can produce misleading results.

6. Kinematics below the still water level can be reasonably predicted from the water surface time series.

10. ACKNOWLEDGEMENT

The author would like to thank Ed Funke, Okey Nwogu, and Etienne Mansard for their support and assistance during this work. The efforts of the staff of the Hydraulics Laboratory at the NRC in Ottawa are gratefully acknowledged.

11. REFERENCES

1. A.M. Cornett: Short-crested Wave forces on a Rigid Segmented Vertical Cylinder, Master's Thesis, Dept. of Civil Engineering, University of British Columbia, Vancouver, Canada.

2. E.R. Funke, and E.P.D. Mansard: The NRCC 'Random' Wave Generation Package, NRCC Hydraulics Laboratory Report No. TR-HY-002, Ottawa, Canada, 1984.

3. Y. Goda: Effect of Wave Tilting on Zero-crossing Wave Heights and Periods, Coastal Engineering in Japan, Vol. 29, pp. 79-90, 1986.

4. IAHR Working Group on Wave Generation and Analysis: List of Sea State Parameters, January, 1986.

5. W.W. Jamieson and E.P.D. Mansard: An Efficient Upright Wave Absorber, ASCE Specialty Conference on Coastal Hydrodynamics, pp. 124-139, Newark, Delaware, 1987.

6. S.P. Kjeldsen: Wave Forces and Kinematics in Deep Water Breaking Waves, Marintek Report No. SF10486.33, Trondheim, Norway, 1987.

7. S.P. Kjeldsen and D. Myrhaug: Kinematics and Dynamics of Breaking Waves, Report No. STF6-A78100, Ships in Rough Seas, Part 4., Norwegian Hydrodynamic Laboratory, Trondheim, Norway, 1978.

8. S.P. Kjeldsen, A. Torum, and R.G. Dean: Wave Forces on Vertical Piles Caused by 2- and 3-Dimensional Breaking Waves, 20th International Conference on coastal Engineering, Taipei, Taiwan, 1986.

9. E.P.D. Mansard, S.E. Sand and P. Klinting: Sub and Superharmonics in Natural Waves, Journal of Offshore Mechanics and Arctic Engineering, Vol 110, pp. 210-217, August, 1988.

10. G.R. Mogridge, E.R. Funke and P.W. Bryce: Wave Simulation for Run-up and Deck Clearance Tests on a Gravity-Base Structure, Proceedings of the Eighth Joint International Conference on Offshore Mechanics and Arctic Engineering, The Hague, Netherlands, March, 1989.

11. D. Myrhaug and S.P. Kjeldsen: Steepness and Asymmetry of Extreme Waves and the Highest Waves in Deep Water, Ocean Engineering, Vol. 13, No. 6, pp. 549-568, 1986.

12. H.J. Reddish and D.R. Basco: Breaking Wave Force Distribution on a Slender Pile, ASCE Specialty Conference on Coastal Hydrodynamics, pp. 184-195, Newark, Delaware, 1987.

13. M.M. Reinecker and J.D. Fenton: A Fourier Approximation for Steady Water Waves, Journal of Fluid Mechanics, vol. 104, pp. 119-137, 1981.

DRAG AND INERTIA LOADS ON CYLINDERS

C.J. GARRISON
Oregon State University
Department of Civil Engineering
Corvallis, Oregon 97331

ABSTRACT. Oscillating cylinder results are compared with wave tank results for both regular and random waves and the agreement is found to be quite good. Also, oscillating cylinder results are found to be well-represented by a simple wake model.

1. INTRODUCTION

Experimental results for cylinders have been presented in the literature (Garrison (1989) and Rodenbush (1983)) which define drag and inertia coefficients for use in Morison's equation which are correlated with the relative displacement (or Keulegan–Carpenter number) and the Reynolds number. Unlike most previous results, these results, when taken together cover a wide Reynolds number range and cover the relative amplitude range from $2a/d = 1.0$ to 10.0 (where a = the motion amplitude and d = diameter). Garrison (1980, 1989) has shown that this sort of viscous oscillatory flow about a cylinder induced by oscillating the cylinder is kinematically identical to oscillated flow past a fixed cylinder. Thus, oscillatory cylinder results may be directly compared with wave–cylinder interaction test results.

Recent results (Garrison (1982, 1984, 1989) and Lambrakos (1987)) indicate that the wake model may hold promise for an improved method for computing loads on small–diameter members. Some recent experimental results are compared with a simple wake model giving more emphasis to the wake model concept.

2. COMPARISON OF OSCILLATING CYLINDER AND WAVE TANK TESTS

Many papers showing oscillatory cylinder data as well as wave tank tests, for example, Sarpkaya (1985) and Chakrabarti (1987) have been presented in the small Re, small K_c range, and typically the effect of Reynolds number is unclear if a factor at all. However, results of both Garrison (1989) and Rodenbush et al. (1983) are presented together (Garrison (1989)) wherein the effect of Reynolds number is found to be extremely important in the case of both rough and smooth cylinders. Figure 1 shows a sample of the combined results covering a large

731

A. Tørum and O. T. Gudmestad (eds.), Water Wave Kinematics, 731–734.
© 1990 Kluwer Academic Publishers.

Reynolds number range for both smooth and rough cylinders. These
results indicate a profound effect of Re and suggest that caution
should be exercised in the Keulegan-Carpenter-number-only scaling of
model test results.

The wake-flow model appears to hold promise for improving the
predictive capability of Morison's equation. In the simple model
proposed here (Garrison (1989)) the wake-encounter velocity is characterized by the well-known model of Schlichting which was developed on
the basis of the mixing length concept for steady flow. On the basis
of this model the drag coefficient may be correlated as shown in Figure
2 provided the effect of Reynolds number is not important. Figure 2,
therefore, contains high Re results of Rodenbush et al. (1983) since it
is presumed that at sufficiently high values of Re the drag coefficient
should become independent of Re. These results which include both
smooth and quite rough cylinders (k/d - 0 and 1/50) show the rather
crude model to predict slightly high, but that the results correlate
well as C_D/C_{DS} vs. K/C_{DS} where C_{DS} - the steady drag coefficient.

The combined results for an oscillating cylinder have also been
compared with the large-scale random wave tank tests of Klopman et al.
(1989) in Figure 3. In the wave tank tests C_m and C_d were evaluated by
a least-squares fit on a wave-by-wave basis. The period was defined as
the time between two successive downward mean level crossings of the
velocity signal. The velocity time-derivative was not measured directly but was computed by applying a Fourier transform to the velocity
signal, multiplying by the frequency response function of the derivative and transforming back to the time-domain. The rather extensive
and indirect approach to obtaining the inertia coefficients may account
for the differences between the results but, in general, the comparison
is very good. The Reynolds number represented by the wave tank data in
Figure 3 extends up to about $5x10^5$ with smaller values of Re corresponding to smaller values of KC. The comparison shown in Figure 3
indicates that the scatter in Klopman's data can be explained by
variations in the Reynolds number.

3. CONCLUSIONS

The combined results of Garrison and Rodenbush (Garrison (1989)) show
the Reynolds number is a very important parameter over a large Re and
KC range. These results also compare very well with random-wave tank
tests and it may be concluded that in the case of C_d the scatter in the
wave tank data may be explained by the Reynolds number dependence. The
wake-flow model appears to hold promise as a basic concept for wave
loads on cylindrical members.

4. REFERENCES

Garrison, C.J., 1980, "A Review of Drag and Inertia Forces on Circular
     Cylinders," *Proceedings*, Offshore Technology Conference, OTC 3760,
     Houston, TX.

Garrison, C.J., 1982, "Forces on Semi-Submerged Structures," Keynote
    Paper, *Proceedings*, Ocean Structural Dynamics Symposium, Oregon
    State University, Corvallis, OR.
Garrison, C.J., 1984, "Wave-Structure Interaction," Keynote Paper,
    Proceedings, Computer Methods in Offshore Engineering, Halifax,
    Nova Scotia, May.
Garrison, C.J., 1989, Drag and Inertia Forces on Cylinders in Harmonic
    Flow, (in press) ASCE.
Klopman, G. and Kostense, J.K., 1989, "The Loading on a Vertical
    Cylinder in Random Waves at High Reynolds Number, *Proc. NATO Water
    Wave Kinematics Workshop*, Molde, Norway.
Lambrakos, K.F., Chao, J.C., Beckman, H., and Brannon, H.R., 1987,
    "Wake Model of Hydrodynamic Forces on Pipelines," *Ocean Engineer-
    ing*, Vol. 14, No. 2, pp. 117-136.
Rodenbush, G. and Gutierrez, C.A., "Forces on Cylinders in Two-Dimen-
    sional Flows," Technical Progress Report, Vol. 1, BRC 13-83, May
    1983, Bellaire Research Center, Shell Development Co., Houston,
    TX.
Sarpkaya, T., 1985, Past Progress and Outstanding Problems in Time-
    Dependent Flows About Ocean Structures, Proc., Separated Flow
    Around Marine Structures, Trondheim, Norway.

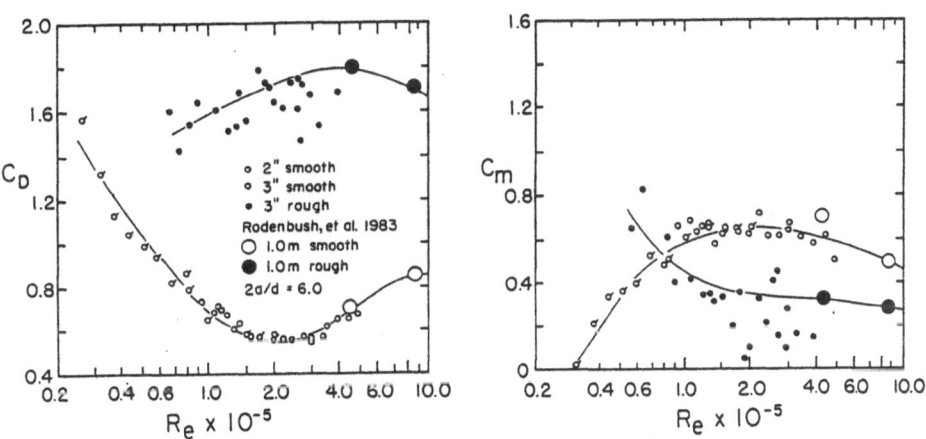

Figure 1   Drag and Inertia Coefficients

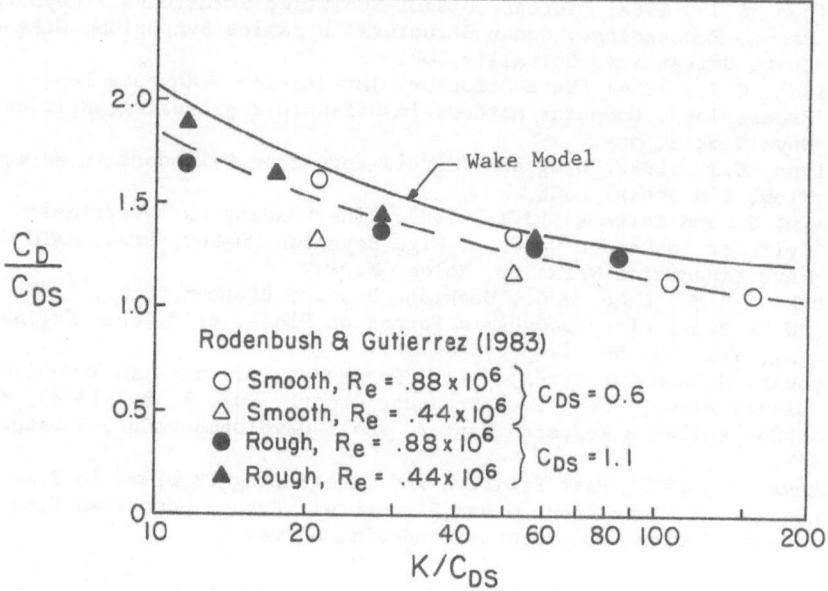

Figure 2   Wake-Flow Model -- Comparison with Data

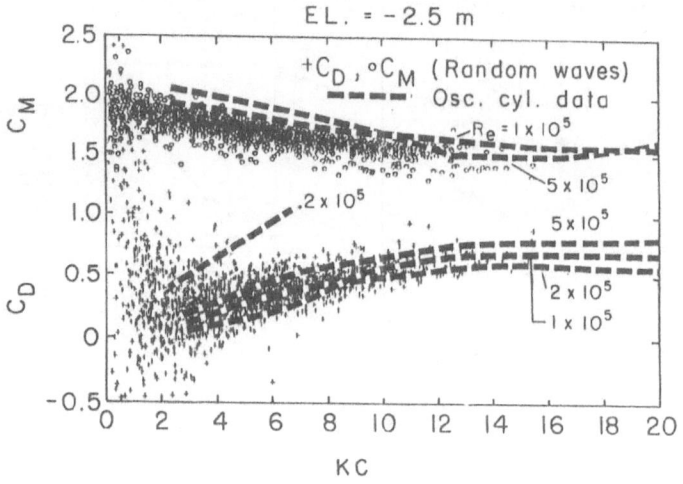

Figure 3   Comparison of Random Wave Tank Results with Oscillating Cylinder Data.

THE VARIATION OF THE WAVE REDUCING EFFICIENCY OF PRISMATIC FIXED SURFACE OBSTACLES WITH RESPECT TO ITS DIMENSIONS

Yalçın ARISOY
Ph.D. Student
Dokuz Eylül University, Dept. of Civil Eng.
Bornova 35100, Izmir, TURKEY

ABSTRACT. Study of the effects of fixed obstacles on incident waves can give considerable insight into the design of offshore structures. With certain assumptions simplifying the phenomena, motions caused by incident waves around the obstacles can be defined as a velocity potential, but this approach may be analytically very complex. Velocity potential around the obstacle can also be numerically determined by means of finite difference methods which are considerably simpler than the analytical techniques.

In this study, the effects of dimensions of prismatic obstacles on the incoming sinusoidal waves with small amplitudes are investigated. It is found that sectional barriers are more effective on reflecting waves than homogenous barriers with the same external dimensions if one evaluates them on the basis of their cross sectional areas and the material required for construction.

1. INTRODUCTION

Breakwaters which are rigid structures block the flow section completely and produce still basins by reducing and reflecting wave energies. Structures partially blocking the flow section reduce the wave energies partially. Since wave energies are concentrated near the surface, floating objects are more efficient in reducing the wave energy. However, the development of mathematical models for this type of breakwaters is quite difficult due to the presence of wave generated horizontal and vertical oscillations. This case can be simplified by disaggregating the system as flows created by incident waves and flows created in still water by oscillations and rotations of the same rigid object. Eventually, these two phenomena can be aggregated or superposed.

There are numerous theoretical and physical type of models investigated to describe the effects of rigid surface obstacles and the resulting velocity profiles. Some researchers made physical modelling studies by using rectangular (Nece & Richey, 1972; Sudko & Haden, 1974), horizontal (Şendil & Graf, 1975), and vertical (Matsson & Cederwall, 1977) type surface obstacles to study the effects of

A. Tørum and O. T. Gudmestad (eds.), Water Wave Kinematics, 735–738.

simple sinusoidal waves whose spectral caracteristics are known. Mei (1978) described this case by velocity potentials and used numerical solution techniques. Akyarlı (1980) investigated the wave reducing efficiency of prismatic surface obstacles and similar breakwaters by solving the velocity potential by means of finite difference methods.

2. APPLIED METHOD

Small amplitude sinusoidal incident waves can be defined by a velocity potential $\Phi(x,y,t)$ which agrees with Laplace equations. But, as seen from Fig.1, if a barrier is placed in this domain, the system is no more homogenous. Therefore, it becomes quite difficult to describe the phenomenon analytically; thus numerical techniques are preferred.

Due to the periodicity of the wave motion, the velocity potential can be defined by space variables which are independent of time and which conform to boundary values and Laplace equations as a complex $Q(x,y)$ function due to the relation of

$$\Phi(x,y,t) = \text{Re} \{ Q(x,y).\exp(-i.s.t)\} \tag{1}$$

where s represents $2\pi/L$ with $i = \sqrt{-1}$. Asymptotical values of the Q function at $\mp\infty$ must satisfy the conditions:

$$Q = Q_I + Q_R \qquad x \longrightarrow -\infty \tag{2}$$

$$Q = Q_T \qquad x \longrightarrow +\infty \tag{3}$$

where Q_I, Q_R and Q_T are incident, reflected and transmitted wave equations, respectively. They can be defined as

$$Q = A.\exp(iK_o x).\text{Cosh } K_o(y-d) \tag{4}$$

where d is water depth. K_o represents the wave number and can be determined by $K_o = 2\pi/L$. The term A represents constants which are complex functions of amplitudes and phase angles of the wave. Asymptotical values of the Q function at $\mp\infty$ are given in equations (2) and (3). Yet, within the obstacle zone, the analytical form of the Q function is different. Here, it is assumed that the Q function assumes the forms defined by (2) and (3) at a distance of 3d from the external surface of the barrier on each side, d representing the water depth. At distances less than 3d towards the obstacle, the analytical form of the function Q is more complex. This fact defines the boundaries of the complex form of the Q function around the barrier. On the other hand, by defining Q function in terms of the Q_I function, which represents the incident wave, and an unknown F function as,

$$Q = Q_I + F \tag{5}$$

the problem of defining Φ velocity potential reduces to the determination of the F function.

F function can be defined as

$$(s^2/g) F + F_y = 0 \qquad |x| < b \qquad y = 0 \qquad (6)$$

at surface. At the bottom,

$$F_y = 0 \qquad -\infty < x < +\infty \qquad y = d \qquad (7)$$

Due to the boundary conditions, given in equation (5) and since $Q_x = 0$ and $Q_y = 0$ at the surfaces of the obstacle (Fig. 1), the following equations can be written.

$$F_x = -Q_{I,x} \qquad x = \mp b \qquad 0 < y < a \qquad (8)$$

$$F_x = -Q_{I,x} \qquad x = \mp b_o \qquad a_o < y < a \qquad (9)$$

$$F_y = -Q_{I,y} \qquad b_o < |x| < b \qquad y = a \qquad (10)$$

$$F_y = -Q_{I,y} \qquad -b_o < x < b_o \qquad y = a_o \qquad (11)$$

Because the boundary value problem given by (6-11) is composed of the F function and the derivatives of this function (F_x, F_y) normal to the obstacle surfaces, this type of problem can be defined as a "mixed type boundary value problem" and can be solved by finite difference techniques.

After the determination of F functions using Q and Φ functions, reflection and transmission coefficients, which are defined as the ratio of reflected and transmitted wave amplitudes to the incident wave amplitude, can be computed together with orher related hydrodynamic parameters.

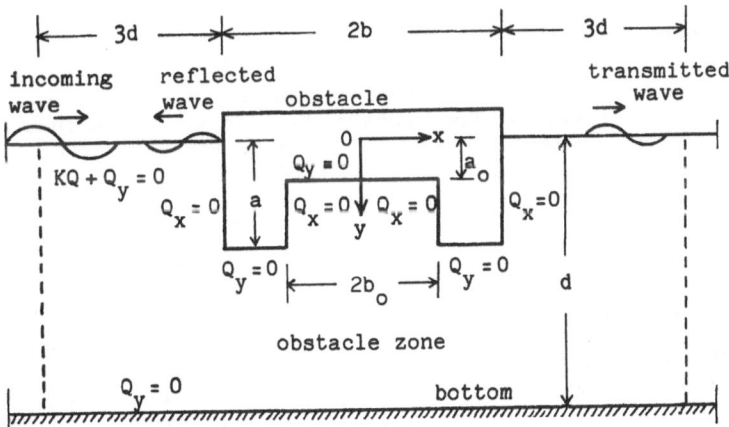

Figure 1 : Schematic description of prismatic sectional fixed surface obstacle.

3. NUMERICAL APPLICATION AND RESULTS

With a computer program developed, characterictics of reflected and transmitted waves can be determined by using the characteristics of approaching waves and surface obstacles. To study the wave reducing efficiency of prismatically partitioned surface obstacles, various wave heights, wave periods and water depths are used. To determine the effects of the dimensions of obstacles on the wave reducing capacity, changes in wave reflection coefficients are related to the dimensionless $2\pi a/L$ parameters which include water depth and the incoming wave caracteristics.

To compare the wave reducing efficiency of prismatic rectangular obstacles and obstacles with partial prismatic sections having the same exterior dimensions, the variation of wave reflection coefficients with recpect to dimensionless parameter $2\pi a/L$ is also investigated. It is found that prismatic obstacles have more reflecting capacity than obstacles with partial prismatic sections having total exterior dimensions equal to those of prismatic obstacles. But, if the cross sectional areas are taken into account, partitioned obstacles may be considered more efficient.

The effects of the dimensions of prismatic sections of obstacles with constant exterior dimensions are also investigated. It is found that an increase in the length of the middle thinner part of the obstacle without any change in the exterior dimensions causes a decrease in reflection coefficients. The case transforms into the optimization problem if reflection is related to the cross sectional area of the obstacle. An important advantage of sectional obstacles, if used for wave reducing, is that, besides savings in contruction material volume, water masses enclosed between the fore and aft parts of the obstacle act as a part of the obstacle, and may cause in increase in the wave reflecting capacity.

REFERENCES

AKYARLI, A. (1980) 'The Analysis of the Wave Reducing Efficiency of Prismatic Floating Breakwaters by Finite Difference Methods', Izmir, Ege University, Thesis for Associate Professorship, 180 p.

MATSSON, B. & CEDERWALL, K. (1977) 'Vertical Barriers as Wave Attenuators', Baden-Baden, Proc. of Seventeenth Cong. of IAHR, Vol. 4, pp. 57-64

MEI, C.C. (1978) 'Numerical Methods in Water Wave Differaction and Radiation', Ann. Rev. Fluid Mech., Vol. 10, pp. 393-416.

NECE, R.E. & RICHEY, E.P. (1972) 'Wave Transmission Tests on Floating Breakwater for Oak Harbor', Univ. of Washington, C.W. Harris Hydraulic Lab., Tech. Rep. 32.

SUTKO, A.A. & HADEN, E.L. (1974) 'The Effect of Surge, Heave and Pitch on the Performance of a Floating Breakwater', New Port, Rhode Island, Proc. of the Floating Breakwaters Conf., pp. 41-53.

ŞENDİL, U. & GRAF, W.H. (1975) 'Transmission of Regular Waves Past Fixed Plates', Sao Paulo, Brasil, Proc. of Sixteenth Cong. of IAHR, Vol. 1, pp. 254-261.

Discussion on Session 4: Forcing

Mr R. M. Webb

After the general criticism of Morisons Equation and non
spectral data analysis methods I wish to confirm that, in
full scale, real sea, measurements there has been a
remarkable degree of agreement between Morison formula
computations of force and measurements on a wave by wave
basis. This is found when the measurements are clearly
in either the drag or the inertia regimes; when the
individual waves are of significant magnitude in a storm,
and the appropriate coefficients are applied taking into
account the Re and Kc numbers. Problems arise in the
generally unimportant overlap phase between drag and
inertia when the load is past its maximum, or in
uncertain flow regimes. It is acknowledged that this can
be of scientific interest and be of engineering concern
in some situations, but it was not generally an important
engineering problem. The fact is that Morisons Equation
does a good job when applied appropriately and should not
be lightly discarded in favour of complex partial
solutions.

A. Tørum and O. T. Gudmestad (eds.), Water Wave Kinematics, 739–740.
© 1990 Kluwer Academic Publishers.

Mr R. M. Webb

Several full scale offshore measurement programmes have been undertaken over the past 20 years. Many of these have been in the North Sea. Much of the data is distributed widely within the offshore industry and, contrary to widespread perception, much is available to interested parties. When I was Manager of Ocean Engineering at BP I had a list of offshore projects compiled. A summary of this was published as table 1 in "Platform Monitoring as a Tool for Cost Reduction" by K. M. Svehla and P. Elliot at the ICE Offshore Cost Reduction Conference, London, February 1988. A copy is attached.

The much fuller list of projects has subsequently been reviewed by Atkins R & D for Mr M. Birkinshaw at MATSU to assess the availability of data and the applicability of the records to analysis for a range of different purposes. I suggest that interested parties contact him direct to see if that information can be passed to them. He can be contacted at:-

    MATSU,
    B 10.69,
    Harwell Laboratory,
    Didcot,
    Oxon OX11 0RA
    UK.

In particular, the BP Forties data is available at Liverpool University (and I have learned at this Workshop that selected data sets are still available at IFREMER. Esso made a public invitation a couple of years ago to share the Odin data and facilities with interested parties. Shell have recently made similar suggestions to suitable bodies for the further analysis of the Tern monitoring data. Prof Dean indicated during these proceedings that the Wave Projects 1 and 2 data may still be available on the NOAA data base in the USA. In short there is a lot of good data about and a willingness by the owners to have it analysed by up to date methods.

Summary Paper

WATER WAVE KINEMATICS: STATE OF THE ART AND FUTURE RESEARCH NEEDS

R. G. DEAN
Coastal and Oceanographic Engineering Department
University of Florida
336 Weil Hall
Gainesville, Florida 32611 U.S.A.

ABSTRACT. A review is presented of the state of the art of water wave kinematic prediction and understanding and areas of needed research. Existing capabilities to predict kinematics in the laboratory under regular wave conditions appear to be within the accuracy (≈ 5%) of inherent errors in the measurements. Recent contributions in numerical techniques for regular waves over a sloping bottom also appear to yield similar accuracy up to the point of wave breaking. Predictive capability of irregular waves within the laboratory environment appears adequate except in the case where strong nonlinearities exist, such as near the free surface. A proven methodology does not exist to address the general problem of interaction of waves with a current of arbitrary vorticity distribution. In nature, waves are much more complicated than can be generated in most laboratory facilities and the conditions much less controlled; thus the state of the art is less satisfying than for laboratory conditions and the associated research needs are greater. Research needs include a nonlinear and directional wave theory to be more representative of extreme conditions in nature. Ultimately, the theory should encompass the presence of an arbitrary three-dimensional current field. The magnitude and kinematic significance of the second-order forced waves, usually most pronounced in narrow spectra of limited directional spreading, appear to be significant and should be incorporated into design procedures. Wave systems generated by concentrated and complex wind fields have a high degree of directionality which may have significance for the extreme waves, both due to the directional content of the kinematics and also due to the local three-dimensionality of the sea surface. Because of the significance of freak waves to design, their causes should be investigated further and design procedures developed. Recommendations are made for the continuation of research in: theory, numerical developments and laboratory and field research. Due to the inherent difficulties in conducting field research and the fact that the first order processes are well-understood, these programs should be well-coordinated with a broad range of interests involved in the planning and assurance that the data are of high quality resulting from frequent instrument calibration and validation of the resulting data as it becomes available.

743

A. Tørum and O. T. Gudmestad (eds.), Water Wave Kinematics, 743–756.
© 1990 Kluwer Academic Publishers.

744

1. Introduction

Understanding of water waves and the associated kinematics has advanced substantially during the last three to four decades. The first comprehensive "wave force projects" conducted in nature from 1954-1963 (Thrasher and Aagaard, 1969) inferred water wave kinematics from wave measurements at one or more locations due to the unavailability of suitable instrumentation for measuring kinematics. Indeed, the first field measurements of water particle kinematics under storm conditions were conducted in 1973 less than 20 years ago (Forristall, et al., 1978). Much of the development in understanding water wave kinematics was spurred predominantly by the interest to ensure that offshore platforms would withstand extreme wave forces. An emphasis on surf zone dynamics that commenced at about the same time but has received emphasis in the last decade has also stimulated investigations in water wave research. The nonlinearity and directionality of water waves are both crucial elements to an adequate understanding of wave forces, especially for the highest waves generated by complex wind fields such as tropical storms.

In any attempt such as this is, it is difficult, if not impossible to encompass adequately all significant research needs. In this summary, I will strive to indicate our status relative to the needs and to identify fruitful future research areas. In this effort, one approach would be to look through the eyes of the practitioner responsible for the design of offshore structures. Concerns are whether there are any "surprises" due to effects that are not now routinely included in the design process. The threshold for these concerns may be at about the 10% level - thus one question we should ask is whether the expectation is great that there are factors above the 10% level? Of greatest concern are overlooked elements that would increase the loads above those used in design; however, from considerations of economy, the designer is also interested in improved understanding that could reduce safely the design loads by more than 10%. My assessment is that with respect to the 10% level threshold, there are both conservative and non-conservative effects that justify study and improved understanding.

We might also ask the research needs question from the standpoint of the researcher with an interest in understanding water wave phenomena. Here, there is even a more commanding justification of continuing, even accelerating, water wave kinematic research.

The discussion presented here will exclude wave forecasting not because our understanding and operational capabilities are complete and adequate, but simply to establish a reasonable scope for this evaluation. Additionally, the problem of wave-induced forces on structures will not be considered directly, although it is recognized that much of the stimulus for understanding water wave kinematics is to develop an improved capability for wave force predictions.

2. The Laboratory Environment

Laboratory conditions provide substantial advantages in water wave research, especially in terms of control of the experiment parameters, the ability to repeat experiments, the relatively low cost and the ability to conduct frequent calibrations of the instruments. Additionally, it is possible to generate essentially two-dimensional conditions in the laboratory which correspond closely to many theoretical formulations.

2.1. REGULAR WATER WAVES; HORIZONTAL BOTTOM

For the case of <u>periodic</u> water waves propagating in uniform depth, the available theories are as accurate as our ability to generate and measure water wave phenomena within the wave tank. It is believed that due to tank reflection and other spurious effects, this limit is approximately 5%. This assessment is believed to apply up to limiting waves and is based, in part, on the studies by Iwagaki, et al. (1971) as reported by Dean (1984).

2.2. REGULAR WAVES; SLOPING BOTTOM

A number of global or local wave properties could be considered in this case which also encompasses conditions up to wave breaking. Numerical techniques first demonstrated by Longuet-Higgins and Cokelet (1976), improved by Vinje and Brevik (1981) and further refined by Dold and Peregrine (1986) appear adequate to represent well the phenomenon of wave shoaling and deformation up to the point of touchdown of the wave nappe. Based on the excellent wave shoaling data base of Hansen and Svendsen (1979), the stream function approach has been shown to agree well (Swift and Dixon, 1987). Hino and Kashiwayanagi (1979) have reported good agreement in kinematics and wave set-down and set-up under shoaling conditions using the irregular stream function method.

2.3. IRREGULAR WATER WAVES; HORIZONTAL BOTTOM

Most of the reported measurements in this category represent either relatively small amplitude water waves or kinematics below the trough elevation. For these cases, available wave theories seem to represent adequately the measurements. Above the trough, especially near the crest region where the nonlinearities are strongest, more accurate measurements and comparison with theory are needed. Vis (1980) presents a comparison of measured and predicted water wave kinematics in the laboratory.

2.4. IRREGULAR WATER WAVES; SLOPING BOTTOM

In this case, experiments are needed defining the transformation and water particle kinematic field as waves propagate up the beach. A comparison of these data with theory should be made to assess the adequacy of existing theories.

3. The Field Environment

Field measurements of water wave kinematics are obviously more difficult to obtain than in the laboratory. Factors include biological growth, instrument calibration and the three dimensionality of the wave system.

Since, to some degree, the wave form is less controlled in the field than under laboratory conditions, present understanding of the wave form in the field requires considerable research. Several specific examples are discussed below.

3.1. NEARLY REGULAR WAVES

This class of waves is likely to occur where the generation area is a great distance from the observation point such that the spectrum has been narrowed through wave dispersion. An examination of the waves over a short period of record may suggest that the waves are regular; however since narrow spectra are "groupy", the wave heights will vary gradually and a forced long wave will occur and may be of importance. Figure 1 presents the characteristics of a two-dimensional group of waves and the forced long wave. Longuet-Higgins and Stewart (1962) have shown the long wave to be a second order wave forced by radiation stress gradients of the primary wave system. This forced wave is a long wave and is always out of phase with the wave group maxima. Thus the associated velocity of this forced wave reduces the maximum velocities and associated forces on, for example, a platform support piling.

Providing support for the significance of this phenomena are the measurements of Forristall, et al. (1978) in Hurricane Delia which occurred in 1973. Figure 2 presents, for the extreme wave in the storm, time variations of the water surface and measured water particle velocities at 10 ft, 35 ft and 55 ft above the bottom. It is seen that the

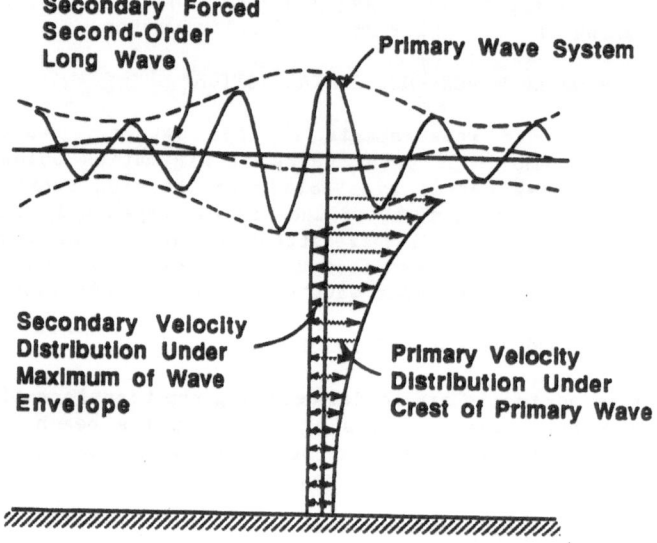

Figure 1. Primary and secondary wave systems and associated velocity distributions at crest phase position of primary wave system when wave envelope is maximum.

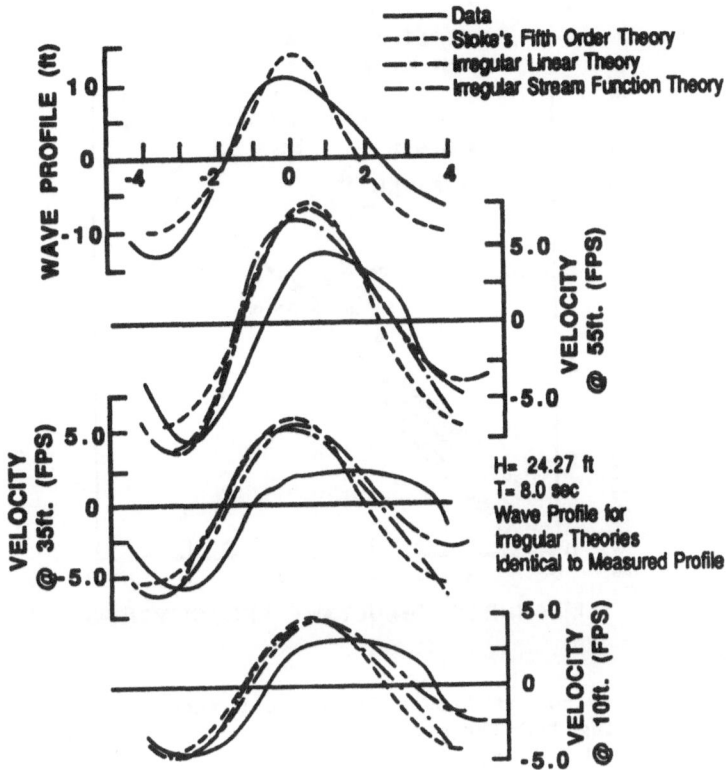

Figure 2. Comparison of Stokes and irregular unidirectional wave theories to measurements made during the largest wave in Hurricane Delia. (after Forristall, et al., 1978).

measured crest velocities are less than the trough velocities. Figure 3 compares, for the maximum waves in the storm, the maximum measured crest and trough velocities and those determined from Stokes Fifth Order Wave Theory at the upper current meter. It is seen that the measured crest velocities are less than calculated from the Stokes theory whereas the measured trough velocities are greater than calculated. These differences are entirely consistent with the presence and phasing of a forced second order wave as discussed earlier.

3.2. WAVE DIRECTIONALITY

There are a number of aspects of wave directionality which can have a significant effect on water wave kinematics and the associated forces. Several of these will be discussed below.

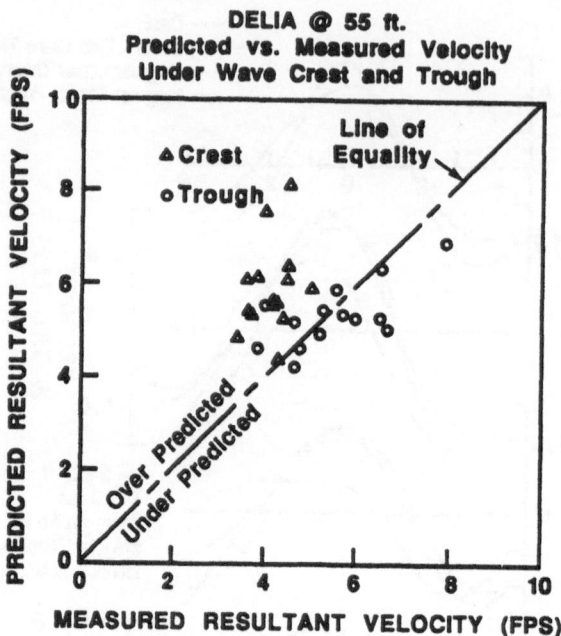

Figure 3. Comparison of Stokes fifth-order theory to measured maximum velocities during the largest waves at the top current meter. (after Forristall, et al., 1978).

3.2.1. *Second Order Forced Wave* - The second order forced wave described above for a two-dimensional sea surface is quite dependent on wave direction. Stig (1982) has shown that for two primary waves propagating in arbitrary directions, the magnitude of the second order forced wave decreases markedly with small differences in direction of wave propagation. Thus, it is not clear whether or not or under what wave conditions it is justified to account for the significant reductions in crest velocities that would result from collinear waves. Clearly, a definitive field measurement program is required to clarify the relationship of directional spread on second order waves and the associated kinematics.

3.2.2. *Kinematics of Directional Waves* - At the linear level, the water surface displacements of the two waves propagating in different directions add as scalars whereas the velocities add vectorially. Thus the maximum velocity associated with two intersecting crests would be less than if the waves were propagating in the same direction. The <u>magnitude</u> of the combined vector $|u_c|$

$$|u_c| = \sqrt{u_1^2 + 2u_1 u_2 \cos(\theta_1 - \theta_2) + u_2^2} \tag{1}$$

where u_1 and u_2 are the magnitudes of the two velocity vectors oriented at directions θ_1 and θ_2. For a small angular difference $(\theta_1 - \theta_2)$, the ratio of the vectorial addition to the scalar addition of velocities can be shown to be approximately

$$\frac{|\vec{u}_1 + \vec{u}_2|}{|\vec{u}_1| + |\vec{u}_2|} \approx 1 - \frac{2u_1 u_2}{(|\vec{u}_1| + |\vec{u}_2|)^2} \sin^2 \frac{1}{2}(\theta_1 - \theta_2) \tag{2}$$

Thus if the included angle $(\theta_1 - \theta_2)$ is 60^0 and the magnitudes of u_1 and u_2 are the same, there is a reduction of 12.5%. The reduction from the complete equation (Eq. (1)) is 13.4%. For structure loadings which are drag-dominant, this would amount to a force reduction of 25%. Dean (1977) has investigated the reduction of distributions of wave direction of the form $\cos^{2n}(\theta)$ and, as an example, has shown for n=1, the peak velocity is reduced by 13% compared to the case of collinear waves.

3.2.3. *Composition of Extreme Waves in Directional Seas* - Longuet-Higgins (1952) first proposed the Rayleigh distribution for application to the population of ocean wave heights. The Rayleigh distribution applies for narrow spectra; for spectra with appreciable widths, the generalized Rice distribution (1944,1945) of which the Rayleigh distribution is a special case should apply. The spectral width parameter, ε, can be shown to have only a small effect on certain wave height parameters, such as the ratio of significant wave height to the root mean square water surface elevation.

There have been a number of field investigations confirming the general validity of the Rayleigh distribution for representing wave heights, for example Goodknight and Russell (1963). However, considering the discussion above of the significance of wave direction on wave kinematics and wave forces, it becomes important to better understand the role of directionality in the composition of the highest waves. In particular, for waves caused by fairly localized and rapidly moving storm systems, the directionality may play a dominant role in forming the larger waves. This would have two significant effects relative to wave forces on multi-legged platforms. The first is the reduction in maximum velocities as discussed previously and as quantified for simple cases in Eqs. (1) and (2). The second and possibly greater effect is due to the possible lack of correlation of maximum wave heights at the various support pilings. For example, if the cause of an extreme wave height at one piling is the combination of all wave components in a spectrum, then in a directional sea, this may occur by all crest phase components coinciding at the piling under consideration. Thus, the associated wave heights and crest heights (allowing for a suitable lag for travel time) at the other support piling may be less than the extreme at the reference piling under discussion and may not be the extreme height at the other support piling for the same storm. In some areas, such as the Gulf of Mexico, this may represent a major difference between two-dimensional seas and directional seas, especially for the extreme waves which can dictate design wave loads.

3.3. FREAK WAVES

Freak waves are waves of large height relative to Rayleigh distribution predictions. As this distribution does not predict a limiting wave height, the classification of a wave height as a freak wave requires judgement. A recommendation of the Working Group on Freak Waves at this Workshop is that all waves satisfying the criterion

$$\frac{H_m}{H_s} > 2.0 \qquad\qquad (3)$$

be considered as freak waves. In Eq. (3), H_m is the maximum wave height in a train of waves and H_s is the significant wave height. Based on the Rayleigh distribution (Figure 4), one would expect, on the average, that a wave record having approximately 2,000 waves would satisfy this criterion. Also, there is a 5% chance that a record having only 200 waves would include a wave height meeting this criterion. Therefore, it appears that one can define a freak wave only in a statistical sense, by examining a population of the largest waves occurring in one or more records and determining whether or not, as an aggregate, they are larger than predicted by the Rayleigh distribution. Because of the significance of freak waves to the design integrity of offshore structures, it is clear that much more research is needed to understand and develop rational design procedures for this phenomenon. Theoretical, laboratory and field studies appear warranted. Nonlinearities and directionality could play a role.

3.4. INTERACTION OF WAVES AND CURRENTS

The kinematics resulting from the interaction of an arbitrary distribution of current over depth with highly nonlinear waves are poorly understood and improved procedures are needed for design purposes. Developments by Dalrymple (1974a,1974b) for linear and bilinear shear currents have demonstrated the significance of the interaction.

At this workshop, Chaplin (1989) reported on a promising procedure for dealing with the problem of an arbitrary distribution of current over depth. The availability of quality data sets would allow evaluation of existing methods and would stimulate theoretical developments.

3.5. NEAR SURFACE KINEMATICS

A characteristic of water waves is that the nonlinearities increase with proximity to the free surface, especially for the crest phases. Additionally, high-quality measurements of kinematics are most difficult within this region. These complexities have resulted in a greater demonstrated capability to calculate accurate kinematics for design below

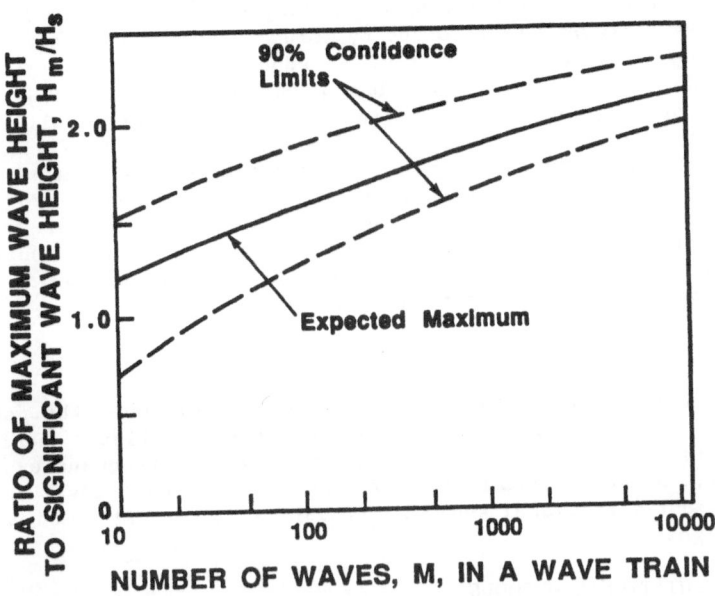

Figure 4. Rayleigh distribution results for expected maximum wave height in a train of M waves. Maximum wave height expressed as a ratio to significant wave height.

the trough level than above it. There should be a concerted effort to develop high quality field data sets of near surface kinematics during energetic wave conditions. As in the case of other research needs, these data sets would provide a basis for assessing existing capabilities and where found deficient, would provide motivation and direction for analytical efforts.

3.6. INTERACTION OF KINEMATICS WITH STRUCTURES

In practice, this facet of the problem is usually represented in terms of drag and inertia force coefficients, which are not the primary subject of this workshop. However, it appears appropriate to discuss the near free-surface interaction case where the forces are poorly understood.

As a wave crest passes a piling, there is run-up on the leading edge of the piling and a draw down or ventilation on the down wave side. The proximity of the free surface and the modifications to the kinematics described above result in near surface forces that are poorly understood. Analysis of wave and force data from Wave Projects I and II by Dean, et al. (1981) concluded that the local wave forces vary about those that would occur without the proximity of the free surface, with an overall reduction in the total force on the piling that could be as large as 20%. Laboratory measurements reported by Torum (1989) which included Reynolds Numbers up to $6x10^4$ did not find a reduction of that magnitude. Torum recommended drag coefficients based on large scale results below a distance of $u^2/2g$ <u>below</u> the instantaneous mean water level. The inertia coefficient was presented as an algebraic function of $(u^2/2g)$. Torum also found that the force was a maximum at $u^2/2g$ below the instantaneous mean water level and recommended that it decrease linearly to zero at a distance of $u^2/2g$ <u>above</u> the instantaneous mean water level.

An improved understanding of the interaction of near surface water particle kinematics with a cylindrical structure would assist in developing more reliable design procedures.

3.7. IMPROVED WAVE THEORIES

Although wave theories have advanced considerably over the past several decades, existing capabilities to represent realistic nonlinear and directional wave systems including a range of frequencies are still deficient. A need exists to develop and verify computational capabilities to represent directional seas with nonlinearity content up to the limiting wave heights.

3.8. COMPUTATIONAL METHODS

Capabilities to compute unsteady and nonuniform two-dimensional free surface flows have advanced remarkably over the last few decades. Much of this progress has been due to the methodology developed by Dold and Peregrine (1986) which allows reliable computations to be extended up to the point at which the wave nappe touches down. Demonstrated capabilities include the presence of internal boundaries such as a circular cylinder (without separation) as demonstrated by Skourup and Jonsson (1989).

The computational advances can be expected to continue with the increasing availability of larger and faster computers with greater memory. The next generation of computation capabilities may address three-dimensional flows.

3.9. SURF ZONE

The surf zone region is one of intensive dissipation of wave energy, turbulent motions and transfer of momentum (Battjes, 1988). Additionally the medium is three phase (water, sediment and air) and the motion is fully three-dimensional with each component comprising scales ranging from quasi-steady to wave-induced to turbulent fluctuations (with the latter two overlapping). Commonly, the mean motion near the surface is shoreward with a seaward undertow near the bottom. Within the surf zone, momentum transfer causes an increase in water level which for the case of a spectrum of incident waves can result in long period motions termed infragravity waves. The geomorphic response to this system can include substantial quantities of longshore and cross-shore sediment transport, depending on wave and sediment conditions. During extreme storms the seaward transport of sediment can cause beach erosion of up to 50 meters or more in several days and the seasonal erosion/depositional cycle can be of the same magnitude. Under some conditions, horizontal circulation cells exist including concentrated seaward flows (rip currents) balanced by more diffuse return flows into the surf zone. An intense interest in kinematics within the surf zone commenced in the early 1940's about the same time as for deeper water. Initially, the surf zone interest was motivated by the need to land troops and equipment through the surf zone during the second world war. More recently the emphasis has shifted to concern to predict reliably sediment transport, including the longshore component and the cross-shore component, primarily during storms.

Progress in understanding and a quantitative description of kinematics within the surf zone has advanced considerably due to the introduction of theoretical concepts, especially radiation stresses and the development of instrumentation such as the electro-magnetic current water capable of operating in this erosive and energetic environment.

Research needs include an emphasis on wave breaking across the surf zone, and the generation of turbulence and transfer of momentum. An understanding of the effects of suspended sediment and entrained air on surf zone kinematics is needed. At least two fairly uniform longshore periodicities are intriguing and challenge a conclusive exposition of the causes and mechanisms. These are the aforementioned rip currents and beach cusps. Edge waves which propagate parallel to shore and may be progressive or standing have been proposed as possible causes. Flow and sedimentary instabilities are a possible alternate explanation for these periodicities.

4. Summary, Conclusions and Recommendations

The last four decades have seen very substantial advances in fundamental understanding and capabilities to design rationally in the marine environment. This progress has resulted from the combined efforts of theoreticians, and laboratory and field experimenters. Numerical advances have contributed significantly due, in part, to the availability of fast computers with large memories. Present capabilities are such that many of the problems are understood conceptually and it is possible to say that, given the wave field, predictions of kinematics can be made within approximately $\pm$ 30% in field conditions and $\pm$ 5% in the laboratory. The associated uncertainties in wave forces may not be acceptable and can result in under design or unnecessarily costly structures.

To improve understanding and existing capabilities will require a coordinated effort including theory and laboratory and field measurements. In all cases identified, quality data sets encompassing broad ranges of parameters would both provide a basis for evaluating procedures and provide guidance for theoretical efforts.

As a recommendation, with the perceived demanding nature of future advances, it is suggested that broad ranges of individual projects continue to be conducted by theoreticians, and numerical analysts and laboratory experimenters. However, field experiments should be coordinated to the degree required to ensure high quality of the data obtained. With further improvements in understanding now focussed on the last 30% or so, it is imperative that the data be of unusually high quality with good maintenance of the instruments and assurances of data validity. As in the past, it is expected that theoretical progress will be stimulated and guided by the availability of quality sets which allow identification of differences with the best theories available.

5. References

Battjes, J.A. (1988) 'Surf zone dynamics', Annual Review of Fluid Mechanics, Vol. 20, pp. 257-293.

Chaplin, J.R. (1989) 'Computation of steep waves on a current of arbitrary profile', Proceedings, NATO Advanced Research Workshop on Water Wave Kinematics, Molde, Norway.

Dalrymple, R.A. (1974a) 'A finite amplitude wave on a linear shear current', Journal of Geophysical Research, Vol. 70, No. 30.

Dalrymple, R.A. (1974b) 'A finite amplitude wave on a bilinear current', Proceedings, Fourteenth Conference on Coastal Engineering, Chapter 36, pp. 626-641.

Dean, R.G. (1977) 'Hybrid method of computing wave loading', Proceedings, Offshore Technology Conference, pp. 483-492.

Dean, R.G. (1984) 'Prediction of water wave and current kinematics, state of the art for offshore design', Proceedings, Ocean Structural Dynamics Symposium '84, Corvallis, Oregon, pp. 1-39.

Dean, R.G., Dalrymple, R.A., and Hudspeth, R.T. (1981) 'Force coefficients from wave project I and II data including free surface effects', Journal, Society of Petroleum Engineers, pp. 779-786.

Dold, J.W. and Peregrine, D.H. (1986) 'An effective boundary-integral method for steep unsteady water waves', K.W. Morton and M.J. Baines (eds.), Numerical Methods for Fluid Dynamics II, Clarendon, Oxford, pp. 671-679.

Forristall, G.Z., Ward, E.G., Borgman, L.E. and Cardone, V.J. (1978) 'Storm wave kinematics', Proceedings, Offshore Technology Conference, OTC Paper No. 3227, pp. 1503-1514.

Goodknight, R.C. and Russell, T.L. (1963) 'Investigation of the statistics of wave heights', ASCE Journal of Waterways and Harbors Division, Vol. 89, No. WW2, pp. 29-54.

Grilli, S. and Svendsen, I.A. (1989) 'Computation of nonlinear wave kinematics during propagation and runup on a slope', Proceedings, NATO Advanced Research Workshop on Water Wave Kinematics, Molde, Norway.

Hansen, J.B. and Svendsen, I.A. (1979) 'Regular waves in shoaling water: experimental data', Institute of Hydrodynamics and Hydraulic Engineering, Technical University of Denmark, Series Paper No. 21.

Hino, M. and Kashiwayanagi, M. (1979) 'Applicability of Dean's stream function method to estimation of wave orbital velocity and wave set-down and set-up', Proceedings, Coastal Engineering in Japan, Vol. 22, pp. 11-20.

Iwagaki, Y.T., Sakai, T., Kainuma, J., and Kawashima, T. (1971) 'Experiments on horizontal water particle velocity at water surface of near breaking waves', Proceedings, Coastal Engineering in Japan, Vol. 14, pp. 15-24.

Longuet-Higgins, M.S. (1952) 'On the statistical distribution of the heights of sea waves', Journal of Marine Research, Vol. XI, No. 3.

Longuet-Higgins, M.S. and Stewart, R.W. (1962) 'Radiation stress and mass transport in gravity waves with application to 'surf beats'', Journal of Fluid Mechanics, Vol. 13, pp. 481-504.

Longuet-Higgins, M.S. and Stewart, R.W. (1964) 'Radiation stresses in water waves: a physical discussion, with applications', Deep Sea Research, Vol. 11, pp. 529-562.

Longuet-Higgins, M.S. and Cokelet, E.D. (1976) 'The deformation of steep surface waves on water. I A numerical method of computation', Proceedings of the Royal Society of London, Vol. a, No. 350, pp. 1-26.

Rice, S.O. (1944,1945) 'The mathematical analysis of random noise', Bell System Technical Journal, No. 23 (1944), No. 24 (1945).

Sand, S.E. (1982) 'Long waves in directional seas', Journal of Coastal Engineering, Vol. 6, pp. 195-208.

Sand, S.E., Ottesen Hansen, N.E., Klinting, P., Gudmestad, O.T., and Sterndorf, M.J. (1989) 'Freak wave kinematics', Proceedings, NATO Advanced Research Workshop on Water Wave Kinematics, Molde, Norway.

Skourup, J., and Jonsson, I.G. (1989) 'Computational modelling of velocities and accelerations in steep waves', Proceedings NATO Advanced Research Workshop on Water Wave Kinematics, Molde, Norway.

Swift, R.H. and Dixon, J.C. (1987) 'Transformation of regular waves', Proceedings, Institution of Civil Engineers, Vol. 83, Part 2, pp. 359--380.

Thrasher, L.W. and Aagaard, P.M. (1969) 'Measured wave force data on offshore platforms', Proceedings, Offshore Technology Conference, OTC Paper No. 1007, pp. 83-94.

Torum, A. (1989) 'Wave forces on pile in surface zone', Journal of Waterway, Port, Coastal, and Ocean Engineering, Vol. 115, No. 4, pp. 547-565.

Vinje, T. and Brevik, P. (1981) 'Numerical simulation of breaking waves', Advances in Water Resources, Vol. 4, pp. 77-82.

Vis, F.C. (1980) 'Orbital velocities in irregular waves', Proceedings, Seventeenth International Conference on Coastal Engineering, Chapter 9, pp. 173-185.

Index List

762

List of Participants

Arisoy, Yalcin, Civ.Eng., M.Sc.

Dokuz Eylol University
Faculty of Eng. & Arc.
Civil Eng.,Bornova/IZMIR
Turkey

Barltrop, Niegel, Dr.

Atkins Oil and Gas Eng.
Woodcote Grove, Ashley Road
Epsom Surrey KT18 5BW
United Kingdom

Barnouin, Bruno, Dr.

IFREMER
Centre de Brest
BP 337, 29273 Brest Cedex
France

Belorgey, Michel, Professor

Université du Havre
Laboratorie de Mécanique des
Fluides, Faculté des Sciences
et Techniques
25, Rue Phillippe-Lebon
B.P. 540
76058 Le Havre Cedex
France

Borekci, Osman, Asst. Prof.

Bogazici University
Civil Engineering Department
80815 Bebek, Istanbul
Turkey

Borgman, Leon, Professor

University of Wyoming
Department of Statistics
P.O.Box 3332, University St.
Laramie, WY 82071
U.S.A.

Brathaug, Hans-Petter, Research Eng.

SINTEF
Div. of Structural Eng.
7034   TRONDHEIM
Norway

Brevik, Iver H., Professor

NTH
Inst. for mekanikk
7034   TRONDHEIM
Norway

765

Burcharth, Hans F., Professor

Aalborg University
Dept. of Civil Engineering
Sohngardsholmsvej.57
9000 Aalborg
Denmark

Chaplin, John R., Professor

City University
Dept. of Civil Engineering
Northampton Square
London EC1V OHB
United Kingdom

Clauss, Gunter, Professor

Techn. Uni. Berlin
Inst. fur schiffs- und
meerestechnik
Salzufer 17-19
1000 Berlin 10
F.R.G.

Cornett, Andrew, Research Officer

NRC Hydraulics Laboratory
Bldg. M-32, Montreal Rd.
Ottawa, ONT, KIA-ORG
Canada

Dean, R.G., Professor

University of Florida
Coastal & Oceanographic Eng.
Weil Hall, Gainesville
Florida 32611
U.S.A.

Durukanoglu, Fehmi, Dr.

Karadeniz Techn. University
Higher School of Marine
Sciences and Technology
61080 Trabzon
Turkey

Dysthe, Kristian, Professor

IMR University of Tromsø
9000 Tromsø
Norway

Frigaard, Peter

Aalborg University
Inst. 5
9000 Aalborg
Denmark

Førland, Even, Overingeniør

Statoil
P.O.Box 300
4001 Stavanger
Norway

Garrison, C.J., Professor

Oregon State University
Dept. of Civil Engineering
Apperson Hall 206
Corvallis, OR 97331-2302
U.S.A.

Germain, Jean Pierre, Professor

Inst. de Mecanique de
Grenoble
B.P. 53X - 38041 Grenoble
Cedex
France

Grilli, Stephan, Dr. (Ass.Prof.)

University of Delaware
Ocean Group
Civil Engineering Department
NEWARK-DE, 19716
U.S.A.

Gudmestad, Ove T., Dr. Scient

Statoil
P.O.Box 300
4001 Stavanger
Norway

Günbak, Ali Riza

Middle East Techn. University
Coastal & Harbour Eng. Lab.
Ankara
Turkey

Jacobsen, Vagner, Dr.

Danish Hydraulic Institute
Agern alle 5
DK-2970 Hørsholm
Denmark

Karabas, Theofanis, Ph.D.Student

Aristotele University of
Thessaloniki,
School of Technology
Dept. of Civil Engineering
Div. of Hydraulics & Environ-
mental Engineering,
Thessaloniki
Greece

Kharif, Christian, Doctor

I.M.S.T.
Campus Universitaire de
Luminy,
Case 903 - 13288 Marseille
Cedex 9
France

Kjeldsen, Peter, Chief Research Eng.

MARINTEK
P.O.Box 4125
7002 TRONDHEIM
Norway

Klinting, Per, Senior Researcher

Danish Hydraulic Institute
Agern alle 5
DK-2970 Hørsholm
Denmark

Klopman, Gert

Delft Hydraulics
Rotterdamsweg 185
P.O.Box 152
8300 ad Emmeloord
The Netherlands

Krogstad, Harald E., Dr.ing.

SINTEF
Section for industrial
mathematics,
7034 TRONDHEIM
Norway

Lambrakos, Kostas, Research Assoc.

Exxon Production Research Co.
P.O.Box 2189
Houston, Texas 77001
U.S.A.

Legras, Jean-Louis, Project Manager

E.T.P.M.
57 avenue Jules Quentin
92002 Nanterre cedex
France

Le Mehaute, Bernard, Prof.

Rosenstiel School of Marine
and Atmospheric Science,
4600 Rickenbacker Causeway
Miami, Florida 33149
U.S.A.

MacKenzie, Neil, Overingeniør

Statoil
P.O.Box 300
4001 Stavanger
Norway

Marcos Rita, Manuel, Civil Engineer

LNEC - DH
Av. Brasil 101, P-1799
Lisboa Codex
Portugal

Massel, Stanislav, Professor

Polish Academy of Sciences
Inst. of Hydroengineering
Gdansk - 5 Ceptersow 11
Poland

Mathiesen, Martin

Norwegian Hydrotechnical Lab.
7034 Trondheim
Norway

Mei, C.C., Professor

Massachussetts Inst. of
Technology,
Room 48-413
Cambridge, Mass. 02139
U.S.A.

Moe, Geir, Professor

NTH
Section of Structural Mech.
7034   TRONDHEIM
Norway

Myrhaug, Dag, Professor

NTH
Div. of Marine Hydrodynamics
7034   TRONDHEIM
Norway

Ochi, Michel K., Professor

University of Florida
Coastal & Oceanographic
Engineering Department,
Weil Hall, Gainesville
Florida 32611
U.S.A.

Olagnon, Michel

IFREMER
Centre de Brest
BP 337
29273 Brest Cedex
France

Pedersen, Bjarke

LIC Engineering A/S
Ehlersvej 24
DK-2900 Hellerup
Denmark

Peregrine, D.H., Dr., Professor

University of Bristol
School of Mathematics
Univ. Walk,
Bristol BS8 1TW
United Kingdom

Sand, Stig E., Dr.

Dansk Operatørselskab i-s
Slotsmarken 10
DK-2970 Hørsholm
Denmark

Schaeffer, Bernard

8, Allee Bonaparte
91080 - Courcouronnes
France

Schäffer, Hemming A., M.Sc.

Technical Univ. of Denmark
Inst. of Hydrodynamics and
Hydraulic Engineering,
Bldg.115, 2800 Lyngby
Denmark

Shaw, Christopher
Chief Oceanographer

Shell Internationale
Petroleum,
EPD/55, P.O.Box 162
2501 AW The Hague
The Netherlands

Skjelbreia, J.

MARINTEK
P.O.Box 4125
7002 Trondheim
Norway

Skourup, Jesper, M.Sc.

Technical Univ. of Denmark
Inst. of Hydrodynamics and
Hydraulic Engineering,
Bldg. 115, 2800 Lyngby
Denmark

Skyner, David, Mr.

Edinburgh University
Physics Dept.
Mayfield Rd.
Edinburgh EH9 3JZ
United Kingdom

Sobey, Rodney J., Professor

University of California
c/o Prof. J.A. Battjes
Delft Univ. of Technology
Dept. of Civil Engineering
P.O.Box 5048
2600 GA Delft
The Netherlands

Soulsby, Richard

Hydraulics Research Ltd.
Wallingford
Oxfordshire OXON OX10 8BA
United Kingdom

Spidsøe, Nils, Division Director

SINTEF
Div. of Structural Eng.
7034 TRONDHEIM
Norway

Stansberg, Carl Trygve, Dr.

MARINTEK A/S
P.O.Box 4125
7002 TRONDHEIM
Norway

Swan, Christopher, Dr.

Cambridge University
Dept. of Engineering
Cambridge CB2 IPZ
United Kingdom

Sætre, Hans Jørgen

A/S Norske Shell
P.O.Box 40
4056 Tananger
Norway

Tallent, James R., Researcher

Kyoto University
Disaster Prevention Res.
Inst., Gokasho 611,
Vjishi, Kyoto
Japan

Tørum, Alf, Professor

NTH/NHL
7034 Trondheim
Norway

Webb, Robin, Mr.

Robin Webb Consulting Ltd.
Ewer House 44-46
Crouch Street
Colchester, Essex, CO3 3MM
United Kingdom

Yeh, Harry, Assist. Professor

University of Washington
Dept. of Civil Engineering
FX-10, Seattle
Washington, 98195
U.S.A.

60 participants